Biology of the Reptilia
Volume 17, Neurology C
Sensorimotor Integration

Biology of the Reptilia

Edited by
Carl Gans

Volume 17, Neurology C
Sensorimotor Integration

Coeditor for this volume
Philip S. Ulinski

The University of Chicago Press
Chicago and London

The University of Chicago Press, Chicago 60637
The University of Chicago Press, Ltd., London
© 1992 by The University of Chicago
All rights reserved. Published 1992
Printed in the United States of America

00 99 98 97 96 95 94 93 92   5 4 3 2 1

ISBN (cloth): 0-226-28118-3
ISBN (paper): 0-226-28119-1

Library of Congress Cataloging-in-Publication Data
(Revised for vol. 17)

Gans, Carl, 1923–
    Biology of the reptilia.
    Vol. 14–15 published: New York : Wiley;
  v. 16–   published: New York : A. R. Liss;
    v. 17 published Chicago : University of Chicago
Press.
    Includes bibliographies and indexes.
    1. Reptiles.   I. Title.
QL641.G3   597.9   68–9113

Contents

Preface

This volume of the *Biology of the Reptilia* is the third dedicated primarily to topics of the nervous system, in many cases involving sensory paths. The volume was begun jointly with R. Glenn Northcutt; however, he found it difficult to complete the sections he had intended to write and to my sincere regret finally resigned both the authorship and the editorship.

In editing the chapters, we again encountered the difficulty of characterizing the diversity of species examined in a logical pattern (cf. Gans, 1989, *Fortschr. Zool.* 35,631–637). For obvious reasons, many students of the nervous system appear to have used two major criteria to select their experimental animals: availability and size. This selection process extends inadequate coverage to the rare and the small and tends to restrict coverage of very large organisms to hatchlings and juveniles. What is more critical is that the assumption remains that any crocodilian is representative of the Crocodilia, and any turtle of the Testudines; adequate comparative biology of "reptiles" then would seem to require sampling only of a single member of each major taxon.

We know that this is not the case; for instance, marine turtles and desert tortoises differ profoundly. We may recognize that some such intraordinal differences are physiology associated; they may reflect saltwater adaptation or the needs associated with distinct body temperature regimes and body volumes. Similarly, we know that the brain volumes of lizards and snakes reflect both absolute size and such questions as limblessness. However, it is essential to base correlations on multiple comparisons; occurrence of a condition in one limbless form does not make it a general correlate of limblessness.

In the present case, we often lack sufficient samples to permit much generalization. The decisions of which species to sample were made by multiple independent investigators who may have been active on different continents some time during the last century. What is needed is to provide the reader with an indication of the robustness of the data base. This should permit evaluation of the reliability of what is offered and, more important, suggest the nature of the next organism to be sampled by those readers who are involved in continuing such studies.

To facilitate this process, we have again asked the authors to tabulate the species that were sampled in the several studies. Use of these tables facilitates interpretation of text statements; for instance, a comparison of conditions in testudines and crocodilians might have been based on data only from *Pseudemys scripta* and *Caiman crocodilus*, respectively. It also reduces redundancy—the need to repeat the names and taxonomic assignments of all of the species on which a particular statement is based. Naturally, this procedure requires that the reader attempt to evaluate what is offered. If there is a physiological component it may be possible to estimate whether, for instance, on ecological grounds one would expect *Pseudemys* and *Caiman* to differ from *Chelonia* and *Crocodylus*. It has been the desirability of such future comparison that has led us again to attempt to standardize nomenclature. I am most appreciative of George Zug's careful check of Latin names, which permits us to refer to species by the currently appropriate terminology. In the text of two chapters we have allowed the authors to list some of the original generic designations, adding the currently accepted name in parentheses. This style should not imply that we accept the old usage, only that this style of citation should facilitate explicit recognition that names referred to in previous studies no longer apply.

I am very grateful to our contributors for their responsiveness to the comments of multiple reviewers and for their patience while awaiting the completion of the last-arrived manuscripts of the volume. It is my pleasure to acknowledge the assistance of my coeditor Philip S. Ulinski and of many ad hoc reviewers. Their efforts represent the key to the generation of volumes intended to remain useful to the herpetological and, indeed, the zoological community. At the risk of omitting some names we would like to mention Mark Braford, Jr., Edward Gruberg, Karl F. Guthe, R. Glenn Northcutt, André Parent, Michael Pritz, Dolores M. Schroeder, and Wilhelmus J. A. J. Smeets, as well as the authors of the several chapters. Ms. Katherine Vernon aided substantially with the correspondence. Finally, we appreciate the support of our institutions toward the ever increasing costs of postage and copying.

CARL GANS

1

Retinal Structure

ELLENGENE H. PETERSON

CONTENTS

I. INTRODUCTION

This review summarizes what is known about the structure of reptilian retinas. It is organized into three sections, which treat the retina at increasing levels of integration. The first section deals with the gross morphology and ultrastructure of retinal elements, that is, photoreceptors and the several varieties of retinal neurons. The second describes the interconnections of these elements to form retinal information channels by which the visual stimulus array is analyzed,

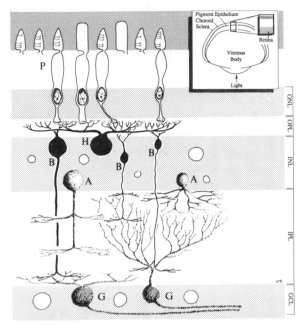

Fig. 1.1. Laminar organization of vertebrate retinas. There are five cell types. *Photoreceptors* (P) have outer segments embedded in the pigment epithelium (dark shading), nuclei in the *outer nuclear layer* (ONL) and axons that extend into the *outer plexiform layer* (OPL). *Horizontal cells* (H) have somata in the outer portion of the *inner nuclear layer* (INL) and processes that ramify in the OPL. *Bipolar cells* (B) have somata in the INL. Their dendrites contact photoreceptor and horizontal cell processes in the OPL, and their axons ramify in the *inner plexiform layer* (IPL). *Amacrine cells* (A) have somata in the inner portion of the INL and processes that extend into the IPL. *Ganglion cells* (G) have somata in the *ganglion cell layer* (GCL) and dendrites that ramify in the IPL, where they contact the process of amacrine and bipolar cells. Inset: Schematic eye illustrating the relation of the retina to the outer layers of the eye. Throughout the text, the terms *inner, proximal,* and *vitread* indicate positions toward the center of the eye. The terms *outer, distal,* and *sclerad* indicate positions toward the periphery of the eye (toward the photoreceptor layer).

coded, and transmitted to the rest of the central nervous system. The third deals with the assembly of these information channels into whole retinas that reflect the physical and ecological constraints under which they have evolved and currently function.

In writing this chapter I have had a number of secondary goals in mind. First, I have attempted to show, albeit in a limited manner, how the structural organization observed in reptiles compares with that of other vertebrates. Second, I have attempted to make this report understandable to readers who may not be especially expert regarding retinal structure, but who nevertheless may want some useful information about the organization of reptilian retinas. Finally, I have attempted to be comprehensive in my treatment of the literature through late 1989, although I will emphasize recent work since reviews of earlier studies are readily available (Verrier, 1935; Walls, 1942a; Detwiler, 1943; Rochon-Duvigneaud, 1943; Duke-Elder, 1958; Tansley, 1965; Underwood, 1970).

II. RETINAL ELEMENTS
A. Overview

The retina of reptiles, like that of all other vertebrates except certain agnathans (Holmberg, 1977), is composed of five primary cell types whose somata are arranged in layers parallel to the choroidal coat of the eye (Fig. 1.1). Processes from these cells enter complex neuropil regions, the outer and inner plexiform layers where they intermingle and form functional contacts. This section describes the structure of these cells, beginning with the photoreceptors, which form the outer (scleral) layer of the retina, and proceeding toward the ganglion cell layer at the vitreal surface. The circuitry of the outer and inner plexiform layers will be described in Section III.

B. Photoreceptors

1. GENERAL The schematic photoreceptor in Fig. 1.2 serves as a guide to this section. Most sclerad is the *outer segment*, with its stacked

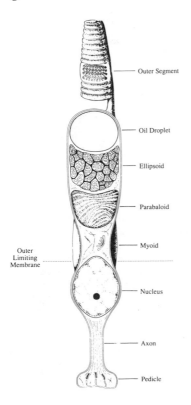

Outer Segment

Oil Droplet

Ellipsoid

Paraboloid

Myoid

Outer Limiting Membrane

Nucleus

Axon

Pedicle

Fig. 1.2. Composition of a reptilian photoreceptor. The *outer segment* bears numerous photosensitive disks. It is connected to the *inner segment* by a ciliary stalk. The inner segment comprises four organelles: an *oil droplet*, an *ellipsoid*, a *paraboloid*, and a *myoid*. In the myoid region, two fins are illustrated. Some reptilian photoreceptors lack one or more of these organelles. A *nucleus* lies proximal to the outer limiting membrane; it is surrounded by a thin rim of cytoplasm and is connected to the synaptic *pedicle* by an *axon*. Three synaptic (ribbon) complexes are illustrated in the pedicle.

disks containing photosensitive pigment molecules. It is attached, sometimes through a connecting cilium, to the *inner segment*, which may exhibit any or all of the following organelles: (1) an oil droplet, (2) an ellipsoid, (3) a paraboloid, (4) a myoid. The inner segment may also bear cytoplasmic extensions commonly known as calyceal processes and fins (Dunn, 1973; Owen, 1985). Proximal to the inner segment and subjacent to the outer limiting membrane is the photoreceptor *nucleus*, which with other such nuclei forms the outer nuclear layer (Fig. 1.1). The cell nucleus is connected by a fiber to the synaptic *pedicle*, which extends into the outer plexiform layer and, directly or through basal processes, contacts other retinal cells.

Fig. 1.2 illustrates a *single* photoreceptor. Many nonmammalian vertebrates, including reptiles, also have multiple visual cells in which two or more single elements appear to be fused (Fig. 1.3B).

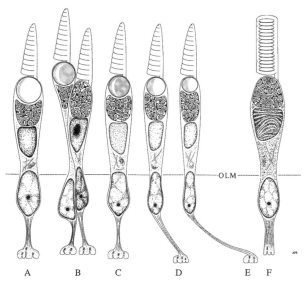

Fig. 1.3. *Pseudemys (Trachemys) scripta*. Photoreceptors. Most of the available data on the structure and physiology of reptilian photoreceptors come from work on this species. The drawings are based on data from Kolb and Jones, 1982, 1987. (A) Red-sensitive single cone. (B) Double cone comprising a principal (left) and an accessory (right) cone. Both are probably red sensitive. (C) Green-sensitive single cone. (D) Oblique cone with clear, fluorescent oil droplet. (E) Small oblique cone, with clear, nonfluorescent oil droplet. The spectral sensitivities of single cones with clear oil droplets are in dispute. (F) Rod; these visual cells account for 5% to 10% of photoreceptors in the retina of *Pseudemys*. Note that the oil droplets of the cones differ in size. They are arranged in order of decreasing size from left to right. Nuclei of single cones and the accessory cone extend distal to the outer limiting membrane (OLM); those of rods and the principal cone lie more proximally. The pedicles form three layers from distal to proximal in the outer plexiform layer: (1) double cones and rods (B and F); (2) red and green single cones (A and C); and (3) oblique single cones (D and E).

Most are paired cells of two broad kinds: (1) *twin* photoreceptors, in which the two members are mirror images of each other, and (2) *double* photoreceptors, in which members of the pair are obviously dissimilar. In double visual cells, one member resembles single photoreceptors in the same retina and is referred to as the chief, or principal, cell to distinguish it from the other, accessory, member of the pair.[1] These cells show morphological specializations that appear to bind them together; they may or may not be electrically coupled (see Section II.B.5.c.3). Occasionally, more than two photoreceptors appear to fuse, forming triplet or quintuplet visual cells. The distribution of single and multiple cell types among the Reptilia has been reviewed by Underwood (1970).

2. Outer Segment

a. Fine Structure and Renewal   In reptiles, as in other vertebrates, outer segments are composed of stacked disks, or saccules, enclosed entirely or in part by a plasma membrane (Fig. 1.4, left). These disks

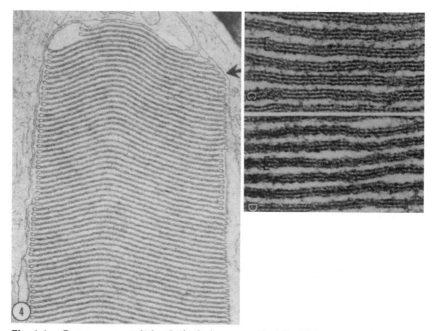

Fig. 1.4.   Outer segment disks. Left: *Sceloporus occidentalis*. Disk structure in cone photoreceptor. Most disks are continuous with the plasma membrane (arrow); this pattern is characteristic of cones. Approximately half way along the outer segment, several disks have separated from the plasma membrane, exhibiting a rod-type configuration. The cilium side of the outer segment is to the right (8064 x). Right: *Crotalus viridis* (upper) and *Eublepharus macularis* (lower). Electron micrograph showing the pentalaminar structure of outer segment disks (204,000 x) (A, from Young, 1977; B, from Dunn, 1973).

represent the site of photic transduction. Under appropriate fixation conditions (Dunn and Adomian, 1971; Dunn, 1973; Cohen, 1972), the complete disk (including proximal and distal leaflets) has a pentalaminar structure (Fig. 1.4, right) consisting of an alternating series of three electron-dense and two electron-lucent bands. Sjostrand and Kreman (1978) attempted to analyze the molecular structure of disk membranes in *Gekko gecko* (also *Rana* and guinea pigs) using freeze-fracture techniques; they proposed that each disk leaflet is composed of two layers: a monolayer of rhodopsin molecules facing the potential intradisk space and an outer, bimolecular layer of low-weight lipids. More recent data suggest that rhodopsin actually spans the disk membrane; that is, it protrudes into the intradisk space, as proposed by Sjostrand and Kreman (1978); but it is also exposed at the cytoplasmic surface of the disk (see reviews in Olive, 1980; Hargrave, 1986). Rhodopsin is also believed to span the plasma membrane of the outer segment (Besharse, 1986), and this has been confirmed in *Pseudemys (Trachemys) scripta* (Gaur et al., 1988).

Immunological probes against cell-specific and species-specific molecular structures have recently been applied to reptilian outer segments. For example, one monoclonal antibody (15–18) raised against *Pseudemys (Trachemys) scripta* photoreceptors labels the outer segments of rods and some single cones, but not double cones (Fig. 1.5), suggesting that the antigenic site on turtle rod opsin (the putative binding site of 15–18) resembles a peptide sequence on some, but not all, turtle cones. The same antibody cross-reacts with retinas from *Anolis carolinensis* and several mammals, suggesting antigenic similarities among photoreceptor pigments in these distantly related groups (Sarthy and Gaur, 1986; Gaur et al., 1988). In contrast, antibodies directed against bovine phosphodiesterase (an outer segment protein thought to be involved in phototransduction; Yau and Baylor, 1989) failed to cross-react with *Emys orbicularis* retina (Etingof et al., 1986). These new techniques enable investigators to assess and compare the molecular structure of retinas, and they will undoubtedly be applied more frequently to reptiles in the coming decade.

Relatively few quantitative data on disk dimensions and spacing are available for reptiles. Dunn (1973) reports an average disk thickness of 13.0 to 14.5 nm in *Crotalus* and several geckos (Fig. 1.4, right); these values are somewhat lower than those given by Sjostrand and Kreman (1978), but they agree well with figures for other vertebrates (Nilsson, 1965; Dunn, 1973). Disk thickness does not differ in rods and cones (Dunn, 1973).

There are, however, marked differences between photoreceptor types in the patterns of disk-plasma membrane continuities. This issue has attracted considerable attention over the past three decades,

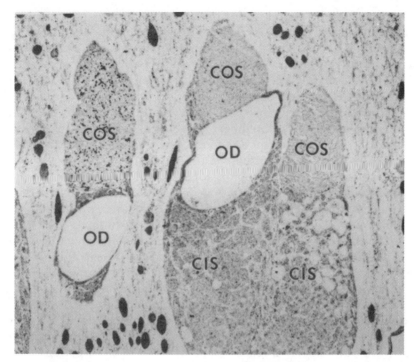

Fig. 1.5. *Pseudemys (Trachemys) scripta.* Electron micrograph showing immunogold labeling with monoclonal antibody (MAb) that binds to a subset of turtle outer segments. One single cone is labeled, but the outer segments of the double cone (right) are not. This MAb probably binds to a photopigment found in rods and some single cones, but not in double cones. COS, cone outer segment; CIS, cone inner segment; OD, oil droplet (630 X). (Reprinted with permission from *Vision Research*, Vol. 28, Gaur et al., A monoclonal antibody that binds to photoreceptors in the turtle retina, 1988, Pergamon Press plc.)

both because it raises important questions about the mechanism of photoreceptor action and because it has sometimes appeared to offer a reliable criterion for distinguishing between visual cell types. Briefly, vertebrate photoreceptor disks develop as elaborations of the plasma membrane, either by successive invaginations of the outer segment membrane (Nilsson, 1964) or as evaginations of the plasma membrane of the connecting cilium (Steinberg et al., 1980; review in Besharse, 1986). Thus, at least in the early stages of disk development, there is some continuity between disk and plasma membrane. In many receptors, however, disks become partially or completely separated from the plasma membrane during development, raising the question of how excitation of photopigment molecules in the disk can be transmitted to the plasma membrane, where it triggers the first electrical sign of visual excitation—suppression of the dark current

(Hagins, 1979; Sjostrand and Kreman, 1978). The pattern of disk membrane continuities appears to differ between photoreceptor types: in classic rods, disks are discontinuous except for a small number at the outer segment base, where disk formation occurs (Steinberg et al., 1980; Cohen, 1972; Moody and Robertson, 1960; see Fig. 1.7); in cones, the basal region of open disks is more extensive and at least some continuities occur throughout the length of the outer segment (Anderson et al., 1978; Steinberg et al., 1980). Thus, disk-plasma membrane continuities may be an important criterion for distinguishing between photoreceptor types (Dunn, 1973).

Among reptiles, most receptors for which ultrastructural evidence exists have a cone-type pattern (Fig. 1.4, left). At one extreme there are outer segments in which virtually all disks are continuous with the plasma membrane on the cilium side. (Testudines: *Pseudemys (Trachemys) scripta*—Bush and Lunger, 1972; Villegas, 1960; *Chrysemys picta*—Dunn, 1973. Snakes: *Leptodeira annulata*—Miller and Snyder, 1977. Lizards: *Sceloporus occidentalis*—Eakin, 1964; Young, 1977.) In nocturnal gecko photoreceptors (Fig. 1.6) there is a marked tendency toward separation of disks from plasma membrane (i.e., toward a rodlike pattern), but extensive regions of continuity exist in the proximal outer segment (Cohen, 1972; Pedler and Tilly, 1964) and scattered along the outer segment shaft (Yoshida, 1976, 1978), a pattern reminiscent of some mammalian cones (Pedler, 1965; Cohen, 1972; Anderson et al., 1978; Steinberg et al., 1980). It is unfortunate that these experiments have done little to ascertain the disk patterns of different receptor types within the same retina. Yoshida does suggest that the pattern he observed in *Gekko japonicus* was primarily characteristic of Class B (double) cells[2] and that single visual cells may have a more rodlike configuration. In the diurnal gecko *Ebenauia inunguis* (*Phelsuma;* Pedler and Tansley, 1963) the disks are described as "two parallel linear densities connected at both ends" (i.e., a rodlike pattern), and, in an explicit comparison between nocturnal and diurnal forms, Pedler and Tilly (1964) report no difference in the ultrastructural appearance of outer segment disks. However, the use of osmium tetroxide fixation without glutaraldehyde in these studies appears to have produced artifactual vesiculation and coalescence of cone disk membranes (Cohen, 1972; Dunn, 1973, Fig. 14; Eakin, 1965; see Kalberer and Pedler, 1963, for a similar problem in *Alligator mississippiensis*), so the data are difficult to interpret. Thus, we do not know whether nocturnal and diurnal geckos exhibit similar patterns of continuity between disks and the plasma membrane. Classic rodlike configurations (all disks free floating except those at the outer segment base) have been described in *A. mississippiensis* (Kalberer and Pedler, 1963) and reported (but not illustrated) in the snake *Leptodeira annulata* (Miller and Snyder, 1977).

A related issue that has attracted attention recently is the mechanism of disk shedding and renewal. This mechanism differs between photoreceptor types, and several investigators have sought to take advantage of the simplex condition in lizard retinas to establish the rodlike or conelike pattern. Briefly, both rods and cones shed packets

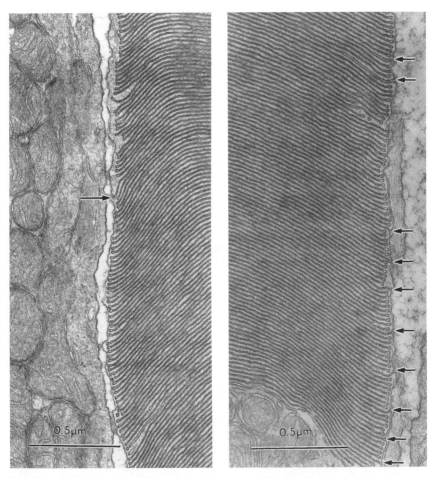

Fig. 1.6. *Gekko japonicus.* This pure "rod" retina exhibits continuities between outer segment disks and plasma membrane (arrows) in Class A single (left) and the accessory member of Class B double (right) visual cells. Such continuities are thought to be characteristic of cones (38,000 x). (Reprinted with permission from *Vision Research,* Vol. 18, Yoshida, Some observations on the patency in the outer segments of photoreceptors of the nocturnal gecko, 1978, Pergamon Press plc.)

of disks from the distal outer segment, which are then phagocytosed by the pigment epithelium (Anderson et al., 1978; Besharse, 1986). In rods, this process is triggered by light onset. In cone photoreceptors of most species, shedding begins with light offset, an observation that may explain early failures to detect disk shedding in cones. Such cone shedding and disk phagocytosis have been demonstrated in *Scelopo-rus* (Young, 1977), in which it exhibits both dark-triggered and circadian components (Bernstein et al., 1984). Tritiated fucose (Bunt and Klock, 1980a) and procion yellow (Laties et al., 1976a, 1976b) incorporation studies indicate that, in the rodlike visual cells of *Gekko gecko*, disks are formed and "pinched off" at the outer segment base (Fig. 1.7). They then travel distally and eventually are shed in a manner typical of vertebrate rods. Dual pulses of procion yellow have been used to monitor the rate of disk movement (Fig. 1.7); the data suggest that renewal rates differ between photoreceptor types. In cone visual cells of *Chrysemys picta, Thamnophis ordinoides,* and a variety of other vertebrates (Bunt and Klock, 1980a, 1980b), ^{3}H fucose is incorporated throughout the outer segment (rather than in a discrete band at the base, as in rods); the reason for this diffuse labeling of cone outer segments is not known (reviewed in Besharse, 1986). Thus, reptilian visual cells exhibit the same patterns of membrane turnover and disk shedding seen in other vertebrates.

Simplex lizard retinas have also been used to study the role of cyclic nucleotides in normal and abnormal retinal function. Cyclic GMP (cGMP) and cGMP phosphodiesterase are now known to mediate phototransduction in rod visual cells of frogs and mammals (Stryer, 1986; Dowling, 1987), and evidence supporting their role in visual excitation has been adduced in the rodlike photoreceptors of *Gekko gecko* (Kawamura and Murakami, 1986). Abnormally elevated cyclic nucleotide levels are also known to cause photoreceptor degeneration in certain inherited diseases of mammals (Farber and Shuster, 1986). Under such conditions, rods typically degenerate before cones, and Williams et al. (1987) have used the pure cone retina of *Sceloporus occidentalis* to demonstrate that cone loss results directly from elevated cyclic nucleotide levels rather than indirectly from rod degeneration.

b. GROSS MORPHOLOGY

Because outer segment size and form among reptilian species vary greatly and little necessary background information is available to interpret these differences, few useful conclusions are possible. Species differ along two dimensions.

1. Outer segments within a given taxon may exhibit a duplex (i.e., some conical and some cylindrical) or a simplex (all outer segments conical or cylindrical) pattern. Most turtles (but see Gillett, 1923) and

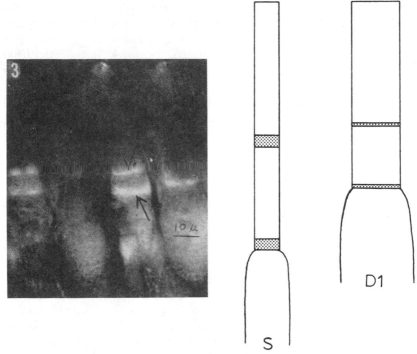

Fig. 1.7. *Gekko gecko*. Migration of newly formed disk leaflets within the outer segment. Left: Two disk leaflets (arrows) within the outer segment have incorporated procion yellow that was injected twice, with a two-day separation between injections. The more distal band represents disks that have formed at the outer segment base and migrated toward the photoreceptor tip. The distance between labeled leaflets indicates the extent of disk migration during the two days between injections. Right: Schematic drawing shows differences between outer segment widths and migration rates in single (S) and D1 double cells. (From Laties et al., 1976a.)

crocodilians for which data exist exhibit a duplex pattern, as do some snakes. Some snakes and all lizards possess a simplex retina. Walls (1934, 1942a, 1942b) and Underwood (1951, 1954, 1967a, 1967b, 1968, 1970) have documented these patterns in some detail; they argue that the duplex condition is primitive, that snakes arose from lizards before the duplex condition was lost in that group, that the simplex pattern in both lizards and snakes (when present) is secondary, and that duplicity once lost cannot be regained.

2. Outer segments may also vary across species in size (length, diameter, or both) and shape. Not surprisingly, these parameters generally correlate with the degree of habitat insolation and the apparent dependence of their owners on visual cues for foraging, sexual, and agonistic behaviors. The retinas of diurnal species such as iguanids and agamids are dominated by photoreceptors with narrow, tapering

outer segments, whereas taxa that operate under low illumination levels tend to adopt some strategy for increasing the light absorptive surface of the outer segment by becoming cylindrical (e.g., *Telescopus* pattern), long (e.g., crotaline pattern), or thick (e.g., nocturnal geckonids; see Tansley, 1959, 1964, for explicit comparisons between nocturnal and diurnal geckos). In nocturnal squamates that depend heavily on vision for foraging, outer segments are especially massive (*Phyllorhynchus, Hypsiglena*). In contrast, amphisbaenians and other burrowing forms with reduced eyes (*Dibamus:* Underwood, 1970; scolecophidian pattern, Underwood, 1967b) have diminutive outer segments. Beyond these simple generalizations, however, we know almost nothing about the functional sequelae of different outer segment patterns; their significance can only be assessed in the context of variables such as eye morphology and optical properties, specific sensory ecology, and visual system organization. For example, outer segments of visual cells in *Gekko gecko* have a larger diameter than those of iguanid cones, suggesting that the former may be less able to resolve stimuli of high spatial frequency. However, calculations of the schematic eye of *G. gecko* and *Iguana iguana* suggest that the retinal image for a given object will be 30% larger (with twice the illuminance) in *Gekko* than in *Iguana* (Citron and Pinto, 1973), and this larger image will help offset the effects of increased photoreceptor diameter. Dunn (1966c, 1969) has calculated that the large, cylindrical outer segments of geckos occupy 53% to 63% of total retinal area and that they absorb 42% to 63% of incident light, depending on the species.

3. INNER SEGMENT

a. GENERAL   The inner segments of vertebrate photoreceptors contain a number of organelles, the primary role of which is optical, metabolic, or both. In reptiles, inner segment organization is especially complex and variable. Two organelles or divisions of the inner segment, the ellipsoid and myoid, are present in all species; two others, the paraboloid and an oil droplet, are variable in occurrence. This section describes each component of these inner segments in the order of its distal-proximal occurrence (Fig. 1.2).

b. OIL DROPLETS

Many reptilian photoreceptors contain a refractive element at the distal tip of the inner segment whose putative function is to concentrate incident light on the outer segments. In all groups except snakes and some nocturnal geckos (see next section), this element takes the form of an oil droplet (Fig. 1.8), an inclusion that is also found in certain fishes, amphibians, birds, and noneutherian mammals. These inert lipid bodies occur in close association with ellipsoidal mitochondria

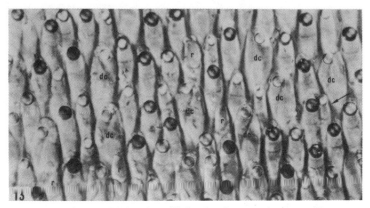

Fig. 1.8. *Mauremys caspica.* Oil droplets fill the distal inner segment. Differences in oil droplet color are reflected in gray level differences in this black and white light micrograph. The arrow indicates a single cone with a small, clear oil droplet. Rods (r) have no oil droplets; dc, double cone. (From Gallego and Perez-Arroyo, 1976.)

(Pedler and Tansley, 1963; Ishikawa and Yamada, 1969) and may be derived from them (see Section II.B.3.c).

In electron micrographs (Figs. 1.9 and 1.16), oil droplets show an amorphous or finely granular internal structure that varies in density as a function of their composition of carotenoids (Kolb and Jones, 1982). Whenever osmium tetroxide is used as a primary fixative, oil droplets show an irregular perimeter (Ishikawa and Yamada, 1969) and no limiting membrane (e.g., *Phelsuma*, Pedler and Tansley, 1963; *Pseudemys (Trachemys) scripta*, Bush and Lunger, 1972; *Rana*, Craig et al., 1963). In such preparations, the oil droplet appears to merge with the surrounding ellipsoidal mitochondria, suggesting to some investigators that oil droplets are derived from these mitochondria by secretion (Pedler and Tansley, 1963; Ishikawa and Yamada, 1969). However, the contours of oil droplets are smooth and sometimes show a limiting membrane in glutaraldehyde fixed tissue (e.g., *Chamaeleo*, Armengal et al., 1981; *Chrysemys*, Dunn, 1973, see Fig. 32 of that paper), suggesting that they may be a form of highly modified mitochondria, as described in Section II.B.3.c.

Oil droplets are found in at least some photoreceptors of turtles, *Sphenodon*, *Amphisbaena* (Underwood, 1951), and many lizards. However, their assignment to different photoreceptor types is a matter of some confusion due to the difficulty in classifying many reptilian visual cells (Pedler, 1965; Underwood, 1968; Cohen, 1972). There is general agreement that oil droplets occur in "classic" single cones and in the chief member of double cones of diurnal reptiles and that they are absent in typical rods and in accessory cones (see Fig. 1.3). Oil droplets are often listed among the distinguishing features of cones for this reason. Problems arise, however, in the case of droplet-bearing species (*Sphenodon* and certain geckos) that have adopted nocturnal

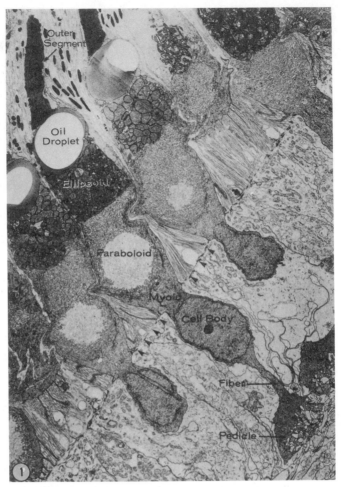

Fig. 1.9. *Pseudemys (Trachemys) scripta.* Electron micrograph illustrating the components of a single cone (compare with Figs. 1.2 and 1.3). Arrowheads indicate the outer limiting membrane (2430 x). (Reproduced from the *Journal of Cell Biology*, 1978, Vol. 79, 802–825, by copyright permission of the Rockefeller University Press.)

habits and in which outer segments are demonstrably enlarged (i.e., rodlike). For example, Walls (1942a) classifies the large visual cells in *Sphenodon* as rods primarily on the basis of the morphology of their outer segments, concluding that this species offers a rare example of droplet-containing rods, whereas Vilter (1951a) argues on the basis of several criteria that these photoreceptors are true cones.[3] Similar questions could be raised about the droplet-containing "rods" in the gecko *Aristelliger* (Underwood, 1951). Thus, oil droplets characteristically occur in cones; wherever they have been reported in "rods" the proper classification of these visual cells may be in doubt.

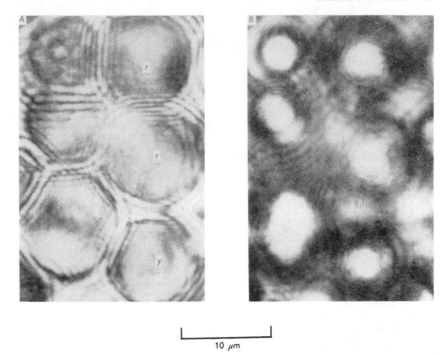

10 μm

Fig. 1.10. *Pseudemys (Trachemys) scripta.* Retinal whole mount showing light collection by oil droplets. Left: The microscope is focused at the level of the ellipsoids. The transmitted light is relatively uniform. r, ellipsoid of cone-bearing red oil droplet. y, ellipsoid of cone-bearing yellow oil droplet. Right: When the microscope is focused just above the oil droplets, illumination of the field is nonuniform. The bright spots are located at the bases of cone outer segments. The larger, dark circles correspond to the (out of focus) ellipsoids. (From Baylor and Fettiplace, 1975.)

Oil droplets are thought to contribute to light collection in reptilian visual cells because of their high refractive indices (Ives et al., 1983). When *Pseudemys (Trachemys)* retinas are whole-mounted with the photoreceptor layer up and transilluminated, the distribution of light at the level of the ellipsoids is relatively uniform, but light is concentrated into an array of bright spots at the outer segment base, that is, above the oil droplets (Fig. 1.10; Baylor and Fettiplace, 1975). Such light funneling, which increases the intensity of light at the outer segments and thus the change in membrane voltage for a given retinal illumination, probably involves inner segment structures in addition to oil droplets, that is, the ellipsoid (Winston and Enoch, 1971; see Section II.B.3.c) and the paraboloid (see Section II.B.3.d). Indeed, a similar concentrating effect has been reported in snake cones (Tansley and Johnson, 1956), which lack both oil droplets and paraboloids; in this case the phenomenon is attributed to the enlarged cone ellipsoids (see Section II.B.3.c). Nevertheless, oil droplets clearly play a significant role in light collection, because incident light is concentrated above the oil droplets even when this light is oriented perpendicular to the cones (i.e., when the cones are viewed from the side) and thus does not pass through the paraboloid and ellipsoid before encountering oil droplets (Baylor and Fettiplace, 1975). Increased light collection by outer segments can be expected to enhance retinal sensitivity as well as contrast and motion detection (Young and Martin, 1984; Ives et al., 1983).

Oil droplets are colorless in the few nocturnal species with droplet-bearing visual cells (e.g., *Sphenodon*, Walls, 1942a; Vilter, 1951a; geckos, Underwood, 1970). In turtles and diurnal lizards, however, most droplets are brightly colored: red, orange, and yellow. Examination of their absorbance spectra suggests that these colors reflect the presence of carotenoid pigments that differ in type and concentration (Liebman and Granda, 1975; reviewed in Lipetz, 1985). In *Pseudemys (Trachemys) scripta* and *Emydoidea blandingii*, for example, red and orange droplets have an absorbance spectrum characteristic of the beta carotene derivative astaxanthin (Granda and Dvorak, 1977; Lipetz, 1984a, 1984b), but the pigment concentration (and thus the peak absorbance) of the red droplets is higher than the orange by a factor of 10. Yellow droplets in *P. scripta* and *E. blandingii* have an absorbance spectrum characteristic of zeaxanthin or, perhaps, xanthophyll. In *Chelonia mydas* the yellow and orange droplets have somewhat different spectra, and their pigments have not been identified.

The net effect of these dense carotenoids is to absorb light at short wavelengths. Thus, they act as powerful cutoff filters, virtually eliminating transmission to the outer segment of any wavelengths below some value that is characteristic for each oil droplet type (Fujimoto et al., 1957; Strother, 1963; Liebman, 1972). Not surprisingly, oil droplets with high (long wavelength) cutoff points are found in photoreceptors with peak sensitivities at slightly longer wavelengths. For example, the pigment of red-sensitive cones in *Pseudemys (Trachemys)* has a peak sensitivity of 620 nm, and these cones contain a red oil droplet that virtually eliminates all wavelengths below 550 nm ($\lambda_{1/2} = 602$ nm; Ohtsuka, 1985a). Thus, wavelengths for which the photopigment is maximally sensitive are transmitted, but virtually all those below its λ_{max} are filtered out. This relationship between the λ_{max} of each cone pigment and the absorbance spectrum of its screening oil droplet leads to an orderly set of photopigment–oil droplet pairings (see next paragraph). The calculated spectral peaks of these photopigment–oil droplet complexes predict the peak spectral sensitivities of the behaving animal (Neumeyer and Jager, 1985).

In *Pseudemys (Trachemys)*, red oil droplets are uniquely associated with red-sensitive (620 nm λ_{max}) single cones, orange oil droplets with the red-sensitive chief member of double cones, and yellow droplets with green-sensitive (518 to 540 nm λ_{max}) single cones (Liebman, 1972; Baylor and Fettiplace, 1975; Granda and Dvorak, 1977; Ohtsuka, 1985a). Similar pairings are found in *Chinemys reevesii* (*Geoclemys*, Ohtsuka, 1978, 1985b) and in *Chelonia mydas* (Granda and Dvorak, 1977). The chief and accessory members of double cones in turtles were originally thought to be red- and green-sensitive, respectively (Granda and Dvorak, 1977), but it now appears that both contain a red-sensitive photopigment similar to that found in single red cones (Lipetz, 1985; Ohtsuka, 1985a, 1985b). Early reports also placed clear oil droplets within blue-sensitive (450 nm λ_{max}) single cones (reviewed in Granda and Dvorak, 1977), but it is now clear that there are two types of clear droplets (Ohtsuka, 1984), each associated with a different spectral type of photoreceptor (Fig. 1.11; Lipetz, 1985; Ohtsuka, 1985a). Small, nonfluorescent droplets and larger, pale green, fluorescent droplets are located in red- or blue-sensitive single cones, but Lipetz (1985) and Ohtsuka (1985a, 1985b) report opposite photopigment–oil droplet pairings. This controversy about photopigment–oil droplet pairings has led to conflicting models of color coding in the outer plexiform layer (see Section III.C.3). Recent morphological evidence suggests that blue-sensitive single cones probably contain the larger, fluorescent (pale green) droplets (Kolb and Jones, 1987, in agreement with Lipetz, 1985). These authors have suggested that single cones containing the small, nonfluorescent, clear droplets may be ultraviolet sensitive, as are some cones in birds and fish. The

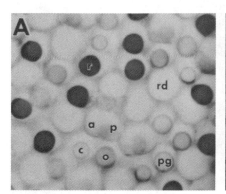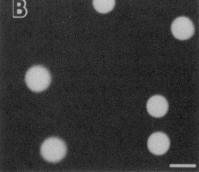

Fig. 1.11. *Chinemys (Geoclemys) reevesii.* Retinal whole mount illustrating oil droplet types. Left: Light micrograph showing red (r), orange (o), pale green (pg), and clear (c) oil droplets. Rods (rd) and the accessory member of double cones (a) have no oil droplets. Oil droplets in the principal member of double cones (p) are located sclerad to the plane of focus and are blurred. Right: The same field photographed with fluorescence microscopy to illustrate that pale green droplets fluoresce upon irradiation with short wavelength (near UV) light. The pale green oil droplets correspond to the large, clear droplets described in the text. Scale: 10 μm. (From Ohtsuka, 1985b.)

relative frequency of different oil droplets has been characterized in adult turtles (Peiponen, 1964; Brown, 1969; Granda and Haden, 1970; Baylor and Fettiplace, 1975; Peterson, 1981a; Ohtsuka, 1985a; Kolb and Jones, 1987) and under different conditions of illumination and development (Pezard, 1957, 1964; Tiemann, 1970).

As noted previously, oil droplets enhance light capture by cone outer segments and so increase the sensitivity of these visual cells (Ives et al., 1983). Some have theorized that oil droplets also filter out glare, reduce chromatic aberration and retinal scatter (especially of short wavelengths), and enhance contrast of objects against the background light. Data bearing on these latter theories have been reviewed repeatedly (e.g., Walls and Judd, 1933; Meyer et al., 1965; Muntz, 1972; Lythgoe, 1979; Bowmaker, 1980) and will not be considered here. More recent evidence (Baylor and Fettiplace, 1975) suggests that oil droplets may increase the directional selectivity of cones (Stiles-Crawford effect of the first kind).

The role of oil droplets in color vision is still unclear. Krause's (1893) idea that different-colored oil droplets could support color vision in the presence of a single photopigment by producing receptors with different peak sensitivities has fallen into disfavor since the demonstration of multiple photopigments in single retinas (Muntz, 1972; Granda and Dvorak, 1977). However, Orlov and Maximova (1964; cited in Muntz, 1972) have claimed that *Agama* responds as a monochromat and *Emys orbicularis* as a dichromat when the retina is illuminated from behind (so that light need not pass through the oil droplets to reach the outer segments), but as dichromats and tetrachromats, respectively, when illuminated normally. These data, if reliable (see Wallman, 1979), are important because they suggest that oil droplets increase the number of spectral channels. In any case, oil droplets clearly are not necessary for color vision because Japanese quail maintained on a carotenoid-free diet to render their oil droplets colorless still exhibit color vision, although of a somewhat altered sort (Wallman, 1979). An interesting recent suggestion is that, by narrowing the spectral range of cones, oil droplets may enhance their spectral resolution (Baylor and Hodgkin, 1973; Barlow, 1982).

c. Ellipsoids

An ellipsoid fills the outer third of all photoreceptors, except the space occupied by oil droplets (i.e., in photoreceptors that contain oil droplets). It is composed of massed mitochondria (Figs. 1.2 and 1.12) and is presumed to play a role in photoreceptor energetics, although its exact role is not known. It is also probable that the ellipsoid of some reptilian photoreceptors is refractile and contributes to the light-collecting properties of the inner segment (Winston and Enoch,

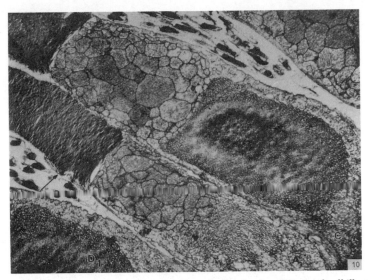

Fig. 1.12. *Coleonyx variegatus.* Electron micrograph of a class D2 visual cell illustrating the mitochondrial composition of its ellipsoids. In the larger, principal cell, mitochondria are enlarged in the center of the ellipsoid. The paraboloid has a granular inner core and a membranous perimeter. An arrow indicates a calyceal process extending from the inner segment of the accessory cell. The ellipsoid of this cell is less differentiated than that of the principal cell, and it has a membranous paraboloid (7340 x). (From Dunn, 1966a.)

1971; Tansley and Johnson, 1956). Recent evidence suggests that the ellipsoid of most cones in the snake, *Thamnophis sirtalis*, contains numerous, small microdroplets packed among the mitochondrial cristae (Fig. 1.13; Bossomaier et al., 1989; Wong, 1989). They are probably lipid droplets that increase the refractive index of the inner segment and thus increase the light-capturing capacity of the photoreceptor. This may be especially important for vertebrates with eyes of low f-number, such as *Thamnophis* (Bossomaier et al., 1989). Interestingly, this central mass of specialized mitochondria is enlarged relative to the outer segment, yielding the highly tapered photoreceptor contour characteristic of an ideal light collector (Winston and Enoch, 1971). When the retina is mounted with the photoreceptor layer up and transilluminated, the outer segments appear as an array of bright spots in an otherwise dimly illuminated surrounding. This suggests that the ellipsoid (snake cones have no oil droplets or paraboloids) must concentrate light on the outer segments, as do the oil droplets of turtles (Tansley and Johnson, 1956).

In the few turtles and crocodilians for which data exist, the ellipsoid consists of densely packed, agranular mitochondria with little internal structure beyond the tendency for central mitochondria to be rel-

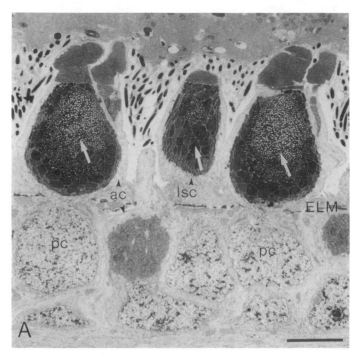

Fig. 1.13. *Thamnophis sirtalis.* Electron micrograph illustrating the accumulation of numerous "microdroplets" in the ellipsoid of large single cones (lsc) and the principal member of double cones (pc). Accessory cones (ac) lack microdroplets, but the light path to their outer segments is shadowed by the enlarged ellipsoid of the principal cell. ELM, external limiting membrane. Scale: 5 μm. (From Wong, 1989.)

atively large in some visual cells of turtles (*Pseudemys (Trachemys)*, Bush and Lunger, 1972; *Mauremys (Clemmys) japonica*, Maekawa and Mizuno, 1976). In *Alligator* (Kalberer and Pedler, 1963), the mitochondria are long and arranged parallel to the photoreceptor long axis, as in mammalian rods. In diurnal lizards (including diurnal geckonids; Pedler and Tansley, 1963; Tansley, 1964), the ellipsoid of most photoreceptor types resembles that of *Pseudemys (Trachemys)* but with a more pronounced central-peripheral or scleral-vitreal (*Chamaeleo*, Armengol et al., 1981; *Phelsuma*, Pedler and Tansley, 1963) gradient of mitochondrial size. Thus, the size and internal structure of the ellipsoid varies substantially, both across species and across visual cell types within a single group, and it is frequently used as a criterion in photoreceptor classification.

One of the more interesting features of the ellipsoid is its relation to photoreceptor oil droplets, which, as noted above, typically lie at the distal end of the inner segment, closely surrounded by a cup of ellipsoidal mitochondria (Fig. 1.2). Several authors have suggested that this spatial relation may reflect a more fundamental developmental relation, namely, that oil droplets are actually radically modified mitochondria (Berger, 1966; Ishikawa and Yamada, 1969; Armengol et al., 1981). The argument runs as follows. In many species, ellipsoidal mitochondria are not structurally uniform but instead exhibit central-peripheral or vitreal-scleral gradients such as those described in the preceding paragraph. In snakes and nocturnal geckos, groups that generally lack oil droplets, this inhomogeneity is especially pronounced. For example, snake single cones have central ellipsoidal mitochondria that are enlarged and contain regular arrays of osmiophilic granules that are Sudan black B positive, suggesting that they are composed of lipids (Ishikawa and Yamada, 1969; these osmiophilic granules probably correspond to the microdroplets of Wong, 1989). In the accessory member of the Class B double cell of nocturnal geckos (*Gekko japonicus*, Ishikawa and Yamada, 1969; *Gekko gecko*, Pedler and Tilly, 1964) the central ellipsoidal mitochondria have double membranes and rudimentary cristae, but they are strikingly swollen with a homogenous matrix similar to that of oil droplets (Fig. 1.14). As in snake ellipsoids, there are transitional forms between these central,

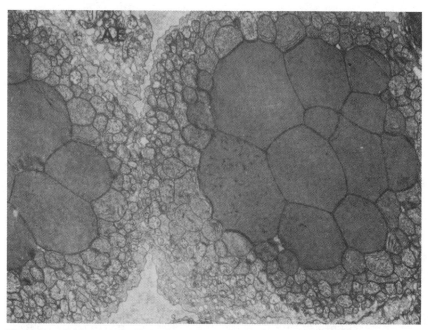

Fig. 1.14. *Gekko japonicus.* Electron micrograph illustrating enlarged central mitochondria of ellipsoids in a Class B double cell. Note that the homogeneous matrix of the central mitochondria resembles that of oil droplets. (From Ishikawa and Yamada, 1969.)

modified mitochondria and the entirely typical mitochondria at the ellipsoidal periphery. A similar pattern has been described in carp (Ishikawa and Yamada, 1969), guppies (Berger, 1966), and frogs (Craig et al., 1963).

One inference (e.g., Berger, 1966; Armengol et al., 1981) is that the central, enlarged mitochondria represent stages in the development of oil droplets and that classic oil droplets of the kind seen in turtles, lizards, and birds represent a further modification or transmutation of those central, specialized mitochondria. In contrast, Ishikawa and Yamada (1969) have argued on morphological and developmental grounds that there are two fundamentally different types of oil droplets. Type I droplets, found in turtles, diurnal lizards (including diurnal geckos), and birds, are membraneless, have a "homogeneous" internal structure, and are present in early neonatal retinas. They are believed to arise by secretion from the ellipsoidal mitochondria (Pedler and Tansley, 1963). Type II droplets, found in nocturnal geckos, snakes, frogs, carp, and guppies, have a limiting membrane, a vesicular or granular internal structure, and are derived after hatching by transmutation of mitochondria at the distal end of the ellipsoid. The value of this distinction is difficult to assess in view of the evident confusion about whether oil droplets are membrane bound (see previous section) and the conflicting use of the terms *homogeneous* and *granular* in the literature. Furthermore, the internal structure of the mitochondria may vary within a single species depending on whether the retina is light or dark adapted (Kunz and Wise, 1978). Thus, it is not clear whether there are distinct categories of oil droplets or whether the different morphologies represent evolutionary or metabolic variants of a single type (see discussion of ellipsosomes in Borwein, 1981, p. 46). The important point seems to be that highly modified ellipsoids occur in species that lack classic oil droplets, suggesting that ellipsoids play a role within the photoreceptor analogous to that of oil droplets.

d. PARABOLOIDS

At least some visual cells of all reptiles except snakes have a paraboloid (Dunn, 1973). It is composed of presumed glycogen granules and a tubular system of membranes that are continuous with the rough endoplasmic reticulum of the myoid region. Photoreceptors vary in the relative proportions and distribution of the two components. At one extreme is the membranous type with a rich system of tubules interspersed with granules; at the other is the granular type in which granules are sequestered into a well-defined central core (Fig. 1.15; Dunn, 1966a; Anh and Anh, 1970; Yamada, 1960a, 1960b; reviewed in Dunn, 1973).

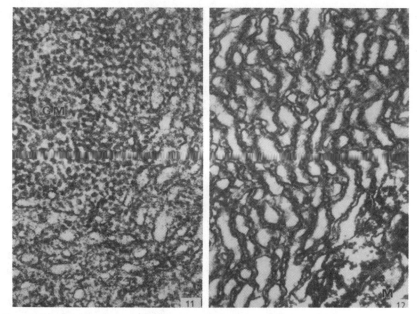

Fig. 1.15. *Coleonyx variegatus*. Electron micrographs illustrating differences in the internal structure of paraboloids. Left: Granular type paraboloid as seen in a vertical section through a single visual cell. The central granular mass (GM) is surrounded by membranous profiles (49,350 x). Right: Membranous-type paraboloid of D2 double cell. A portion of the myoid region (M) can be seen at the bottom of this longitudinal section (52,640 x). (From Dunn, 1966a.)

Some form of paraboloid probably occurs in all single visual cells and in at least one member of all paired photoreceptors (excluding those of snakes; Yamada et al., 1966). Its glycogen content and continuity with the cytoplasmic rough endoplasmic reticulum suggest a role in cellular metabolism and macromolecular transport. Recently, a monoclonal antibody that is specific for a protein in the paraboloid of *Pseudemys (Trachemys) scripta* has been described (Gaur et al., 1989). The molecular weight of the antigen (40 KDa) resembles that of glycogenin, a protein involved in glycogen synthesis. This is consistent with reports that photoreceptor paraboloids are involved in glycogenesis as well as glycogen storage (see Gaur et al., 1989).

The distribution of paraboloids in paired photoreceptors suggests that this organelle may also play a role in light collection. Specifically, a paraboloid is present in both members of all paired cells in which the two members are of equal diameter or nearly so (*Gekko* D2 twin and triplet cells; Dunn, 1966a). It is, in fact, only absent in the chief member of double cells (*Gekko* Type B or D1 cells and in double cones in turtles and diurnal lizards) in which the very large paraboloid of the accessory cell shadows the light path to the outer segments of both members (Figs. 1.3 and 1.16). Thus, a paraboloid lies in the light path of virtually all reptilian photoreceptors except those of snakes.

e. MYOIDS: PHOTOMECHANICAL CHANGES AND TIERED RETINAS

The myoid region of each photoreceptor lies just distal to the outer (external) limiting membrane (see Fig. 1.2). It contains free ribosomes, granular and agranular reticulum, and Golgi apparatus (e.g., Bush and Lunger, 1972; Schaeffer and Raviola, 1978; Kalberer and Pedler, 1963; Pedler and Tansley, 1963; Kolb and Jones, 1982; Terada et al., 1983) and is thought to be involved in the production, transport, and secretion of membrane-bound proteins such as those used in outer segment renewal (Fliesler and Anderson, 1983; Besharse, 1986). The cell membrane in this region often forms lateral extensions, or "fins," that project radially from the perimeter of the myoid and interdigitate with processes of Müller cells (Fig. 1.17; see, e.g., Carasso, 1956; Anh, 1968; Dunn, 1973; Owen, 1985). Interestingly, ouabain binding studies in *Pseudemys (Trachemys) scripta* have localized the cone sodium-potassium pump to this portion of the inner segment (Stirling and Sarthy, 1985). The density of putative pump sites is significantly higher here than elsewhere in the retina, perhaps because this region must sustain the large photoreceptor dark current.

The myoid may also play a significant, if indirect, role in image formation. Its shape, which can vary from long and narrow to short and stout, helps determine the intraretinal position of the light-absorbing outer segment. This, in turn, has important consequences for the optical characteristics of the photoreceptor layer as a whole and the light-collecting ability of different visual cell types. Myoid length can vary in two ways. In some species of reptiles, photoreceptors have no known photomechanical capabilities, but the myoids of various visual cell types differ in length, yielding a tiered photoreceptor arrangement (Fig. 1.18). This is evident in drawings and photographs of visual cells from several reptilian taxa, but is best documented in certain snakes (Miller and Snyder, 1977; see later discussion of *Leptodeira annulata*). Underwood has used this tiering as a diagnostic feature for certain ophidian visual cell patterns (Underwood, 1967a, 1967b, 1970).

In other species of reptiles, myoid shape varies as a function of light level: cones shorten in the light and elongate in the dark, whereas rods undergo opposite configurational changes. Thus, both cones and rods are shortest under light conditions for which they are maximally sensitive. This will, of course, change their space/time constants and, thus, the electrotonic conduction properties of the photoreceptors (Rodieck, 1973). It will also place the proximal elements first in the light path and so increase the probability of their intercepting

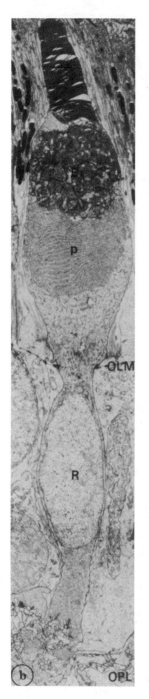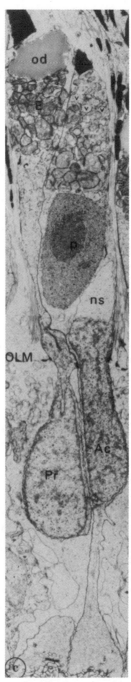

Fig. 1.16. *Pseudemys (Trachemys) scripta.* Electron micrographs illustrating the internal structure of a rod (R; left) and a double cone (right). Compare with Fig. 1.3. The rod lacks and oil droplet. It has a thick outer segment, a greatly enlarged inner segment, and a thick axon. The accessory member of the double cone (ac) also lacks an oil droplet. od, oil droplet; E, ellipsoid; v, small, pigmented vesicles in ellipsoid of accessory cell; p, paraboloid; ns, nuclear sac, an inclusion that is sometimes observed in cone inner segments and may be an artifact; Pr, principal member of double cell; OLM, outer limiting membrane (4116 x). (From Kolb and Jones, 1982.)

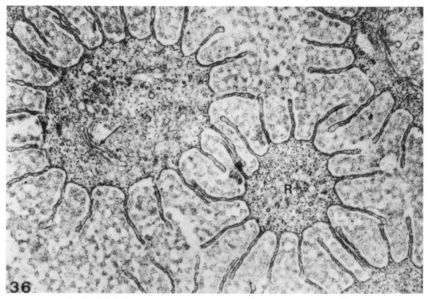

Fig. 1.17. *Sceloporus occidentalis.* Electron micrograph of a vertical section through the myoid region of a cone and a rod (R). Cytoplasmic fins project radially from the inner segment and interdigitate with the processes of Müller cells. The arrow indicates one of the membrane complexes that are common in this region (compare with Figs. 1.2 and 1.3) (11,844 x). (From Dunn, 1973.)

incident photons (see below). Such photomechanical changes have been reported in turtle (Detwiler, 1916) and lizard (Detwiler, 1923a) cones and in rods and cones of *Alligator mississippiensis* (Laurens and Detwiler, 1921; see Detwiler, 1943, for review). However, the amplitudes of these changes are rather small compared with those seen in some birds, amphibians and, especially, teleosts (Ali, 1974; Rodieck, 1973; Burnside, 1988). In teleosts, some photomechanical changes (e.g., cone contraction) are known to be mediated by filamentous (F-) actin (reviewed in Burnside and Dearry, 1986; Burnside, 1988), and appropriately oriented F-actin has been localized to the myoid region of several other vertebrates, including *Sceloporus* (Vaughan and Fisher, 1987). These longitudinal actin bundles are better developed in species with clear retinomotor capacities (turtle, frog, fish) than in *Gekko*, mouse, and rat, which show few photomechanical changes (Drenckhahn and Wagner, 1985). Myoid length is known to be modulated by dopamine in at least some vertebrates (Daw et al., 1989).

The immediate consequence of both tiering and photomechanical changes is that two different photoreceptor types each come to occupy almost the entire light capture area, but at different levels. In the snake *Leptodeira annulata*, for example, rods and cones form 80% and 20% of the photoreceptor population respectively (Miller and Snyder,

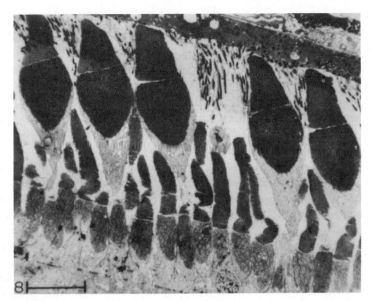

Fig. 1.18. *Leptodeira annulata*. Tiering in the photoreceptor layer. An array of numerous, small-diameter rods occupies the light path in the proximal part of the photoreceptor layer. Light that passes through the layer of rods is captured by an array of enlarged cone ellipsoids that are presumed to act as light guides, funneling light to the cone outer segments. Scale: 10 μm. (Reprinted with permission from *Vision Research*, Vol. 17, Miller and Snyder, The tiered vertebrate retina, 1977, Pergamon Press plc.)

1977). In the first tier (Fig. 1.18) rods occupy approximately 80% of the light capture area. As their outer segments end, the cone ellipsoids (which act as optical guides and so define the light collection area of the cones; Enoch and Tobey, 1981; Bossomaier et al., 1989) expand to occupy far more of the photoreceptor layer than their low relative numbers would suggest. Thus, light that is not absorbed by the outer segments of the proximal cells is intercepted by the sheet of cone ellipsoids. Furthermore, optical experiments indicate that unabsorbed light from the first tier is not collimated by the outer segments of the proximal cells and beamed toward the pigment epithelium; instead, it spreads out from the tips of the outer segments into the distal layer of cones, where it is captured by their ellipsoids and outer segments (Miller and Snyder, 1977). Thus, the layered arrangement and optical properties of photoreceptors in *Leptodeira* collectively produce a condition in which both photoreceptor levels intercept (and potentially absorb) all the available light.

4. NUCLEUS AND CONNECTING FIBER
Photoreceptor nuclei lie immediately subjacent to the external limiting membrane and collectively form the outer nuclear layer (see Fig. 1.1). Typically, the nucleus contains a nucleolus and an extended net-

work of chromatin aggregates (see Figs. 1.3 and 1.16; Dunn, 1973; Bush and Lunger, 1972; Kolb and Jones, 1982; Pedler and Tilly, 1964; Hibbard and Lavergne, 1972), which gives these nuclei a reticulated appearance under the light microscope (see, for example, drawings of Walls, 1942a, 1942b). In the cones of *Pseudemys (Trachemys)*, the nucleus is surrounded by a very thin rim of cytoplasm so that it sits at the vitreal end of the cell body "like a ball into a chalice or even a cork in a bottle," a condition of potential significance for cytoplasmic resistivity and, thus, information flow in these visual cells (Lasater et al., 1989). In snakes, the accessory member of the Class B double cell bears a cluster of mitochondria (Yamada et al., 1966)—the paranuclear body (Walls, 1942a)—between the cell nucleus and the outer limiting membrane. Thus, the accessory nucleus is displaced vitread relative to that of the chief cell. A similar "tiering" of photoreceptor nuclei has been reported in several species (e.g., Vilter, 1951a; Kalberer and Pedler, 1963; Underwood, 1966). Although the significance of this layering is not known, it appears that nuclear position varies with photoreceptor type. For example (see Fig. 1.3), rod nuclei lie most vitread in the outer plexiform layer in turtles (Odashima et al., 1979; Schwartz, 1973; Kolb and Jones, 1982), and the nucleus of the accessory cell is more distal than that of the chief cell in double cones (Walls, 1942a; Kolb and Jones, 1982).

The nucleus is connected to the synaptic pedicle by an axon that varies in thickness and orientation (Figs. 1.3, 1.16, and 1.20). In photoreceptors with low-lying nuclei, for example, rods of turtles (e.g., Kolb and Jones, 1982; Schaeffer and Raviola, 1978; Lasansky, 1971; Owen, 1985) and certain visual cells of alligators (Kalberer and Pedler, 1963), axons are short and thick, whereas they are slender and long in most cone-type visual cells (e.g., Hibbard and Lavergne, 1972; Kolb and Jones, 1982; Cajal, 1893). For example, axons of turtle cones are 1 to 1.5 μm thick and 3 to 60 μm long (Kolb and Jones, 1987; Lasater et al., 1989). For single cones, axons are longer in the visual streak (see Section IV.C) than in the retinal periphery, and in all retinal regions axon length varies systematically with cone spectral sensitivity. Indeed, differences in axon length between cones with clear oil droplets have been used to support the position that the larger clear (fluorescent) droplets are found in blue- rather than red-sensitive cones (Kolb and Jones, 1987; see Section II.B.3.b). These differences in axon caliber can be expected to affect the space-time constants of the photoreceptors, but their precise consequences are not known. Recent patch clamp experiments (Lasater et al., 1989) indicate that the relatively short axons of double cones and red and green single cones do not electrically isolate the synaptic pedicles from their respective somata.

In lizards with foveae, the axons of foveal or parafoveal cones run

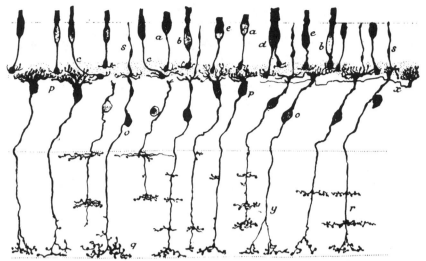

Fig. 1.19. *Lacerta viridis.* Cones and bipolar cells visualized by Golgi impregnation. a, Cone cell body situated near the outer limiting membrane; b, more elongated cell body situated in the middle of the outer nuclear layer; c, oblique cone fiber; d, twin cones; e, cone cell body displaying clear transverse bands; o, fine or inner bipolar cells; p, larger or outer bipolar cells; q, terminal arborization situated above the ganglion cells; r, descendant process of a bipolar cell, giving rise to collateral arborizations; s, Landolt's club. Original drawings and descriptions of cells are from Cajal, 1893. (From *The Vertebrate retina: principles of structure and function.* By R. W. Rodieck. Copyright (c) 1973 by W. H. Freeman and Company. Reprinted with permission of W. H. Freeman and Company.)

obliquely for long distances (up to 500 μm) before reaching their pedicles (*Anolis carolinensis,* Makaretz and Levine, 1980; *Chamaeleo,* Armengol et al., 1988; see Fig. 1.36). Presumably this reflects the greater packing constraints in more vitreal layers of the retina. Outside these specialized retinal regions, most axons are oriented parallel to the photoreceptor axis, but in putative blue single cones and putative ultraviolet-sensitive cones of *Pseudemys (Trachemys) scripta* (Leeper, 1978b; Kolb and Jones, 1982, 1987; Lasater et al., 1989) and in *Sphenodon* "fine cones" (Vilter, 1951a), axons run obliquely to their pedicles. Such oblique cones have also been described in *Lacerta viridis* (Fig. 1.19; Cajal, 1893). It is not known whether the length or orientation of the oblique connecting fibers vary systematically with retinal locus.

5. Synaptic Pedicles: Structural Organization and Plasticity

a. General   Synaptic pedicles present a number of features that may shed light on photoreceptor function. First, differences in pedicle morphology and laminar position provide a marker for functional classes of visual cells at both light microscopic and ultrastructural levels. Second, pedicles often contact other retinal elements at morphologically distinct junctions and so provide important information about outer plexiform layer circuitry. Finally, pedicle membranes are labile and can be used to study mechanisms of synaptic vesicle formation and release. Virtually all the available data on synaptic pedicles come from work on turtles.

b. Pedicle Structure and Lamination

In turtles, cone- and rod-type cells are easy to distinguish; they differ, *inter alia*, in the structure of their pedicles. In radial sections, the cone pedicle erupts from a narrow connecting fiber and its synaptic surface is wide (5 to 8 μm), usually as wide as the inner segment, which gives it a characteristic inverted bowl appearance (Figs. 1.3 and 1.20). It may bear five to seven 10- to 25-μm basal processes that radiate hori-

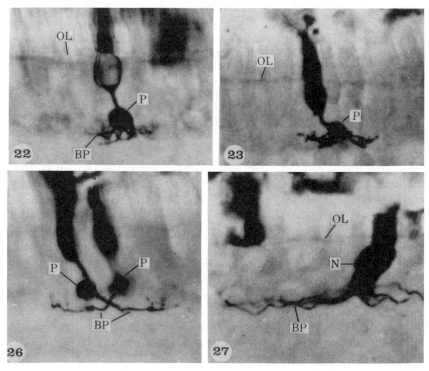

Fig. 1.20. *Pseudemys (Trachemys) scripta.* Structural variation in pedicles and connecting fibers (axons) of photoreceptors. Upper: Cone photoreceptors have a narrow axon and a wide pedicle (P) that bears numerous, short basal processes (BP; i.e., telodendria). Lower left: Some cones have a single, long basal process (BP) that terminates in a spray of short telodendria. Lower right: A rod has a thick axon that extends from the nucleus (N) to a pedicle that bears many long telodendria (BP). OL, Outer limiting membrane (800 x). (From Lasansky, 1971.)

zontally in all directions and terminate in minute swellings or a single, longer (30 to 40 μm) process with a terminal cluster of teloden-dria (Fig. 1.20; Lasansky, 1971; Kolb and Jones, 1985). Such basal pro-cesses have been described in representatives of all vertebrate orders, including reptiles (e.g., *Alligator*, Kalberer and Pedler, 1963; *Lacerta*, Cajal, 1893; *Chamaeleo*, Armengol et al., 1988; *Pseudemys (Trachemys)*, Lasansky, 1971; *Chelydra*, Owen, 1985). The structure of basal pro-cesses varies with retinal locus; in central retina they are shorter, more branched, and end in smaller terminal varicosities than in the periphery (Armengol et al., 1988). The pedicle itself bears vacuoles, cisternae, agranular reticulum, microtubules, and synaptic com-plexes (Figs. 1.21, 1.22, 1.24, and 1.27). These complexes consist of (1) a synaptic ribbon or lamella that may have one to five visible layers, depending on the species and fixation conditions (Dunn, 1973); (2) an arciform density, which is a puff of flocculent material of unknown function between the ribbon and the adjacent synaptic membrane (Lasansky, 1971); and (3) coated and smooth vesicles, the latter often clustered around the synaptic ribbon. In *Pseudemys (Trachemys)* there are approximately 12 such complexes per cone pedicle (Lasansky, 1972).

In rods, the pedicle is relatively narrow (Figs. 1.16, 1.20), and its 5 to 10 or more basal processes are longer than those of most cones (35 to 40 μm), with both *en passant* and terminal swellings (Copenhagen and Owen, 1976a, 1976b; Ohtsuka, 1978; Mariani and Lasansky, 1984; Owen, 1985). Perhaps because of its narrower synaptic surface, rod pedicles appear less complex; they have fewer invaginating processes and ribbon complexes than do cones (Dunn, 1973; Lasansky, 1971; Odashima et al., 1979). Furthermore, synaptic vesicles are somewhat larger (approximately 65 nm versus 58 to 59 nm) and more numerous in rods than in cones (Bush and Lunger, 1972; Odashima et al., 1979; Lasansky, 1971; Kolb and Jones, 1982).

Thus, in turtles, rod- and cone-type cells differ in pedicle structure. Most other species of reptiles for which we have data exhibit simplex retinas. Not surprisingly, pedicle structure shows little variation within such species. All photoreceptors in these species, including the "rods" of nocturnal geckonids, bear complex pedicles similar to those of turtle cones (Tansley, 1964; Pedler and Tilly, 1964; Pedler, 1965). Two known exceptions are *Alligator mississippiensis* and the liz-ard *Tiliqua rugosa*, both with duplex retinas (Walls, 1942a; Laurens and Detwiler, 1921; Kalberer and Pedler, 1963; Braekevelt, 1989) that bear two types of synaptic pedicles (Kalberer and Pedler, 1963; Braekevelt, 1989): a "simple" variety, usually with one very long synaptic ribbon (compare Fig. 44 in Dunn, 1973), and a wide "complex" pedicle with one to three ribbon synapses, numerous invaginating processes, and

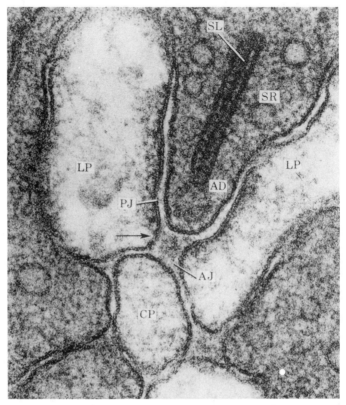

Fig. 1.21. *Pseudemys (Trachemys) scripta.* Electron micrograph illustrating invaginated processes that form a cone triad. The cytoplasm of the cone pedicle is dark and bears vesicles, a synaptic lamella (SL), and an arciform density (AD). The pedicle forms a synaptic ridge (SR) that is flanked on both sides by lateral processes (LP), probably from horizontal cells. The lateral processes bear prominent membrane densities opposite the synaptic ridge, suggesting that these proximal junctions (PJ) may be sites of synaptic contact. Opposite the apex of the ridge, a central process (CP) forms an apical junction (AJ) with the cone pedicle. The central process probably belongs to a bipolar cell, but no membrane specializations suggestive of a synaptic contact have been identified at these apical junctions (157,000 x). (From Lasansky, 1971.)

basal filaments that contact other pedicles (see next section). The larger pedicles with multiple synaptic sites are characteristic of cones in *Tiliqua*. Thus, rod and cone visual cells probably differ in pedicle structure throughout the Reptilia, much as they do in turtles; conversely, variation in pedicle structure within a single retina is correlated with the presence of both rod- and cone-type visual cells.

In addition to structural variation, synaptic pedicles may also vary in their laminar distribution as a function of photoreceptor type. In *Lacerta viridis* (Cajal, 1893), the pedicles form two layers: those of the oblique cones and the chief member of double cones lie more proximal than those of other visual cells (see Fig. 1.19). In *Pseudemys (Trachemys) scripta*, pedicles are segregated into three laminae according to photoreceptor type (see Fig. 1.3): pedicles of oblique cones are most vitreal; those of red and green single cones occupy the middle of the outer plexiform layer, and rod plus double cone pedicles are outermost (Kolb and Jones, 1982). As noted by Cajal, this suggests that different classes of receptors may contact separate populations of bipolar cells.

c. Pedicle Junctions

Photoreceptor pedicles contact other retinal elements through several morphologically distinct junctions. These contacts can be grouped into three broad types based on the identity of the postsynaptic elements.

1. *Contacts between photoreceptors and bipolar cells* occur in two forms: invaginating and noninvaginating (basal or distal) junctions. In some invaginating junctions (Fig. 1.21), the central process of a triad almost contacts the apex of a synaptic ridge but remains separated from it by a 30- to 80-nm cleft containing electron opaque material. Such invaginating junctions are sometimes called ribbon synapses because of their proximity to the photoreceptor ribbon. The postjunctional membrane and cytoplasm at these contacts are unspecialized; there are no membrane densities, vesicles, or intramembrane particles (Lasansky, 1969, 1971, 1972, 1974, 1977; Schaeffer et al., 1982; Kolb and Jones, 1984). Furthermore, the prejunctional membrane shows no features that are specifically related to the apical process, that is, the P face particles[4] at the synaptic ridge apex also occur in dyads, suggesting that they are not specifically related to the invaginating central process. Thus, the apical contact shows no evidence of an active zone either in thin sections or in freeze-fracture material, and the functional significance of this contact is unclear. Kolb et al. (1986) have also described a form of "semiinvaginating" photoreceptor to bipolar contact in which the invaginating process approaches the apical ridge but remains separated from it by lateral processes (see Fig. 1.24). These junctions, like the ribbon junctions described previously, bear no obvious synaptic specializations. They are formed by processes that make "distal" synaptic contacts with photoreceptor pedicles (see below), and the authors suggest that bipolar cells are postsynaptic to photoreceptors at the distal, and not the semiinvaginating, contact.

Noninvaginating basal junctions are more variable, and the classification schemes used by different authors sometimes do not correspond. Basal junctions (Fig. 1.22) may be symmetrical or asymmetrical (i.e., with little or no postjunctional membrane thickening;

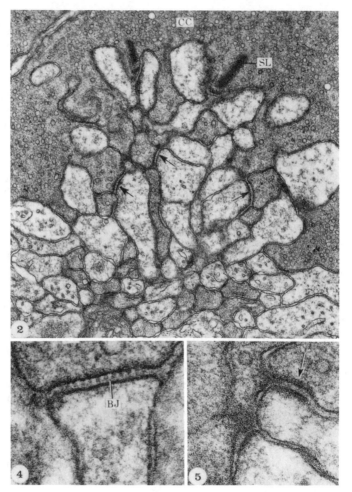

Fig. 1.22. *Pseudemys (Trachemys) scripta.* Basal junctions made by cone pedicles. Upper: Low-power electron micrograph showing contacts of numerous processes with a cone pedicle (CC). Some processes invaginate the cone pedicle; others form basal junctions (arrows). SL, synaptic lamella (32,000 x). Lower left: High-power micrograph of a basal junction (BJ) showing the wide synaptic cleft and regular vertical striations that characterize these contacts (120,000 x). Lower right: These striations are sometimes seen within the cytoplasm of the cone pedicle (arrow) (108,000 x). (From Lasansky, 1971.)

Schaeffer et al., 1982) and slightly invaginating or superficial (Lasansky, 1971). The width of the intercellular cleft can vary substantially, often within the same junction, yielding "wide-gap" or "narrow-gap" contacts (Lasansky, 1971). A simplifying suggestion (Kolb and Jones, 1984) is that there is a single type of wide-gap basal junction (between photoreceptors and bipolar cells; see Section III.B.2) with a 30- to 40-

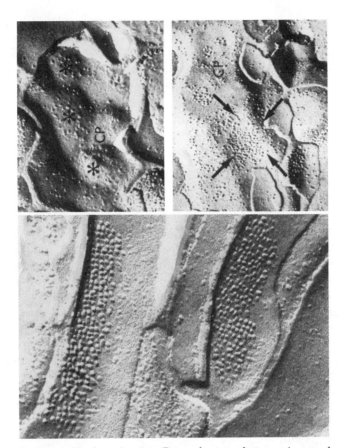

Fig. 1.23. *Pseudemys (Trachemys) scripta.* Freeze-fracture electron micrographs of basal junctions. Upper left: The E face of a cone pedicle (CP) bearing three basal junctions (asterisks). Membrane particles are randomly distributed at these contacts (67,200 x). Upper right: The P face of a cone pedicle (CP) bearing a basal junction (arrows). The indentation is characterized by an accumulation of particles, suggesting that this region of the pedicle membrane may be specialized (61,320 x). Lower: E face particles in the membranes of bipolar cell dendrites that form basal junctions with a cone pedicle. The particles are arranged in a square lattice (110,880 x). (From Schaeffer et al., 1982.)

nm cleft and somewhat asymmetrical membrane densities. In thin sections, the cleft of these basal junctions exhibits electron-dense material that often forms periodic striations (Fig. 1.22); viewed *en face*, this material may be arranged in an orthogonal lattice. Thus, these contacts are called wide-gap striated junctions (Kolb and Jones, 1984). In freeze-fracture material the pedicle P face bears clusters of particles but no vesicle sites (Fig. 1.23). This is consistent with observations that in thin sections vesicles do not appear to cluster around basal junctions and in fact may be more scarce there than elsewhere in the pedicle (Lasansky, 1971). On the postjunctional (bipolar cell) E face there are aggregations of 10-nm particles, often arranged in a square lattice array reminiscent of the cleft material seen in thin sections (Fig. 1.23; Schaeffer et al., 1982). Thus, the freeze-fracture appearance of basal junctions resembles that of central excitatory synapses, whereas the unspecialized membrane of apical processes described above is similar to that of the postsynaptic element in a central inhibitory contact (Landis et al., 1974; Landis and Reese, 1974). These issues become important in attempts to understand the structural basis of sign-preserving and sign-inverting synapses between photoreceptors and bipolar cells (see Section III.B.2).

Noninvaginating distal junctions resemble wide-gap basal junctions (contacts 1 and 5, in Fig. 1 of Lasansky, 1971), but the synaptic clefts are narrower and the presynaptic membrane is characterized by puncta of dense material (Fig. 1.24; Kolb and Jones, 1984). Thus, distal junctions have been called *narrow-cleft punctate basal junctions* (Kolb and Jones, 1984). The primary difference between basal and distal junctions (of Lasansky, 1971; wide-cleft striated basal junctions and narrow-cleft punctate basal junctions, respectively, of Kolb and Jones, 1984) seems to be that distal junctions (but not basal junctions) are made by bipolar dendrites that *also* invaginate the photoreceptor pedicles (e.g., to form the semiinvaginating contacts described above; Kolb et al., 1986). This invaginating process may or may not become the central element of a triad (Fig. 1.24). Thus, there are two or more classes of bipolar cells: one that only makes wide gap basal junctions and one or more that invaginates the receptor pedicles and makes distal synaptic contacts *en route* (Lasansky, 1971; Kolb and Jones, 1984; Kolb et al., 1986).

2. *Contacts between photoreceptors and horizontal cells* occur either as invaginated (ribbon) or distal junctions. In the former, two horizontal cell processes lie on either side of the synaptic ridge, separated from it by a 15-nm cleft (Figs. 1.21 and 1.22). Such a configuration is called a dyad or, whenever there is an additional central process, a triad. In both cases, the lateral processes that form ribbon junctions derive from horizontal cells (Lasansky, 1971; Kolb and Jones, 1984). For example, some of these lateral processes have been traced to H1 horizontal cell somata or axon terminals (Kolb and Jones, 1984; see Section II.C.1 for types of horizontal cells). At invaginated junctions, the horizontal cell membrane is markedly thickened, but the pedicle membrane remains unspecialized; thus, the contact is asymmetrical. Because the photoreceptor pedicle contains synaptic vesicles and the horizontal cell process often does not, the presumption has been that the photoreceptors are presynaptic to horizontal cells at these contacts. This is consistent with freeze fracture data described in the following paragraph. The site of horizontal cell feedback to photoreceptors is less clear (see following paragraph and Section II.C.2.b).

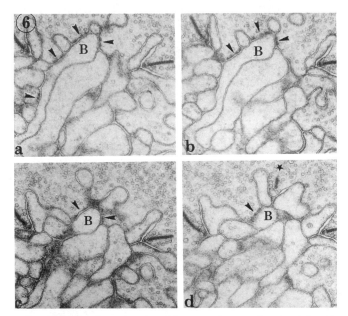

Fig. 1.24. *Pseudemys (Trachemys) scripta.* Electron micrographs of serial sections through a bipolar cell dendrite (B) that makes numerous contacts with a cone pedicle. In the first three sections, the process makes narrow-cleft, punctate junctions (arrowheads) with the cone pedicle. In the last section, the same bipolar cell process becomes the central element in a triad (star). Note that the bipolar cell process approaches the apex of the synaptic ridge but remains separated from it by the adjacent lateral processes (21,000 x). (From Kolb and Jones, 1984, by permission of the publishers, Chapman and Hall.)

Ribbon junctions of both dyads and triads exhibit an array of two to three hundred 9- to 11-nm P face particles within the pedicle membranes, along the apex of the synaptic ridge (Schaeffer et al., 1982). On the ridge slope, opposite the horizontal cell processes, there are 10-30-nm vesicle sites associated with the E face. These vesicle sites often occur in linear arrays and appear to correspond to the rows of vesicles seen along the synaptic ribbon in appropriately sectioned material. Further lateral (away from the apex) there may be 40-60-nm protrusions suggesting the formation of coated vesicles. Each of these intramembranous specializations occurs in both dyads and triads, suggesting that none are specifically related to the central (apical) contact made by bipolar cell processes. The postjunctional (horizontal cell) membrane of the invaginating junction shows a cluster of 9-11-nm particles opposite the apex of the synaptic ridge, but there are no E face specializations except for an occasional, and apparently random, small cluster of particles. Thus, the horizontal cell membrane resembles the postsynaptic element of an excitatory peripheral (cholinergic) synapse (Schaeffer et al., 1982).

Photoreceptors also contact horizontal cells at distal junctions (contact 3 in Fig. 1 of Lasansky, 1971), similar to those between photoreceptors and the central process of a triad (contact 5). Lasansky (1972) suggested that, at these junctions, horizontal cells may be presynaptic to cone pedicles but later (1973) denied this on the grounds that, in the salamander *Ambystoma*, some horizontal cells engaging in distal junctions lack any synaptic vesicles. More recent work (Kolb and Jones, 1984) indicates that photoreceptor telodendria are the presynaptic elements at these "distal junctions of lateral (horizontal cell) processes." No turtle distal junctions have been identified in freeze-fracture material (Schaeffer et al., 1982). Indeed, there are no known intramembranous specializations characteristic of presynaptic active sites in horizontal cells, either at distal junctions or elsewhere (See Section II.C.2.b).

3. *Interreceptor contacts* can occur between adjacent pedicles or through basal processes. They can be chemical or gap junctions. Gap junctions between pedicles (Fig. 1.25) typically include a wide-cleft segment characterized by a 20-nm gap containing moderately opaque material and symmetrical thickenings of the opposed membranes. Immediately adjacent to the wide-gap segment is a focal region of close (2 nm) membrane apposition (Lasansky, 1971). Such "punctate" gap junctions are especially frequent in the visual streak of *Pseudemys (Trachemys)*; here they appear to link neighboring photoreceptors irrespective of spectral type (Kolb and Jones, 1984, 1985; see Section III.C.2.b). In freeze-fracture material these latter, narrow-gap, contacts are characterized by rows or circular arrays of 9- to 10-nm P face particles on an almost particle-free elevation (Schaeffer et al., 1982). Similar freeze-fracture configurations have been observed at interreceptor contacts in mammals, in which they correspond to regions of close membrane apposition seen in thin sections and are continuous with hexagonal arrays of P face particles characteristic of vertebrate gap junctions (Raviola and Gilula, 1973). The adjacent wide-gap segments may simply play a mechanical ("anchoring") role, since they exhibit no vesicle accumulations or intramembrane particles characteristic of an active zone. Wide and punctate, narrow-gap contacts between adjacent photoreceptor pedicles are rare or absent in *Chelydra* retina (Owen, 1985).

More extensive gap junctions have been documented between photoreceptor telodendria (unidentified as to spectral type) in peripheral retina of *Pseudemys (Trachemys) scripta* (Schaeffer et al., 1982; Kolb and Jones, 1984, 1985) and in several other testudines (Fig. 1.25; Davydova, 1983; Owen, 1985). They resemble the junctions between basal processes of photoreceptors in monkey peripheral retina (Raviola and

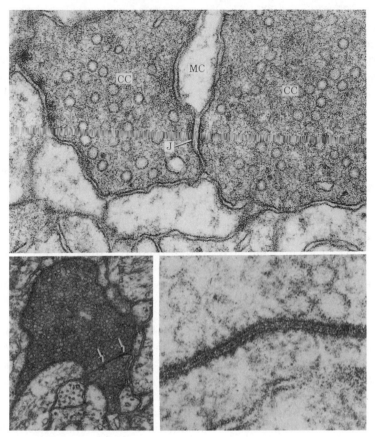

Fig. 1.25. Gap junctions between photoreceptors. Upper: *Pseudemys (Trachemys) scripta*. Junctions (J) between adjacent cone pedicles (CC) bear wide and narrow (2 nm) cleft segments. The narrow cleft contact may mediate electrical interactions between adjacent photoreceptors (81,000 x). (From Lasansky, 1971.) Lower left: *Chelydra serpentina*. Large gap junction between two telodendria (arrows) (25,200 x). The gap junction is shown at higher magnification at the lower right (226,800 x). (From Owen, 1985.)

Gilula, 1973). Kolb and Jones (1984, 1985) have suggested that these contacts may mediate the electrical coupling between spectrally similar cones that has been documented in *Pseudemys (Trachemys)* peripheral retina (Baylor and Hodgkin, 1973; Detwiler and Hodgkin, 1979; see Section III.C.2.b).

Chemical interactions between photoreceptors are mediated by telodendria that make basal contacts on neighboring pedicles, either terminating or en route to becoming the invaginating element of a dyad or triad (Lasansky, 1971; Mariani and Lasansky, 1984; Kolb and Jones,

1984, 1985; Owen, 1985). The latter telodendria are probably postsynaptic to the invaginated photoreceptor at their basal contacts and not at the apical ribbon junction (Kolb and Jones, 1985). These interreceptor basal contacts have a narrow (20 nm) cleft and symmetrical membrane densities (Kolb and Jones, 1985; they correspond to the narrow-cleft, symmetrical contacts between cones and possible bipolar cells of Kolb and Jones, 1984). Thus, their structure differs from the wide-cleft (30 to 40 nm) contacts that link photoreceptors to bipolar cells (see Owen, 1985, for an opposing view). Because chemical synapses link photoreceptors of different spectral sensitivities (as inferred from their oil droplets), they may mediate the color mixing that has been demonstrated in turtle cones (Normann et al., 1984, 1985a, 1985b). The two types of basal contacts between photoreceptors resemble cone-to-bipolar cell contacts in that some postsynaptic elements terminate at the base of receptor pedicles, whereas others contact photoreceptor bases en route to a dyad or triad. It is unclear whether the invaginating telodendria interact in some way with adjacent horizontal cell processes within the dyad/triad. Kolb and Jones (1985) have proposed that they are postsynaptic neither to horizontal cells, as originally proposed by Lasansky (1971), nor to photoreceptor ribbons. However, rod telodendria in *Chelydra* appear to be postsynaptic to rod or cone pedicles as lateral elements at a ribbon synapse (Mariani and Lasansky, 1984). Thus, the role of these invaginating photoreceptor telodendria is unclear.

Perhaps the most intimate contact between photoreceptors is that between the principal and accessory members of double cones (see Figs. 1.3 and 1.16). Several authors have noted that their inner-segment membranes are closely apposed, with wide- and narrow-gap segments (Fig. 1.26; Anh, 1968; Nishimura and Shimai, 1980; Nishimura et al., 1980; Ogura and Arispe, 1967; Hara and Miyoshi, 1984; Kolb and Jones, 1985; Owen, 1985; Lasater et al., 1989). The latter may have a cleft width as low as 2 nm, suggesting that it might mediate electrical signaling between the two elements. Early physiological studies yielded opposing data about electrical coupling between double cone elements (Richter and Simon, 1974; Detwiler and Hodgkin, 1979), but recent patch clamp recordings suggest that these visual cells are electrically independent (Lasater et al., 1989). Principal and accessory elements also make chemical contacts with each other through their telodendria (Kolb and Jones, 1984, 1985). This was originally interpreted as a contact between spectrally dissimilar cones (Kolb and Jones, 1984), but more recent evidence suggests that both members of the pair have red-sensitive photopigments (reviewed in Lipetz, 1985; Ohtsuka and Kouyama, 1986b; see Section II.B.3.b).

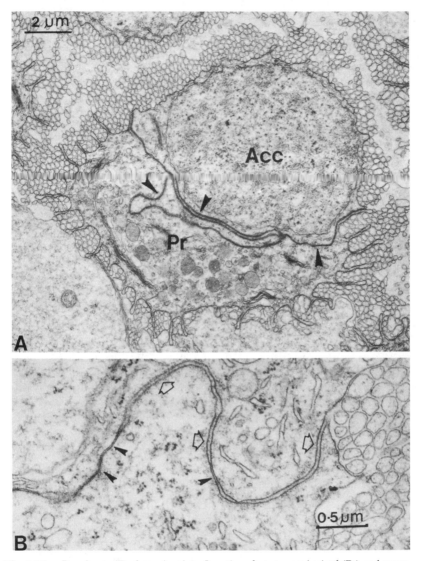

Fig. 1.26. *Pseudemys (Trachemys) scripta.* Junctions between principal (Pr) and accessory (Acc) members of a double cone. Upper: Tangential section just distal to the outer limiting membrane. The section has passed through the myoid region of the principal cell and through the distal tip of the nucleus of the accessory cell (see Fig. 1.3). Fins project radially from the myoid of the principal cell. The membranes of the two cells are thrown into folds (arrowheads), which increase their contact area. Lower: Higher-power micrograph of the junction between members of a double cone illustrating wide (open arrow) and narrow (arrowheads) cleft contacts. The narrow cleft contacts resemble gap junctions, but simultaneous recordings from the two cells suggest that they are not electrically coupled. (From Lasater et al., *Signal integration at the pedicle of turtle cone photoreceptors: an anatomical and electrophysiological study,* 1989, by permission of the publishers, Cambridge University Press.)

d. Pedicle Plasticity: Changes with Light and Temperature

In vertebrate photoreceptor pedicles, transmitter is released from vesicles in the dark and, upon illumination, transmitter release is inhibited (review in Kaneko, 1979; Dowling, 1987). Simultaneously, new vesicles must be formed and readied for the next cycle of transmitter release from the pedicle. The mechanisms underlying this process of vesicle formation and release have been studied in *Pseudemys (Trachemys)* (Schaeffer and Raviola, 1975, 1978), a form in which the in vitro retina is very robust and will allow manipulation for several hours (see Holtzman and Mercurio, 1980).

The amount of presynaptic membrane increases in the dark and decreases when the retina is light adapted (Fig. 1.27). This is consistent with the hypothesis that, in the dark, vesicles release their contents by fusing with the presynaptic membrane and that this transmitter release is inhibited by illumination. Retrieval of membrane from the presynaptic terminal continues in both light- and dark-adapted retinas, as evidenced by uptake of exogenous horseradish peroxidase (HRP) under both conditions (see Evans, 1966). Apparently, vesicle release and retrieval are in equilibrium both in light-adapted and dark-adapted retinas because the characteristic membrane configuration under both conditions (Fig. 1.27) is stable once reached. However, the rate of membrane exchange is higher in the dark. The rate of vesicle release decreases upon illumination, with uptake levels changing more slowly to the new equilibrium level; this results in a net decrease of presynaptic membrane.

Membrane retrieval can be inhibited by cold, leading to a condition in which the amount of presynaptic membrane is greatly increased. When such a retina is illuminated and warmed in the presence of exogenous HRP, the synaptic surface decreases, accompanied by an increase in the number of HRP-labeled vacuoles and coated vesicles within the synaptic terminal (Fig. 1.28). However, these inclusions disappear within 2 hours and are succeeded by labeled vesicles. These data suggest that vesicle formation involves retrieving membrane from the presynaptic terminal by coated vesicles and/or vacuoles from which synaptic vesicles are eventually formed. A similar pattern of vesicle recycling probably occurs at other central synapses of vertebrates (review in Landis, 1988).

In addition to these daily cycles of presynaptic membrane deposition and uptake, *Pseudemys (Trachemys) scripta* shows a diurnal change in the structure of rod synaptic ribbons (Fig. 1.29; Abe and Yamamoto, 1984). Starting at midnight, rod ribbons increase in complexity from a single linear density to a multilayered ribbon complex. After about 12 hours, these complexes begin to degenerate, gradually returning to the single-layered configuration. This diurnal cycle ap-

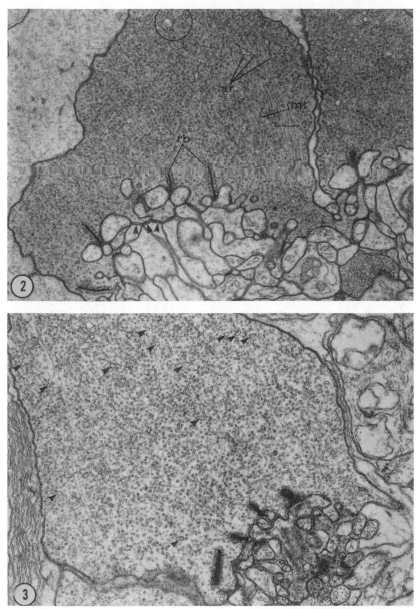

Fig. 1.27. *Pseudemys (Trachemys) scripta.* Light-induced changes in the structure of cone pedicles. Upper: Dark-adapted retina. The base of the cone pedicle shows numerous infoldings. Processes of unidentified cells make basal (arrowheads) and ribbon (rb) contacts with the photoreceptor; ar, agranular reticulum; mt, microtubules; circle, vacuole surrounded by coated vesicles (see Fig. 1.28, inset) (15,738 x). Lower: Light-adapted retina. The base of the pedicle is relatively narrow and flat, with few invaginations, suggesting that its total membrane area is less than under the dark-adapted condition. Arrowheads indicate round vacuoles in the cytoplasm (11,135 x). (Reproduced from the *Journal of Cell Biology,* 1978, Vol. 79, 802–825, by copyright permission of the Rockefeller University Press.)

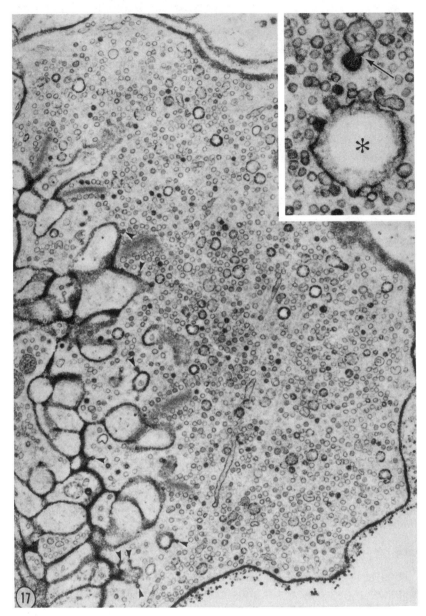

Fig. 1.28. *Pseudemys (Trachemys) scripta.* A dark-adapted retina has been exposed to light for 45 min with horseradish peroxidase (HRP) in the bathing medium. Numerous large vacuoles are labeled. Forming coated vesicles filled with reaction product are indicated by arrowheads. Such coated vesicles are shown at higher power in the inset. A few HRP-filled synaptic vesicles are visible; some are aligned with the synaptic ribbon at the lower left of the figure. These data are consistent with the interpretation that, in the light, membrane is pinched off from the base of the synaptic pedicle and recycled to form new synaptic vesicles (35,190 x; inset, 72,590 x). (Reproduced from the *Journal of Cell Biology,* 1978, Vol. 79, 802–825, by copyright permission of the Rockefeller University Press.)

0:00   5:00   7:00

12:00   17:00   19:00

Fig. 1.29. *Pseudemys (Trachemys) scripta.* Schematic illustration of the daily changes in the synaptic ribbon complex of rod visual cells. 0.00 = midnight. (From Abe and Yamamoto, 1984.)

pears to depend on photic stimulation, not on an internal circadian rhythm (Abe and Yamamoto, 1988). Its possible effect on visual function remains unknown.

C. Horizontal Cells

1. CELL STRUCTURE AND CLASSIFICATION   The somata of horizontal cells lie along the outermost boundary of the inner nuclear layer, and their primary dendrites form a rich plexus in the proximal outer plexiform layer (see Fig. 1.1; Yamada and Ishikawa, 1965; Bush and Lunger, 1972; Gallego and Perez-Arroyo, 1976). Distal segments of primary dendrites and secondary dendrites may run horizontally at more scleral levels of the outer plexiform layer (Leeper, 1978a; Witkovsky et al., 1983). Fine processes ascend from them and from the cell bodies to the level of the photoreceptor pedicles, where they end in minute knobs (turtles: *Emys orbicularis*—Borovyagin, 1966; *Mauremys (Clemmys) japonica*—Gallego and Perez-Arroyo, 1976; *Chelydra serpentina*—Leeper, 1978a; *Pseudemys (Trachemys) scripta*—Lasansky, 1971; Leeper, 1978a; Kolb and Jones, 1984; *Geoclemys*—Ohtsuka, 1983; Ohtsuka and Kouyama, 1982, 1986a. Lizards: *Lacerta viridis*—Cajal, 1893. Snakes: *Pelamis platurus* and *Natrix (Nerodia) valida*—Hibbard and Lavergne, 1972).

Cajal (1893) examined horizontal cells in *Lacerta* and concluded that there was more than one morphological type ("brush" and "flattened stellate" cells; Fig. 1.30). He further predicted that each type contacted different classes of photoreceptor on the grounds that their ascending processes terminate at different levels within the outer plexiform layer, as do photoreceptor pedicles. This has proven to be the case, at least in turtles (see Section III.C.3). Unfortunately, we know

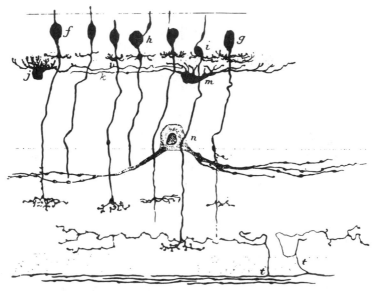

Fig. 1.30. *Lacerta viridis.* Retinal neurons visualized by Golgi impregnation. f, g, h, i, various types of displaced bipolar cells; j, brush-shaped horizontal cell with a fine axon; m, stellate horizontal cell with an axon (k) (this cell has been stained with methylene blue); n, large mitral amacrine cell, stained with methylene blue; t, fine processes arising from a fiber of the optic nerve; these fine processes were impregnated in the retina of an embryonic *Lacerta agilis.* Original drawings and descriptions of cells are from Cajal, 1893. (From *The vertebrate retina: principles of structure and function,* by R. W. Rodieck. Copyright (c) 1973 by W. H. Freeman and Company. Reprinted with permission by W. H. Freeman and Company.)

almost nothing about horizontal cells in any other orders of reptiles.

The most complete account of reptilian horizontal cells is that of Leeper (1978a, 1978b) for *Pseudemys (Trachemys) scripta.* This species has five morphologically distinct horizontal cell types. One (H1AT) is not a true cell, but rather is the axon terminal of a second horizontal cell type (H1CB). It is described separately because it differs from its parent cell in important ways and functions as an independent physiological unit (Ohtsuka, 1983; review in Leeper and Copenhagen, 1982). Most available evidence suggests that turtles lack the laminar segregation of horizontal cell types described in other vertebrates (Cajal, 1893).

a. H1CB Cells
The H1CB type is the cell body portion of the H1 horizontal cell type (Fig. 1.31), and it differs in three ways from other nucleated horizontal cells. First, the dendritic arbor is richly branched with overlapping processes. Numerous fine, ascending fibers, especially in the region

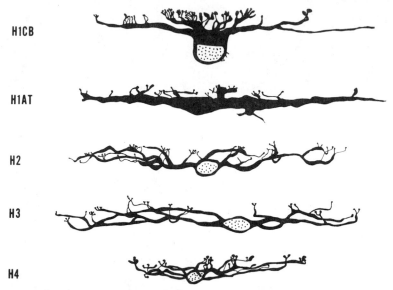

H1CB

H1AT

H2

H3

H4

Fig. 1.31. *Pseudemys (Trachemys) scripta.* Horizontal cell types visualized by Golgi impregnation. H1CB and H1AT are the cell body and axon terminal, respectively, of the H1 horizontal cell type. Scale, 20 μm. (From Leeper, 1978, with permission.)

of the soma, give it a "bushy" appearance (Saito et al., 1974; Leeper, 1978a; Ohtsuka and Kouyama, 1986a, 1986b). Second, the prominent nucleus lies proximal to the dendrites. Third, H1CB cells give rise to an axon.

The H1CB type is characterized ultrastructurally by numerous organelles (Golgi apparatus, endoplasmic reticulum, and small, rounded mitochondria); its few microtubules and neurofilaments appear randomly dispersed within the cytoplasm (Fig. 1.32; Yamada and Ishikawa, 1965; Witkovsky et al., 1983; Kolb and Jones, 1984). A fine axon (0.5 to 1.0 μm) leaves the soma or a primary dendrite and follows an erratic course for 200 to 300 μm before ending in an axon terminal (Fig. 1.33; Leeper, 1978a; Ohtsuka, 1983). Thus, axonal length does not predict the linear distance between cell body and axon terminal. The length and small caliber of this axon electrically isolates the two components of H1 horizontal cells (Ohtsuka, 1983).

The H1CB type appears to correspond to the procion yellow–injected L2 (luminosity) cells of Simon (1973) and Saito et al. (1974). These cells hyperpolarize to light of any wavelength (reviews in Drujan and Laufer, 1982; Ohtsuka and Kouyama, 1986b; Lipetz, 1985) and differ from the second variety of luminosity cell (L1) in the size of their receptive field (Simon, 1973; Saito et al., 1974; Ohtsuka, 1983; see Section III.C.2.c). The H1CB type is similar to the "brush" horizontal cell

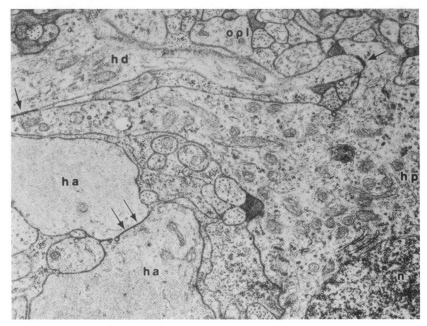

Fig. 1.32. *Pseudemys (Trachemys) scripta.* Low-power electron micrograph illustrating contacts between H1AT horizontal cells and between H1CB horizontal cells. A gap junction between H1AT profiles (ha) is indicated by double arrows. Note that the cytoplasm of these cells is electron lucent and bears numerous filaments. Two arrows mark contacts between an H1 perikaryon (hp) and dendrite (hd). The cytoplasm of these profiles is rich in organelles. The nucleus of the horizontal cell soma (n) is visible in the lower right corner. opl, outer plexiform layer (13,860 x). (From Witkovsky et al., 1983.)

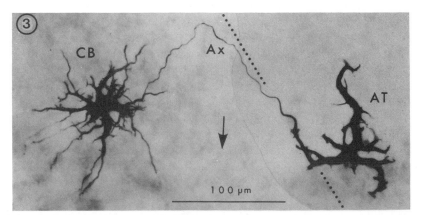

Fig. 1.33. *Chinemys (Geoclemys) reevesii.* The cell body (CB) and axon terminal (AT) of an H1 horizontal cell have been visualized by intracellular filling with horseradish peroxidase. The microelectrode (tract indicated by dotted line) penetrated the axon terminal. HRP diffused through the narrow connecting axon (Ax) to fill the cell body. Arrow indicates the direction of the optic disk. (From Ohtsuka, 1983.)

of Lasansky (1971) and, perhaps, of Cajal (1893). A similar cell has been described in *Mauremys japonica* (Gallego and Perez-Arroyo, 1976), *Chelydra serpentina* (Leeper, 1978a; Leeper and Copenhagen, 1979), and *Chinemys (Geoclemys) reevesii* (Ohtsuka, 1983; Ohtsuka and Kouyama, 1986a, 1986b).

b. H1AT CELLS
The axon terminal portion of H1 horizontal cells (H1AT) has a characteristic "tuberous" morphology with a primary trunk up to 10 μm thick and three to five small (5 μm long), terminal dendrites that leave the trunk in an irregular pattern (Figs. 1.31 and 1.33; Leeper, 1978a; Simon, 1973; Ohtsuka, 1983). Ultrastructurally, H1AT cells differ from the H1CB type in that they contain relatively few organelles except for prominent bundles of neurofilaments (Witkovsky et al., 1983; Kolb and Jones, 1984). Roughly similar descriptions of horizontal cell ultrastructure have been reported by earlier authors (*Mauremys japonica*, Yamada and Ishikawa, 1965; *Emys orbicularis*, Borovyagin, 1966; *Pseudemys (Trachemys) scripta*, Bush and Lunger, 1972). These studies do not distinguish between horizontal cell types, but it seems likely that they are describing the H1AT type because their accounts emphasize large processes with an accumulation of filaments. Saito et al. (1974) have argued that H1CB and H1AT cells can be distinguished by their laminar position, but other evidence (Simon, 1973; Kolb and Jones, 1984; Ohtsuka, 1983), including a quantitative analysis of horizontal cell laminar position (Leeper, 1978a), suggests that both the H1 cell body and axon terminal occupy similar retinal strata. Interestingly, a horizontal cell type has been described in snakes with a probable axon that descends through the thickness of the inner nuclear layer (Hibbard and Lavergne, 1972), suggesting that its axon terminal lies near the inner plexiform layer.

H1AT cells appear to correspond to the L-type cells of Miller et al. (1973; Hashimoto et al., 1973), the L1 cells of Simon (1973) and Saito et al. (1974), and the luminosity-type axon terminal of Ohtsuka (1983). They are probably the "flat stellate" variety of Lasansky (1971) and Cajal (1893), which occasionally were seen to bear a fine axon. Similar cells have been reported in *Mauremys japonica* (Gallego and Perez-Arroyo, 1976) and *Chelydra serpentina* (Leeper, 1978a; Leeper and Copenhagen, 1979). Thus, the H1 horizontal cells correspond to the luminosity cells identified in physiological experiments, whereas the cell body and axon terminal portions of the H1 type correspond to the L2 and L1 physiological varieties, respectively. These cells differ in the size of their receptive fields in a way that cannot be accounted for by their relative dendritic tree sizes; the explanation appears to lie in the contacts they make with horizontal cells of the same type (see Sections II.C.2.c and III.C.2.c). In addition, the flash response of the

H1AT type exhibits a slow component not seen in H1CB cells, which reflects their different patterns of photoreceptor contacts (Leeper and Copenhagen, 1979; see Section III.C.3).

c. H2 CELLS

The H2 cell type, like the H3 and H4 varieties, can be distinguished from the H1CB cells because its soma and dendrites occupy a single level of the outer plexiform layer (Fig. 1.31; Leeper, 1978a). It is characterized by five to six relatively thick primary dendrites (Saito et al., 1974; Leeper, 1978a) that branch repeatedly but tend not to cross each other as in the H1CB type; they give rise to numerous fine, ascending processes. Ultrastructurally, putative H2 cells have relatively pale cytoplasm, round nuclei, and fewer primary dendrites than H1 cells (Kolb and Jones, 1984).

H2 cells appear to correspond to a type of chromaticity (C) horizontal cell that depolarizes to red light and hyperpolarizes to green and shorter wavelength light (Leeper, 1978a). This assignment is based on the morphological similarity between H2 cells and R/G chromaticity-type cells injected with procion yellow (Miller et al., 1973; Hashimoto et al., 1973; Saito et al., 1974), and it is supported by the pattern of photoreceptor inputs to H2 cells (Leeper, 1978b; see Section III.C.3). A horizontal cell type with similar morphology has been identified in *Geoclemys* (biphasic chromaticity type of Ohtsuka and Kouyama, 1986a, 1986b; see Fig. 1.56), but these authors present a different interpretation of its photoreceptor afferents (see Section III.C.3).

d. H3 CELLS

H3 horizontal cells can be distinguished from the H2 type by their relatively delicate dendrites, which branch less frequently than those of H2 cells (Figs. 1.31 and 1.55). Furthermore, H3 cells give rise to relatively few ascending processes. Structurally, they resemble the triphasic chromaticity cells of Ohtsuka (1986a; see Fig. 1.56). H3 cells have a characteristic pattern of photoreceptor inputs (Leeper, 1978b; see Section III.C.3), which suggests that they correspond to the green depolarizing, blue hyperpolarizing (G/B) chromaticity-type horizontal cell of Fuortes and Simon (1974) and Yazulla (1976a). A similar horizontal cell has been described in Golgi impregnated retinas of *Mauremys caspica* (see Fig. 6 in Gallego and Perez-Arroyo, 1976).

e. H4 CELLS

H4 cells are distinguished from the H3 variety by their size; at a given retinal locus they are smaller than H3 cells (Fig. 1.31). Leeper (1978a) quotes a case in which an H3 and an H4 cell were 250 μm apart and had dendritic field diameters of 180 μm and 100 μm, respectively.

Only two cases of the H4 type have been observed, and neither was in whole mounted retinas, so the dendritic architecture of H4 cells is poorly understood. Nevertheless, their pattern of photoreceptor contacts suggests that H4 cells are a distinct horizontal cell type (see Section III.C.3).

2. Morphology of Synapses Made by Horizontal Cells

a. General   Horizontal cells engage in contacts with photoreceptors, other horizontal cells, and bipolar cells. The available data suggest that each type of contact has a characteristic morphology, and this section reviews these patterns. Most of our knowledge about horizontal cell contacts in reptilian retinas comes from work on the retina of turtles, especially *Pseudemys (Trachemys) scripta*.

b. Contacts with Photoreceptors
Junctions between horizontal cells and photoreceptors have been described in Section II.B.5.c. Briefly, horizontal cells contact photoreceptors at ribbon synapses, where they are always the lateral processes of a dyad or, more rarely, a triad. Such junctions have been described in many reptiles, but the identity of the lateral processes has only been confirmed in turtles (Lasansky, 1971; Yamada and Ishikawa, 1965; Borovyagin, 1966; Kolb and Jones, 1984; Ohtsuka and Kouyama, 1985, 1986a). Only rarely are the lateral processes seen to contain clusters of synaptic vesicles (see Figs. 1.15 and 1.16 in Odashima et al., 1978, 1979; Kolb and Jones, 1984; Fig. 14 in Lasansky, 1971). Their appearance in thin sections and freeze-fracture material suggests that photoreceptors are presynaptic to horizontal cells at these junctions (Schaeffer et al., 1982). Thus, we do not know the structural substrate of the horizontal cell feedback to photoreceptors that has been demonstrated physiologically in turtles (Tachibana and Kaneko, 1984; Kaneko et al., 1985; Kaneko and Tachibana, 1986a, 1986b). A second type of photoreceptor to horizontal cell contact occurs at distal junctions (Lasansky, 1971; see Section II.B.5.c); we do not know its function.

c. Contacts with Other Horizontal Cells
Evidence based on freeze-fracture material and transmission electron microscopy suggests that H1 horizontal cells contact each other through gap junctions (Schaeffer et al., 1982; Witkovsky et al., 1983; Kolb and Jones, 1984). Specifically, H1CB cells contact other H1CB cells, and H1AT cells contact other H1ATs (Fig. 1.32). The contacts linking these two cell types differ morphologically in ways that may explain their dissimilar receptive field sizes (see below and Section III.C.2.c).

H1CB cells contact each other at gap junctions in which only short

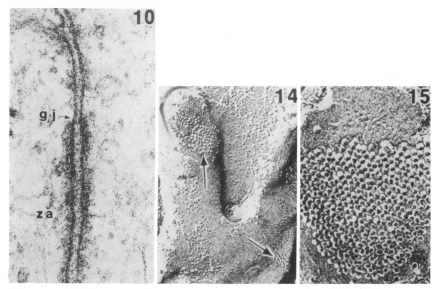

Fig. 1.34. *Pseudemys (Trachemys) scripta.* Gap junctions between horizontal cell somata and dendrites. Left: Transmission electron micrograph illustrating a gap junction (gj) with an adjacent zonula adhaerens (za) (99,000 x). Middle: Freeze-fracture
micrograph showing clusters of P face particles marking two gap junctions (arrows)
(78,000 x). Right: Higher-power micrograph illustrating gap junction particles packed
in an hexagonal array (198,000 x). (From Witkovsky et al., 1983, with permission.)

patches of membrane are opposed (0.02 to 0.07 μm²; Witkovsky et al.,
1983; Kolb and Jones, 1984). Adjacent to the gap junction there is usually a classic zonula adhaerens, which gives these contacts a configuration similar to that linking photoreceptor pedicles (Fig. 1.34; Lasansky, 1971). The internal structure of the membrane at these contacts is
characteristic of gap junctions, that is, a patch of 9-μm P face particles, often with a central dimple, in a roughly hexagonal array (Fig.
1.34). H1CB cells are also presynaptic to putative wide-field dendrites
(i.e., dendrites of H2 or H3 cells) at chemical synapses (Kolb and
Jones, 1984).

H1ATs can be distinguished by their high-density neurofilaments
and low complement of other intracellular organelles (Fig. 1.32). Profiles with these characteristics contact each other by gap junctions
that differ from those linking H1CB cells primarily by the greater extent of the opposed membranes (0.1 to 1 μm²; Witkovsky et al., 1983;
Fig. 1.35). Similar stretches of opposed membrane link H1AT processes as they ascend toward the photoreceptor pedicles. In freeze-
fracture material, H1AT contacts appear to correspond to large
patches of P face particles with complementary E face pits similar to
those described previously; however, these particles do not form a
crystalline array (Fig. 1.35). H1AT cells have also been observed as the

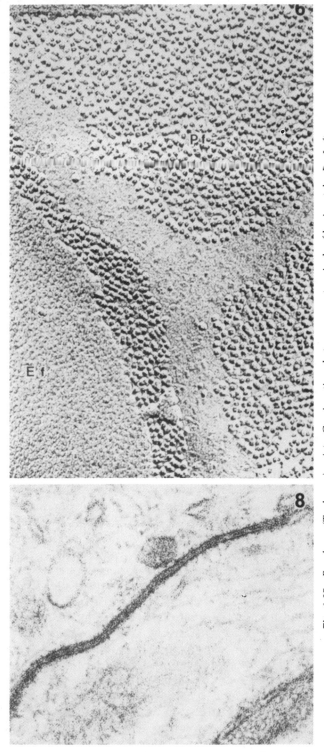

Fig. 1.35 *Pseudemys (Trachemys) scripta.* Gap junctions between axon terminals of horizontal cells. Left: Transmission electron micrograph of large gap junction between two axon terminals (110,880 x). Right: Freeze-fracture preparation showing the extensive areas of P face (Pf) and E face (Ef) particles that characterize gap junctions between axon terminals (166,320 x). (From Witkovsky et al., 1983.)

presynaptic element at a chemical synapse with H1CB dendrites (Kolb and Jones, 1984). Earlier reports of contacts between large, fibril-bearing horizontal cell processes (Yamada and Ishikawa, 1965; Schaeffer et al., 1982; Lasansky, 1972) may refer to H1AT junctions, although it is impossible to be certain, as the ultrastructure of horizontal cells other than the H1 type is unknown.

d. CONTACTS WITH BIPOLAR CELLS

H1CB cells engage in chemical synapses with profiles that are probably bipolar cells (Witkovsky et al., 1983; Kolb and Jones, 1984). At these contacts there are symmetrical membrane densities and an adjacent cluster of vesicles in the horizontal cell process suggesting that H1 cells are the presynaptic element. Similar contacts have been observed between bipolar cells and putative wide-field (H2 or H3) horizontal cells (Kolb and Jones, 1984).

D. Bipolar Cells

Bipolar cells in reptiles resemble those of other vertebrates, at least in broad outline. They have a small cell body (3 to 30 μm in *Pseudemys* (*Trachemys*) *scripta*, Bush and Lunger, 1972; Dacheux, 1982; Kolb et al., 1986; Schutte and Weiler, 1987) that is generally located in the middle of the inner nuclear layer (Figs. 1.1, 1.19, and 1.36). The nucleus is heavily invested with chromatin, which accounts for the characteristic dark staining of bipolar cells in Nissl material, and it is surrounded by a thin margin of cytoplasm containing smooth endoplasmic reticulum, scattered ribosomes, and small mitochondria (Meller and Eschner, 1965; Bush and Lunger, 1972; Kolb and Jones, 1984). A primary dendrite erupts from the soma and ascends, sometimes obliquely (Fig. 1.36), toward the photoreceptor layer. Just proximal to the rod and cone pedicles, a number of secondary dendrites arise from this ascending process and spread out horizontally for a distance that varies with bipolar cell type and with retinal locus (Kolb, 1982; Schutte and Weiler, 1987; see Fig. 1.39). From them, or from the ascending process itself, projects a robust, vertical Landolt's club, which reaches as far as the outer limiting membrane (Kolb, 1982; Detwiler and Sarthy, 1981). The secondary dendrites give rise to fine vertical processes that terminate in minute swellings at the level of the photoreceptor pedicles (see Fig. 1.19; *Lacerta viridis, Chamaeleo*, Cajal, 1893; *Pseudemys* (*Trachemys*), Lasansky, 1971; Kolb, 1982; Dacheux, 1982; Kolb et al., 1986). In thin sections of *Pseudemys* (*Trachemys*) retina, bipolar cell dendrites can be distinguished from photoreceptor pedicles by their electron-lucent cytoplasm and from horizontal cell processes by their rich complement of axially oriented microtubules (Witkovsky et al., 1983; Kolb and Jones, 1984). Some dendrites bear

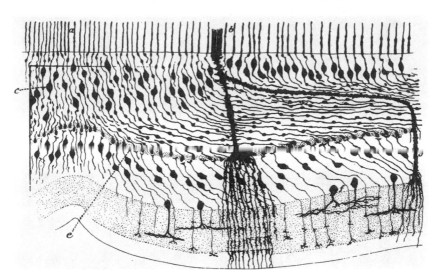

Fig. 1.36. *Chamaeleo vulgaris.* Vertical section at the level of the fovea illustrating oblique process of cones and bipolar cells. Golgi impregnated retina. a, Thin cones; b, thicker cones; c, cone cell bodies; e, tiny dendrites of a bipolar cell; f, amacrine cells; g, lateral process of a Müller cell. Original drawings and descriptions of cells are from Cajal, 1893. (From *The vertebrate retina: principles of structure and function,* by R. W. Rodieck. Copyright (c) 1973 by W. H. Freeman and Company. Reprinted with permission by W. H. Freeman and Company.)

lamellar structures of unknown function (Owen, 1985). The ultrastructure of bipolar contacts with photoreceptors and horizontal cells has already been described (Sections II.B.5.c and II.C.2.d, respectively).

A central process (axon) passes from the vitreal pole of the bipolar soma toward the inner plexiform layer (see Fig. 1.19). It may run obliquely, like the primary dendrite, but it turns normal to the retinal laminae at the border of the inner plexiform layer and descends for a variable distance toward the ganglion cell layer (Fig. 1.36). It terminates as one or more planar arborizations within the inner plexiform layer. Ultrastructurally, these terminals bear synaptic ribbons and numerous round vesicles (Fig. 1.37; Guiloff et al., 1988). Some may release glutamate as a neurotransmitter (in *Pseudemys (Trachemys) scripta;* Ehinger et al., 1988; Copenhagen and Jahr, 1989). The oblique course of the ascending and descending processes means that the axon terminal is often separated from the dendritic field by a considerable linear distance (150 μm in *Anolis carolinensis,* Makaretz and Levine, 1980; *Chamaeleo,* Cajal, 1893; *Pseudemys (Trachemys),* Lasansky, 1971; Kolb, 1982; Dacheux, 1982; Schutte and Weiler, 1987). Moreover, the length and orientation of these oblique processes is not random, reflecting instead the underlying pattern of photoreceptor density

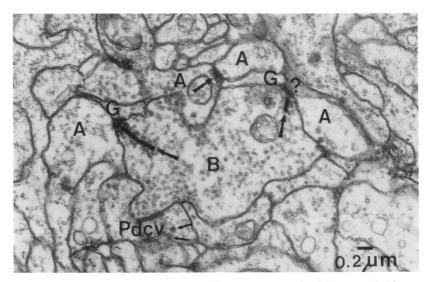

Fig. 1.37. *Pseudemys (Trachemys) scripta.* Electron micrograph of the inner plexiform layer in peripheral retina. A bipolar cell (B) with two ribbon complexes (large arrows) is contacting several amacrine (A) and ganglion (G) cells and one unidentified process (?). A chemical synapse between two amacrine cells (large arrow) and a profile bearing two large, dense-cored vesicles (Pdcv) are also visible. (From Guiloff et al., 1988.)

(Schutte and Weiler, 1987; see Fig. 1.39). Thus, amacrine and ganglion cells at a given retinal locus may receive their primary input from photoreceptors some distance away.

The morphology of bipolar cells varies in three ways. First, the somata vary in size and location within the inner nuclear layer (Figs. 1.19 and 1.38). At least some of this variation may reflect retinal locus (Fig. 1.39; Weiler and Schutte, 1985a; Schutte and Weiler, 1987). Some somata may be displaced to the proximal outer nuclear layer (Fig. 1.30); one-third of profiles in the outer nuclear layer of *Pseudemys (Trachemys) scripta* may belong to these displaced bipolar cells (Kouyama and Ohtsuka, 1985). In such cells the horizontally arranged dendrites arise from the descending processes, just below the soma (Cajal, 1893; Davydova, 1981; Kolb, 1982). Second, dendritic arbors vary in diameter, depending on retinal locus, and in the density of fine, ascending processes (Figs. 1.38 and 1.39). However, they do not appear to be stratified within the outer plexiform layer, as they are in teleosts and birds (Cajal, 1893). Finally, axonal arbors range from delicate and tendrillike to robust and varicose; they vary conspicuously in the number and stratification levels of their planar axon fields.

There have been several attempts to group bipolar cells into morphological types. Cajal (1893) described these classes of bipolar cells in *Lacerta viridis* on the basis of soma location and size: displaced bi-

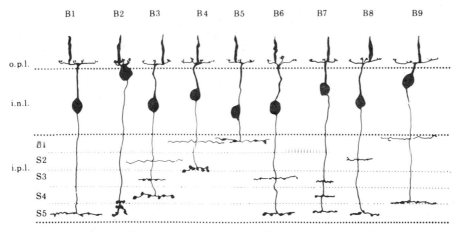

Fig. 1.38. *Pseudemys (Trachemys) scripta.* Bipolar cell types visualized in Golgi impregnated retinas. These neurons have dendrites in the outer plexiform layer (opl) and a Landolt's club that ascends through the photoreceptor layer. Their somata occupy different levels of the inner nuclear layer (inl), and their axons descend through the inner plexiform layer (ipl) to terminate as tangentially oriented arbors in one or more strata of the ipl. (From Kolb, 1982.)

polars with cell bodies in the outer nuclear layer (see Fig. 1.30), large "outer" bipolars with somata in the distal inner nuclear layer, and small "inner" bipolars with cell bodies scattered throughout the inner nuclear layer (see Fig. 1.19). Similar cells have also been observed in experiments on bipolar cells of *Pseudemys (Trachemys) scripta* that were identified physiologically and then stained with procion yellow (Yazulla, 1976b; Weiler, 1981). These data suggest that both center-hyperpolarizing (center-off; see introduction to Section III for definitions) and center-depolarizing (center-on) bipolar cells can have somata of virtually any size or location. Thus, somatal characteristics are not correlated with the polarity of center responses, at least in non-color-coded bipolar cells (Yazulla, 1976b) or those stimulated with white light (Weiler, 1981). Perhaps they correspond to differences in spectral profiles, as they appear to do in teleosts (Ishida et al., 1980).

Kolb (1982) has described nine morphological types of bipolar cells in *Pseudemys (Trachemys) scripta* based primarily on the stratification levels of their axon terminals (Fig. 1.38). Such interest in axonal strata within the inner plexiform layer derives from two findings. First, the inner plexiform layer of reptiles and other vertebrates has been divided into five sublaminae based on the levels at which the dendrites of different amacrine and ganglion cell types arborize (Cajal, 1893; Kolb, 1982; Guiloff et al., 1988). Thus, the levels at which bipolar cell axons ramify may provide clues to their functional contacts in the inner plexiform layer. Second, neurons synapsing in the outer and in-

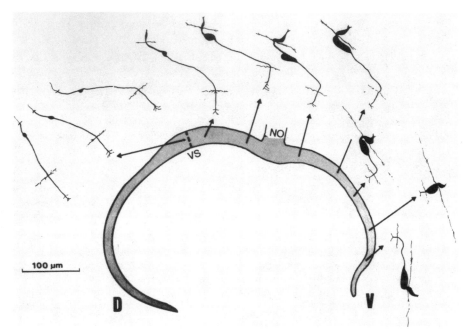

Fig. 1.39. *Pseudemys (Trachemys) scripta.* Setotoninergic bipolar cells visualized in vertical sections at different distances from the visual streak (VS). Somata and dendritic arbors are larger in ventral (V) than in dorsal (D) retina. Both increase with distance from the visual streak. In contrast, the size of axonal ramifications changes very little with retinal eccentricity. NO, Optic nerve. Scale applies to cells. (From Schutte and Weiler, 1987.)

ner zones of the inner plexiform layer have distinct physiological characteristics in teleost fish (Famiglietti et al., 1977) and mammals (Nelson et al., 1978). In these species, bipolar and ganglion cells with processes terminating in the distal inner plexiform layer (strata 1 to 2; review in Guiloff et al., 1988) exhibit center-off responses, whereas center-on cells form synaptic contacts closer to the ganglion cell layer (strata 3 to 5). Thus, it is of interest to know if reptilian bipolar cells can be similarly classified.

Interpretation of these stratification levels in reptiles is complicated by the fact that most reptilian bipolar cells ramify in two or even three retinal strata (Fig. 1.38). Thus, no simple pairings between bipolar and amacrine or ganglion cells are possible on the basis of matched stratification levels. Furthermore, many reptilian bipolar cells ramify in *both* outer and inner sublaminae of the inner plexiform layer; thus, they cannot be readily classified morphologically as off-center and on-center types, in contrast to the bipolar cells of fish and mammals, which typically stratify at only one level of the inner plexiform layer.

In spite of these complications, the available data suggest that a functional division of the inner plexiform layer into outer and inner zones is also characteristic of reptiles (Marchiafava and Weiler, 1980; Weiler, 1981; Weiler and Marchiafava, 1981; Marchiafava, 1982; Ashmore and Copenhagen, 1983; see Section III.B.3). Briefly monostratified bipolar cells arborize in either the outer or inner half of the inner plexiform layer, depending on their response to center stimulation (hyperpolarizing or depolarizing, respectively). For bi- and tristratified bipolar cells the center response is best predicted by the level of the most distal (vitread) axonal arbor. Center hyperpolarizing cells always have one ramification in the outer sublaminae of the inner plexiform layer, whereas the second arbor (and third, where it exists) can be found anywhere within the inner plexiform layer. In contrast, center-depolarizing cells have both arbors in the inner sublayers, the most distal one usually at about the midpoint between inner nuclear and ganglion cell layers (layer 3). Interestingly, many published pictures of reptilian bipolar cells indicate that the most distal arbor, which predicts the center response, has a wider spread and more delicate morphology than more proximal axonal arbors of the same cell (e.g., Weiler, 1981; Kolb, 1982; see Fig. 1.38).

Recent attempts to characterize retinal neurons using immunofluorescence and molecular probes have revealed that bipolar cells also differ in their cytochemistry. Almost all bipolar cells of *Pseudemys* are strongly immunoreactive for glutamate, suggesting that these neurons may use glutamate as a neurotransmitter (Ehinger et al., 1988). A subset of bipolar cells in *Pseudemys (Trachemys)* are also serotoninergic (Figs. 1.39 and 1.40; Witkovsky et al., 1984; Weiler and Schutte, 1985a, 1985b; Schutte and Weiler, 1987). Distal to the inner plexiform layer, the structure of these neurons varies with retinal locus, but their axonal arbors are everywhere the same (Fig. 1.39). They arborize distally in layer 1, suggesting that they are off-center bipolars, at the 4–5 border, and in lamina 5 (Fig. 1.40; Schutte and Weiler, 1987). They do not clearly resemble any of Kolb's nine bipolar types (Kolb, 1982) but are closest to the B9 variety (Witkovsky et al., 1984).

E. Amacrine Cells

Amacrine cells typically have somata in the proximal inner nuclear layer (see Fig. 1.1), where they may be arranged in layers according to cell type (*Pseudemys (Trachemys) scripta*, e.g., Eldred and Karten, 1983; Weiler and Ball, 1984). Amacrine cell perikarya are occasionally displaced to the inner plexiform layer or even to the ganglion cell layer (e.g., *Lacerta*, Cajal, 1893; *Pseudemys (Trachemys)*, Eldred and Karten, 1983; Weiler and Ball, 1984; Eldred and Cheung, 1989; Hurd and Eldred, 1989; Williamson and Eldred, 1989; *Thamnophis*, Wong,

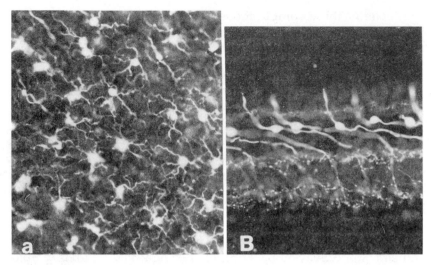

Fig. 1.40. *Pseudemys (Trachemys) scripta.* Serotoninergic bipolar cells visualized by immunofluorescence. Left: Whole mount illustrating tangentially oriented dendritic processes. The bright spots are Landolt's clubs that mark the center of each cell's dendritic field. The diameter of these dendritic fields increases with distance from the visual streak (see Fig. 1.39). Right: Vertical section taken ventral to the optic nerve. The axons arborize in three layers: 1, 4/5, and 5. This ramification pattern does not change with retinal locus. (From Schutte and Weiler, 1987.)

1989; see Fig. 1.44), but for *Pseudemys (Trachemys)*, at least, it is unlikely that they account for a high proportion of profiles in the ganglion cell layer (Peterson and Ulinski, 1979), as they do in birds (Bingelli and Paule, 1969) and mammals (Hughes and Vaney, 1980).

Amacrine cell processes enter the inner plexiform layer, where they arborize in a rich variety of patterns (Fig. 1.41). Most arbors appear planar in vertical sections, delineating five sublaminae within the inner plexiform layer (Cajal, 1893; Boycott and Dowling, 1969). The processes of individual amacrine cells ramify in one or more of these sublaminae or, less frequently, they spread diffusely throughout the inner plexiform layer. All attempts to classify reptilian amacrine cells have focused on these vertical arborization patterns (see Section E, last paragraph).

In whole-mount preparations it is clear that amacrine cell dendritic fields can also vary in their tangential shape and orientation. Some have radiate or elliptical dendritic fields in which the soma may be eccentric or centrally placed, and others are markedly anisotropic with elongated dendritic fields oriented parallel or orthogonal to the nasotemporal axis of the retina (Figs. 1.42 and 1.43; *Lacerta viridis*, Cajal, 1893; *Pseudemys (Trachemys) scripta*, Kolb, 1982; Eldred and Karten, 1983; Williamson and Eldred, 1989; Ammermuler and Weiler, 1988;

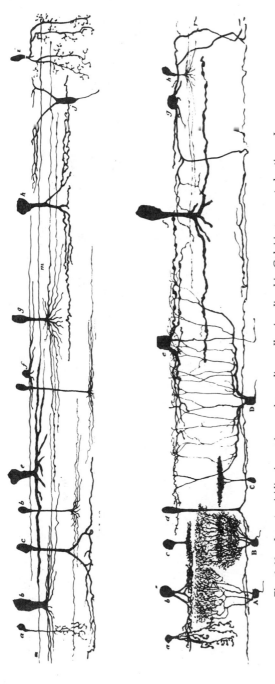

Fig. 1.41. *Lacerta viridis.* Amacrine and ganglion cells visualized in Golgi impregnated retinas. Upper: a, Nonstratified amacrine cell; b, radiating amacrine cell of the third level; c, amacrine cell with sinuous branches at the fifth level; e, giant amacrine cell of the second level, whose branches, thick at first, become fine like axons; f, radiating amacrine cell of the first level; g, radiating amacrine cell of the third level; h, amacrine cell with sinuous branches at the fourth level; i, nonstratified amacrine cell with multiple processes; j, displaced ganglion cell; m, fine fiber coming from the giant amacrine cell (e). Lower: A and B, Two ganglion cells with a rich, delicate arborization; C, ganglion cell of the second level; D, ganglion cell whose fine, ascendant branches terminate at the first level; a, nonstratified amacrine cell; b, amacrine cells with sinuous branches at the second level; d, radiating amacrine cell of the fifth level; e, amacrine cell; f, large amacrine cell of the third level; g, amacrine cell whose fine branches seem to ramify at the first and fifth levels; h, radiating amacrine cell of the second level. Original drawings and descriptions of cells are from Cajal, 1893. (From *The vertebrate retina: principles of structure and function.* by R. W. Rodieck. Copyright (c) 1973 by W. H. Freeman and Company. Reprinted with permission by W. H. Freeman and Company.)

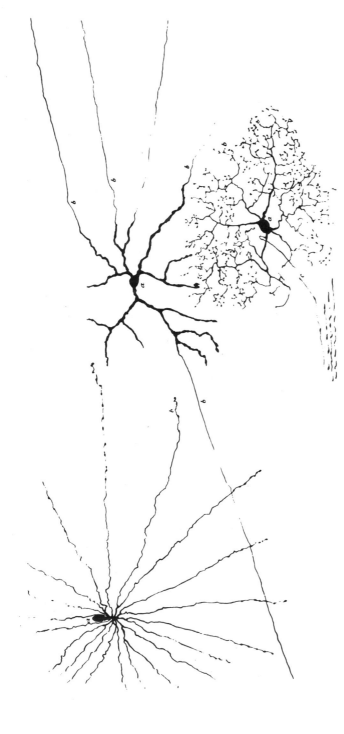

Fig. 1.42. *Lacerta viridis.* Amacrine and ganglion cells visualized in whole mounts of Golgi impregnated retinas. Left: Radiating amacrine cell from the inner plexiform layer; a, pyriform cell body; b, swellings at the end of a radiating fiber. Upper right: Giant amacrine cell; a, cell body; b, fine axon-like branchlets (this cell is of the same kind as (e) in Fig. 1.41, upper half). Lower right: Ganglion cell; a, cell body; b, varicose branchlets terminating freely; c, axon that becomes a fiber of the optic fiber layer. Original drawings and descriptions of cells are from Cajal, 1893. (From *The vertebrate retina: principles of structure and function,* by R. W. Rodieck. Copyright (c) 1973 by W. H. Freeman and Company. Reprinted with permission by W. H. Freeman and Company.)

Mauremys caspica, Kolb et al., 1988). Dendritic fields also vary in diameter, from 30 μm to several millimeters. In general, it is not clear how much of this morphological variation is a function of retinal locus and how much reflects structural differences between distinct functional types. For example, two amacrine cell types with enkephalin-like immunoreactivity in *Pseudemys (Trachemys)* have the same soma size, laminar position in the inner nuclear layer, and stratification pattern (Eldred and Karten, 1983), but they differ in the varicosity of their processes, and their dendritic fields have radically dissimilar shapes. One type has an elongated dendritic arbor and is found only in the visual streak (Fig. 1.43); the other is found outside the visual streak and has an isotropic dendritic field. Thus, it is not clear whether these two variants are best considered as a single amacrine cell type that changes the shape of its dendritic field with retinal locus or as two distinct cell classes. A similar argument could be raised regarding amacrine cells that differ primarily in size (e.g., Cajal, 1893; see Fig. 1.41, lower half, cells b and c). At least some amacrine cells clearly vary in size as a function of retinal locus (Kolb et al., 1987).

Morphological classifications of reptilian amacrine cells have relied on Golgi preparations, immunological or autoradiographic probes for putative neurotransmitters, or dye injections following physiological characterization. Cajal (1893) described at least 18 morphological varieties of amacrine cell in *Lacerta viridis* using Golgi material (Fig. 1.41), and more recent Golgi studies of *Pseudemys (Trachemys) scripta* (Kolb, 1982) and *Mauremys caspica* (Kolb et al., 1988) retinas list 27 and 30 amacrine cell types, respectively. In each case, the cells are distinguished from each other on the basis of soma and arbor size, process morphology, and stratification patterns within the inner plexiform layer. Kolb and colleagues have grouped the 27 to 30 amacrine cell types of *Pseudemys* and *Mauremys* into four broader categories on the basis of dendritic field size (narrow, small, medium, and wide fields). Unfortunately, there has been no systematic attempt to determine whether any of these different amacrine cell "types" are actually regional variants.

Amacrine cells in reptiles can also be grouped according to their putative neurotransmitters, as they have been for other vertebrates (reviewed in Karten and Brecha, 1983; Brecha, 1983; Brecha et al., 1984; Weiler, 1985; Marc, 1986; Massey and Redburn, 1987; Dowling, 1987; Daw et al., 1989). At present, the primary value of these transmitter-specific probes is to partition the variance in amacrine cell morphology; the role of these transmitters in visual information coding is still largely unknown (Daw et al., 1989, review recent evidence about the function of retinal transmitters). Studies using autoradiographic

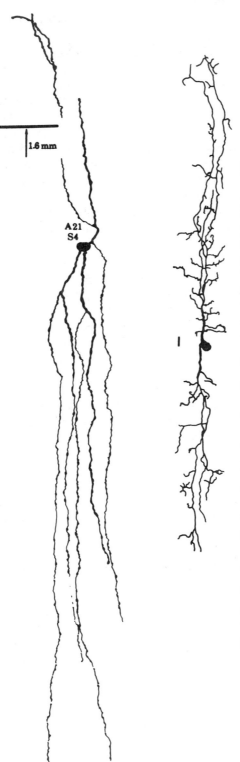

Fig. 1.43. *Pseudemys (Trachemys) scripta.* Amacrine cells with anisotropic dendritic fields. Upper: A21 amacrine cell visualized in whole mount of Golgi impregnated retina. The processes of this monostratified cell ramify in the fourth stratum (S4) of the inner plexiform layer. They are oriented orthogonal to the visual streak. The orientation of the visual streak is indicated by the black bar, which is approximately 70 μm long. This amacrine cell was located 1.6 mm from the streak. Lower: Amacrine cell showing enkephalinlike immunoreactivity. Its processes are oriented parallel to the visual streak. Scale: 10 μm. (Upper, from Kolb, 1982; Lower, from Eldred and Karten, 1983.)

or immunocytochemical probes for transmitters and their metabolic precursors have identified reptilian amacrine cells that are positive for *glutamate* (Ehinger et al., 1988), *glycine* (Marc, 1985; Weiler and Ball, 1984; Eldred and Cheung, 1989), *γ-aminobutyric acid* (*GABA*) (Brandon, 1985; Mosinger et al., 1986; Zucker and Adolph, 1988; Hurd and Eldred, 1989), *dopamine* (Scheie and Laties, 1971; Ehinger, 1982; Osborne et al., 1982; Engbretson and Battelle, 1987; Witkovsky et al., 1984; Nguyen-Legros et al., 1985; Witkovsky et al., 1987; Kolb et al., 1987), *serotonin* (Osborne et al., 1982; Witkovsky et al., 1984, 1987; Weiler and Schutte, 1985a; Engbretson and Battelle, 1987), and the neuropeptides *enkephalin* (Eldred and Karten, 1983, 1985; Zucker and Adolph, 1988), *Y* (Isayama and Eldred, 1988; Isayama et al., 1988), *substance P* (Reiner et al., 1984; Cuenca and Kolb, 1989), *glucagon* (Eldred and Karten, 1983), *neurotensin* (Eldred and Karten, 1983; Weiler and Ball, 1984; Eldred and Carraway, 1987), *LANT-6* (Eldred et al., 1987, 1988), *somatostatin* (Adolph and Zucker, 1988), and *corticotropin releasing factor* (*CRF*) (Williamson and Eldred, 1989). Many of these chemically identified neurons are also morphologically distinct.

Attempts to correlate chemically identified amacrine cells with the morphological classes seen in Golgi material have met with varying degrees of success (reviewed in Guiloff et al., 1988; Kolb et al., 1988). Several of the putative neurotransmitters have been localized to more than one morphological "type" of amacrine cell (e.g., Eldred and Cheung, 1989; Hurd and Eldred, 1989; Weiler and Schutte, 1985a; Eldred and Karten, 1983; Isayama and Eldred, 1988; Weiler and Ball, 1984; Eldred et al., 1987; Williamson and Eldred, 1989), but it seems likely that at least some of these types represent regional or random variations (Rowe and Stone, 1980) within a single amacrine cell class, rather than functionally distinct amacrine cell classes. With rare exceptions (e.g., Kolb et al., 1987) there have been no systematic attempts to assess local or regional variation in any chemically identified type of amacrine cell.

Colocalization of neurotransmitters has been demonstrated in a few amacrine cell types. One is immunoreactive for neurotensin and also accumulates glycine (Fig. 1.44; Weiler and Ball, 1984). Two amacrine cell types are immunoreactive for both neurotensin and LANT-6 (Eldred et al., 1987, 1988), and approximately 10% of amacrine cells in *Pseudemys* (*Trachemys*) *scripta* are immunoreactive for both enkephalin and glutamic acid decarboxylase (GAD; a key enzyme in the synthesis of GABA; Zucker and Adolph, 1988). In addition, there is indirect, ultrastructural evidence for such colocalization in enkephalin (Eldred and Karten, 1985), neuropeptide Y (Isayama et al., 1988), and neurotensin (Eldred and Carraway, 1987) immunoreactive amacrine cells because the large, dense-cored vesicles that contain these peptides are accompanied by small, unlabeled vesicles that are presumed to store "conventional" neurotransmitters.

Several laboratories have attempted to group amacrine cells in the retina of *Pseudemys (Trachemys) scripta* based on their physiological responses and subsequent visualization by dye injections. Probably the best-understood type, first described by Schwartz (1973), gives a transient response to both light onset and offset, and these "on" and "off" responses are differently affected by changes in stimulus diameter and intensity (Marchiafava and Torre, 1978; Marchiafava and Weiler, 1980, 1982; Marchiafava, 1982; Jensen and DeVoe, 1982), suggesting that they may be mediated by distinct classes of retinal elements, perhaps by two bipolar cell types (Marchiafava and Torre, 1978). In dye-injected material, these cells often have a flask-shaped soma with a single primary dendrite that extends to a point approximately half way between the inner nuclear layer and ganglion cell layer and there gives rise to a planar, characteristically radiate, arbor (Fig. 1.45, left; Jensen and Devoe, 1982). This cell may correspond to a type of amacrine cell that shows neurotensinlike reactivity (Fig. 1.45, right; Eldred and Karten, 1983; Eldred and Carraway, 1987; Weiler and Ball, 1984) and Kolb's cell A22 (see Table 2 in Kolb et al., 1989). Other amacrine cells that respond similarly to photostimulation are also monostratified in the middle inner plexiform layer but have a more irregular branching pattern (Marchiafava and Weiler, 1982) resembling that of Kolb's A20. This irregularly branching cell type may be immunoreactive for substance P (see Fig. 22 in Guiloff et al., 1988, and Table 2 in Kolb et al., 1988; but see Cuenca and Kolb, 1989, for a different pairing of substance P immunoreactive and Golgi impregnated amacrine cell types).

At present, the most comprehensive treatment of reptilian amacrine cells seeks to relate the time course of their membrane depolarization to the vertical distribution of their processes in the inner plexiform layer (Weiler and Marchiafava, 1981; Marchiafava and Weiler, 1982; Weiler and Ball, 1984; Ammermuller and Weiler, 1988). In *Pseudemys (Trachemys) scripta,* as in other vertebrates (Dowling, 1987), amacrine cells respond to center illumination with transient or sustained depolarizations (including center-on, center-off, and center-on-off variants), and the time course of a given amacrine cell response is positively correlated with the (vertical) thickness of its processes. This classification contains three broad categories of amacrine cells. (1) The *fast-transient class* gives a phasic response with rapid decay time (< 100 msec) and has narrowly monostratified dendritic arbors. Its response polarity can be predicted by the laminar distribution of its processes. One fast-transient subtype that ramifies near the *middle* of

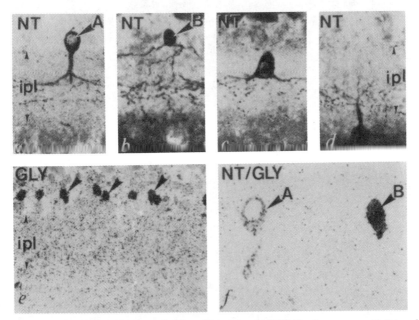

Fig. 1.44. *Pseudemys (Trachemys) scripta.* Neurotensin (NT)-like immunoreactivity and
³H-glycine uptake in turtle amacrine cells. Upper: Four morphological types of amacrine cells that exhibit NT-like immunoreactivity. Large Type A and small Type B amacrines are observed most frequently. Some immunoreactive cells have somata in the
inner plexiform layer (ipl; third cell in row) or ganglion cell layer (last cell in row) (318
x). Lower left: Autoradiograph showing amacrine cells that have accumulated ³H-
glycine (arrowheads). Lower right: Combined immunocytochemistry-autoradiog-
raphy showing NT-like immunoreactivity and ³H-glycine uptake. Both Type A and
Type B amacrines show NT-like immunoreactivity, but only the Type B amacrine
shows colocalization of ³H-glycine (650 x). (Reprinted by permission from Weiler and
Ball, *Nature* Vol. 311, pp. 759–761. Copyright (c) 1984 Macmillan Magazines Ltd.)

the inner plexiform layer gives an on-off response. These are probably
the amacrine cells described in the preceding paragraph (Schwartz,
1973; Marchiafava and Torre, 1978; Marchiafava and Weiler, 1980; Jensen and Devoe, 1982; see Fig. 1.45, left). A second fast-transient subtype (off-center) is monostratified within the *distal* inner plexiform
layer (i.e., laminae 1 or 2; Ammermuller and Weiler, 1988). Three variants of this second subtype have been identified (stellate, giant, and
"starburst") that differ structurally and in their photoresponses. A
stellatelike amacrine that ramifies in the *proximal* inner plexiform layer
may represent a third, on-center, variant of the fast-transient amacrine cell class (Ammermuller and Weiler, 1988, citing Kolb, 1982;
they do not identify the amacrine type). (2) *Slow-transient* amacrine
cells give an on-off response with a decay time up to 1 sec, and they
are either broadly monostratified (extending across approximately

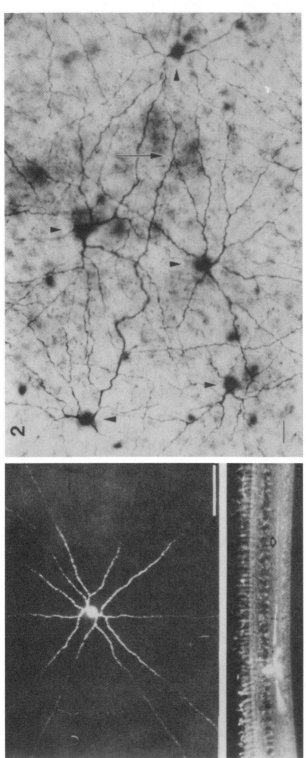

Fig. 1.45. *Pseudemys (Trachemys) scripta.* Monostratified amacrine cell. Left: On-off amacrine cell filled with lucifer yellow after physiological characterization of its receptive field. Its processes ramify in stratum 3 of the inner plexiform layer. The cell is shown in whole mounted (above) and vertically sectioned (below) retinas. Scale: 100 μm. Right: Type A amacrine cells (arrowheads) exhibiting neurotensinlike immunoreactivity. This cell type may correspond to the on-off amacrine cell of Jensen and DeVoe (1982). Arrowhead marks apparent contact between processes of two cells. Scale: 10 μm. (Left, from Jensen and DeVoe, 1982; right, reprinted with permission from *Neuroscience,* Vol. 21, Eldred and Carraway, Neurocircuitry of two types of neurotensin-containing amacrine cells in the turtle retina, 1987, Pergamon Press plc.)

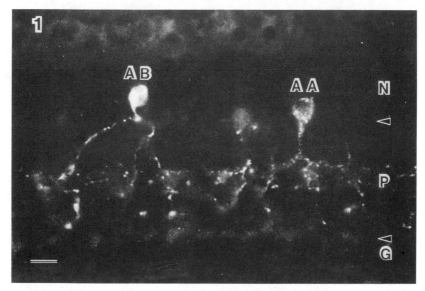

Fig. 1.46. *Pseudemys (Trachemys) scripta.* Two types of amacrine cells that exhibit neurotensinlike reactivity, visualized by immunofluorescence. Type A (AA) has a soma in the inner nuclear layer (N) and a single primary dendrite that arborizes in the middle of the inner plexiform layer (P). Type B (AB) has several fine primary dendrites that arborize in the proximal inner plexiform layer. This cell may correspond to an amacrine type that is immunoreactive for neurotensin and also accumulates ³H-glycine (see Fig. 1.44). Arrowheads mark the boundaries of the inner plexiform layer. Scale: 10 μm. (Reprinted with permission from Neuroscience, Vol. 21, Eldred and Carraway, Neurocircuitry of two types of neurotensin-containing amacrine cells in the turtle retina, 1987, Pergamon Press plc.)

50% of the inner plexiform layer) or bistratified. (3) *Sustained* amacrine cells were originally thought to stratify diffusely throughout the entire inner plexiform layer (Marchiafava and Weiler, 1982), but recent evidence indicates that one sustained, on-center amacrine cell arborizes diffusely within the proximal inner plexiform layer only. This cell corresponds to the Type B amacrine that is immunoreactive for neurotensin (Fig. 1.46; Eldred and Karten, 1983; Eldred and Carraway, 1987) and LANT-6 (Eldred et al., 1987) and probably to Kolb's A10 amacrine. Thus, within the three response groups of amacrine cells (fast-transient, slow-transient, and sustained), the morphology and lateral extent of their processes can vary widely; neither parameter predicts the time course of the amacrine cell response to center illumination as well as the vertical distribution of its processes does. Therefore, vertical coverage of the inner plexiform layer is a second nested variable, like inner and outer inner plexiform layer stratification types, that may be useful in partitioning the variance seen in the reptilian inner plexiform layer.

F. Interplexiform Cells

Classic descriptions of vertebrate retinas describe five cell types (see Fig. 1.1), but over the past 20 years it has become clear that a sixth class exists in some species (Dowling, 1987). These interplexiform cells resemble amacrine cells, except that their processes also extend into the outer plexiform layer where they are presynaptic to bipolar and horizontal cells. Thus, they are believed to mediate interactions between the two layers of retinal neuropil. Interplexiform cells have been identified in teleosts and mammals, where they are usually dopaminergic or GABAergic. Putative glycinergic interplexiform cells have also been described in fish and some amphibians.

Dopaminergic interplexiform cells in teleosts are known to modulate electrical coupling between horizontal cells. Dopamine also uncouples the L1 horizontal cell syncytium in turtles (see Section III.C.2.c), and several investigators have sought a dopaminergic interplexiform cell that might mediate this effect. To date, all available evidence suggests that turtle retinas contain no dopaminergic interplexiform cells (Witkovsky et al., 1984, 1987; Nguyen-Legros et al., 1985; Kolb et al., 1987), and these authors have suggested that any modulation of horizontal cell coupling by the one dopaminergic amacrine cell in turtle retinae (A28 of Kolb et al., 1988) must occur "at a distance" because the processes of this neuron ramify exclusively within the inner plexiform layer. Such humoral effects of retinal dopamine have been described in other vertebrates where they may mediate retinomotor movements (reviewed in Daw et al., 1989; see Section II.B.3.e). A putative glycinergic interplexiform cell has recently been described in *Pseudemys (Trachemys) scripta* (Eldred and Cheung, 1989); alternatively, these glycinergic neurons may be bipolar cells.

G. Ganglion Cells

Ganglion cell perikarya form the most vitreal layer of retinal elements (see Fig. 1.1), although they may occasionally be displaced to the inner plexiform layer or the proximal inner nuclear layer (e.g., Dogiel, 1888; Cajal, 1893; Kolb, 1982; Reiner, 1981; Eldred et al., 1988). Their dendritic processes enter the inner plexiform layer, where they make synaptic contact with bipolar and amacrine cells (Guiloff et al., 1988), as they do in other vertebrates (Rodieck, 1973; Dowling, 1987). Ultrastructurally, ganglion cell cytoplasm is more electron lucent than that of bipolar and amacrine cells. It is heavily invested with organelles such as rough endoplasmic reticulum, free ribosomes, fibrils, and mitochondria, but it lacks synaptic vesicles (*Pseudemys (Trachemys) scripta*, Villegas, 1960; Bush and Lunger, 1972; Guiloff et al., 1988; *Thamnophis sirtalis*, Chandler, 1974; Wong, 1989). Indeed, in analyses of inner plexiform layer circuitry, profiles that lack synaptic vesicles are generally presumed to arise from ganglion cells (e.g., Guiloff et al., 1988). The axons of ganglion cells leave the retina at the optic nerve and distribute terminals to a number of central visual targets. Thus, ganglion cells are the "final common path" by which the products of retinal analysis are transmitted to the rest of the central nervous system.

The number of ganglion cells characteristic of a given species varies dramatically, from iguanids in which the ganglion cell layer may be four or five cells thick (*Anolis* species, Makaretz and Levine, 1980; Fite and Lister, 1981; *Dipsosaurus dorsalis, Iguana iguana*, Peterson, unpublished), to *Thamnophis sirtalis* (Wong, 1989), *Crotalus* (Peterson, unpublished), and sea snakes (*Pelamis platurus*, Hibbard and Lavergne, 1972), in which ganglion cell density is much lower. Direct counts of ganglion cell numbers are only available for the turtle *Pseudemys (Trachemys) scripta*. Retinas 14 mm in diameter (116.5 mm² total area) have approximately 364,500 total ganglion cells (Peterson and Ulinski, 1979); this agrees well with estimates of axon numbers in the optic nerve (395,900; Geri et al., 1982). Integration of retinal isodensity maps suggests that there are approximately 127,500 ganglion cells in *Thamnophis* (retinal diameter, 5.8 mm; Wong, 1989). Estimates (Davydova et al., 1982) of ganglion cell numbers based on axon counts in the optic nerve are available for *Emys orbicularis* (483,000), *Mauremys (Clemmys) caspica* (302,000), *Testudo (Agrionemys) horsfieldi* (384,000), and *Testudo graeca* (213,000), but such figures are difficult to interpret without corresponding data on eye size.

Attempts to partition the total ganglion cell population into morphological groups have proceeded along three lines. Some investigators have grouped ganglion cells on the basis of structure, as visualized in Golgi material, with special emphasis on dendritic stratification patterns, since this presumably sets boundary conditions for the afferent contacts that can be made by each type. Others have identified ganglion cell types on the basis of their immunoreactivity to putative neurotransmitters. A third group of studies has focused on the physiological properties of ganglion cells, using intracellular injections of fluorescent dyes or HRP to characterize the morphology of ganglion cells with known receptive field characteristics. The result has been a wealth of descriptive data that so far lack a compelling organizational framework.

Detailed morphological descriptions of ganglion cells based on Golgi data are available for *Lacerta viridis* (Cajal, 1893), *Pseudemys (Trachemys) scripta* (Kolb, 1982), and *Mauremys caspica* (Kolb et al., 1988); there is also a brief account of two or three ganglion cell types in the sea snake *Pelamis platurus* (Hibbard and Lavergne, 1972). Cajal described nine ganglion cell varieties in *Lacerta*, each with a distinctive dendritic architecture, remarking that the "reptilian retina is rich in ganglion cell types" (Fig. 1.47). Twenty-one types have been described in *Pseudemys (Trachemys)* (Kolb, 1982) and *Mauremys* (Kolb et al., 1988). In all three species, a majority of ganglion cells ramify in more than one stratum of the inner plexiform layer. Thus, ganglion cells differ from amacrine cells (most of which are monostratified; Kolb, 1982; Kolb et al., 1988) in that they probably sample inner plexiform activity at more than one level. Most ganglion cells have highly branched "bushy" dendrites and, viewed tangentially, their dendritic fields tend to be circular or elliptical (Kolb, 1982; Kolb et al., 1988). The cell soma may be placed eccentrically in the dendritic field. In a few anisotropic types of ganglion cells, the long axis is oriented parallel or orthogonal to the horizontal meridian of the eye (G7, G12, G16), but no ganglion cells have the strikingly oriented processes seen in several amacrine cell types (A14, A18, A20, A21, A27; Kolb, 1982; Eldred and Karten, 1983; Williamson and Eldred, 1989).

Attempts to identify classes of ganglion cells with immunocytochemical probes have so far met with limited success. Ganglion cells in *Pseudemys (Trachemys)* have been identified that are immunoreactive to glycine (Eldred and Cheung, 1989), CRF (Williamson and Eldred, 1989), GABA (Mosinger et al., 1986; Hurd and Eldred, 1989), LANT-6 (Eldred et al., 1987, 1988), substance P (Cuenca and Kolb, 1989), and serotonin (Weiler and Ammermuller, 1986). In several cases, labeled profiles in the ganglion cell layer have identified as ganglion cells (rather than displaced amacrines) by backfilling from the optic tectum (Eldred et al., 1988; Williamson and Eldred, 1989; Hurd and Eldred, 1989) or physiological analysis (Weiler and Ammermuller, 1986). In each case, however, labeling of processes is poor, so it is impossible adequately to assess the dendritic structure of these immunoreactive neurons. As a result, no immunoreactive ganglion cells clearly correspond to types identified in Golgi preparations (Kolb, 1982; Eldred et al., 1988). Several classes of ganglion cells that are immunoreactive to LANT-6 have been described, including several large, monostratified "types" that ramify in distal, middle, and proximal inner plexiform layer. One or more of these types may give rise to the large-caliber axons that terminate in the nucleus of the basal optic root (Reiner, 1981; Eldred et al., 1988). The serotoninergic ganglion cells have an 18-μm soma and dendrites that ramify "mainly in layers 2 and 3" (Weiler and Ammermuller, 1986). They show on-off type responses to center illumination and are wavelength sensitive. Unfortunately, they do not correspond well to either of the physiological types that have been more completely characterized in turtle retina (Types A and B; see below).

Whenever ganglion cells are characterized physiologically and then

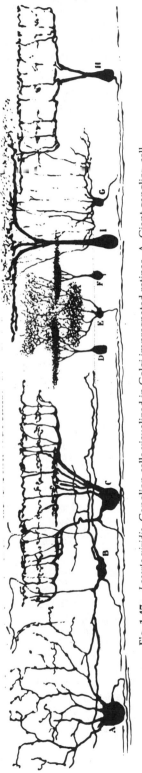

Fig. 1.47. *Lacerta viridis.* Ganglion cells visualized in Golgi impregnated retinas. A, Giant ganglion cell with diffuse branches. B, Horizontal ganglion cell, most of whose branches become lost at the fifth level. C, Polystratified ganglion cell forming plexuses at the second, third, and fourth levels. D and F, Two ganglion cells with granular arborizations at the fourth level. E, ganglion cell whose delicate branches fill the third, fourth, and upper half of the fifth levels. G, Ganglion cell whose branches terminate at the first level. H, A Ganglion cell similar to C, but smaller. Original drawings and descriptions of cells are from Cajal, 1893. (From *The vertebrate retina: principles of structure and function,* by R. W. Rodieck. Copyright (c) 1973 by W. H. Freeman and Company. Reprinted with permission by W. H. Freeman and Company.)

visualized with intracellular dye injections, relatively few types are described. This is probably because of electrode sampling biases and because cell labeling with fluorescent dyes or immunological probes seldom reveals the kind of morphological detail obtainable in Golgi material, which may give two neurons a sufficiently dissimilar appearance to earn them descriptions as separate cell types (e.g., G8/G9; G20/G21 in Kolb, 1982). Two attempts to describe the morphology of physiologically identified ganglion cells in *Pseudemys (Trachemys) scripta* have yielded eight and two types, respectively. Jensen and Devoe (1983) characterized cells on the basis of physiological parameters (interspike interval, center responses to spots of different sizes and directional relativity) and of gross morphology in vertical sections and whole mounts. They described eight dendritic patterns (Fig. 1.48). Ganglion cells resembling three of these types are directionally selective (Type I and, less clearly, Types II and VIII), but they differ from each other in dendritic morphology and stratification levels. Furthermore, not all Type I ganglion cells show directional properties. Thus, directionally selective ganglion cells do not constitute a unique morphological type. Only two types (I and VI) were monostratified, which is consistent with findings in Golgi material that a majority of ganglion cells ramify in more than one stratum of the inner plexiform layer. These two cells arborize in the distal (Type I) and proximal (Type VI) inner plexiform layer and, as might be expected, give hyperpolarizing and depolarizing responses, respectively, to center illumination. One bistratified type (III) is physiologically indistinguishable from a class of transient on-off amacrine cells (Schwartz, 1973), except that its latencies are 3 to 6 msec longer (Jensen and Devoe, 1982), suggesting that these two on-off retinal cells may be synaptically coupled. Probable correspondences between the ganglion cell types of Jensen and Devoe (1983) and Golgi impregnated types are given in Kolb et al. (1988, Table 2).

Marchiafava, Weiler, and their colleagues have investigated the properties of two ganglion cell types in detail (Marchiafava and Weiler, 1980; Marchiafava and Wagner, 1981; Weiler and Marchiafava, 1981; Marchiafava, 1982; 1983). Their Type A ganglion cells are encountered relatively infrequently (33%). They exhibit center-surround antagonism and may produce either transient or sustained responses as a function of stimulus conditions, but they are most often sustained. Type A cells appear to be largely bipolar driven and have color opponent organization with one peak at 646 nm and a second peak at one of several shorter wavelengths, suggesting that all Type A ganglion cells receive input from red-dominated bipolar cells and from one other single cone type. They have high firing rates and fast conduction velocities, indicating that Type A cells have large-

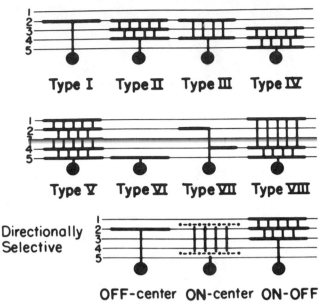

Type I Type II Type III Type IV

Type V Type VI Type VII Type VIII

Directionally Selective

OFF-center ON-center ON-OFF

Fig. 1.48. *Pseudemys (Trachemys) scripta.* Dendritic arborizations of physiologically characterized ganglion cells. Directionally selective ganglion cells exhibit one of three dendritic stratification patterns. Each of the eight physiological types that are not directionally selective have distinctive dendritic morphologies. Type I is very similar to the off-center directionally select ganglion cell. (From Jensen and DeVoe, 1983.)

caliber axons. This is consistent with observations that they have relatively large cell bodies and dendritic fields. Their large cell size means that Type A cells will probably be encountered preferentially in electrode penetrations and, thus, their relative frequency in the retina is probably lower than that suggested by their encounter rate (33%) in physiological experiments.

Type A ganglion cells are monostratified and have two variants, an off-center cell that ramifies in the outermost inner plexiform layer and an on-center cell with dendrites in the proximal inner plexiform layer (Figs. 1.49 and 1.50). Their large, relatively straight, unbranched dendrites and stratification patterns suggest that they probably correspond to Kolb's G20 and G21 ganglion cells, respectively, and perhaps to the Type I and Type VI cells of Jensen and Devoe (Peterson, 1982; Jensen and Devoe, 1983). Kolb seems to have first opposed (1982) and then supported (Kolb et al., 1988) this view. These cells are always the largest profiles at any retinal locus and have circular or elliptical dendritic fields as large as 1 mm in diameter. They probably account for less than 5% of the total ganglion cell population (Peterson and Ulinski, 1982). Cells resembling off-center Type A cells are labeled following application of HRP to the nucleus of the basal optic root (nBOR;

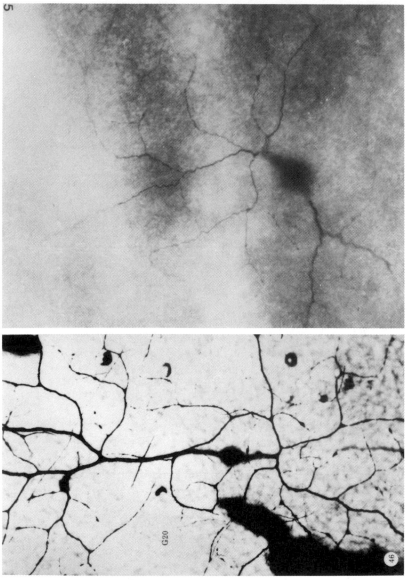

Fig. 1.49. *Pseudemys (Trachemys) scripta.* Monostratified ganglion cell visualized in whole mounted retinas. Upper: This Type A cell gave a sustained off-center response to light. It ramifies in the outermost portion of the inner plexiform layer (ipl). The ganglion cell has been injected with horseradish peroxidase (330 x). Lower: G20 ganglion cell visualized in Golgi impregnated retina. Its processes ramify at the border between strata 1 and 2 of the ipl. This ganglion cell may correspond to the off-center Type A cell of Marchiafava (1983). (370 x) (Upper, reprinted with permission from *Vision Research*, Vol. 23, Marchiafava, The organization of inputs establishes two functional and morphologically identifiable classes of ganglion cells in the retina of the turtle, 1983, Pergamon Press plc; lower, from Kolb, 1982.)

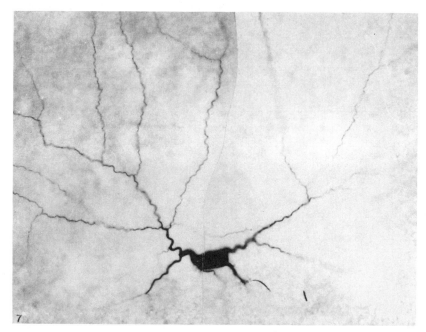

Fig. 1.50. *Pseudemys (Trachemys) scripta.* A Type A ganglion cell that gave a sustained on-center response to light. Its processes ramify in the innermost portion of the inner plexiform layer. This HRP-injected cell may correspond to the G21 type of Kolb (1982) (330 x). (Reprinted with permission from *Vision Research,* Vol. 23, Marchiafava, The organization of inputs establishes two functional and morphologically identifiable classes of ganglion cells in the retina of the turtle, 1983, Pergamon Press plc.)

Reiner, 1981), which is consistent with observations that Type A ganglion cells have large-caliber axons and that nBOR typically is supplied by the largest-caliber retinal efferents. This suggests that Type A ganglion cells may be involved in optomotor (optokinetic and eye-head) coordination through the accessory optic system (Reiner and Karten, 1978; Fite et al., 1979). As noted previously, these ganglion cells probably are immunoreactive for LANT-6.

Type B ganglion cells (Fig. 1.51), like Type A cells, can give either transient or sustained responses depending on stimulus conditions (Marchiafava and Weiler, 1980), but most often they show a transient on-off response to center illumination (Marchiafava, 1983). Physiological analysis suggests that they receive mixed bipolar and amacrine cell input and have a single peak sensitivity at 646 nm; thus they are not color coded. Some are directionally selective for a limited range of stimulus velocities. Type B ganglion cells appear to have slightly smaller somata and dendritic field diameters and much slower conduction velocities (0.4 ± 0.3 m/sec) than Type A cells. Their smaller

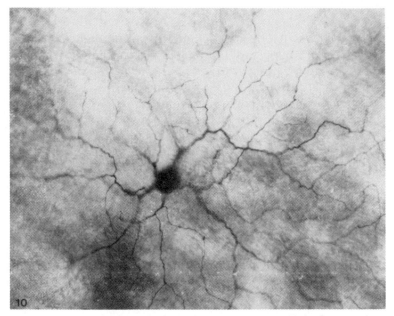

Fig. 1.51. *Pseudemys (Trachemys) scripta.* Type B ganglion cell processes appear more highly branched than those of Type A ganglion cells when injected with HRP and visualized in whole mounted retinas (compare Figs. 1.49 and 1.50). These dendrites ramify diffusely throughout the inner plexiform layer. Type B ganglion cells give a transient, on-off response to light (330 x). (Reprinted with permission from *Vision Research*, Vol. 23, Marchiafava. The organization of inputs establishes two functional and morphologically identifiable classes of ganglion cells in the retina of the turtle, 1983, Pergamon Press plc.)

cell bodies and higher encounter rates (67% of total ganglion cells) suggest that their relative frequency is more than twice that of Type A cells. Many Type B cells exhibit a brief depolarization on the falling limb of their spikes (post-spike depolarization) that has been ascribed to the action of recurrent axon collaterals from the same or adjacent ganglion cells (Marchiafava and Torre, 1977).

In vertical sections, Type B cells are either multistratified or diffuse (Marchiafava and Weiler, 1980). The size and dendritic architecture of the multistratified Type B cells suggest that they may correspond to Type II ganglion cells, as suggested by Jensen and Devoe (1983). This is consistent with observations that some Type B cells are directionally selective (Marchiafava, 1983). Diffusely stratifying Type B cells resemble Kolb's G13 cell most closely (Kolb et al., 1988). Unfortunately, correspondences between Type B ganglion cells and those visualized with other techniques are extremely difficult to determine in the absence of accurate data on retinal location, cell size, and stratification levels in the inner plexiform layer.

III. CONSTRUCTION OF INFORMATION CHANNELS:
RETINAL CIRCUITRY
A. Overview

The previous section described the structure of neurons in reptilian retinas. These cells interact through chemical and electrical synaptic contacts in the two plexiform layers (see Fig. 1.1) to form channels for the analysis and transmission of information about the visual field. Such interactions constitute the retinal circuitry. We know relatively little about these interconnections, but it may nevertheless be useful to consider the available data and attempt to develop a picture of information channels in reptilian retinas. Virtually all the relevant data come from work on the retinas of turtles (usually *Pseudemys (Trachemys) scripta*).

Before proceeding, it is necessary to define certain terms that describe the physiological properties of retinal cells. Briefly, the region of retinal or visual space within which changes in illumination can influence the output of a retinal cell is called that cell's *receptive field*. Such a receptive field can be characterized in several ways. For example, retinal neurons respond to local illumination of their receptive fields with depolarization or hyperpolarization. *On-cells* depolarize to light on and hyperpolarize to light off, whereas *off-cells* depolarize to light off and hyperpolarize to light on. In other words, on-cells are excited by an increase in illumination and off-cells are excited by a decrease.[5] If the response is maintained throughout the stimulus presentation, it is described as tonic or *sustained;* if the membrane potential (or spike frequency, in the case of ganglion cells and some amacrine cells) declines significantly during presentation of a constant stimulus, the cell is described as *phasic* or *transient*. Some amacrine and ganglion cells give phasic depolarizations at both onset and offset of a light stimulus (*on-off cells*). The receptive fields of many retinal neurons also have an internal spatial organization, responding with potential changes of opposite polarity, depending on whether the stimulus falls in the center or periphery of their receptive field. Such cells are said to exhibit *center-surround antagonism* (sometimes called spatial opponency or spatial contrast sensitivity), and the available data suggest that the center and surround responses are generated by different retinal mechanisms. The receptive fields of on-, off- or on-off retinal cells can also be characterized in terms of their shape, spectral and movement sensitivity, direction or orientation selectivity, temporal contrast sensitivity, and spatial frequency selectivity (size). These receptive field properties are believed to encode information about the visual image, although the relation between the receptive field characteristics of a cell and the information it encodes is not always clear (see Dowling, 1987, for a summary of receptive field generation in retinal neurons).

B. Vertical Organization: Structure of the On and Off Pathways

1. GENERAL The minimum path by which light-generated electrical activity can reach the intracranial CNS involves three retinal elements: photoreceptor → bipolar cell → ganglion cell. This pathway is responsible for generating one of the most fundamental features of bipolar and ganglion cell activity, namely, that they respond to focal increases in retinal illumination with either depolarizing (center-on) or hyperpolarizing (center-off) changes in membrane potential in spite of the fact that the photoreceptor response to light is always a hyperpolarization (Kaneko, 1979; see footnote 5). Thus, a first step in constructing retinal information channels in reptiles is to outline what we know about this straight signal pathway and the generation of on and off responses. It is convenient to divide the relevant circuitry into two parts: (1) photoreceptor-bipolar cell contacts, at which the photoreceptor response polarity may be either preserved (yielding hyperpolarizing bipolar cells) or reversed (to yield depolarizing bipolar cells), and (2) inner plexiform layer contacts, at which the two resulting bipolar cell types impart their response properties to more proximal retinal elements (amacrine cells and ganglion cells).

2. PHOTORECEPTOR-BIPOLAR JUNCTIONS

Bipolar cells contact photoreceptors overlying their dendritic fields, and it is generally accepted that these junctions account for the properties of bipolar cell receptive field centers (Kaneko, 1979; Miller, 1979; Dowling, 1987). The size of the dendritic field varies with bipolar cell type and retinal locus by almost a factor of 10, from less than 15 μm to 120 μm (Richter and Simon, 1975; Kolb, 1982; Copenhagen et al., 1983; Schutte and Weiler, 1987), but it is always smaller than the physiological receptive field center of the same bipolar cell due to electrical coupling between the presynaptic photoreceptors (see Section III.C.2.b). Because the bipolar cells of turtles also differ in their center response polarities (Schwartz, 1974; Yazulla, 1976b; Ashmore and Copenhagen, 1980; Weiler, 1981; Copenhagen et al., 1983), it is likely that their contacts with the overlying photoreceptors also differ. No direct morphological data on the afferent contacts of physiologically identified center-hyperpolarizing and center-depolarizing bipolar cells in reptiles account for their different response polarities; however, some inferences can be drawn from ultrastructural and freeze-fracture data on the contacts between bipolar cells and photoreceptor pedicles.

In turtles, as in other vertebrates, bipolar cells are postsynaptic to photoreceptors at basal and invaginating (ribbon) junctions (see Figs. 1.21, 1.22, 1.24, and 1.52, and Section II.B.5.c). At ribbon junctions, the tip of the bipolar cell process is lodged at the apex of a synaptic ridge, where it forms the central element of a triad. The intramembrane structure of these apical ribbon junctions resembles that of central inhibitory (sign inverting) synapses (*Pseudemys (Trachemys)*, Raviola, 1976; Schaeffer et al., 1982). In contrast, the freeze-fracture appearance of basal junctions resembles that of excitatory contacts elsewhere in the central nervous system, suggesting that it may function as a sign-preserving synapse. Because center-hyperpolarizing and center-depolarizing bipolar cells are linked to photoreceptors by sign-preserving and sign-inverting synapses, respectively, an appealing inference is that center-hyperpolarizing bipolar cells make basal junctions with photoreceptors and that center-depolarizing bipolar cells form ribbon synapses. Indeed, this seems to be the case in mammals (reviewed in Sterling, 1983; Dowling, 1987) and some teleost fish (Stell et al., 1977).

Unfortunately, this association between contact type, intramembrane structure, and response polarity is less predictable in turtles. Certain bipolar cells make basal junctions exclusively (Lasansky, 1971; Kolb et al., 1986). At least some of these (B4 bipolars of Kolb, 1982) resemble bipolar cells that are known from physiological analyses to be hyperpolarizing (Marchiafava and Weiler, 1980; Ashmore and Copenhagen, 1980, 1983). As might be expected, the axons of these neurons arborize in the distal inner plexiform layer. They contact cones exclusively, and always through wide-cleft basal junctions. In contrast, neurons resembling depolarizing (on) bipolars (B6 bipolar cells of Kolb, 1982) contact cones through "narrow-cleft, semiinvaginating" contacts, that is, through processes that approach but do not reach the presynaptic apex and make basal/distal synapses en route (Kolb et al., 1986; see Section II.B.5.c). These bipolar cells ramify exclusively in strata 3 to 5 of the inner plexiform layer. Thus, two bipolar types that arborize exclusively in distal (off) and proximal (on) zones of the inner plexiform layer form photoreceptor contacts resembling those seen in mammals and teleost fish, suggesting that wide-cleft basal junctions may be sign-preserving (excitatory) contacts and narrow-cleft semiinvaginating contacts may be sign-reversing (inhibitory) contacts.

However, some bipolar cell types present a different pattern. Serial reconstructions of bipolar cell dendritic trees reveal that some small bipolars make *both* ribbon and basal contacts with the same photoreceptors (Fig. 1.52; Dacheux, 1982). These small bipolars have "axon terminal collaterals occupying two sublaminae of the middle" inner plexiform layer (Dacheux, 1982). They do not clearly correspond to any of Kolb's nine bipolar types (Kolb, 1982) or to any physiologically identified bipolars; however, if ribbon and basal contacts simply mediate synaptic linkages of different polarity, then Dacheux's small bi-

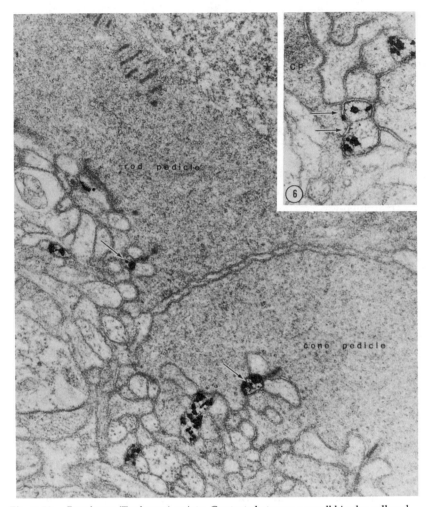

Fig. 1.52. *Pseudemys (Trachemys) scripta.* Contacts between a small bipolar cell and photoreceptor pedicles visualized by Golgi impregnated, gold-toned retina. One dendritic process terminates as the central element of a cone triad (arrow). A second process from the same dendrite ends as the central element of a rod triad (arrow, left). Upper right: The process that ended as the central element of a cone triad also makes a basal contact (arrows) with the same cone pedicle. The labeled processes in these two electron micrographs were traced through contiguous serial sections to the same small bipolar cell (10,500 x). (From Dacheux, 1982.)

polar cells will be driven in opposite directions by activation of over-lying photoreceptors.

The complexities and apparent contradictions that characterize photoreceptor-bipolar contacts in nonmammals have received wide attention (reviewed in Kolb et al., 1986; Dowling, 1987; Daw et al., 1989) and have not yet been resolved. In this context, it is worth emphasizing that postsynaptic response polarity probably depends on the nature of the receptors that control postsynaptic membrane ion conductances, and several forms of conductance change have been identified at vertebrate photoreceptor contacts (reviewed in Miller and Slaughter, 1986; Dowling, 1987). Thus, the morphological heterogeneity of photoreceptor-bipolar contacts may reflect the diversity of postsynaptic ion channel mechanisms rather than the (binary) polarity of the postsynaptic response (Kolb et al., 1986).

3. BIPOLAR–GANGLION CELL JUNCTIONS

Output from the outer plexiform layer, whether sign preserving or sign reversing, is transmitted to amacrine and ganglion cells in the inner plexiform layer by the central processes of bipolar cells. There is abundant physiological evidence that reptilian ganglion cells (and amacrine cells) exhibit center-on and center-off responses (e.g., Baylor and Fettiplace, 1977a, 1977b; Jensen and Devoe, 1983; Marchiafava, 1983; Bowling, 1980; Granda and Fullbrook, 1989). The presumption is that this property is conferred by contacts with appropriate bipolar cells, but the morphological substrate for these interactions is not well understood. We know that mono- or multistratified bipolar cells with axonal arbors restricted to roughly the outer half of the inner plexiform layer are usually center hyperpolarizing, whereas bipolar cells with axons restricted to the proximal inner plexiform layer are depolarizing (Weiler, 1981; Marchiafava and Weiler, 1980; Ashmore and Copenhagen, 1983). Ganglion cells contacting these bipolar types can be expected to be center hyperpolarizing or depolarizing, respectively. The response polarities of bipolar cells that ramify in *both* the distal and proximal inner plexiform layer are less predictable. As described in Section II.D., there is a general tendency for the most distal process of such bipolars to correlate with the response polarity of the cell, but there is considerable overlap. For example, Weiler (1981) has described two bistratified bipolar cells that gave opposite responses to center illumination but apparently have identical stratification patterns in the inner plexiform layer. We do not know how, if at all, such widely stratified bipolar cells confer response polarity on retinal ganglion cells.

At least one ganglion cell type appears to be dominated by activity in the straight signal pathway. These Type A ganglion cells receive their primary input from bipolar cells (Marchiafava and Weiler, 1980; Marchiafava, 1983; see Section II.G and Figs. 1.49 and 1.50). They probably correspond to G20 and G21 ganglion cells (Kolb, 1982), and they are monostratified, typically in either the distal (strata 1/2) or proximal (stratum 4) inner plexiform layer, respectively (Kolb et al., 1988). The type or types of bipolar cell they contact is not known. Presumably, off-center Type A cells contact bipolars with at least one distally ramifying axon terminal (e.g., B3-B5, B8, B9), whereas on-center Type A cells should contact bipolars having all axonal arbors confined to the proximal inner plexiform layer (e.g., B1, B2, B6, B7).

These contacts between Type A ganglion cells and bipolar cells, if confirmed, may represent a substantial proportion of total bipolar-to-ganglion-cell contacts in the retina of turtles. A recent ultrastructural analysis of the inner plexiform layer of *Pseudemys (Trachemys) scripta* suggests that only a small fraction of bipolar cell synapses are made directly with ganglion cells, especially in the visual streak (Guiloff et al., 1988). Thus, the response polarity of most ganglion cells must be conferred indirectly through the more numerous bipolar contacts with amacrine cells. Interestingly, ganglion cells that receive *direct* bipolar input make these contacts almost exclusively in layers 2 and 4 (see Figs. 13 and 14 in Guiloff et al., 1988). This is consistent with physiological data described previously that Type A ganglion cells, which are unistratified either at the border between layers 1 and 2, or in layer 4, receive direct bipolar cell input. Such ganglion cells are thought preferentially to monitor and transmit outer plexiform layer processing (Marchiafava, 1983). Thus, they will tend to signal spatial and wavelength information (stimulus properties coded in the outer plexiform layer) rather than the temporal stimulus parameters that generally are the province of the inner plexiform layer (Dowling, 1987).

C. Tangential Organization: Further Elaboration of Receptive Field Properties

1. GENERAL   The previous section described how information about local increases and decreases in retinal flux is transmitted vertically from photoreceptors to ganglion cells. However, electrical activity generated at a point in the photoreceptor layer also spreads tangentially across the retina, and this lateral flow of activity allows extraction of additional information about the visual array. Lateral spread occurs in two ways: (1) Retinal elements of the same kind may be interconnected in electrical networks, and (2) two cell types that have primarily a lateral orientation (horizontal cells and amacrine cells; see Fig. 1.1) interconnect relatively distant points on the retina by chemical synapses and perform local operations on information flowing in the straight signal pathways. This section summarizes what we know about these two types of tangential interactions in reptilian retinas.

2. ELECTRICAL NETWORKS

a. GENERAL   Abundant evidence indicates that at least three classes of vertebrate retinal cells are interconnected by electrical networks (Kaneko, 1979). Some of these interactions, especially those between photoreceptors and between horizontal cells, are particularly well documented in reptiles (turtles: *Pseudemys (Trachemys) scripta* and *Chelydra serpentina*), and they are summarized in Sections III.C.2.b and III.C.2.c. In addition there are isolated reports of electrical coupling between turtle amacrine cells (Jensen and Devoe, 1981) and ganglion cells (Stewart, 1978). Many of these phenomena have also been described in nonreptiles, suggesting that they may be a general design feature of vertebrate retinas (Kaneko, 1979; Dowling, 1987). Taken together, these phenomena suggest that the outer and inner plexiform layers are built up of superimposed syncytial networks that mediate a tangential flow of information over retinal distances far greater than has hitherto been supposed.

b. PHOTORECEPTORS

Dye injected into a single turtle cone can move into an adjacent cone, suggesting that they are connected by low-resistance junctions (Stewart, 1978). Consistent with this are observations that current injected into a turtle cone can be detected by an electrode in neighboring cones (Baylor et al., 1971; Detwiler and Hodgkin, 1979). This coupling is electrical and bidirectional. It is thought to mediate the spatial characteristics of adaptation (Copenhagen and Green, 1987) and to improve the signal-to-noise ratio when cones are stimulated in dim light. Both direct and indirect physiological evidence indicates that cones are electrically coupled only to cones of the same spectral type (Baylor and Hodgkin, 1973; Lamb and Simon, 1976; Detwiler and Hodgkin, 1979). These data suggest that red- and green-sensitive cones form independent electrical networks.

Two candidates for this cone-cone electrical coupling are the gap junctions that have been observed between cone pedicles and between basal processes in several vertebrates, including turtles (see Section II.B.5.c). In *Pseudemys (Trachemys) scripta,* as in primates (Raviola and Gilula, 1973), gap junctions between cone pedicles are found in regions of high photoreceptor density (Kolb and Jones, 1984, 1985). The function of these contacts is not known. They probably do not mediate coupling between spectrally similar cones, because they appear to link all contiguous cone pedicles, irrespective of spectral type (Kolb and Jones, 1984, 1985).

A more likely substrate for electrical coupling of spectrally similar cones is the gap junction between basal processes that link more widely separated cones (Lasansky, 1971; Kolb and Jones, 1984, 1985). The lengths of these basal processes are reasonably consistent with data on effective coupling distances (up to 40 μm, Baylor et al., 1971; 40 to 61 μm, Detwiler and Hodgkin, 1979) if we assume that the effects of coupling are limited to contiguous cells (i.e., current from a polarized photoreceptor cannot flow across adjacent cells and into more distal photoreceptors) and that basal processes can contact either other basal processes or the cone pedicles themselves. Unfortunately, there are no direct data on the spectral properties of cones linked by gap junctions between basal processes because the basal process of red and green cones are apparently indistinguishable (Kolb and Jones, 1985). This structural similarity is surprising, given physiological observations that the effective coupling distances between green cones (39 μm) are longer than those for red cones (26 μm; Detwiler and Hodgkin, 1979; but note that the green cone figures are based on only two pairs) and that the less numerous green cones are necessarily more widely separated than are red cones. One might expect the basal process of green cones to be longer than those of red cones if they link photoreceptors of like spectral type.

The preceding data suggest that turtle cones will sum inputs over an area 50 to 80 μm in diameter, and this is consistent with the available physiological evidence (e.g., Baylor and Hodgkin, 1973). However, we do not know whether a single cone contacts all photoreceptors of the same spectral type within the range of its basal processes. The most widely accepted explanation of photoreceptor coupling is that it acts to average intrinsic noise (random fluctuations in the membrane potential of visual cells, which is most pronounced at low illumination levels) and so increases the signal-to-noise ratio in dim light (see Lamb and Simon, 1976). If this is so, it would seem advantageous for a photoreceptor to contact as many other photoreceptors as is consistent with the spatial frequency demands of that cone system. Thus, we would expect basal processes of red and green cones to contact all appropriate photoreceptors within their range.

Rods, like cones, are interconnected by gap junctions between telodendria (Owen, 1985) that probably mediate the rod-rod electrical coupling reported by several authors (Schwartz, 1976; Copenhagen and Owen, 1976b; Detwiler et al., 1980). However, the behavior of the rod network differs from that seen in coupled cones in several respects for which there is currently no good explanation (see Detwiler et al., 1980, p. 214). One of the most striking differences is that the effective coupling distance for rods (120 to 200 μm) is much greater than for cones (40 μm). This is true whether coupling distances are calculated by simultaneously recording from rod pairs or by measuring the distance over which a single rod sums inputs. Most of the summation distances measured for cones can be accounted for by assuming that the cells are interconnected by basal processes (10 to 40 μm long in cones) and that current flow in one cone will affect the membrane potential of only the immediately adjacent cones. However, the summation distances of rods (120 to 150 μm, Copenhagen and Owen, 1976a, 1976b; 200 to 250 μm, Schwartz, 1973, 1974) are greater than twice the length of the longest basal processes (35 to 45 μm, Lasansky, 1971; Owen, 1985; see Fig. 1.20). Thus, coupling between rods must be tighter than between cones, so current flow set up in one rod by light or electrical stimulation will spread beyond the adjacent receptors. Unfortunately, there is no morphological information about differences between the gap junctions joining rods and those joining cones. Owen (1985) has calculated that the size of gap junctions linking rod telodendria is sufficient to mediate the observed coupling resistance between rods; there has been no comparable quantitative analysis for gap junctions between cones in *Chelydra* or any other reptile.

It is not clear how electrical coupling between photoreceptors, which increases their receptive field size, is compatible with high-resolution vision. Interestingly, Schwartz (1976) has calculated that voltage spread in the rod syncytium evoked by a point source is not much greater than the light scatter from the same source. Thus, electrical coupling in this system will have little degrading effect on spatial resolution. Presumably the effect of cone networks on resolution will be even less, given their weaker coupling characteristics.

c. Horizontal Cells

When a fluorescent dye is injected into the soma of a luminosity-type horizontal cell body (H1CB, Leeper, 1978a), the dye diffuses through the fine axon and into the axon terminal of the injected cell. In addition, a small number of neighboring H1 cell bodies are labeled. If the same dye is injected into the *axon terminal* of a horizontal cell (Leeper's H1AT), adjacent axon terminals are labeled, and the dye spreads extensively, yielding a meshwork of filled axonal processes that can extend laterally for several hundred micrometers (Fig. 1.53; Stewart, 1978; Piccolino et al., 1982, 1984; Byzov and Shura-Bura, 1985). These findings are consistent with several physiological findings (Simon, 1973; Piccolino et al., 1982): (1) both types of luminosity horizontal cells (H1CB, H1AT) are synaptically linked to cells of their same type; (2) the receptive fields of these cells are larger than can be accounted for by their dendritic arborization, and (3) H1CB and H1AT cells have very different receptive field sizes even though they differ little in morphological extent (see Figs. 1.31 and 1.33). Specifically, H1CB cells have dendritic spreads of 120 to 160 μm in peripheral retina (Leeper, 1978a; Normann and Kolb, 1981), but they sum responses to retinal illumination over twice that distance (300 μm; Simon, 1973). H1AT cells have dendritic spreads similar to that of H1CB cells at the same retinal locus (Normann and Kolb, 1981), but their summation areas average 750 μm, approximately five times the size of their dendritic

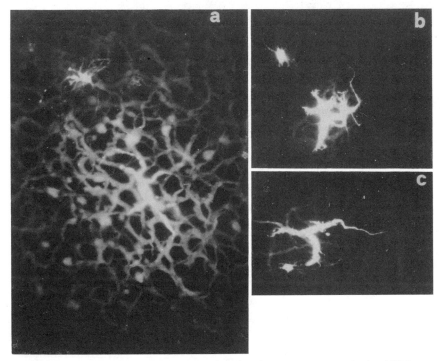

Fig. 1.53. *Pseudemys (Trachemys) scripta.* Coupling between axon terminals of H1 horizontal cells as visualized by intracellular injection of Lucifer Yellow. Left: Injection of one axon terminal leads to extensive spread of the dye into neighboring axon terminals. Numerous horizontal cell somata (small round profiles) are also labeled; the well-filled perikaryon at the upper left probably corresponds to the injected axon terminal. Upper right: The spread of Lucifer Yellow is decreased after a bath application of 10 μm forskolin, which stimulates adenylate cyclase activity and thus increases intracellular cyclic AMP concentrations. The effect resembles that induced by dopamine. Lower right: Dye spread is reduced with 10 μm dopamine in the bathing medium. This treatment also narrows the receptive field of the horizontal cell axon terminal. (From Piccolino et al., 1984.)

arbors and two to three times the receptive field size of the H1CB type.

Differences between the dendritic and receptive field sizes of H1 horizontal cells are probably mediated by the gap junctions that link these retinal neurons (see Section II.C.2.c); that is, electrical coupling between H1 somata or between axon terminals extends the distance over which these cells respond to light beyond the boundaries of their dendrites. Differences in the coupling characteristics of H1 somata and H1 axons may be mediated by structural differences in their respective gap junctions (Witkovsky et al., 1983). Freeze-fracture studies suggest that the junctions linking H1AT cells are larger (i.e., more P face particles) and perhaps more numerous than those linking the

H1CB type (see Figs. 1.34 and 1.35). Since the intramembrane specializations in the two cases do not otherwise differ, the different summation areas of H1AT and H1CB cells may reflect the number of channels interconnecting cells of each type.

Interestingly, the permeability of these gap junctions is modulated by dopamine (Fig. 1.53; Gerschenfeld et al., 1982; Piccolino et al., 1982, 1984, 1985, 1987; Weiler et al., 1988), which acts on D1 receptors in the horizontal cell membranes to decrease gap junction permeability by a mechanism that probably involves cyclic AMP as a second messenger. The effect apparently involves changes in the density of gap junction connexons, since dopamine-depleted retinas exhibit a significant increase in these membrane particles (Fig. 1.54; Weiler et al., 1988). This light-activated dopaminergic system will therefore decrease the spatial extent of horizontal cell coupling, with consequent effects on photoreceptor surround inhibition and adaptation levels (see Section III.C.3). The source of retinal dopamine in turtles is unclear. Dopaminergic interplexiform cells, which mediate these effects in teleosts (Dowling, 1987), are not found in *Pseudemys (Trachemys) scripta* (see Section II.F). Indeed, no dopaminergic processes of any kind contact turtle horizontal cells (Piccolino et al., 1987), suggesting that coupling between these visual neurons is modulated by humoral effects of dopamine, perhaps from dopaminergic amacrine cells (see Section II.E and F).

3. LATERAL EFFECTS MEDIATED BY CHEMICAL TRANSMISSION

a. INTERACTIONS BETWEEN HORIZONTAL CELLS AND PHOTORECEPTORS   Two cell classes in vertebrate retinas mediate lateral interactions that differ from the electrical network effects described above. Horizontal cells contact photoreceptors, other horizontal cells, and bipolar cells in the outer plexiform layer, where they contribute to the receptive field structures and response dynamics of these cells (e.g., center-surround antagonism, regulation of sensitivity levels, color coding). Amacrine cells contact bipolar and ganglion cells in the inner plexiform layer, where they mediate lateral interactions and localized operations that are believed to underlie complex response characteristics such as motion, direction, and orientation selectivity. A substantial amount of information is now available on the circuitry of horizontal cells in reptiles, but with one significant exception (Guiloff et al., 1988), we know little about amacrine cell connectivity in the inner plexiform layer. Accordingly, this section focuses on interactions mediated by horizontal cells.

There are four morphologically distinct horizontal cell types in *Pseudemys (Trachemys) scripta*, one of which (the H1 or luminosity cell) has a cell body (L1 cell) and a physiologically independent axon ter-

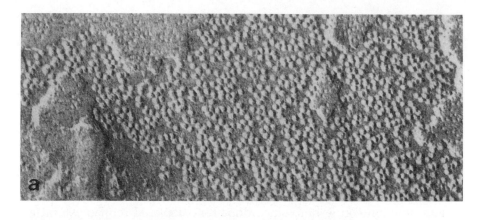

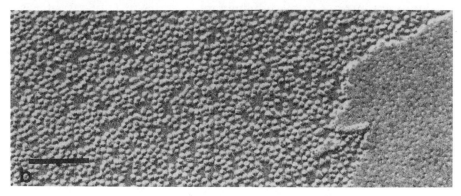

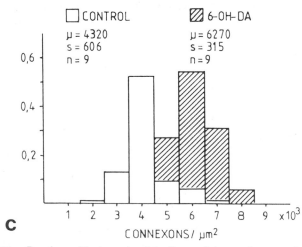

Fig. 1.54. *Pseudemys (Trachemys) scripta.* Freeze-fracture electron micrographs of gap junctions between horizontal cell axon terminals in control (upper) and dopamine-depleted (lower) retinas. The density of gap junction particles is higher in retinas that have been treated with 6-hydroxydopamine (6-OHD) to deplete dopamine. These differences are quantified in the histogram. Scale: 0.1 μm. (From Weiler et al., 1988.)

minal (L2 cell; see Section II.C.1). Leeper (1978a, 1978b) used serial reconstructions of 1-μm sections through identified, Golgi-impregnated horizontal cells to characterize their photoreceptor contacts at the light microscope level, relying on oil droplet size to distinguish different spectral types of cones. He concluded that each horizontal cell type engages in an idiosyncratic pattern of contacts with photoreceptors (Fig. 1.55). H1CB cells contact red- and green-sensitive single cones and the chief and accessory members of double cones. There is some indication that chief cones make most of the contacts onto distal H1CB processes, whereas contacts with the other three photoreceptor types occur closer to the soma, probably in a region corresponding to the bushy cluster of ascending processes overlying the soma (see Fig. 1.31). H1AT cells contact single red cones, the chief member of double cones, and rods. H2 cells contact green single cones and cones with the smallest (clear) oil droplets, which at the time were thought to be exclusively blue sensitive (see following paragraph and Section II.B.3.b). H3 and H4 cells contact only these "blue-sensitive" single cones and accessory cones, respectively. Leeper used these data on the characteristic morphology and pattern of horizontal cell connections with photoreceptors to formulate hypotheses about the elaboration of receptive field properties in these neurons.

Shortly after Leeper's model was published, new data from microspectrophotometry (Lipetz, 1985) and intracellular recording (Ohtsuka and Kouyama, 1982, 1985, 1986a, 1986b) established that some of the original photopigment–oil droplet pairings of Liebman and Granda (1975) were incorrect. There is now general agreement that the accessory member of double cones is red sensitive, not green sensitive as previously supposed. Thus, Leeper's H4 chromaticity cells probably receive red (not green) cone input. However, the same authors disagree about the chromatic input to H2 and H3 horizontal cells, which contact cones with clear oil droplets (Fig. 1.56; compare with Fig. 1.55 and with Fig. 1 in Leeper, 1978b). These cones were originally considered a single, blue-sensitive class. However, it is now clear that there are two populations of clear oil droplets: large fluorescent, and smaller, nonfluorescent droplets that are found in different spectral classes of cones (see Section II.B.3.b). Lipetz holds that the larger clear droplet is found in blue-sensitive cones, whereas the smaller clear droplet is found in red cones. Ohtsuka and Kouyama report exactly opposite pairings. Thus, the chromatic input to H2 and H3 horizontal cells is unclear. Because cones bearing the larger clear droplets are much more numerous (13% of total cones; Kolb and Jones, 1987) than cones with small clear droplets (5%), the model of Lipetz for the chromatic input to H2 and H3 horizontal cells is similar

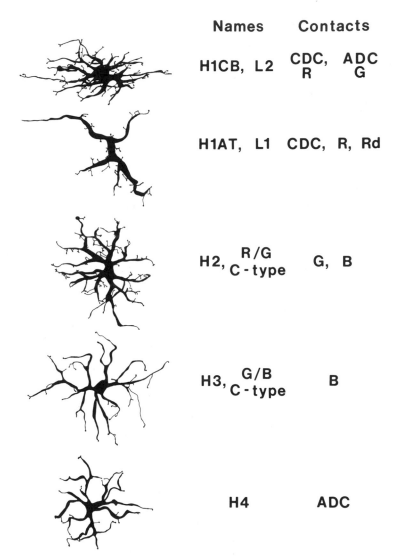

	Names	Contacts
	H1CB, L2	CDC, ADC R G
	H1AT, L1	CDC, R, Rd
	H2, R/G C-type	G, B
	H3, G/B C-type	B
	H4	ADC

Fig. 1.55. *Pseudemys (Trachemys) scripta.* Photoreceptor contacts made by different types of horizontal cells. To the left, the structure of each horizontal cell type is shown as visualized in Golgi impregnated, whole mounted retinas. Under "Names," the author's terminology is given (H1CB, H3, etc.) with probable correspondences to physiologically identified horizontal cell types. Under "Contacts," the types of photoreceptors contacting each horizontal cell type are listed. Rd, rod; CDC, chief member of double cone; ADC, accessory member of double cone; R, red-sensitive single cone; G, green-sensitive single cone; B, "blue-sensitive" single cone. In this scheme, the spectral sensitivities of single cones were inferred from their oil droplet types. Recent evidence has cast doubt on the spectral sensitivities of cones with clear droplets (blue-sensitive cones in this figure). Thus, the connectivities of H2 and H3 horizontal cells are still in doubt. (From Leeper, 1978b.)

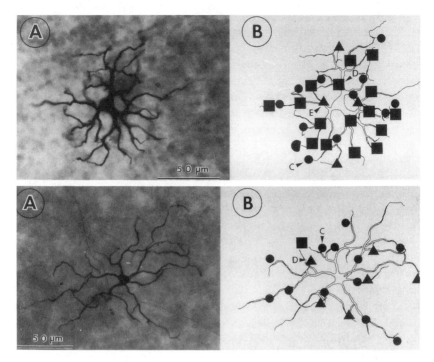

Fig. 1.56. *Chinemys (Geoclemys) reevesii.* Photoreceptor contacts made by chromaticity horizontal cells. The morphology of each cell is shown to the left. The cells have been injected with horseradish peroxidase and visualized in whole mounted retinas. Photoreceptor contacts made by each cell are shown to the right. These horizontal cells may correspond, respectively, to the H2 and H3 horizontal cells of Leeper (1978a, 1978b; compare with Fig. 1.55). Upper: Biphasic chromaticity cell. Lower: Triphasic chromaticity cell. Squares, green sensitive single cones; circles, cones with large, clear (pale green, fluorescent) oil droplets; triangles, cones with small, clear oil droplets; incomplete circle, accessory member of double cone. Capital letters refer to electron micrographs in original figures. (From Ohtsuka and Kouyama, 1986.)

to that of Leeper (who assumed that all clear droplets are found in blue cones). Kolb and Jones (1987) have reported light and electron microscopic evidence supporting the pairings of Lipetz. Therefore, the following sections are based on the models of Leeper and Lipetz (but see Ohtsuka and Kouyama, 1986a, 1986b, for an opposing view.

b. CONE RECEPTIVE FIELDS
When a cone is stimulated with a small spot of white light, it responds with a graded hyperpolarization. If the spot is very small (2 to 4 μm), it effectively stimulates only the cone over which it is centered. If the size of the spot is increased up to 80 to 100 μm, it also stimulates surrounding cones that are electrically coupled to the central cone,

and the response of the latter cell is enhanced (faster rise time and peak amplitude) but still hyperpolarizing. This "near periphery" effect is probably mediated by gap junctions between cone basal processes as described previously. With further increases in spot size, a delayed depolarizing inflection is superimposed on the direct hyperpolarizing response and this "far periphery" effect is believed to be mediated by feedback from surrounding horizontal cells through a chemical, sign-inverting synapse (Leeper and Copenhagen, 1982; Kaneko et al., 1985).

This negative feedback from horizontal cells to cones was first demonstrated in reptiles (*Pseudemys (Trachemys) scripta*, Baylor et al., 1971) and has profoundly influenced the development of thought about receptive field formation in retinal neurons. It is therefore interesting that no consistent morphological substrate for this feedback has been demonstrated. Horizontal cells contact photoreceptors at ribbon and distal junctions. However, neither of these contacts shows the typical cluster of synaptic vesicles, images of vesicle exocytosis, or intramembrane particles characteristic of a presynaptic active site (see Section II.C.2.b). Occasionally, small clusters of E face particles are observed on the lateral (horizontal cell) processes adjacent to a synaptic ridge, but they are not reliably associated with membrane specializations on either cone pedicles or the horizontal cell P face. Since the effect of horizontal cell activity on cones is probably to increase transmitter vesicle exocytosis by altering Ca^{++} conductance in the pedicles (Piccolino and Neyton, 1982), it seems likely that horizontal cells act near the slope of the synaptic ridge where vesicle exocytosis probably occurs. Perhaps horizontal cells modulate photoreceptor activity by some form of nonvesicular transmitter release (Yazulla and Kleinschmidt, 1983; Dacheux, 1982; reviewed in Dowling, 1987). In any case, it is clear that horizontal cells generate a negative feedback signal to cones that can adjust their sensitivity and provide them with an antagonistic surround.

This negative feedback creates spectral antagonism in cones that can be explained, at least in part, by the connections that horizontal cells make with specific photoreceptors. Red and green cones are interconnected by luminosity horizontal cells (Fig. 1.57). When a green cone is stimulated with a large spot or annulus of dim red light, it responds with a depolarization (Fuortes et al., 1973) that is probably mediated by the pathway: red cones → H1CB → green cones (Fig. 1.57; Leeper, 1978b; Leeper and Copenhagen, 1982). In red cones, feedback is more difficult to demonstrate because these cones are very sensitive to both red (680 nm) and green (550 nm) light, and the direct hyperpolarizing response of the central red cones tends to mask feedback effects from photoreceptors in the far periphery. Nevertheless,

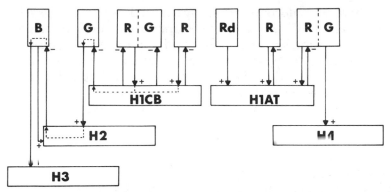

Fig. 1.57. *Pseudemys (Trachemys) scripta.* Interconnections between horizontal cell types and photoreceptors. The model is an attempt to account for the spectral properties of photoreceptor and horizontal cell receptive fields. Some of its features have been modified by more recent data (see text). R, G, B single elements, red-, green-, and blue-sensitive single cones. The spectral sensitivities of these "blue" cones are still in question. R and G in double elements, chief and accessory members of double cones. Accessory cones were originally thought to be green sensitive (G), but recent evidence suggests that both members of the double cone are red sensitive (R). +, noninverting synapse; − inverting synapse. (From Leeper, 1978b.)

very large spots shorten the hyperpolarizing response of red cones, as would be expected if a late depolarizing inflection were superimposed on the primary, hyperpolarizing response, and this effect is greater for stimuli at 550 nm (green; Fuortes et al., 1973). Thus, the spatial antagonism of turtle cones is also color dependent in a way that is consistent with the known connections of luminosity horizontal cells.

c. COLOR DEPENDENCE OF HORIZONTAL CELL RESPONSES
Luminosity cells, both H1CB and H1AT types, respond with hyperpolarization to light of any wavelength but are maximally sensitive to red. This is readily explained given the spectral sensitivity of red single cones (Baylor and Hodgkin, 1973), which provide much of the input to luminosity horizontal cells through sign-preserving synapses. Rods also contact H1AT cells (Fig. 1.57), but the broad spectral sensitivity of rods (Baylor and Hodgkin, 1973) is not reflected in the spectral curves of H1AT cells (to which they are connected, Figs. 1.55 and 1.57) except under special conditions (Leeper and Copenhagen, 1979), probably because the slow, low-amplitude component contributed by rods is masked by the more robust cone-driven response.

Chromaticity horizontal cells respond to lights of different wavelength with either depolarization or hyperpolarization, depending on stimulus wavelength (review in Leeper and Copenhagen, 1982). Two types have been documented in turtles: red depolarizing/green hyperpolarizing (R/G), and green depolarizing/blue hyperpolarizing (G/B). Fig. 1.57 summarizes Leeper's model of these horizontal cell response profiles (compare with Fig. 56 and Fig. 9 in Ohtsuka and Kouyama, 1986a). Briefly, H2 and H3 cells correspond to R/G and G/B horizontal cells, respectively. Their hyperpolarizing responses are mediated by direct cone input through sign-conserving synapses and their depolarizing responses by the path: red cones → H1CB → green cones → H2 (R/G type), and green cones → H2 → blue cones → H3 (G/B type). Since in this scheme the feedback from H1CB and H2 cells is through a sign-inverting synapse, the green (R/G path) or blue (G/B path) cones will be depolarized, as will the H2 and H3 cells they contact. Thus, the polarity of H2 and H3 cell responses reflects the wavelength of impinging light.

This model makes certain assumptions about the synaptic relations between horizontal cells and photoreceptors. Such assumptions are necessary to reconcile data on the pattern of contacts seen at the light microscopic level and the physiological properties of these cells, namely, that at least some of these contacts are not reciprocal (i.e., only the pedicle or the horizontal cell process is presynaptic). Without such assumptions it is difficult to understand, for example, why there is apparently no horizontal cell feedback to rods (Copenhagen and Owen, 1976b) and no direct input to H1CB cells from green single cones (Leeper and Copenhagen, 1982). Thus, uncertainties about synaptic organization and polarity in the outer plexiform layer (see Section II.C.2.b) need to be resolved before we can understand the chromatic properties of cones and horizontal cells.

IV. ASSEMBLY OF INFORMATION CHANNELS: THE RETINA AS A WHOLE
A. General

The previous section dealt with one way the structure of retinas can be characterized, namely by the composition of their information channels. But these channels must also be assembled to form a complete retina, and it is clear that this assembly is not random. Rather, patterns can be distinguished at two levels: (1) at any given point on the retina the cellular elements may exhibit orderly spatial arrangements, and (2) different regions of the retina may have distinctive compositions. This section summarizes what we know about these more global aspects of retinal structure in reptiles.

B. Local Order: Photoreceptor Mosaics

Regular spatial arrangements of retinal cells have been demonstrated in several vertebrates (see review in Sterling, 1983), but none are more striking than the photoreceptor mosaics of teleost fish and sauropsids

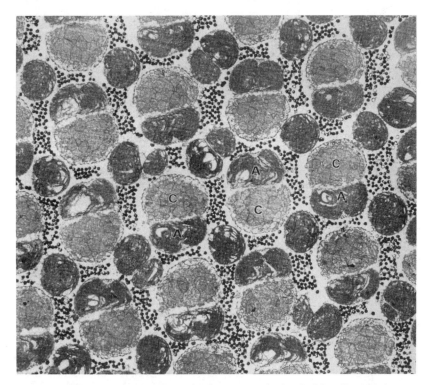

Fig. 1.58. *Coleonyx variegatus.* Tangential section at the level of the ellipsoid showing the highly structured photoreceptor mosaic. Note alternating positions of chief (C) and accessory (Λ) members of D1 cell. The smaller profiles represent single, twin, and D2 receptors (2840 x). (From Dunn, 1966b.)

(reviewed in Dunn, 1973; Underwood, 1970). In some lizards, Type B double cells are aligned in rows (Fig. 1.58; Dunn, 1966b). Within each row the polarity of the double cells (as seen in tangential sections) is not random; rather, they are arranged in an alternating pattern, the details of which vary with each species (*Coleonyx variegatus,* Dunn, 1966b, 1973; *Gekko swinhonis,* Detwiler, 1923a; *Aristelliger praesignis,* Underwood, 1951; *Aprasia pulchella,* Underwood, 1957; *Anolis lineatopus,* Underwood, 1970). Each row of Type B doubles is separated from the next by a row containing the remaining photoreceptor types (singles, Type C doubles, triplets, and so on). Although the arrangement of these intercalated cells often appears less regular to the eye, quantitative techniques reveal a degree of underlying order (Dunn, 1966b). In *Gekko gecko,* each photoreceptor type fluoresces at a different wavelength, revealing its own subpattern within the total receptor array (Liebman and Leigh, 1969). The net effect of these regularly alternating rows is a mosaic with repeating square or rectangular subunits (See Fig. 10 in Dunn, 1966b).

With one exception (Land and Snyder, 1985), photoreceptor mosaics have not been reported in other reptiles; it is unclear whether they are absent or have not been looked for. Even in species that lack striking rectilinear arrays (e.g., *Sphenodon*, which has a pattern of three single:one double cone; Vilter, 1951a), statistical techniques may disclose an orderly spatial arrangement of photoreceptors (Morris, 1970). The functional role of these mosaics is unknown. In fish, they have been supposed to mediate motion detection because they are best developed in species that feed on fast-moving prey (Borwein, 1981). In both geckos and fish, photoreceptor patterns are often most regular in "central" retina (Dunn, 1973; Borwein, 1981), suggesting that they are not simply a general property of the photoreceptor layer but reflect a functional role that is localized to a part of the retina. The mosaics in geckos appear to be arranged for maximum separation between photoreceptors of a given type; perhaps they represent a compromise between the need for high packing density and the necessity of minimizing optical cross-talk (Kirschfeld, 1983) between receptors of the same spectral type that would tend to degrade images of high spatial frequency.

C. Retinal Topography

It has been known since at least the 18th century that the structure of vertebrate retinas is not spatially uniform. There are, instead, regional differences in the size, morphology, density, and relative frequency of retinal elements. The most fundamental reason for this inhomogeneity is that vertebrates are polarized along two axes. On the one hand, the encephalized body plan and characteristic locomotor patterns of vertebrates require that they be polarized in a rostrocaudal direction. The significance of objects in the forward visual field often differs from that of objects behind an animal, and this difference is reflected in regional variations along the nasotemporal axis of the retina. At the same time, vertebrates are also polarized along a vertical axis in response to physical features of the environment. Gravitational forces and the position of the sun ensure that the distribution of light intensities and of objects in the environment is not random but follows orderly vertical gradients. For example, the amount of light coming from below an animal can generally be expected to be less than that along an upward line of sight. Similarly, at an air-water interface, water will be below and air above. Thus, there are vertical gradients in the environment, and this is often reflected in regional variation along the dorsoventral axis of the retina. Further constraints on visual space are imposed by the morphology, behavioral ecology, and specific habitat of each species. In short, the visual space of any given animal is not equipotential in terms of the input it is likely to afford, and it should not be surprising that the functional organization of vertebrate retinas differs with retinal locus.

Regional variation in the structure of reptilian retinas has been noted repeatedly, but virtually all the available data deal with gross features of retinal morphology such as local increases in cell density (*areae retinae*) or the occurrence of more or less well-developed foveas. Probably the most extensive treatment is that of Chievitz (1889, 1890, 1891). Many of the earlier studies (see Appendix and reviews in Slonaker, 1897; Franz, 1913; Rochon-Duvigneaud, 1943; Duke-Elder, 1958) are probably confounded by poor fixation and preparation techniques that introduce mechanical distortions of the retina, by inadequate sampling procedures, and perhaps by excessive enthusiasm for discovering "foveas" in nonhuman vertebrates. Thus, their significance is primarily historical. More recently, undoubted foveas have been described in *Sphenodon* (Walls, 1942a; Vilter, 1951c) and many lizards (*Phrynosoma cornutum*, Detwiler and Laurens, 1920; *Eremias argus*, Detwiler, 1923b; *Amphibolurus barbatus* and *Leiolopisma entrecasteauxii*, O'Day, 1939; *Anolis* species, Polyak, 1957; Underwood, 1951, 1970; Makaretz and Levine, 1980; Fite and Lister, 1981; *Phelsuma* species, Tansley, 1961, 1964; *Podarcis (Lacerta) muralis*, Vilter, 1949b; *Chamaeleo*, Cajal, 1893; Johnson, 1927; *Iguana iguana*, Meneghini and Hamasaki, 1967; *Dipsosaurus dorsalis*, Peterson, 1981b; *Gonatodes* and *Sphaerodactylus*, Underwood, 1970; *Agama tournevillii*, Verrier, 1933). In these foveas, photoreceptors become attenuated, double cones may be absent (*Eremias*, Detwiler, 1923b; *Iguana iguana*, Meneghini and Hamasaki, 1967), and oil droplets lost (*Phrynosoma*, Detwiler and Laurens, 1970), although the latter two observations may simply reflect the difficulty in resolving structural detail in the very slender foveal cones. Interestingly, foveal receptors are not always elongated, even in species with well-developed foveas, and comparative studies of these cells in anoles suggest that receptor length in the fovea may be correlated with some aspects of foraging strategy (Fite and Lister, 1981). A large number of anoline species have been examined (Underwood, 1970; Makaretz and Levine, 1980; Fite and Lister, 1981), and all have both central (laterally directed) and temporal (forward directed) foveas that differ in size, shape, photoreceptor and ganglion cell densities, and probably convergence ratios (Fig. 1.59). This is the best-known example of regional differences along the horizontal retinal axis in reptiles. Horizontally elongated regions of high cell density have been described in turtles (see below), crocodilians (Chievitz, 1889, 1890), and lizards (*Phrynosoma*, Slonaker, 1897; *Dipsosaurus*, *Iguana*, Peterson, 1981b). These lizards also have a fovea embedded in the horizontal band.

The significance of these regions of high cell density is unclear. In the older literature (see Slonaker, 1897) they were evidently considered specializations for high-acuity vision, but with growing awareness that there are different functional types of cells within any pop-

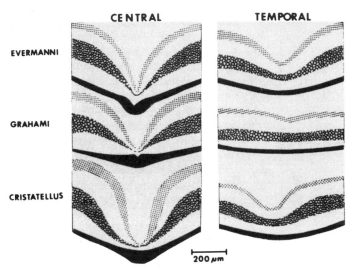

Fig. 1.59. Variation in the structure of central and temporal foveas across anoline species. The three shaded layers represent, from proximal to distal retina, the optic axon layer, the inner nuclear layer, and the photoreceptor-pigment epithelial layer. (From Fite and Lister, 1981, by permission of the publisher, S. Karger AG, Basel.)

ulation of retinal elements (e.g., within ganglion cells, bipolar cells, horizontal cells; see review in Rowe and Stone, 1977; Kaneko, 1979; Sterling, 1983; Dowling, 1987), it has become clear that only some cells of each major group may mediate high visual acuity. Thus, local increases in packing density coupled with decreased soma or nuclear size probably represent some form of functional specialization but are not necessarily related to high-resolution vision. Such data cannot be interpreted in isolation. The same can be said for reports of regional variation in convergence ratios (e.g., between photoreceptors and bipolar cells or ganglion cells), a phenomenon that received much attention in the earlier literature (e.g., Vilter, 1949a, 1951b; reviewed in Underwood, 1970). They are difficult or impossible to interpret without knowing which functional classes of cells are involved in the observed frequency changes.

In addition to these gross variations in retinal structure there are limited data on the relative frequencies of different functional types of photoreceptors (*Coleonyx variegatus*, Dunn, 1966a; *Thamnophis sirtalis*, Wong, 1989; *Pseudemys (Trachemys) scripta*, Peterson, 1981a; Ohtsuka, 1985a; Kouyama and Ohtsuka, 1985; Kolb and Jones, 1987; *Geoclemys reevesii*, Ohtsuka, 1985b; *Alligator mississippiensis*, Laurens and Detwiler, 1921). In *Alligator*, such effects are especially marked along the dorsoventral axis of the retina. In dorsal retina, rods dominate the photoreceptor population; cones are infrequent and, whenever present, they are usually of the double cone type (possibly specialized for vision in dim light; see below). The frequency of cones increases toward the ventral retina; thus, the composition of the photoreceptor layer appears to reflect differences in luminous intensity along upward and downward lines of sight.

Retinal topography (in the sense of local features of retinal organization) is best understood in *Pseudemys (Trachemys) scripta*. In the photoreceptor layer it is possible to map the size and density of the various oil droplets. Because each droplet type is associated with a single cone type (see Section II.B.3.b), and because oil droplets completely fill the inner segment (Kolb and Jones, 1982) and—with the ellipsoid—define the collecting area of the photoreceptor (Baylor and Fettiplace, 1975), data on oil droplet density and size reveal the regional characteristics of the cone population. These cones account for more than 90% of photoreceptors in *Pseudemys (Trachemys)*. Such an analysis reveals that the retina is dominated by a horizontal band—the visual streak, which extends from nasal to temporal periphery and in which cones are small and densely packed (Figs. 1.60 and 1.61; Brown, 1969; Granda and Haden, 1970; Peterson, 1981a; Normann and Kolb, 1981). They are maximally concentrated in a small region of the streak at the approximate center of the retina (peak density area; Peterson, 1981a). In most of the streak, the relative frequency of red oil droplets (red single cones) is increased (compared with the dorsal and ventral periphery; Fig. 1.62), and in the peak density area of the streak, the relative frequency of yellow droplets (green-sensitive single cones) is especially high. Visual streaks dominated by red and orange oil droplets are found only in certain surface feeding sea birds, where they have been supposed to enhance vision through the air-water interface (Muntz, 1972) or to enable the birds to see long distances through atmospheric haze (Lythgoe, 1979). On the other hand, yellow filters of various sorts (including oil droplets) have traditionally been associated with high-resolution pattern vision because they absorb short wavelength light, which is most likely to scatter and so degrade a visual image (Walls and Judd, 1933; Muntz, 1972).

The density gradients that define the visual streak in the photoreceptor layer are extremely sharp. For example, 78% of the total density increase from far dorsal periphery to visual streak occurs within 250 μm (4% of 12-mm diameter retina, or 3% visual space) of the streak (Peterson, unpublished). Outside the streak, the dorsal and ventral hemiretinae are relatively homogeneous in the size and relative frequency of different oil droplet (and presumably cone) types (see Granda and Haden, 1970, for a dissenting view). The primary differences between dorsal and ventral retina are that (1) for a given distance from the streak, photoreceptor density is higher in ventral

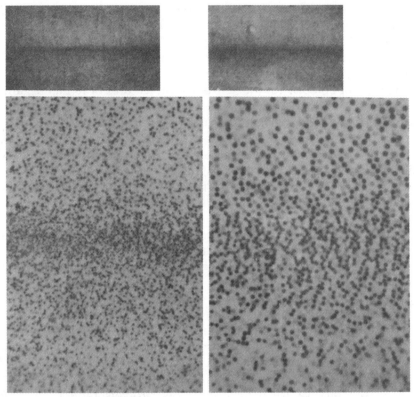

Fig. 1.60. *Pseudemys (Trachemys) scripta.* Density gradients in the photoreceptor layer. The retinas have been whole mounted with dorsal retina toward the top and the visual streak parallel to earth horizontal. The position of photoreceptors is inferred from their oil droplets. Upper: Low-power views of the visual streak in peripheral (left) and central (right) retina. The visual streak is thickest at the midpoint of the visual streak (right micrograph, slightly to the left of center; compare Fig. 1.61) (15 x). Lower: Higher-magnification views of the visual streak showing the steep falloff in density with distance from the streak, especially in dorsal retina. Oil droplet size is larger in dorsal retina than at a comparable distance from the streak in ventral retina (left, 128 x; right 315 x). (Reprinted with permission from *Vision Research* Vol. 9, Brown, A linear area centralis extending across the turtle retina and stabilized to the horizon by non-visual cues, 1969, Pergamon Press plc.)

than in dorsal retina and (2) there are more yellow droplets (green single cones) than orange droplets (double cones) in ventral retina, whereas the converse is true in the dorsal retina (Fig. 1.62). The dorsal retina in *Pseudemys (Trachemys)* sees the relatively dim upwelling space light, and the high frequency of double cones in dorsal retina is consistent with observations in fish (Boehlert, 1978) that double cones form in response to low illumination levels. Interestingly, orange oil droplets occupy a stratum distal to that of single cones (Kolb and Jones, 1982). Thus, they represent a second sheet of light collectors

TOTAL

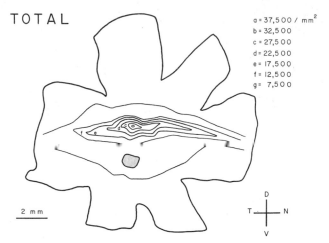

a = 37,500 / mm²
b = 32,500
c = 27,500
d = 22,500
e = 17,500
f = 12,500
g = 7,500

2 mm

D
T ─┼─ N
V

Fig. 1.61. *Pseudemys (Trachemys) scripta.* Distribution of total cone population as inferred from the distribution of oil droplets in a whole mounted retina. The isodensity lines form a horizontally oriented visual streak that is somewhat broader in temporal (T) than in nasal (N) retina. Values for isodensity lines are given at upper right. Total cone density falls off steeply above the streak. D, dorsal; V, ventral; shading, optic disk.

arranged to maximally intercept light not absorbed by the layer of single cones, much like the distal elements in "tiered" retinas described previously (Section II.B.3.e).

In the inner nuclear layer, isodensity maps of amacrine cells labeled with different antisera have distinctive distributions (e.g., Eldred and Karten, 1983; Kolb et al., 1987; Isayama and Eldred, 1988; Williamson and Eldred, 1989; Cuenca and Kolb, 1989). Some amacrine types with enkephalinlike or CRF-like immunoreactivity are found only in the visual streak (Fig. 1.63), and other types have a more radial distribution. A second CRF immunoreactive amacrine type is found only in ventral retina (Fig. 1.63), but neuropeptide Y–immunoreactive amacrine cells are found preferentially outside the visual streak. Similarly, one substance P immunoreactive amacrine cell type is found throughout the retina, whereas a second type is found only in the periphery (Fig. 1.64; Cuenca and Kolb, 1989). These data indicate that the relative frequency of different amacrine cell types varies with retinal locus. Thus, at least in *Pseudemys (Trachemys) scripta*, there are qualitative as well as quantitative differences in the retinal circuitry that analyzes different parts of the visual field.

The visual streak is also represented in the ganglion cell layer by a local increase in the density of neuronal profiles, at least 95% of which are ganglion cells (Peterson and Ulinski, 1979). Ganglion cells are small within the streak; there is some evidence that this is due, at least in part, to a relative increase in one or more populations of small cells

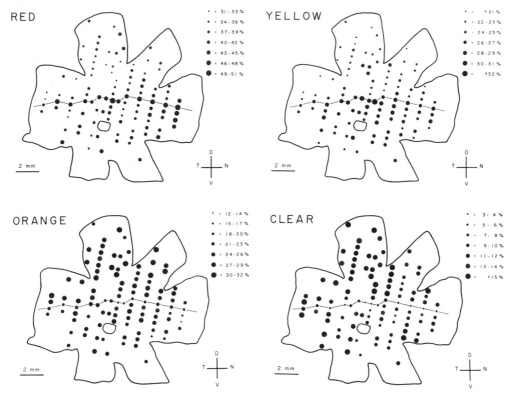

Fig. 1.62. *Pseudemys (Trachemys) scripta.* Distribution of cones with different colors of oil droplets. Dot sizes represent the relative frequency of oil droplet types at various points on the retina. Each oil droplet type has a characteristic spatial profile. Red and yellow droplets are most numerous in the visual streak and in ventral retina. Yellow droplets reach peak density in the center of the retina. Orange droplets (double cones) are infrequent in the visual streak and are most numerous in dorsal retina. Clear droplets are relatively rare in the visual streak. (This figure reflects the distribution of all clear oil droplets combined.)

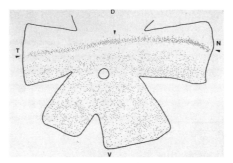

Fig. 1.63. *Pseudemys (Trachemys) scripta.* Distribution of amacrine cells with CRF-like immunoreactivity. Type A amacrines are found in the visual streak (arrowheads). Type B amacrines are found in ventral (V) retina. No CRF immunoreactive amacrine cells are found in dorsal (D) retina. T, temporal; N, nasal. (From Williamson and Eldred, 1989.)

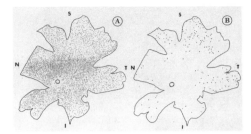

Fig. 1.64. *Pseudemys (Trachemys) scripta.* Distribution of Substance P–immunoreactive amacrine cells. Left: All immunoreactive amacrine cells in one retina are represented by dots. Their density is highest in the visual streak. Right: Distribution of Type A immunoreactive amacrine cells. They are absent from the visual streak and are most numerous in peripheral retina. S, superior; I, inferior; N, nasal; T, temporal. (From Cuenca and Kolb, 1989.)

and not simply to a size decrease in the ganglion cell population as a whole (Peterson and Ulinski, 1982). The shape of the streak outlined by ganglion cells resembles that seen in the photoreceptor layer, but the gradients are not so sharp (Fig. 1.65). This is because central (and sometimes peripheral) processes of bipolar cells radiate away from the visual streak so that most ganglion cells receive their straight signal path input from cones that are closer to the horizontal meridian of the eye than they are. Thus, in species like *Pseudemys (Trachemys) scripta* that have oblique bipolar cell processes, maps of ganglion cell density are only a rough index of spatial sampling in the retina; photoreceptor maps provide a more accurate picture. Indeed, a recent study of regional variation in serotoninergic bipolar cells indicates that the visual streak formed by photoreceptors is enormously magnified at the ganglion cell layer (Schutte and Weiler, 1987).

Visual streaks occur in many vertebrates, but their role is unknown (see review in Peterson and Ulinski, 1979). They have been supposed to mediate some form of movement detection (Rowe and Stone, 1976) or high-resolution vision (Hughes, 1977) in a manner analogous to primate foveas or the cat area centralis (thus the expression *linear area centralis;* Brown, 1969; Kolb, 1982), and the limited data from *Pseudemys (Trachemys) scripta* cannot distinguish between the two hypotheses. Densely packed cells with small receptive field centers are probably necessary to detect high spatial frequencies, and this condition is satisfied in the photoreceptor layer at least, where cones have diameters of 1.5 to 3.0 μm and reach densities of over 40,000/mm^2. Electrical coupling between these cones would appear troublesome, but it must not prevent high-acuity vision since it occurs in primate foveas (Raviola and Gilula, 1973). Coupling distances between red cones in dorsal periphery average 40 μm (Section III.C.2). In the streak, cones are smaller by a factor of 3 (Peterson, 1981a); thus, if coupling were

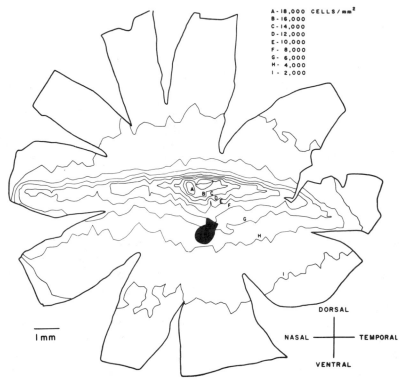

Fig. 1.65. *Pseudemys (Trachemys) scripta.* Isodensity map showing the distribution of ganglion cells as visualized in a whole mounted retina. Ganglion cells are concentrated in a horizontally elongated band, the visual streak. They are most dense at the midpoint of the streak, at the approximate center of the retina. Note that the density gradient falls off much more sharply in dorsal than in ventral retina. (From Peterson and Ulinski, 1979.)

linearly related to cone size, effective coupling distances should be very small (average 13 to 14 μm in the streak or 0.13° visual field, assuming 12° of visual field per millimeter of retina; Peterson and Ulinski, 1979). Some horizontal cell types become smaller in the streak (Figs. 1.66 and 1.67; Leeper, 1978a; Normann et al., 1979), but they are never less than 30 × 30 μm mean orthogonal diameters and, furthermore, cells of this type are widely coupled. Their role in high-resolution vision, if any, is unknown. Certain bipolar, amacrine and ganglion cell types may become smaller in the streak (e.g., Kolb, 1982; Peterson, 1982; Ammermuller and Weiler, 1988; Isayama and Eldred, 1988; Weiler and Ammermuller, 1986) but, with the exceptions noted above (Schutte and Weiler, 1987; Kolb et al., 1987), we lack any quantitative analyses of these cell types that might shed light on their ability to mediate vision at high spatial frequencies.

Some retinal elements in *Pseudemys (Trachemys) scripta* that seem unlikely to be involved in high-acuity vision because of their size or cou-

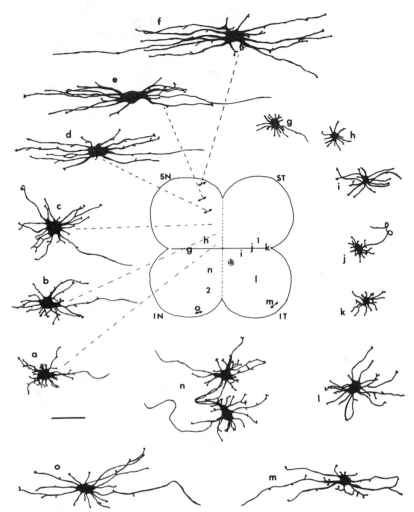

Fig. 1.66. *Pseudemys (Trachemys) scripta.* Variation in the size and dendritic architecture of H1CB horizontal cells with retinal locus. The retina has been whole mounted, and the horizontal cells are Golgi impregnated. The somata and dendritic fields get larger with distance from the visual streak. The dendritic fields are always radially oriented except in mid- to far-peripheral retina. Scale: 50 μm. (From Leeper, 1978a.)

pling characteristics nevertheless exhibit morphological specializations associated with the visual streak. This is especially pronounced among interneurons of the outer and inner plexiform layers that intersect the straight signal pathways. Many of these horizontal and amacrine cells have markedly anisotropic dendritic fields oriented parallel or orthogonal to the streak. For example, the axon terminals of luminosity horizontal cells (H1AT) in the streak (Fig. 1.67) have elongated, horizontally oriented, dendritic fields (20 × 50 μm; Nor-

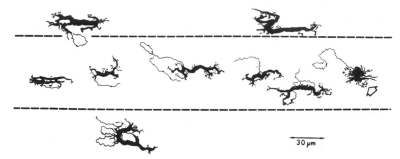

Fig. 1.67. *Pseudemys (Trachemys) scripta.* Orientation of H1 axon terminals parallel to the visual streak (dashed lines). In the same region, the dendrites of H1 cell bodies (open arrow) have a radial distribution. Axon terminals are smaller within the streak than in more peripheral retina. (Reprinted by permission from Normann et al., NATURE Vol. 280, pp. 60–62. Copyright (c) 1979 Macmillan Magazines Ltd.)

mann et al., 1979). Interestingly, their cell bodies (H1CB) maintain radial dendritic fields throughout the central retina, including the visual streak (Figs. 1.66 and 1.67; Leeper, 1978a; Normann et al., 1979); this suggests that, in the visual streak, their functional role differs from that of the H1AT cells. Many large-field amacrine cells also have elongated, oriented, dendritic fields (see Fig. 1.43; Kolb, 1982), and some of these occur only in the streak (see Figs. 1.43 and 1.63; Eldred and Karten, 1983; Williamson and Eldred, 1989). Oriented cells such as these have been proposed to mediate directional or orientational sensitivity (Normann et al., 1979; Normann and Kolb, 1981; Kolb, 1982; Williamson and Eldred, 1989), and a model has been proposed in which horizontal cells code stimulus orientation relative to the visual streak (Normann and Kolb, 1981). This visual streak is maintained at earth horizontal by nonvisual (presumably vestibular) reflexes of the head and neck (Fig. 1.68; Brown, 1969). Thus, coding of orientation relative to the visual streak refers direction on the retina to an external coordinate frame.

V. CONCLUSION

Following publication of the duplicity theory of Schultz (1866), the early literature on reptilian retinas was dominated by the search for "rods" and "cones," and later by the transmutation theories of Walls (1934, 1942a, 1942b), who responded to the growing evidence of intermediate photoreceptor types by suggesting that rods and cones might become modified by changing environmental demands over evolutionary time. With the development of more powerful analytical techniques and a growing understanding of the need for multiple criteria in neuronal classification schemes (Tyner, 1975; Rowe and Stone, 1977), attention has shifted toward understanding the properties of

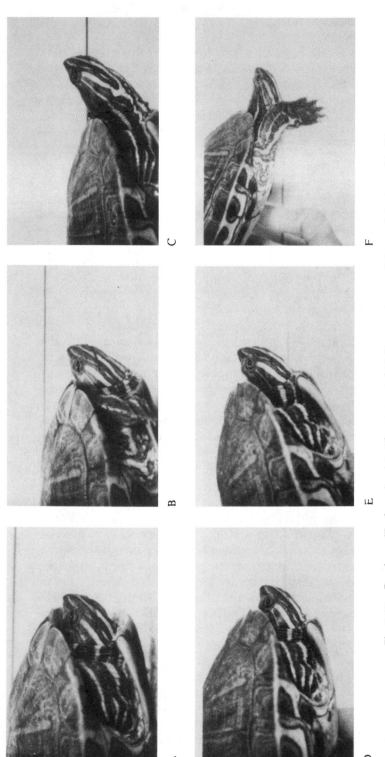

Fig. 1.68. *Pseudemys (Trachemys) scripta.* The pigmented iris line remains parallel to earth horizontal (solid line) as head position changes. The animal was held in the dark and photographed under stroboscopic illumination. Therefore, this optomotor reflex is probably mediated by non-visual cues. The iris line is oriented parallel to the visual streak. Thus, this reflex maintains the visual streak parallel to earth horizontal. Turtles can override these reflexes during active feeding and exploration (Peterson and Ulinski, 1979). (Reprinted with permission from *Vision Research*, Vol. 9, Brown, A linear area centralis extending across the turtle retina and stabilized to the horizon by non-visual cues, 1969, Pergamon Press plc.)

retinal elements: their functional architecture, connectivities, bio-chemistry, and physiological characteristics. Relatively little attention has been paid to the third level of organization described in this' paper, that is, properties of the retinas as a whole, or to correlations between retinal organization and behavioral ecology of the kind that have been studied so successfully in teleosts (Lythgoe, 1979; Levine and MacNichol, 1979; Crescitelli et al., 1985; Loew and Lythgoe, 1985).

The greatest progress has been made in understanding the organization of the distal retina in turtles (especially *Pseudemys (Trachemys) scripta*). This is primarily because the number of cell types is limited, the classes of photoreceptors and some horizontal cells are morpho-logically distinct at both the light and electron microscopic levels, and there is a substantial body of information available on both the photochemistry and the physiological properties of photoreceptors and horizontal cells in turtles. Much less is known about more proximal retinal elements and about the organization of the inner plexiform layer. One problem is that the number of cell types at this level is evidently very great. A second and perhaps related problem is that data on bipolar, amacrine, and ganglion cell types have, with few exceptions, consisted of simple descriptive catalogues of cell "types" with no attempt to determine whether the different types may actually be regional or random variants of a single cell class (Rowe and Stone, 1977). Moreover, these studies have relied heavily on dendritic stratification patterns, even though the sublamination of the inner plexiform layer is poorly understood. It is not clear, for example, whether the five layers of Cajal are equal in thickness; some authors assume they are (Kolb, 1982), and some do not (Jensen and DeVoe, 1983). More important, we do not know whether the relative thickness of the sublaminae changes with retinal locus. Because different amacrine and ganglion cell types stratify preferentially in different layers, and because their relative frequencies vary across the retina, there is good reason to suppose that the relative thickness of the five sublaminae will not be everywhere the same. Until we know whether this is so, descriptions of stratification levels for different cells will be difficult to compare or interpret. Perhaps the critical step in understanding the morphology of mammalian ganglion cells was the realization that the properties of a given cell type vary with retinal locus (Sterling, 1983), but such issues have received little notice in the literature on reptilian retinas.

Finally, it should be pointed out that over 90% of all the studies published since the structure of reptilian retinas was last reviewed (Underwood, 1970) have been on turtles, usually on *Pseudemys (Trachemys) scripta*. This is due, ultimately, to its well-known robustness as a preparation for retinal recording; the resulting physiological data created a demand to understand their structural substrates. These data have made an important contribution to our understanding of vertebrate vision and so must be welcome. Regrettably, however, the functional morphology of other reptilian retinas, at least beyond the photoreceptor layer, remains almost entirely unknown.

APPENDIX A: REVIEWS AND SURVEY ARTICLES

Reviews and survey articles that give information on the structure of reptilian retinas are listed below. They are grouped according to their primary focus.

1. Vertebrates: Brecha, 1983; Brecha et al., 1984; Cajal, 1893; Chievitz, 1889, 1890, 1891; Cohen, 1972; Detwiler, 1943; Dogiel, 1888; Dowling, 1987; Duke-Elder, 1958; Dunn, 1973; Ehinger, 1982; Franz, 1913; Gallego, 1983; Garten, 1907; Hannover, 1840; Heinemann, 1877; Lasansky, 1977; Lund and von Barsewisch, 1975; Marc, 1985, 1986; Meyer et al., 1965; Müller, 1857; Muntz, 1972; Pedler, 1965; Polyak, 1957; Putter, 1912; Ranvier, 1889; Rochon-Duvigneaud, 1943; Rodieck, 1973; Schiefferdecker, 1886; Schultze, 1866, 1873; Slonaker, 1897; Tansley, 1965; Verrier, 1935; Walls, 1942a, 1942b; Walls and Judd, 1933.

2. Reptiles: Hess, 1910; Hulke, 1863–1865, 1867; Johnson, 1927; Krause, 1893; Leydig, 1853; O'Day, 1939; Rochon-Duvigneaud, 1917; Underwood, 1951, 1968, 1970; Verrier, 1933; Verriest et al., 1959; Walls, 1934.

3. Turtles: Granda and Dvorak, 1977.

4. Lizards: Anh, 1969; Knox, 1823; Tansley, 1964; Underwood, 1954, 1957.

5. Snakes: Baumeister, 1908; Underwood, 1967a, 1967b.

6. Crocodilians: Abelsdorff, 1898; Tafani, 1883.

APPENDIX B: REPTILIAN SPECIES DISCUSSED

(This list does not include all species referred to in reviews and surveys of retinal structure [Appendix A].)

TESTUDINES

Agrionemys horsfieldii
  Davydova, 1981, 1983
  Davydova et al., 1982
Chelonia mydas
  Granda and Dvorak, 1977
  Granda and Haden, 1970
  Liebman and Granda, 1975
Chelopus species
  Detwiler, 1916
Chelydra serpentina
  Ashmore and Copenhagen, 1980, 1983
  Baylor and Fettiplace, 1975, 1977a, 1977b
  Copenhagen and Green, 1987
  Copenhagen and Owen, 1976a, 1976b
  Copenhagen et al., 1983

Detwiler et al., 1980
Kolb and Jones, 1985
Leeper, 1978a, 1978b
Leeper and Copenhagen, 1979, 1982
Mariani and Lasansky, 1984
Owen, 1985
Pedler, 1965
Sarthy and Gaur, 1986
Schwartz, 1974, 1976
Chinemys reevesii
  Fujimoto et al., 1957
  Gallego and Perez-Arroyo, 1976
  Kaneko and Tachibana, 1986a, 1986b
  Kaneko et al., 1985
  Kouyama and Ohtsuka, 1985
  Nishimura and Shimai, 1980

Nishimura et al., 1980
Ohtsuka, 1978, 1983, 1985b
Ohtsuka and Kouyama, 1982,
    1985, 1986a, 1986b
Tachibana and Kaneko, 1984
Terada et al., 1983
Chrysemys picta
    Bunt and Klock, 1980a
    Detwiler, 1916
    Dunn, 1973
    Fite et al., 1979
    Reiner, 1981
    Reiner and Karten, 1978
    Reiner et al., 1984
    Yamada, 1960b
Clemmys insculpta
    Strother, 1963
Emydoidea blandingii
    Lipetz, 1984a, 1984b
Emys orbicularis
    Borovyagin, 1966
    Byzov and Shura-Bura, 1985
    Cajal, 1893
    Davydova, 1981, 1983
    Davydova et al., 1982
    Etingof et al., 1986
    Orlov and Maximova, 1964
    Pezard, 1964
    Vilter, 1949a
Mauremys caspica
    Davydova, 1981, 1983
    Davydova et al., 1982
    Gallego and Perez-Arroyo, 1976
    Kolb et al., 1988
Mauremys japonica
    Fujimoto et al., 1957
    Gallego and Perez-Arroyo, 1976
    Ishikawa and Yamada, 1969
    Maekawa and Mizuno, 1976
    Odashima et al., 1978, 1979
    Yamada and Ishikawa, 1965
Pseudemys Trachemys scripta
    Abe and Yamamoto, 1984, 1988
    Adolph and Zucker, 1988
    Ammermuller and Weiler, 1988
    Baylor and Fettiplace, 1975, 1977a,
        1977b
    Baylor et al., 1971

Baylor and Hodgkin, 1973
Bowling, 1980
Brandon, 1985
Brecha et al., 1984
Brown, 1969
Bush and Lunger, 1972
Copenhagen and Jahr, 1989
Cuenca and Kolb, 1989
Dacheux, 1982
Detwiler and Hodgkin, 1979
Detwiler and Sarthy, 1981
Ehinger et al., 1988
Eldred and Carraway, 1987
Eldred and Cheung, 1989
Eldred et al., 1987, 1988
Eldred and Karten, 1983, 1985
Fuortes et al., 1973
Fuortes and Simon, 1974
Gaur et al., 1988, 1989
Geri et al., 1982
Gerschenfeld et al., 1982
Granda and Fulbrook, 1989
Granda and Haden, 1970
Guiloff et al., 1988
Hashimoto et al., 1973
Hurd and Eldred, 1989
Isayama and Eldred, 1988
Isayama et al., 1988
Ives et al., 1983
Jensen and DeVoe, 1982, 1983
Kolb, 1982
Kolb et al., 1986, 1987
Kolb and Jones, 1982, 1984, 1985,
    1987
Kouyama and Ohtsuka, 1985
Lamb and Simon, 1976
Lasansky, 1969, 1971, 1972, 1974,
    1977
Lasater et al., 1989
Leeper, 1978a, 1978b
Leeper and Copenhagen, 1982
Liebman and Granda, 1975
Lipetz, 1985
Marchiafava, 1983
Marchiafava and Torre, 1977, 1978
Marchiafava and Wagner, 1981
Marchiafava and Weiler, 1980,
    1982

Miller et al., 1973
Mosinger et al., 1986
Neumeyer and Jager, 1985
Nguyen-Legros et al., 1985
Normann and Kolb, 1981
Normann et al., 1979, 1984, 1985a
Ohtsuka, 1984, 1985a
Ohtsuka and Kouyama, 1985, 1986b
Peiponen, 1964
Peterson, 1981a, 1982
Peterson and Ulinski, 1979, 1982
Piccolino et al., 1982, 1984, 1987
Raviola, 1976
Raviola and Gilula, 1973
Reiner, 1981
Reiner et al., 1984
Richter and Simon, 1974, 1975
Saito et al., 1974
Schaeffer and Raviola, 1975, 1978
Schaeffer et al., 1982
Schutte and Weiler, 1987
Schwartz, 1973
Simon, 1973
Stewart, 1978
Stirling and Sarthy, 1985
Strother, 1963
Villegas, 1960
Weiler, 1981, 1985

Weiler and Ammermuller, 1986
Weiler and Ball, 1984
Weiler et al., 1988
Weiler and Marchiafava, 1981
Weiler and Schutte, 1985a, 1985b
Williamson and Eldred, 1989
Witkovsky et al., 1983, 1984, 1987
Yazulla, 1976a, 1976b
Pelodiscus sinensis
  Davydova, 1981, 1983
Sternotherus odoratus
  Ernst et al., 1970
Testudo graeca
  Davydova, 1981, 1983
  Davydova et al., 1982
  Evans, 1966
  Peiponen, 1964
  Pezard, 1964
"Soft-shelled turtle"
  Gillett, 1923
"Tortoise"
  Okabe, 1972
  Okuda, 1961
"Turtle"
  Drenckhahn and Wagner, 1985
  Marchiafava, 1982
  Yamada, 1960a
  Zucker and Adolph, 1988

CROCODILIA

Alligator mississippiensis
  Dunn, 1973
  Kalberer and Pedler, 1963

  Laurens and Detwiler, 1921
  Pedler, 1965
  Tafani, 1883

RHYNCHOCEPHALIA

Sphenodon punctatus
  Vilter, 1951a, 1951b, 1951c

SAURIA

Agama caucasia
  Orlov and Maximova, 1964
Agama tournevilli
  Verrier, 1933
Amphibolurus barbatus
  O'Day, 1939
Amphisbaena
  Underwood, 1970

Anguis fragilis
  Anh, 1969
  Anh and Anh, 1970
Anolis species
  Fite and Lister, 1981
Anolis carolinensis
  Brecha et al., 1984
  Gaur et al., 1988

Makaretz and Levine, 1980
Sarthy and Gaur, 1988
Aprasia pulchella
Underwood, 1957
Aristelliger praesignis
Pedler, 1965
Underwood, 1951
Chamaeleo chamaeleo
Armengol et al., 1981, 1988
Chamaeleo lateralis
Anh, 1969
Anh and Anh, 1970
Chamaeleo vulgaris
Cajal, 1893
Coleonyx variegatus
Dunn, 1966a, 1966b, 1966c, 1973
Dibamus
Underwood, 1970
Dipsosaurus dorsalis
Peterson, 1981b
Ebenauia inunguis
Pedler, 1965
Pedler and Tansley, 1963
Pedler and Tilly, 1964
Tansley, 1964
Eremias argus
Detwiler, 1923b
Eublepharis macularis
Dunn, 1969, 1973
Gekko gecko
Brecha et al., 1984
Bunt and Klock, 1980a
Citron and Pinto, 1973
Dunn, 1969, 1973
Kawamura and Murakami, 1986
Laties et al., 1976a, 1976b
Liebman and Leigh, 1969
Maneghini and Hamasaki, 1967
Pedler, 1965
Pedler and Tilly, 1964
Scheie and Laties, 1971
Sjostrand and Kreman, 1978
Gekko japonicus
Ishikawa and Yamada, 1969
Yamada, 1976
Yoshida, 1976, 1978
Gekko swinhonis
Detwiler, 1923a

Gekko species
Dunn and Adomian, 1971
Gonatodes
Underwood, 1970
Gonatodes vittatus
Ogura and Arispe, 1967
Hemidactylus flaviviridis
Tansley, 1964
Hemidactylus turcicus
Anh, 1969
Anh and Anh, 1970
Pedler, 1965
Pedler and Tilly, 1964
Tansley, 1959, 1964
Iguana iguana
Citron and Pinto, 1973
Meneghini and Hamasaki, 1967
Lacerta ocellata
Pezard, 1957, 1964
Lacerta viridis
Anh, 1968, 1969
Anh and Anh, 1970
Cajal, 1893
Pedler, 1965
Pezard, 1957, 1964
Tiemann, 1970
Lacerta vivipara
Peiponen, 1964
Leiolopisma entrecastauxii
O'Day, 1939
Lygodactylus coloratus
Pedler and Tilly, 1964
Oplurus cyclurus
Anh, 1969
Anh and Anh, 1970
Phelsuma madagascariensis
Ahn, 1969
Pedler, 1965
Tansley, 1961, 1964
Phelsuma quadriocellata
Dunn, 1973
Phrynosoma cornutum
Detwiler and Laurens, 1920
Slonaker, 1897
Podarcis muralis
Anh and Anh, 1970
Cajal, 1893
Meller and Eschner, 1965

Pedler, 1965
Vilter, 1949b
Sceloporus occidentalis
 Bernstein et al., 1984
 Dunn, 1973
 Eakin, 1964
 Vaughan and Fisher, 1987
 Williams et al., 1987
 Young, 1977
Sceloporus undulatus
 Detwiler, 1916
Tarentola mauritanica
 Carasso, 1956
 Pedler, 1965

Pedler and Tilly, 1964
Tansley, 1959, 1964
Tiliqua rugosa
 Braekevelt, 1989
Uta stansburiana
 Brecha et al., 1984
 Engbretson and Battelle, 1987
"Gekko"
 Drenckhahn and Wagner, 1985
"Lizard"
 Okuda, 1961
 Osborne et al., 1982
Pygopodid species
 Underwood, 1957

SERPENTES

Agkistrodon halys
 Ishikawa and Yamada, 1969
Crotalus species
 Dunn and Adomian, 1971
Crotalus viridis
 Dunn, 1973
Elaphe climacophora
 Ishikawa and Yamada, 1969
 Yamada, 1976
 Yamada et al., 1966
Elaphe quadrivirgata
 Hara and Miyoshi, 1984
Enhydris pakistanica
 Underwood, 1966
Hypsiglena
 Underwood, 1970
Leptodeira annulata
 Miller and Snyder, 1977
Natrix maura
 Pedler, 1965

Natrix natrix
 Pedler, 1965
 Tansley and Johnson, 1956
Nerodia valida
 Hibbard and Lavergne, 1972
Pelamis platurus
 Hibbard and Lavergne, 1972
Phyllorhynchus
 Underwood, 1970
Telescopus
 Underwood, 1967b
Thamnophis species
 Chandler, 1974
 Land and Snyder, 1985
Thamnophis ordinoides
 Bunt and Klock, 1980a
Thamnophis sirtalis
 Bossomaier et al., 1989
 Wong, 1989

ACKNOWLEDGMENTS
I thank Phil Ulinski, Carl Gans, and two anonymous reviewers for their helpful comments, and Mike Rowe for stimulating discussions. Alan Brichta made the drawings in Figs. 1.1, 1.2, and 1.3. I am most grateful to him and to the Photographic Unit at the College of Osteopathic Medicine. Finally, I thank the authors and publishers who granted permission for their illustrations to be reproduced. Some of the work described in this paper was supported by NIH grant EYNS 0370001 and the Australian Research Grants Committee.

NOTES

1. This is the usage adopted by most authors (e.g., Walls, 1942a; Rodieck, 1973). Unfortunately, these same terms are occasionally used with the reverse meaning (e.g., Carasso, 1956; Kalberer and Pedler, 1963; Pedler and Tilly, 1964). In such cases I have reported the data using the nomenclature described previously rather than the original terminology of the authors. Underwood (1968, 1970) has suggested calling the more robust cell of a Type B double (the only double cell in all groups except geckos) an *axial cell* and the more slender member that wraps itself around it a *peripheral cell*. These terms have not been widely adopted, and they will not be used here.

2. See Dunn (1966a), Table I, and Underwood (1970) for useful comparisons of terms used to describe visual cells in certain squamates.

3. Vilter argues that Walls' "cones" are really developing photoreceptors. Consistent with this idea is the observation (Underwood, 1970) that these small cells usually occur near the ora serrata, where photoreceptors may be produced as the retina develops (Johns, 1977).

4. In freeze-fracture preparations, the fracture plane passes between inner and outer membrane leaflets, exposing a P face (the outer face of the internal, protoplasmic, leaflet) or an E face (the inner face of the external leaflet that faces the cytoplasm). The intramembrane particles thus exposed typically adhere preferentially to the P or the E face, depending on the particular membrane being fractured. In studies of photoreceptor disk membranes, the internal disk leaflet is considered to be the external membrane leaflet (in the sense described previously) because it is exposed to the extracellular matrix, or was once, during early stages of disk development. Thus, the E face of a disk membrane faces the interdisk space and the P face looks toward the disk lumen.

In freeze-fracture-etch experiments, one exposed face is "nibbled" away by the etching process, revealing the underlying tissue. At a synapse, for example, the exposed E face of one cell can be etched away to reveal the true outer surface of the plasmalemma of the opposed cell.

In the earlier literature, the P and E faces were referred to as the A and B faces, respectively.

5. All vertebrate photoreceptors are hyperpolarized by light captured by their own outer segments and would, therefore, be classified as off-cells by these criteria.

REFERENCES

Abe, H., and Yamamoto, T. Y. (1984). Diurnal changes in synaptic ribbons of rod cells of the turtle. *J. Ultrastruct. Res.* 86, 246–251.

Abe, H., and Yamamoto, T. Y. (1988). Modification of diurnal changes in the ultrastructure of synaptic ribbons of the turtle. *Tohoku J. Exp. Med.* 156, 381–393.

Abelsdorff, G. (1898). Physiologische Beobachtungen am Auge der Krokodile. *Arch. Anat. Physiol. (Physiol. Abt.).* 1898, 155–167.

Adolph, A. R., and Zucker, C. L. (1988). Exclusive localization of somatostatin-containing amacrine cell bodies within the visual streak of the turtle retina. *Invest. Ophthalmol. Vis. Sci.* Suppl. 29, 196.

Ali, M. A. (1974). Retinomotor responses. In *Vision in Fishes* (M. A. Ali, ed.). Plenum Press, New York, pp. 313–355.

Ammermuller, J., and Weiler, R. (1988). Physiological and morphological characterization of OFF-center amacrine cells in the turtle retina. *J. Comp. Neurol.* 273, 137–148.

Anderson, D. H., Fisher, S. K., and Steinberg, R. H. (1978). Mammalian cones: disc shedding, phagocytosis, and renewal. *Invest. Ophthalmol. Vis. Sci.* 17, 117–133.

Anh, J. N. H. (1968). Ultrastructure des récepteurs visuels de la retine de *Lacerta viridis. Bull. Assoc. Anat.* 53, 1247–1259 [English summary].

Anh, J. N. H. (1969). Ultrastructure des récepteurs visuels chez les vertébrés. *Arch. Ophthalmol.* (Paris) 29, 795–822 [English summary].

Anh, J. N. H., and Anh, N. H. (1970). Aspects de la synthèse du glycogène dans le paraboloide des cellules visuelles de la rétine. *C. R. Assoc. Anat.* 148, 468–474 [English summary].

Armengol, J. A., Prada, F., Ambrosiani, J., and Genis-Galvez, J. M. (1988). The photoreceptors of the chameleon retina (*Chamaeleo chamaeleo*). *J. Hirnforsch.* 29, 403–409.

Armengol, J. A., Prada, F., and Genis-Galvez, J. M. (1981). Oil droplets in the chameleon (*Chamaeleo chamaeleo*) retina. *Acta Anat.* 110, 35–39.

Ashmore, J. F., and Copenhagen, D. R. (1980). Different postsynaptic events in two types of retinal bipolar cell. *Nature* (London) 288, 84–86.

Ashmore, J. F., and Copenhagen, D. R. (1983). An analysis of transmission from cones to hyperpolarizing bipolar cells in the retina of the turtle. *J. Physiol.* (London) 340, 569–597.

Barlow, H. B. (1982). What causes tricromacy? A theoretical analysis using comb-filtered spectra. *Vision Res.* 22, 635–643.

Baumeister, L. (1908). Beiträge zur Anatomie und Physiologie der Rhinophiden. *Zool. Jb., Anat.* 26, 423–526.

Baylor, D. A., and Fettiplace, R. (1975). Light path and photon capture in turtle photoreceptors. *J. Physiol.* (London) 248, 433–464.

Baylor, D. A., and Fettiplace, R. (1977a). Transmission from photoreceptors to ganglion cells in turtle retina. *J. Physiol.* (London) 271, 391–424.

Baylor, D. A., and Fettiplace, R. (1977b). Kinetics of synaptic transfer from receptors to ganglion cells in turtle retina. *J. Physiol.* (London) 271, 425–448.

Baylor, D. A., Fuortes, M. G. F., and O'Bryan, P. M. (1971). Receptive fields of cones in the retina of the turtle. *J. Physiol.* (London) 214, 265–294.

Baylor, D. A., and Hodgkin, A. L. (1973). Detection and resolution of visual stimuli by turtle photoreceptors. *J. Physiol.* (London) 234, 163–198.

Berger, E. R. (1966). On the mitochondrial origin of oil drops in the retinal double cone inner segments. *J. Ultrastruct. Res.* 14, 143–157.

Bernstein, S. A., Breding, D. J., and Fisher, S. K. (1984). The influence of light on cone disk shedding in the lizard, *Sceloporus occidentalis. J. Cell Biol.* 99, 379–389.

Besharse, J. C. (1986). Photosensitive membrane turnover: differentiated membrane domains and cell-cell interaction. In *The Retina: a Model for Cell*

Biology Studies (R. Adler and D. Farber, eds.). Academic Press, New York, pp. 297–352.

Bingelli, R. L., and Paule, W. J. (1969). The pigeon retina: Quantitative aspects of the optic nerve and ganglion cell layer. *J. Comp. Neurol.* 137, 1–18.

Boehlert, G. W. (1978). Intraspecific evidence for the function of single and double cones in the teleost retina. *Science* (N.Y.) 202, 309–311.

Borovyagin, V. L. (1966). Submicroscopic morphology and structural connection of the receptor and horizontal retinal cells of the retina of a number of lower vertebrates. *Biophysics* 11, 930–940 [translated from *Biofizika* 11, 810–817].

Borwein, B. (1981). The retinal receptor: a description. In *Vertebrate Photoreceptor Optics* (J. M. Enoch and F. L. Tobey, eds.). Springer-Verlag, New York, pp. 11–81.

Bossomaier, T. R. J., Wong, R. O. L., and Snyder, A. W. (1989). Stiles-Crawford effect in garter snake. *Vision Res.* 29, 741–746.

Bowling, D. B. (1980). Light responses of ganglion cells in the retina of the turtle. *J. Physiol.* (London) 299, 173–196.

Bowmaker, J. K. (1980). Colour vision in birds and the role of oil droplets. *Trends Neurosci.* 3, 196–199.

Boycott, B. B., and Dowling, J. E. (1969). Organization of the primate retina: light microscopy. *Phil. Trans. Roy. Soc. Lond.* B255, 109–184.

Braekevelt, C. R. (1989). Photoreceptor fine structure in an Australian reptile, the bobtail goanna (*Tiliqua rugosa*). *Anat. Rec.* 223, 17A.

Brandon, C. (1985). Retinal GABA neurons: localization in vertebrate species using an antiserum to rabbit brain glutamate decarboxylase. *Brain Res.* 344, 286–295.

Brecha, N. (1983). Retinal neurotransmitters: histochemical and biochemical studies. In *Chemical Neuroanatomy* (P. C. Emson, ed.). Raven Press, New York, pp. 85–130.

Brecha, N. C., Eldred, W., Kuljis, R. O., and Karten, H. J. (1984). Identification and localization of biologically active peptides in the vertebrate retina. In *Progress in Retinal Research* (N. Osborne and J. Chandler, eds.). Pergamon Press, New York, pp. 185–226.

Brown, K. T. (1969). A linear area centralis extending across the turtle retina and stabilized to the horizon by non-visual cues. *Vision Res.* 9, 1053–1062.

Bunt, A. H., and Klock, I. B. (1980a). Comparative study of 3H-fucose incorporation into vertebrate photoreceptor outer segments. *Vision Res.* 20, 739–747.

Bunt, A. H., and Klock, I. B. (1980b). Fine structure and radioautography of retinal cone outer segments in goldfish and carp. *Invest. Ophthalmol. Vis. Sci.* 19, 707–719.

Burnside, B. (1988). Photoreceptor contraction and elongation: calcium and cyclic adenosine 3′,5′ monophosphate regulation of actin- and microtubule-dependent changes in cell shape. In *Intrinsic Determinants of Neuronal Form and Function*. Alan R. Liss, New York, pp. 323–359.

Burnside, B., and Dearry, A. (1986). Cell motility in the retina. In *The Retina* (R. Adler and D. Farber, eds.). Academic Press, New York, pp. 151–206.

Bush, W. G., and Lunger, P. D. (1972). Ultrastructural survey of the retina of *Pseudemys scripta elegans*. *Brain Behav. Evol.* 5, 114–123.

Byzov, A. L., and Shura-Bura, T. M. (1985). Local nonuniformities in the horizontal cell syncytium of the turtle retina. *Neurofiziologiya* 17, 239–245.

Cajal, S. R. (1893). La rétine des vertébrés. *La Cellule* 9,17–257.

Carasso, N. (1956). Mise en evidence de prolongements cytoplasmiques inframicroscopiques au niveau du segment interne des cellules visuelles du *Gecko* (Reptile). *C. Rend. hebd. Seanc. Acad. Sci.* (Paris) 242, 2988–2991.

Chandler, J. P. (1974). Electronmicroscopic observations of the retina of the garter snake, *Thamnophis*. *Anat. Rec.* 178, 325.

Chievitz, J. H. (1889). Untersuchungen über die Area centralis retinae. *Arch. Anat. (Physiol.)*, Suppl. 1, 139–196

Chievitz, J. H. (1890). Untersuchungen über die Entwickelung der Area und Fovea centralis retinae. *Arch. Anat. (Physiol.)*, S. 332–366.

Chievitz, J. H. (1891). Ueber das Vorkommen der Area centralis retinae in den vier höheren Wirbelthierklassen. *Arch. Anat. Entw.* 311–334.

Citron, M. C., and Pinto, L. H. (1973). Retinal image: larger and more illuminous for a nocturnal than for a diurnal lizard. *Vision Res.* 13, 873–876.

Cohen, A. I. (1972). Rods and cones. In *Handbook of Sensory Physiology* (M. G. F. Fuortes, ed.). Springer-Verlag, Berlin, VII/2, pp. 63–110.

Copenhagen, D. R., and Green, D. G. (1987). Spatial spread of adaptation within the cone network of turtle retina. *J. Physiol.* (London) 393, 763–776.

Copenhagen, D. R., and Jahr, C. E. (1989). Release of endogenous excitatory amino acids from turtle photoreceptors. *Nature* (London) 341, 536–538.

Copenhagen, D. R., and Owen, W. G. (1976a). Coupling between rod photoreceptors in a vertebrate retina. *Nature* (London) 260, 57–59.

Copenhagen, D. R., and Owen, W. G. (1976b). Functional characteristics of lateral interactions between rods in the retina of the snapping turtle. *J. Physiol.* (London) 259, 251–282.

Copenhagen, D. R., Ashmore, J. F., and Schnapf, J. K. (1983). Kinetics of synaptic transmission from photoreceptors to horizontal and bipolar cells in turtle retina. *Vision Res.* 23, 363–369.

Craig, E. L., Eglits, J. A., and McConnell, D. G. (1963). Observations on the oil droplets of the principal cone cells of the frog retina. *Exp. Eye Res.* 2, 268–271.

Crescitelli, F., McFall-Ngai, M., and Horwitz, J. (1985). The visual pigment sensitivity hypothesis: further evidence from fishes of varying habitats. *J. Comp. Physiol.* A157, 323–334.

Cuenca, N., and Kolb, H. (1989). Morphology and distribution of neurons immunoreactive for substance P in the turtle retina. *J. Comp. Neurol.* 290, 391–411.

Dacheux, R. F. (1982). Connections of the small bipolar cells with the photoreceptors in the turtle: an electron microscope study of Golgi-impregnated, gold-toned retinas. *J. Comp. Neurol.* 205, 55–62.

Davydova, T. V. (1981). The ultrastructure of displaced bipolars in the turtle retinae. *Tsitologhiya* 23, 12–16 [in Russian with English summary].

Davydova, T. V. (1983). Contacts of the cone cell basal processes in the external plexiform layer of the tortoise tunica sensory bulbi. *Archives of Anatomy, Histology, and Embryology* (Leningrad) 85, 49–55 [in Russian with English summary].

Davydova, T. V., Goncharova, N. V., and Boyko, V. P. (1982). Correlation between the morpho-functional organization of some portions of the visual analyser of *Chelonia* and their ecology. I. Normal morpho-functional characteristics of the optic nerve and the tectum opticum. *J. Hirnforsch.* 23, 271–286.

Daw, N. W., Brunken, W. J., and Parkinson, D. (1989). The function of synaptic transmitters in the retina. *Ann. Rev. Neurosci.* 12, 205–225.

Detwiler, S. R. (1916). The effect of light on the retina of the tortoise and the lizard. *J. Exp. Zool.* 20, 165–189.

Detwiler, S. R. (1923a). Studies on the retina: an experimental study of the gecko retina. *J. Comp. Neurol.* 36, 125–141.

Detwiler, S. R. (1923b). Photomechanical responses in the retina of *Eremias argus*. *J. Exp. Zool.* 37, 89–99.

Detwiler, S. R. (1943). *Vertebrate Photoreceptors*. Macmillan, New York.

Detwiler, P. B., and Hodgkin, A. L. (1979). Electrical coupling between cones in turtle retina. *J. Physiol.* (London) 291, 75–100.

Detwiler, P. B., Hodgkin, A. L., and McNaughton, P. A. (1980). Temporal and spatial characteristics of the voltage response of rods in the retina of the snapping turtle. *J. Physiol.* (London) 300, 213–250.

Detwiler, S. R., and Laurens, H. (1920). Studies on the retina: the structure of the retina of *Phrynosoma cornutum*. *J. Comp. Neurol.* 32, 347–356.

Detwiler, P. B., and Sarthy, P. V. (1981). Selective uptake of Lucifer yellow by bipolar cells in the turtle retina. *Neurosci. Lett.* 22, 227–232.

Dogiel, A. S. (1888). Ueber das Verhalten der nervösen Elemente in der Retina der Ganoiden, Reptilien, Vögel und Säugethiere. *Anat. Anz.* 3, 133–143.

Dowling, J. E. (1987). *The Retina: an Approachable Part of the Brain*. Harvard University Press, Cambridge, Mass.

Drenckhahn, D., and Wagner, H.-J. (1985). Relation of retinomotor responses and contractile proteins in vertebrate retinas. *Eur. J. Cell Biol.* 37, 156–168.

Drujan, B. D., and Laufer, M. (eds.) (1982). *The S-Potential*. Alan R. Liss, New York.

Duke-Elder, S. (1958). *System of Ophthalmology*, Vol. I: *The Eye in Evolution*. The C. V. Mosby Co., St. Louis.

Dunn, R. F. (1966a). Studies on the retina of the gecko *Coleonyx variegatus*. I. The visual cell classification. *J. Ultrastruct. Res.* 16, 651–671.

Dunn, R. F. (1966b). Studies on the retina of the gecko *Coleonyx variegatus*. II. The rectilinear visual cell mosaic. *J. Ultrastruct. Res.* 16, 672–684.

Dunn, R. F. (1966c). Studies on the retina of the gecko *Coleonyx variegatus*. III. Photoreceptor cross-sectional area relationships. *J. Ultrastruct. Res.* 16, 685–692.

Dunn, R. F. (1969). The dimensions of rod outer segments related to light absorption in the gecko retina. *Vision Res.* 9, 603–609.

Dunn, R. F. (1973). The ultrastructure of the vertebrate retina. In *The Ultrastructure of Sensory Organs* (I. Friedman, ed.). North-Holland/American Elsevier, New York, pp. 153–265.

Dunn, R. F., and Adomian, G. E. (1971). The pentalaminar configuration of the vertebrate outer segment discs. In *Electron Microscopy 1971* (C. J. Arceneaux, ed.). Claitor's Pub. Division, Baton Rouge, La., pp. 268–269.

Eakin, R. M. (1964). The effect of vitamin A deficiency on photoreceptors in the lizard *Sceloporus occidentalis*. *Vision Res.* 4, 17–22.

Eakin, R. M. (1965). Differentiation of rods and cones in total darkness. *J. Cell Biol.* 25, 162–165.

Ehinger, B. (1982). Neurotransmitter systems in the retina. *Retina* 2, 305–321.

Ehinger, B., Ottersen, O. P., Storm-Mathisen, J., and Dowling, J. E. (1988). Bipolar cells in the turtle retina are strongly immunoreactive for glutamate. *Proc. Natl. Acad. Sci. USA* 85, 8321–8325.

Eldred, W. D., and Carraway, R. E. (1987). Neurocircuitry of two types of neurotensin-containing amacrine cells in the turtle retina. *Neurosci.* 21, 603–618.

Eldred, W. D., and Cheung, K. (1989). Immunocytochemical localization of glycine in the retina of the turtle (*Pseudemys scripta*). *Visual Neurosci.* 2, 331–338.

Eldred, W. D., Isayama, T., Reiner, A., and Carraway, R. (1988). Ganglion cells in the turtle retina contain the neuropeptide LANT-6. *J. Neurosci.* 8, 119–132.

Eldred, W. D., and Karten, H. J. (1983). Characterization and quantification of peptidergic amacrine cells in the turtle retina: enkephalin, neurotensin and glucagon. *J. Comp. Neurol.* 221, 371–381.

Eldred, W. D., and Karten, H. J. (1985). Ultrastructure and synaptic contacts of enkephalinergic amacrine cells in the retina of turtle (*Pseudemys scripta*). *J. Comp. Neurol.* 232, 36–42.

Eldred, W. D., Li, H. B., Carraway, R. E., and Dowling, J. E. (1987). Immunocytochemical localization of LANT-6-like immunoreactivity within neurons in the inner nuclear and ganglion cell layers in vertebrate retinas. *Brain Res.* 424, 361–370.

Engbretson, G. A., and Battelle, B. A. (1987). Serotonin and dopamine in the retina of a lizard. *J. Comp. Neurol.* 257, 140–147.

Enoch, J. M., and Tobey, F. L. (1981). Waveguide properties of retinal receptors: techniques and observations. In *Vertebrate Photoreceptor Optics.* (J. M. Enoch and F. L. Tobey, eds.). Springer-Verlag, New York, pp. 169–218.

Ernst, C. H., Soenarjo, S., and Hamilton, H. F. (1970). The retinal histology of the stinkpot, *Sternotherus odoratus. Herpetologica* 26, 222–223.

Etingof, R. N., Kalinina, S. N., and Dumler, I. L. (1986). Immunochemical study of the cyclic nucleotide system of retinal photoreceptor membranes: antibodies raised to phosphodiesterase, its protein inhibitor and GTP-binding proteins. *Vision Res.* 26, 415–423.

Evans, E. M. (1966). On the ultrastructure of the synaptic region of visual receptors in certain vertebrates. *Z. Zellforsch. Mikrosk. Anat.* 71, 499–516.

Famiglietti, E. V., Kaneko, A., and Tachibana, M. (1977). Neuronal architecture of on and off pathways to ganglion cells in carp retina. *Science* (N.Y.) 198, 1267–1269.

Farber, D. B., and Shuster, T. A. (1986). Cyclic nucleotides in retinal function and degeneration. In *The Retina: a Model for Cell Biology Studies* (R. Adler and D. Farber, eds.). Academic Press, New York, pp. 239–296.

Fite, K. V., and Lister, B. C. (1981). Bifoveal vision in anolis lizards. *Brain Behav. Evol.* 19, 144–154.

Fite, K. V., Reiner, A., and Hunt, S. P. (1979). Optokinetic nystagmus and the accessory optic system of pigeon and turtle. *Brain Behav. Evol.* 16, 192–202.

Fliesler, S. J., and Anderson, R. E. (1983). Chemistry and metabolism of lipids in the vertebrate retina. *Prog. Lipid Res.* 22, 79–131.

Franz, V. (1913). Sehorgan. In *Oppel's Lehrbuch der vergleichenden mikroskopischen Anatomie der Wirbeltiere.* G. Fischer, Jena, Vol. 7, i–x, 1–417.

Fujimoto, K., Yanase, T., and Hanaoka, T. (1957). Spectral transmittance of retinal colored oil globules re-examined with microspectrophotometer. *Jap. J. Physiol.* 7, 339–346.

Fuortes, M. G. F., Schwartz, E. A., and Simon, E. J. (1973). Colour-dependence of cone responses in the turtle retina. *J. Physiol.* (London) 234, 199–216.

Fuortes, M. G. F., and Simon, E. J. (1974). Interactions leading to horizontal cell responses in the turtle retina. *J. Physiol.* (London) 240, 177–198.

Gallego, A. (1983). Organization of the outer plexiform layer of the tetropoda retina. In *Progress in Sensory Physiology 4* (H. Autrum, D. Ottoson, E. R. Perl, R. F. Schmidt, H. Shimazu, W. D. Willis, eds.). Springer-Verlag, New York, pp. 83–114.

Gallego, A., and Perez-Arroyo, M. (1976). Photoreceptors and horizontal cells of the turtle retina (*Clemmys caspica*). In *The Structure of the Eye* (E. Yamada and S. Mishima, eds.). *Jap. J. Ophthalmol* 3, 311–317.

Garten, S. (1907). Die Veränderungen der Netzhaut durch Licht. In *Graefe-Saemisch Handbuch der gesamten Augenheilkunde,* Part 13(12), Appendix, pp. 1–130, Wilhelm Engelmann, Leipzig.

Gaur, V. P., Adamus, G., Arendt, A., Eldred, W., Possin, D. E., McDowell, J. H., Hargrave, P. A., Sarthy, P. V. (1988). A monoclonal antibody that binds to photoreceptors in the turtle retina. *Vision Res.* 28, 765–776.

Gaur, V. P., Eldred, W., Possin, D. E., and Sarthy, P. V. (1989). A monoclonal antibody marker for the paraboloid region of cone photoreceptors in turtle retina. *Cell Tissue Res.* 257, 497–503.

Geri, G. A., Kimsey, R. A., and Dvorak, C. A. (1982). Quantitative electron microscopic analysis of the optic nerve of the turtle, *Pseudemys. J. Comp. Neurol.* 207, 99–103.

Gerschenfeld, H. M., Neyton, J., Piccolino, M., and Witkovsky, P. (1982). L-Horizontal cells of the turtle: network organization and coupling modulation. In *Neurotransmitters in the Retina and the Visual Centers* (A. Kaneko, N. Tsukahara, and K. Uchizono, eds.). Biomedical Research Foundation, Tokyo, pp. 21–34.

Gillett, W. G. (1923). The histologic structure of the eye of the soft-shelled turtle. *Am. J. Ophthalmol.* 6, 955–973.

Granda, A. M., and Dvorak, C. A. (1977). Vision in turtles. In *Handbook of Sensory Physiology* (H. Autrum, R. Jung, W. R. Loewenstein, D. M. Mackay, H. L. Teuber, eds.). Springer-Verlag, New York, Vol. VII/5, pp. 451–495.

Granda, A. M., and Fulbrook, J. E. (1989). Classification of turtle retinal ganglion cells. *J. Neurophysiol.* 62, 723–737.

Granda, A. M., and Haden, K. W. (1970). Retinal oil globule counts and distributions in two species of turtles: *Pseudemys scripta elegans* (Wied) and *Chelonia mydas mydas* (Linnaeus). *Vision Res.* 10, 79–84.

Guiloff, G. D., Jones, J., and Kolb, H. (1988). Organization of the inner plexiform layer of the turtle retina: an electron microscopic study *J. Comp. Neurol.* 272, 280–292.

Hagins, W. A. (1979). Excitation in vertebrate photoreceptors. In *The Neurosciences: Fourth Study Program* (F. O. Schmitt, F. G. Worden, eds.). The MIT Press, Cambridge, Mass., pp. 183–191.

Hannover, A. (1840). Ueber die Netzhaut und ihre Gehirnsubstanz bei Wirbelthieren mit Ausnahme des Menschen. *Arch. Anat. Physiol. wiss. Med.* 1840, 320–345.

Hara, S., and Miyoshi, M. (1984). Intercellular junctions between visual cells in the snake retina. *J. Electron Microsc.* 33, 39–45.

Hargrave, P. A. (1986). Molecular dynamics of the rod cell. In *The Retina* (R. Adler and D. Farber, eds.). Academic Press, New York, pp. 207–237.

Hashimoto, Y., Saito, T., Miller, W. H., and Tomita, T. (1973). Morphological and physiological identification of retinal cells in the turtle. In *Intracellular Staining in Neurobiology* (S. B. Kater and C. Nicholson, eds.). Springer-Verlag, New York, pp. 181–188.

Heinemann, C. (1877). Beiträge zur Anatomie der Retina. *Arch. Mikroskop. Anat.* 14, 409–441.

Hess, C. (1910). Untersuchungen über den Lichtsinn bei Reptilien und Amphibien. *Pflüger's Archiv. ges. Physiol.* 132, 255–295.

Hibbard, E., and Lavergne, J. (1972). Morphology of the retina of the seasnake, *Pelamis platurus. J. Anat.* 112, 125–136.

Holmberg, K. (1977). The cyclostome retina. In *Handbook of Sensory Physiology* (H. Autrum, R. Jung, W. R. Loewenstein, D. M. MacKay, and H. L. Teuber, eds.). Springer-Verlag, New York, VII/5, pp. 47–66.

Holtzman, E., and Mercurio, A. M. (1980). Membrane circulation in neurons and photoreceptors: some unresolved issues. *Int. Rev. Cytol.* 67, 1–67.

Hughes, A. (1977). The topography of vision in mammals of contrasting life style: comparative optics and retinal organisation. In *Handbook of Sensory Physiology* (H. Autrum, R. Jung, W. R. Loewenstein, D. M. MacKay, and H. L. Teuber, eds.). Springer-Verlag, New York, VII/5, pp. 613–756.

Hughes, A., and Vaney, D. I. (1980). Coronate cells: displaced amacrines in the rabbit retina. *J. Comp. Neurol.* 189, 169–189.

Hulke, J. W. (1863–1865). A contribution to the anatomy of the amphibian and reptilian retina. *Roy. Lond. Ophthalmol. Hosp. Rep.* 4, 243–268.

Hulke, J. W. (1867). On the retina of amphibia and reptiles. *J. Anat. Physiol.* 1, 94–106.

Hurd, L. B., and Eldred, W. D. (1989). Localization of GABA- and GAD-like immunoreactivity in the turtle retina. *Visual Neurosci.* 3, 9–20.

Isayama, T., and Eldred, W. D. (1988). Neuropeptide Y-immunoreactive amacrine cells in the retina of the turtle *Pseudemys scripta elegans. J. Comp. Neurol.* 271, 56–66.

Isayama, T., Polak, J., and Eldred, W. D. (1988). Synaptic analysis of amacrine cells with neuropeptide Y-like immunoreactivity in turtle retina. *J. Comp. Neurol.* 275, 452–459.

Ishida, A. T., Stell, W. K., and Lightfoot, D. O. (1980). Rod and cone inputs to bipolar cells in goldfish retina. *J. Comp. Neurol.* 191, 315–335.

Ishikawa, T., and Yamada, E. (1969). Atypical mitochondria in the ellipsoid of the photoreceptor cells of vertebrate retinas. *Invest. Ophthalmol. Vis. Sci.* 8, 302–316.

Ives, J. T., Normann, R. A., and Barber, P. W. (1983). Light intensification by cone oil droplets: electromagnetic considerations. *J. Opt. Soc. Am.* 73, 1725–1731.

Jensen, R. J., and DeVoe, R. D. (1982). Ganglion cells and (dye-coupled) amacrine cells in the turtle retina that have possible synaptic connection. *Brain Res.* 240, 146–150.

Jensen, R. J., and DeVoe, R. D. (1983). Comparisons of directionally-selective with other ganglion cells of the turtle retina: intracellular recording and staining. *J. Comp. Neurol.* 217, 271–287.

Johns, P. R. (1977). Growth of the adult goldfish eye. III. Source of the new retinal cells. *J. Comp. Neurol.* 176, 343–358.

Johnson, G. L. (1927). Contributions to the comparative anatomy of the reptilian and the amphibian eye, chiefly based on ophthalmological examination. *Phil. Trans. Roy. Soc. Lond.* B215, 315–353.

Kalberer, M., and Pedler, C. H. (1963). The visual cells of the alligator: an electron microscopic study. *Vision Res.* 3, 323–329.

Kaneko, A. (1979). Physiology of the retina. *Ann. Rev. Neurosci.* 2, 169–191.

Kaneko, A., Ohtsuka, T., and Tachibana, M. (1985). GABA sensitivity in solitary turtle cones: evidence for the feedback pathway from horizontal cells to cones. In *Neurocircuitry of the Retina: a Cajal Memorial.* (A. Gallego and P. Gouras, eds.). Elsevier Science Publishing Co., New York, pp. 89–98.

Kaneko, A., and Tachibana, M. (1986a). Effects of gamma-aminobutyric acid on isolated cone photoreceptors of the turtle retina. *J. Physiol.* (London) 373, 443–461.

Kaneko, A., and Tachibana, M. (1986b). Blocking effects of cobalt and related ions on the gamma-aminobutyric acid-induced current in turtle retinal cones. *J. Physiol.* (London) 373, 463–479.

Karten, H. J., and Brecha, N. (1983). Localization of neuroactive substances in the vertebrate retina: evidence for lamination in the inner plexiform layer. *Vision Res.* 23, 1197–1205.

Kawamura, S., and Murakami, M. (1986). In situ cGMP phosphodiesterase and photoreceptor potential in gecko retina. *J. Gen. Physiol.* 87, 737–759.

Kirschfeld, K. (1983). Are photoreceptors optimal? *Trends Neurosci.* 6, 97–101.

Knox, R. (1823). On the discovery of the foramen centrale of the retina in the eyes of reptiles. *Edinburgh Philosophical J.* 9, 358–359.

Kolb, H. (1982). The morphology of the bipolar cells, amacrine cells and ganglion cells in the retina of the turtle, *Pseudemys scripta elegans. Phil. Trans. Roy. Soc. Lond.* B298, 355–393.

Kolb, H., Cline, C., Wang, H. H., and Brecha, N. (1987). Distribution and morphology of dopaminergic amacrine cells in the retina of the turtle (*Pseudemys scripta elegans*). *J. Neurocytol.* 16, 577–588.

Kolb, H., and Jones, J. (1982). Light and electron microscopy of the photoreceptors in the retina of the red-eared slider. *Pseudemys scripta elegans. J. Comp. Neurol.* 209, 331–338.

Kolb, H., and Jones, J. (1984). Synaptic organization of the outer plexiform layer of the turtle retina: an electron microscope study of serial sections. *J. Neurocytol.* 13, 567–591.

Kolb, H., and Jones, J. (1985). Electron microscopy of golgi-impregnated photoreceptors reveals connections between red and green cones in the turtle retina. *J. Neurophysiol.* 54, 304–317.

Kolb, H., and Jones, J. (1987). The distinction by light and electron microscopy of two types of cone containing colorless oil droplets in the retina of the turtle. *Vision Res.* 27, 1445–1458.

Kolb, H., Perlman, I., and Normann, R. A. (1988). Neural organization of the retina of the turtle *Mauremys caspica:* a light microscope and golgi study. *Visual Neurosci.* 1, 47–72.

Kolb, H., Wang, H. H., and Jones, J. (1986). Cone synapses with golgi-stained bipolar cells that are morphologically similar to a center-hyperpolarizing and a center-depolarizing bipolar cell type in the turtle retina. *J. Comp. Neurol.* 250, 510–520.

Kouyama, N., and Ohtsuka, T. (1985). Quantitative morphological study of the outer nuclear layer in the turtle retina. *Brain Res.* 345, 200–203.

Krause, W. (1893). Die Retina der Amphibien und Reptilien. *Mon. internat. J. Anat. Physiol.* 10, 12–84.

Kunz, Y. W., and Wise, C. (1978). Structural differences of cone 'oil-droplets' in the light and dark adapted retina of *Poecilia reticulata P. Experientia* 34, 246–248.

Lamb, T. D., and Simon, E. J. (1976). The relation between intercellular coupling and electrical noise in turtle photoreceptors. *J. Physiol.* (London) 263, 257–286.

Land, M. F., and Snyder, A. W. (1985). Cone mosaic observed directly through natural pupil of live vertebrate. *Vision Res.* 25, 1519–1523.

Landis, D. M. D. (1988). Membrane and cytoplasmic structure at synaptic junctions in the mammalian central nervous system. *J. Electron Microsc. Tech.* 10, 129–151.

Landis, D. M. D., and Reese, T. S. (1974). Differences in membrane structure between excitatory and inhibitory synapses in the cerebellar cortex. *J. Comp. Neurol.* 155, 93–126.

Landis, D. M. D., Reese, T. S., and Raviola, E. (1974). Differences in membrane structure between excitatory and inhibitory components of the reciprocal synapse in the olfactory bulb. *J. Comp. Neurol.* 155, 67–92.

Lasansky, A. (1969). Basal junctions at synaptic endings of turtle visual cells. *J. Cell Biol.* 40, 577–581.

Lasansky, A. (1971). Synaptic organization of cone cells in the turtle retina. *Phil. Trans. Roy. Soc. Lond.* B262, 365–381.

Lasansky, A. (1972). Cell junctions at the outer synaptic layer of the retina. *Invest. Ophthalmol. Vis. Sci.* 11, 265–275.

Lasansky, A. (1973). Organization of the outer synaptic layer in the retina of the larval tiger salamander. *Phil. Trans. Roy. Soc. Lond.* 265, 471–489.

Lasansky, A. (1974). Synaptic actions on retinal photoreceptors: structural aspects. *Fed. Proc.* 33, 1069–1073.

Lasansky, A. (1977). Synaptic organization of retinal receptors. In *Vertebrate Photoreception* (H. B. Barlow and P. Fatt, eds.). Academic Press, New York, pp. 275–290.

Lasater, E. M., Normann, R. A., and Kolb, H. (1989). Signal integration at the pedicle of turtle cone photoreceptors: an anatomical and electrophysiological study. *Visual Neurosci.* 2, 553–564.

Laties, A., Liebman, P., and Bok, D. (1976a). Effects of procion yellow on *Gekko gecko* rod outer segments. In *The Structure of the Eye* (E. Yamada and S. Mishima, eds.). *Jap. J. Ophthalmol* 3, 247–259.

Laties, A. M., Bok, D., and Liebman, P. (1976b). Procion yellow: a marker dye for outer segment disc patency and for rod renewal. *Exp. Eye Res.* 23, 139–148.

Laurens, H., and Detwiler, S. R. (1921). Studies on the retina: the structure of the retina of *Alligator mississippiensis* and its photomechanical changes. *J. Exp. Zool.* 32, 207–234.

Leeper, H. F. (1978a). Horizontal cells of the turtle retina. I. Light microscopy of Golgi preparations. *J. Comp. Neurol.* 182, 777–793.

Leeper, H. F. (1978b). Horizontal cells of the turtle retina. II. Analysis of interconnections between photoreceptor cells and horizontal cells by light microscopy. *J. Comp. Neurol.* 182, 795–809.

Leeper, H. F., and Copenhagen, D. R. (1979). Mixed rod-cone responses in horizontal cells of snapping turtle retina. *Vision Res.* 19, 407–412.

Leeper, H. F., and Copenhagen, D. R. (1982). Horizontal cells in turtle retina: structure, synaptic connections, and visual processing. In *The S-Potential* (B. D. Drujan and M. Laufer, eds.). Alan R. Liss, New York, pp. 77–104.

Levine, J. S., and MacNichol, E. F. (1979). Visual pigments in teleost fishes: effects of habitat, microhabitat, and behavior on visual system evolution. *Sensory Proc.* 3, 95–131.

Leydig, F. (1853). *Anatomisch-Histologische Untersuchungen über Fische und Reptilien.* Georg Reimer, Berlin.

Liebman, P. A. (1972). Microspectrophotometry of photoreceptors. In *Handbook of Sensory Physiology* (H. J. A. Dartnall, ed.). Springer-Verlag, New York, VII/1, 481–528.

Liebman, P. A., and Granda, A. M. (1975). Super dense carotenoid spectra resolved in single cone oil droplets. *Nature* (London) 253, 370–372.

Liebman, P. A., and Leigh, R. A. (1969). Autofluorescence of visual receptors. *Nature* (London) 221, 1249–1251.

Lipetz, L. E. (1984a). A new method for determining peak absorbance of dense pigment samples and its application to the cone oil droplets of *Emydoidea blandingii. Vision Res.* 24, 597–604.

Lipetz, L. E. (1984b). Pigment types, densities and concentrations in cone oil droplets of *Emydoidea blandingii. Vision Res.* 24, 605–612.

Lipetz, L. E. (1985). Some neuronal circuits of the turtle retina. In *The Visual System.* Alan R. Liss, New York, pp. 107–132.

Loew, E. R., and Lythgoe, J. N. (1985). The ecology of color vision. *Endeavour* 14, 170–174.

Lund, O. E., and von Barsewisch, B. (1975). Die Macula in der Tierreihe. *Ber Dtsch. Ophthalmol. Ges.* 73, 11–17.

Lythgoe, J. N. (1979). *The Ecology of Vision.* Oxford University Press, New York.

Maekawa, N., and Mizuno, K. (1976). Cryoultramicroscopy of lower vertebrate retina. *Albrecht von Graefes Arch. Klin. exp. Ophthalmol.* 201, 1–10.

Makaretz, M., and Levine, R. L. (1980). A light microscopic study of the bifoveate retina in the lizard *Anolis carolinensis:* general observations and convergence ratios. *Vision Res.* 20, 679–686.

Marc, R. E. (1985). The role of glycine in retinal circuitry. In *Retinal Transmitters and Modulators: Models for the Brain* (R. E. Marc, ed.). CRC Press, Boca Raton, Fla., pp. 119–158.

Marc, R. E. (1986). Neurochemical stratification in the inner plexiform layer of the vertebrate retina. *Vision Res.* 26, 223–238.

Marchiafava, P. L. (1982). Electrophysiological and structural studies of the organization of inputs to ganglion cells in the turtle retina. *Arch. Ital. Biol.* 120, 271–282.

Marchiafava, P. L. (1983). The organization of inputs establishes two functional and morphologically identifiable classes of ganglion cells in the retina of the turtle. *Vision Res.* 23, 325–338.

Marchiafava, P. L., and Torre, V. (1977). Self-facilitation of ganglion cells in the retina of the turtle. *J. Physiol.* (London) 268, 335–351.

Marchiafava, P. L., and Torre, V. (1978). The responses of amacrine cells to light and intracellularly applied currents. *J. Physiol.* (London) 276, 83–102.

Marchiafava, P. L., and Wagner, H. G. (1981). Interactions leading to colour opponency in ganglion cells of the turtle retina. *Proc. Roy. Soc. Lond. (Biol.)* 211, 261–267.

Marchiafava, P. L., and Weiler, R. (1980). Intracellular analysis and structural correlates of the organization of inputs to ganglion cells in the retina of the turtle. *Proc. Roy. Soc. Lond. (Biol.)* 208, 103–113.

Marchiafava, P. L., and Weiler, R. (1982). The photoresponses of structurally identified amacrine cells in the turtle retina. *Proc. Roy. Soc. Lond. (Biol.)* 214, 403–415.

Mariani, A. P., and Lasansky, A. (1984). Chemical synapses between turtle photoreceptors. *Brain Res.* 310, 351–354.

Massey, S. C., and Redburn, D. A. (1987). Transmitter circuits in the vertebrate retina. *Prog. Neurobiol.* 28, 55–96.

Meller, K., and Eschner, J. (1965). Vergleichende Untersuchungen über die Feinstruktur der Bipolarzellschicht der Vertebratenretina. *Z. Zellforsch. Mikrosk. Anat.* 68, 550–567 [English summary].

Meneghini, K. A., and Hamasaki, D. I. (1967). The electroretinogram of the iguana and Tokay gecko. *Vision Res.* 7, 243–251.

Meyer, D. B., Cooper, T. G., and Gernez, C. (1965). Retinal oil droplets. In *The Structure of the Eye; Symposium* (J. W. Rohen, ed.). Schattauer-Verlag, Stuttgart, Vol. 2, pp. 521–533.

Miller, R. F. (1979). The neuronal basis of ganglion-cell receptive-field organization and the physiology of amacrine cells. In *The Neurosciences: Fourth Study Program* (F. O. Schmidt and F. G. Worden, eds.). The MIT Press, Cambridge, Mass., pp. 227–245.

Miller, R. F., and Slaughter, M. M. (1986). Excitatory amino acid receptors of the retina: diversity of subtypes and conductance mechanisms. *Trends Neurosci.* 9, 211–218.

Miller, W. H., Hashimoto, Y., Saito, T., and Tomita, T. (1973). Physiological and morphological identification of L- and C-type S-potentials in the turtle retina. *Vision Res.* 13, 443–447.

Miller, W. H., and Snyder A. W. (1977). The tiered vertebrate retina. *Vision Res.* 17, 239–255.

Moody, M. F., and Robertson, J. D. (1960). The fine structure of some retinal photoreceptors. *J. Biophys. Biochem. Cytol.* 7, 87–92.

Morris, V. B. (1970). Symmetry in a receptor mosaic demonstrated in the chick from the frequencies, spacing and arrangement of the types of retinal receptor. *J. Comp. Neurol.* 140, 359–398.

Mosinger, J. L., Yazulla, S., and Studholme, K. M. (1986). GABA-like immunoreactivity in the vertebrate retina: a species comparison. *Exp. Eye Res.* 42, 631–644.

Müller, H. (1857). Anatomisch-physiologische Untersuchungen über die Retina bei Menschen und Wirbelthieren. *Z. Wiss. Zool.* 8, 1–122.

Muntz, W. R. A. (1972). Inert absorbing and reflecting pigments. In *Handbook of Sensory Physiology* (H. J. A. Dartnall, ed.). Springer-Verlag, Berlin, VII/1, pp. 529–565.

Nelson, R., Famiglietti, E. V., and Kolb, H. (1978). Intracellular staining reveals different levels of stratification for on- and off-center ganglion cells in cat retina. *J. Neurophysiol.* 41, 472–483.

Neumeyer, C., and Jager, J. (1985). Spectral sensitivity of the freshwater turtle *Pseudemys scripta elegans:* evidence for the filter-effect of colored oil droplets. *Vision Res.* 25, 833–838.

Nguyen-Legros, J., Versaux-Botteri, C., Vigny, A., and Raoux, N. (1985). Tyrosine hydroxylase immunohistochemistry fails to demonstrate dopaminergic interplexiform cells in the turtle retina. *Brain Res.* 339, 323–328.

Nilsson, S. E. G. (1964). Receptor cell outer segment development and ultrastructure of the disk membranes in the retina of the tadpole (*Rana pipiens*). *J. Ultrastruct. Res.* 11, 581–620.

Nilsson, S. E. G. (1965). The ultrastructure of the receptor outer segment in the retina of the leopard frog. *J. Ultrastruct. Res.* 12, 207–231.

Nishimura, Y., and Shimai, K. (1980). Clasping structure in the double cone of the retina of the turtle (*Geoclemys reevesii*). *Cell Tissue Res.* 209, 391–398.

Nishimura, Y., Terada, S., Sekiguchi, M., and Shimai, K. (1980). Clasping structure in the double cone of the retina in vertebrates. *J. Electron Microsc.* (Tokyo), 29, 403–405.

Normann, R. A., and Kolb, H. (1981). Anatomy and physiology of the horizontal cells of the visual streak region of the turtle retina. *Vision Res.* 21, 1585–1588.

Normann, R. A., Kolb, H., Hanani, M., Pasino, E., and Holub, R. (1979). Orientation of horizontal cell axon terminals in the streak of the turtle retina. *Nature* (London) 280, 60–62.

Normann, R. A., Perlman, I., and Daly, S. J. (1985a). Mixing of color signals by turtle cone photoreceptors. *J. Neurophysiol.* 54, 293–303.

Normann, R. A., Perlman, I., and Kolb, H. (1985b). Chromatic interactions between cones of differing spectral classes: anatomical and electrophysiological studies in turtle. In *Neurocircuitry of the Retina: a Cajal Memorial* (A. Gallego and P. Gouras, eds.). Elsevier Science Publishing Co., New York, pp. 19–33.

Normann, R. A., Perlman, I., Kolb, H., Jones, J., and Daly, S. J. (1984). Direct excitatory interactions between cones of different spectral types in the turtle retina. *Science* (N.Y.) 224, 625–627.

Odashima, S., Tazawa, Y., and Yamauchi, A. (1978). Ultrastructure of visual cells in the turtle retina. *Internat. Cong. Series #450,* Elsevier North-Holland, pp. 687–689.

Odashima, S., Tazawa, Y., and Yamauchi, A. (1979). Ultrastructure of visual cells in the turtle retina [author's translation]. *Nippon Ganka Gakkai Zasshi* 83, 472–484.

O'Day, K. J. (1939). The visual cells of Australian reptiles and mammals. *Trans. Ophthalmol. Soc. Aust.* 1, 12–20.

Ogura, M., and Nelson Arispe, A. (1967). Special structure of the apposition zone of double cones. *Acta Biologica Venezuelica* 5, 107–122.

Ohtsuka, T. (1978). Combination of oil droplets with different types of photoreceptor in a freshwater turtle, *Geoclemys reevesii. Sensory Processes* 2, 321–325.

Ohtsuka, T. (1983). Axons connecting somata and axon terminals of luminosity-type horizontal cells in the turtle retina: receptive field studies and intracellular injections of HRP. *J. Comp. Neurol.* 220, 191–198.

Ohtsuka, T. (1984). Fluorescence from colorless oil droplets: a new criterion for identification of cone photoreceptors. *Neurosci. Lett.* 52, 241–245.

Ohtsuka, T. (1985a). Relation of spectral types to oil droplets in cones of turtle retina. *Science* (N.Y.) 229, 874–877.

Ohtsuka, T. (1985b). Spectral sensitivities of seven morphological types of photoreceptors in the retina of the turtle, *Geoclemys reevesii. J. Comp. Neurol.* 237, 145–154.

Ohtsuka, T., and Kouyama, N. (1982). Rod input to luminosity-type horizontal cells in the turtle retina. *Biomed. Res.* (Suppl.) December, 1–7.

Ohtsuka, T., and Kouyama, N. (1985). Synaptic contacts between red-sensitive cones and triphasic chromaticity horizontal cells in the turtle retina. *Brain Res.* 346, 374–377.

Ohtsuka, T., and Kouyama, N. (1986a). Electron microscopic study of synaptic contacts between photoreceptors and HRP-filled horizontal cells in the turtle retina. *J. Comp. Neurol.* 250, 141–156.

Ohtsuka, T., and Kouyama, N. (1986b). Physiological and morphological studies of cone-horizontal cell connections in the turtle retina. *Neurosci. Res.,* Suppl. 4, S69–S84.

Okabe, S. (1972). Electron microscopic studies on the inner plexiform layer of the retina. 1. Comparative anatomy of synapses. *Acta Soc. Ophthalmol. Jap.* 76, 970–976 [in Japanese with English summary].

Okuda, K. (1961). Electron microscope observations of the vertebrate retina. *Acta Soc. Ophthalmol. Jap.* 65, 2126–2151 [in Japanese with English summary].

Olive, J. (1980). The structural organization of mammalian retinal disc membrane. *Int. Rev. Cytol.* 64, 107–169.

Orlov, O., and Maximova, E. M. (1964). The role of intracone light filters

(Mechanism of color vision of lizard and tortoise). *Dokl. Biophys.* 154, 11–14 [translation of Dokl. Acad. Nauk SSR 154, 463–466].

Osborne, N. N., Nesselhut, T., Nicholas, D. A., Patel, S., and Cuello, A. C. (1982). Serotonin-containing neurones in vertebrate retinas. *J. Neurochem.* 39, 1519–1528.

Owen, W. G. (1985). Chemical and electrical synapses between photoreceptors in the retina of the turtle, *Chelydra serpentina. J. Comp. Neurol.* 240, 423–433.

Pedler, C. (1965). Duplicity theory and microstructure of the retina. Rods and cones: a fresh approach. In *Ciba Foundation Symposium on Physiology and Experimental Psychology of Color Vision* (C. E. Wolstenholme and J. Knight, eds.). J. and A. Churchill, London, pp. 52–58.

Pedler, C., and Tansley, K. (1963). The fine structure of the cone of a diurnal gecko (*Phelsuma inunguis*). *Exp. Eye Res.* 2, 39–47.

Pedler, C., and Tilly, R. (1964). The nature of the gecko visual cell: a light and electron microscopic study. *Vision Res.* 4, 499–510.

Peiponen, V. A. (1964). Zur Bedeutung der Oelkugeln im Farbensehen der Sauropsiden. *Ann. Zool. Fenn.* 1, 281–302.

Peterson, E. H. (1981a). Regional variation in the photoreceptor layer of the turtle, *Pseudemys scripta. Soc. Neurosci. Abstr.* 7, 275.

Peterson, E. H. (1981b). Regional specialization in retinal ganglion cell projection to optic tectum of *Dipsosaurus dorsalis* (Iguanidae). *J. Comp. Neurol.* 196, 225–252.

Peterson, E. H. (1982). Morphology of retinal ganglion cells in a turtle, *Pseudemys scripta. Soc. Neurosci. Abstr.* 8, 48.

Peterson, E. H., and Ulinski, P. S. (1979). Quantitative studies of retinal ganglion cells in a turtle, *Pseudemys scripta elegans.* I. Number and distribution of ganglion cells. *J. Comp. Neurol.* 186, 17–42.

Peterson, E. H., and Ulinski, P. S. (1982). Quantitative studies of retinal ganglion cells in a turtle, *Pseudemys scripta elegans.* II. Size spectrum of ganglion cells and its regional variation. *J. Comp. Neurol.* 208, 157–168.

Pezard, A. G. (1957). Influence de la lumière et de l'activité spermatogénétique sur les boules colorées de la rétine des Oiseaux et des Sauriens. *C. r. Soc. Biol. Paris* 151, 840–842.

Pezard, A. G. (1964). Les boules colorees des cellules visuelles de certains vertébrés et la relation de leur couleur avec les fonctins sexuelles. *Arch. d'Anat. Micro. Morphol. Exp.* 53, 45–55.

Piccolino, M., and Neyton, J. (1982). The feedback effect from luminosity horizontal cells to cones in the turtle retina: a key to understanding the response properties of the horizontal cells. In *The S-Potential* (B. D. Drujan and M. Laufer, eds.). Alan R. Liss, New York, pp. 161–179.

Piccolino, M., Neyton, J., and Gerschenfeld, H. M. (1984). Decrease of gap junction permeability induced by dopamine and cyclic adenosine 3′:5′-monophosphate in horizontal cells of turtle retina. *J. Neurosci.* 4, 2477–2488.

Piccolino, M., Neyton, J., Witkovsky, P., and Gerschenfeld, H. M. (1982). Gamma-aminobutyric acid antagonists decrease junctional communication between L-horizontal cells of the retina. *Proc. Natl. Acad. Sci. USA* 79, 3671–3765.

Piccolino, M., Witkovsky, P., Neyton, J., Gerschenfeld, H. M., and Trimarchi,

C. (1985). Modulation of gap junction permeability by dopamine and GABA in the network of horizontal cells of the turtle retina. In *Neurocircuitry of the Retina: a Cajal Memorial* (A. Gallego and P. Gouras, eds.). Elsevier Science Publishing Co., New York, pp. 66–76.

Piccolino, M., Witkovsky, P., and Trimarchi, C. (1987). Dopaminergic mechanisms underlying the reduction of electrical coupling between horizontal cells of the turtle retina induced by d-amphetamine, bicuculline, and veratridine. *J. Neurosci.* 7, 2273–2284.

Polyak, S. (1957). *The Vertebrate Visual System.* University of Chicago Press, Chicago.

Putter, A. (1912). *Organologie des Auges.* Wilhelm Engelmann, Leipzig.

Ranvier, L. A. (1889). *Traité Technique d'Histologie.* 2nd ed. F. Savy, Paris.

Raviola, E. (1976). Intercellular junctions in the outer plexiform layer of the retina. *Invest. Ophthalmol. Vis. Sci.* 15, 881–895.

Raviola, E., and Gilula, N. B. (1973). Gap junctions between photoreceptor cells in the vertebrate retina. *Proc. Natl. Acad. Sci. USA* 70, 1677–1681.

Reiner, A. (1981). A projection of displaced ganglion cells and giant ganglion cells to the accessory optic nuclei in turtle. *Brain Res.* 204, 403–409.

Reiner, A., and Karten, H. J. (1978). A bisynaptic retinocerebellar pathway in the turtle. *Brain Res.* 150, 163–169.

Reiner, A., Krause, J. E., Keyser, K. T., Eldred, W. D., and McKelvy, J. F. (1984). The distribution of substance P in turtle nervous system: a radioimmunoassay and immunohistochemical study. *J. Comp. Neurol.* 226, 50–75.

Richter, A., and Simon, E. J. (1974). Electrical responses of double cones in the turtle retina. *J. Physiol.* (London) 242, 673–683.

Richter, A., and Simon, E. J. (1975). Properties of centre-hyperpolarizing red-sensitive bipolar cells in the turtle retina. *J. Physiol.* (London) 248: 317–334.

Rochon-Duvigneaud, A. (1917). Les fonctions des cones et des batonnets. Indications fournies par la physiologie comparée. *Ann. D'Oculist.* 154, 633–648.

Rochon-Duvigneaud, A. (1943). *Les Yeux et la Vision des Vertébrés.* Masson et Cie., Paris.

Rodieck, R. W. (1973). *The Vertebrate Retina.* W. H. Freeman, San Francisco.

Rowe, M. H., and Stone, J. (1976). Properties of ganglion cells in the visual streak of the cat's retina. *J. Comp. Neurol.* 169, 99–126.

Rowe, M. H., and Stone, J. (1977). Naming of neurones: classification and naming of cat retinal ganglion cells. *Brain Behav. Evol.* 14, 185–216.

Rowe, M. H., and Stone, J. (1980). The interpretation of variation in the classification of nerve cells. *Brain Behav. Evol.* 17, 123–151.

Saito, T., Miller, W. H., and Tomita, T. (1974). C- and L-type horizontal cells in the turtle retina. *Vision Res.* 14, 119–123.

Sarthy, P. V., and Gaur, V. P. (1986). Monoclonal antibodies to turtle retinal cells. *Invest. Ophthalmol. Vis. Sci.* (Suppl.) Vol. 27, No. 3, 330.

Schaeffer, S. F., and Raviola, E. (1975). Ultrastructural analysis of functional changes in the synaptic endings of turtle cone cells. *Cold Spring Harbor Symp. Quant. Biol.* 40, 521–528.

Schaeffer, S. F., and Raviola, E. (1978). Membrane recycling in the cone cell endings of the turtle retina. *J. Cell Biol.* 79, 802–825.

Schaeffer, S. F., Raviola, E., and Heuser, J. E. (1982). Membrane specializations in the outer plexiform layer of the turtle retina. *J. Comp. Neurol.* 204, 253–267.

Scheie, E., and Laties, A. M. (1971). Catecholamine-containing cells in the retinas of *Gekko gecko* and *Rana pipiens*. *Herpetologica* 27, 77–80.

Schiefferdecker, P. (1886). Studien zur vergleichenden Histologie der Retina. *Arch. Mikrosk. Anat.* 28, 305–396.

Schultze, M. (1866). Zur Anatomie und Physiologie der Retina. *Arch. Mikrosk. Anat.* 2, 175–286.

Schultze, M. (1873). The retina. In *Manual of Human and Comparative Histology* (S. Sticker, ed.). New Sydenham Society, London, 3, 218–298.

Schutte, M., and Weiler, R. (1987). Morphometric analysis of serotoninergic bipolar cells in the retina and its implications for retinal image processing. *J. Comp. Neurol.* 260, 619–626.

Schwartz, E. A. (1973). Organization of on-off cells in the retina of the turtle. *J. Physiol.* (London) 230, 1–14.

Schwartz, E. A. (1974). Responses of bipolar cells in the retina of the turtle. *J. Physiol.* (London) 236, 211–224.

Schwartz, E. A. (1976). Electrical properties of the rod syncytium in the retina of the turtle. *J. Physiol.* (London) 257, 379–406.

Simon, E. J. (1973). Two types of luminosity horizontal cells in the retina of the turtle. *J. Physiol.* (London) 230, 199–211.

Sjostrand, F. S., and Kreman, M. (1978). Molecular structure of outer segment disks in photoreceptor cells. *J. Ultrastruct. Res.* 65, 195–226.

Slonaker, J. R. (1897). A comparative study of the area of acute vision in vertebrates. *J. Morphol.* 13, 445–503.

Steinberg, R. H., Fisher, S. K., and Anderson, D. H. (1980). Disc morphogenesis in vertebrate photoreceptors. *J. Comp. Neurol.* 190, 501–519.

Stell, W. K., Ishida, A. T., and Lightfoot, D. O. (1977). Structural basis for on- and off-center responses in retinal bipolar cells. *Science* (N.Y.) 198, 1269–1271.

Sterling, P. (1983). Microcircuitry of the cat retina. *Ann. Rev. Neurosci.* 6, 149–187.

Stewart, W. W. (1978). Functional connections between cells as revealed by dye-coupling with a highly fluorescent naphthalimide tracer. *Cell* 14, 741–759.

Stirling, C. E., and Swarthy, P. V. (1985). Localization of the Na-K pump in turtle retina. *J. Neurocytol.* 14, 33–47.

Strother, G. K. (1963). Absorption spectra of retinal oil globules in turkey, turtle and pigeon. *Exp. Cell. Res.* 29, 349–355.

Stryer, L. (1986). Cyclic GMP cascade of vision. *Ann. Rev. Neurosci.* 9, 87–119.

Tachibana, M., and Kaneko, A. (1984). Gamma-aminobutyric acid acts at axon terminals of turtle photoreceptors: difference in sensitivity among cell types. *Proc. Natl. Acad. Sci. USA* 81, 7961–7964.

Tafani, A. (1883). Parcours et terminason du nerf optique dans la retine des crocodiles (*Champsia lucius*). *Arch. Ital. Biol.* T.4, 210–233.

Tansley, K. (1959). The retina of two nocturnal geckos, *Hemidactylus turcicus* and *Tarentola mauritanica*. *Pflügers Arch. ges. Physiol.* 268, 213–220.

Tansley, K. (1961). The retina of a diurnal gecko. *Pflügers Arch. ges. Physiol.* 272, 262–269.

Tansley, K. (1964). The gecko retina. *Vision Res.* 4, 33–37.

Tansley, K. (1965). *Vision in Vertebrates.* Chapman and Hall Ltd., London.

Tansley, K., and Johnson, B. K. (1956). The cones of the grass snake's eye. *Nature* (London) 178, 1285–1286.

Terada, S., Nishimura, Y., Sekiguchi, M., and Shimai, K. (1983). Three-dimensional reconstruction of a single cone in the turtle retina. *Okajimas Folia Anat. Japan.* 60, 125–146.

Tiemann, G. (1970). Untersuchungen über die Entwicklung und Bedeutung der farbigen Ölküglein in der Retina von *Lacerta vivipara. Experientia* 26, 1274.

Tyner, C. F. (1975). The naming of neurons: applications of taxonomic theory to the study of cellular populations. *Brain Behav. Evol.* 12, 75–96.

Underwood, G. (1951). Reptilian retinas. *Nature* (London) 167, 183–185.

Underwood, G. (1954). On the classification and evolution of geckos. *Proc. Zool. Soc.* (London) 124, 469–492.

Underwood, G. (1957). On lizards of the family Pygopodidae. *J. Morphol.* 100, 207–268.

Underwood, G. (1966). On the visual-cell pattern of a homalopsine snake. *J. Anat.* 100, 571–575.

Underwood, G. (1967a). A comprehensive approach to the classification of higher snakes. *Herpetologica* 23, 161–168.

Underwood, G. (1967b). *A Contribution to the Classification of Snakes.* British Museum (Natural History), London.

Underwood, G. (1968). Some suggestions concerning vertebrate visual cells. *Vision Res.* 8, 483–488.

Underwood, G. (1970). The eye. In *Biology of the Reptilia* (C. Gans and T. S. Parsons, eds.). Academic Press, London, Vol. 2B, pp. 1–97.

Vaughan, D. K., and Fisher, S. K. (1987). The distribution of F-actin in cells isolated from vertebrate retinas. *Exp. Eye Res.* 44, 393–406.

Verrier, M. L. (1933). Recherches sur la vision des reptiles. *Bull. Biol. France Belg.* 67, 350–370.

Verrier, M. L. (1935). Recherches sur l'histophysiologie de la retine des vertebres et les problemes qu'elle souleve. *Bull. Biol. France Belg.*, Suppl. 20, 1–140.

Verriest, G., de Rouck, A., and Rabaey, M. (1959). Etude comparative de l'histologie rétinienne et de l'electroretinogramme chez les batraciens et chez les reptiles. *Biol. Jaarb.* 27, 102–191 [English summary].

Villegas, G. M. (1960). Electron microscopic study of the vertebrate retina. *J. Gen. Physiol.* 43, 15–43.

Vilter, V. (1949a). Relations synaptiques dans la retine de la Tortue. *Comptes Rendus Soc. Biol.* 143, 781–783.

Vilter, V. (1949b). Nouvelle conception de relations neuronales dans la fovea. *Comptes Rendus Soc. Biol.* 143, 784–785.

Vilter, V. (1951a). Valeur morphologique des photorecepteurs retiniens chez la hatterie (*Sphenodon punctatus*). *Comptes Rendus Soc. Biol.* 145, 20–23.

Vilter, V. (1951b). Organisation generale de la retine nerveuse chez le *Spheno-don punctatus*. *Comptes Rendus. Soc. Biol.* 145, 24–26.

Vilter, V. (1951c). Recherches sur les structures foveales dans la retine du *Sphenodon punctatus*. *Comptes Rendus, Soc. Biol.* 145, 26–29.

Wallman, J. (1979). Role of the retinal oil droplets in the color vision of Japanese quail. In *Neural Mechanisms of Behavior in the Pigeon* (A. M. Granda and J. H. Maxwell, eds.). Plenum Press, New York, pp. 327–351.

Walls, G. L. (1934). The reptilian retina. *Am. J. Ophthalmol.* 17, 892–915.

Walls, G. L. (1942a). *The Vertebrate Eye and its Adaptive Radiation*. Cranbrook Press, Bloomfield Hills, Mich.

Walls, G. L. (1942b). The visual cells and their history. In *Visual Mechanisms* (H. Kluver, ed.). Jacques Cattell Press, Lancaster, Pa., pp. 203–251.

Walls, G. L., and Judd, H. D. (1933). The intra-ocular colour filters of vertebrates. *Br. J. Ophthalmol.* 17, 641–675.

Weiler, R. (1981). The distribution of center-depolarizing and center-hyperpolarizing bipolar cell ramifications within the inner plexiform layer of turtle retina. *J. Comp. Physiol.* 144, 459–464.

Weiler, R. (1985). Afferent and efferent peptidergic pathways in the turtle retina. In *Neurocircuitry of the Retina* (A. Gallego and P. Gouras, eds.). Elsevier, New York, pp. 245–256.

Weiler, R., and Ammermuller, J. (1986). Immunocytochemical localization of serotonin in intracellularly analyzed and dye-injected ganglion cells of the turtle retina. *Neurosci. Lett.* 72, 147–152.

Weiler, R., and Ball, A. K. (1984). Co-localization of neurotensin-like immunoreactivity and ^{3}H-glycine uptake system in sustained amacrine cells of turtle retina. *Nature* (London) 311, 759–761.

Weiler, R., Kohler, K., Kolbinger, W., Wolburg, H., Kurz-Isler, G., and Wagner, H-J. (1988). Dopaminergic neuromodulation in the retinas of lower vertebrates. *Neurosci. Res.,* Suppl. 8, S183–S196.

Weiler, R., and Marchiafava, P. L. (1981). Physiological and morphological study of the inner plexiform layer in the turtle retina. *Vision Res.* 21, 1635–1638.

Weiler, R., and Schutte, M. (1985a). Morphological and pharmacological analysis of putative serotonergic bipolar and amacrine cells in the retina of a turtle, *Pseudemys scripta elegans. Cell Tissue Res.* 241, 373–382.

Weiler, R., and Schutte, M. (1985b). Kainic acid-induced release of serotonin from OFF-bipolar cells in the turtle retina. *Brain Res.* 360, 379–383.

Williams, D. S., Colley, N. J., and Farber, D. B. (1987). Photoreceptor degeneration in a pure-cone retina. *Invest. Ophthalmol. Vis. Sci.* 28, 1059–1069.

Williamson, D. E., and Eldred, W. D. (1989). Amacrine and ganglion cells with corticotropin-releasing-factor-like immunoreactivity in the turtle retina. *J. Comp. Neurol.* 280, 424–435.

Winston, R., and Enoch, J. M. (1971). Retinal cone receptor as an ideal light collector. *J. Opt. Soc. Am.* 61, 1120–1121.

Witkovsky, P., Alones, V., and Piccolino, M. (1987). Morphological changes induced in turtle retinal neurons by exposure to 6-hydroxydopamine and 5,6-dihydroxytryptamine. *J. Neurocytol.* 16, 55–67.

Witkovsky, P., Eldred, W., and Karten, H. J. (1984). Catecholamine- and indoleamine-containing neurons in the turtle retina. *J. Comp. Neurol.* 228, 217–225.

Witkovsky, P., Owen, W. G., and Woodworth, M. (1983). Gap junctions among the perikarya, dendrites, and axon terminals of the luminosity-type horizontal cell of the turtle retina. *J. Comp. Neurol.* 216, 359–368.

Wong, R. O. L. (1989). Morphology and distribution of neurons in the retina of the American garter snake *Thamnophis sirtalis. J. Comp. Neurol.* 283, 587–601.

Yamada, E. (1960a). The fine structure of the paraboloid in the turtle retina as revealed by electron microscopy. *Anat. Rec.* 137, 172.

Yamada, E. (1960b). Observations on the fine structure of photoreceptive elements in the vertebrate eye. *J. Electron Microscopy* 9, 1–14.

Yamada, E. (1976). Structure analysis of the retina with electron microscopy. *Jap. J. Clip. Med.* 34, 1603–1609 [in Japanese].

Yamada, E., and Ishikawa, T. (1965). The fine structure of the horizontal cells in some vertebrate retinae. *Cold Spring Harbor Symposium* 30, 383–392.

Yamada, E., Ishikawa, T., and Hatae, T. (1966). Some observations on the retinal fine structure of the snake *Elaphe climacophora. Sixth Internat. Congr. Electron Microscopy, Kyoto* 2, 495.

Yau, K.-W., and Baylor, D. A. (1989). Cyclic GMP-activated conductance of retinal photoreceptor cells. *Ann. Rev. Neurosci.* 12, 289–327.

Yazulla, S. (1976a). Cone input to horizontal cells in the turtle retina. *Vision Res.* 16, 727–735.

Yazulla, S. (1976b). Cone input to bipolar cells in the turtle retina. *Vision Res.* 16, 737–744.

Yazulla, S., and Kleinschmidt, J. (1983). Carrier-mediated release of GABA from retinal horizontal cells. *Brain Res.* 263, 63–75.

Yoshida, M. (1976). Unusual arrangement of lamellae in the outer segment of the photoreceptors in the gecko retina. In *The Structure of the Eye* (E. Yamada and S. Mishima, eds.). *Jap. J. Ophthalmol.* 3, 223–228.

Yoshida, M. (1978). Some observations on the patency in the outer segments of photoreceptors of the nocturnal gecko. *Vision Res.* 18, 137–143.

Young, R. W. (1977). The daily rhythm of shedding and degradation of cone outer segment membranes in the lizard retina. *J. Ultrastruct. Res.* 61, 172–185.

Young, S. R., and Martin, G. R. (1984). Optics of retinal oil droplets: a model of light collection and polarization detection in the avian retina. *Vision Res.* 24, 129–137.

Zucker, C. L., and Adolph, A. R. (1988). Interactions between enkephalin and GABA in the turtle retina. *Invest. Ophthalmol. Vis. Sci.* (Suppl.) 29, 196.

2

Retinal Function in Turtles

ALAN M. GRANDA
AND
DAVID F. SISSON

CONTENTS

I. INTRODUCTION

The retina is brain made peripheral, and many of the functions normally reserved to central brain structure in higher mammals are operative at peripheral retinal sites in lower vertebrates. That principle is well exemplified in the retina of turtles, where there is sufficient experimental data to show that fundamental parameters of color, space, and time have their limits set at that level. The retina of turtles is organized on the vertebrate plan with a complement of photoreceptor, bipolar, horizontal, amacrine, and ganglion cells (see Chapter 1); functional information is available on each of these cell types. This chapter reviews the physiology, neuropharmacology, and psychophysics of turtle retinas. Granda and Dvorak (1977) previously reviewed retinal physiology in turtles. Dowling (1987) provides a recent review of vertebrate retinal physiology, and Naka et al. (1987) review the physiology of turtle photoreceptors.

II. PHOTOPIGMENTS AND OIL DROPLETS

Vertebrate vision initially derives its informational content from the discrete sampling of spatially continuous images. Sampling of this kind occurs in the photoreceptor mosaic, which in *Pseudemys scripta elegans* consists of red-, green-, and blue-sensitive single cones (Liebman and Granda, 1971). (The stated colors conform to perceived stimulus wavelengths at 640 nm (red), 520 nm (green), and 460 nm (blue) by human observers.) A separate class of double cones comprises red-sensitive principal and green-sensitive accessory members (Liebman and Granda, 1971), although there is strong evidence that both members contain identical red-sensitive photopigments (Lipetz and MacNichol, 1982; Lipetz, 1985; Ohtsuka, 1985b). Finally, rods can be easily distinguished from any of the cones by their distinct morphology. In turtles, cone photoreceptors generally are tied closely to colored oil droplets that are contained in the distal portions of their inner segments. The droplets are located so that all of the light destined for photopigment absorption in the outer segments must pass through them. The droplets consist of neutral lipids and carotenoid pigments of high optical density. When colored, as they are in the freshwater turtle, *P. scripta elegans*, and in the eastern box turtle, *Terrapene carolina carolina*, droplets appear to be red, orange, and yellow, although there are also colorless droplets (Granda and Haden, 1970; Kolb and Jones, 1982) that are clear or fluoresce to ultraviolet light (Ohtsuka, 1984). The sea turtle, *Chelonia mydas mydas,* has only orange, yellow, and colorless droplets. Liebman and Granda (1975) showed these optical densities to be extremely high, a general finding since confirmed by Lipetz and MacNichol (1979), and Lipetz (1984a, 1984b).

The photopigments of turtles were measured by microspectrophotometry in *Pseudemys scripta elegans* and *Chelonia mydas mydas* (Liebman and Granda, 1971) and in *Emydoidea blandingii* (Lipetz, 1984b). The different photopigments are housed singly in separate receptors. The several photopigments, characterized by their peak absorptions (λ_{max}), and the oil droplets conjoined with them are listed in Table 2.1.

Table 2.1. Absorption spectra of visual pigments found in retina of *Pseudemys scripta elegans* corrected for filtration by oil droplets.[a]

Photopigment Sensitivity	λ_{max} (nm)	#	Color	$\lambda_{1/2}$ (nm)	Diameter (μm)	Population (%)	Corrected λ_{max} (nm)
Red	620	60	Red	602	8.5 (0.3)	30	640
	620	6	Fluor		6.1 (0.3)	9	623
	620	19	Orange	563	7.3 (0.3)	17	626
	620	19	None			17	623
Green	540	16	Yellow	537	7.1 (0.2)	18	563
	518	2	None			4	518
Blue	460	3	Clear		5.2 (0.3)	5	462

From T. Ohtsuka, *Science* 229 (1985), 874–877, with permission of author and publisher.
[a] Oil droplet diameters are mean values with standard deviations in parentheses.

Red oil droplets appear to be exclusively associated with a red-sensitive photopigment (λ_{max} = 620 nm), orange droplets are associated with chief members of double cones that also contain the same red-sensitive photopigment, and yellow droplets are associated with a green-sensitive photopigment (λ_{max} = 518 nm) (Liebman and Granda, 1971, 1975; Baylor and Fettiplace, 1975; Ohtsuka, 1978). Accessory members of double cones are thought to contain green- or red-sensitive pigments. The rods, which are also sensitive to green light, contain no droplets and can be easily identified. Small, clear droplets are associated with blue-sensitive cones that contain a photopigment of λ_{max} = 450 nm; somewhat larger, colorless droplets that fluoresce to ultraviolet light have been reported to occur in receptors containing red-sensitive photopigments (Ohtsuka, 1984). Photopigments that are conjoined to particular colored oil droplets have their regions of greatest absorption shifted to longer wavelengths, that is, toward the red end of the visible spectrum. That shift is evident in red-sensitive and green-sensitive single cones and in chief members of double cones, which are also red-sensitive. Single, blue-sensitive cones that contain colorless droplets, as do some of the red-sensitive single cones, and the accessory members of double cones and rods that contain none, are all unaffected by the selective action of colored droplets. The described characteristics then make it relatively easy to identify photoreceptor contributions by deriving spectral sensitivity functions under appropriate conditions. This approach has been used with great success in the analysis of visual functions.

III. ELECTROPHYSIOLOGY
A. Photoreceptor Cells

1. GENERAL The photoreceptors in turtles, as in those of all vertebrates examined so far, hyperpolarize to light by suppressing a steady sodium current that enters the outer segments in darkness (Baylor and Fuortes, 1970). The hyperpolarizations recorded from intracellular penetrations of single photoreceptors show graded voltages that depend on the numbers of photons caught (Baylor and Hodgkin, 1973). Baylor and Fuortes (1970) demonstrated that photoreceptors in *Pseudemys scripta elegans* obey the principle of univariance, which makes the very sensible statement that for a particular pigment system, effectively absorbed photons will have equal visual effects (Naka and Rushton, 1966), Visual effects are taken to mean action spectra, for which the reciprocal of light intensity at each wavelength evokes a response of some defined magnitude. In the determination of such spectra, the curves of voltage-amplitude responses to changes in stimulus intensity, defined from noise level to response saturation, can be described by the sigmoidal Michaelis-Menton function (Michaelis and Menton, 1913):

$$U/U_{max} = I/(I + U_k), \qquad (1)$$

in which U is response amplitude, U_{max} = response amplitude to a saturating flash, I = flash intensity of U, and U_k = flash intensity required to produce half the maximum amplitude. For such curves, changes in stimulus wavelength in a given photoreceptor cause it to be displaced along the abscissa, with the shapes of the curves remaining constant (Baylor and Hodgkin, 1973). By calculating the displacements necessary to superimpose intensity response curves to a wide range of stimulus wavelengths, relative spectral sensitivity curves for single cones and rods (Baylor and Hodgkin, 1973), as well as double cones (Richter and Simon, 1974; Baylor and Fettiplace, 1975), have been determined. Peak sensitivities are near 460 nm for blue-sensitive cones, 550 nm for green-sensitive cones, 630 nm for red-sensitive cones, and 520 nm for rods (Fig. 2.1). Double cones exhibit two spectral peaks, one near 550 nm and the other near 630 nm.

The sensitivity of photoreceptors to light may be defined by the metric flash sensitivity, S_F (cf. Baylor and Hodgkin, 1973). This function is determined electrophysiologically from the amplitude of intracellular responses according to the equation:

$$S_F = U_{peak} / (I_F \cdot \Delta t), \qquad (2)$$

in which U_{peak} is the amplitude of peak hyperpolarization in response to a brief quantity of light of unit area, $I_F \cdot \Delta t$, at optimal wavelength. S_F has units of $\mu V \cdot \phi^{-1} \mu m^2$ (Baylor and Hodgkin, 1973). Sensitivity at the optimal wavelength for the photoreceptor is called absolute sensitivity, which differs somewhat among the various cones and considerably more between cones and rods (Table 2.2).

Codified in this manner, flash sensitivity also relates to the number of photoisomerizations of individual chromophores, that is, the mo-

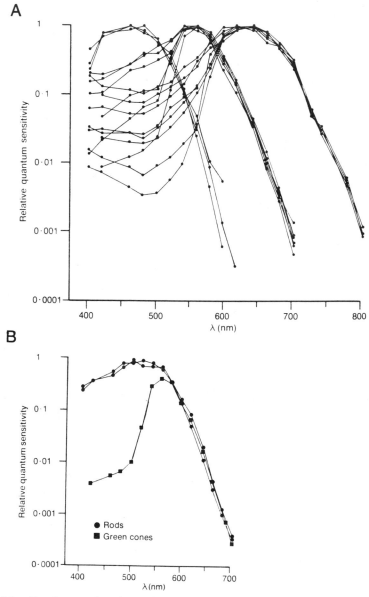

Fig. 2.1. *Pseudemys scripta elegans.* Spectral sensitivities of single retinal photoreceptors. The ordinates are relative quantum sensitivity; the abscissae, wavelength. (A) Three cells have a peak sensitivity at about 460 nm, six have maxima at about 550 nm, and eight have maxima at about 630 nm. (B) Spectral sensitivity of two rods compared with that of a green-sensitive cone showing a large drop in short wavelength sensitivity. The curves agree at long wavelengths but diverge at shorter wavelengths because of the presence of a colored oil droplet in the cone. (From Baylor and Hodgkin, 1973.)

Table 2.2. Flash sensitivities of cones and rods in *Pseudemys scripta elegans*.[a]

Photo-receptors	References[b]						
	1	2	3	4	5	6	7
Red-sensitive cones	72.4 [8] (14.6–246)			32.4 [5] (9–55)	217 [6] (50–334)		
Green-sensitive cones	74.4 [6] (10.7–247)			21 [1]	77.3 [4] (30–135)		
Double cones			3.7 [2] (0.85–6.74)		15.9 [1]		
Rods	680.7 [2] (71.4–1290)	1920 [1]		1000 [1]		9800 [1]	19,600 [107] (3,000–77,000)

[a] Electrophysiologically determined flash sensitivities in the dark of retinal photoreceptors. Each entry consists of a mean value followed by the range of values in parentheses on which the mean is based. Values are reported in $\mu V \cdot \phi^{-1} \cdot \mu m^2$. Figures in brackets are the number of photoreceptors examined.
[b] Key to References: (1) Baylor and Hodgkin, 1973; (2) Schwartz, 1973b; (3) Richter and Simon, 1974; (4) Baylor and Hodgkin, 1974; (5) Baylor and Fettiplace, 1975; (6) Copenhagen and Owen, 1976; (7) Detwiler et al., 1980.

lecular changes in photopigment that are contingent upon the absorption of individual photons. Values of 25 μV per isomerization have been obtained for red- and green-sensitive cones, 12 μV for blue-sensitive cones (although that figure is not particularly reliable), and the much higher value of 1000 μV for rods, in each case after corrections for factors that affect those absorptions (Baylor and Hodgkin, 1973; Baylor and Fettiplace, 1975). Considering that in red-sensitive cones the maximum hyperpolarization to a bright flash is about 25 mV, a single isomerization gives about 0.1% of the maximum light response. In rods, a higher figure of 5% of the maximal light response is obtained, a finding consistent with the common experience that rods are more sensitive than cones at low levels of light (Baylor et al., 1974a).

2. ADAPTATION TO LIGHT
It is well known in psychophysics that background lights or sustained periods of bright lights decrease sensitivity and increase temporal resolution. The same effects occur in the cones of *Pseudemys scripta elegans*, the photoreceptor responses of which become both desensitized and accelerated (Baylor and Hodgkin, 1974; Normann and Anderton, 1983). The desensitization can be expressed as the reciprocal of flash sensitivity plotted against background light. This is the equivalent of

threshold detection on lighted backgrounds in psychophysical determinations, namely

$$1/S_F = 1/S_F^D + \zeta \cdot I_B \qquad (3)$$

for which S_F is sensitivity to short flashes dependent on a steady background field of intensity I_B, S_F^D is the flash sensitivity in the dark ($I_B = 0$), and ζ is a parameter with units of $sec \cdot V^{-1}$, the electrophysiological correlate of the Weber fraction (Baylor and Hodgkin, 1974). The Weber fraction is a psychophysical ratio that measures minimally noticeable differences to stimulus intensity, which is presumed to be constant for different stimulus values. Relative discriminability in this way can be assessed over fairly large ranges.

Rods have very much slower temporal responses than do cones. Background lights speed up both rod and cone responses considerably. These responses have been treated theoretically in a model advanced by Baylor et al. (1979). They propose that a series of time delays occurs between the absorption of a photon and the actual response of the photoreceptor. For cones, a series of six delays ranging from 20 to 60 msec fit the data remarkably well, whereas a series of four delays of 500 msec each works well for rods. These delays cannot be accounted for by the delays or time constants of the underlying photochemical reactions because these reactions are much too fast. Rather it appears that the accelerational changes in the presence of background lights take place in the transduction process itself, most probably in the outer segments very close to where the photons are absorbed (Lamb et al., 1981).

3. Electrical Coupling

The responses of photoreceptors contain fairly large amounts of electrical noise that limit their sensitivities. Of particular interest is that the noise is much larger in dark-adapted photoreceptors than in those illuminated with bright lights (Simon et al., 1975; Lamb and Simon, 1976). At issue is the problem of how a red-sensitive cone, for example one with a nominal response of 25 μV per photon capture, can be detected in a peak-to-peak noise of some 3 mV, with a standard deviation of variability of about 600 μV. One possible way is to average responses, because noise sources are statistically uncorrelated with each other and therefore independent, whereas the response signals are correlated and hence summate. In practice, photoreceptors can be shown to average responses over a number of elements; in effect they average over space rather than time. Averaging in time is at the expense of temporal resolution, a tactic likely inimical for survival. Spatial averaging, on the other hand, entails a loss of acuity so that large objects are perceived at low light levels even though there is difficulty in resolving detail. The tradeoff is more advantageous in comparison to temporal averaging. Spatial averaging is accomplished by the photoreceptors themselves mostly through electrical coupling between neighboring cells of like type. In the retina of *Pseudemys scripta elegans*, red-sensitive cones are seemingly coupled to red-sensitive cones, green-sensitive cones to green-sensitive cones, and rods to rods (Detwiler and Hodgkin, 1979; Detwiler et al., 1980). Such averaging is logical if spectral information is to be preserved, as it appears to be at other levels of the visual pathway.

There is, however, the added complication that not all photoreceptors, at least not all cones, are coupled exclusively to cones of the same chromatic sensitivity. There is evidence from *Pseudemys scripta elegans* and from *Chinemys reevesi*, both anatomical (Kolb and Jones, 1985; Owen, 1985; Ohtsuka and Kawamata, 1990) and functional (Normann et al., 1985; Normann et al., 1984), that telodendria radiating from synaptic terminals connect red- and green-sensitive cones. Ohtsuka and Kawamata have very good evidence to show that green-sensitive cones make specific telodendrial contacts with red- and blue-sensitive cones, blue-sensitive cones with red- and green-sensitive cones, and red-sensitive cones connect with both red- and green-sensitive cones. In turtles, the three chromatic classes of cones are thus linked selectively by basal junctions at the extreme ends of telodendrial processes, although the functional properties of these selective contacts are not yet defined. Rods and cones, on the other hand, do not directly make synaptic contacts with each other.

B. Horizontal Cells

1. GENERAL   In *Pseudemys scripta elegans*, horizontal cells have been classified functionally by a number of investigators (Miller et al., 1973; Simon, 1973; Saito et al., 1974; Fuortes and Simon, 1974; Yazulla, 1976a). In general, two broad classes of horizontal cells have been described: luminosity (L) cells and chromaticity (C) cells, the one (L-cells) responding to light falling on their receptive fields with graded hyperpolarizations, the other (C-cells) responding to some light wavelengths with hyperpolarizations and to other wavelengths with depolarizations. Each of these major categories can be further subdivided on the basis of more specific properties. Receptive field diameters of L-cells are distributed bimodally (Simon, 1973; Saito et al., 1974). Those horizontal cells with receptive fields greater than 750 μm in diameter are classified as L_1-cells, whereas those with the smaller receptive fields of about 300 μm in diameter are classified as L_2-cells. C-cells have been subdivided according to their responses to different color stimuli (Fuortes and Simon, 1974; Yazulla, 1976a). Cells that depolarize to long wavelength light and hyperpolarize to intermediate

and short wavelengths are classified as R/G-B cells, whereas those that depolarize to intermediate wavelengths and hyperpolarize to both long and short wavelengths are classified as G/B-R cells.

Injection of procion yellow dye into single horizontal cells that have been functionally classified permits correlation of functional categories with morphological categories (Leeper, 1978a). L_1-cells possess tuberous dendrites, lack a cell body, and lie near or on the border between the inner nuclear and outer plexiform layers; they correspond to the H1AT (axon terminal) cells of Leeper. L_2-cells possess a well-defined cell body and radially dispersed dendrites and appear to correspond to Leeper's H1CB (cell body) cells. These two cells actually represent a single complex consisting of two main regions; namely the dendritosomatic region and the axonal terminal expansion. The two regions are connected by an axon fiber that is sufficiently long and thin to be electrically inexcitable and therefore it functionally separates the two regions. Leeper (1978b) has shown that H1AT cells make synaptic contacts with rods and several classes of cones, whereas H1CB cells make synaptic contacts only with cones. Saito et al. (1974) reported similar selective synapses between L_1- and L_2-cells, respectively, using morphological differences between cone pedicles and rod spherules as criteria (Lasansky, 1971). H2, H3, and H4 cells appear to be the morphological correlates of chromaticity cells. The dendritic arborizations of H2, H3, and H4 cells show more subtle differences than do those between H1CB and H1AT and cannot be reliably differentiated with procion yellow dye.

2. Feedback

Horizontal cells participate in the transfer of information by a process of feedback; Baylor et al. (1971) first demonstrated this mechanism in the retina of *Pseudemys scripta elegans*. Hyperpolarization of horizontal cells, evoked by light or by extrinsic current, causes depolarization in cones. The feedback pathway cannot be electrotonic, as the pre- and postsynaptic potentials are opposite in polarity and not synchronized in time. The underlying mechanism, probably chemical, remains uncertain. The benefits of feedback are spatial and chromatic. Wherever light falls on peripheral areas beyond 80 μm in diameter, a cone can exhibit a center-surround receptive field organization with a hyperpolarizing center surrounded by a depolarizing region. The depolarization results from feedback from L-cells.

The wavelength of light is also involved in the following feedback: red light hyperpolarizes red-sensitive cones, which in turn hyperpolarize luminosity (L_2) cells. L_2-cells feed back onto green-sensitive cones, causing these cells to depolarize (Fuortes et al., 1973). Green-sensitive cones then feed forward onto chromaticity (R/G-B) cells, causing them to depolarize to stimulation with red light. The hyperpolarization of R/G-B cells to green and blue stimuli is most easily attributed to direct input from green- and blue-sensitive cones. The spectral responses of G/R-B cells might be generated by the signals in R/G-B cells being fed back to blue-sensitive cones, which in turn feed forward to G/R-B cells.

3. ELECTRICAL COUPLING

The intercommunication of horizontal cells via gap junctions is diagnostic in their organization. The contacts are quite specific and relate only to horizontal cells of the same type and in particular regions of that type. The coupling ranges from very tight, supporting an extensive spread of signal into a network of coupled cells, to being loose or absent. The coupling can be visualized by injection of lucifer yellow (Fig. 2.2), as this dye is able to diffuse across gap junctions (Piccolino et al., 1982; Piccolino et al., 1984). Whereas horizontal cell processes only extend over areas of 100 to 200 μm in diameter, coupling can extend receptive fields to diameters of more than 5 mm.

C. Bipolar Cells

The receptive fields of bipolar cells have center-surround properties, with the diameter of the center field being approximately 200 μm and that of the surround over 1,000 μm (Schwartz, 1974; Richter and Simon, 1975; Yazulla, 1976b). Stimulation of the center field produces effects antagonistic to those of the surround. Studies on carp and salamanders show two functional organizations. The first will hyperpolarize in response to small light flashes centered over the cell body, and the second will depolarize in response to the same stimulus configuration. Chromatic properties are apparently of three types (Yazulla, 1976b): (1) red-dominated cells respond to central stimulation by either hyperpolarization or depolarization, (2) central hyperpolarizing cells exhibit a broad, spectral sensitivity in the central portion, and (3) a class of cells show a central, red/green color opponency. The surrounds of these opponent cells are broadly sensitive to spectral light (Fig. 2.3).

The absolute sensitivity of bipolar cells is greater than that of the photoreceptors afferent to them. In a study of bipolar cells that hyperpolarize to stimulation of the red-sensitive central field, Richter and Simon (1975) showed a gain of three- to fourfold over the red-sensitive cones; the receptor-bipolar synapse amplifies the signal in ways that are not readily apparent.

In *Pseudemys scripta elegans*, central illumination of the bipolar receptive field is conveyed directly by the cones. The surrounds, however, stem from horizontal cell inputs. Marchiafava (1978) simulta-

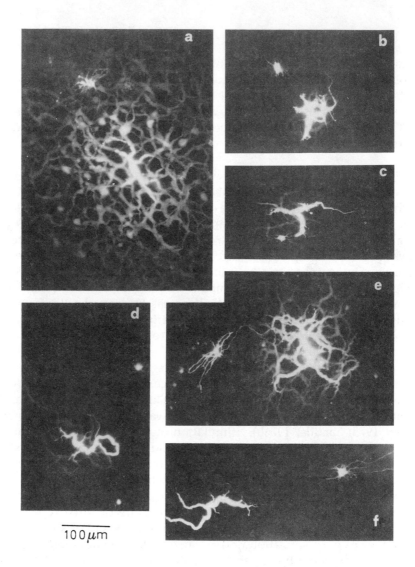

100 μm

Fig. 2.2. Diffusion of lucifer yellow in the H1AT axonal terminal network of horizontal cells in *Pseudemys scripta elegans*. (A) Control. Note the dye diffusion among several cell bodies and their associated network. (B) After application of 10 μM of forskolin in the perfusion bath. Forskolin is a diterpene compound known to stimulate the activity of adenylate cyclase independently from the activation of dopamine receptors. (C) Dye injected after 10-min application of 10 μM dopamine in the extracellular medium. Dye is confined to few H1ATs beyond the injected one. (D) Dye injected after application of 50 μM IBMX, which increases cyclic AMP activity by inhibiting its catabolic enzyme, phosphodiesterase. (E) Dye injected in the presence of 3μM cobalt chloride in the perfusion bath, which produces a mild but consistent decrease in the diffusion of lucifer yellow in the H1AT network. (F) In the continuous presence of cobalt, 10 μM dopamine was added to the extracellular medium. Note that dopamine was still able to reduce dye diffusion in the network. (From Piccolino et al., 1984; reprinted by permission of the *Journal of Neuroscience*.)

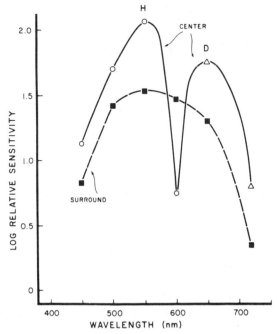

Fig. 2.3. Spectral sensitivities of the center (solid line) and surround (dashed line) of a bipolar cell. The center both hyperpolarizes (H) and depolarizes (D); the surround only hyperpolarizes. (Reprinted with permission from *Vision Research*, Vol. 16, Yazulla, Cone input to bipolar cells in the turtle retina, 1976, Pergamon Press plc.)

neously recorded from a center-hyperpolarizing bipolar cell while stimulating a neighboring L-type horizontal cell electrically; the bipolar cell reflected an opposite polarity to the passed current; that is, a hyperpolarizing current that was passed through the horizontal cell membrane resulted in a sustained depolarization from the center-hyperpolarizing bipolar cell. Center-depolarizing bipolar cells, on the other hand, were depolarized or hyperpolarized by passed currents of the same polarity. The results are expected but nevertheless important, as there is no anatomical evidence of chemical synapses in this circuitry. There is nothing to exclude, of course, the possible feedback of horizontal cells through cones (see preceding discussion) to bipolar cells, but that influence is not thought to play a role in establishing the surround.

It may be argued that bipolar cells have the main task of comparing two signals (Witkovsky, 1980), namely, a central response derived from photoreceptor input via a chemical synapse and a related surround response derived from horizontal cells through means not clearly understood. That comparison can play a critical role in shaping spatial border and object definition.

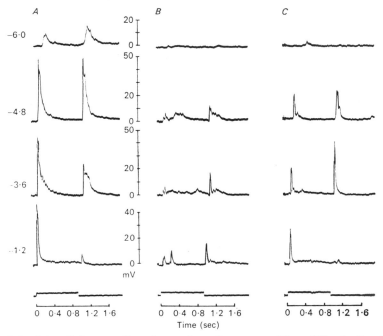

Fig. 2.4. Intracellular recording of amacrine cell responses. Time of illumination is shown on the lower line. Stimuli consisted of a 120-μm radius spot in (A), an annulus with 200 and 800 μm internal and external radii, respectively, in (B), and a spot of 800 μm radius in (C). (From Marchiafava and Torre, 1978.)

D. Amacrine Cells

Amacrine cells depolarize to light stimulation and exhibit the first action or spike potentials that appear in the retinal network (Fig. 2.4). In *Pseudemys scripta elegans,* the responses occur in transient bursts at the onset and offset of light, independent of the retinal area stimulated or the wavelength involved (Schwartz, 1973a; Marchiafava, 1976; Marchiafava and Torre, 1978). There are also graded depolarizations in response to the onset and offset of the light, as well as to shocks of the optic nerve on which a few action potentials may be superimposed. The antagonistic center-surround arrangement is also preserved; responses in some cells, optimally evoked by bright lights of 120 μm in diameter, can be reduced or suppressed by larger, intense light spots of 1,600 μm in diameter.

The spectral characteristics of amacrine cells have been measured so far in only three cells. Each cell appears to be dominated by red-sensitive inputs but also appears to receive green- and blue-sensitive inputs (Marchiafava and Torre, 1978).

To these functional properties of the amacrine cells may be added Golgi impregnated silhouettes accomplished in *Pseudemys scripta ele-*

gans and *Mauremys caspica* by Kolb and co-workers (Kolb, 1982; Kolb et al., 1988). In the first paper, they report a group of 27 different cell types and add two new ones in a second report. Both species are remarkably similar morphologically. Of course, it would be desirable to associate structure and function; regrettably, this has yet to be accomplished.

A definite role for amacrine cells is hard to assess. The cells are connected to bipolar cells as well as to adjacent amacrine cells that determine their receptive field properties. There is also centrifugal input from more central visual levels (Marchiafava, 1976; Cervetto et al., 1976; Marchiafava and Torre, 1978). Amacrine cells have action potentials that are thought to encode information over long distances. The limited available data leave open the question whether the observed response properties are representative.

E. Ganglion Cells

1. GENERAL   At the ganglion cell layer, signals originally initiated by quantal absorption in the photoreceptors are converted from slow analogue responses to digital action potentials that are transmitted to the central nervous system. The receptive fields of ganglion cells have center-surround organizations. The response characteristics of the central portion permit them to be classified into three groups: (1) on-cells, which respond to the onset of light falling on the centers of their receptive fields, (2) off-cells, which respond to the offset of light falling on the centers of their receptive fields, and (3) on-off-cells, which combine the response characteristics of on- and off-cells (Baylor and Fettiplace, 1976; Baylor and Fettiplace, 1977a; Baylor and Fettiplace, 1979). The surround portion of a receptive field is generally antagonistic to the center (Schwartz, 1973a; Marchiafava, 1979; Bowling, 1980).

On the basis of intracellular recording in *Pseudemys scripta elegans*, Marchiafava and Weiler (1980) described two classes of ganglion cells differing in the reversal potentials of their photoresponses. The first class has higher conduction velocities and response frequencies and receives inputs only from bipolar cells: its dendrites are monostratified in the outer part of the inner plexiform layer (off-Type A) or in the middle part of this same layer (on-Type A), which only contains cell processes from bipolar cells. The second class (Type B ganglion cells) has dendritic ramifications spread throughout the entire inner plexiform layer and makes synapses with both bipolar and amacrine cells (Weiler and Marchiafava, 1981). How these inputs resolve themselves into spatially antagonistic center-surround arrangements in both Type A and Type B ganglion cells remains unknown. Fig. 2.5 shows the firing patterns of the two groupings.

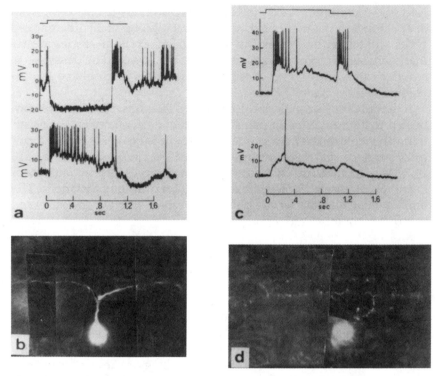

Fig. 2.5. Types A and B ganglion cell responses to spots and annuli. Responses of the Type A cell to a small spot and to an annulus are shown in the upper and lower traces of (A), respectively. The soma of a Type A cell filled with procion yellow dye is shown in a montage in (B). Responses of a Type B cell to spots and annuli are shown in (C) in the same order as in (A). The soma of a Type B cell is shown in a montage in (D). (Reprinted with permission from *Vision Research*, Vol. 21, Physiological and morphological study of the inner plexiform layer in the turtle retina, 1981, Pergamon Press plc.)

Signal transmissions from photoreceptor to ganglion cell reflect long delays that include a differentiation to prevent steady responses. Presumably the delays are due to a variety of factors such as cell time constants, transmitter release and diffusion delays, as well as action kinetics. Differentiation, on the other hand, is more difficult to place, and it is not easy to estimate where it might occur (Baylor and Fettiplace, 1977a; Baylor and Fettiplace, 1977b). Cones and rods operate along different time scales, with the cones responding electrically to a dim flash with a delay and signal duration of about 175 msec (Baylor et al., 1974b); the rods respond severalfold more slowly (Copenhagen and Owen, 1976; Schwartz, 1976). These differences suggest independence of transmission before convergence onto ganglion cells. The duration of voltage change produced in a receptor appears to match its transmission time; voltage changes are temporally different in the two kinds of receptors.

Whereas it is difficult to quantify the input to ganglion cells, because of the increased degrees of freedom involved and the accompanying uncertainties, it is possible to estimate the requirement for the photoreceptor to induce a response in a ganglion cell. The amount of stimulating current needed to be injected into the most sensitive pair described by Baylor and Fettiplace (1977a; 1977b) was about 2×10^{-11} amperes. If the input resistance of an isolated cone is estimated as 170 MΩ, and the voltage produced by one photoisomerization as 25 μV (see Section III.A), the net current resulting from one isomerization is about 1.5×10^{-13} amperes. About 130 photoisomerizations from red-sensitive cones would be required to provoke a ganglion cell; as the photon photocurrent of rods would be larger, perhaps 50 photoisomerizations would be required.

2. Descriptive Classification

In the same papers that reported the several amacrine cell types in *Pseudemys scripta elegans* and *Mauremys caspica* (see Section III.D), Kolb described 21 morphological varieties of ganglion cell common to both species on the basis of Golgi impregnation. Unfortunately this morphological diversity shows little correspondence to particular sets of functional properties.

The ganglion cell responses of the eastern box turtle, *Terrapene carolina carolina*, were initially classified by Lipetz and Hill (1970). They described several response categories and noted their suitability for movement detection. A more definitive classification by Bowling (1980), in *Pseudemys scripta elegans*, grouped response properties into color, orientation, motion sensitivity, and directional selectivity. It is of particular interest that a large proportion of these ganglion cells received inputs from both rods and red-sensitive cones.

Granda and Fulbrook (1982; 1989) used the complexity of receptive field organization to define eight categories, with response properties mutable for stimulus parameters of wavelength, light and dark adaptation, and velocity (Table 2.3).

All ganglion cells in *Pseudemys scripta elegans* are sensitive to movement, although they differ in their responses to movement parameters. Some cells respond only to target movement in one direction, being quiescent to movement in any other direction; other cells are directionally preferring, with vigorous responses to one direction of target movement but lower responses to movement in the otherwise null direction. In addition, other cells are sensitive to bidirectional movement on an axis orthogonal to the long axis of the receptive field.

Table 2.3. Classification of RGC types in *Pseudemys scripta elegans*.[a]

Cell Class	Response Type	RF Shape Size Range	Directional Sensitivity	# of Cells
Simple	+, −, +	C, O 3°–22°	non-D, B-D D-P	16
ON-sustained	+ sustained	C 3°–5°	non-D	6
Annular	+	A 8°–15°	non-D	3
Wavelength sensitive	+, −, ±	C, O 3°–10°	non-D, B-D D-P	4
Directionally selective	+, −, ±	C, O, B 3°–22°	D-S	23
Bar-shaped	+, −, ±	B 5°–50°	D-S, B-D	9
Large-field	+, ± + sustained	B, O, Q 25°–75°	non-D, B-D D-P, D-S	5
Velocity	±	C, O 8°–20°	non-D, B-D	12

[a] Source: Granda & Fulbrook, 1989.
Abbreviations: +, ON; ±, ON-OFF; −, OFF; RF, receptive field; C, circular; O, oval; B, bar-shaped; A, annular; Q, crescent-shaped; non-D, nondirectional; D-P, directionally preferred; B-D, bidirectional; D-S, directionally selective.

Finally, some cells are sensitive to any kind of movement with no preference to its direction.

Most ganglion cells of *Pseudemys scripta elegans* are sensitive to red light, hardly surprising since red-sensitive cones predominate in number (Granda and Dvorak, 1977). Dark-adapted ganglion cells often show broader spectral responses that depend on additional input from rods (Leeper, 1978a; Leeper, 1978b; Bowling, 1980). Under light adaptation, responses to red light are more vigorous and those to blue and green lights practically disappear (Fig. 2.6).

Receptive field maps of ganglion cells can be rather elaborate, with areal profiles that are circular, oval, or crescent shaped. Circular or oval maps, described by Granda and Fulbrook (1989), range from 3° to 22° in diameter, with some especially large fields measuring up to 75°. Generally, the smallest receptive fields lie close to the area centralis, a region of greater cellular density. Figure 2.7 shows several of these configurations.

Because so many ganglion cells are sensitive to movement, it is natural to inquire as to their preference for particular stimulus velocities. Not all units are sensitive to this. Those of *Pseudemys scripta elegans* encounter a range that extends more than a thousandfold, from 0.2° to 320°·sec^{-1}. The units can be classified into one of five velocity response types, depending on their sensitivity ranges: slow-pass;

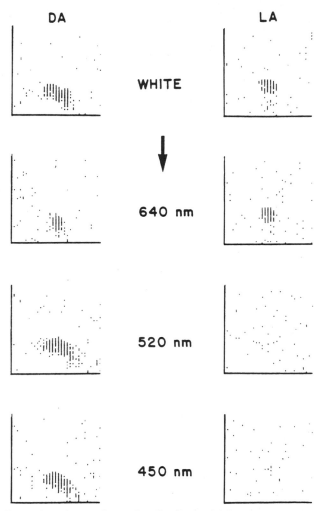

Fig. 2.6. Responses of a ganglion cell to l° stimuli of different wavelengths under dark (DA) and light (LA) adaptation. A field of 45° × 45° was scanned at 9° sec⁻¹ in a downward direction signified by the central arrow. The spot stimuli were equated for numbers of photons at the several wavelengths. The number of photons for *white* light was based on an integrated value between wavelength limits of 400 and 700 nm. (From Spatial organization of retinal ganglion cells in turtle, Granda and Fulbrook, *Color Research and Application,* copyright 1982, John Wiley & Sons, Inc.)

broad-band; narrow-band, slow-pass; midrange-pass; and velocity-tuned, fast-pass cells. This classification is one of convenience modified from the classification of velocity response categories proposed for cat striate cortex by Orban et al. (1981).

It is easy to surmise that ganglion cells in turtles must be complex and that the number and variety of their receptive field organizations possess intricate retinal integrative capacities, among the most elabo-

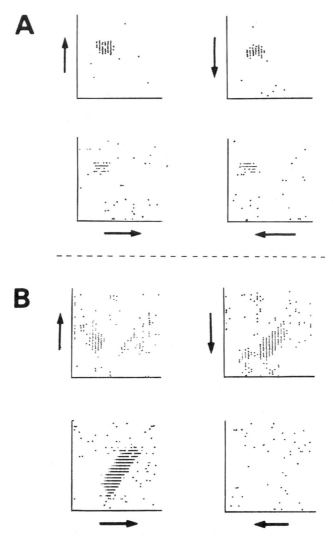

Fig. 2.7. Receptive fields of two ganglion cells scanned at a speed of 9°·sec⁻¹ in the directions indicated by arrows. (A) Nondirectional ON cell. (B) Bar-shaped, directionally selective OFF cell. The overall visual field being scanned is 45° × 45°. The receptive field is clearly defined by a horizontal scan to the right, with its null response to a scan in the opposite direction. Up and down scans show intermediate responses. (From Spatial organization of retinal ganglion cells in turtle, Granda and Fulbrook, *Color Research and Application*, copyright 1982, John Wiley & Sons.)

rate found in vertebrates. These functions in ganglion cells of turtles are relegated in part to later steps in the visual pathway of mammals for which there is an extensive and developed anatomical network. It is astonishing that turtles, with far fewer cells and fewer elaborations, have retinal machinery that is efficient and sophisticated and yet contains the visual mechanisms described.

IV. NEUROPHARMACOLOGY
A. General

The development of the retina confirms that it is part of the central nervous system (CNS). Its use as a model that is more easily accessible for the study of central functions was first suggested by Cajal. Of the 40 or so chemicals thought to be neurotransmitters, retinal cells have been found to respond to almost two dozen. The retina of turtles is particularly well suited for these studies. First, the cells are robust in that they survive in vitro for extended periods, functionally intact. This robustness enables experimenters more easily to control their chemical milieu. Second, because the retinal cells are large, their activity can be easily recorded during the application of putative neurotransmitters, their agonists and antagonists. Third, as can be seen from the two preceding sections, the anatomy and physiology of the turtle retina are fairly well understood, which establishes a framework to which the pharmacological data can be related. Finally, the response properties of ganglion cells in turtle retina are complex; for example, they have the orientation and velocity specific properties associated with visual cortical cells in higher vertebrates. Questions concerning the formation of these receptive field properties can be answered using peripheral structure in turtles.

B. Excitatory Amino Acids

In *Pseudemys scripta elegans*, as in most vertebrates, an excitatory amino acid (EAA), probably L-glutamate, mediates chemical transmission among photoreceptors and horizontal cells and between photoreceptors and bipolar cells in the outer plexiform layer (Cervetto and MacNichol, 1972; Copenhagen and Jahr, 1989). L-glutamate is released in the dark by depolarized photoreceptors, and the release is decreased by light-mediated hyperpolarizations. This results in one of two responses in second-order neural cells: (1) a graded, sign-conserving hyperpolarization in horizontal cells and off-center bipolar cells or (2) a graded, sign-reversing depolarization in on-center bipolar cells. The opposing effects of the same transmitter are mediated by the postsynaptic receptor. Receptors sensitive to the L-glutamate agonist, kainate, occur in horizontal cells and off-center bipolar cells (Slaughter and Miller, 1983a, 1983b, 1983c); receptors sensitive to 2-amino-4-phosphonobutyric acid occur in on-center bipolar cells (Slaughter and Miller, 1981; Bloomfield and Dowling, 1985).

Bipolar cells communicate with ganglion cells in the inner plexiform layer with an EAA as a transmitter. As is the case in photoreceptors, L-glutamate is the most likely transmitter candidate. On-center bipolar cells contribute to the central receptive fields of on-center ganglion cells, and off-center bipolar cells contribute to the central receptive fields of off-center ganglion cells. Unlike segregated inputs to bipolar receptive fields, the centers of which are mediated by direct photoreceptor inputs—whereas antagonistic surrounds are mediated by horizontal cells, both central and surround portions of ganglion cell receptive fields receive significant inputs from bipolar and amacrine cells. Whenever amacrine cells depolarize the central portions of ganglion cell receptive fields, an EAA is likely to be the transmitter involved.

C. Inhibitory Amino Acids

The two inhibitory amino acids are γ-aminobutyric acid (GABA) and glycine. GABA is the most completely characterized transmitter in the retina. Both GABA and its synthetic enzyme occur in presynaptic terminals of certain retinal neurons. In the presence of Ca^{++}, GABA is released from these presynaptic terminals upon stimulation; there are specific uptake and catabolic mechanisms for GABA at the synapse. Finally, GABA, applied iontophoretically, produces the same response in the postsynaptic cell as does the endogenous transmitter (Wu et al., 1989).

Horizontal cells mediate the antagonistic surrounds of bipolar cells through GABAergic synapses (Cervetto and MacNichol, 1972; Yazulla, 1986). Horizontal cells also use GABAergic synapses to feed back onto photoreceptors. GABA is also found in certain amacrine cells that mediate specific characteristics of ganglion cell receptive fields. Picrotoxin, a GABA antagonist, degrades directional and orientation specificity of ganglion cells (Wyatt and Daw, 1976; Ariel and Adolph, 1985).

Many amacrine cells contain glycine (Marc, 1985), and their dendrites are broadly stratified in the inner plexiform layer. Both of these observations suggest widespread inhibitory effects of glycine on receptive field properties of ganglion cells. Application of strychnine, a glycine antagonist, confirms this view by eliciting a general increase in ganglion cell responsiveness through disinhibition. More specifically, glycine appears to sharpen the transient activity of on and off ganglion cells as strychnine broadens the temporal response profiles of these cells (Marc, 1985).

D. Catecholamines

Dopamine is the most common catecholamine in the retina. In *Pseudemys scripta elegans*, it is located exclusively in amacrine cells (Witkovsky et al., 1984; Nguyen-Legros et al., 1985; Witkovsky et al., 1987), although in several other species dopamine is also localized in inter-plexiform cells, which feed back from the inner to the outer plexiform layer (Dowling, 1987). Dopamine has three main sites of action: (1) It acts on ganglion cells, in which it activates the inhibitory surround of ganglion cell receptive fields through D1 receptors (Jensen and Daw, 1986); (2) it acts on horizontal cells, in which it decouples the exten-sive gap-junctional networks between similar horizontal cells by act-ing through D1 receptors; and (3) finally, it affects photoreceptors and pigment granule motility in the pigment epithelium, where it acts via D2 receptors.

A common thread through each of these properties is that dopa-mine appears to drive light adaptation. Retinal dopamine metabolism also increases in the light. During light adaptation, rods, which me-diate dark-adapted vision, elongate so that their outer segments become further embedded in the pigment epithelium, effectively shielding them from light. At the same time, cones, which mediate photopic vision, shorten, so that their outer segments are partially retracted from the pigment epithelium and the melanin granules in the pigment epithelial cells disperse into the processes that surround each photoreceptor. In this configuration, cone outer segments can only be reached by light that travels radially from the pupil to the retina; at the same time, these outer segments will be effectively shielded from light scattered within the eye. The net effect of these physical changes is to trade sensitivity, which is counter-productive at high light levels, for spatial acuity.

Dopamine also sharpens the inhibitory surrounds of bipolar and ganglion cells. In ganglion cells, light adaptation and dopamine es-sentially turn on the surround portion of the receptive field. In bipo-lar cells, the surround portion of the receptive field is produced by horizontal cells. In the dark and in the absence of dopamine, horizon-tal cells are strongly coupled by gap junctions that allow ions, and thereby electrical information, to move freely among cells. In this state, horizontal cells respond best to large diffuse lights. Light adap-tation and the presence of dopamine decrease this electrical coupling. In this state, horizontal cells respond best to spatially smaller lights, and their response actually decreases for large, diffuse lights. Dopa-mine and light adaptation act on bipolar cells by sharpening the spa-tial characteristics of the surround. Because antagonistic center-surround receptive fields serve as contrast detectors, sharpening of the spatial properties of the surround improves the ability of bipolar cells to detect local contrasts. Again, the net effect of dopamine and light adaptation improve visual acuity at both the bipolar and gan-glion cell levels.

Turtles lack interplexiform cells and thus have no obvious mechanism that can deliver dopamine directly to the outer plexiform layer. Dopamine, however, is released from amacrine cells, and there is evidence that it can diffuse from the inner to outer plexiform layers in an active form (Piccolino et al., 1984; Piccolino et al., 1985).

In opposition to the idea that dopamine mediates the light-adaptive response in the retina, Dowling and co-workers (Yang et al., 1988a, 1988b; Tornqvist et al., 1988) showed that prolonged dark adaptation of the retina in teleosts reduces horizontal cell responsiveness and decreases the electrical coupling between horizontal cells that appear to be mediated by dopamine. They argue that the sensitivity of the retina to light is increased by the loss of the inhibitory surrounds of bipolar cells. It is possible, though highly speculative, that interplexiform cells provide a means whereby a sufficient quantity of dopamine can be delivered to horizontal cells after prolonged dark adaptation, essentially to shut down their inhibitory influence on bipolar cells. At the same time, dopamine would have to be sequestered from the rest of the retina in order to prevent it from eliciting the remainder of its light-adapting influences.

E. Indolamines

The two indolamines found in the retina are melatonin and serotonin. Along with the EAA l-glutamate, photoreceptors also secrete melatonin in the dark (Cardinali and Rosner, 1971). Exposure to light disrupts this secretion by inactivating N-acetyltransferase, a key enzyme in the synthesis of melatonin (Binkley et al., 1980). Melatonin inhibits dopamine release from amacrine cells (Dubocovich, 1983; Dubocovich, 1985; Dubocovich et al., 1985; Dubocovich, 1986) and in this way is implicated in the regulation of dark and light adaptation; this role is complementary to the one that melatonin plays in circadian rhythms. Melatonin also has effects that are antagonistic to those mediated by dopaminergic D2 receptors; namely, melatonin causes the aggregation of melanin granules in the cells of pigment epithelia (Pang and Yew, 1979). It also causes cones to elongate and rods to shorten (Pierce and Besharse, 1985). These are all dark-adaptive mechanisms.

The indolamine serotonin has also been implicated in dark adaptation. In the dark-adapted retina, serotonergic amacrine cells form a positive feedback loop with bipolar cells. The result of this loop is to enhance the signal-to-noise ratio of transmission from bipolar cells to ganglion cells in the dark. The positive feedback is mediated by HT_{1A} receptors. It is limited by HT_2 receptors, which are less sensitive to serotonin than the HT_{1A} receptors but which have stronger effects once activated (Brunken and Daw, 1987).

F. Acetylcholine

Unlike the amino acid transmitters involved in the definition of the trigger features of ganglion cells, acetylcholine (Ach) alters the efficacy with which a given stimulus causes a ganglion cell to respond. Agonists for Ach increase the slope of the intensity-response curves of the ganglion cell. Ach may then be involved in setting the gain of retinal output (Masland and Ames, 1976; Ariel and Daw, 1982; Schmidt et al., 1987).

G. Neuropeptides

In *Pseudemys scripta elegans*, various peptides have been implicated in synaptic transmission at the amacrine cell level. These include somatostatin, neurotensin, met-enkephalin (Adolph, 1989), glucagon (Eldred and Karten, 1983), neuropeptide Y (Isayama et al., 1988), substance P (Kolb et al., 1988), and corticotropin-releasing factor (Williamson and Eldred, 1989). The function of these neuropeptides in the retina is not well defined. Most are localized in amacrine cells that are interneurons within the inner plexiform layer; that is, these cells form amacrine-amacrine synapses and do not directly influence ganglion cell responses. Application of the neuropeptides neurotensin and substance P causes general excitation; application of the neuropeptide enkephalin causes inhibition of ganglion cells without altering more specific receptive field properties.

Unlike the situation in the CNS, most neuropeptides in the retina do not appear to be co-localized with more traditional transmitter agents. Possible exceptions are met-enkephalin, which may be co-localized with GABA (Zucker and Adolph, 1988), and neurotensin, which may share the same amacrine cells with glycine (Weiler and Ball, 1984); the physiological implications of a single amacrine cell containing both an excitatory (neurotensin) and an inhibitory (glycine) transmitter are not readily apparent. Whereas substance P and Ach are not co-localized in amacrine cells, they have parallel functions in that they increase retinal gain (Glickman et al., 1982b). The main difference between the two transmitters is in their kinetics; Ach acts rapidly and briefly, whereas the effects of substance P have latencies of several seconds and last for several minutes.

V. PSYCHOPHYSICS
A. General

The sum total of the outputs of visual cells is reflected in behavior. When that behavior is displayed under conditions controlled for particular functions, the validity of the processing is that much better supported. In *Pseudemys scripta elegans*, the spectral, spatial, and temporal properties of the visual system have been studied under differing conditions of adaptation (Maxwell and Granda, 1975; Granda and Dvorak, 1977). The methodology is one that utilizes conditioned shock avoidance to light by head withdrawal. In *P. scripta elegans*, a shy animal, shock applied to the jaw immediately elicits head withdrawal. The withdrawal controls automatic programming equipment that then tracks thresholds for light contingent on that response in a modified up-and-down sequence. The value of each stimulus is determined by the behavioral response to the preceding one. This method has the virtue that it concentrates behavioral responses near threshold. Fig. 2.8 shows a diagram of the apparatus.

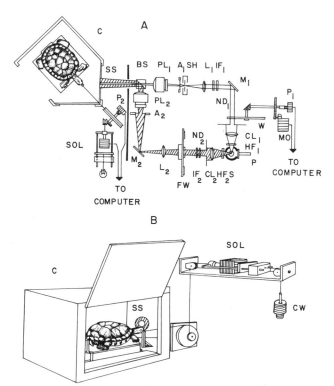

Fig. 2.8. Apparatus for measuring visual thresholds in turtles. (A) Optical pathway. (B) Apparatus used to control and monitor turtle head position. Abbreviations: A, aperture; BS, beam splitter; C, experimental chamber; CL, condensing lens; CW, counterweight; FW, neutral density filter wheel; HF, heat filter; IF, interference filter; L, collecting lens; M, mirror; MO, wedge-drive motor; ND, neutral density filter; P, potentiometer; PL, projection lens; S, light source; SH, shutter; SOL, solenoid; SS, stimulus screen; W, neutral density wedge. (Reprinted with permission from *Vision Research*, Vol. 29, Psychophysically derived visual mechanisms in the turtle. II. Spectral properties, 1989, Pergamon Press plc.)

B. Adaptation

The time course of dark adaptation is wavelength dependent in *Pseudemys scripta elegans* (Granda et al., 1972). Minimum thresholds for colored stimuli are seen within one or two minutes of dark adaptation. For intermediate and short wavelength stimuli, dark-adaptation curves also exhibit a threshold plateau within approximately two minutes but then again begin to decrease after several more minutes in the dark. These second decreases in threshold continue for over an hour. During this time course, the peak spectral sensitivity of the visual system changes from approximately 650 nm to 550 nm (Granda et al., 1972). *Chrysemys picta* shows a similar increase in sensitivity to shorter wavelengths (Sokol and Muntz, 1966).

C. Increment Threshold

Signals in individual cones at behavioral threshold can be derived from information about quantal input. For example, the sensitivity of red-sensitive cones varies with background intensity according to Equation 3 (see Section III.A). At threshold, the receptor response peak amplitude, ΔV_T, equals the product of flash sensitivity, S_F, and increment flash intensity, ΔI_T. Consequently, the voltage response at threshold may be expressed as

$$\Delta V_T = S_F \cdot \Delta I_T. \tag{4}$$

If the receptor response at threshold remains constant with background intensity, Equations 3 and 4 can be combined to give the Weber relation for psychophysical threshold in terms of the electrical properties of the photoreceptors involved:

$$\Delta I_T = \Delta V_T / S_F^D + \zeta \cdot \Delta V_T \cdot I_B, \tag{5}$$

where the Weber fraction, k, is equal to $\zeta \cdot \Delta V_T$, and $IO = \Delta V_T / S_F^D$.

Using a value for k of $10^{-4} \cdot \text{flash}^{-1}$ with this derivation, Fain et al. (1977) derived a value of 5 to 10 μV as the receptor response at visual threshold in red-sensitive cones. The voltage, of course, is small and indicates that the signaling properties among photoreceptors and other cells can be modified by very little change.

Stiles (1978) developed a methodology whereby the intensities of the increment threshold for one monochrome can be determined against background intensities of another. Data obtained by this two-color threshold technique are plotted as log threshold intensity at one wavelength against log background intensity at another. Whenever a single mechanism is involved, the resultant curve is composed of (1) a portion obtained at low background intensity near absolute thresh-

Table 2.4. Parameters of psychophysically derived visual mechanisms in *Pseudemys scripta elegans*.[a]

Mechanism	k (sec)	I_0 ($\phi \cdot \mu m^{-2}$)	ζ (sec $\cdot$ V^{-1})	ΔV_r (μV)	S_F^{DM} (μV $\cdot \phi^{-1} \cdot \mu m^2$)
1	0.0533	0.0136	1580	33.7	2480
2	0.0533	0.0697			
3	0.0055	0.7430	20	273.0	367
4	0.0034	0.6998	20	170.0	242

Source: Reprinted with permission from *Vision Research*, Vol. 29, Psychophysically derived visual mechanisms in the turtle. I. Spectral properties, 1989, Pergamon Press plc.)
[a] Psychophysically determined flash sensitivities in the dark of visual mechanisms calculated from Equation 6. I_0 is in units of irradiance falling on the retina, assuming a critical duration of 125 msec for red-sensitive cones, 145 msec for green-sensitive cones, and 500 msec for rods (Granda and Dvorak, 1977).

old for that mechanism, (2) a central portion that conforms to Weber's law, and (3) a portion at high background intensity that departs from Weber's law and may reflect saturation of the mechanism (Aguilar and Stiles, 1954).

The definition of the psychophysical parameters of k (the Weber fraction) and I_O (absolute threshold) in terms of electrophysiological parameters (Fain et al., 1977; Granda and Sisson, 1989) permits the calculation of the flash sensitivity of a visual mechanism (S_F^{DM}) as

$$S_F^{DM} = k/\zeta \cdot I_O = \Delta V_T/I_O, \tag{6}$$

and is expressed in the same units as S_F^P. Four visual mechanisms have been isolated in *Pseudemys scripta elegans* using this approach (Granda and Sisson, 1989). The dark-adapted flash sensitivities of these mechanisms have been calculated from Equation 6, and are listed in Table 2.4.

One visual mechanism appears to be mediated by rods and red-sensitive cones active at low intensities of background light. A second mechanism involves coupled green- and red-sensitive cones that predominate at intermediate light levels. The third and fourth process, mediated by red- and green-sensitive single cones, act alone at high intensities of background light (Fig. 2.9).

The spread of visual mechanisms across the visible spectrum and over a log unit in flash sensitivities permits *Pseudemys scripta elegans* to effectively sample its environment. Spectral visual mechanisms obtained through behavioral psychophysics accord well with data from intracellular electrophysiology (Baylor and Hodgkin, 1973; Baylor and Hodgkin, 1974) and also from absorption microspectrophotometry (Liebman and Granda, 1971; Ohtsuka, 1985). This approach to visual mechanisms through visual behavior allows an integrative assessment of cellular inputs, their combinations, their transmission properties, and their ultimate displays.

D. Spatial Factors

The spatial characteristics for these color mechanisms have also been investigated in terms of Ricco areas (Sisson and Granda, 1989). Up to a critical value, threshold intensity for a given light stimulus is proportional to the inverse of stimulus area (Riggs, 1972). Increase of stimulus area beyond this critical size has no further effect on threshold levels. In terms of the metrics employed, for those areas of less than critical size, stimulus threshold is determined by radiant flux, photons per unit time; for those areas greater than this critical size, stimulus threshold is determined by stimulus radiance, photons per unit area per unit time. Whenever spatial summation is graphed, with abscissal values of log stimulus area versus ordinal values of log threshold-stimulus intensity, the slopes are -1 for areas less than the critical size and 0 for areas greater than that value.

Critical areas for the several mechanisms isolated in *Pseudemys scripta elegans* have been determined. For similar adaptive states, the values are at least an order of magnitude less than the relative areas of the central portions of the receptive fields of either bipolar cells (Schwartz, 1974; Yazulla, 1976b) or ganglion cells (Bowling, 1980; Granda and Fulbrook, 1982). It is very likely, therefore, that critical area is determined early in the visual pathway, probably at the level of the photoreceptors. In support of this conclusion, the diameter of the critical area of the dark-adapted mechanism (about 140 µm) approximates the spatial spread of information among rods (150 to 200 µm (Copenhagen and Owen, 1976; Schwartz, 1973b); the diameter of the light-adapted mechanisms (about 50 µm) approximates the spatial spread of information among cones (50 to 100 µm) (Baylor and Hodgkin, 1973; Copenhagen and Owen, 1976) (Fig. 2.10).

E. Temporal Factors

In *Pseudemys scripta elegans*, visual mechanisms exhibiting long critical durations, large critical areas, and high sensitivity are generally attributed to the activity of rods; cones show shorter critical durations, smaller critical areas, and lower sensitivities. All these properties have been determined by psychophysical studies. By varying the duration of test stimuli, the shortest duration capable of threshold detection can be determined in a paradigm reflecting Bloch's law (Granda and Dvorak, 1977). Stimulus duration and intensity then are integrated to some critical value of stimulus duration. Thereafter integra-

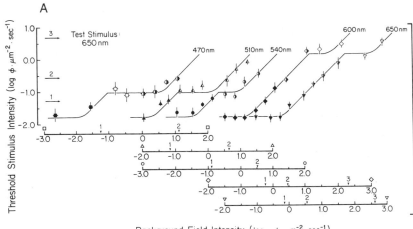

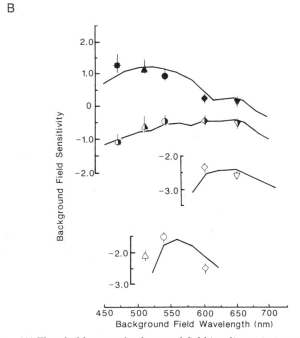

Fig. 2.9. (A) Threshold versus background field irradiance (tvi) curves. The test stimulus for each curve was a 650-nm spot, with a diameter of 4.0 mm projected onto 25-mm diameter background fields at indicated wavelengths. The symbol shape for the data points of each tvi curve is also used to mark the appropriate abscissa for that curve, for example, the 650-nm increment threshold curve relates to the bottom scale and is so indicated by the symbols in common. Superimposed on these shapes is a secondary code denoting visual mechanism: filled symbols are attributed to Mechanism 1, half-filled symbols to Mechanism 2, and open symbols to Mechanism 3. Criterion-threshold stimulus intensities, set at 0.5 log unit above the respective I_o val-

ues for the visual mechanism, are used to calculate background field sensitivity and are indicated by numbered arrows along the ordinate; associated background field intensities are indicated by corresponding numbered arrowheads on the appropriate abscissa. The inverse of these values are plotted in (B) as spectral sensitivity points. Consistent symbols are maintained in both figures. Error bars indicate 95% confidence intervals. (B) Spectral sensitivity curves derived from the data displayed in (A). Mechanism 1 is displayed in the topmost curve, Mechanism 2 in the middle curve, and Mechanism 3 (against its own ordinate) at the right of the figure. Mechanism 4, derived from a similar plot as in (A), but with a 540-nm spot stimulus, is shown at the bottom left (also against its own ordinate). Mechanism 1 has been fit to action spectra for red-sensitive cones and is do, Mechanism 2 to action spectra for red- and green-sensitive cones, Mechanism 3 to the action spectrum for red-sensitive cones alone, and Mechanism 4 to the action spectrum for green-sensitive cones alone (see text). The dashed line in the middlemost curve indicates required absorption by the yellow oil droplets associated with green-sensitive single cones. Background field sensitivity is in units of inverse retinal irradiance. (Reprinted with permission from *Vision Research*, Vol. 29, Psychophysically derived visual mechanisms in the turtle. II. Spectral properties, 1989, Pergamon Press plc.)

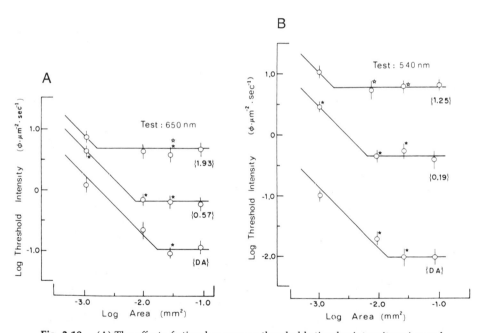

Fig. 2.10. (A) The effect of stimulus area on threshold stimulus intensity using red (650 nm) stimuli. The 650-nm test stimuli were presented against 600-nm background fields. Background intensity, the value in braces to the right of each curve (log $\phi \cdot \mu m^{-2} \cdot sec^{-1}$), was chosen to isolate one of the mechanisms. (B) The effect of stimulus area on threshold stimulus intensity using green (540 nm) stimuli. The 540-nm test stimuli were projected against 510-nm background fields. Background intensity, the value in braces to the right of each curve (log $\phi \cdot \mu m^{-2} \cdot sec^{-1}$), was chosen to isolate one of the mechanisms. (Reprinted with permission from *Vision Research*, Vol. 29, Psychophysically derived visual mechanisms in the turtle. II. Spatial properties, 1989, Pergamon Press plc.)

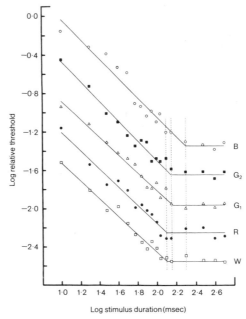

Fig. 2.11. Critical duration as a function of wavelength: 125 msec for 620 nm (R) and white (W) light, 150 msec for 566 (G₁) and 533 (G₂) nm light, and 200 msec for 466 nm (B) light. (From Granda and Dvorak, 1977.)

tion fails, and threshold is determined solely by stimulus intensity. The values under dark adaptation are long, about 500 msec, reflecting the activity of rods. Light adaptation discloses three different critical durations that depend on the color of the test stimulus, selected to reflect the peak absorptions of the underlying cone photopigments. Blue or short wavelength stimuli result in critical durations of 200 msec; green or intermediate wavelength stimuli generate critical durations of 150 msec; red or long wavelength stimuli produce the shortest values of all, 125 msec. Critical duration thresholds as a function of wavelength are shown in Fig. 2.11. These figures are in keeping with what is already known about the activity of rods and cones in the retina (see Section III.A).

VI. CONCLUSIONS

The aspects of vision described in this chapter are all encompassed by the properties of retinal cells. The major boundary conditions in perception are surely set by the photoreceptors, for they afford spectral analysis, kinetics of signal production, noise filtering through spatial averaging, and the intricacies of light and dark adaptation. At the outer plexiform layer, horizontal and bipolar cells interact to afford an elaborate spatial analysis of visual space by means of center-surround, receptive-field arrangements. The organization permits enhancement of object contour and contrast, properties that are important and necessary for determining figure-ground discriminations. Such discriminations play an obvious role in survival.

Recodification of signals takes place at the inner plexiform layer, where amacrine and ganglion cells digitize visual messages for transmittal to successive way stations in the central nervous system. Ganglion cells, in turtles at least, then elaborate sensitivities of the receptive field to the speed and direction of movement; the degree and range of sensitivity allows prediction of object placement, useful to the animal in affording a quick and reliable assay of the visual environment. Turtles differ from higher mammals in that they accomplish these things in the retina without the extensive brain development of mammals.

APPENDIX: REPTILIAN SPECIES DISCUSSED

TESTUDINES

Chelonia mydas mydas
  Granda and Haden, 1970
  Liebman and Granda, 1971
Chelydra serpentina
  Copenhagen and Jahr, 1989
  Copenhagen and Owen, 1976
  Detwiler and Hodgkin, 1979
  Detwiler et al., 1980
  Fuortes et al., 1973
  Fuortes and Simon, 1974
  Lamb et al., 1981
  Lamb and Simon, 1976
  Owen, 1985
  Richter and Simon, 1974, 1975
  Schwartz, 1973a, 1973b, 1974, 1976
  Simon, 1973
  Simon et al., 1975
Chinemys reevesii
  Ohtsuka, 1978, 1984, 1985
  Ohtsuka et al., in press
Chrysemys picta
  Sokol and Muntz, 1966
Emydoidea blandingii
  Lipetz, 1984a, 1984b, 1985
  Lipetz and Hill, 1970
Mauremys caspica
  Kolb et al., 1988

Terrapene c. carolina
  Granda and Halen, 1970
  Kolb and Jones, 1982
  Lipetz and Hill, 1970
Trachemys scripta (elegans) = (ex *Pseudemys*)
  Adolph, 1989
  Ariel and Adolph, 1985
  Baylor and Fettiplace, 1975, 1976, 1977a, 1977b, 1979
  Baylor and Fuortes, 1970
  Baylor et al., 1971
  Baylor and Hodgkin, 1973, 1974
  Baylor et al., 1974a, 1974b
  Baylor et al., 1979
  Bowling, 1980
  Cervetto and MacNichol, 1972
  Cervetto et al., 1976
  Eldred and Karten, 1983
  Fain, Granda and Maxwell, 1977
  Granda and Fulbrook, 1982, 1989
  Granda and Haden, 1970
  Granda et al., 1972
  Granda and Sisson, 1989
  Isayama et al., 1988
  Kolb, 1982
  Kolb and Jones, 1982, 1985

Lasansky, 1971
Leeper, 1978a, 1978b
Liebman and Granda, 1971, 1975
Marchiafava, 1976, 1978, 1979
Marchiafava and Torre, 1978
Marchiafava and Weiler, 1980
Maxwell and Granda, 1975
Normann et al., 1985
Normann et al., 1984
Piccolino et al., 1984
Piccolino et al., 1982
Piccolino et al., 1985
Sisson and Granda, 1989

Weiler and Ball, 1984
Weiler and Marchiafava, 1981
Williamson and Eldred, 1989
Witkovsky et al., 1987
Witkovsky et al., 1984
Yazulla, 1976a, 1976b
Zucker and Adolph, 1988
Turtle
Miller et al., 1973
Naka et al., 1987
Nguyen-Legros et al., 1985
Saito et al., 1974

REFERENCES

Adolph, A. R. (1989). Pharmacological actions of peptides and indolamines on turtle retinal ganglion cells. *Visual Neurosci.* 3, 411–423.

Aguilar, M., and Stiles, W. S. (1954). Saturation of rod mechanism of the retina at high levels of stimulation. *Optica Acta* 1, 59–65.

Ariel, M., and Adolph, A. R. (1985). Neurotransmitter inputs to directionally sensitive turtle retinal ganglion cells. *J. Neurophysiol.* 54, 1123–1143.

Ariel, M., and Daw, N. W. (1982). Pharmacological analysis of directionally sensitive rabbit retinal ganglion cells. *J. Physiol.* (London) 324, 161–185.

Baylor, D. A., and Fettiplace, R. (1975). Light path and photon capture in turtle photoreceptors. *J. Physiol.* (London) 248, 433–464.

Baylor, D. A., and Fettiplace, R. (1976). Transmission of signals from photoreceptors to ganglion cells in the eye of the turtle. *Cold Spring Harbor Symp. Quant. Biol.* 40, 529–536.

Baylor, D. A., and Fettiplace, R. (1977a). Transmission from photoreceptors to ganglion cells in the turtle retina. *J. Physiol.* (London) 271, 391–424.

Baylor, D. A., and Fettiplace, R. (1977b). Kinetics of synaptic transfer from receptors to ganglion cells in the turtle retina. *J. Physiol.* (London) 271, 425–448.

Baylor, D. A., and Fettiplace, R. (1979). Synaptic drive and impulse generation in ganglion cells of turtle retina. *J. Physiol.* (London) 288, 107–127.

Baylor, D. A., and Fuortes, M. G. F. (1970). Electrical responses of single cones in the retina of the turtle. *J. Physiol.* (London) 207, 77–92.

Baylor, D. A., Fuortes, M. G. F., and O'Bryan, P. M. (1971). Receptive fields of cones in the retina of the turtle. *J. Physiol.* (London) 214, 265–294.

Baylor, D. A., and Hodgkin, A. L. (1973). Detection and resolution of visual stimuli by turtle photoreceptors. *J. Physiol.* (London) 234, 163–198.

Baylor, D. A., and Hodgkin, A. L. (1974). Changes in time scale and sensitivity in turtle photoreceptors. *J. Physiol.* (London) 242, 729–758.

Baylor, D. A., Hodgkin, A. L., and Lamb, T. D. (1974a). The electrical response of turtle cones to flashes and steps of light. *J. Physiol.* (London) 242, 685–727.

Baylor, D. A., Hodgkin, A. L., and Lamb, T. D. (1974b). Reconstruction of the electrical responses of turtle cones to flashes and steps of light. *J. Physiol.* (London) 242, 759–791.

Baylor, D. A., Lamb, T. D., and Yau, K.-W. (1979). The membrane current of single rod outer segments. *J. Physiol.* (London) 288, 589–611.

Binkley, S., Reilly, K. B., and Hryshchyshyn, M. (1980). N-Acetyltransferase in the chick retina: I. Circadian rhythms controlled by environmental lighting are similar to those in the pineal gland. *J. Comp. Physiol.* 189B, 103–108.

Bloomfield, S. A., and Dowling, J. E. (1985). Roles of aspartate and glutamate in synaptic transmission in rabbit retina: I. Outer plexiform layer. *J. Neurophysiol.* 53, 699–713.

Bowling, D. B. (1980). Light responses of ganglion cells in the retina of the turtle. *J. Physiol.* (London) 299, 173–196.

Brunken, W. J., and Daw, N. W. (1987). The actions of serotonergic agonists and antagonists on the activity of brisk ganglion cells in the rabbit retina. *J. Neurosci.* 7, 4054–4065.

Cardinali, D. P., and Rosner, J. M. (1971). Metabolism of serotonin by the rat retina *in vitro*. *J. Neurochem.* 18, 1769–1770.

Cervetto, L., and MacNichol, E. F., Jr. (1972). Inactivation of horizontal cells in turtle retina by glutamate and aspartate. *Science* (N.Y.) 178, 767–768.

Cervetto, L., Marchiafava, P. L., and Pasino, E. (1976). Influence of efferent retinal fibers on responsiveness of ganglion cells to light. *Nature* (London) 260, 56–57.

Copenhagen, D. R., and Jahr, C. E. (1989). Release of endogenous excitatory amino acids from turtle photoreceptors. *Nature* (London) 341, 536–539.

Copenhagen, D. R., and Owen, W. G. (1976). Functional characteristics of lateral interactions between rods in the retina of the snapping turtle. *J. Physiol.* (London) 259, 251–282.

Detwiler, P. B., and Hodgkin, A. L. (1979). Electrical coupling between cones in turtle retina. *J. Physiol.* (London) 291, 75–100.

Detwiler, P. B., Hodgkin, A. L., and McNaughton, P. A. (1980). Temporal and spatial characteristics of the voltage response of the rods in the retina of the snapping turtle. *J. Physiol.* (London) 300, 213–250.

Dowling, J. E. (1987). *The Retina.* Harvard University Press, Cambridge, Mass.

Dubocovich, M. L. (1983). Melatonin is a potent modulator of dopamine release in the retina. *Nature* (London) 306, 782–784.

Dubocovich, M. L. (1985). Characterization of a retinal melatonin receptor. *J. Pharmacol. Exp. Ther.* 234, 395–401.

Dubocovich, M. L. (1986). Modulation of dopaminergic activity by melatonin in retina. In *Pineal and Retinal Relationships.* (P. J. O'Brian and D. C. Klein, eds.). Academic Press, New York, pp. 239–252.

Dubocovich, M. L., Lucas, R. C., and Takahashi, J. S. (1985). Light-dependent regulation of dopamine receptors in mammalian retina. *Brain Res.* 335, 321–325.

Eldred, W. D., and Karten, H. J. (1983). Characterization and quantification of peptidergic amacrine cells in the turtle retina: enkephalin, neurotensin, and glucagon. *J. Comp. Neurol.* 221, 371–381.

Fain, G. L., Granda, A. M., and Maxwell, J. H. (1977). Voltage signals of photoreceptors at visual threshold. *Nature* (London) 265, 181–183.

Fuortes, M. G. F., Schwartz, E. A., and Simon, E. J. (1973). Colour-dependence of cone responses in the turtle retina. *J. Physiol.* (London) 234, 199–216.

Fuortes, M. G. F., and Simon, E. J. (1974). Interactions leading to horizontal cell responses in the turtle retina. *J. Physiol.* (London) 240, 177–198.

Glickman, R. D., Adolph, A. R., and Dowling, J. E. (1982b). Inner plexiform circuits in the carp retina: effects of cholinergic agonists, GABA and substance P on the ganglion cells. *Brain Res.* 234, 81–99.

Granda, A. M., and Dvorak, C. A. (1977). Vision in turtles. In *Handbook of Sensory Physiology: the Visual System in Vertebrates*, Vol. VII/5 (F. Crescitelli, ed). Springer-Verlag, Heidelberg, pp. 451–495.

Granda, A. M., and Fulbrook, J. E. (1982). Spatial organization of retinal ganglion cells in turtle, *Pseudemys*. *Color Res. Appl.* 7, 173–177.

Granda, A. M., and Fulbrook, J. E. (1989). Classification of turtle retinal ganglion cells. *J. Neurophysiol.* 62, 723–737.

Granda, A. M., and Haden, K. W. (1970). Retinal oil globules counts and distributions in two species of turtles: *Pseudemys scripta elegans* (Wied) and *Chelonia mydas mydas* (Linnaeus). *Vision Res.* 10, 79–84.

Granda, A. M., Maxwell, J. H., and Zwick, H. (1972). The temporal course of dark adaptation in the turtle, *Pseudemys*, using a behavioral avoidance paradigm. *Vision Res.* 12, 653–672.

Granda, A. M., and Sisson, D. F. (1989). Psychophysically derived visual mechanisms in the turtle. I. Spectral properties. *Vision Res.* 29, 93–105.

Isayama, T., Polak, J., and Eldred, W. D. (1988). Synaptic analysis of amacrine cells with neuropeptide Y-like immunoreactivity in turtle retina. *J. Comp. Neurol.* 275, 452–459.

Jensen, R. J., and Daw, N. W. (1986). Effects of dopamine and its agonists and antagonists on the receptive field properties of ganglion cells in the rabbit retina. *Neuroscience* 17, 837–855.

Kolb, H. (1982). The morphology of the bipolar cells, amacrine cells and ganglion cells in the retina of the turtle *Pseudemys scripta elegans*. *Philos. Trans. Roy. Soc. Lond.* 298, 355–393.

Kolb, H., and Jones, J. (1982). Light and electron microscopy of the photoreceptors in the retina of the red-eared slider, *Pseudemys scripta elegans*. *J. Comp. Neurol.* 209, 331–338.

Kolb, H., and Jones, J. (1985). Electron microscopy of Golgi-impregnated photoreceptors reveals connections between red and green cones in the turtle retina. *J. Neurophysiol.* 54, 304–317.

Kolb, H., Perlman, I., and Normann, R. A. (1988). Neural organization of the retina of the turtle *Mauremys caspica*: a light microscope and Golgi study. *Vis. Neurosci.* 1, 47–72.

Lamb, T. D., McNaughton, P. A., and Yau, K.-W. (1981). Spatial spread of activation and background sensitization in rod outer segments. *J. Physiol.* (London) 319, 463–496.

Lamb, T. D., and Simon, E. J. (1976). The relation between intracellular coupling and electrical noise in turtle photoreceptors. *J. Physiol.* (London) 263, 257–286.

Lasansky, A. (1971). Synaptic organization of cone cells in the turtle retina. *Philos. Trans. Roy. Soc. Lond.* 262, 365–381.

Leeper, H. F. (1978a). Horizontal cells of the turtle retina. I. Light microscopy of Golgi preparations. *J. Comp. Neurol.* 182, 777–794.

Leeper, H. F. (1978b). Horizontal cells of the turtle retina. II. Analysis of interconnections between photoreceptor cells and horizontal cells by light microscopy. *J. Comp. Neurol.* 182. 795–810.

Liebman, P. A., and Granda, A. M. (1971). Microspectrophotometric measurements of visual pigments in two species of turtle, *Pseudemys scripta* and *Chelonia mydas. Vision Res.* 11, 105–114.

Liebman, P. A., and Granda, A. M. (1975). Super dense carotenoid spectra resolved in single cone oil droplets. *Nature* (London) 253, 370–372.

Lipetz, L. E. (1984a). A new method for determining peak absorbance of dense pigment samples and its application to the cone oil droplets of *Emydoidea blandingii. Vision Res.* 24, 597–604.

Lipetz, L. E. (1984b). Pigment types, densities and concentrations in cone oil droplets of *Emydoidea blandingii. Vision Res.* 24, 605–612.

Lipetz, L. E. (1985). Some neuronal circuits of the turtle retina. In *The Visual System: a Symposium to honor Edward F. MacNichol, Jr.* (A. Fein, ed.). Alan R. Liss, New York, pp. 107–132.

Lipetz, L. E., and Hill, R. M. (1970). Discrimination characteristics of the turtle's retinal ganglion cells. *Experientia* 26, 373–374.

Lipetz, L. E., and MacNichol, E. F., Jr. (1979). A survey of oil droplets in the cones of turtles' retinas. *Invest. Ophthalmol. Vis. Sci.*, Suppl., April, 118.

Lipetz, L. E., and MacNichol, E. F., Jr. (1982). Photoreceptors of freshwater turtles: cell types and visual pigments. *Biol. Bull.* 163, 396.

Marc. R. E. (1985). The role of glycine in retinal circuitry. In *Retinal Transmitters and Modulators: Models for the Brain* (W. W. Morgan, ed.). CRC Press, Boca Raton, Fla., pp. 119–158.

Marchiafava, P. L. (1976). Centrifugal actions on amacrine and ganglion cells in the retina of the turtle. *J. Physiol.* (London) 255, 137–155.

Marchiafava, P. L. (1978). Horizontal cells influence membrane potential of bipolar cells in the retina of the turtle. *Nature* (London) 275, 141–142.

Marchiafava, P. L. (1979). The responses of retinal ganglion cells to stationary and moving stimuli. *Vision Res.* 19, 1203–1211.

Marchiafava, P. L., and Torre, V. (1978). The responses of amacrine cells to light and intracellularly applied currents. *J. Physiol.* (London) 276, 83–102.

Marchiafava, P. L., and Weiler, R. (1980). Intracellular analysis and structural correlates of the organization of inputs to ganglion cells in the retina of the turtle. *Proc. Roy. Soc. Lond.* B208, 103–113.

Masland, R. H., and Ames, A. (1976). Responses to acetylcholine of ganglion cells in an isolated mammalian retina. *J. Neurophysiol.* 39, 1220–1235.

Maxwell, J. H., and Granda, A. M. (1975). An automated apparatus for the determination of visual thresholds in turtles. *Physiol. Behav.* 15, 131–132.

Michaelis, L., and Menton, M. L. (1913). Die Kinetic der Invertinwirkung. *Biochem. Zeit.* 49, 333–339.

Miller, W. H., Hashimoto, Y., Saito, T., and Tomita, T. (1973). Physiological and morphological identification of L- and C-type S-potentials in the turtle retina. *Vision Res.* 13, 443–447.

Naka, K. I., Itoh, M. A., and Chappell, R. L. (1987). Dynamics of turtle cones. *J. Gen. Physiol.* 89, 321–337.

Naka, K. I., and Rushton, W. A. H. (1966). An attempt to analyse colour reception by electrophysiology. *J. Physiol.* (London) 185, 556–586.

Nguyen-Legros, J., Versaux-Botteri, C., Vigny, A., and Raoux, N. (1985). Tyrosine hydroxylase immunohistochemistry fails to demonstrate dopaminergic interplexiform cells in the turtle retina. *Brain Res.* 339, 323–328.

Normann, R. A., and Anderton, P. J. (1983). The incremental sensitivity curve of turtle cone photoreceptors. *Vision Res.* 23, 1731–1733.

Normann, R. A., Perlman, I., and Daly, S. J. (1985). Mixing of color signals by turtle cone photoreceptors. *J. Neurophysiol.* 54, 293–303.

Normann, R. A., Perlman, I., Kolb, H., Jones, J., and Daly, S. J. (1984). Direct excitatory interactions between cones of different spectral types in the turtle retina. *Science* (N.Y.) 224, 625–627.

Ohtsuka, T. (1978). Combinations of oil droplets with different types of photoreceptor in a freshwater turtle, *Geoclemys reevesii. Sens. Proc.* 2, 321–325.

Ohtsuka, T. (1984). Fluorescence from colorless oil droplets: a new criterion for identification of cone photoreceptors. *Neurosci. Lett.* 52, 241–245.

Ohtsuka, T. (1985). Relation of spectral types to oil droplets in cones of turtle retina. *Science* (N.Y.) 229, 874–877.

Ohtsuka, T., and Kawamata, K. (1990). Telodendrial contact of HRP-filled photoreceptors in the turtle retina: pathways of photoreceptor coupling. *J. Comp. Neurol.* 292, 599–613.

Orban, G. A., Kennedy, H., and Maes, H. (1981). Response to movement of neurons in areas 17 and 18 of the cat: velocity sensitivity. *J. Neurophysiol.* 45, 1043–1058.

Owen, W. G. (1985). Chemical and electrical synapses between photoreceptors in the retina of the turtle, *Chelydra serpentina. J. Comp. Neurol.* 240, 423–433.

Pang, S. F., and Yew, D. T. (1979). Pigment aggregation by melatonin in the retinal epithelium and choroid of guinea pigs, *Cavia porcellus. Experientia* 35, 231–233.

Piccolino, M., Neyton, J., and Gerschenfeld, H. M. (1984). Decrease of gap junction permeability induced by dopamine and cyclic adenosine 3':5'-monophosphate in horizontal cells of turtle retina. *J. Neurosci.* 4, 2477–2488.

Piccolino, M., Neyton, J., Witkovsky, P., and Gerschenfeld, H. M. (1982). γ-Aminobutyric acid antagonists decrease junctional communication between L-horizontal cells of the retina. *Proc. Nat. Acad. Sci.* USA 79, 3671–3675.

Piccolino, M., Witkovsky, P., Neyton, J., Gerschenfeld, H. M., and Trimarchi, C. (1985). Modulation of gap junction permeability by dopamine and GABA in the network of horizontal cells of the turtle retina. In *Neurocircuitry of the Retina: a Cajal Memorial* (A. Gallego and P. Gouras, eds.). Elsevier, New York, pp. 66–75.

Pierce, M. E., and Besharse, J. C. (1985). Circadian regulation of retinomotor movements. I. Interaction of melatonin and dopamine in the control of cone length. *J. Gen. Physiol.* 86, 671–689.

Richter, A., and Simon, E. J. (1974). Electrical responses of double cones in the turtle retina. *J. Physiol.* (London) 242, 673–683.

Richter, A., and Simon, E. J. (1975). Properties of centre-hyperpolarizing, red-sensitive bipolar cells in the turtle retina. *J. Physiol.* (London) 248, 317–334.

Riggs, L. A. (1972). Vision. In *Woodworth and Schlosberg's Experimental Psychology* (J. W. Kling and L. A. Riggs, eds.). Holt, Rinehart and Winston, New York, pp. 273–314.

Saito, T., Miller, W. H., and Tomita, T (1974) C- and L-type horizontal cell responses in the turtle retina. *Vision Res.* 14, 119–123.

Schmidt, M., Humphrey, M. F., and Wassle, H. (1987). Action and localization of acetylcholine in the cat retina. *J. Neurophysiol.* 58, 997–1015.

Schwartz, E. A. (1973a). Organization of on-off cells in the retina of the turtle. *J. Physiol.* (London) 230, 1–14.

Schwartz, E. A. (1973b). Responses of single rods in the retina of the turtle. *J. Physiol.* (London) 232, 503–514.

Schwartz, E. A. (1974). Responses of bipolar cells in the retina of the turtle. *J. Physiol.* (London) 236, 211–224.

Schwartz, E. A. (1976). Electrical properties of the rod syncytium in the retina of the turtle. *J. Physiol.* (London) 257, 379–406.

Simon, E. J. (1973). Two types of luminosity horizontal cells in the retina of the turtle. *J. Physiol.* (London) 230, 199–211.

Simon, E. J., Lamb, T. D., and Hodgkin, A. L. (1975). Spontaneous voltage fluctuations in retinal cones and bipolar cells. *Nature* (London) 256, 661–662.

Sisson, D. F., and Granda, A. M. (1989). Psychophysically derived visual mechanisms in the turtle. II. Spatial properties. *Vision Res.* 29, 107–114.

Slaughter, M. M., and Miller, R. F. (1981). 2-Amino-4-phosphonobutyric acid: a new pharmacological tool for retinal research. *Science* (N.Y.) 211, 182–185.

Slaughter, M. M., and Miller, R. F. (1983a). The role of excitatory amino acid transmitters in the mudpuppy retina: an analysis with kainic acid and N-methyl aspartate. *J. Neurosci.* 3, 1701–1711.

Slaughter, M. M., and Miller, R. F. (1983b). Bipolar cell in the mudpuppy retina uses an excitatory amino acid transmitter. *Nature* (London) 303, 537–538.

Slaughter, M. M., and Miller, R. F. (1983c). An excitatory amino acid antagonist blocks cone input to sign-conserving second-order retinal neurons. *Science* (N.Y.) 219, 1230–1232.

Sokol, S., and Muntz, W. R. A. (1966). The spectral sensitivity of the turtle *Chrysemys picta picta*. *Vision Res.* 6, 285–292.

Stiles, W. S. (1978). *Mechanisms of Colour Vision.* Academic Press, New York.

Tornqvist, K., Yang, X.-L, and Dowling, J. E. (1988). Modulation of cone horizontal cell activity in the teleost fish retina. III. Effects of prolonged darkness and dopamine on electrical coupling between horizontal cells. *J. Neurosci.* 8, 2279–2288.

Weiler, R., and Ball, A. K. (1984). Colocalization of neurotensin-like immunoreactivity and 3H-glycine uptake system in sustained amacrine cells of turtle retina. *Nature* (London) 311, 759–761.

Weiler, R., and Marchiafava, P. L. (1981). Physiological and morphological study of the inner plexiform layer in the turtle retina. *Vision Res.* 21, 1635–1638.

Williamson, D. E., and Eldred, W. D. (1989). Amacrine and ganglion cells with corticotropin-releasing-factor-like immunoreactivity in the turtle retina. *J. Comp. Neurol.* 280, 424–435.

Witkovsky, P. (1980). Excitation and adaptation in the vertebrate retina. *Curr. Top. Eye Res.* 2, 1–66.

Witkovsky, P., Alones, V., and Piccolino, M. (1987). Morphological changes induced in turtle retinal neurons by exposure to 6-hydroxydopamine and 5-6-dihydroxytryptamine. *J. Neurocytol.* 16, 55–67.

Witkovsky, P., Eldred, W., and Karten, H. J. (1984). Catecholamine- and indoleamine-containing neurons in the turtle retina. *J. Comp. Neurol.* 228, 217–225.

Wu, J.-Y., Lin, C. T., Liu, J. Y., Yan, G. H., Evans, D., and Liao, C. C. (1989). Amino acid transmitters in the vertebrate retina. In *Neurology and Neurobiology: Extracellular and Intracellular Messengers in the Vertebrate Retina* (D. A. Redburn and H. Pasantes-Morales, eds.). Alan R. Liss, New York, pp. 177–190.

Wyatt, H. J., and Daw, N. W. (1976). Special effects of neurotransmitter antagonists on ganglion cells in rabbit retina. *Science* (N.Y.) 191, 204–205.

Yang, X.-L., Tornqvist, K., and Dowling, J. E. (1988a). Modulation of cone horizontal cell activity in the teleost fish retina. I. Effects of prolonged darkness and background illumination on light responsiveness. *J. Neurosci.* 8, 2259–2268.

Yang, X.-L. Tornqvist, K., and Dowling, J. E. (1988b). Modulation of cone horizontal cell activity in the teleost fish retina. II. Role of interplexiform cells and dopamine in regulating light responsiveness. *J. Neurosci.* 8, 2269–2278.

Yazulla, S. (1976a). Cone input to horizontal cells in the turtle retina. *Vision Res.* 16, 727–735.

Yazulla, S. (1976b). Cone input to bipolar cells in the turtle retina. *Vision Res.* 16, 737–744.

Yazulla, S. (1986). GABAergic mechanisms in the retina. In *Progress in Retinal Research* (N. Osborne and J. Chader, eds.). Pergamon Press, New York, pp. 1–52.

Zucker, C. L., and Adolph, A. R. (1988). Interaction between enkephalin and GABA in turtle retina. *Invest. Ophthalmol. Vis. Sci.*, Suppl. 29, 196.

3

Comparative Analysis of the Primary Visual System of Reptiles

JACQUES REPÉRANT, JEAN-PAUL RIO, ROGER WARD,
STÉPHANE HERGUETA, DOM MICELI, AND MICHEL LEMIRE

CONTENTS

I. INTRODUCTION

The ganglion cells of the retina (see Chapters 1 and 2) are the main source of information about the visual world. Their axons leave the posterior surface of the eyeball as the two optic nerves. These nerves intersect in the optic chiasm, which lies on the ventral surface of the diencephalon. The axons of the retinal ganglion cells continue into the brain as the optic tracts and ultimately form synaptic endings in a series of brain structures known as the primary visual centers. This chapter is a comparative analysis of the organization of the primary visual pathways in reptiles; the Appendix lists species studied by different investigators.

The first studies of the primary visual pathways in reptiles relied solely on descriptive methods. Stieda (1875) believed the contralateral optic tectum to be the only brain structure that received retinal information. Some years later, Bellonci (1888), Ramón (1896), and Edinger (1899) identified up to five primary visual centers in the thalamopretectal region of the brain, and Edinger (1899) identified the accessory optic system of reptiles. During the first 40 years of the 20th century, many authors confirmed the earlier findings (Gisi, 1908; de Lange, 1913; Beccari, 1923; Huber and Crosby, 1926, 1933a; Cairney, 1926; Shanklin, 1930; Durward, 1930; Frederikse, 1931; Kuhlenbeck, 1931; Papez, 1935; Ariëns-Kappers et al., 1936; Ströer, 1939; Warner, 1931, 1942).

Armstrong (1950, 1951) was the first to apply experimental methods to the study of reptilian primary visual pathways. The Nonidez technique (1933) was used to demonstrate the degeneration of visual fibers following ablation of the retina. In addition to the superficial layers of the optic tectum and the basal optic nucleus, Armstrong also described five or six primary visual structures that receive projections from the contralateral retina. He neglected the first suggestion of the occurrence of an ipsilateral visual contingent. Later works (Burns and Goodman, 1967; Kosareva, 1967; Knapp and Kang, 1968a, 1968b), using Nauta methods (1951, 1954, 1957), only partially confirmed Armstrong's observations. For example, Burns and Goodman (1967) found three contralateral retinorecipient structures at the thalamo-pretectal level, whereas Kosareva (1967) identified eight. Knapp and Kang (1968a, 1968b) found no ipsilateral visual projections, but Burns and Goodman (1967) mentioned that *Caiman crocodilus* has contralateral retinal fibers that cross through the posterior commissure.

More recent studies have shown that these contradictions arose principally from technical difficulties (Halpern and Frumin 1973; Repérant, 1973a, 1978). The Nauta method is prone to produce a number of artifacts, such as an inadequate suppression of normal fibers (Burns and Goodman, 1967; Kosareva, 1967) or a lack of impregnation of degenerating visual terminals (Knapp and Kang 1968a, 1968b). The introduction of Fink-Heimer techniques (1967) resolved many of these difficulties. They are more sensitive and easier to interpret than the method of Nauta, consequently their use has allowed considerable clarification of the organization of the visual system in many species of reptiles (Ebbesson, 1970, 1972; Hall and Ebner, 1970a, 1970b; Butler and Northcutt, 1971; Repérant, 1972, 1973a; Braford, 1973; Halpern, 1973; Halpern and Frumin, 1973; Northcutt and Butler, 1974a, 1974b; Butler, 1974; Platel et al., 1975; Repérant, 1978; Ebbesson and Karten, 1981; Peterson 1981). Even more recently, radioautography and the orthograde axonal transport of horseradish peroxidase (HRP) have been used to trace the retinofugal projections in many reptiles (Cruce and Cruce, 1975, 1978; Repérant, 1978; Repérant and Rio, 1976; Hall et al., 1977; Repérant et al., 1978; Northcutt, 1978; Butler and Northcutt, 1978; Bass and Northcutt, 1981a, 1981b; Schroeder, 1981a; Künzle and Schnyder, 1983; Rio et al., 1983; Dacey and Ulinski 1986d; de la Calle et al., 1986; Kenigfest et al., 1986; Sjöström and Ulinski, 1985; Rio, 1987; Torcol, 1988; Lemire et al., 1988; Ulinski and Nautiyal, 1988).

These studies have allowed the determination of a general pattern underlying the contralateral retinal projections in reptiles. This pattern is outlined in Section II. Section III considers the principal variations of this pattern, including those regarding the extent of ipsilateral retinal projections, the reduction of the visual system in microphthalmic reptiles, and the special variations seen in snakes. Throughout, the discussion emphasizes phylogenetic considerations and comparative aspects.

II. GENERAL PLAN OF PRIMARY VISUAL CENTERS IN MACROPHTHALMIC SURFACE-DWELLING REPTILES: CONTRALATERAL RETINAL PROJECTIONS

A. The Optic Tracts

The decussation of optic nerve fibers at the chiasm does not take place in a single compact bundle, but in alternating fascicles, and can be either partial or complete (Fig. 3.1). The contralateral optic tract consists of three components: (1) a small and variable anterior optic fascicle (tractus opticus anterior) to the hypothalamus and thalamus; (2) a small basal optic root (tractus opticus basalis, TOB) to the ventral mesencephalic tegmentum; and (3) the principal fascicle, the marginal optic tract (tractus opticus marginalis, TOM), which divides at the anterior margin of the tectum into lateral and medial fascicles. The TOM comprises the axons that are distributed to the different thalamic, pretectal, and tectal visual relays (Fig. 3.2).

B. Primary Visual Centers

1. HYPOTHALAMIC NUCLEI   No classic author working with normal material described a retinohypothalamic projection in reptiles. Some experimental studies found this projection (Armstrong, 1950, 1951; Burns and Goodman, 1967; Kosareva, 1967; Butler and Northcutt, 1971; Repérant, 1973a), whereas others either did not, or demonstrated it with less certainty (Knapp and Kang, 1968a, 1968b; Halpern and Frumin, 1973; Butler, 1974; Northcutt and Butler, 1974a, 1974b; Repérant, 1975; Repérant and Rio, 1976). The nucleus supraopticus (Knapp and Kang, 1968a, 1968b; Northcutt and Butler, 1974a; Repérant, 1975; Cruce and Cruce, 1978), the nucleus suprachiasmaticus

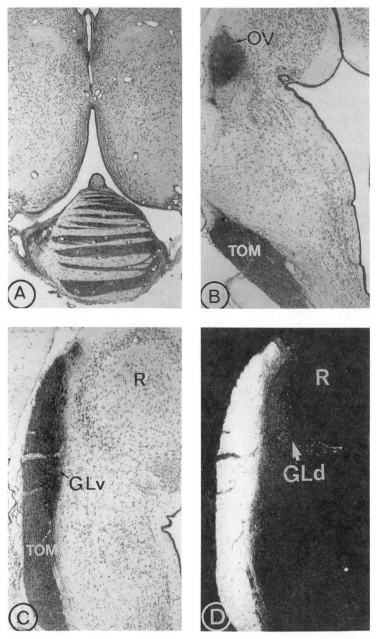

Fig. 3.1. Radioautograms of transverse sections showing the projection to the contralateral thalamus 48 hours following right intraocular injection of 3H proline (25HμCi) in *Cyclemys dentata* (A) and *Chinemys reevesi* (B, C, and D). (A) Chiasm. Note the crossing of fibers in alternative fascicles. (B) Rostral thalamus. (C) Intermediate thalamus. (D) The same level photographed under dark-field illumination (40 x). Abbreviations: GLd, nucleus geniculatus pars dorsalis; GLv, nucleus geniculatus pars ventralis; OV, nucleus ovalis; R, nucleus rotundus; TOM, tractus opticus marginalis.

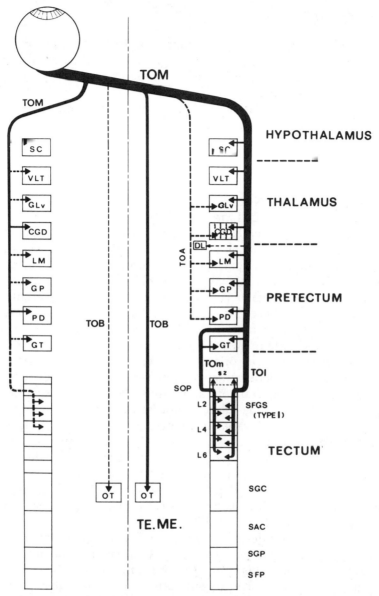

Fig. 3.2. Diagram of primary visual system organization in reptiles; contralateral (right) and ipsilateral (left) projections. Inconstant projections are indicated by dashed lines and the midline by the vertical broken line. Abbreviations: CGD, corpus geniculatus pars dorsalis; DL, nucleus dorsolateralis; Glv, nucleus geniculatus pars ventralis; LM, nucleus lentiformis mesencephali; OT, nucleus opticus tegmenti; PD, nucleus posterodorsalis; TE.ME., tectum mesencephali; TOA, tractus opticus anterioris; TOB, tractus opticus basalis; TOl, tractus opticus lateralis; TOm, tractus opticus medialis; TOM, tractus opticus marginalis; SC, nucleus suprachiasmaticus; SAC, stratum album centrale; SFGS, stratum fibrosum et griseum superficiale; SFP, stratum fibrosum periventriculare; SGC, stratum griseum centrale; SGP, stratum griseum periventriculare; SOP, stratum opticum; VLT, nucleus ventrolateralis thalami.

(Butler, 1974; Repérant, 1978; Repérant et al., 1978; Ebbesson and Karten, 1981), or both (Halpern and Frumin 1973; Northcutt and Butler, 1974b; Repérant and Rio, 1976) have been described as receiving retinal projections (Fig. 3.2).

Radioautographic and electron-microscopical observations in *Vipera* demonstrate unequivocally that retinal fibers course toward the hypothalamus and always terminate in the suprachiasmatic nucleus (Repérant, 1978).

2. THALAMIC VISUAL NUCLEI

a. NUCLEUS GENICULATUS LATERALIS PARS VENTRALIS   The lentiform nucleus geniculatus lateralis pars ventralis (GLv) lies in the convexity of the marginal optic tract along almost the entire extent of the thalamus. It may comprise a lateral region of neuropil, the pars molecularis of Beccari (1923), and a medial cellular region, the pars magnocellularis of Kuhlenbeck (1931). The two regions are clearly differentiated in the Testudines, Crocodilia, and Lacertilia but are more or less reduced in snakes (Repérant, 1978). In turtles, the well-defined pars magnocellularis contains two to four rows of predominantly fusiform cells with their long axes oriented perpendicular to the optic tract, and the pars molecularis contains a few vertically oriented, fusiform cells (Repérant, 1978; Franzoni and Fasolo, 1982). In snakes, on the other hand, the pars molecularis is not (sensu stricto) a neuropil but contains many cells of variable orientation. Furthermore, the pars magnocellularis is not clearly demarcated from the former region nor from the medial N. ventrolateralis partes ventralis et dorsalis; its organization is imprecise, and the neurons do not show any particular preferential orientation (Repérant, 1978). The visual projection, which is generally dense in the pars molecularis, arises principally from the TOM (Figs. 3.1C, 3.2, 3.3A, 3.5A & 4).

Bellonci (1888) first described the ventral lateral geniculate nucleus as the corpus geniculatum thalamicum. Neuroanatomists of the classic period redescribed this structure using terms such as N. genicolato inferior (Ramón 1896), corpus subopticum (Edinger, 1899), corpus geniculatum laterale (Edinger, 1899; Herrick, 1910; de Lange, 1913; Huber and Crosby, 1926; Ströer, 1939), N. anterior thalami (Gisi, 1908), N. genicolato laterale (Beccari, 1923), corpus geniculatus lateralis pars interna and pars inferior (Shanklin, 1930), corpus geniculatus (or geniculatum) lateralis pars ventralis (Cairney 1926; Durward, 1930), N. geniculatus lateralis and N. ventralis (Frederikse, 1931), and N. geniculatus lateralis pars ventralis (Kuhlenbeck, 1931; Huber and Crosby, 1933a; Papez, 1935; Warner 1931, 1942, 1955). The last name has been adopted by the majority of contemporary authors (Kosar-

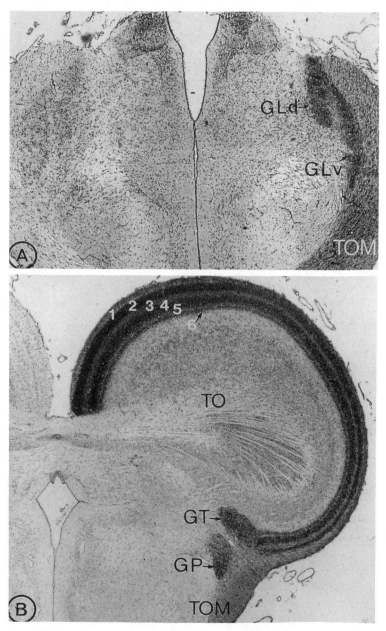

Fig. 3.3. *Uromastyx acanthinurus.* Radioautograms of retinal projections at the level of
the diencephalon (A) and intermediate mesencephalon (B) 72 hours following contralateral intraocular injection of a mixture of 3H-fucose and 3H-proline (25μCi). Transverse sections. In this species, the absence of ipsilateral labeling should be noted. (A,
40 x; B, 29 x). Abbreviations: GLd, nucleus geniculatus pars dorsalis; GLv, nucleus
geniculatus pars ventralis; GP, nucleus geniculatus pretectalis; GT, nucleus griseus
tectalis; TO, tectum opticum; TOM, tractus opticus marginalis; numbers refer to layers of the stratum fibrosum et griseum superficiale.

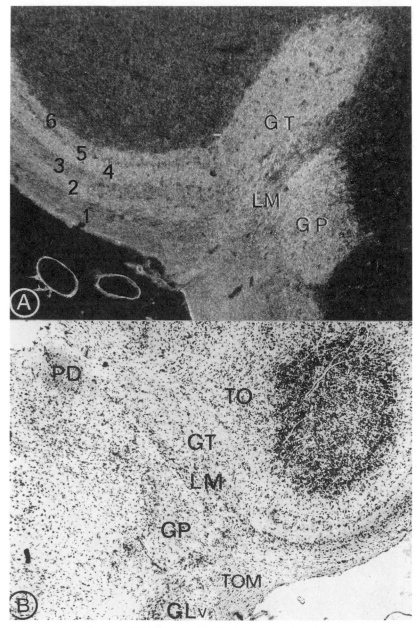

Fig. 3.4. *Uromastyx acanthinurus.* (A) Dark-field radioautograph of retinal projections at the level of the anterior mesencephalon 48 hours following contralateral intraocular injection of a mixture of 3H-fucose and 3H-proline (25μli). (B) Cresyl violet staining of the same region on the ipsilateral side showing the architectonic organization of the pretectal primary visual centers. Transverse sections (160 x). Abbreviations: GLv, nucleus geniculatus pars ventralis; GP, nucleus geniculatus pretectalis; GT, nucleus griseus tectalis; LM, nucleus lentiformis mesencephali; PD, nucleus posterodorsalis; TO, tectum opticum; TOM, tractus opticus marginalis; numbers refer to layers of the stratum fibrosum et griseum superficiale.

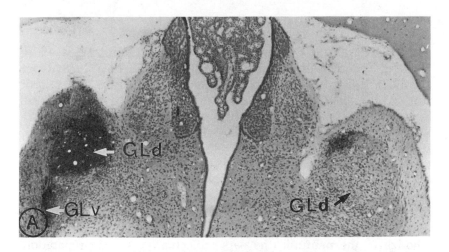

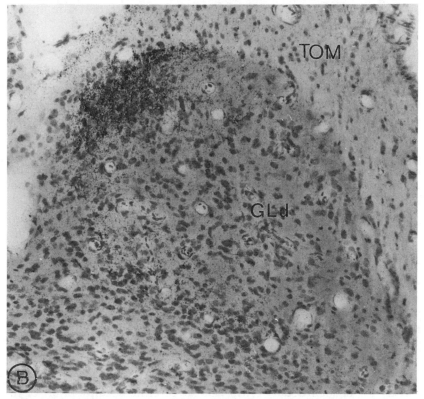

Fig. 3.5. *Vipera berus*. Radioautograms of the retinothalamic projections 24 hours following the right intraocular injection of 3H-proline (20μli). Transverse sections. (A) Low magnification showing the radioautographic labeling at the contralateral and ipsilateral levels. (B) Detail of radioautographic labeling in the ipsilateral nucleus geniculatus pars dorsalis (GLd). Note the accumulation of silver grains in the dorsomedial region of the structure. (A, 46 x; B, 190 x). Abbreviations: GLd, nucleus geniculatus pars dorsalis; GLv, nucleus geniculatus pars ventralis; TOM, tractus opticus marginalis.

eva, 1967; Burns and Goodman, 1967; Knapp and Kang, 1968a, 1968b; Hall and Ebner, 1970a; Butler and Northcutt, 1971, 1973, 1978; Repérant, 1972, 1973a, 1975; Braford, 1973; Butler, 1974; Cruce, 1974; Northcutt and Butler 1974a, 1974b; Cruce and Cruce, 1975; Northcutt, 1978; Repérant, 1978; Bass and Northcutt, 1981a, 1981b; Schroeder, 1981a; Künzle and Schnyder, 1983; de la Calle et al., 1986; Kenigfest et al., 1986; Torcol, 1988; Lemire et al., 1988; Ulinski and Nautiyal, 1988). However, Cruce and Cruce (1978) and Ebbesson and Karten (1981) refer to it respectively as the N. geniculatus lateralis ventralis and N. geniculatus ventrolateralis pars ventralis. All classic authors recognize this structure as a primary visual center, and experimental studies (see references below) confirm this observation. The retinal fibers arborize in the neuropil (the pars molecularis). Golgi preparations show that a large proportion of these optic arborizations are formed by collateral branches of fibers of the TOM (Repérant, 1978).

b. Nucleus Ventrolateralis
The nucleus ventrolateralis (VL) forms a large mass of cells lying against the medial face of the GLv. It can be divided into dorsal and ventral components (Northcutt, 1978). The ventrolateral nucleus (Cruce, 1974), also known as the area ventrolateralis (Huber and Crosby, 1926), area ventrolateralis anterior thalami (Shanklin, 1930), and N. suprapeduncularis (Frederikse, 1931), was never recognized as a primary visual center by classic authors. Only recently have the techniques of experimental neuroanatomy made it possible to document a discrete retinal projection to this nucleus, particularly to its pars ventralis (Fig. 3.2). This projection has been described in many lizards (*Iguana iguana, Anolis carolinensis, Gekko gecko, Xantusia vigilis, Podarcis hispanica;* Butler, 1974; Butler and Northcutt, 1971; Northcutt and Butler, 1974b; Northcutt, 1978; de la Calle et al., 1986). Cruce and Cruce (1978), Ebbesson et al. (1972), and Ebbesson and Karten (1981) also observed this projection in *Tupinambis teguixin,* but they described the nucleus itself as an integral part of the GLv. This retino-thalamic projection was also observed in *Caiman crocodilus* (Braford, 1973) and in the turtles *Chrysemys picta* and *Caretta caretta* (Bass and Northcutt, 1981a, 1981b). Most observers have not identified it in snakes (Halpern and Frumin, 1973; Repérant, 1973a; Northcutt and Butler, 1974a; Schroeder, 1981a), but a careful reexamination of radioautograms in several kinds of snakes has demonstrated the occurrence of retinal terminals in this projection (Repérant, 1978). It is possible that most authors tended to confuse the N. ventrolateralis with the pars magnocellularis of the ventral lateral geniculate nucleus because the two structures cannot be easily distinguished in snakes.

c. DORSAL LATERAL GENICULATE COMPLEX

The dorsal lateral geniculate complex (CGD) is an architecturally heterogeneous structure in which cellular aggregations and areas of neuropil vary from one reptilian group to another. It is therefore not surprising that different authors have provided divergent and sometimes contradictory descriptions of this structure, reporting from one to four cytoarchitectural subdivisions.

The following description adopts the nomenclature proposed by Repérant (1978). The CGD can be divided rostrocaudally into two components: the subnuclei ovalis (OV) and geniculatus lateralis pars dorsalis (GLd). In snakes and turtles, the OV consists of a region of neuropil immediately internal to the optic tract with a plate of densely packed somata forming the medial face of the subnucleus (Fig. 3.1B). In crocodilians and lizards, OV is better differentiated and is demarcated medially, dorsally, and ventrally by a multicellular shell surrounding a zone of neuropil. Huber and Crosby defined the name and cytoarchitectural characteristics of this structure for the first time in 1926 (in *Alligator mississippiensis*, although Ramón, 1896, described a N. genicolato anterior in *Chamaeleo*, which possibly corresponds to the N. ovalis). The nucleus was subsequently identified and named as such in other crocodilians (Addens, 1938; Repérant, 1975), lizards (Kuhlenbeck 1931; Ströer, 1939; Senn, 1966, 1968a), snakes (Armstrong 1951; Senn, 1966; Repérant, 1972, 1973a; Halpern and Frumin 1973; Northcutt and Butler, 1974a), and turtles (Papez, 1935; Kosareva, 1967; Knapp and Kang, 1968a; Künzle and Schnyder, 1983; Sjöström and Ulinski, 1985; Rainey and Ulinski, 1986; Ulinski and Nautiyal, 1988). It appears that the oval nucleus of Butler and Northcutt (1973 in *Iguana*) and that of Cruce and Cruce (1975, in *Tupinambis*) do not correspond to the differently located N. ovalis described by the preceding authors; it is probably the equivalent of the structure described as the pars dorsalis of the lateral geniculate nucleus in different turtles (Bass and Northcutt, 1981a, 1981b; Torcol, 1988) and lizards (Butler and Northcutt 1978; Northcutt, 1978). Finally, some authors either do not report a N. ovalis (Burns and Goodman, 1967, in *Caiman crocodilus*) or include it in the GLd (Hall and Ebner 1970a, 1970b; Hall et al., 1977, in *Pseudemys scripta*).

Several authors describe the OV as a primary visual center on the basis of normal material (*Chamaeleo vulgaris*, Ramón, 1896; *Sphenodon punctatus*, Cairney, 1926; *Alligator mississippiensis*, Huber and Crosby, 1926; *Crocodilus porosus*, Addens, 1938; *Chelonia mydas*, Papez, 1935; *Lacerta agilis*, Kuhlenbeck, 1931, Ströer 1939; *Podarcis sicula*, Senn, 1968a). The existence of a retinal projection has been confirmed many times by experimental techniques, most notably by Kosareva (1967,

Emys orbicularis), Knapp and Kang (1968, *Chelydra serpentina*), Repérant (1972, 1973a, *Vipera aspis*), Repérant (1975, *Caiman crocodilus*), and by Künzle and Schnyder (1983), Sjöström and Ulinski (1985), and Ulinski and Nautiyal (1988, *Pseudemys scripta elegans* and *Chrysemys picta belli*).

The GLd is the principal cellular aggregation of the CGD. It lies ventrally on the GLv, laterally facing the tractus opticus marginalis. Anteromedially, it is bounded by the N. dorsolateralis anterior and further caudomedially by the N. rotundus. In squamates (Figs. 3.3A and 3.5) and crocodilians, it is composed of zones of neuropil usually placed laterally and cellular aggregations that vary interspecifically in size and number (Repérant, 1978). The GLd appears to be larger in snakes than in other reptiles (Northcutt and Butler, 1974b; Repérant, 1978). Lizards show at least three architectural subdivisions (the dorsal optic nucleus, the intercalated optic nucleus, and the dorsolateral nucleus, Butler and Northcutt, 1978; Northcutt, 1978). The GLd of turtles is formed by an external region of neuropil that lies laterally against the marginal optic tract and an internal cellular band (Figs. 3.1C and D). In *Pseudemys* and *Chrysemys*, this band could be subdivided into a medial parvocellular zone and a lateral magnocellular zone (Rainey and Ulinski, 1986). The neurons, of highly variable morphology, are exclusively of Type 1 (Rainey and Ulinski, 1986).

Ramón (1896, as the nucleo anterior) and Gisi (1908, as the N. geniculatum laterale) first identified this nucleus. However, the first serious cytoarchitectural investigation was that of Cairney (1926), who named it the corpus geniculatus lateralis pars dorsalis. Since 1930 it has been identified in the majority of species examined.

Some indication of a retinal projection to this nucleus is given by Ramón's (1896) illustrations of Golgi preparations, but it was not until the experimental work of Belekhova and Kosareva (1967) and Kosareva (1967) that the retinal projection was confirmed. Subsequent investigations have demonstrated its existence in a wide variety of species (Knapp and Kang, 1968a, 1968b; Hall and Ebner, 1970a, 1970b; Ebbesson, 1970, 1972; Butler and Northcutt 1971; Repérant, 1972, 1973a, 1975, 1978; Halpern and Frumin 1973; Braford, 1973; Butler, 1974; Northcutt and Butler, 1974a, 1974b; Northcutt et al., 1974; Cruce and Cruce, 1975, 1978; Ebbesson and Karten, 1981; Bass and Northcutt, 1981a, 1981b; Künzle and Schnyder 1983; Sjöström and Ulinski, 1985; Kenigfest et al., 1986; de la Calle et al., 1986; Torcol, 1988; Lemire et al., 1988; Ulinski and Nautiyal, 1988). Use of Fink-Heimer techniques to investigate the effect of restricted retinal lesions in *Pseudemys* and *Chrysemys* has documented a single, topographically organized, projection of the retinal surface to the CGD (Ulinski and Nautiyal, 1988). The nasotemporal axis of the retina projects along the rostrocaudal axis of the CGD, and the dorsoventral axis of the retina corresponds to the dorsoventral axis of the CGD. At least three patterns of retinogeniculate axon terminals, based on the size of the arborization and the number and size of varicosities on its branches, have been reported (Sjöström and Ulinski, 1985, *Pseudemys* [Fig. 3.6]). All three types occur throughout the mediolateral extent of the complex, and each has a preferential laminar distribution.

Electron microscopy of the CGD in *Vipera aspis* (Repérant, 1978) and *Pseudemys* (Ulinski, 1986b) shows that the visual fibers frequently make synaptic contacts in a glomerular configuration. The central

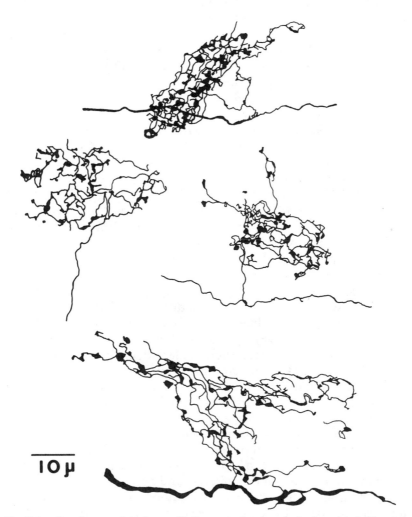

Fig. 3.6. *Pseudemys scripta elegans*. Retinogeniculate terminal arbors labeled by orthograde filling from extracellular injection of HRP into the optic tract. (From Sjöström and Ulinski, 1985.)

core of these synaptic zones is encapsulated by a glial envelope and generally contains boutons with spheroidal synaptic vesicles that make contacts (Gray Type I) with dendritic profiles. These postsynaptic profiles may contain synaptic vesicles and are thus part of an optodendrodendritic series. Two other types of axonal terminals, one containing pleomorphic synaptic vesicles, the other containing small round synaptic vesicles, also occur in these glomeruli.

The relationship of the CGD to the dorsal thalamus has been the subject of some long-standing controversies. Embryological investigations (Kuhlenbeck, 1931, *Lacerta vivipara*; Warner, 1942, *Nerodia sipedon*) led to consideration of the CGD as a dorsal thalamic structure. However, Senn (1968a, 1979) has argued that the CGD may derive from the ventral thalamus. The retrograde degeneration produced in the diencephalon by telencephalic lesions (*Lacerta vivipara*, Powell and Kruger, 1960; *Alligator mississippiensis*, Kruger and Berkowitz, 1960) led Kruger and his colleagues to conclude that the CGD does not project to the telencephalon and thus is part of the ventral thalamus. In contrast, more recent work has shown that the CGD very likely serves as a thalamic relay for visual information destined for the telencephalon, the thalamofugal optic tract; therefore, the DGC is a dorsal thalamic structure (Belekhova, 1970, 1979; Belekhova and Kosareva, 1967; Hall and Ebner, 1970a, 1970b; Hall, 1972; Hall et al., 1977; Wang and Halpern, 1977; Lohman and van Woerden-Verkley, 1978; Bass et al., 1983; Bruce and Butler, 1984; Ouimet et al., 1985; Rainey and Ulinski, 1986; Heller and Ulinski, 1987; Desan, 1988). However, some important differences exist among these later findings. In turtles, the projection to the telencephalon arises throughout the CGD (OV and GLd) and terminates in the dorsal cortex; most if not all of the cells of the CGD project to this structure (Rainey and Ulinski, 1986). In squamates, the results are slightly different and somewhat discordant. In *Thamnophis* (Wang and Halpern, 1977) and *Tupinambis* (Lohman and van Woerden-Verkley, 1978) only the GLd, but in its entirety, projects to the external part of the dorsal ventricular ridge. On the other hand, *Iguana iguana* and *Gekko gecko* have this projection arising from a subdivision of the GLd (the intercalated optic nucleus), which projects to the pallial thickening, a rostral extension of the dorsolateral cortex (Bruce and Butler, 1984).

In conclusion, the dorsal lateral geniculate complex is a constant and important site of retinal projections in reptiles (Fig. 3.2). Considered as a whole, it appears to be homologous in all species, but interspecific variation of its cellular subunits and neuropil makes more detailed comparisons difficult. It is to be hoped that further detailed studies of afferent and efferent connections will improve our understanding of the significance of this variation.

d. NUCLEUS DORSOLATERALIS ANTERIOR
The N. dorsolateralis anterior (DL) and the N. dorsomedialis constitute the rostral pole of the dorsal thalamus. The DL lies medially to the CGD and extends to the midthalamic level (Repérant, 1978).

Several authors have described a retinal projection to the DL (Fig. 3.2). Huber and Crosby (1926) had already described the retinal terminals in normal *Alligator mississippiensis*. Degeneration techniques applied respectively to *Emys orbicularis* (Kosareva, 1967) and *Chelydra serpentina* and *Podocnemis unifilis* (Knapp and Kang, 1968a, 1968b) reported evidence of a retinal projection. More recently, Cruce and Cruce (1978) provided radioautographic data of a retinal projection to DL in *Tupinambis teguixin*. A variety of tracing techniques (radioautography, degeneration, and HRP) also demonstrated this projection in *Ophisaurus apodus* (Kenigfest et al., 1986). Whereas Ulinski and Nautiyal (1988) do not include the DL among the primary visual centers of *Pseudemys* and *Chrysemys*, their illustrations show numerous visual terminals in the neuropil of this structure.

Several studies in a variety of species have documented massive projections of the DL onto the hippocampal cortex (Lohman and van Woerden-Verkley, 1978; Belekhova and Kenigfest, 1983; Gaidaenko, 1983; Bruce and Butler, 1984; Belekhova and Nemova, 1987; Hoogland and Vermeulen-van der Zee, 1988; Martinez-Garcia and Olucha, 1988), and others have shown that the archicortex (in particular the mediodorsal archipallium) is electrophysiologically responsive to visual stimulation (Gusel'nikov and Supin, 1964; Andry and Northcutt, 1976; Ivazov and Belekhova, 1982; Belekhova and Ivazov, 1983; Belekhova et al. 1983). These results jointly indicate the probable existence, in reptiles, of a retinothalamohippocampal pathway relayed through the DL, a pathway also described in amphibians (Vesselkin et al. 1971, 1978; Northcutt and Kicliter, 1980) and suspected to occur in mammals (Conrad and Stumpf, 1975; Itaya et al., 1981, 1986; Pakhamova et al., 1987; Repérant, Weidner, Pakhamova, et al., 1987).

3. PRETECTAL NUCLEI
a. NUCLEUS LENTIFORMIS MESENCEPHALI   The fusiform anterior portion of the nucleus lentiformis mesencephali (LM) lies immediately caudal to the habenular nuclei, inclined medially, and bounded dorsally by the marginal optic tract and ventrally by the N. geniculatus pretectalis. Caudally, its medial portion becomes progressively thinner and is eventually reduced to a triangular mass of cells adhering externally to the optic tract and enveloped internally by the N. griseus tectalis. The LM is composed principally of large, fusiform, deeply stained cells mingled with a few smaller and less basophilic cells (see Fig. 3.4B). It is particularly stable in all the macrophthalmic reptiles

and receives a pronounced contingent of optic fibers arising, for the most part, from the lateral optic bundle (see Figs. 3.2 and 3.4A). It is largest in snakes, and in some families of lizards (Lacertidae, Scincidae, Cordylidae, Gekkonidae), but poorly developed in most turtles (Repérant, 1978).

Gisi (1908) first described the LM in *Sphenodon punctatus*, and most authors subsequently found it. Although Huber and Crosby (1933a) suggested that the optic tract and the mesencephalic lentiform nucleus of *Anolis* are connected, there are no similar comments in the classic literature. Nevertheless, experimental methods adopted after 1950 have demonstrated that retinal ganglion cells terminate in this structure (see Fig. 3.2).

b. Nucleus Geniculatus Pretectalis

The nucleus geniculatus pretectalis (GP) spreads out beneath the N. lentiformis mesencephali. Caudally, as the central part of the LM disappears, the GP merges dorsally into the N. griseus tectalis. Its structure is comparable to that of GLv, with a medial pars magnocellularis poorly supplied with visual terminals and a lateral pars molecularis, which receives the majority of the visual projections (see Figs. 3.3A and 3.4). These two subdivisions are, however, more difficult to establish in snakes (Repérant, 1978).

Bellonci (1888) first identified the nucleus as the N. thalami posterior; Ramón (1896) subsequently identified it as the nucleo medio; Edinger (1899), Gisi (1908), and de Lange (1913) identified it as the nucleus pretectalis. Beccari (1923) termed it the nucleo genicolato pretectale, and Huber and Crosby (1926) described it as the nucleus geniculatus pretectalis. Other neuroanatomists have defined cytoarchitectural entities in the pretectum that topographically resemble the GP: Papez (1935), N. pretectalis dorsalis; Curwen and Miller (1939), N. pretectalis dorsomedialis; or Warner (1955) and Kosareva (1967), N. pretectalis partes dorsalis et ventralis. Cairney (1926, *Sphenodon punctatus*) was the first to consider the GP as a primary visual center, and numerous experimental studies beginning with Armstrong's (1950, *Lacerta vivipara*) investigation have confirmed this conclusion (see Fig. 3.2).

c. Nucleus Griseus Tectalis

The N. griseus tectalis (GT) appears first as a rectangular band of widely dispersed small cells behind the main body of the lentiform nucleus (see Fig. 3.4B). It lies obliquely across the anterior margin of the tectum. More caudally, the dorsomedial part of the nucleus progressively disappears, leaving a ventral ovoid mass of cells (see Figs. 3.3A and 3.4A). This structure, clearly seen in *Caiman crocodilus* and in all lizards, is noticeably less differentiated in snakes. The visual fibers, arising predominantly from the medial optic bundle, arborize throughout the nucleus. It is possible that this nucleus corresponds to the N. superior of Ramón (1896), but it has always been considered as a zone through which fibers from the medial fascicle of the optic tract pass without termination.

One of us (Repérant, 1972, 1973a) has described this nucleus as the N. pretectalis, referring to a pretectal visual center as previously defined by Butler and Northcutt (1971) in *Iguana* and *Anolis*. Cruce and Cruce (1978) and Schroeder (1981a) later adopted the same term to describe the nucleus in *Tupinambis teguixin* and *Crotalus viridis*, respectively. However, this nucleus is more comparable to the N. griseus tectalis described in a variety of avian species (Repérant, 1973b; Raffin, 1974, 1976, 1977; Repérant and Raffin, 1974; Raffin et al., 1977). The latter term is based on subsequent investigations (Repérant, 1975, 1978; Repérant and Rio, 1976), as well as on a consideration of topographical, cytoarchitectural, and hodological findings and a reevaluation of the data of Butler and Northcutt (1971). Furthermore, it appears that the N. pretectalis, which Butler and Northcutt (1971) describe as being devoid of visual terminals, is situated more medially and more caudally than the visual N. griseus tectalis. We have described this latter nucleus in *Vipera aspis* (Repérant and Rio, 1976) and in several species of turtles, lizards, and snakes (Repérant, 1978). The nucleus, which has occasionally been either confused with or included within the N. lentiformis mesencephali, appears to be a constant pretectal visual structure in reptiles (Fig. 3.2).

d. NUCLEUS POSTERODORSALIS

The N. posterodorsalis (PD) is small and rounded, rich in neuropil, and contains a small group of poorly stained cells. The nucleus lies in the anterodorsal pretectal region at the level of the posterior commissure (Fig. 3.7). The PD has been identified in the chameleon (Shanklin, 1926), as the N. dorsalis commissurae posterior and in *Lacerta agilis* (Kuhlenbeck, 1931) as the N. posterodorsalis. The structure has since been described, generally as the PD, in most of the species examined. However, Papez (1935), Curwen and Miller (1939), and Hall and Ebner (1970a, 1970b) call it the area pretectalis. The N. posterodorsalis of Curwen and Miller (1939) and of Warner (1955) corresponds to the N. lentiformis of Beccari (1923).

Huber and Crosby (1926, *Anolis carolinensis*) described a small bundle of retinal fibers that appear to arborize in the PD. The first experimental findings only partially confirmed this observation. Using Nonidez preparations of *Natrix natrix*, Armstrong (1951) identified terminal degeneration in the contralateral PD; however, this nu-

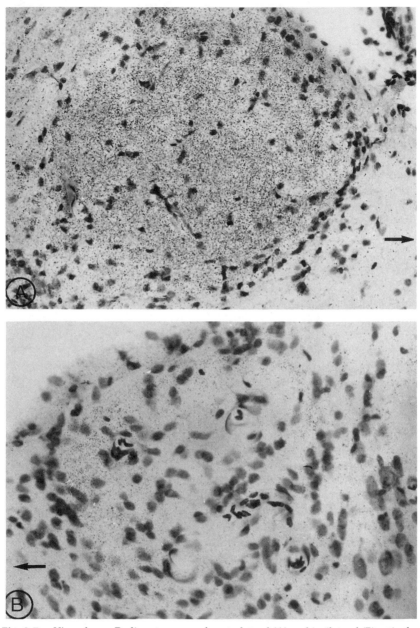

Fig. 3.7. *Vipera berus.* Radioautograms of contralateral (A) and ipsilateral (B) retinal projections in the nucleus posterodorsalis following right intraocular injection of 3H-proline (20μli). Transverse sections; arrows indicate the medial side (A, 200 x; B, 400 x). Abbreviations: SFGS, stratum griseum et fibrosum superficiale; SGC, stratum griseum centrale; SOP, stratum opticum.

cleus lacked any trace of degenerating visual fibers in *Lacerta vivipara* (Armstrong, 1950). Similarly, Nauta preparations of *Caiman crocodilus* (Burns and Goodman, 1967) and of *Chelydra serpentina* and *Podocnemis unifilis* (Knapp and Kang, 1968a, 1968b) failed to reveal any terminal degeneration of visual fibers. However, following retinal ablation, Fink-Heimer techniques show terminal degeneration in the PD in *Pseudemys scripta* (Hall and Ebner, 1970a, 1970b), *Iguana iguana, Anolis carolinensis* (Butler and Northcutt, 1971), *Tupinambis teguixin* (Ebbesson et al., 1972), *Gekko gecko* (Northcutt and Butler, 1974a, 1974b), *Xantusia vigilis* (Butler, 1974), *Thamnophis sirtalis* (Halpern and Frumin, 1973), *Nerodia sipedon* (Northcutt and Butler, 1974a, 1974b), and *Vipera aspis* (Repérant, 1973a). The radioautographic technique and the orthograde transport of HRP in optic fibers have confirmed a retinal projection to the PD in several species (Repérant, 1975; Cruce and Cruce, 1975, 1978; Repérant and Rio, 1976; Butler and Northcutt, 1978; Northcutt, 1978; Repérant, 1978; Bass and Northcutt, 1981a, 1981b; Schroeder, 1981; Kenigfest et al., 1986; Lemire et al., 1988; Torcol, 1988, Fig. 3.2).

4. TECTUM OPTICUM

a. GENERAL   Cell bodies and fibers in the reptilian optic tectum are arranged in concentric layers or laminae between the pial and ventricular surfaces of the brain. Reptilian taxa differ markedly in the relative development of individual laminae and in the number of laminae (Huber and Crosby, 1933a, 1933b; Leghissa, 1962; Senn, 1966, 1970b; Senn and Northcutt, 1973; Northcutt, 1978, 1984a). These differences have compounded the problem of describing tectal anatomy, and several systems of nomenclature have been proposed (see reviews in Anthony, 1970, and Northcutt, 1984a). Two nomenclatures proposed by Ramón (1896) and Huber and Crosby (1933a) have received particular attention.

Ramón (1896, *Chamaeleo*) recognized 14 tectal layers, which he numbered from the ventricular to the pial surface. This nomenclature was subsequently adopted by many investigators (de Lange, 1913; Huber and Crosby, 1926, 1933a, 1933b; Shanklin, 1930; Senn 1966, 1968a, 1968b, 1970a, 1970b; Butler and Northcutt, 1971, 1973; Senn and Northcutt, 1973; Butler, 1974; Northcutt and Butler, 1974a, 1974b; Butler and Ebbesson, 1975; Cruce and Cruce, 1975, 1978; Foster and Hall 1975; Quiroga, 1978; Northcutt, 1978, 1984b; Bass and Northcutt 1981a, 1981b; Ebbesson and Karten, 1981; Peterson, 1981).

Huber and Crosby (1933a) proposed regrouping these 14 layers in six strata, ordering them from the pia to the ventricle: (1) a stratum opticum (SO; Layer 14); (2) a stratum fibrosum et griseum superficiale

(SFGS; Layers 13 to 8); (3) a stratum griseum centrale (SGC; Layer 7); (4) a stratum album centrale (SAC; Layer 6); (5) a stratum griseum periventriculare (SGP; Layers 5 and 4); and (6) a stratum fibrosum periventriculare (SFP; Layers 3 to 1). This nomenclature, which has the advantage of being readily transposable to other vertebrate groups (Huber and Crosby, 1933b; Repérant, 1978), has also been widely adopted (Ariëns-Kappers et al., 1936; Armstrong, 1950, 1951; Burns and Goodman, 1967; Kosareva, 1967; Knapp and Kang, 1968a, 1968b; Davydova and Gonchareva 1970; Davydova, 1971; Repérant, 1972, 1973a, 1975, 1978; Davydova and Smirnov, 1973; Halpern and Frumin, 1973; Davydova et al., 1976; Repérant et al., 1978, 1981; Schechter and Ulinski 1979; Schroeder, 1981a, 1981b; Dacey and Ulinski, 1986a, 1986b, 1986c, 1986d, 1986e; Kenigfest et al., 1986; Lemire et al., 1988; Torcol, 1988).

The retinotectal projection was clearly established in reptiles by the end of the 19th century (Bellonci, 1888; Ramón 1896); most authors describe it as occurring essentially in the superficial layers. Whereas this region differs substantially among the different reptilian groups, Repérant (1978) proposed a system of nomenclature inspired by that of Huber and Crosby (1933a) that may be applied to the different layers and strata of the tectum in all reptiles.

b. Stratum Zonale (Layer 1')

The stratum zonale layer is always extremely well differentiated in snakes (Huber and Crosby, 1933a; Ariëns-Kappers et al., 1936; Armstrong, 1951; Senn, 1966, 1970a, 1970b; Repérant, 1972, 1973a, 1978; Halpern and Frumin, 1973; Senn and Northcutt, 1973; Repérant and Rio, 1976; Repérant, Peyrichoux, and Rio, 1981; Schroeder 1981a, 1981b; Northcutt, 1984; Dacey and Ulinski, 1986a, 1986b, 1986c, 1986d, 1986e) and receives a rich retinal projection in these species (Repérant and Rio, 1976; Repérant, 1978; Schroeder, 1981a; Dacey and Ulinski, 1986d; Fig. 3.8B). Whereas a very thin s. zonale has been described in some lizards, turtles, and crocodilians (Huber and Crosby, 1933a; Armstrong, 1950; Kosareva, 1967; Burns and Goodman, 1967; Knapp and Kang, 1968a, 1968b; Repérant, 1978), Senn (1966, 1970b) and Northcutt (1984a) consider this layer to be a tectal structure characteristic of snakes.

c. Stratum Opticum (Layer 1)

The stratum opticum (SO) corresponds to the principal incoming layer of visual fibers to the tectum and appears to be most highly organized in snakes (see Figs. 3.3B, 3.4A, 3.8, 3.9, and 3.10). Recalling the fascicular organization of the optic nerve (Ward et al., 1987), the SO of these reptiles is formed by 140 to 250 (depending on the spe-

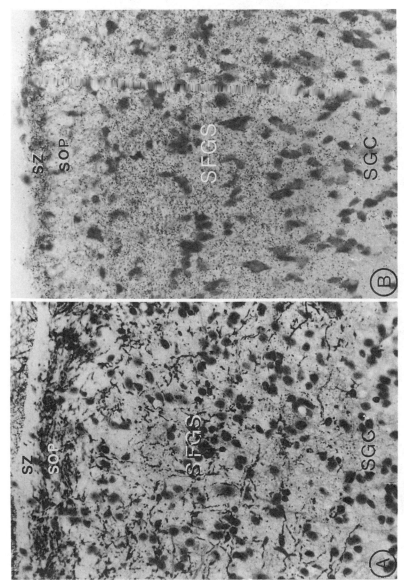

Fig. 3.8. *Vipera aspis.* (A) Contralateral retinotectal projections demonstrated by the Fink-Heimer technique 26 days following retinal ablation. Note the absence of staining of degenerated fibers and boutons in the stratum zonale (SZ). (B) Radioautograph of the contralateral retinotectal projections 24 hours following the intraocular injection of 3H-proline (20μli). Note the accumulation of silver grains in the stratum zonale (SZ) and the stratum fibrosum et griseum superficiale and the weak labeling of the stratum zonale. Transverse sections (400 x).

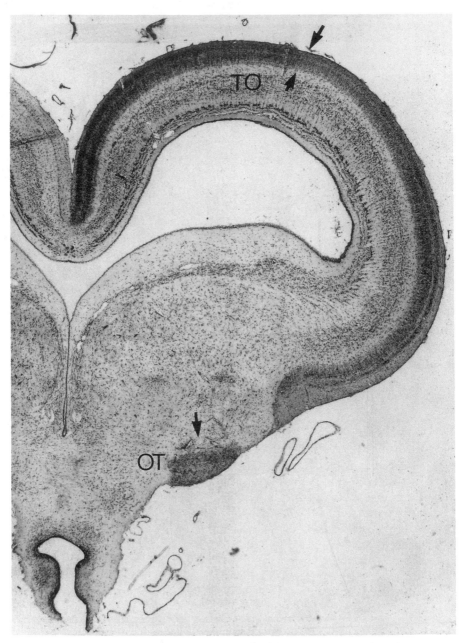

Fig. 3.9. *Uromastyx acanthinurus*. Radioautographic labeling at the contralateral posterior mesencephalic level of the retinotectal projection (zone indicated by two arrows) and the retinotegmentomesencephalic (arrow) 72 hours following the left intraocular injection of a mixture of 3H-fucose and 3H-proline (25μli). Transverse section (29 x). Abbreviations: OT, nucleus opticus tegmenti; TO, tectum opticum.

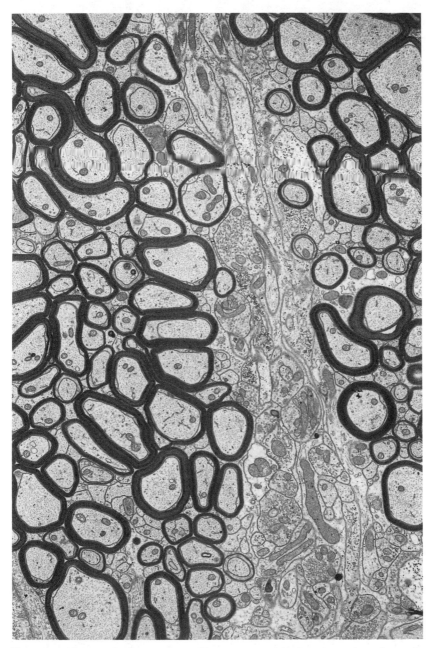

Fig. 3.10. *Vipera aspis.* Optic tectum. Ultrastructure of the stratum opticum (normal material) at the level of two fascicles. Most of the optic fibers are myelinated. The bridge of neuropil separating the two fascicles connecting the stratum fibrosum et griseum superficiale contains numerous axonal terminals making synapse. Transverse section (10,000 x).

cies) bundles of axons, which are for the most part myelinated and clearly separated by bridges of neuropil formed mainly by dendritic prolongations and axonal ramifications of the cells situated in underlying layers (Schroeder, 1981a, 1981b; Repérant, Peyrichoux, and Rio, 1981; Dacey and Ulinski 1986d).

d. Stratum Fibrosum et Griseum Superficiale

Two major architectural types can be distinguished in the stratum fibrosum et griseum superficiale (SFGS). One of these (Type I)—typical of turtles, crocodilians, and lizards—is characterized by a pronounced lamination in which plexiform and cellular layers alternate. The other (Type II), typical of snakes, is poorly organized and lacks precise stratification.

e. Type I

The narrow, plexiform layer L2 extends beneath the s. opticum; in some species this layer contains a considerable number of cells. The thickness of the cellular layer L3 varies both among species and tectal regions. Layer L4 is wide and plexiform, whereas layer L5 forms a narrow (occasionally a single row) band of cells. The last layer, L6, is thin and plexiform (see Figs. 3.3A, 3.4A, and 3.9).

Investigations carried out with Golgi techniques, mainly in lizards, have demonstrated the morphological diversity of the neuronal types found in SFGS (Ramón, 1896; Ramón y Cajal, 1911; de Lange, 1913; Huber and Crosby, 1933a; Leghissa, 1962; Butler and Ebbesson, 1975; Quiroga, 1978; Schechter and Ulinski, 1979). It appears that the Type I SFGS in all reptiles examined is characterized by horizontal cells that are exclusively intrinsic (interneurons) to the superficial tectal zone. In addition, vertically oriented fusiform, pyramidal, and polygonal neurons form short axon connections within the superficial zone or project to the central zone. In some lizard species, some visual fibers penetrate the tectum by way of the layer L3 (Quiroga, 1978; Northcutt, 1984a).

In Golgi impregnated pieces of optic tecta of *Chamaeleo*, Ramón (1896) identified three sheets of retinal arborizations in layers L2, L4, and L6 of the chameleon. With reduced silver and Weigert stains, Huber and Crosby (1926, 1933b) described retinal projections to all layers between L2 and the outermost layer of the SGC (L7) in different species.

The experimental studies of the late 1950s are contradictory. Whereas some authors (Butler and Northcutt, 1971; Butler, 1974; Northcutt and Butler, 1974b; Cruce and Cruce, 1975) agree with Huber and Crosby (1926, 1933a, 1933b), others (Armstrong, 1950; Repérant, 1978; Repérant et al., 1978; Quiroga, 1978; Bass and Northcutt, 1981b; Kenigfest et al., 1986; Torcol, 1988) report that retinal fibers terminate only in layers L2 to L6. Knapp and Kang (1968a, 1968b) note that retinotectal fibers in *Chelydra* and *Podocnemis* arborize mainly in layers L2, L3, and L4, whereas Kosareva (1967) and Davydova and Smirnov (1973) report retinal terminals in *Emys orbicularis* even in deeper laminae of the central and periventricular zones. The axonal fragments observed by Kosareva (1967) were probably artifacts due to the inadequate suppression of normal fibers in Nauta-paraffin preparations. Other hodological techniques (HRP and radioautography) fail to show any retinal projection to layers deeper than L6 in this species (Torcol, 1988).

f. TYPE II
The five subdivisions of the SFGS are not apparent, but three more or less well-characterized zones can be identified (see Fig. 3.8). A wide and more or less plexiform layer, containing cell bodies, extends beneath the s. opticum; it possibly corresponds to a particularly well-developed layer L2. Beneath this zone lies a thick band of cells in which the neurons are generally packed in pairs or triads. Under the electron microscope (Fig. 3.11), they display many cell membrane appositions without any obvious synaptic differentiation (Repérant, Peyrichoux, and Rio, 1981). These cells, generally fusiform and of variable size, typically give off radially oriented apical and basal dendritic trunks (Repérant, Peyrichoux, and Rio, 1981; Dacey and Ulinski 1986b). Most of these cells are efferent and project to the N. rotundus, the GLv, the N. isthmi, and the brainstem (in *Thamnophis sirtalis;* Dacey and Ulinski, 1986b). Stellate cells and horizontal interneurons also occur (Repérant, Peyrichoux, and Rio, 1981; Dacey and Ulinski, 1986c). This central cellular region most likely corresponds to an agglomeration of the layers L3, L4, and L5 described previously and is succeeded, without any clear boundary, by a relatively narrow plexiform layer containing cell bodies; the latter layer is probably equivalent to the layer L6 (see Fig. 8.8). The three-layered organization of the SFGS is most clearly seen in modern snakes (Caenophidians), whereas in primitive snakes (Henophidians) the SFGS appears much less well organized (Repérant, 1978).

All experimental studies carried out in snakes (Repérant, 1972, 1973a, 1978; Halpern and Frumin, 1973; Northcutt and Butler, 1974a; Repérant and Rio, 1976; Schroeder, 1981a; Dacey and Ulinski, 1986d) report that layers 1 and 3 of the triple-layered SFGS contain most terminals. In *Thamnophis sirtalis*, the optic arborizations labeled with HRP (Fig. 3.12) are ellipsoidal, with their long axes oriented mediolaterally and their short axes oriented rostrocaudally (Dacey and Ulinski, 1986d). These arborizations vary both in their overall size (from

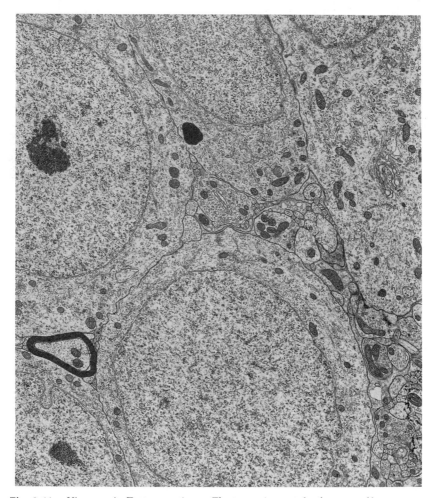

Fig. 3.11. *Vipera aspis.* Tectum opticum. Electron micrograph of stratum fibrosum et griseum superficiale showing five clustered small somata with scanty cytoplasm. Note direct apposition of membranes between them and almost complete absence of glial covering of these cells, which are bordered by thin neural profiles (9,800 x).

45 to 150 μm) and in the diameter of their parent axons (from less than 0.5 to 3.0μm). In *Vipera aspis,* the neuropil of the stratum fibrosum et griseum superficiale is made up of small nerve elements, including three types of profiles containing synaptic vesicles:

1. Boutons with pleomorphic synaptic vesicles (P), representing over 47% of the total population of profiles containing synaptic vesicles and comprising three subgroups (P_1, P_2, and P_3).

2. Boutons with spheroidal synaptic vesicles (S), forming more than 29% of the total population of profiles containing synaptic vesicles and comprising two categories, S_1 and S_2 (S_2, the more numerous,

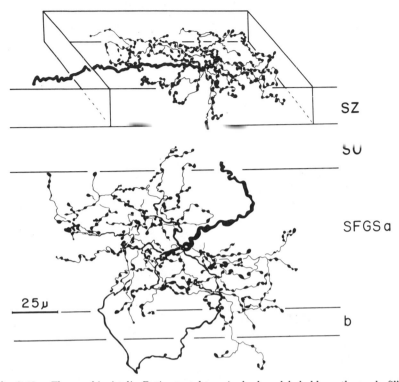

Fig. 3.12. *Thamnophis sirtalis*. Retinotectal terminal arbors labeled by orthograde filling from extracellular injection of HRP into the optic tract. Abbreviations: SFGSa, sublayer a of the stratum griseum et fibrosum superficiale; SO, stratum opticum; SZ, stratum zonale. (From Dacey and Ulinski, 1986).

represents the optic boutons, which make up 22% of the total population of profiles containing synaptic vesicles).

3. Dendrites with pleomorphic vesicles, accounting for approximately 23% of the total population of profiles containing synaptic vesicles (Figs. 3.13 and 3.14). A study of synaptic patterns revealed a large number of serial synapses and a lesser number of triplets or triadic synapses. The presynaptic components are boutons containing spheroidal (S_1, S_2) or pleomorphic (P_1, P_2, P_3) synaptic vesicles. The intermediate profile was always a dendrite with synaptic vesicles, which frequently belonged to an interneuron (Repérant, Peyrichoux, and Rio, 1981). An immunochemical study has shown that a large proportion of P boutons and dendrites containing synaptic vesicles are GABAergic (Rio et al., 1988; Fig. 3.14B).

Though little information is available regarding the electrophysiological properties of the diencephalic and tegmental visual centers, those related to the optic tectum are well documented in several spe-

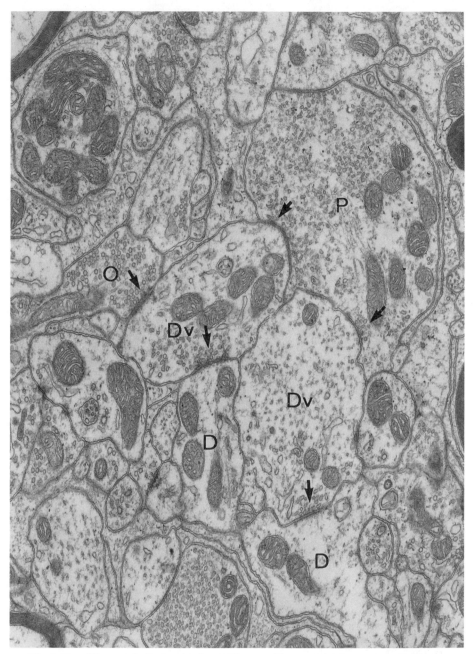

Fig. 3.13. *Vipera aspis.* Tectum opticum. Stratum fibrosum et griseum superficiale. The center of this electron micrograph exhibits three serial arrangements (arrows). Dendrites containing synaptic vesicles (Dv) are contacted by a bouton containing pleomorphic vesicles (P) and an optic bouton (O). This Dv synapses on a dendritic profile (25,500 x). Abbreviations: D, dendrites; Dv, dendrite containing synaptic vesicles.

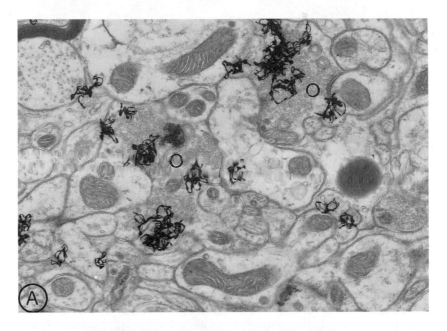

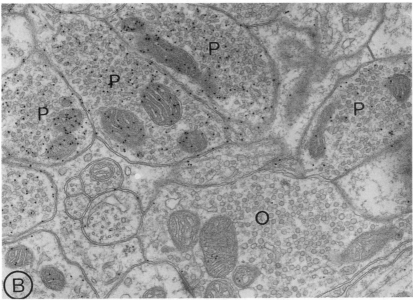

Fig. 3.14. *Vipera aspis*. Tectum opticum. Stratum fibrosum et griseum superficiale. (A) High-resolution radioautograph showing labeled optic boutons (O) 24 hours following contralateral intraocular injection of 3H-proline. (B) Post embedding immunoelectron microscopic photograph depicting the presence of densely GABA-immunopositive profiles containing pleomorphic synaptic vesicles. Note the mitochondrial GABA vesicles for the most labeled profiles. The bottom part of the picture is occupied by an irregularly shaped optic terminal devoid of labeling. (A, 12,000 x; B, 28,000 x). Abbreviations: O, optic boutons; P, bouton containing pleomorphic vesicles.

cies (*Alligator mississippiensis*, Heric and Kruger, 1965; turtles—*Emys orbicularis*, Zagorul'ko, 1967; Belekhova and Kosareva, 1971; Gusel-'nikov et al., 1970; Boiko and Davydova, 1973; Boiko and Gonchareva, 1976; *Agrionemys horsfieldii*, Belekhova and Kosareva, 1971; Boiko and Gonchareva, 1976; *Chelonia mydas mydas*, Granda and Stirling, 1965; Granda and O'Shea, 1972; and *Pseudemys scripta*, Robbins, 1972. Snakes: *Crotalus viridis*—Hartline et al., 1978; Kaas et al., 1978 and *Agkistrodon sp.*, Terashima and Goris, 1975. Lizards: *Iguana iguana*—Gaither and Stein, 1979; Stein and Gaither, 1981, 1983). The nasotemporal axis of the retina is oriented along the rostrocaudal axis of the tectum, and the dorsoventral axis corresponds to the mediolateral tectal axis as shown by electrophysiological studies in the alligator (Heric and Kruger, 1965), turtle (Gusel'nikov et al. 1970), and crotaline snakes (Terashima and Goris, 1975; Hartline et al., 1978). In *Iguana*, on the other hand (Gaither and Stein 1979; Stein and Gaither, 1981), these axes are rotated through a right angle so that the nasotemporal retinal axis corresponds to the lateromedial tectal axis, and the dorsoventral visual axis is oriented along the rostrocaudal tectal surface.

Electrophysiological investigations in *Iguana*, *Crotalus*, and *Alligator* reveal that the tectal maps of the retina show varying degrees of nonuniform retinal magnification (Gaither and Stein 1979; Terashima and Goris, 1975; Heric and Kruger, 1965). In *Iguana*, the central 10° of the visual field occupies approximately 20% of the tectal surface; in *Crotalus*, the frontal quadrant of the visual field occupies a disproportionally large fraction of the tectal surface; and in *Alligator*, the magnified region lies just below the equator and posterior to the visual axis.

It has been reported that visually sensitive units in the reptilian tectum are sensitive to changes in intensity or wavelength of the stimulus (Heric and Kruger, 1966; Zagorul'ko 1968; Granda and O'Shea, 1972; Robbins, 1972). The functional properties of visually sensitive cells in different tectal layers may be inferred by comparing the depth of the electrode from the tectal surface and the histological measurement of the depths of various laminae. In turtles, lizards, and snakes, the neurons of the SFGS give only excitatory responses to visual stimuli—the size of their receptive fields varying from 5° to 20° (Boiko and Davydova, 1973; Boiko and Gonchareva, 1976; Robbins, 1972; Kaas et al., 1978; Gaither and Stein, 1979). In deeper layers of the tectum, the response changes with increasing depth. As the electrode is advanced farther into the tectum, the receptive fields become larger, the sensitivity to movement increases, the habituation of the response becomes more pronounced, a greater proportion of units is binocularly sensitive, and progressively more units are multimodally sensitive (Granda and Stirling, 1965; Belekhova and Kosareva, 1971; Robbins, 1972; Boiko and Davydova, 1973; Boiko and Gonchareva, 1976; Hartline et al., 1978; Gaither and Stein, 1979; Stein and Gaither, 1981).

In lizards and snakes, the tectal receptive field units are typically round or elliptical (Hartline et al., 1978; Stein and Gaither, 1981); some interesting data concerning the electrophysiology of tectal units are provided by Boiko and Davydova (1973) and Boiko and Gonchareva (1976), who compared two species of turtles. In the aquatic species *Emys orbicularis*, many tectal units have highly elongated receptive fields, uncommon in the terrestrial *Agrionemys horsefieldii*. The latter chows many units with large receptive fields sensitive to stationary stimuli. Furthermore, *Emys* has a greater proportion of tectal units with small receptive fields that are directionally sensitive than does *Agrionemys*. It has been suggested that the preferential sensitivity to large, stationary regions of contrast might be related to the vegetarian habits of the terrestrial *Agrionemys*, whereas the differential sensitivity to elongated, moving objects might be related to the aquatic habitat and predatory mode of life of *Emys*.

5. MESENCEPHALIC TEGMENTUM (NUCLEUS OPTICUS TEGMENTI)

The basal or accessory optic system has long been known. The retinal terminal zone in the ventral mesencephalic tegmentum, the N. opticus tegmenti (OT) (also known as the N. ectomammilaris or the N. opticus basalis), was first described in *Varanus* (Edinger, 1899). Thereafter, neuroanatomists working with normal material rediscovered this system in many reptiles (de Lange, 1913; Beccari, 1923; Cairney, 1926; Huber and Crosby, 1926, 1933a; Shanklin, 1930, 1933; Frederikse, 1931; Papez, 1935; Ariëns-Kappers et al., 1936). Experimental studies performed after the 1950s confirm these earlier observations and demonstrate a prominent basal optic system in most macrophthalmic species (Armstrong, 1950, 1951; Burns and Goodman, 1967; Kosareva, 1967; Knapp and Kang, 1968a, 1968b; Hall and Ebner, 1970a; Butler and Northcutt, 1971; Repérant, 1972, 1973a, 1975, 1978; Ebbesson et al., 1972; Braford, 1973; Halpern and Frumin, 1973; Butler, 1974; Northcutt and Butler, 1974a, 1974b; Platel et al., 1975; Repérant and Rio, 1976; Cruce and Cruce, 1978; Kubie and Allen, 1978; Repérant et al., 1978; Bass and Northcutt, 1981a, 1981b; Ebbesson and Karten 1981; Schroeder, 1981a; Rio et al., 1983; de la Calle et al., 1986; Kenigfest et al., 1986; Rio, 1987; Lemire et al., 1988; Torcol 1988).

The tractus opticus basalis (TOB) emerges from the TOM at the posterior thalamic level, courses caudally on the ventrolateral surface of the brain, and terminates in the nucleus opticus tegmenti (see Figs. 3.7 and 3.9). The TOB is mainly composed of large-diameter myelinated fibers (Shanklin, 1933; Burns and Goodman, 1967; Ebbesson and Karten, 1981; Reiner, 1981; Rio et al. 1983; Rio, 1987). The OT located on the ventrolateral surface of the mesencephalic tegmentum, rostrally to the third nerve, lies between the nucleus of Beccari (1923), anteriorly, and the nucleus lemniscus lateralis of Huber and Crosby

(1926), posteriorly. Three cell types have been described (Beccari, 1923; Shanklin, 1933; Reiner and Karten, 1978, Rio et al., 1983; Rio, 1987): (1) small ovoid cells (27% in *Vipera aspis*); (2) medium-size elongated cells (35% in *Vipera aspis*); and (3) large multipolar cells (44% in *Vipera aspis*). A large proportion of small cells are GABAergic (*Vipera aspis*, Rio, 1987). A detailed electron microscopic analysis has been carried out on the neuropil of the OT in *Vipera aspis* (Rio et al. 1983; Rio, 1987). Six types of profiles containing synaptic vesicles (PCSVs) have been identified. Those containing spheroidal synaptic vesicles (S) comprise two classes: S_1, optic terminals, which are the most numerous (48.5% of the total PCSVs), and S_2, axon terminals (19%). On the basis of density of flattened synaptic vesicles (F), FPCVs have been subdivided into three types: F_1 (9%), F_2 (8%), and F_3 (15%). Whereas F_1 and F_2 are axonal in nature, F_3 correspond to dendrites containing synaptic vesicles. These latter profiles participate in serial or, more rarely, triadic arrangements within glomerulus-like structures in which they are postsynaptic to optic terminals. A very low proportion of axon terminals (0.5%) represents a population of mixed large granular vesicles. Occasionally, gap junctions have been observed between axon terminals and dendrites. Following partial lesions of the basal optic tract, the turtle *Chrysemys picta* shows reduced optokinetic nystagmus (OKN) at high pattern velocities (Fite et al., 1979). These results lend some support to the hypothesis that the basal optic nucleus is involved in visuomotor behavior associated with the OKN.

C. Conclusions

This brief review of the literature has shown that the macrophthalmic reptiles share a common plan of organization of the primary visual system, which consists in a single hypothalamic center (N. suprachiasmaticus); certainly three thalamic centers (NN. ventrolateralis and geniculatus lateralis pars ventralis, together with the dorsal lateral geniculate complex) and possibly a fourth one (N. dorsolateralis anterior); four pretectal centers (NN. geniculatus pretectalis, lentiformis mesencephali, posterodorsalis and griseus tectalis); the superficial layers of the optic tectum; and the N. opticus tegmenti (see Fig. 3.2).

III. VARIATIONS OF THE FUNDAMENTAL PLAN AND THEIR IMPLICATIONS
A. Overview

There are two major departures from the visual plan thus described. The plan may be complicated by the addition of an ipsilateral projection or more or less drastically simplified (in microphthalmic reptiles).

The structural detail of certain primary visual centers shows more subtle variations of the fundamental plan; the optic tectum of snakes, for example, shows an architecture differing from the general pattern. These variations have led to a wide variety of interpretations.

B. The Ipsilateral Visual Projections

Classic authors, working on normal specimens, have generally agreed that the optic chiasm crosses completely in reptiles (Ramón y Cajal, 1911; Cairney, 1926; Huber and Crosby 1926; Shanklin, 1930; Ariëns-Kappers et al., 1936; Warner, 1947). However, in most squamates (*Uromastyx acanthinurus* is an exception; Repérant et al., 1978), experimental methods have shown an ipsilateral projection (Armstrong, 1950, 1951; Ebbesson, 1970; Butler and Northcutt, 1971; Ebbesson et al., 1972; Repérant, 1972, 1973a, 1978; Halpern, 1973; Halpern and Frumin, 1973; Northcutt and Butler, 1974a, 1974b; Butler, 1974; Cruce and Cruce, 1975, 1978; Platel et al., 1975; Repérant and Rio, 1976; Repérant et al., 1978; Ebbesson and Karten, 1981; Schroeder, 1981a; Kenigfest et al., 1986; de la Calle et al., 1986). An ipsilateral projection has been observed neither in the Rhynchocephalia (Northcutt et al., 1974) nor in crocodilians (Braford, 1973; Repérant, 1975). The situation in turtles is unclear. Some authors consider the decussation to be complete (Knapp and Kang, 1968a, 1968b; Hall and Ebner, 1970a), whereas others consider it to be partial (Kosareva, 1967; Bass and Northcutt, 1981a, 1981b; Repérant, 1978; Künzle and Schnyder, 1983; Lemire et al., 1988; Torcol, 1988; Ulinski and Nautiyal, 1988).

The ipsilateral projection in reptiles is generally of modest importance and varies with the size and location of the projection zones. In most squamates and many turtles, the ipsilateral visual fibers project to the dorsal lateral geniculate complex and the N. posterodorsalis (see references below and Figs. 3.2, 3.5, and 3.7). These projection sites, always more extensive in snakes (see references below) than in lizards, remain in burrowing snakes (Halpern, 1973; Repérant, 1978; Repérant et al., 1987a). Sometimes, there are ipsilateral projections to the N. geniculatus lateralis pars ventralis (Armstrong, 1950; Butler and Northcutt, 1971; Halpern and Frumin 1973; Butler, 1974; Northcutt and Butler, 1974a, 1974b; Cruce and Cruce 1975, 1978; Repérant, 1978; Bass and Northcutt, 1981a, 1981b; Lemire et al., 1988), the N. geniculatus pretectalis (Butler, 1974; Northcutt and Butler, 1974a, 1974b; Cruce and Cruce, 1978), the N. lentiformis mesencephali (Repérant, 1973; Butler, 1974; Northcutt and Butler, 1974a, 1974b; Cruce and Cruce, 1978; Repérant, 1978), the N. griseus pretectalis (Schroeder, 1981a), the N. ventrolateralis pars ventralis (Cruce and Cruce, 1978), and to the superficial layers of the optic tectum (Repérant,

1972, 1973a, 1978; Butler, 1974; Northcutt and Butler, 1974a, 1974b; Repérant and Rio, 1976; Cruce and Cruce, 1975, 1978; Bass and Northcutt, 1981a; Lemire et al., 1989). More rarely, ipsilateral optic fibers have been reported in the basal optic nucleus (Repérant, 1973a; Repérant and Rio, 1976; Repérant, 1978; Schroeder, 1981a; Bass and Northcutt, 1981a; Rio et al., 1983; Rio, 1987; Torcol, 1988).

The absence of an ipsilateral contingent in *Sphenodon* and crocodilians raises obvious questions about the origin of this system. Does it represent a recent development that appeared in certain reptilian radiations (squamates and turtles), or is it a primitive character that has been maintained in certain lines? All amphibian species that have been examined experimentally possess well-developed ipsilateral visual projections (Lazar, 1969, 1978; Jakway and Riss, 1972; Scalia and Fite, 1974; Scalia, 1976; Guillery and Updyke, 1976; Repérant, 1978). An ipsilateral component has also been described in other anamniote groups: dipnoans (Northcutt, 1980), selachians (Northcutt and Wathey, 1980; Repérant et al., 1986), teleosteans (Repérant et al., 1976; Ebbesson and Ito, 1980), holosteans (Northcutt and Butler, 1976), chondrosteans (Repérant et al., 1982), polypterids (Repérant et al., 1981), and cyclostomes (Kennedy et al., 1977; Kosareva, 1980; Vesselkin et al., 1980). This suggests that the ipsilateral visual projections reflect a pattern characteristic of the earliest vertebrates; their absence in *Sphenodon* and crocodilians is probably a secondary condition.

It has recently been proposed that lizards may be subdivided into two groups on the basis of some neuroanatomical specializations (Senn, 1970a; Northcutt, 1972, 1978; Platel, 1975). Group I are lacertomorphs (lacertids, scincids, cordylids, gerrhosaurids, gekkonids, xantusids, and anguids), and Group II are the dracomorphs (agamids, iguanids, chameleonids, teiids, and varanids). Most lizards show a moderate ipsilateral visual contingent, the major targets of which are the dorsal lateral geniculate complex and the N. posterodorsalis. The nocturnal predatory lizards of Group I also have a feeble ipsilateral retinotectal projection (*Gekko gecko*, Northcutt and Butler, 1974b; *Xantusia vigilis*, Butler, 1974; *Tarentola mauritanica*, Repérant et al., 1978). Northcutt and Butler (1974b) consider this ipsilateral retinotectal projection possibly to be specific to Group I lizards or to be an adaptive character related to predation in dim light. However, all Group I lizards do not share this ipsilateral retinotectal projection; it is absent in the lacertids *Lacerta vivipara* (Armstrong, 1950) and *Acanthodactylus boskianus* (Repérant et al., 1978), the scincid *Scincus scincus* (Platel et al., 1975; Repérant et al., 1978), and the gerrhosaurid *Zonosaurus madagascarensis* (Repérant, 1978). This projection has also been described in the Group II teiid *Tupinambis teguixin* (Cruce and Cruce, 1978). Furthermore, whereas some nocturnal predators (*Gekko, Xan-*

tusia, and *Tarentola*) have ipsilateral retinotectal projections, other cre-
puscular forms lack it (*Sphenodon punctatus,* Northcutt et al., 1975), as
does the euryphotic *Scincus scincus* (Platel et al., 1975). It has also been
reported in *Tupinambis teguixin,* which is a diurnal predator (Cruce
and Cruce, 1978). Hence, the presence or absence of this projection
does not correspond either to the distinction between Group I and
Group II lizards or to the distinction between nocturnal and diurnal
predators; its significance in lizards remains obscure.

In mammals, the proportion of uncrossed visual fibers increases in
parallel with the size of the binocular visual field and is thus related
to the development of stereoscopic vision (Walls, 1942; Duke-Elder,
1958). This does not seem to be the case for reptiles (Repérant, 1978).
For example, in *Podarcis muralis* the ipsilateral projection is well developed and the binocular field extends through 18°, whereas the binocular field of the turtle *Chelydra serpentina* covers 38° and its ipsilateral
contingent is weak. At least crocodilians do not possess an ipsilateral
visual contingent, although their binocular fields range between 24°
and 26°, depending on the species (Walls, 1942).

C. Microphthalmia and Adaptation
Microphthalmia represents an adaptation to underground life. In the
squamates, it is found in the most diverse families (Walls, 1942; Angel
and Rochon-Duvigneaud, 1942; Rochon-Duvigneaud, 1943, 1970;
Senn, 1966; Underwood, 1970, 1977; Senn and Northcutt, 1973). With
the exception of amphisbaenians, the more or less substantial reduction of the peripheral optic tract is rarely accompanied by degenera-
tive changes in the retina (Angel and Rochon-Duvigneaud, 1942;
Rochon-Duvigneaud, 1943, 1970; Bonin, 1965; Underwood, 1977).
Different investigators have studied the brains of microphthalmic reptiles (Senn, 1966, 1968b, 1968c, 1969, 1970a; Senn and Northcutt, 1973;
Masai, 1973). In general, these authors report a more or less severe
atrophy of the primary visual centers and in particular the superficial
layers of the optic tectum, which shows a thinning and loss of lamination. In contrast, the intermediate and deeper layers of the optic
tectum always retain a relatively remarkable development. The N. rotundus, which is the visual thalamic relay for the tectofugal tract, also
tends to disappear. Whatever the systematic position of the species
under question, the degree of atrophy is related to the reduction in
size of the eyeball. These observations of normal material are confirmed by the experimental findings of Halpern (1973), Repérant
(1978), and Repérant et al. (1987a), whose results add the necessary
details to discern the organization of the visual pathways of microphthalmic reptiles.

The African *Calabaria reinhardti* (Repérant, Miceli, Rio, and Weidner,

1987b), a henophidian boid, lives mostly in burrows. Its eyes are somewhat reduced in size but perfectly visible and are covered by a bulging ocular scale. The optic nerve is small (200 μm) and composed of 150 to 170 tiny fascicles, each containing 30 to 50 axons, most of which are unmyelinated. Radioautography shows contralateral retinal projections to four thalamopretectal centers (GLv, GLd, GP, and PD), to the superficial layers of the tectum (SZ, SO, SFGS), and to the mesencephalic tegmentum (OT). With the exception of PD, these centers show pronounced signs of atrophy. The GLv and GP are poorly developed and lack cytoarchitectural details; neither the pars magnocellularis nor the pars molecularis can be discerned. Moreover, the GLv fades medially without any distinct transition into the VL. The GLd is reduced and shows no precise cytoarchitectural differentiation. The pretectal LM and GT, present in macrophthalmic reptiles, have disappeared in *Calabaria*. The superficial layers of the tectum, to which the retina projects, are very thin and imprecisely laminated. The intermediate tectal layers (SGC and SAC) are also proportionally less thick than in surface-dwelling snakes. The thickness of the superficial (SZ, SO, SFGS) and intermediate (SGC, SAC) layers represent respectively 17% and 27% of the total tectum thickness, whereas in the surface-dwelling *Naja haje* these are 30% and 43%, respectively. In contrast, the deeper strata (SGP, SFP) are relatively more pronounced in *Calabaria*. In addition, in this species the intermediate and deep strata are well separated from one another, whereas in macrophthalmic snakes they are more or less fused. Finally, in the latter snakes the SGP is small and forms a single broad layer, whereas it is thick and multilayered in *Calabaria*. The nucleus opticus tegmenti in this species shows marked atrophy and appears as a small mass containing a few cell bodies (Fig. 3.15).

In addition, *Calabaria* has a relatively dense ipsilateral contingent of visual fibers that project to the GLd, PD and the superficial layers of the optic tectum (Fig. 3.15).

The markedly microphthalmic scolecophidian *Typhlops vermicularis* (Halpern, 1973) shows an even greater regression of the primary visual system in Fink-Heimer preparations. Contralateral retinal projections are restricted to three levels, localized respectively in the thalamus (probably equivalent to the GLd), pretectum (PD), and tectum (the superficial layers). The PD is relatively well developed; the optic thalamic area is small and poorly differentiated cytoarchitecturally. The superficial tectal layers, which receive a weak projection, are considerably thinned and lack lamination; they represent approximately 8% of the total thickness of the tectum. *Typhlops* has an ipsilateral retinal projection to the visual thalamic area and to the PD.

It thus appears that in reptiles, microphthalmia induces the regres-

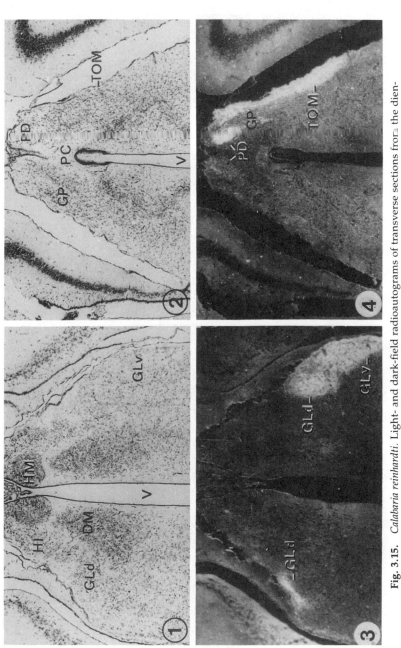

Fig. 3.15. *Calabaria reinhardti:* Light- and dark-field radioautograms of transverse sections from the diencephalon (1) to the mesencephalon (8) showing the retinal projections 48 hours following left intraocular injection of 3H-proline (18μli) (50 x). Abbreviations (A): DM, nucleus dorsomedialis; GLd, nucleus geniculatus pars dorsalis; GP, nucleus geniculatus pretectalis; Hl, nucleus habenularis lateralis; Hm, nucleus habenularis medialis; PD, nucleus posterodorsalis; TOM, tractus opticus marginalis. (B): nu-

Fig. 3.15 *(cont.)*

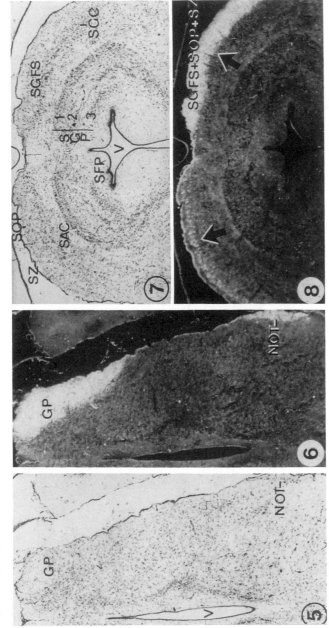

cleus geniculatus pretectalis; SAC, stratum album centrale; SFP, stratum fibrosum periventriculare; SGC, stratum griseum centrale; SFGS, stratum fibrosum et griseum superficiale; SGP, stratum griseum periventriculare; SOP, stratum opticum; SZ, stratum zonale.

sion or disappearance of the retinotectal and retinotegmental systems, whereas its effects on the retinothalamic and retinopretectal systems (especially PD) are less apparent. Investigations of the primary visual system in other naturally microphthalmic species (agnathan, *Eptatretus burgeri*, Kusunoki and Amemiya, 1983; teleost, *Astyanax hubbsi*, Voneida and Sligar 1976; caecilians, *Typhlonectes compressicauda*, Clairambault et al., 1980, and *Ichthyophis kohtaoensis*, Himstedt and Manteuffel 1985; mammal, *Talpa europaea*, Lund and Lund, 1965) have provided similar results. Finally, it is interesting to note that microphthalmia, at least in *Typhlops* and *Calabaria*, is not associated with any reduction of the ipsilateral visual component.

The modification of the fundamental reptilian pattern of organization of the primary visual system observed in microphthalmic reptiles is difficult to interpret. On the one hand, very few behavioral data concerning the visual capacities of these animals are available; on the other hand, the data obtained in macrophthalmic species give only fragmentary and sometimes contradictory evidence as to the precise functional role of the different components of the visual system (for a review see Repérant, 1978; Belekhova, 1979; Peterson, 1980). In spite of these difficulties, two hypotheses can be advanced to account for the alterations associated with microphthalmia in reptiles. The first possibility is that the functions of the retinothalamic and retinopretectal systems are more important and more useful to these subterranean species than are those of the retinotectal and retinotegmental systems. Alternatively, the retinothalamic and retinopretectal systems may be no longer functional and are thus vestigial components of the basic macrophthalmic organization. In this case, the reduction of the retinotectal and retinotegmental systems in subterranean forms may reflect their importance in surface-dwelling, macrophthalmic reptiles.

IV. EVOLUTIONARY SIGNIFICANCE OF SOME ASPECTS OF THE PRIMARY VISUAL SYSTEM OF SNAKES

Whereas the overall organization of the primary visual system of macrophthalmic snakes agrees with the general plan described previously, some morphological features are characteristic of this suborder. While some primary visual centers (NN. lentiformis mesencephali, posterodorsalis, griseus tectalis, and opticus tegmenti) are in all aspects comparable to their homologues in other reptilian suborders, other primary visual centers show remarkable structural particularities. Thus, the dorsal geniculate complex of snakes is clearly more developed than that of other reptiles, and one can observe a relative dedifferentiation of the GLv and GP, the partes molecularis and magnocellularis no longer appearing very distinct. Most strikingly, the architecture of the ophidian optic tectum is extremely idiosyncratic. The pronounced five-layered stratification of the SFGS typical of turtles, crocodilians, and lizards (Type I architecture) is replaced in all surface-dwelling snakes by a Type II architecture in which the SFGS is poorly differentiated and only three layers can be recognized. This impoverishment of the superficial tectal layering of snakes, which has been noted by many authors, is interpreted by some (Senn, 1966, 1970a; Senn and Northcutt, 1973; Northcutt, 1984a) as the result of the disappearance of Ramón's Layer 14. In addition, both SGC and SAC have a tendency to merge into a single tectal layer in snakes, whereas they are clearly separated in other reptiles. Finally, in snakes the N. rotundus, the thalamic relay of the tectofugal visual pathway, is displaced toward the posterior thalamus and appears less well developed than in other reptiles (Dacey and Ulinski, 1983). The question arises whether these specializations in snakes reflect a primitive state or a more recent acquisition; one might expect that the prehistory of snakes might answer this question.

The phylogenetic origin and history of snakes have provided the grounds for abundant controversy. One of the major difficulties arises from the paucity of Mesozoic fossils. The abundant Tertiary fossils are also too close to extant forms to be informative, and Cretaceous specimens are rare except for a few ribs and vertebrae (Hoffstetter, 1955, 1962, 1968; Romer, 1968; Rage, 1976, 1984). Prior to the Cretaceous, palaeontologists resort to speculation. In the absence of clear palaeontological evidence, many authors have attempted to formulate theories of the origin of snakes. Most agree that snakes, lizards, and amphisbaenians have a common ancestor. However, the snakes show a number of special features (see Hoffstetter, 1968; Guibé, 1970; Gasc, 1973; Rage, 1976, 1984; Estes and Pregill, 1988), the diverse interpretations of which provide the basis for continuing controversy.

It is possible that snakes derive from Jurassic lacertilians. Some authors (Cope, 1869, 1875; Boulenger, 1893; Nopsca, 1923; Camp, 1923; McDowell and Bogert, 1954; Porter, 1972) have argued that they separated from the varanids or platynotids of the Secondary epoch, whereas Underwood (1957) rejects this theory and brings out affinities between snakes and certain gekkonids (pygopodids). Finally, Senn and Northcutt (1973) argue for an origin among a group of burrowing lizards related to the dibamids, the feylinids, or perhaps the anelytropsids.

Several authors (Mahendra, 1938; Walls, 1942; Bellairs, 1949; Bellairs and Underwood, 1951; Underwood, 1957, 1970) have suggested that snakes may derive from burrowing lizards. After adaptation to subterranean life, the ancestral forms would have become progressively snakelike, perhaps similar in form to modern scolecophidians (typhlopids and leptotyphlopids), and showed a similar degree of involution of the peripheral visual system.

For reasons that are far from clear, these subterranean reptiles re-colonized on the surface, and this change of habitat was accompanied by a number of morphological changes, the most notable of which in this context was the redevelopment of macrophthalmic eyes. The earliest surface-dwelling species, placed with the henophidians, are assumed to have been little specialized and gave rise to the larger and more diverse caenophidians.

This theory is best illustrated by Walls (1942) on account of the differences between the eyes of snakes on the one hand and lizards on the other (see also Baird, 1970). In snakes, the eyeball is spherical, and there is no inflexion at the junction of the sclera and the cornea, no pericorneal bony ring being present. They lack a cartilaginous layer, such as the lacertilian sclera, in which the envelope of the eyeball is entirely fibrous. The ophidian lens is spherical and yellow, and the ciliary muscles have disappeared. Schlemm's canal lies in the sclera of lizards, but in the cornea of snakes. The muscles of the ophidian iris have a mesodermal origin; those of lizards have an ectodermal one and are involved both in pupillary adjustment and accommodation by hernia of the lens. The papillary cone, an ectodermal structure of lizards, is either absent in snakes or replaced by a rudimentary papillary cone of mesodermal origin or a hyaloid vascular network similar to that of teleosts and amphibians. The mobile eyelids of most lizards have disappeared and are replaced by a single fixed transparent scale.

The ophidian retina also differs from that of lizards. Whereas lizards typically have a cone retina, that of snakes may contain only cones (as in the diurnal elapids), rods (as in *Hypsiglena* and *Trimorphodon*), or a mixture of both (as in boids). The rods and cones may be single or double, and the cones contain neither an oil droplet nor a paraboloid (Walls, 1942; Underwood, 1970).

Beginning with these anatomical peculiarities, Walls (1942) envisaged a series of transformations that led from the atrophied eye of the "preserpents," analogous to the modern amphisbaenians, to the macrophthalmic eye of modern snakes. Initially, as the ancestral forms began a subterranean life, the pupils shrank and the photoreceptor cells lost their oil droplets and paraboloids. After adaptation to life in the dark, the apparatus of accommodation disappeared, together with the pericorneal bony ring and the cartilaginous component of the sclera. As the eyeball shrank, it became spherical, the iris lost its motility, and the Schlemm's canal became blocked. Simultaneously, a protective brille covered the cornea. In their return to the surface, these animals needed to reconstruct a peripheral visual system. The vestigial photoreceptor cells differentiated once more into rods and cones. As the eyeball enlarged, the ciliary body and the lens, which retained their initial microphthalmic features, were no longer in contact and a mechanism of accommodation was developed by the migration of the vestigial ciliary body into the iris, which henceforth played the double role of a pupil and a means of accommodation. The retina regained its principal role, its vasculature redeveloped, the mesodermal papillary cone reappeared and was replaced by a more efficient hyaloid system. A new canal of Schlemm, localized in the cornea, reestablished relations with the venous system and contributed to the irrigation of the iris.

Provocative as it might appear, this theory was well received both by palaeontologists (e.g., Romer, 1956, 1968) and a number of anatomists (Bellairs, 1949, 1972; Bellairs and Underwood 1951; Underwood, 1957, 1970; Senn and Northcutt, 1973), who sought additional evidence in its favor. Thus Bellairs (1972), for example, mentions additional morphological characteristics of snakes that are compatible with a subterranean history, such as the loss of the limbs, the closed and rigid cranial capsule, the disappearance of structures such as the temporal arch, the interorbital septum and the external ear, and the exaggerated development of the olfactory bulb, the Jacobson's organ, and the Harderian gland.

Nevertheless, other authors (Hoffstetter, 1955; Dowling, 1959; Gasc, 1973) do not accept this point of view and argue that snakes originated much earlier, near the period of divergence of the lacertilian and amphisbaenian lines of the Squamata. In this view, some characteristics of snakes, such as the platytraby of the chondrocranium (similar to that of some fossil crossopterygians; Hoffstetter, 1955), the pattern of articulation among ribs and vertebrae, and similarity of the structure of the eyeball of amphibians, are regarded as primitive characteristics that were fixed early in the divergence of the ophidian group. In addition, Gasc (1967, 1968, 1970, 1974) has argued that snakes may not have been derived from subterranean forms but from surface-dwelling tetrapods. Gasc and Renous (1976) point out that the loss of the limbs is seen in other squamates, and different theories have attempted to account for the mechanism that leads to the rudimentation of limbs. It is thus possible that the absence of limbs in snakes is not necessarily pertinent to Walls' (1942) hypothesis. Some other evidence that he presents can also be used against his hypothesis. For example, the brille equivalent to that found in snakes is also seen in some surface-dwelling lizards (Rochon-Duvigneaud, 1970).

It has therefore been argued that the earliest legless ophidians colonized either the surface of grassland or the superficial strata of the litter lying on the surface (Hoffstetter, 1968; Gasc, 1973). This group, corresponding to the henophidians, gave rise to the more specialized caenophidians, some of which (xenopeltids, aniliids, uropeltids) adapted a burrowing mode of life. The scolecophidians diverged early from this ancestral trunk and also took to a burrowing way of life. The subterranean forms, in this view, represent a recent highly specialized adaptation, rather than the ancestral forms of the adaptive radiation of surface-dwelling snakes (Gasc, 1973).

Whatever their origin, either recently from a subterranean group of lizards or from the earliest squamate stock, snakes are undeniably squamates and their relation to the lizards is clear. Nevertheless, the marked differences in structure of the eyeball and central visual projections between snakes and lizards are sufficiently pronounced that they do not suggest close relationship.

The architecture of the lacertilian eyeball and primary visual centers exhibits many similarities to those of turtles and crocodilians; in all three groups the delineation between the partes magnocellularis and molecularis of the GLv and GP is clear-cut, and the optic tectum is always well laminated. The genealogy of the reptiles suggests that these characteristics were fixed early, very likely in the forms ancestral to these three lines. The peculiarities of snakes are thus either a new tendency seen in a single reptilian line and maintained since its divergence from the lacertilian squamates, or, as Walls (1942) proposed, they result from a slow process of adaptation. If the snake condition represents a new tendency, we should recall that a new type of eyeball appeared with the snakes; the new architecture of some of the primary visual centers then developed in correlation with it. However, the structural differences between the retinae of surface-dwelling snakes and lizards do not appear to be sufficiently profound to have led to so pronounced a series of changes in the projection centers.

Could the structural peculiarities of snakes be interpreted in terms of Walls' (1942) theory? Senn (1966, 1968c, 1969, 1970a), Senn and Northcutt (1973), Masai (1973), Repérant (1978), Repérant, Miceli, Rio, and Weidner, (1987), and Quiroga (1979) provide a large sample covering the entire range of squamate species from the surface-dwelling to the most obligate subterranean forms. Within this sample, whatever the systematic position of the species, the more the animal is adapted to subterranean life, the more pronounced are the changes in the primary optic tract. The optic tectum is the first visual center that begins to adapt to subterranean life. One finds a gradual thinning of the tectum and a progressive loss of its initial lamination, most noticeable in the superficial SFGS. The deeper layers, on the other hand, increase in their relative size. In the thalamus and pretectum, the separation between the partes magnocellularis and molecularis of the GLv and GP becomes progressively less distinct. Extreme microphthalmia, whatever the species in question, is associated with the disappearance of the LM, GP, GT and, above all, the N. opticus tegmenti. In contrast, the lateral geniculate and the posterodorsalis nuclei remain unaffected. Finally, the N. rotundus, a thalamic relay of the tectofugal visual pathway, shrinks to the point of disappearance in some obligate subterranean forms.

This series of material provides numerous examples of the effect of partial sensory deprivation on central visual structures and raises questions about the morphogenetic influence of visual fibers on central visual structures. Experimental embryological investigations, mainly carried out in birds, provide pertinent evidence (La Vail and Cowan, 1971a, 1971b; Kelly and Cowan, 1972; Raffin, 1972, 1974, 1976, 1977; Crossland et al., 1974, 1975; Raffin and Repérant, 1975; Crossland, 1979; Peduzzi and Crossland, 1983a, 1983b). Early ablation of the optic vesicle leads both to a thinning of the tectum and the virtually total disappearance of lamination of its superficial layers. Moreover, the nucleus of the basal optic root does not differentiate, and the molecular and magnocellular parts of GLv and GP fuse (Raffin and Repérant, 1975; Raffin, 1977). In addition, the volume of N. rotundus decreases (Raffin, 1976, 1977).

In surface-dwelling snakes, the structural organization of GLv and GP, together with the reduction in size of N. rotundus, both resemble some of the changes seen in cavernicolous forms and are also comparable to the preceding observations. However, macrophthalmic snakes differ from subterranean forms with respect to other characteristics such as the importance of the NN. opticus tegmenti, lentiformis mesencephali, and the overdevelopment of the dorsal geniculate complex. Walls' (1942) hypothesis thus forces us to accept that after a pronounced modification of some structures (TO, GLd, GLv, GP) and the disappearance of others (GT, LM, OT) consecutive to microphthalmia, the reconstruction of the eyeball led to the redeployment of the first set of centers and the reappearance of the second set. This problematical hypothesis has been reconsidered by Senn and Northcutt (1973), who consider that the Scolecophidians (particularly microphthalmic forms) are the true archetypes of surface-dwelling snakes. Their main argument comes from the fact that the deeper layers of the optic tectum of typhlopids and leptotyphlopids show typically lizardlike architectural features. The plurilaminar SGC of the Scolecophidians, seen also in all surface-dwelling and subterranean lizards, is in their view a primitive character that they interpret as testimony to the descent of these burrowing snakes from lizards. This may, however, be no more than the result of convergence; more "recent" subterranean snakes (henophidians and boids, Repérant, 1978; caenophidians and elapids, Masai, 1973) show the same arrangement.

These contradictory arguments lead us to the following hypothesis. The ophidian eyeball is not a reconstruction of an initially microphthalmic, involuted eyeball resembling that of the amphisbaenians but corresponds to the appearance, probably very early in phylogeny, of a new reptilian ocular architecture. Nevertheless, on the basis of the modification of some of the primary visual centers, it is highly likely that these reptiles passed through a period of subterranean life. This led to a reduction in the size of the eyeball and the consequent involution of centers such as the optic tectum. The return to the surface involved a redevelopment of the eyeball and an expansion of the primary visual centers, with the thalamofugal pathway rather than the tectofugal pathway hypertrophying. The imprecise architecture of the ophidian optic tectum and the GLv and GP remain as evidence of the period of semiunderground existence of earlier snakes. As Hoffstetter (1968) has suggested, support for Walls' (1942) hypothesis may come from Nopsca's (1923) suggestion that the first snakes burrowed in mud.

V. CONCLUDING REMARKS

The present review covers over a century of research on the visual system of reptiles, and it shows how the long and difficult morphological analysis of this system has been influenced by various technical advances. For more than 50 years the use of traditional histological methods produced no more than modest results. However, the recent availability of experimental hodological techniques has led to considerable advances in our knowledge of the reptilian visual system. The comparative analysis has revealed a plan of organization that is common to all presently existing groups of surface-dwelling reptiles (turtles, crocodilians, lacertilians, and snakes). Whatever the taxonomic position of the species in question, the retina projects contralaterally to one hypothalamic center (N. suprachiasmaticus); three thalamic centers (NN. geniculatus lateralis pars ventralis and ventralis lateralis, together with the dorsal lateral geniculate complex) and possibly to a fourth (N. dorsolateralis anterior); four pretectal centers (NN. geniculatus pretectalis, lentiformis mesencephali, posterodorsalis, and griseus tectalis); the superficial layers of the optic tectum; and to one tegmentomesencephalic center (N. opticus tegmenti).

The establishment of this common plan from anatomical data depends on the observation of structural similarities, according to a wide variety of criteria. This leads immediately to the question of whether these structural similarities result from a common ancestry, convergent evolution, or random accident. In other words, is the considerable structural similarity of the primary visual centers among groups of reptiles due to homology (inheritance from a common ancestry) or homoplasy (the development of similar characters from different preexisting characters; see Wiley, 1981; Northcutt, 1984b; Gans, 1985)?

Before attempting to answer this question, it should be borne in mind that the species examined, although taxonomically diverse, belong to a single phylogenetic category deriving from the palaeozoic amphibians. The different radiations probably arose from a common ancestral reptilian stock, the captorhinomorph cotylosaurs of the Carboniferous (see, in particular, Olson, 1977).

Assuming the optic tracts and visual centers to be homoplasic, one argues that a single pattern has been repeatedly derived from different morphological archetypes, or that convergent evolution has led to the adoption of essentially the same pattern of organization in all living reptiles. The classic explanation of convergent evolution invokes a selective process imposed by comparable environments. It is evident that the selective pressures that have been applied to the visual system differ greatly among reptilian groups. Under such conditions one would expect initial ancestral differences to be amplified rather than converging toward a final common pattern.

The possibility that the similarity described is merely accidental can therefore be discarded, as can the possibility of convergent evolution. We thus conclude that the similarities observed were not acquired secondarily, but rather reflect the retention of ancestral traits. It follows that the organization of the contralateral pathways and centers of living macrophthalmic reptiles corresponds to a basic plan that probably arose in the precursor organisms of the Upper Carboniferous.

We have also seen that the general plan of organization can be modified in two ways. It may be complicated by the addition of an ipsilateral visual contingent, the size and the distribution of which vary considerably from one group to another, or it may be simplified to the point of partial disappearance in highly microphthalmic, cavernicolous reptiles.

Whereas the study of the variation of the components of the primary visual system has provided us with a wealth of data, a more detailed and exhaustive analysis of this variation in a larger sample of species remains to be undertaken. Such an analysis could provide a better understanding of how each of the primary visual centers has evolved in the different groups of reptiles or, in the vocabulary of cladistic analysis, the degree to which these structures show the retention of ancestral or plesiomorphic characteristics or disclose newly derived apomorphic characteristics. In this context, the comparative analysis of the optic tectum has already yielded some information that has led, notably in snakes, to the formulation of interesting hypotheses concerning the evolution of this structure.

One comparative analysis that has yet to be undertaken in reptiles is that of the dorsal lateral geniculate complex (CGD), the relay of the thalamofugal visual pathway. Most recent studies indicate that this structure is extremely variable in reptiles. On the other hand, this structure has acquired considerable importance in mammals with the development of the visual cortex. It is evident that the CGD is a key structure in the evolution of the visual system in those reptiles that gave rise to mammals. Whereas no living forms are closely related to the therapsids, the detailed comparative analysis of the reptilian CGD should enable us to acquire a better understanding of the evolutionary history of the mammalian thalamofugal visual relay.

ACKNOWLEDGMENTS

This work was supported by INSERM, CNRS, MINHN (Actions Spécifiques), FCAR, and cooperative France-Québec research grants. The authors thank S. Arnold, D. Le Cren, M. Médina, and F. Roger for their excellent technical assistance.

APPENDIX: REPTILIAN SPECIES DISCUSSED

(Studies using experimental techniques are indicated by an asterisk.)

TESTUDINES

Agrionemys horsfieldii
  Belekhova, 1971*
  Davydova et al., 1976*
  Belekhova, 1979*
  Gaidaenko, 1983*
  Boiko and Gonchareva, 1976*
Amyda cartilaginea
  Torcol, 1988*
Caretta caretta
  Bass and Northcutt, 1981a*
Chelonia mydas
  Papez, 1935
  Granda and Stirling, 1965*
  Granda and O'Shea, 1972*
Chelydra serpentina
  Huber and Crosby, 1933
  Knapp and Kang, 1968a*
  Senn, 1971
Chinemys reevesii
  Torcol, 1988*
Chrysemys picta picta
  Hall et al., 1977*
  Reiner and Karten, 1978*
  Repérant, 1978*
  Fite et al., 1979*
  Bass and Northcutt, 1981b*
  Reiner, 1981*
  Heller and Ulinski, 1987*
Chrysemys picta bellii
  Rainey and Ulinski, 1986*

Chrysemys picta marginata
  Huber and Crosby, 1933
Cyclemys dentata
  Torcol, 1988*
Dermochelys coriacea
  Torcol, 1988*
  Lemire et al., 1988*
Emys orbicularis
  Belekhova, 1967*
  Bellonci, 1888
  Kosareva, 1967*
  Davydova and Gonchareva, 1970*
  Gusel'nikov et al., 1970*
  Belekhova, 1971*
  Boiko and Davydova, 1973*
  Davydova and Smirnov, 1973*
  Boiko and Davydova, 1975*
  Boiko and Gonchareva, 1976*
  Repérant, 1978*
  Belekhova, 1979*
  Gaidaenko, 1983*
  Stieda, 1875
  Torcol, 1988*
Graptemys pseudogeographica
  Huber and Crosby, 1933
Platysternon megacephalum
  Torcol, 1988*
Podocnemis unifilis
  Knapp and Kang, 1968b*

Sternotherus minor
  Senn, 1968b
Sternotherus odoratus
  Huber and Crosby, 1933
Testudo graeca
  Stieda, 1875
  Leghissa, 1962
  Repérant, 1978*
  Rio, 1987*
  Torcol, 1988*
Trachemys scripta (= Pseudemys)
  Curwen and Miller, 1939
  Hall and Ebner, 1970a
  Robbins, 1972*

Foster and Hall, 1975*
Hall et al., 1977*
Schechter and Ulinski, 1979
Reiner, 1981*
Künzle and Schnyder, 1983*
Ouimet et al., 1985*
Ulinski, 1986a
Heller and Ulinski, 1987*
Trachemys scripta elegans
  Huber and Crosby, 1933
  Rainey and Ulinski, 1986*
  Ulinski and Nautiyal, 1988*
Trachemys scripta scripta
  Sjöström and Ulinski, 1985*

CROCODILIA

Alligator mississippiensis
  Huber and Crosby, 1926
  Huber and Crosby, 1933b
  Heric and Kruger, 1965*
Caiman crocodilus
  Braford, 1973*
  Rio, 1987*
  Repérant, 1975*

Burns and Goodman, 1967*
Repérant, 1978*
Crocodylus niloticus
  Leghissa, 1962
Crocodylus porosus
  Addens, 1938

RHYNCHOCEPHALIA

Sphenodon punctatus
  Gisi, 1908
  Cairney, 1926

Durward, 1930
Northcutt et al., 1974*
Northcutt, 1978

SAURIA

Ablepharus kitaibeli
  Senn, 1970b
  Senn and Northcutt, 1973
  Northcutt, 1978
Acanthodactylus boskianus
  Repérant, 1978*
  Repérant et al., 1978*
Agama jayakari
  Northcutt, 1978
  Senn, 1970b
Agama mutabilis pallida
  Senn, 1970b
  Northcutt, 1978
Ameiva ameiva
  Northcutt, 1978
Anelytropsis papillosus
  Senn and Northcutt, 1973
  Northcutt, 1978

Anguis fragilis
  Senn, 1970b
  Senn and Northcutt, 1973
  Northcutt, 1978
Anniella pulchra
  Senn, 1968c
  Northcutt, 1978
  Senn, 1970b
Anolis carolinensis
  Huber and Crosby, 1933b
  Senn, 1970b
  Butler and Northcutt, 1971*
  Northcutt, 1978
Bachia intermedia
  Northcutt, 1978
Bachia trisanale (Ophignomon abendro-
    thi)
  Northcutt, 1978

Basiliscus basiliscus
  Northcutt, 1978
Callisaurus draconoides
  Northcutt, 1978
Calotes versicolor
  Northcutt, 1978
Chalcides chalcides
  Leghissa, 1962
  Senn, 1966
  Senn, 1970b
  Senn and Northcutt, 1973
  Northcutt, 1978
Chalcides ocellatus
  Senn, 1970b
  Senn and Northcutt, 1973
  Northcutt, 1978
Chamaeleo bitaeniatus
  Northcutt, 1978
Chamaeleo chamaeleon
  Ramón, 1896
  Shanklin, 1930
  Senn, 1970b
  Northcutt, 1978
Chamaeleo cristatus
  Repérant, 1978
Chamaeleo jacksonii
  Northcutt, 1978
Cnemidophorus sexlineatus
  Northcutt, 1978
Cnemidophorus tigris
  Northcutt, 1978
Coleonyx variegatus
  Northcutt, 1978
Cordylus cordylus
  Northcutt, 1978
Crotaphytus collaris
  Northcutt, 1978
Ctenosaura hemilopha
  Northcutt, 1978
Dibamus bourreti
  Senn and Northcutt, 1973
  Northcutt, 1978
Dipsosaurus dorsalis
  Northcutt, 1978
  Peterson, 1981*
Dracaena guianensis
  Northcutt, 1978
Draco sp.
  Northcutt, 1978

Draco volans
  de Lange, 1913
Elgaria multicarina
  Senn, 1970b
  Senn and Northcutt, 1973
  Northcutt, 1978
Enyaliosaurus quinquecarinatus
  Northcutt, 1978
Eumeces algeriensis
  Northcutt, 1978
Euspondylus rahmi
  Northcutt, 1978
Feylinia currori
  Senn and Northcutt, 1973
Gambelia wislizeni
  Northcutt, 1978
Gekko gecko
  Senn, 1970b
  Northcutt and Butler, 1974a*
  Andry and Northcutt, 1976*
  Butler and Northcutt, 1978*
  Northcutt, 1978
  Bruce and Butler, 1984*
Gerrhosaurus validus
  Northcutt, 1978
Gymnophthalmus speciosus
  Northcutt, 1978
Heloderma horridum
  Senn and Northcutt, 1973
  Northcutt, 1978
Heloderma suspectum
  Huber and Crosby, 1933b
  Northcutt, 1978
Holbrookia sp.
  Huber and Crosby, 1933b
Iguana iguana
  Senn, 1970b
  Butler and Northcutt, 1971*
  Foster and Hall, 1975*
  Butler and Northcutt, 1978*
  Northcutt, 1978
  Bruce and Butler, 1984*
Lacerta agilis
  de Lange, 1913
  Kuhlenbeck, 1931
Lacerta lepida
  Repérant, 1978*
Lacerta viridis
  Bellonci, 1888

Leghissa, 1962
Senn, 1970b
Senn and Northcutt, 1973
Northcutt, 1978
Lacerta vivipara
Frederikse, 1931
Ströer, 1939
Armstrong, 1950*
Northcutt, 1978
Lanthanotus borneensis
Senn and Northcutt, 1973
Northcutt, 1978
Leposoma parietale
Northcutt, 1978
Lialis burtonis
Senn and Northcutt, 1973
Northcutt, 1978
Mabuya vittata
Northcutt, 1978
Neusticurus ecpleopus
Northcutt, 1978
Ophiomorus latastii
Senn and Northcutt, 1973
Northcutt, 1978
Ophisaurus apodus
Ivazov and Belekhova, 1982*
Belekhova and Ivazov, 1983*
Belekhova and Kenigfest, 1983*
Belekhova et al., 1983*
Kenigfest et al., 1986*
Belekhova and Nemova, 1987*
Ophisaurus compressus
Northcutt, 1978
Ophisaurus ventralis
Senn, 1970b
Senn and Northcutt, 1973
Northcutt, 1978
Pantodactylus schreibersii
Quiroga, 1978
Pholidobolus montium
Northcutt, 1978
Phrynosoma cornutum
Northcutt, 1978
Phrynosoma coronatum
Huber and Crosby, 1933b
Phrynosoma platyrhinos
Northcutt, 1978
Physignathus lesueurii
Northcutt, 1978

Platysaurus intermedius
Northcutt, 1978
Podarcis hispanica
de la Calle et al., 1986*
Podarcis muralis
Bellonci, 1888
Beccari, 1923
Leghissa, 1962
Repérant, 1978*
Podarcis sicula
Senn, 1966
Senn, 1968b
Senn, 1968c
Senn, 1970b
Senn and Northcutt, 1973
Podarcis viridis
Northcutt, 1978
Franzoni and Fasolo, 1982
Prionodactylus argulus
Northcutt, 1978
Proctoporus bolivianus
Northcutt, 1978
Ptyodactylus hasselquisti
Northcutt, 1978
Repérant, 1978*
Sceloporus magister
Northcutt, 1978
Sceloporus undulatus
Northcutt, 1978
Scincella ladacense
Northcutt, 1978
Scincus scincus
Platel et al., 1975*
Repérant, 1978*
Repérant et al., 1978*
Sphaerodactylus cinereus
Northcutt, 1978
Tarentola mauritanica
Senn, 1966
Senn, 1970b
Northcutt, 1978
Repérant, 1978*
Repérant et al., 1978*
Rio, 1987*
Tropiocolotes tripolatanus
Repérant, 1978*
Tupinambis nigropunctatus
Northcutt, 1978
Senn, 1970b

Butler and Ebbesson, 1975
Cruce and Cruce, 1975*
Cruce and Cruce, 1978*
Lohman and Van Woerden-
  Verkley, 1978*
Repérant, 1978*
Ebbesson and Karten, 1981*
Typhlosaurus cregoi
  Senn and Northcutt, 1972
Uromastyx acanthinurus
  Repérant, 1978*
  Repérant et al., 1978*
  Rio, 1987*
Varanus bengalensis
  Northcutt, 1978
Varanus flavescens
  Senn, 1966
  Senn, 1970b
  Senn and Northcutt, 1973
  Northcutt, 1978
Varanus griseus
  Huber and Crosby, 1933b
  Repérant, 1978*

Varanus indicus
  Senn, 1966
  Senn, 1970b
  Senn and Northcutt, 1973
  Northcutt, 1978
Varanus niloticus
  Northcutt, 1978
Varanus salvator
  de Lange, 1913
Varanus sp.
  Edinger, 1899
Xantusia vigilis
  Butler, 1974*
  Northcutt, 1978
Xenosaurus grandis
  Northcutt, 1978
Zonosaurus madagascariensis
  Repérant, 1978*
  Repérant et al., 1978*

AMPHISBAENIA

Amphisbaena alba
  Repérant, 1978
Amphisbaena darwini heterozonota
  Quiroga, 1979

Agamodon anguliceps
  Bonin, 1965*

SERPENTES

Achalinus spinalis
  Masai, 1973
Acrochordus javanicus
  Senn and Northcutt, 1973
Agkistrodon contortrix mokeson
  Huber and Crosby, 1933b
Agkistrodon blomhoffi brevicaudatus
  Terashima and Goris, 1975*
Agkistrodon caliginosus
  Terashima and Goris, 1975*
Amphiesma vibakari
  Masai, 1973
Anomalepis aspinosus
  Senn and Northcutt, 1973
  Senn, 1969
Boa constrictor
  Huber and Crosby, 1933
  Senn, 1966

  Senn and Northcutt, 1973
  Repérant, 1978*
  Rio, 1987*
Brachyophidium rhodogaster
  Senn and Northcutt, 1973
  Senn, 1966
Bungarus fasciatus
  Senn and Northcutt, 1973
Calabaria reinhardtii
  Repérant, 1978*
  Repérant et al., 1987*
Cerastes cerastes
  Repérant, 1978*
  Rio, 1987*
Coluber gemonensis
  Senn and Northcutt, 1973
Crotalus adamanteus
  Warner, 1947

Crotalus viridis helleri
  Hartline et al., 1978*
  Kass et al., 1978*
  Schroeder, 1981*
Cylindrophis rufus
  Senn and Northcutt, 1973
  Senn, 1966
Dinodon orientalis
  Masai, 1973
Dinodon rufozonatus
  Masai, 1973
Elaphe climacophora
  Masai, 1973
Elaphe conspicillata
  Masai, 1973
Epicrates cenchria
  Senn, 1966
  Senn and Northcutt, 1973
Hypsiglena
  Walls, 1942
Leptotyphlops humilis
  Senn, 1966
  Senn and Northcutt, 1973
  Repérant, 1978
Liotyphlops albirostris
  Senn, 1969
  Senn and Northcutt, 1973
  Repérant, 1978
Malpolon monspessulana
  Senn, 1966
  Senn and Northcutt, 1973
Naja haje
  Repérant, 1978
Natrix natrix
  Bellonci, 1888
  Warner, 1942
  Armstrong, 1951*
  Warner, 1955
  Senn, 1966
  Senn, 1970a
  Senn and Northcutt, 1973
  Repérant, 1978
  Rio, 1987*
Natrix tesselata
  Senn and Northcutt, 1973
Natrix sp.
  Huber and Crosby, 1933

Nerodia sipedon
  Northcutt and Butler, 1974b*
Philothamnus semivariegatus
  Repérant, 1978*
Python molurus
  Huber and Crosby, 1933
Python reticularis
  Rio, 1987*
  Senn and Northcutt, 1973
Rhabdophis tigrina
  Masai, 1973
Rhamphotyphlops braminus
  Masai, 1973
Telescopus fallax
  Repérant, 1978*
Thamnophis sirtalis
  Huber and Crosby, 1933b
  Halpern and Frumin, 1973*
  Dacey and Ulinski, 1983*
  Dacey and Ulinski, 1986a–e*
Trimorphodon
  Underwood, 1970
Typhlops sp.
  Halpern, 1973*
Typhlops vermicularis
  Senn, 1966
  Senn and Northcutt, 1973
  Repérant, 1978*
Vipera aspis
  Senn, 1966
  Repérant, 1972*
  Repérant, 1973a*
  Senn and Northcutt, 1973
  Repérant and Rio, 1976*
  Repérant, 1978*
  Repérant et al., 1981*
  Rio et al., 1983*
  Rio, 1987*
  Ward et al., 1987
  Rio et al., 1988*
Vipera berus
  Repérant, 1978*
  Rio, 1987*
Xenopeltis unicolor
  Senn, 1966
  Senn and Northcutt, 1973

REFERENCES

Addens, J. L. (1938). The presence of a nucleus of Bellonci in reptiles and mammals. *Proc. Kon. Ned. Akad. Wet. Amst.* 41, 1134–1145.

Andry, M. L., and Northcutt, R. G. (1976). Telencephalic visual responses in the lizard *Gekko gecko. Neurosci. Abstr.* 2, 176.

Angel, F., and Rochon-Duvigneaud, A. (1942). Contribution à l'étude des yeux chez les Sauriens et les Ophidiens fouisseurs. *Bull. Mus. Hist. Nat.*, Ⴌ ᴍᴍ 3, 14, 163–166 (1re note) and 255–260 (?me note),

Anthony, J. (1970). Le névraxe des reptiles. In *Traité de Zoologie* (P. P. Grassé, ed.). Masson, Paris, 14, Vol. 2, pp. 203–332.

Ariëns Kappers, C. U., Huber, G. C., and Crosby, E. C. (1936). *The Comparative Anatomy of the Nervous System of Vertebrates, including Man.* Macmillan, New York.

Armstrong, J. A. (1950). An experimental study of the visual pathways in a reptile (*Lacerta vivipara*). *J. Anat.* (London) 84, 146–167.

Armstrong, J. A. (1951). An experimental study of the visual pathways in a snake (*Natrix natrix*). *J. Anat.* (London) 85, 275–288.

Baird, I. L. (1970). The anatomy of the reptilian ear. In *Biology of the Reptilia* (C. Gans and T. S. Parson, eds.). Academic Press, London, Vol. 2, pp. 193–275.

Bass, A. H., and Northcutt, R. G. (1981a). Primary retinal targets in the Atlantic loggerhead sea turtle, *Caretta caretta. Cell. Tiss. Res.* 218, 253–264.

Bass, A. H., and Northcutt, R. G. (1981b). Retinal recipient nuclei in the painted turtle, *Chrysemys picta:* an autoradiographic and HRP study. *J. Comp. Neurol.* 199, 97–112.

Bass, A. H., Andry, M. L., and Northcutt, R. G. (1983). Visual activity in the telencephalon of the painted turtle, *Chrysemys picta. Brain Res.* 263, 201–210.

Beccari, N. (1923). Il centro tegmentale o interstiziale ed altre formazioni poco note nel mesencefalo e nel diencefalo di un Rettile. *Arch. Ital. Anat. Embriol.* 20, 560–619.

Belekhova, M. G. (1970). Peculiarities of the organization of the thalamo-cortical system in turtles. In *Electrophysiological Studies on the Central Nervous System of Vertebrates* (A. I. Karamian, ed.). *Zh. Evol. Biok. Fiziol.* Suppl. Nauka, Leningrad, pp. 137–146 [in Russian].

Belekhova, M. G. (1979). Neurophysiology of the forebrain. In *Biology of the Reptilia* (C. Gans, R. G. Northcutt, and P. S. Ulinski, eds.). Academic Press, London, Vol. 10, 287–359.

Belekhova, M. G., and Ivazov, N. I. (1983). Analysis of conduction of visual, somatic and audiovibratory sensory information to the hippocampal cortex in lizard. *Neurofiziol.* 15(2), 153–160 [in Russian].

Belekhova, M. G., and Kenigfest, N. B. (1983). A study of the hippocampal mediodorsal cortex connections in lizard by means of horseradish peroxidase axonal transport. *Neurofiziol.* 15 (2), 145–152 [in Russian].

Belekhova, M. G., and Kosareva, A. A. (1967). Characteristics of visual representation in the brain of the turtle. *Zh. Evol. Biok. Fiziol.* 3, 248–258 [in Russian].

Belekhova, M. G., and Kosareva, A. A. (1971). Organization of the turtle thalamus: visual, somatic and tectal zones. *Brain Behav. Evol.* 4, 337–375.

Belekhova, M. G., and Nemova, G. V. (1987). Study of connections of supposed limbic diencephalic nuclei in lizards using axonic HRP transport. *Neurofiziol.* 19 (1), 110–120 [in Russian].

Belekhova, M. G., Ivazov, N. I., and Safarov, Kh. M. (1983). New data on sensory representation in the telencephalon of the lizard *Ophisaurus apodus*. *Zh. Evol. Biok. Fiziol.* 19 (5), 478–485 [in Russian].

Bellairs, A. d'A. (1949). The anterior braincase and interorbital septum of Sauropsida, with a consideration of the origin of snakes. *J. Linn. Soc. (Zool.)* 41, 482–512.

Bellairs, A. d'A. (1972). Comments on the evolution and affinities of the origin of snakes. In *Studies in Vertebrate Evolution* (K. A. Joysey and T. S. Kemp, eds.). Oliver and Boyd, Edinburgh, pp. 157–172.

Bellairs, A. d'A., and Underwood, G. (1951). The origin of snakes. *Biol. Rev.* 26, 193–237.

Bellonci, J. (1888). Ueber die centrale Endigung des Nervus opticus bei den Vertebraten. *Z. Wiss. Zool.* 47, 1–46.

Boiko, V. P., and Davydova, T. V. (1973). Comparative study of functional and morphological characteristics of mesencephalic neurons of the tortoise *Emys orbicularis*. *J. Evol. Biok. Fiziol.* 9, 476–478 [in Russian].

Boiko, V. P., and Gonchareva, N. V. (1976). Morphological and functional organisation of the tectal visual center in the turtles *Agrionemys horsfieldi* and *Emys orbicularis*. *J. Evol. Biok. Fiziol.* 12, 399–404 [in Russian].

Bonin, J. J. (1965). The eye of *Agamodon anguliceps* Peters (Reptilia, Amphisbaenia). *Copeia* 1965, 324–331.

Boulenger, G. A. (1893). On some newly described Jurassic and Cretaceous lizards and rhynchocephalians. *Ann. Mag. Nat. Hist.* 11, 204–210.

Braford, M. R., Jr. (1973). Retinal projections in *Caiman crocodilus*. *Am. Zool.* 13, 1345.

Bruce, L. L., and Butler, A. B. (1984). Telencephalic connections in lizards. I. Projections to cortex. *J. Comp. Neurol.* 229, 585–601.

Burns, A. H., and Goodman, D. C. (1967). Retinofugal projections in *Caiman sclerops*. *Exp. Neurol.* 18, 105–115.

Butler, A. B. (1974). Retinal projections in the night lizard, *Xantusia vigilis* Baird. *Brain Res.* 80, 116–121.

Butler, A. B., and Ebbesson, S. O. E. (1975). A Golgi study of the optic tectum of the tegu lizard, *Tupinambis nigropunctatus*. *J. Morphol.* 146, 215–228.

Butler, A. B., and Northcutt, R. G. (1971). Retinal projections in *Iguana iguana* and *Anolis carolinensis*. *Brain Res.* 26, 1–13.

Butler, A. B., and Northcutt, R. G. (1973). Architectonic studies of the diencephalon of *Iguana iguana* (Linnaeus). *J. Comp. Neurol.* 149, 439–462.

Butler, A. B., and Northcutt, R. G. (1978). New thalamic visual nuclei in lizards. *Brain Res.* 149, 469–476.

Cairney, J. (1926). A general survey of the forebrain of *Sphenodon punctatus*. *J. Comp. Neurol.* 42, 255–348.

Camp, C. L. (1923). Classification of lizards. *Bull. Am. Mus. Nat. Hist.* 48, 289–481.

Clairambault, P., Cordier-Picouet, M. J., and Pairault, C. (1980). Premières données sur les projections visuelles d'un Amphibien Apode (*Typhlonectes compressicauda*). *C. R. Acad. Sci. Paris* 291, 283–286.

Conrad, C., and Stumpf, W. (1975). Direct visual input to the limbic system: crossed retinal projections to the nucleus anterodorsalis thalami in the tree shrew. *Exp. Brain Res.* 23, 141–149.

Cope, E. D. (1869). On the reptilian orders Pythonomorpha and Streptosauria. *Proc. Boston Soc. Nat. Hist.* 12, 250–266.

Cope, E. D. (1875). The vertebrata of the Cretaceous formations of the West Rep. *U.S. Geol. Surv.* 3, 1–1009.

Crossland, W. J. (1979). Retinal afferents affect the formation of the avian ectomammillary nucleus. *Brain Res.* 165, 127–132.

Crossland, W. J., Cowan, W. M., and Rogers, L. A. (1975). Studies on the development of the chick optic tectum. IV. An autoradiographic study of the development of retinotectal connections. *Brain Res.* 91, 1–24.

Crossland, W. J., Cowan, W. M., Rogers, L. A., and Kelly, J. P. (1974). The specification of the retinotectal projections in the chick. *J. Comp. Neurol.* 155, 127–165.

Cruce, J. A. F. (1974). A cytoarchitectonic study of the diencephalon of the tegu lizard, *Tupinambis nigropunctatus*. *J. Comp. Neurol.* 153, 215–238.

Cruce, J. A. F., and Cruce, W. L. R. (1978). Analysis of the visual system in a lizard, *Tupinambis nigropunctatus*. In *Behavior and Neurology of Lizards* (N. Greenberg and P. D. MacLean, eds.). U.S. Department of Health, Education and Welfare Public Health Service, DHEW Publication No. (ADM) 77–491, NIH, Rockville, Md., pp. 79–90.

Cruce, W. L. R., and Cruce, J. A. F. (1975). Projections from the retina to the lateral geniculate nucleus and mesencephalic tectum in a reptile (*Tupinambis nigropunctatus*): a comparison of anterograde transport and anterograde degeneration. *Brain Res.* 85, 221–228.

Curwen, A. O., and Miller, R. N. (1939). The pretectal region of the turtle *Pseudemys scripta troostii*. *J. Comp. Neurol.* 71, 99–120.

Dacey, D. M., and Ulinski, P. S. (1983). Nucleus rotundus in a snake, *Thamnophis sirtalis*: an analysis of a nonretinotopic projection. *J. Comp. Neurol.* 216, 175–191.

Dacey, D. M., and Ulinski, P. S. (1986a). Optic tectum of the eastern garter snake, *Thamnophis sirtalis*. I. Efferent pathways. *J. Comp. Neurol.* 245, 1–28.

Dacey, D. M., and Ulinski, P. S. (1986b). Optic tectum of the eastern garter snake, *Thamnophis sirtalis*. II. Morphology of efferent cells. *J. Comp. Neurol.* 245, 198–237.

Dacey, D. M., and Ulinski, P. S. (1986c). Optic tectum of the eastern garter snake, *Thamnophis sirtalis*. III. Morphology of intrinsic neurons. *J. Comp. Neurol.* 245, 283–300.

Dacey, D. M., and Ulinski, P. S. (1986d). Optic tectum of the eastern garter snake, *Thamnophis sirtalis*. IV. Morphology of afferents from the retina. *J. Comp. Neurol.* 245, 301–318.

Dacey, D. M., and Ulinski, P. S. (1986e). Optic tectum of the eastern garter snake, *Thamnophis sirtalis*. V. Morphology of brainstem afferents and general discussion. *J. Comp. Neurol.* 245, 423–453.

Davydova, T. V. (1971). The ultrastructure of nerve terminals in the tectum opticum of the turtle (*Emys orbicularis*). *Cytologia* 13, 433–440 [in Russian].

Davydova, T. V., and Gonchareva, N. V. (1970). Cytoarchitectonics and neuronal composition of midbrain tectum in the turtle (*Emys orbicularis*). *Arch. Anat. Histol. Embriol.* 59, 53–59 [in Russian].

Davydova, T. V., Gonchareva, N. V., and Boiko, V. P. (1976). Retinotectal system of the tortoise, *Testudo horsfieldi* Gray (Morpho-functional study in the normal and after enucleation). *J. Hirnforsch.* 17, 463–488.

Davydova, T. V., and Smirnov, G. D. (1973). Retinotectal connections in the tortoise: an electron microscope study of degeneration in optic nerve and midbrain tectum. *J. Hirnforsch.* 14, 473–492.

de la Calle, A., Davilla, J. C., Guirado, S., and Marin Giron, F. (1986). Retinal projection in the lizard *Podarcis hispanica*. *J. Hirnforsch.* 27, 707–713.

de Lange, S. J. (1913). Das Zwischenhirn und das Mittelhirn der Reptilien. *Folia Neurobiol.* 7, 67–138.

Desan, P. H. (1988). Organization of the cerebral cortex in turtle. In *The Forebrain of Reptiles: Current Concepts of Structure and Function* (W. K. Schwerdtfeger and W. J. A. J. Smeets, eds.). Karger, Basel, pp. 1–11.

Dowling, H. G. (1959). Classification of the serpents: a critical review. *Copeia* 1959 (1), 38–52.

Duke-Elder, S. (1958). *The Eye in Evolution*. Kimpton, London.

Durward, A. (1930). The cell masses in the forebrain of *Sphenodon punctatum*. *J. Anat.* (London) 65, 8–44.

Ebbesson, S. O. E. (1970). On the organization of the central visual pathways in vertebrates. *Brain Behav. Evol.* 3, 178–194.

Ebbesson, S. O. E. (1972). A proposal for a common nomenclature for some optic nuclei in vertebrates and the evidence for a common origin of two such cell groups. *Brain Behav. Evol.* 6, 75–91.

Ebbesson, S. O. E., and Ito, H. (1980). Bilateral retinal projections in the black piranha (*Serralamus niger*). *Cell. Tiss. Res.* 213, 483–495.

Ebbesson, S. O. E., Jane, J. A., and Schroeder, D. M. (1972). A general overview of major interspecific variations in thalamic organization. *Brain Behav. Evol.* 6, 92–130.

Ebbesson, S. O. E., and Karten, H. J. (1981). Terminal distribution of retinal fibers in the tegu lizard (*Tupinambis nigropunctatus*). *Cell Tiss. Res.* 215, 591–601.

Edinger, L. (1899). Untersuchungen über die vergleichende Anatomie des Gehirns. 4. Studien über das Zwischenhirn der Reptilien. *Abh. Senck. Natur. Ges.* 20, 161–202.

Estes, R., and Pregill, G., eds. (1972). *Phylogenetic Relationships of the Lizard Families: Essays Commemorating Charles L. Camp*. Stanford University Press, Stanford, Calif.

Fink, R. P., and Heimer, L. (1967). Two methods for selective silver impregnation of degenerating axons and their synaptic ending in the central nervous system. *Brain Res.* 4, 369–374.

Fite, K. V., Reiner, A., and Hunt, S. P. (1979). Optokinetic nystagmus and the accessory optic system of pigeon and turtle. *Brain Behav. Evol.* 16, 192–202.

Foster, R. E., and Hall, W. C. (1975). The connections and laminar organisation of the optic tectum in a reptile (*Iguana iguana*). *J. Comp. Neurol.* 163, 397–426.

Foster, R. E., Hall, J. A., and Ebner, F. F. (1976). Visual cortex in the turtle (*Pseudemys scripta* and *Chrysemys picta*). *Anat. Rec.* 184, 405.

Franzoni, M., and Fasolo, A. (1982). A study on neuronal morphology of the ventral lateral geniculate nucleus of the lizard. *J. Hirnforsch.* 23, 245–256.

Frederikse, A. (1931). *The Lizard's Brain: an Investigation of the Histological Structure of the Brain of Lacerta vivipara. Acad. Proefschr. Callen Boch, Nijkerk*

Gaidaenko, G. V. (1983). Connections of the dorsomedial hippocamp in tortoises. *Arch. Anat. Histol. Embriol.* 1, 36–43 [in Russian].

Gaither, N. S., and Stein, B. E. (1979). Reptiles and mammals use similar sensory organization in the midbrain. *Science* (N.Y.) 205, 595–597.

Gans, C. (1985). Differences and similarities: Comparative methods in mastication. *Am. Zool.* 25, 291–301.

Gasc, J. P. (1967). Introduction à l'étude de la musculature axiale des Squamates serpentiformes. *Mém. Mus. Nat. Hist. Nat.*, sér. A, 48, 69–124.

Gasc, J. P. (1968). Contribution à l'ostéologie et à la myologie de *Dibamus novaguinae* (Sauria, Reptilia). Discussion systématique. *Ann. Soc. Nat. Zool.* 10, 127–150.

Gasc, J. P. (1970). Réflexions sur le concept de "régression" des organes. *Rev. Questions Scientif.* 141, 175–195.

Gasc, J. P. (1973). *L'interprétation fonctionnelle de l'appareil musculo-squelettique de l'axe vertébral chez les Serpents (Reptilia)*. Thèse, Paris.

Gasc, J. P. (1974). L'interprétation fonctionnelle de l'appareil musculo-squelettique de l'axe vertébral des Serpents (Reptilia). *Mém. Mus. Nat. Hist. Nat.*, sér. A, 83, 1–182.

Gasc, J. P., and Renous, S. (1976). La céphalisation et le problème de l'intégration organique. *Cah. Etud. Biol.* 13–15, 115–152.

Gisi, J. (1908). Das Gehirn von *Hatteria punctata*. *Zool. Jahrhb. Anat. Ontog.* 25, 71–236.

Granda, A. M., and O'Shea, P. J. (1972). Spectral sensitivity of the green turtle (*Chelonia mydas mydas*) determined by electrical responses to heterochromatic light. *Brain Behav. Evol.* 5, 143–154.

Granda, A. M., and Stirling, C. E. (1965). Differential spectral sensitivity in the optic tectum and eye of the turtle. *J. Gen. Physiol.* 48, 901–917.

Guibé, J. (1970). La systématique des reptiles actuels. In *Traité de Zoologie* (P. P. Grassé, ed.). Masson, Paris, 14(2), 1054–1160.

Guillery, R. W., and Updyke, B. V. (1976). Retinofugal pathways in normal and albino axolotls. *Brain Res.* 109, 235–244.

Gusel'nikov, V. I., and Supin, A. Y. (1964). Visual and auditory regions and hemispheres of lizard forebrain. *Trans. Suppl. Fed. Ame. Soc. Exp. Biol.* 23, 641–646.

Gusel'nikov, V. I., Morenkov, E. D., and Pivovarov, A. S. (1970). On functional organization of the visual system of the tortoise (*Emys orbicularis*). *Fiziol. Zh. SSSR Sechenova* 56, 1377–1385 [in Russian].

Hall, W. C. (1972). Visual pathways to the telencephalon in reptiles and mammals. *Brain Behav. Evol.* 5, 95–113.

Hall, W. C., and Ebner, F. F. (1970a). Parallels in the visual afferent projections of the thalamus in the hedgehog (*Paraechinus hypomolas*) and the turtle (*Pseudemys scripta*). *Brain Behav. Evol.* 3, 135–154.

Hall, W. C., and Ebner, F. F. (1970b). Thalamotelencephalic projections in the turtle (*Pseudemys scripta*). *J. Comp. Neurol.* 140, 101–122.

Hall, J. A., Foster, R. E., Ebner, F. F., and Hall, W. C. (1977). Visual cortex in a reptile, the turtle (*Pseudemys scripta* and *Chrysemys picta*). *Brain Res.* 130, 197–216.

Halpern, M. (1973). Retinal projections in blind snakes. *Science* (N.Y.) 182, 390–391.

Halpern, M., and Frumin, N. (1973). Retinal projections in a snake, *Thamnophis sirtalis*. *J. Morphol.* 141, 359–382.

Hartline, P. H., Kass, L., and Loop, M. S. (1978). Merging of modalities in the optic tectum: Infrared and visual integretion in rattlesnakes. *Science* (N.Y.) 199, 1225–1229.

Heller, S. B., and Ulinski, P. S. (1987). Morphology of geniculocortical axons in turtles of the genera *Pseudemys* and *Chrysemys*. *Anat. Embryol.* 175, 505–515.

Heric, T. M., and Kruger, L. (1965). Organisation of the visual projections upon the optic tectum of a reptile (*Alligator mississippiensis*). *J. Comp. Neurol.* 124, 101–122.

Heric, T. M., and Kruger, L. (1966). The electrical response evoked in the reptilian optic tectum by afferent stimulation. *Brain Res.* 2, 187–199.

Herrick, C. J. (1910). The morphology of the forebrain in Amphibia and Reptilia. *J. Comp. Neurol.* 20, 413–547.

Himstedt, W., and Manteuffel, G. (1985). Retinal projections in the caecilian *Ichthyophis kohtaoensis* (Amphibia, Gymnophiona). *Cell Tiss. Res.* 239, 689–692.

Hoffstetter, R. (1955). Squamates de type moderne. In *Traité de paléontologie* (J. Piveteau, ed.). Masson, Paris, 5, 602–662.

Hoffstetter, R. (1962). Revue des récentes acquisitions concernant l'histoire et la systématique des Squamates. In *Problèmes actuels de Paléontologie, Evolution des Vertébrés*. Editions du C. N. R. S., Paris, pp. 243–279.

Hoffstetter, R. (1968). Review of: A contribution to the classification of snakes (G. Underwood). *Copeia* 1968(1), 201–213.

Hoogland, P. V., and Vermeulen-van der Zee, E. (1988). Intrinsic and extrinsic connections of the cerebral cortex of lizards. In *The Forebrain of Reptiles: Current Concepts of Structure and Function* (W. K. Schwerdtfeger and W. J. A. J. Smeets, eds.). Karger, Basel, pp. 20–29.

Huber, G. C., and Crosby, E. C. (1926). On the thalamic and tectal nuclei and fiber paths in the brain of the American alligator. *J. Comp. Neurol.* 40, 94–227.

Huber, G. C., and Crosby, E. C. (1933a). A phylogenetic consideration of the optic tectum. *Proc. Philadelphia Acad. Nat. Sci.* 19, 15–22.

Huber, G. C., and Crosby, E. C. (1933b). The reptilian optic tectum. *J. Comp. Neurol.* 57, 57–163.

Itaya, S. K., Van Hoesen, C. E., and Benevento, L. A. (1986). Direct retinal input to limbic thalamus of the monkey. *Exp. Brain Res.* 61, 607–613.

Itaya, S. K., Van Hoesen, C. E., and Jeng, C. B. (1981). Direct retinal input to limbic system of the rat. *Brain Res.* 226, 33–42.

Ivazov, N. I., and Belekhova, M. G. (1982). Electrophysiological studies on afferent organisation on the thalamus in the lizard *Ophisaurus apodus*. *Zh. Evol. Biok. Fiziol.* 1, 76–84 [in Russian].

Jakway, J. S., and Riss, W. (1972). Retinal projections in the tiger salamander *Ambystoma tigrinum*. *Brain Behav. Evol.* 5, 401–442.

Kass, L., Loop, M. S., and Hartline, P. H. (1978). Anatomical and physiological localization of visual and infrared cell layers in the tectum of pit vipers. *J. Comp. Neurol.* 182, 811–820.

Kelly, J. P., and Cowan, W. M. (1972). Studies on the development of the chick optic tectum. III. Effects of early eye removal. *Brain Res.* 42, 263–289.

Kenigfest, N. B., Repérant, J., and Vesselkin, N. P. (1986). Retinal projections in the lizard *Ophisaurus apodus*, revealed by autoradiographic and peroxidase methods. *Zh. Evol. Biok. Fiziol.* 2, 181–187 [in Russian].

Kennedy, M. C., Rubinson, K., and Stone, D. J. (1977). Retinal projections in larval, transforming and adult lamprey, *Petromyzon marinus*. *J. Comp. Neurol.* 171, 465–480.

Knapp, H., and Kang, D. S. (1968a). The visual pathways of the snapping turtle (*Chelydra serpentina*). *Brain Behav. Evol.* 1, 19–42.

Knapp, H., and Kang, D. S. (1968b). The retinal projections of the side-necked turtle (*Podocnemis unifilis*). *Brain Behav. Evol.* 1, 369–404.

Kosareva, A. A. (1967). Projection of the optic fibers to visual centers in a turtle (*Emys orbicularis*). *J. Comp. Neurol.* 130, 263–276.

Kosareva, A. A. (1980). Retinal projections in lamprey, *Lampetra fluviatilis*. *J. Hirnforsch.* 21, 243–256.

Kruger, L., and Berkowitz, E. C. (1960). The main afferent connections of the reptilian telencephalon as determined by degeneration and electrophysiological methods. *J. Comp. Neurol.* 115, 125–140.

Kubie, J. L., and Allen, J. O. (1978). Deoxyglucose mapping of the visual system of the garter snake. *Neuroscience* 4, 100.

Kuhlenbeck, H. (1931). Ueber die Grundbestandteile des Zwischenhirnbauplans bei Reptilien. *Gegenb. Morph. Jb.* 66, 244–317.

Künzle, H., and Schnyder, H. (1983). Do retinal and spinal projections overlap within turtle thalamus? *Neuroscience* 10, 161–168.

Kusunoki, T., and Amemiya, F. (1983). Retinal projections in the hagfish, *Eptatretus burgeri*. *Brain Res.* 262, 295–298.

La Vail, J. H., and Cowan, W. M. (1971a). The development of the chick tectum. I. Normal morphology and cytoarchitectonic development. *Brain Res.* 28, 391–419.

La Vail, J. H., and Cowan, W. M. (1971b). The development of the chick tectum. II. Autoradiographic studies. *Brain Res.* 28, 421–441.

Lazar, G. (1969). Distribution of optic terminals in the different optic centers in the frog. *Brain Res.* 16, 1–14.

Lazar, G. (1978). Application of cobalt-filling technique to show retinal projections in the frog. *Neuroscience* 3, 725.

Leghissa, S. (1962). L'evoluzione del tetto ottico nei bassi vertebrati. *Arch. Ital. Anat. Embriol.* 67, 343–413.

Lemire, M., Torcol, N., and Repérant, J. (1988). Les voies visuelles primaires chez la Tortue Luth, *Dermochelys coriacea:* analyse par marquage autoradiographique et peroxydasique. *Mésogée* 48, 51–57.

Lohman, A. H. M., and Van Woerden-Verkley, I. (1978). Further studies on the cortical connections of the tegu lizard. *Brain Res.* 103, 9–28.

Lund, R. D., and Lund, J. S. (1965). The visual system of the mole *Talpa europaea. Exp. Neurol.* 13, 302–316.

McDowell, S. B., Jr., and Bogert, C. M. (1954). The systematic position of *Lanthanotus* and the affinities of the anguinomorphan lizards. *Bull. Am. Mus. Nat. Hist.* 105, 1–142.

Mahendra, B. C. (1938). Some remarks on the phylogeny of the Ophidia. *Anat. Anz.* 86, 347–356.

Martinez-Garcia, F., and Olucha, F. (1988). Afferent projections to the Timm-positive cortical areas of the telencephalon of lizards. In *The Forebrain of Reptiles: Current Concepts of Structure and Function* (W. K. Schwerdtfeger and W. J. A. J. Smeets, eds.). Karger, Basel, pp. 30–40.

Masai, H. (1973). Structural patterns of the optic tectum in Japanese snakes of the family Colubridae in relation to habit. *J. Hirnforsch.* 14, 367–374.

Nauta, W. J. H. (1957). Silver impregnation of degenerating axon. In *New Research Techniques of Neuroanatomy* (W. F. Windle, ed.). Charles C. Thomas, Springfield, Ill.

Nauta, W. J. H., and Gygax, P. A. (1951). Silver impregnation of degenerating axon terminals in the central nervous system: (1) Technique, (2) Chemical notes. *Stain Technol.* 26, 5–11.

Nauta, W. J. H., and Gygax, P. A. (1954). Silver impregnation of degenerating axons in the central nervous system: a modified technique. *Stain Technol.* 29, 91–93.

Nonidez, J. F. (1933). Quoted by R. D. Lillie in *Histopathologic Technic* (1948). Blakiston Co., Philadelphia.

Nopsca, F. (1923). *Eidolosaurus* und *Pachyophis.* Zwei neue Neucom-Reptilien. *Paleontographica* 65, 99–154.

Northcutt, R. G. (1972). The teiid prosencephalon and its bearing on squamate systematics. 52nd Annual Meet. Am. Soc. Ichthyol. Herpetol., Abst., pp. 75–79.

Northcutt, R. G. (1978). Forebrain and midbrain organization in lizards and its phylogenetic significance. In *Behavior and Neurology of Lizards* (N. Greenberg and P. D. MacLean, eds.). U.S. Department of Health, Education and Welfare, Public Health Service, DHEW Publication No. (ADM) 77–491, National Institutes of Health, Rockville, Md., pp. 11–64.

Northcutt, R. G. (1980). Retinal projections in the Australian lungfish. *Brain Res.* 185, 85–90.

Northcutt, R. G. (1984a). Anatomical organization of the optic tectum in reptiles. In *Comparative Neurology of the Optic Tectum* (H. Vanegas, ed.). Plenum Press, New York & London, pp. 547–600.

Northcutt, R. G. (1984b). Evolution of vertebrate central nervous system: patterns and processes. *Am. Zool.* 24, 701–716.

Northcutt, R. G., Braford, M. R., and Landreth, G. E. (1974). Retinal projections in the tuatara *Sphenodon punctatus:* an autoradiographic study. *Anat. Rec.* 178, 428.

Northcutt, R. G., and Butler, A. B. (1974a). Retinal projections in the northern water snake, *Natrix sipedon sipedon* (L.). *J. Morphol.* 142, 117–136.

Northcutt, R. G., and Butler, A. B. (1974b). Evolution of reptilian visual system: retinal projections in a nocturnal lizard, *Gekko gecko* (L.). *J. Comp. Neurol.* 157, 453–466.

Northcutt, R. G., and Butler, A. B. (1976). Retinofugal pathways in the longnose gar *Lepisosteus osseus* (Linnaeus). *J. Comp. Neurol.* 166, 1–16.

Northcutt, R. G., and Kicliter, E. (1980). Organization of the amphibian telencephalon. In *Comparative Neurology of the Telencephalon* (S. O. E. Ebbesson, ed.). Plenum Press, New York & London, pp. 203–255.

Northcutt, R. G., and Wathey, J. C. (1980). Guitarfish possess ipsilateral as well as contralateral retinofugal projections. *Neurosci. Lett.* 20, 237–242.

Olson, E. C. (1977). The history of the vertebrates. In *Handbook of Sensory Physiology: the Visual System in Evolution of Vertebrates* (F. Crescitelli, ed.). Springer-Verlag, Berlin, Heidelberg, & New York, Vol. VII/5, pp. 1–45.

Ouimet, C. C., Patrick, R. L., and Ebner, F. F. (1985). The projection of three extrathalamic cell groups to the cerebral cortex of the turtle *Pseudemys*. *J. Comp. Neurol.* 237, 77–84.

Pakhamova, A. S., Weidner, C., Repérant, J., and Vesselkin, N. P. (1987). The visual input to the hippocampal is relayed in the anterior thalamic nuclei in the rat. *Neurofiziol.* 19, 278–280 [in Russian].

Papez, J. W. (1935). Thalamus of turtles and thalamic evolution. *J. Comp. Neurol.* 61, 433–475.

Peduzzi, J. D., and Crossland, W. J. (1983a). Anterograde transneuronal degeneration in the ectomammillary nucleus and ventral lateral geniculate nucleus of the chick. *J. Comp. Neurol.* 213, 287–300.

Peduzzi, J. D., and Crossland, W. J. (1983b). Morphology of normal and deafferented neurons in the chick ectomammillary nucleus. *J. Comp. Neurol.* 213, 301–309.

Peterson, E. H. (1980). Behavioral studies of telencephalic functions in reptiles. In *Comparative Neurology of Telencephalon* (S. O. E. Ebbesson, ed.) Plenum Press, New York & London, pp. 343–388.

Peterson, E. H. (1981). Regional specialization in retinal ganglion cell projection to optic tectum of *Dipsosaurus dorsalis* (Iguanidae). *J. Comp. Neurol.* 196, 225–252.

Platel, R. (1975). Nouvelles données sur l'encéphalisation des Reptiles Squamates. *J. Zool. Syst. Evol. Forsch.* 13, 65–87.

Platel, R., Raffin, J. P., and Repérant, J. (1975). The primary visual system in a lizard of the family Scincidae *Scincus scincus* (L): an experimental study. *Exp. Brain Res.* Suppl. 3, 162.

Porter, K. R. (1972). *Herpetology.* W. B. Saunders, Philadelphia.

Powell, T. P. S., and Kruger, L. (1960). The thalamic projection upon the telencephalon in *Lacerta viridis*. *J. Anat.* (London) 94, 528–542.

Quiroga, J. C. (1978). The tectum opticum of *Pantodactylus schreiberii* Weigmann (Teiidae, Lacertilia, Reptilia). *J. Hirnforsch.* 19, 109–131.

Quiroga, J. C. (1979). The cell masses in the diencephalon of *Amphisbaena darwinii heterozonata* Burmeister (Amphisbaenia, Squamata, Reptilia). *Acta Anat.* 104, 198–210.

Raffin, J. P. (1972). Quelques effets de l'ablation uni- ou bilatérale de l'ébauche optique sur le tectum du poulet (*Gallus domesticus*). *Acta Embryol. Exp.* 1, 45–83.

Raffin, J. P. (1974). L'architecture des centres visuels diencéphaliques et prétectaux du poussin de *Gallus domesticus* L. Etude expérimentale. *J. Embryol. Exp. Morph.* 32, 763–781.

Raffin, J. P. (1976). La cytoarchitecture des principaux centres visuels diencéphaliques chez l'embryon et le poussin de *Gallus domesticus* L. Etude expérimentale. *J. Hirnforsch.* 17, 581–597.

Raffin, J. P. (1977). "Morphogenèse expérimentale des principaux centres visuels thalamiques, prétectaux et tegmentaux du Poulet." Thèse, Paris.

Raffin, J. P., and Repérant, J. (1975). Etude expérimentale de la spécificité des projections visuelles d'embryons et de poussins de *Gallus domesticus* L. microphtalmes et monophtalmes. *Arch. Anat. Micr. Morph. Exp.* 64 (2), 93–111.

Raffin, J. P., Repérant, J., and Miceli, D. (1977). Analyse radioautographique des projections visuelles primaires d'un rapace diurne (*Buteo buteo* L.). *C. R. Acad. Sci. Paris* 284, 945–947.

Rage, J.-C. (1976). *Paléontologie, Phylogénie et Paléobiogéographie des Serpents.* Thèse, Paris.

Rage, J.-C. (1984). Serpentes. In *Handbuch der Paläoherpetologie,* Part 11. Gustav Fischer Verlag, Stuttgart.

Rainey, W. T., and Ulinski, P. S. (1986). Morphology of neurons in the dorsal lateral geniculate complex in turtles of the genera *Pseudemys* and *Chrysemys*. *J. Comp. Neurol.* 253, 440–465.

Ramón, P. (1896). Estructura del encéfalo del Cameleon. *Rev. Trimest. Micrograf.* 1, 46–82.

Ramón y Cajal, S. (1911). *Histologie du Système Nerveux de l'Homme et des Vertébrés.* Maloine, Paris.

Reiner, A. (1981). A projection of displaced ganglion cells and giant ganglion cells to the accessory optic nuclei in turtle. *Brain Res.* 204, 403–409.

Reiner, A., and Karten, H. J. (1978). A bisynaptic retinocerebellar pathway in the turtle. *Brain Res.* 150, 163–169.

Repérant, J. (1972). Etude expérimentale des projections visuelles chez la Vipère (*Vipera aspis*). *C. R. Acad. Sci. Paris* 275, 695–698.

Repérant, J. (1973a). Les voies et les centres optiques primaires chez la Vipère (*Vipera aspis*). *Arch. Anat. Micr. Morphol. Exp.* 62, 323–352.

Repérant, J. (1973b). Nouvelles données sur les projections visuelles chez le pigeon (*Columba livia*). *J. Hirnforsch.* 14, 152–187.

Repérant, J. (1975). Nouvelles données sur les projections rétiniennes chez *Caiman sclerops:* Etude radioautographique. *C. R. Acad. Sci. Paris* 280, 2881–2884.

Repérant, J. (1978). *Organisation Anatomique du Système Visuel des Vertébrés: Approche Évolutive.* Thèse, Paris.

Repérant, J., Lemire, M., Miceli, D., and Peyrichoux, J. (1976). A radioautographic study of the visual system in fresh water teleost, following intra-ocular injection of tritiated fucose and proline. *Brain Res.* 118, 123–131.

Repérant, J., Miceli, D., Rio, J. P., Peyrichoux, J., Pierre, J., and Kirpitchnikova, E. (1986). The anatomical organization of retinal projections in the shark *Scyliorhinus canicula* with special reference to the evolution of the selachian primary visual system. *Brain Res. Rev.* 11, 227–248.

Repérant, J., Miceli, D., Rio, J. P., and Weidner, C. (1987a). The primary optic system in a microphthalmic snake (*Calabaria reinhardti*). *Brain Res.* 408, 233–238.

Repérant, J., Peyrichoux, J., and Rio, J. P. (1981). Fine structure of the superficial layers of the viper optic tectum: a Golgi and electron-microscopic study. *J. Comp. Neurol.* 199, 393–417.

Repérant, J., and Raffin, J. P. (1974). Les projections visuelles chez le Goéland argenté (*Larus argentatus argentatus* Pontopp) et le Goéland brun (*Larus fuscus graellsii* Brehm). *C. R. Acad. Sci. Paris* 278, 2335–2338.

Repérant, J., and Rio, J. P. (1976). Retinal projections in *Vipera aspis*: a reinvestigation using light radioautographic and electron microscopic degeneration techniques. *Brain Res.* 107, 603–609.

Repérant, J., Rio, J. P., Miceli, D., Amouzou, M., and Peyrichoux, J. (1981). The retinofugal pathways in the primitive bony fish *Polypterus senegalus* (Cuvier 1829). *Brain Res.* 217, 225–243.

Repérant, J., Rio, J. P., Miceli, D., and Lemire, M. (1978). Radioautographic study of the retinal projections in the type I and type II lizard. *Brain Res.* 14, 401–411.

Repérant, J., Vesselkin, N. P., Ermakova, T. V., Rustamov, E. K., Rio, J. P., Palatnikov, G. K., Peyrichoux, J., and Kasimov, R. V. (1982). The retinofugal pathways in a primitive actinopterygian, the chondrostean *Acipenser güldenstädti*: an experimental study using degeneration, radioautographic and HRP methods. *Brain Res.* 251, 1–23.

Repérant, J., Weidner, C., Pakhamova, A., Desroches, A. M., Vesselkin, N. P., and Lemire, M. (1987b). Mise en évidence d'une voie rétino-thalamo-hippocampique chez le rat. *C. R. Acad. Sci. Paris* 305, série 3, 601–604.

Rio, J. P. (1987). *Neuroanatomie comparée du système optique basal chez les Vertébrés*. Thèse, Paris.

Rio, J. P., Repérant, J., and Peyrichoux, J. (1983). A light and electron microscope study of the basal optic system in the reptile *Vipera aspis*. *J. Hirnforsch.* 24, 447–469.

Rio, J. P., Repérant, J., and Peyrichoux, J. (1988). Ultrastructural changes in GABAergic localization in *Vipera* optic tectum following retinal deafferentation. In 11th Annual Meet. of the Europ. Neurosci. Assoc. *Europ. J. Neurosci.* (Suppl.) Abst., p. 41.

Robbins, D. O. (1972). Coding of intensity and wavelength in optic tectal cells in the turtle. *Brain Behav. Evol.* 5, 124–142.

Rochon-Duvigneaud, A. (1943). *Les Yeux et la Vision des Vertébrés*. Masson, Paris.

Rochon-Duvigneaud, A. (1970). L'oeil et la vision. In *Traité de Zoologie* (P. P. Grassé, ed.). Masson, Paris, 14(2): 382–428.

Romer, A. S. (1956). *Osteology of the Reptiles.* University of Chicago Press, Chicago.

Romer, A. S. (1968). *Notes and Comments on Vertebrate Paleontology.* University of Chicago Press, Chicago.

Scalia, F. (1976). The optic pathway of the frog, nuclear organization and connections. In *Frog Neurobiology: a Handbook* (R. Llinas, and W. Precht, eds.). Springer Verlag, Berlin, Heidelberg, & New York, pp. 407–734.

Scalia, F., and Fite, K. V. (1974). A retinotopic analysis of the central connections of the optic nerve in the frog. *J. Comp. Neurol.* 158, 455–478.

Schechter, P. B., and Ulinski, P. S. (1979). Interactions between tectal radial cells in the red-eared turtle, *Pseudemys scripta elegans:* an analysis of tectal modules. *J. Morphol.* 162, 17–36.

Schroeder, D. M. (1981a). Retinal afferents and efferents of an infrared sensitive snake, *Crotalus viridis. J. Morphol.* 170, 29–42.

Schroeder, D. M. (1981b). Tectal projections of an infrared sensitive snake, *Crotalus viridis. J. Comp. Neurol.* 195, 477–500.

Senn, D. G. (1966). Ueber das optische System im Gehirn squamater Reptilien. Eine vergleichend-morphologische Untersuchung, unter besonderer Berücksichtigung einiger Wühlschlangen. *Acta Anat.* Suppl. 52, 65, 1–88.

Senn, D. G. (1968a). Bau und Ontogenese von Zwischen und Mittelhirn bei *Lacerta sicula* (Rafinesque). *Acta Anat.* Suppl. 55, 71, 1–150.

Senn, D. G. (1968b). Der Bau des Reptiliengehirns im Licht neuer Ergebnisse. *Verh. Natur. Ges. Basel* 79, 25–43.

Senn, D. G. (1968c). Ueber den Bau von Zwischen und Mittelhirn von *Anniella pulchra* Gray. *Acta Anat.* 69, 230–261.

Senn, D. G. (1969). Ueber das Zwischen und Mittelhirn von zwei typhlopiden Schlaugen *Anomalepis aspinosus* und *Liotyphlops albirostris. Verh. Natur. Ges. Basel* 88, 32–48.

Senn, D. G. (1970a). Die Zusammenhäge von Grosshirnstriatum dorsalen thalamus und tectum opticum bei Echsen. *Verh. Natur. Ges. Basel* 80, 209–225.

Senn, D. G. (1970b). Zur Ontogenese des Tectum opticum von *Natrix natrix* (L.). *Acta Anat.* 76, 545–563.

Senn, D. G. (1979). Embryonic development of the central nervous system. In *Biology of the Reptilia* (C. Gans, R. G. Northcutt, and P. S. Ulinski, eds.). Academic Press, London, Vol. 9, pp. 173–244.

Senn, D. G., and Northcutt, R. G. (1973). The forebrain and midbrain of some squamates and their bearing on the origin of snakes. *J. Morphol.* 140, 135–141.

Shanklin, W. M. (1930). The central nervous system of *Chameleon vulgaris. Acta Zool.* 11, 426–487.

Shanklin, W. M. (1933). The comparative neurology of the nucleus opticus tegmenti with special reference to *Chameleon vulgaris. Acta Zool.* 14, 163–184.

Sjöström, A. M., and Ulinski, P. S. (1985). Morphology of retinogeniculate terminals in the turtle *Pseudemys scripta elegans. J. Comp. Neurol.* 238, 107–120.

Stein, B. E., and Gaither, N. S. (1981). Sensory representation in reptilian optic tectum: Some comparisons with mammals. *J. Comp. Neurol.* 202, 69–87.

Stein, B. E., and Gaither, N. S. (1983). Receptive field properties in reptilian optic tectum: Some comparisons with mammals. *J. Neurophysiol.* 50, 102–124.

Stieda, L. (1875). Ueber den Bau des centralen Nervensystems der Schild-kroete. *Z. Wiss. Zool.* 25, 29–77.

Ströer, W. F. H. (1939). Zur vergleichenden Anatomie des primären optischen Systems bei Wirbeltieren. *Z. Anat. Entw. Gesch.* 110, 301–321.

Terashima, S. I., and Goris, R. C. (1975). Tectal organization of pit viper infrared reception. *Brain Res.* 83, 490–494.

Torcol, N. (1988). *Analyse Radioautographique et Peroxydasique du Système Visuel Primaire des Chéloniens.* Diplôme d'Etudes Approfondies, Université Paris, VII, 33pp.

Ulinski, P. S. (1986a). Organization of the corticogeniculate projections in the turtle *Pseudemys scripta. J. Comp. Neurol.* 254, 529–542.

Ulinski, P. S. (1986b). Ultrastructure of neurons in the dorsal lateral geniculate complex of turtles. *Brain Behav. Evol.* 29, 117–142.

Ulinski, P. S., and Nautiyal, J. (1988). Organization of retinogeniculate projections in the turtles of the genera *Pseudemys* and *Chrysemys. J. Comp. Neurol.* 276, 92–112.

Underwood, G. (1957). On lizards of the family Pygopodidae: a contribution to the morphology and phylogeny of the Squamata. *J. Morphol.* 100, 207–268.

Underwood, G. (1970). The eye. In *Biology of the Reptilia* (C. Gans and T. S. Parsons, eds.). Academic Press, London, 2, 1–97.

Underwood, G. (1977). Simplification and degeneration in the course of evolution of Squamate Reptiles. In *Colloques Internationaux du C. N. R. S.: Mécanismes de la Rudimentation des Organes chez les Vertébrés.* Edit. C. N. R. S., Paris, 266, 341–351.

Vesselkin, N. P., Agayan, A. L., and Monokonova, L. M. A. (1971). A study of thalamo-telencephalic afferent systems in frogs. *Brain Behav. Evol.* 4, 295–306.

Vesselkin, N. P., and Ermakova, T. V. (1978). Study of the connections of the thalamus in the frog, *Rana temporaria,* with the peroxidase method. *Zh. Evol. Biok. Fiziol.* 14, (2), 117–122 [in Russian].

Vesselkin, N. P., Ermakova, T. V., Repérant, J., Kosareva, A. A., and Kenigfest, N. B. (1980). The retinofugal and retinopetal systems in *Lampetra fluviatilis:* an experimental study using radioautographic and HRP methods. *Brain Res.* 195, 453–460.

Voneida, T. J., and Sligar, C. M. (1976). A comparative neuroanatomic study of retinal projections in two fishes: *Astyanax hubbsi* (the blind cave fish) and *Astyanax mexicanus. J. Comp. Neurol.* 165, 89–106.

Walls, G. L. (1942). The vertebrate eye and its adaptive radiation. *Bull. Cranbrook Inst. Sci.,* 19, 1–785.

Wang, R. T., and Halpern, M. (1977). Afferent and efferent connections of thalamic nuclei of the visual system of garter snakes. *Anat. Rec.* 187, 741–742.

Ward, R., Repérant, J., Rio, J. P., and Peyrichoux, J. (1987). Etude quantitative du nerf optique chez la Vipère Aspic (*Vipera aspis*). *C. R. Acad. Sci. Paris* 304, sér. III, 12, 331–336.

Warner, F. J. (1931). The cell masses in the telencephalon of the rattle snake, *Crotalus atrox. Proc. Kon. Ned. Acad. Wet. Amst.* 34, 1156–1163.

Warner, F. J. (1942). The development of the diencephalon of the American water snake (*Natrix natrix*). *Trans. Roy. Soc. Canada* (Montréal) 36, 53–70.

Warner, F. J. (1947). The diencephalon and midbrain of the American rattle snake, *Crotalus adamenteus. Proc. Zool. Soc.* (London) 116, 331–550.

Warner, F. J. (1955). The development of the pretectal cell masses of the American water snake (*Natrix sipedon*). *J. Comp. Neurol.* 103, 83–104.

Wiley, E. O. (1981). *Phylogenetics.* J. Wiley and Sons, New York.

Zagorul'ko, T. M. (1967). Functional connections of the general cortex and visual center of the tectum mesencephali of the turtle. *Zh. Evol. Biok. Fiziol.* 3, 342–351 [in Russian].

Zagorul'ko, T. M. (1968). Effects of intensity and wavelength of photic stimulus on evoked responses of general cortex and optic tectum in turtles. *Fiziol. Zh. SSSR Sechenova* 54, 436–446 [in Russian].

4

Optic Tectum

PHILIP S. ULINSKI, DENNIS M. DACEY, AND MARTIN I. SERENO

CONTENTS

I. INTRODUCTION

The optic tectum is made up of a pair of dome-shaped lobes that lie on the dorsal surface of the mesencephalon (Fig. 4.1A). All vertebrates have a tectum, but the term *optic tectum* is generally reserved for nonmammalian vertebrates, whereas the term *superior colliculus* is used for the homologous structure in mammals. The older term *optic lobes* is no longer in general usage. The size and shape of the optic tectum differs among species, being particularly large in birds. There is some correlation between the size of the tectum and the size of the eyes; within reptiles the tectum is relatively largest in diurnal lizards and crocodilians and smallest in burrowing forms with reduced eyes.

Some clue to the function of the tectum can be gleaned from a simple dissection of the brain, which shows that the optic tracts wind over the lateral surface of the brainstem and cover the dorsal aspect of the tectum (Fig. 4.1B). More sophisticated anatomical and physiological tracing techniques confirm that retinal ganglion cells carry visual information to the tectum and demonstrate that other sensory modalities are also represented in the tectum. They also demonstrate the existence of descending projections from the tectum to the brainstem reticular formation and spinal cord; by the early decades of this century (e.g., Huber and Crosby, 1943), studies using this technique led to the conclusion that the tectum is an important sensorimotor correlation center.

Functional observations on a variety of species have supported this conclusion. Experiments in which electrical currents are used to stimulate tectal neurons produce coordinated movements of the animal (Schapiro and Goodman, 1969; Distel, 1978). Lesions of the tectum produce defects or errors in visual tasks (Bass et al., 1973). The tectum thus appears to be important in the orientation of the animal to visual stimuli in the environment, as well as a wide range of orienting movements that include the turning of the pinnae toward sounds in cats (Stein and Clamann, 1981), snapping toward prey in frogs (Ingle, 1982; Grobstein, 1988), and eye movements in primates (e.g., Wurtz and Albano, 1980; Sparks, 1986; Sparks and Nelson, 1987).

The importance of orienting movements in behavior has prompted a number of anatomical and physiological studies of the optic tectum.

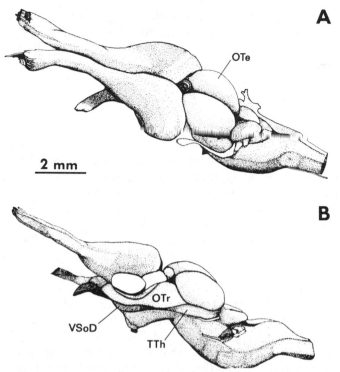

Fig. 4.1. *Thamnophis sirtalis.* (A) A dorsolateral view of the brain showing the optic tectum (OTe). (B) Removal of the telencephalon by cutting the cerebral peduncle reveals the relation of the optic tectum to the optic tract (OTr), tectothalamic tract (TTh), and ventral supraoptic decussation (VSoD).

The earliest studies used cell body stains to characterize the distribution of somata (e.g., Huber and Crosby, 1933); they showed that the tectum is a laminated structure composed of alternating layers of somata and axons. Application of the Golgi technique, notably by Santiago Ramón y Cajal (1911) and his brother Pédro Ramón (1896), to the tectum of lizards revealed that the tectum contains several morphologically distinct classes of neurons. Electrophysiological recording techniques showed that the retinal surface is mapped in a point-to-point or topological fashion onto the tectal surface (e.g., Kruger, 1969). Nauta and Fink-Heimer axon tracing techniques showed that the tectum gives rise to a large set of efferent projections, including some that ascend to the thalamus (e.g., Altman and Carpenter, 1961; Butler and Northcutt, 1971a; Ebbesson and Vanegas, 1976; Martin, 1969; Rubinson, 1968; Ulinski, 1977). Electron microscopic techniques provided an overall impression of the synaptic organization of the tectum and demonstrated the presence of complex forms of interactions between neurons (e.g., Ito, 1970; Lund, 1969, 1972; Sterling,

1971; Szekely et al., 1973; Tigges et al., 1973; Valverde, 1973; Graham and Casagrande, 1980). Recently, electrophysiological and behavioral studies have been combined to obtain information on the roles that individual neurons play in naturally occurring behaviors (e.g., Ewert, 1976; Ingle, 1976; Wurtz and Albano, 1980).

The introduction of horseradish peroxidase (HRP) and other intracellular markers as neuronal tracing and marking agents during the past few years (e.g., Mesulam, 1982) has permitted major advances. It is now possible to use these markers to correlate the morphology of neurons with the destinations of their axons and to visualize neurons that have been physiologically characterized (e.g., Gilbert, 1983). HRP procedures have recently been applied to the optic tectum in several taxa of reptiles to produce an overall picture of tectal organization that identifies the major populations of tectal neurons and outlines the ways in which tectal neurons are related to the fiber systems afferent to the tectum. This chapter will review the organization of the optic tectum in reptiles, emphasizing this new information and its impact on concepts of tectal organization.

The earliest studies of the tectum are summarized in Ramón y Cajal's 1911 monograph. The period up to 1936 is treated in the handbook by Ariëns-Kappers, Huber, and Crosby (1936). The role of the tectum in sensorimotor integration is reviewed by Ingle and Sprague (1975). The tectum in fishes (Vanegas, 1975), amphibians (Szekely and Lazar, 1976), reptiles (Northcutt, 1984), and mammals (Goldberg and Robinson, 1978; Wurtz and Albano, 1980; Huerta and Harting, 1984; Sparks, 1986) has been recently reviewed. The development of the tectum in reptiles has been reviewed by Senn (1979), and its neurophysiology has been reviewed by Belekhova (1979) and Hartline (1984). Ulinski (1983) has discussed the relationship of the tectum to the forebrain in reptiles and birds.

II. GENERAL ANATOMY
A. General

The following three sections will consider in some detail the morphology of the neurons that are efferent from the tectum (Section III), of those neurons that appear to be intrinsic to the tectum (Section IV), and of the fiber systems that are afferent to the tectum (Section V). However, it will be useful to provide a general survey of tectal structure, including the nomenclature of tectal layers and neuronal populations, before turning to more extensive treatments.

B. Tectal Lamination

Routine histological procedures such as cell body and fiber stains give the impression that the tectum consists of alternating layers of neuronal somata and axons (Fig. 4.2). This is illustrated in Figs. 4.3 and

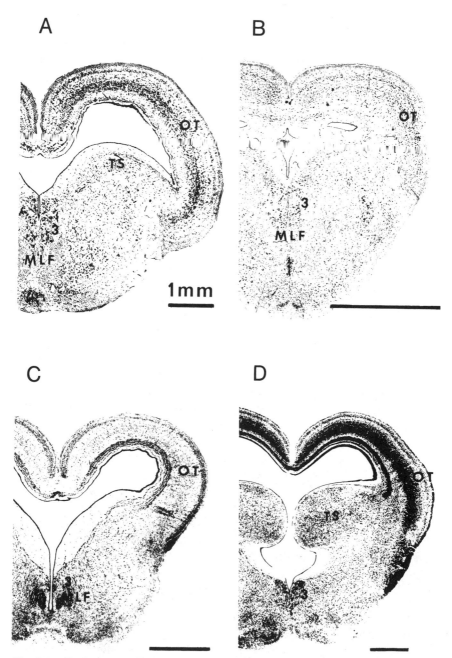

Fig. 4.2. Optic tectum. The general anatomy of the optic tectum is shown in transverse sections of the midbrain through the brains of representatives of each major taxon. (A) *Tupinambis tequixin*. Tegu lizard. (B) *Nerodia sipedon*. Water snake. (C) *Pseudemys scripta*. Red-eared turtle. (D) *Alligator mississippiensis*. American Alligator.

4.4, which show cross-sections through the midbrains of a turtle (Fig. 4.3) and a snake (Fig. 4.4). The cell body stain demonstrates the tendency for layers of densely packed neuronal somata to be separated by clear spaces that are composed largely of myelinated and unmyelinated fiber systems.

Several nomenclatorial systems have been applied to these layers in reptiles. The first stems from the work of Pedro Ramón (1896) on the lizard *Lacerta* and recognizes 14 layers (see Fig. 4.5). Layer 1 consists of the cell bodies of tanycytes, the processes of which extend throughout the depth of the tectum to end as subpial endfeet. The remaining layers consist of alternating bands of somata and fibers. Although this system proved entirely adequate for *Lacerta*, it has led to some confusion when applied to other species because there is enough interspecific variation in lamination patterns to make the identification of some of Ramón's layers difficult in other species. There is also some potential confusion with other, numerically based, nomenclatures. For example, Santiago Ramón y Cajal (1911) recognized 14 layers in the optic tectum of birds but numbered them in the opposite order (so that the ependymal cell bodies were situated in Layer 14). Huber and Crosby (1933) sought to allay confusion by proposing an alternate nomenclature in which the various layers were given Latin names. This nomenclature has the virtue of being more or less applicable to all vertebrates; however, it is generally necessary to subdivide some of the layers in a particular species. More recently, Senn (1968, 1979) recognized periventricular, central, and superficial groups of cells. The pros and cons of various schemes for naming the tectal layers have been discussed by Northcutt (1983).

The nature and clarity of layers among orders of reptiles and among species within orders vary substantially (see Fig. 4.2). In lizards, tectal cytoarchitecture has been described in most detail in the tegu, *Tupinambis teguixin* (Butler and Ebbesson, 1975) and the desert iguana, *Dipsosaurus dorsalis* (Peterson, 1981). In addition, Senn (1968a, 1968b), Senn and Northcutt (1973), and Northcutt (1978, 1983) have surveyed patterns of tectal organization in a large number of lizards. The tectum in lizards usually contains clearly defined laminae (see Fig. 4.3). There are generally several thin periventricular layers and as many as seven sharply defined layers of somata and fibers in the superficial tectum. Two major variants of this general pattern have been recognized in lizards (Northcutt, 1978). The first is seen in agamids, chamaeleonids, iguanids, teiids, and varanids and is called the iguanid pattern. In this variant, Ramón's lamina 14 is always the major layer that contains the axons of retinal ganglion cells. Lamina 12 is at least half the thickness of lamina 14. Laminae 8 and 10 contain many densely packed cells. Lamina 7 is also subdivided into inner and outer

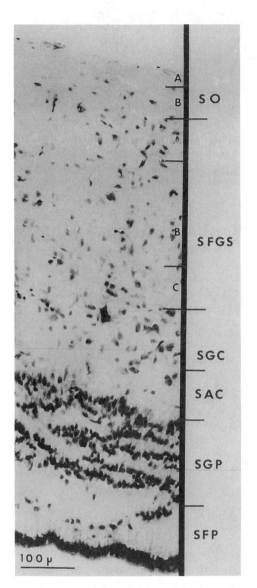

A

B   S O

B   SFGS

C

SGC

SAC

SGP

SFP

100 μ

Fig. 4.3. *Pseudemys scripta*. Tectal cytoarchitecture. The lamination pattern of the tectum is shown in a transverse section through the brain of a red-eared turtle. (From Schechter and Ulinski, 1979.)

subdivisions with the neurons of the inner subdivision scattered within lamina 6. The periventricular layers are relatively reduced. The second pattern is seen in anguids, cordylids, gekkonids, gerrhosaurids, helodermatids, lanthanotids, lacertids, pygopodids, scincids, xantusiids, and xenosaurids and is called the lacertid pattern. In this variant, the superficial layers are relatively poorly developed. Lamina 12 is the principal optic layer and is always better developed than lamina 14. Laminae 8 and 10 are not nearly as well developed as in other

Fig. 4.4. *Thamnophis.* Tectal cytoarchitecture. The lamination pattern of the tectum is shown in a transverse section through the brain of a garter snake. (From Dacey and Ulinski, 1986a.)

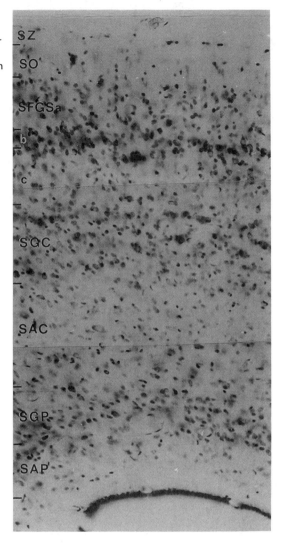

lizards, and lamina 7 consists of single groups of cells with little if any overlap with lamina 6. By contrast, the periventricular laminae are well developed, with lamina 5 being twice the thickness of the comparable lamina in the iguanid pattern. The tectum in amphisbaenians and several groups of lizards with relatively reduced eyes (anelytropsids, *Anniella*, dibamids, feyliniids, some microteids, and some scincids) fits neither the iguanid nor the lacertid pattern. Instead, the tectum in these groups shows a reduction or even loss of the superficial layers and the periventricular layers are diffuse.

In addition to these interspecific differences, Peterson (1981) has pointed out that the lamination pattern of any given species also shows regional variations. She describes five areas within the optic tectum of the desert iguana, *Dipsosaurus dorsalis,* based on variations in the development and clarity of the various layers. She correlates this regional variation in lamination pattern to the pattern of retinal afferents as demonstrated in Fink-Heimer and autoradiographic cases.

Huber and Crosby (1933) have described the tectum in a crocodilian, *Alligator mississippiensis.* Its lamination pattern generally resembles that of lizards with the lacertid pattern. Lamina 5 is poorly developed and, by contrast, lamina 7 is relatively thick. Whereas all the superficial layers are present, they are poorly defined.

Northcutt (1978, 1983) has described the tectum in *Sphenodon.* Its periventricular layers are clearly laminated and well developed, lamina 7 is reduced, and the superficial layers are moderately well developed. As in lizards with the lacertid pattern, lamina 12 is the principal optic layer.

The tectum in turtles has been described by Huber and Crosby (1933) and Schechter and Ulinski (1979). It contains several distinct periventricular layers. Both the intermediate layers 6 and 7, and the superficial layers are relatively poorly developed.

The tectum of snakes has been described by Huber and Crosby (1933), Senn (1969), Senn and Northcutt (1973), Halpern and Frumin (1973), Schroeder (1981b), and Dacey and Ulinski (1986a). As a group, the snakes show relatively poorly developed tectal laminae. Their periventricular layers are poorly developed. The intermediate layers contain a relatively large number of cells, especially in the crotalid snakes and those boids that have infrared systems (Schroeder, 1981a; Auen, 1976). The basic pattern of superficial layers is present, but these are relatively difficult to discern. Snakes are unique among the reptiles in that the layer carrying retinal ganglion cell axons, the stratum opticum, is separated from the pia. Senn (1979) studied the development of the optic tectum in the grass snake, *Natrix natrix,* and discovered that the lamination pattern is more distinct in embryonic snakes than in adults. The periventricular layer in adults, for example, represents a fusion of several distinct embryonic layers.

C. Neuronal Populations
Until the recent development of the HRP technology, essentially all of our information on the morphology of tectal neurons was derived from the use of Golgi preparations. These studies all report the existence of several classes of neurons in the tectum as distinguished by the morphology of their somata and dendrites. Figs. 4.5 to 4.7 illustrate the major types of tectal neurons, which derive from Pedro Ramón's work on lizards.

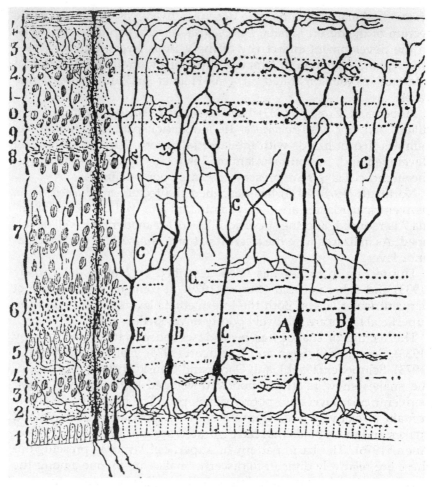

Fig. 4.5. Radial neurons. This and the following four illustrations are taken from Pedro Ramón's 1896 study of the tectum in the lizards *Lacerta* and *Chamaeleo*. They are all drawings done from Golgi preparations. This drawing illustrates Ramón's nomenclature (layers 1 through 14) for the tectal layers. Ependymal cells (not labeled) and radial neurons (A through E) are shown.

The most obvious tectal neurons are those with dendrites that extend radially across the tectal laminae (Figs. 4.5 to 4.7). These neurons have been called pyramidal, piriform, fusiform, or radial cells. They are found in all layers of the tectum but are most common in the strata griseum centrale and the griseum periventriculare. Superficially located radial neurons (Figs. 4.6 and 4.7) have smaller somata, shorter apical dendrites, and more extensive basal dendritic trees. More deeply situated radial neurons (Fig. 4.5) have larger somata, longer apical dendrites—often extending into the stratum opticum—and

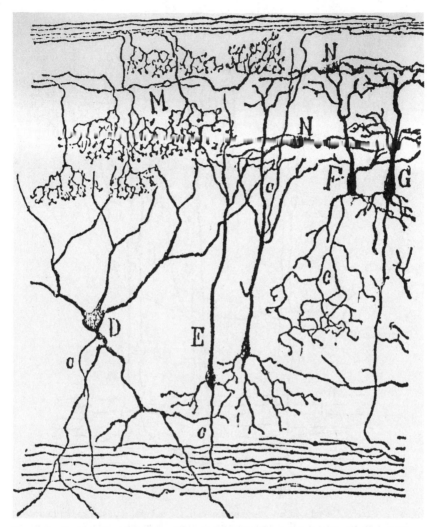

Fig. 4.6. *Lacerta.* Tectal neurons. A ganglion cell (D), radial neurons (E, F, and G), horizontal cells (N), and putative retinotectal arbors (M). (From Ramón, 1896.)

basal dendritic trees that are reduced or absent. The axons of radial neurons originate from their somata, from proximal dendrites, or occasionally from positions on the apical dendrites quite remote from the somata. In the last case, the radial neurons are sometimes called shepherd's crook cells.

A second type of tectal neuron is the ganglion cell. These are neurons with large somata that are often triangular and with extensive dendrites that extend toward the pial surface in a cone-shaped configuration (Figs. 4.6 and 4.7). The axon typically originates from the

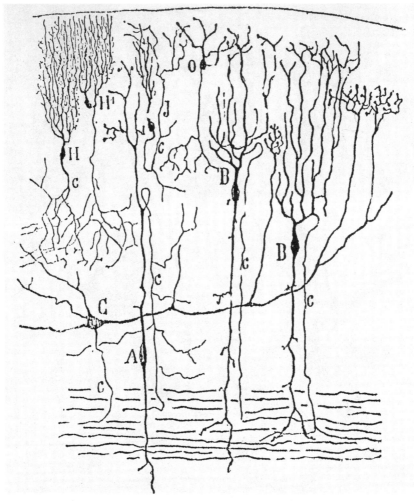

Fig. 4.7. *Lacerta.* Tectal neurons. A ganglion cell (C), several types of radial neurons (A, B, H, and J), and a small neuron with descending axon (0). (From Ramón, 1896.)

soma. Ganglion cells can be identified in Nissl preparations because of their size and shape. Such preparations suggest that they are relatively rare.

Stellate cells are a relatively frequent and variable population of tectal neurons (Fig. 4.8). They have small somata located in the superficial layers of the tectum. Their dendrites extend out in a star-shaped pattern, often bearing elaborate appendages. The axons may either ascend or descend within the tectum, but typically arborize extensively.

A further type of tectal neuron is the horizontal cell, which is usually found in the superficial layers (Fig. 4.9). These neurons have dendrites that extend relatively long distances concentric with the curva-

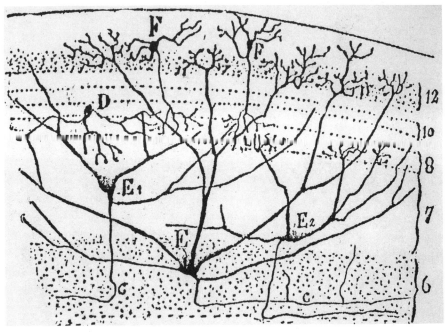

Fig. 4.8. *Lacerta.* Tectal neurons. Ganglion cells (E, E$_1$, and E$_2$) and small neurons with descending axons (F). (From Ramón, 1896.)

ture of the tectum. The primary dendrites are smooth, whereas the secondary dendrites may have dendritic specializations. The axons usually originate from the somata and travel relatively long distances.

Classification of tectal neurons into such general groups provides a convenient vocabulary with which to describe morphological results obtained by a variety of techniques. However, it can be misleading in that it greatly obscures the extent to which tectal neurons vary in morphology. Schechter and Ulinski (1979), for example, attempted to deal with the variation in dendritic morphology that is seen in radial neurons in *Pseudemys* by arranging the apical dendrites of radial neurons along two continua, one in the extent of branching and a second in the density of dendritic specializations and excrescences. Similarly, most authors (e.g., Butler and Ebbesson, 1975; Ramón, 1886; Northcutt, 1984) have recognized several types of radial neurons based on variations in the shape of their soma and the arborization patterns of their dendrites and axons. The central issue here is whether radial neurons, for example, should be viewed as a single population of neurons that vary in morphology, or whether the variation reflects the existence of several types of radial neurons that should be explicitly recognized. This is a general problem, which arises in describing the morphology of any group of neurons and has been discussed extensively in the case of retinal ganglion cells (e.g., Rowe and Stone,

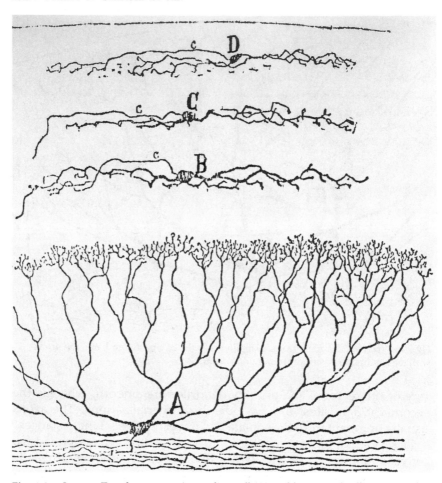

Fig. 4.9. *Lacerta*. Tectal neurons. A ganglion cell (A) and horizontal cells (B, C, and D). (From Ramón, 1896.)

1977). Although the best approach remains a matter of controversy, most authors agree that a classification should be based upon as many criteria as is possible. The ideal is to relate the different populations of tectal neurons just discussed to the pattern of their projections either within or out of the tectum. It is now possible to do this using HRP techniques, and in the following sections tectal neurons will be divided into several distinct subclasses based upon their intrinsic and extrinsic connections.

III. EFFERENT PROJECTIONS
A. General Pattern of Tectal Efferent Projections

The first studies of the efferent projections of the reptilian tectum were based on careful observations of serial sections stained to visualize myelinated axons (e.g., Huber and Crosby, 1933). These studies show that both ipsilateral and contralateral pathways connect the tectum with the brainstem reticular formation. It was suspected that these pathways include tectal axons that descend to the brainstem and ascend to the diencephalon, but the difficulty in determining the polarity of axons in myelin preparations left the issue open. The ambiguity has been resolved with the application of Nauta and Fink-Heimer techniques to the study of tectal efferents (turtles: Foster and Hall, 1975; lizards: Butler and Northcutt, 1971b; snakes: Ulinski, 1977; crocodilians: Braford, 1972). More recently, the reptilian tectum has been studied with tracing techniques based on the anterograde transport of tritiated amino acids (Schroeder, 1981b) and HRP (Auen, 1976; Kass et al., 1978; Dacey and Ulinski, 1986a; Sereno, 1985).

All of the experimental studies, regardless of species, demonstrate a single pattern of efferent projections. This pattern is illustrated by HRP investigation (Sereno, 1985) of tectal efferents in *Pseudemys scripta* (Fig. 4.10). The HRP injection site is seen as a region of stipple in sections F through K. The pathways issuing from the injection site are charted onto sections spaced through the brainstem and diencephalon. The chartings show that several fascicles of axons course laterally from the injection site and leave the tectum. These can be divided into descending and ascending projections.

The first of the descending projections consists of axons that course ventrally and terminate in the mesencephalic tegmentum within the nucleus mesencephalicus profundus (PMr) and the nucleus of the medial longitudinal fasciculus (mlf). As the fibers are traced caudally (sections I and H), they sort out into four fascicles. The most dorsal fascicle terminates just caudal to the tectum in the magnocellular part of nucleus isthmi (lmc). The second is formed by fibers that cross the midline in the mesencephalic tegmentum and turn to descend in the contralateral tegmentum as the predorsal bundle or dorsal tectobulbar tract (Tbd). These axons reach a variety of cell groups in the caudal brainstem. The third fascicle is composed of fibers that descend in the ipsilateral tegmentum as the intermediate tectobulbar path (Tbi). Finally, a contingent of fibers course along the ventrolateral surface of the ipsilateral brainstem as the ventral tectobulbar path (Tbv). These latter two paths reach several cell groups in the caudal brainstem. The particular chartings shown in Fig. 4.10 do not include the spinal cord; however, tectal axons have been traced in the predorsal bundle as far as the cervical levels of the spinal cord in turtles (Foster and Hall, 1975), lizards (Foster and Hall, 1975), and snakes (Gruberg et al., 1979; Dacey and Ulinski, 1986a).

Ascending projections from the tectum are illustrated in sections K through M. Some fibers course ventrorostrally from the tectum and terminate bilaterally in the pretectum in nuclei such as the nucleus

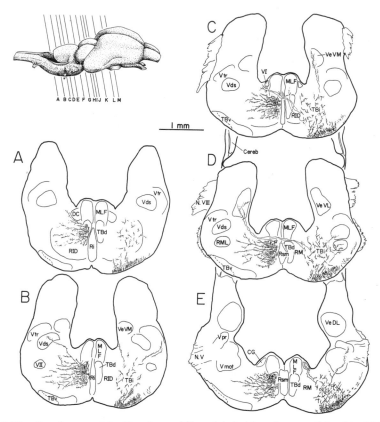

Fig. 4.10. *Pseudemys scripta.* Drawings of fibers labeled by a large unilateral tectal injection of HRP. Retrogradely labeled tracts are marked with an asterisk. The planes of sections A through M are indicated on the brain in the inset. (From Sereno, 1985.)

lentiformis mesencephali and the pretectal nucleus. Other fibers turn and run rostrally in the tectothalamic tract (TTh). These terminate in nucleus rotundus in the dorsal thalamus, in several ventral thalamic nuclei (e.g., the suprapeduncular nucleus, SP), and in the hypothalamus. The projections tend to be bilateral. They reach the contralateral diencephalon principally through the ventral supraoptic decussation, but some fibers cross through the posterior and habenular commissures.

Tracing experiments of this sort adequately display the overall pattern of tectal efferents but provide little or no information on which types of tectal neurons contribute to the various pathways. Does each set of efferent projections originate from a different population of neurons? Can an individual neuron contribute to more than one efferent pathway or terminate in multiple regions within a given pathway? These questions have been addressed directly by intracellular injec-

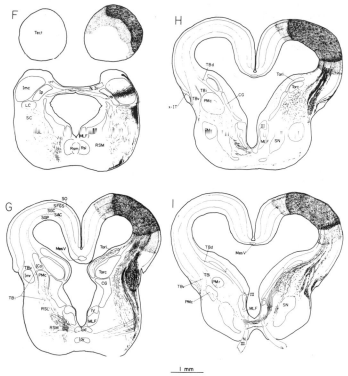

Fig. 4.10. *(cont.)*

tion techniques to demonstrate the complete morphology of individual neurons (e.g., May et al., 1982; Berson and Hartline, 1988; Rhoades et al. 1989). The limitation of these experiments is the technical one of successfully injecting a large number of relatively small neurons. However, some progress has also been made by combining two types of experiments. The first is to inject HRP into the ascending and descending pathways and then look for retrogradely labeled neurons in the tectum so as to demonstrate which populations of tectal neurons contribute to each pathway. The second type of experiment is to trace individual axons from small injections of HRP in the tectum in order to show the morphology of axons in each tectal pathway. In combination, these experiments can provide a relatively complete account of the organization of tectal efferents. Results from procedures of this sort have been obtained in snakes (Dacey and Ulinski, 1986a, 1986b) and turtles (Sereno, 1985; Sereno and Ulinski, 1985).

B. Tectobulbar Neurons

1. GENERAL Tectobulbar neurons are those tectal neurons whose primary terminations are in the brainstem reticular formation. They

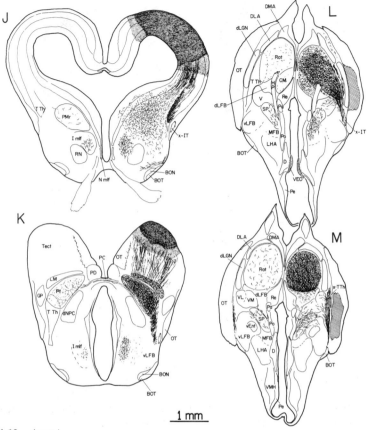

Fig. 4.10. (cont.)

can be identified by injecting HRP into one of the three tectobulbar paths. Experiments of this sort show that tectobulbar neurons are distributed throughout the central and periventricular layers of the tectum both in turtles (Sereno and Ulinski, 1985) and snakes (Dacey and Ulinski, 1986b), but the distribution and morphology of the tectobulbar neurons that project contralaterally and ipsilaterally vary.

2. Contralateral Tectobular Neurons

Tectobulbar neurons that have axons crossing the midline generally resemble the ganglion cells seen in Golgi preparations. Their somata generally lie in the central layers of the tectum. They have large, multipolar somata and four to six thick spiny dendrites. Some of the dendrites descend into the stratum album centrale, whereas most extend toward the pial surface. The extent of the ascending dendrites differs between snakes and turtles. The ascending dendrites stop quite abruptly at the upper boundary of the stratum griseum centrale of snakes (Fig. 4.11) but extend well into the stratum fibrosum et gri-

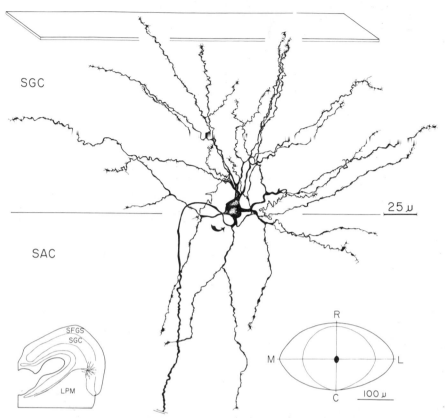

Fig. 4.11. *Thamnophis sirtalis.* Crossed tectobulbar cell. The axon of this neuron (arrow) descends vertically through the central white without branching or collateralizing and makes a sharp lateral turn at the border of the central white with the periventricular gray. The position of this cell and its axon in the tectum is shown in the inset at the lower left. The soma of this cell is situated at the border of the central gray and central white. The dendritic field spread is 320 × 180 μm. (From Dacey and Ulinski, 1986b.)

seum superficiale in turtles (Fig. 4.12). This difference is significant because it makes it unlikely that the crossed tectobulbar neurons of snakes can receive direct retinal input. However, the dendritic fields are extensive in both cases, ranging up to 400 μm in diameter in *Thamnophis* and 700 μm in diameter in *Pseudemys*.

The axons of the crossed tectobulbar neurons are relatively thick, measuring from 2.5 to 3.0 μm in diameter. They course laterally following the ventral border of the stratum album centrale and leave the tectum at its ventrolateral border. The overall configuration of the crossed axons and their collaterals are illustrated in diagrammatic form in Fig. 4.13, which shows a sketch of a crossed tectobulbar cell in *Pseudemys* reconstructed through serial sections. The axons give

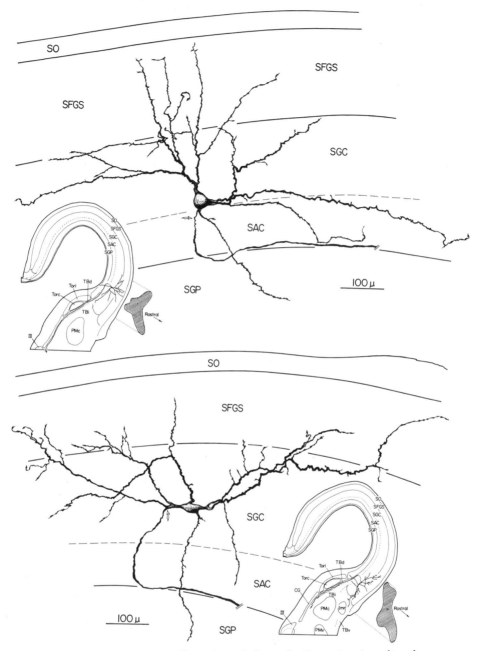

Fig. 4.12. *Pseudemys scripta.* Crossed tectobulbar cells. Reconstruction of two large dorsal tectobulbar pathway (TBd) neurons. They were labeled after an HRP injection into the contralateral medial reticular formation. The open arrows indicate the axon origins. A schematic *en face* view is given alongside each low-power drawing. The dendrites of these neurons were covered with very fine spicules. (From Sereno and Ulinski, 1985.)

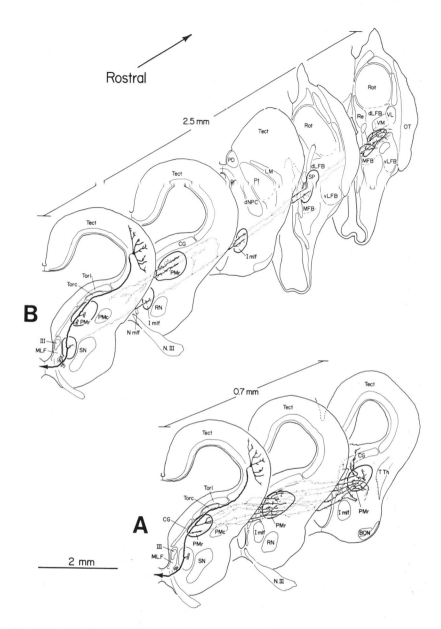

Fig. 4.13. *Pseudemys scripta.* Tectobulbar axon. Schematic diagram of the ipsilateral collaterals of a dorsal pathway (TBd) axon. (A) The first branches arborize in the midbrain reticular nucleus profundus mesencephali rostralis (PMr). (B) The PMr branches have been removed here to better illustrate a robust collateral that courses under PMr, eventually arborizing in the ventral thalamic suprapeduncular nucleus (SP). It gives off terminals throughout its course through the interstitial nucleus of the medial longitudinal fasciculus (lmlf) as well as several branches that reach PMr from below. The main trunk of this axon crosses the midline to run in the contralateral predorsal bundle. (From Sereno and Ulinski, 1985.)

rise to ipsilateral collaterals that run rostrally into the midbrain tegmentum and the diencephalon. The collateral reconstructed in Fig. 4.13A extends principally into the nucleus mesencephali profundus. It consists of several branches that run parallel to each other through the rostrocaudal extent of the nucleus. Other ascending branches are shown in Fig. 4.13B. They run rostrad through the midbrain and terminate in the ventral thalamus.

The main branch of the axon continues ventromedially through the tegmentum, crosses the midline ventral to the medial longitudinal fasciculus, and enters the contralateral predorsal bundle. Its caudal trajectory is best illustrated in horizontal sections. Fig. 4.14 shows a diagram of the major branches of a crossed tectobulbar axon as it courses caudally through the brainstem. They run in the predorsal bundle and issue collaterals at regular intervals. These extend into the medial components of the brainstem reticular formation. The anatomy of one such collateral, from the mesencephalic reticular formation, is illustrated in Fig. 4.15 from an experiment in which a predorsal axon was filled by a tectal injection in *Thamnophis*. It is typical in that the stem branch is relatively thick and gives rise to a larger number of thin collaterals that fan out into the reticular formation. Each collateral bears many varicosities that presumably correspond to synaptic boutons. It has not been possible to label completely crossed tectobulbar axons in turtles, but axons in the predorsal bundle in *Thamnophis* continue caudally into the spinal cord (Fig. 4.16). They run in the ventral funiculus and have axons that turn dorsally into the gray matter of the cervical cord.

3. Ipsilateral Tectobulbar Neurons

Tectobulbar neurons, the axons of which do not cross the midline, resemble the ganglion cells, radial cells, and "plumed cells" (see Fig. 4.9) of Golgi preparations. They have somata positioned in the stratum griseum centrale, stratum album centrale, and stratum griseum periventriculare. They are rather uniform in morphology in *Thamnophis* (Fig. 4.17). Their fusiform somata are 12 to 20 μm long and fusiform in shape. Several dendrites originate from each soma, but the neuron is dominated by a stout primary dendrite that arises from its pole and ascends radial to the pial surface so that it can be classified as a radial cell. These dendrites issue several thin branches that bear a moderate density of short spicules or hairlike protrusions and end abruptly at the interface of the stratum griseum centrale and the stratum fibrosum et griseum superficiale. The morphology of ipsilateral tectobulbar neurons in *Pseudemys scripta* is more variable. Some, like those in *Thamnophis*, have a generally radial cell configuration (Fig.

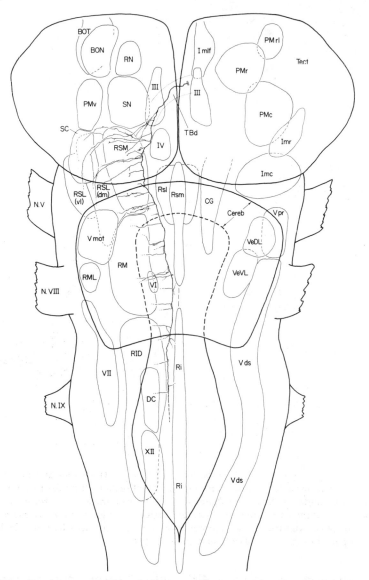

Fig. 4.14. *Pseudemys scripta.* Tectobulbar axon. Reconstruction from horizontal serial sections of the contralateral course of a dorsal pathway (TBd) axon traveling in the predorsal bundle. This axon was labeled after a small HRP injection into the right tectum. It had ipsilateral collaterals similar to those schematically illustrated in Fig. 4.15. Most of the several thousand synaptic boutons supported by this axon were located in the medial half of the reticular formation. A few branches also entered the raphe caudally. The axon was not completely filled past the level of the trigeminal motor nucleus.

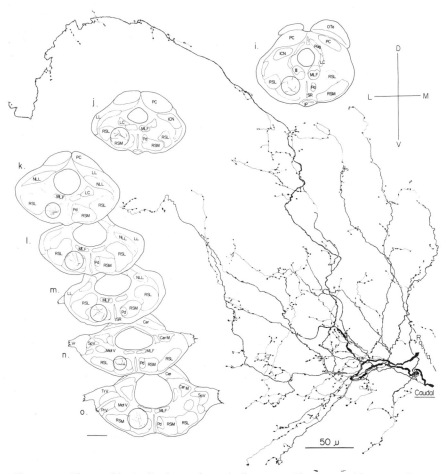

Fig. 4.15. *Thamnophis sirtalis.* Crossed tectobulbar axon. This large-caliber axon (3 μm in diameter) was reconstructed through 25 serial sections from the rostral midbrain to the caudal medulla. This figure shows the seven most caudal sections (i through o) and the part of the axon that terminates in section j. It forms a rostrocaudally elongated cylindrical zone of termination in reticularis medialis and reticularis inferior pars dorsalis. (From Dacey and Ulinski, 1986a.)

4.18), whereas others (Figs. 4.19 and 4.20) have more widespread dendritic fields. The major difference between the two species, however, is that the dendrites of the ipsilateral cells in *P. scripta* extend well into the stratum fibrosum et griseum superficiale, whereas those in *Thamnophis* do not reach the superficial layer.

Axons of ipsilateral cells course dorsally or ventrally in the tectum, depending on the positions of their somata, and enter the stratum album centrale. They may collateralize in this layer (Fig. 4.21) but eventually exit the ventrolateral border of the tectum. Descending col-

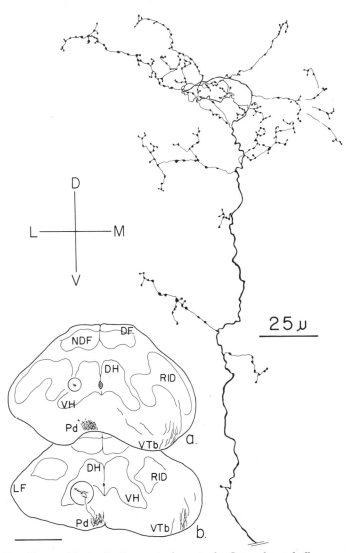

Fig. 4.16. *Thamnophis sirtalis.* Tectospinal terminals. Crossed tectobulbar axons were traced to the cervical spinal cord, where they ascended vertically from the predorsal bundle to terminate in the ventral horn of the spinal cord. These axons ended in a single terminal arbor just dorsal to the region of the motoneuron somata. (From Dacey and Ulinski, 1986a.)

laterals run caudally in the brainstem tegmentum in two tracts. An intermediate tract runs through the central region of the tegmentum, and a ventral tract runs along the ventrolateral surface of the brainstem. The main branches of the axons give off collaterals at regular intervals throughout their trajectory. The collaterals arborize in the lateral regions of the brainstem. Fig. 4.22 shows one such collateral

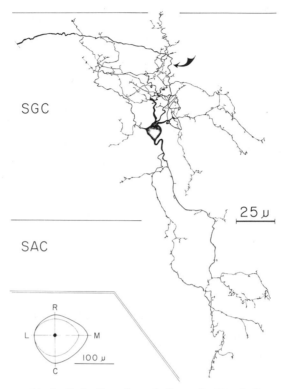

Fig. 4.17. *Thamnophis sirtalis.* Ipsilateral tectobulbar cells. Tectobulbar neurons are multipolar but radially elongate. Their dendritic fields occupy the central gray and extend into the central white but never enter the superficial gray. The inset at the bottom of this figure is an approximate projection of the cell's dendritic field spread in the horizontal plane. The black dot represents the position of the soma. The inner circular profile represents a primary dendritic zone, encompassing the bulk of the dendritic arborization, whereas the outer circle outlines the extension of this primary zone by a few tertiary dendrites. (From Dacey and Ulinski, 1986b.)

filled by an injection of ipsilateral axons running in the intermediate tract in *Thamnophis.* These collaterals bear a general resemblance to the collaterals on contralateral tectobulbar axons in that several thin branches, bearing varicosities, radiate into the reticular formation. Because of their relative positions in the brainstem, the contralateral and two ipsilateral pathways terminate in somewhat separate but overlapping territories throughout the reticular formation caudal of the brainstem.

4. General Features
Each tectobulbar cell has an extensive axonal system that includes both ascending and descending branches. Each individual cell therefore projects to several nuclei in the brainstem, and the projection to each individual target structure involves a highly collateralized pro-

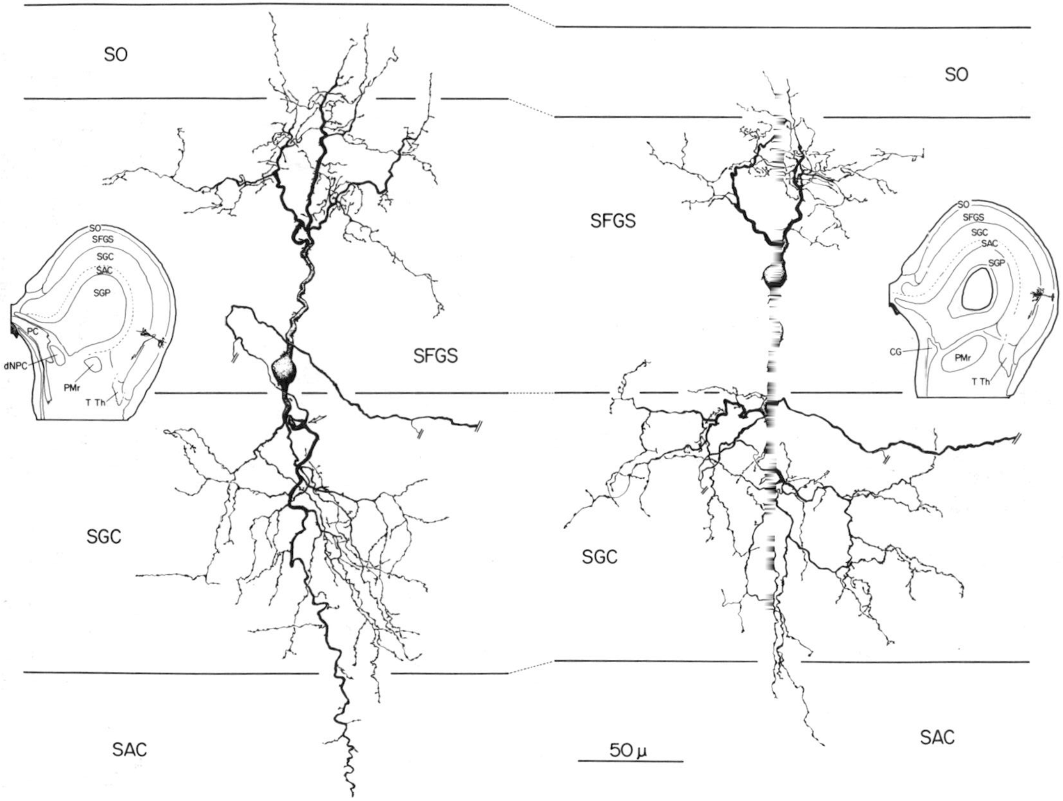

Fig. 4.18. *Pseudemys scripta.* Ipsilateral tectobulbar cells. Reconstruction of two small ventral pathway neurons in the SFGS. They were labeled after ipsilateral HRP injections into the lateral reticular formation. Both have a radially oriented ascending dendrite that gives rise to a plexus of filamentous dendritic appendages just beneath the SO as well as a radial descending dendrite that emits appendages in the SGC. Their thin axons (origins at open arrows) emit "local" collaterals into the SGC (see Fig. 4.23) before leaving the tectum to enter the small-caliber (1 μm) component of the ventral tectobulbar pathway, TBv(sm). (From Sereno and Ulinski, 1985.)

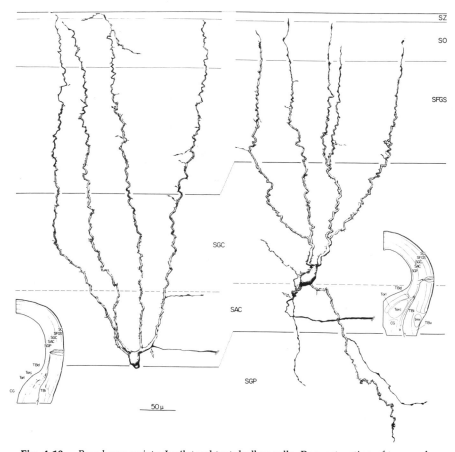

Fig. 4.19. *Pseudemys scripta*. Ipsilateral tectobulbar cells. Reconstruction of two probable intermediate pathway (TBi) neurons. They were labeled after an ipsilateral HRP injection into the lateral reticular formation. Both have radial ascending dendrites covered with very fine spicules similar to those on TBd neurons. Their medium-caliber axons (open arrows) could be traced far enough out of the tectum to suggest that they entered the TBi (rather than the TBv). No collaterals were observed in the tectum. (From Sereno and Ulinski, 1985.)

jection that effectively fills the target. This implies that there may be a convergence of information from many tectal neurons to a single region of the brainstem. This premise has been investigated explicitly in experiments (Fig. 4.23) in which two disjunct HRP injections were made in the tectum in *Pseudemys scripta* (Sereno, 1985). Labeled axons could then be followed from the two injection sites into the brainstem. Although the fascicles were initially separate, the individual axons eventually intermingled in the brainstem so that two adjacent axons may originate from disjunct tectal loci. Morphologically similar tectobulbar cells have recently been described in squirrel monkeys

Fig. 4.20. *Pseudemys scripta*. Ipsilateral tectobulbar cell. Reconstruction of a probable ventral pathway (TBv) neuron. It was labeled after an ipsilateral HRP injection into the lateral reticular formation. The primary dendrites initially travel horizontally before turning upward, generating a mediolaterally elongated dendritic field (see *en face* view in low-power drawing). Many filamentous dendritic appendages are given off in the central gray (SGC). (From Sereno and Ulinski, 1985.)

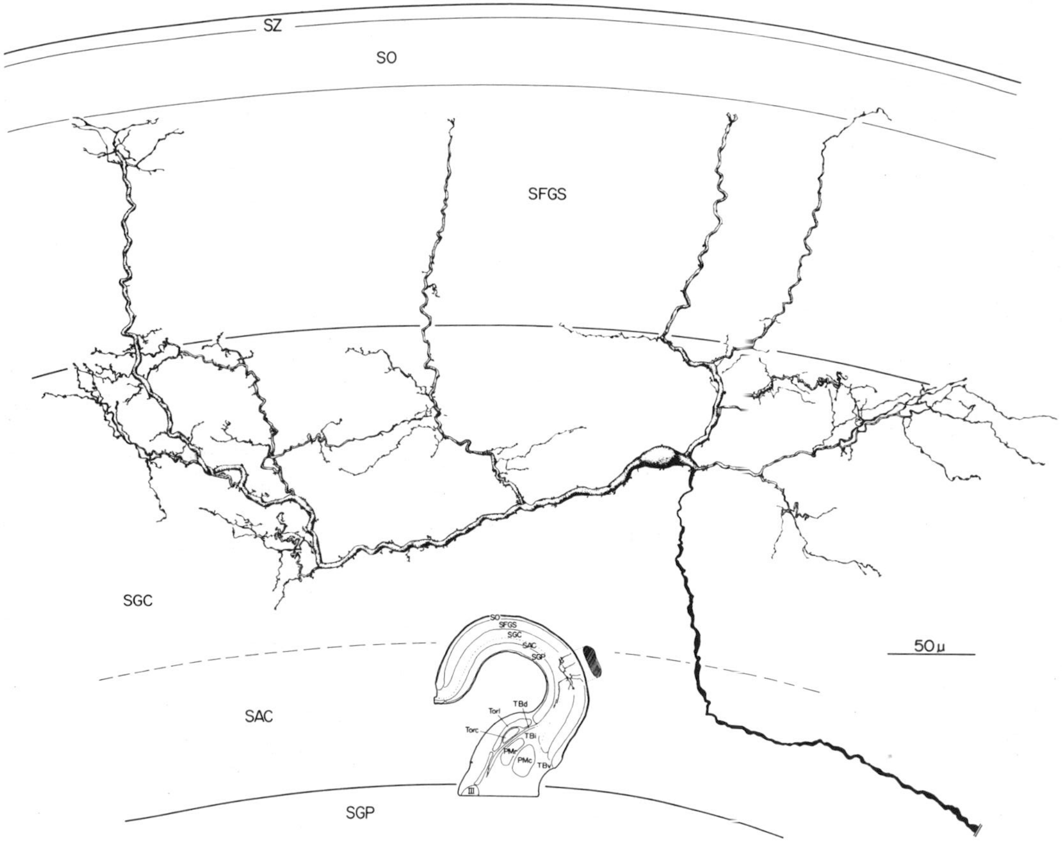

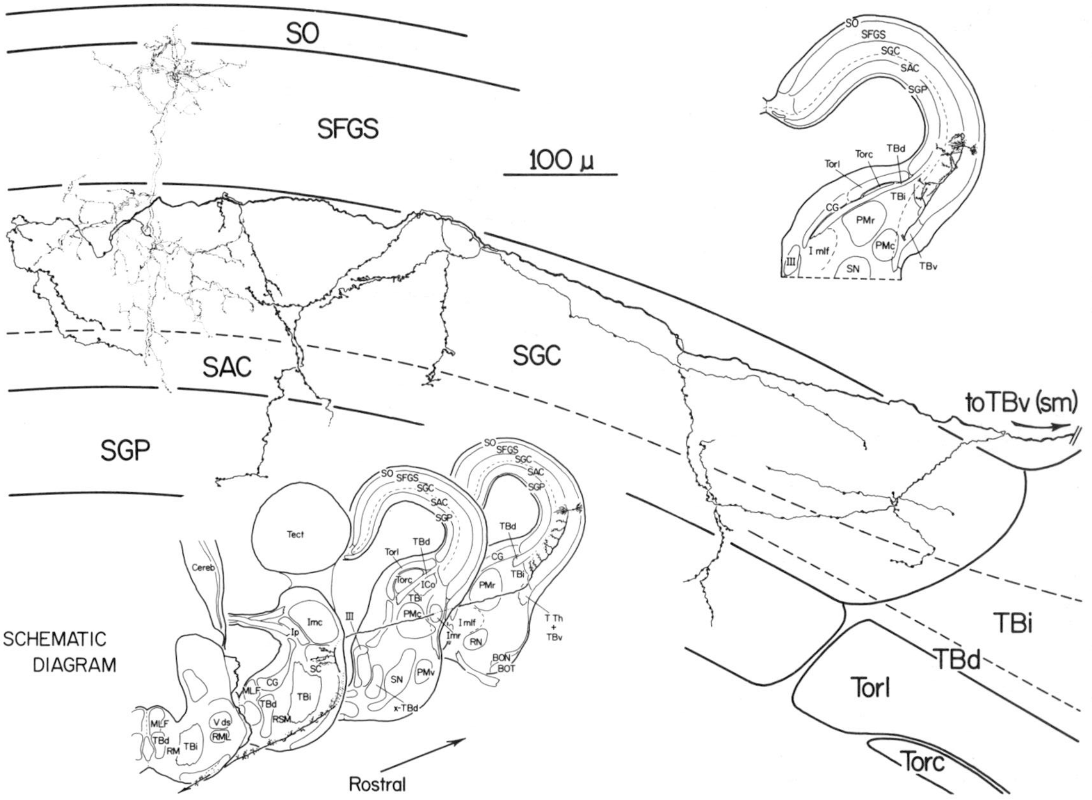

Fig. 4.21. *Pseudemys scripta.* Reconstruction of the intratectal collaterals of one of the TBv (sm) neurons illustrated in Fig. 4.20. The axon arises from the descending radial dendrite and immediately begins emitting strings of synaptic boutons downward into the SGC. Some penetrate the SGP. The main trunk then leaves the tectum (see schematic diagram) to enter the small-caliber component of the TBv. It passes through lmr without branching, emits several long collaterals into a small-celled nucleus ventral to lmc, and finally turns ventrally to run along the ventrolateral surface of the brainstem, where many short branches arise. (From Sereno and Ulinski, 1985.)

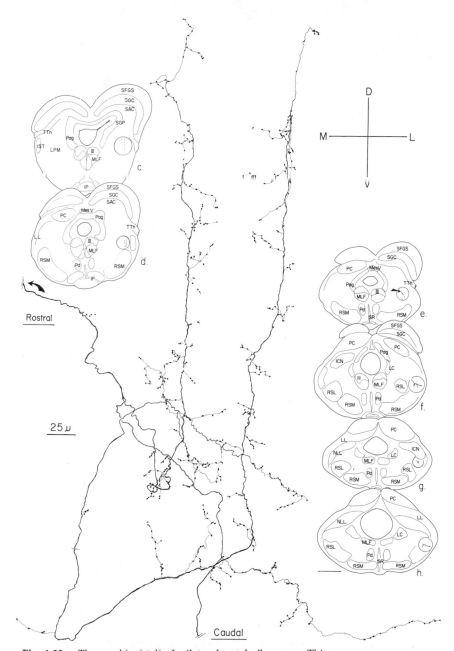

Fig. 4.22. *Thamnophis sirtalis.* Ipsilateral tectobulbar axon. This axon was reconstructed through 16 serial sections from an HRP injection site in the tectum. The figure shows six middle sections from the series. It illustrates a series of terminal collaterals in reticularis superior pars lateralis (RSL) and reticularis medialis (RM) (B and C). Circled areas in the 80-μm coronal sections show the portion of the axon as it descends through the brainstem. Axons from two adjacent sections were collapsed in sections a, h, and i for illustrative purposes. (From Dacey and Ulinski, 1986a.)

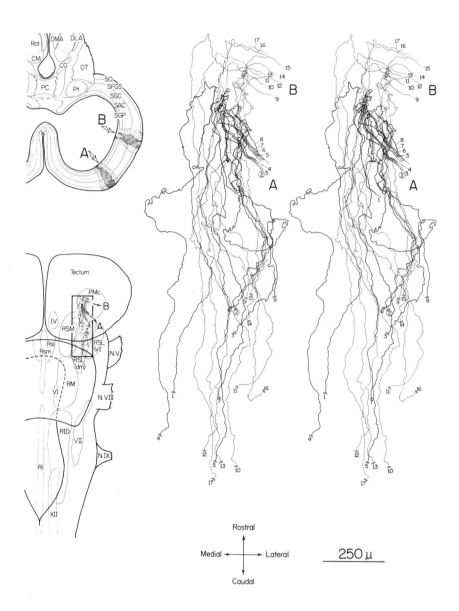

Fig. 4.23. *Pseudemys scripta.* Stereoscopic view of the main trunks of 17 TBi axons emerging from two punctate injection sites and coursing into the lateral pontine reticular formation. Two small HRP injections at A and B labeled eight and nine TBi axons, respectively. They were traced through 17 110-μm serial horizontal sections into the dorsomedial segment of reticularis superioris lateralis, RSL (dm), where they formed two spatially distinct clumps, each, however, containing a mixture of axons from the two injection sites. The axons emerging from site A are drawn thicker to distinguish them from site B axons. The stereodiagram was made by hand as described by Glenn and Burke (1981) and can be viewed by ocular deviation or by using a stereoviewer. (From Sereno, 1985.)

(Moschovakis et al., 1988a, 1988b) and cats (Grantyn and Grantyn, 1982; Grantyn et al., 1984).

C. Tectoisthmi Neurons

The nucleus isthmi is situated in the isthmus region of the brainstem, immediately caudal and ventral to the optic tectum. Tectoisthmi projections have been demonstrated by experimental procedures in the green iguana *Iguana iguana* (Foster and Hall, 1975), the red-eared turtle *Pseudemys scripta* (Foster and Hall, 1975; Sereno, 1983), and garter snakes *Thamnophis sirtalis* (Dacey and Ulinski, 1986a).

The tectoisthmi neurons in *Thamnophis* have been identified following small HRP injections in nucleus isthmi (Fig. 4.24). They fall in the general class of radial cells and have small somata distributed throughout the superficial and central gray layers. Each spherical or pear-shaped soma bears a single, relatively thick radial dendrite that ascends into the superficial gray layers. The overall shape of the dendritic tree is roughly cylindrical and 60 to 80 μm in diameter. The individual dendrites bear thin branchlets and arborize profusely in the stratum zonale, forming nests of horizontally flattened appendages that bear a variety of complex swellings. Many of the somata also bear thin descending dendrites that arborize sparsely.

The axons of tectoisthmi neurons originate from a thin basal dendrite or from the lower pole of a soma. They bear varicose collaterals in the superficial layers. The parent axons exit the tectum in the ventral tectobulbar tract and terminate in the ipsilateral nucleus isthmi (Fig. 4.25). They are of fine caliber, about 0.5 μm in diameter, and descend vertically into the isthmi cell plate to terminate as single arbors. Each arbor is spherical or oval in shape and 10 to 30 μm in diameter. A few tectoisthmi neurons have axons that continue past nucleus isthmi into the brainstem reticular formation after issuing a collateral that terminates in nucleus isthmi. The axons of these tectoisthmibulbar neurons continue in the ventral tectobulbar pathway and issue thin collaterals at irregular intervals.

Like the other tectobulbar axons, the reticular components of the tectoisthmibulbar axons appear to lack any obvious topography in their organization. However, the small terminals of the tectoisthmi collaterals appear to embody a retinotopic, or topographic, map between the tectum and nucleus isthmi. Since nucleus isthmi projects bilaterally to the tectum, the tectoisthmi projections are one limb of a reciprocal set of connections that interconnects the two tectal lobes. This point will be considered in more detail in Section V.B.3.

D. Tectorotundal Neurons

The nucleus rotundus is a large and prominent structure in the dorsal thalamus of reptiles (Rainey, 1979). There are bilateral projections from the tectum to nucleus rotundus in lizards (Butler and Northcutt,

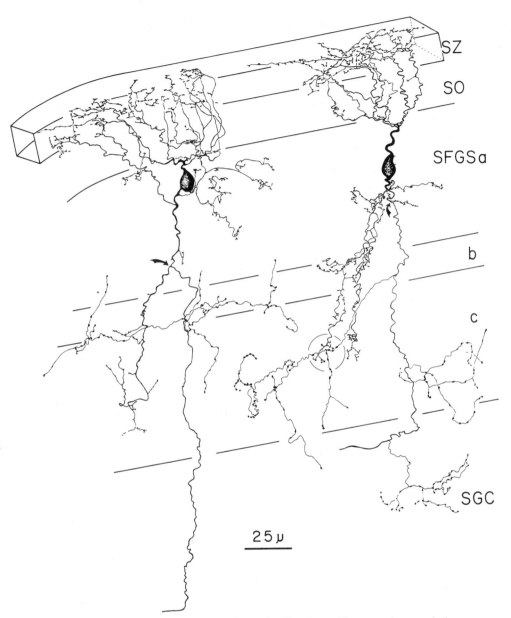

Fig. 4.24. *Thamnophis sirtalis.* Tectoisthmi cells. This figure illustrates the morphology and laminar distribution of some cells that could be backfilled from injections of HRP into the ventral tectobulbar tract rostral but not caudal to nucleus isthmi. Somata in the upper part of the superficial gray have short radial dendrites that ascend through the optic fascicles in the stratum opticum and arborize in the stratum zonale. Thin descending dendrites extend for a variable distance into the SFGS. The axons (arrow) issue local collaterals in the superficial and central gray. (From Dacey and Ulinski, 1986a.)

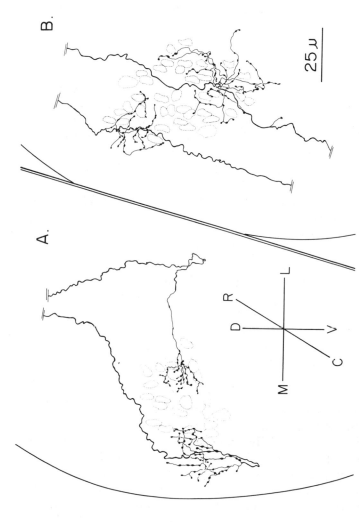

Fig. 4.25. *Thamnophis sirtalis.* Tectoisthmi axons. Fine-caliber axons descend vertically into the nucleus isthmi (Ist) and terminate in extremely small, spherical arbors 10 to 30 μm in diameter (A) or course through the nucleus giving rise to a similarly shaped arbor (B). The somata of a few counterstained isthmi cells are outlined by the dotted lines. (From Dacey and Ulinski, 1986a.)

1971b; Butler, 1978; Hoogland, 1982), turtles (Hall and Ebner, 1970a; Rainey and Ulinski, 1982a), and crocodilians (Braford, 1972). The situation in snakes is less clear. Early cytoarchitectural (Warner, 1947) and Fink-Heimer (Northcutt and Butler, 1974b; Ulinski, 1977) studies failed to unequivocally demonstrate tectorecipient nucleus in the central thalamus of water snakes, *Nerodia sipedon*. Autoradiographic studies (Schroeder, 1981b) were similarly unable to demonstrate a nucleus rotundus in a rattlesnake, *Crotalus*. However, Dacey and Ulinski (1983) identified nucleus rotundus in garter snakes, *Thamnophis sirtalis*, using the orthograde transport of HRP following tectal injections. Nucleus rotundus projects to the ipsilateral anterior dorsal ventricular ridge in each group of reptiles (snakes: Dacey and Ulinski, 1983; Berson and Hartline, 1988; lizards: Distel and Ebbesson, 1975; Butler, 1976; Lohman and van Woerden-Verkley, 1978; Butler and Ebner, 1972; Bruce and Butler, 1979; Bruce, 1982; turtles: Hall and Ebner, 1970b; Kosareva, 1974; Balaban and Ulinski, 1981a, 1981b; crocodilians: Pritz, 1975).

Tectorotundal neurons have been identified in *Thamnophis* following injections of HRP into nucleus rotundus (Fig. 4.26). They belong in the general class of radial neurons. Their somata are positioned throughout the central and superficial gray layers but show a slight tendency to concentrate in the middle of the stratum griseum centrale. They are fusiform, with long axes of 15 to 20 μm. The upper and lower poles of the somata give rise to stout, primary dendrites, and thin branchlets arise from the somata and dendrites. The upper dendrites ascend into the superficial layers and reach the ventral border of the stratum zonale. The lower dendrites descend through the central gray to the upper border of the stratum album centrale. The overall dendritic field is hourglass in shape with a maximum diameter of 75 to 100 μm.

The thick axons of tectorotundal neurons arise from somata or dendrites and turn ventrally to descend to the lower half of the stratum griseum centrale (Fig. 4.27). They form a cluster of collaterals and terminal boutons just adjacent to the lower dendritic tree of the cell. The parent axon emerges from this tangle, descends obliquely into the stratum griseum centrale, and courses laterally into the stratum album centrale. The axons turn into the ipsilateral nucleus rotundus when they reach the thalamus, but at least some have collaterals that continue rostrally to cross the midline in the ventral supraoptic decussation and recurve to terminate in the contralateral nucleus rotundus. Butler (1978) showed that the distribution of terminal degeneration produced in nucleus rotundus in the tokay gecko, *Gekko gecko*, is not uniform. Unilateral tectal lesions produce rod-shaped patches of degeneration. The degeneration is bilateral, with the ipsilateral projection being densest. Bilateral lesions produce evenly distributed de-

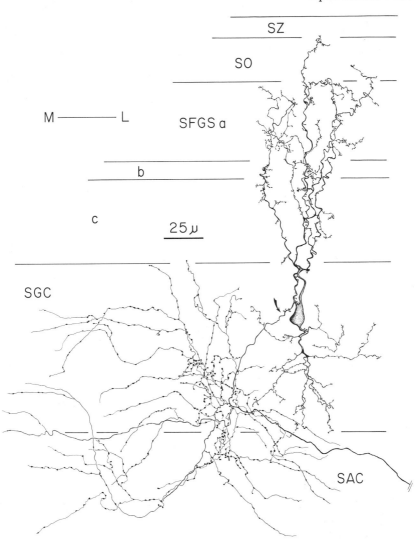

Fig. 4.26. *Thamnophis sirtalis.* Tectorotundal cell. This neuron has its soma positioned in the middle of the central gray. Dendrites bear threadlike, elongate appendages laden with clusters of complexly indented swellings. The axon forms a dense terminal plexus in the vicinity of the cell's lower dendrites and then gives rise to a large number of terminal collaterals that extend away from the cell's dendritic domain in the horizontal plane. The origin of the axon is indicated by the arrow. M-L, Medial-lateral. (From Dacey and Ulinski, 1986b.)

generation bilaterally in rotundus. Degeneration techniques do not provide an unambiguous interpretation of these experiments; however, experiments in which the morphology of tectorotundal axons is demonstrated using the orthograde transport of HRP clarify the organization of the tectorotundal projection. These axons course rostrally in the tectothalamic tract in both *Pseudemys* (Rainey and Ulinski,

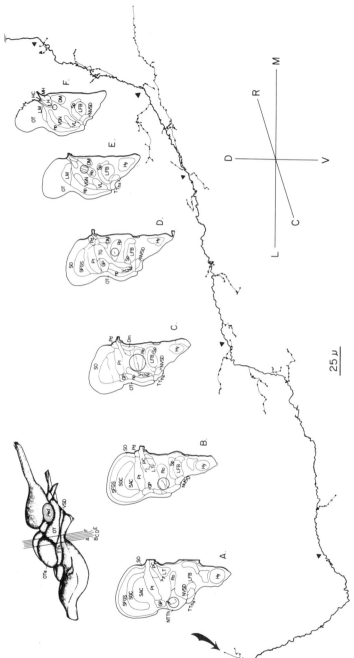

Fig. 4.27. *Thamnophis sirtalis.* The tectorotundal axon. This axon was reconstructed through seven serial sections from the nucleus of the tectothalamic tract (A) to the rostral pole of rotundus (F). After issuing small-terminal collaterals as the nucleus of the tectothalamic tract (NTTh), the axon courses in a straight line from the caudolateral to the dorsal pole of rotundus. The axon gives rise to several fine-caliber collaterals, forming a sheetlike terminal field that extends mediolaterally across the nucleus but is flattened in the dorsoventral axis. (From Dacey and Ulinski, 1983.)

1982b) and *Thamnophis* (Dacey and Ulinski, 1983). As they approach the ipsilateral rotundus, collaterals leave the parent axons and fan into the nucleus. These enter rotundus from its ventrolateral aspect in *Pseudemys* and from its caudal pole in *Thamnophis*. The experiments in *Thamnophis* show explicitly that each parent axon and its collaterals form a sheetlike distribution of terminals that extends rostrocaudally through the nucleus. It is likely that a similar organization obtains in *Pseudemys*. There is no apparent order to the manner in which the axons enter the nucleus, so that each particular region in rotundus is likely to contain collaterals of axons originating from neurons located in several different regions of the tectum. Detailed examination of the terminal collaterals of tectorotundal axons in *Pseudemys* suggests that several individual axons converge to form fascicles of varicose terminals. It seems likely that the rodlike terminal degeneration seen in *Gekko* reflects these fascicles.

The morphology of rotundal neurons is quite constant across species (Rainey, 1979). The central core of rotundus is occupied by multipolar neurons, the dendrites of which extend away from somata in an isodendritic pattern (Fig. 4.28). Neurons situated around the periph-

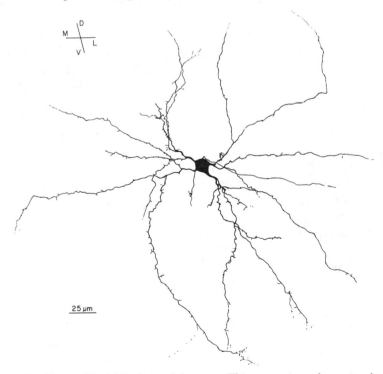

25 μm

Fig. 4.28. *Thamnophis sirtalis*. Rotundal neuron. This neuron in nucleus rotundus was retrogradely labeled following an injection of HRP in the lateral forebrain bundle. (From Dacey and Ulinski, 1983.)

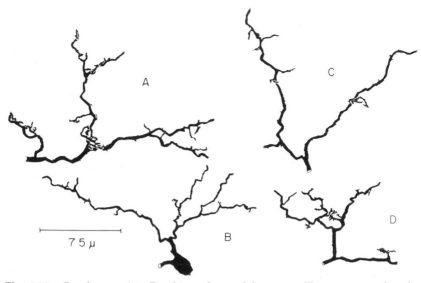

Fig. 4.29. *Pseudemys scripta.* Dendrites of rotundal neurons. These are examples of the dendrites of rotundal neurons, which show complex appendages. They are drawn from neurons that were retrogradely labeled with HRP following an injection in the forebrain bundle. (From Rainey and Ulinski, 1982a.)

ery of the nucleus have dendrites that extend into the core. The diameters of the dendritic fields of rotundus neurons are quite large, so that an individual neuron can span as much as two thirds the diameter of the nucleus. Electron microscopic experiments show that the synapses of tectorotundal axons preferentially contact the distal dendrites of rotundal neurons. They occur on the shafts of the dendrites as well as on complex appendages that are situated on the distal dendrites (Fig. 4.29). The appendages are surrounded by nests of synapses suggesting that each appendage is embedded in a fascicle of terminal axons, each of which contains axons from several tectorotundal neurons (Fig. 4.30).

The available anatomy thus suggests that the tectorotundal projection lacks a simple point-to-point relationship between the tectal surface and nucleus rotundus. Each tectorotundal neuron will potentially synapse on neurons throughout much of the volume of nucleus rotundus. Conversely, any neuron in rotundus is likely to receive input from neurons in many regions of the tectal surface. The results of electrophysiological investigations of the properties of rotundal units in turtles (Morenkov and Pivavarov, 1973, 1975) and pigeons (De Britto et al., 1975; Revzin, 1970, 1979; Granda and Yazulla, 1971; Crossland, 1972; Maxwell and Granda, 1979) are consistent with the anatomy. Rotundal units typically have wide-field receptive fields

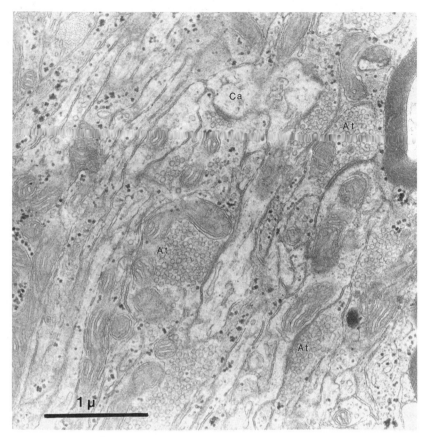

Fig. 4.30. *Pseudemys.* Rotundal synaptic nest. This electronmicrograph shows a detail of the neuropil in nucleus rotundus. The irregular shaped and electron-lucent profiles (Ca) are probably the complex appendages of rotundal neurons. Degeneration experiments suggest that the presynaptic elements (At) are the terminals of tecto-rotundal neurons. (From Rainey and Ulinski, 1981b.)

that respond to stimuli placed throughout much of the contralateral visual field and often through the entire bilateral visual field. They respond well to small, moving stimuli, sometimes with preferred directions, velocities, or colors.

E. Tectogeniculate Neurons

The geniculate complex occupies the lateral diencephalon in all reptiles. It consists of several cytoarchitecturally delineated nuclei that have in common the receipt of direct projections from retinal ganglion cells. The diencephalic structures that receive direct retinal projections have been determined in snakes (Armstrong, 1951; Halpern and Frumin, 1973; Repérant, 1973; Repérant and Rio, 1976; Northcutt and

Butler, 1974b), lizards (Armstrong, 1950; Butler and Northcutt, 1971a; Butler, 1974; Northcutt and Butler, 1974a; Cruce and Cruce, 1975, 1978); turtles (Kosareva, 1967; Knapp and Kang, 1968a, 1968b; Hall and Ebner, 1970a; Belekhova and Kosareva, 1971; Bass and Northcutt, 1981a, 1981b), crocodilians (Burns and Goodman, 1967; Braford, 1973; Repérant, 1975) and *Sphenodon* (Northcutt et al., 1974). The organization of the geniculate complex is still poorly understood, but it is usually possible to recognize a ventral lateral geniculate nucleus that has reciprocal connections with the optic tectum and a dorsal thalamic nucleus that projects to the telencephalon. The latter structure is sometimes called the dorsal lateral geniculate nucleus. It has been claimed that the tectum projects to the dorsal lateral geniculate nucleus (see Ulinski, 1977), but recent work (Ulinski et al., 1983) reaches the conclusion that there is at present no evidence for a tectal projection to the components of the geniculate complex that project to the telencephalon.

Tectogeniculate projections have been best studied in *Thamnophis*. In this species, the geniculate complex is divided into a neuropil that lies internal to the optic tract and contains loosely packed neurons and a cell plate that contains densely packed neurons. The cell plate is divided into a rostrally placed nucleus ovalis and dorsal, dorsomedial, and ventral parts. The tectum projects to the ipsilateral ventral and dorsal parts of the cell plate (Fig. 4.31). The tectal neurons that give rise to this projection have been identified following small HRP injections in the geniculate (Fig. 4.32). They fall in the general class of radial cells. The somata of the tectogeniculate neurons are present in moderate numbers in the central gray and the lower half of the superficial gray, with the largest number occurring at the border between the two layers. Their somata are spherical or pear shaped in diameter and slightly smaller than the somata of the tectorotundal neurons. Each soma gives rise to a single thick dendrite that ascends into the superficial gray layers and one or two extremely thin basal dendrites that descend into the deeper tectal layers. The ascending dendrites occupy a narrow cylindrical space, 60 to 80 μm in diameter, extending from the border of the central and superficial gray layers to the stratum opticum. Many thin and complex branchlets issue from both the primary and secondary ascending dendrites.

Axons of tectogeniculate neurons ascend vertically into sublamina c of the superficial gray. They bifurcate into branches of unequal thickness at about the middle of this sublayer. The thinner branch remains in sublayer c and gives rise to a collateral arbor that overlaps the dendritic field of the neuron. These collaterals have several branches laden with varicosites that form nearly spherical arbors. The thicker branch ascends from sublamina c and turns rostrad as it

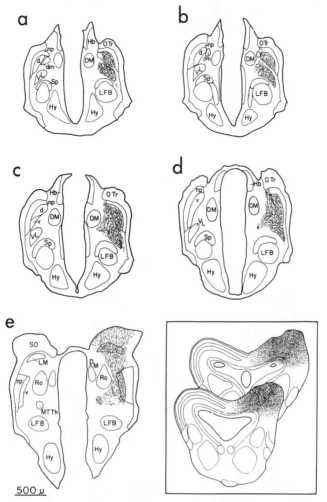

Fig. 4.31. *Thamnophis sirtalis.* Tectogeniculate projections. The inset shows the locus of a tectal HRP injection. The main part of the figure (a–e) shows the pattern of orthograde labeling in the geniculate. Labeled fibers distribute to the dorsal (d) and ventral (v) parts of the geniculate complex and to the ventrolateral nucleus (VL). (From Dacey and Ulinski, 1986a.)

reaches the lower border of the stratum opticum. As it passes ventrorostrally into the pretectal nucleus, it issues collaterals that terminate within the pretectal and geniculate pretectal nuclei of the pretectal complex (Fig. 4.33). The parent axon descends through the pretectum medial to the ventral part of the geniculate cell plate within a region of low cell density that is designated as the tectogeniculate pathway. A collateral extends laterally into the cell plate, forming a highly

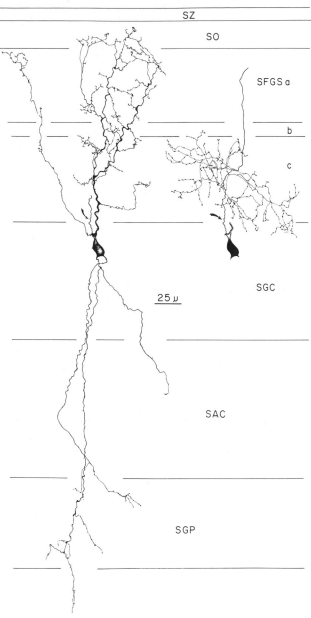

SZ

SO

SFGSa

b

c

SGC

25 μ

SAC

SGP

Fig. 4.32. *Thamnophis sirtalis.* Tectogeniculate cell. This figure illustrates the morphology of neurons whose axons could be traced into the tectogeniculate pathway. The axon collateral system and dendritic trees of this cell were traced separately; the same position in each tracing is indicated by arrows. Tectogeniculate neurons are characterized by a radial dendritic arbor in the superficial gray layers, by extremely thin, unarborized descending dendrites that can often be traced to the deep tectal layers, and by an axon that forms a collateral projection restricted to sublamina c of the SFGC and then ascends to the upper border of SFGSa, where it courses rostrally below the stratum opticum. (From Dacey and Ulinski, 1986b.)

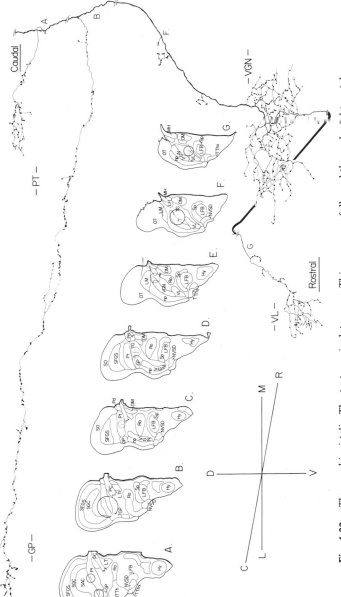

Fig. 4.33. *Thamnophis sirtalis.* The tectogeniculate axon. This axon was followed through eight serial sections (circled areas in A through G). Only the parts of the axon bearing collaterals (circles A, B, F, and G) are illustrated at higher magnification. Single, topographically organized terminal arbors are formed in the geniculate pretectal nucleus (GP), the pretectal nucleus (Pt), the cell plate of the ventral geniculate nucleus (VGN), and the ventrolateral nucleus of the thalamus (VL). The geniculate pretecta, the pretectal, and the ventral geniculate nuclei all receive a topographic retinal input. (From Dacey and Ulinski, 1986a.)

branched arbor that is shaped like a flattened cylinder 50 to 70 μm in diameter. Some geniculate axons extend into the ventrolateral nucleus of the ventral thalamus. Small injections indicate that the projections to the pretectal complex, the ventral part of the geniculate complex, and the ventrolateral nucleus are topologically organized and in register with the direct retinal input to these structures.

F. Summary

All of the studies of the optic tectum based on Golgi preparations recognized that several types of neurons in the tectum have axons that can be traced out of the tectum. The problem has been that it is generally not possible to trace the axons to their ultimate destination in Golgi preparations because single axons have to be followed through serial sections. However, the use of HRP as a neuronal marker in *Thamnophis* and in *Pseudemys* has permitted the correlation of the morphology of various populations of tectal neurons with the ultimate target of their axons. The results for *Thamnophis* are the most complete and are summarized in Fig. 4.34. Each class of tectal neuron, defined on the basis of its connections, has a characteristic morphology and is localized within a specific lamina or set of laminae. The HRP fills show that several of the efferent neurons also have extensive collateral systems within the tectum; beyond the intrinsic interactions within the tectum, they give rise to extrinsic projections. Many of the earlier Golgi studies show these collateral systems, but their extent has been generally underestimated. Most of the efferent neurons in *Pseudemys*, and probably all in *Thamnophis*, have dendrites that extend into the superficial layers and are therefore in potential receipt of retinal inputs. All of the efferent cells have axons that terminate in more than one target structure. Thus, the tectogeniculate neurons have collaterals in the pretectum, some tectoisthmi neurons have collaterals that continue to the reticular formation, and apparently all tectoreticular neurons have widely ranging axonal systems that include both ascending and descending branches.

IV. INTRINSIC NEURONS
A. General

Intrinsic tectal neurons are those that have synaptic interactions only with other tectal neurons. In practice, it is difficult to unequivocally recognize them because of the difficulty in assuring that a neuron with axon collaterals within the tectum really lacks efferent projections. Golgi preparations often impregnate only a fraction of the axon system of an individual neuron and lead to the false impression that a neuron forms only intrinsic connections. Also, intracellular injections of HRP may only partially fill an axon system. Thus, neurons in

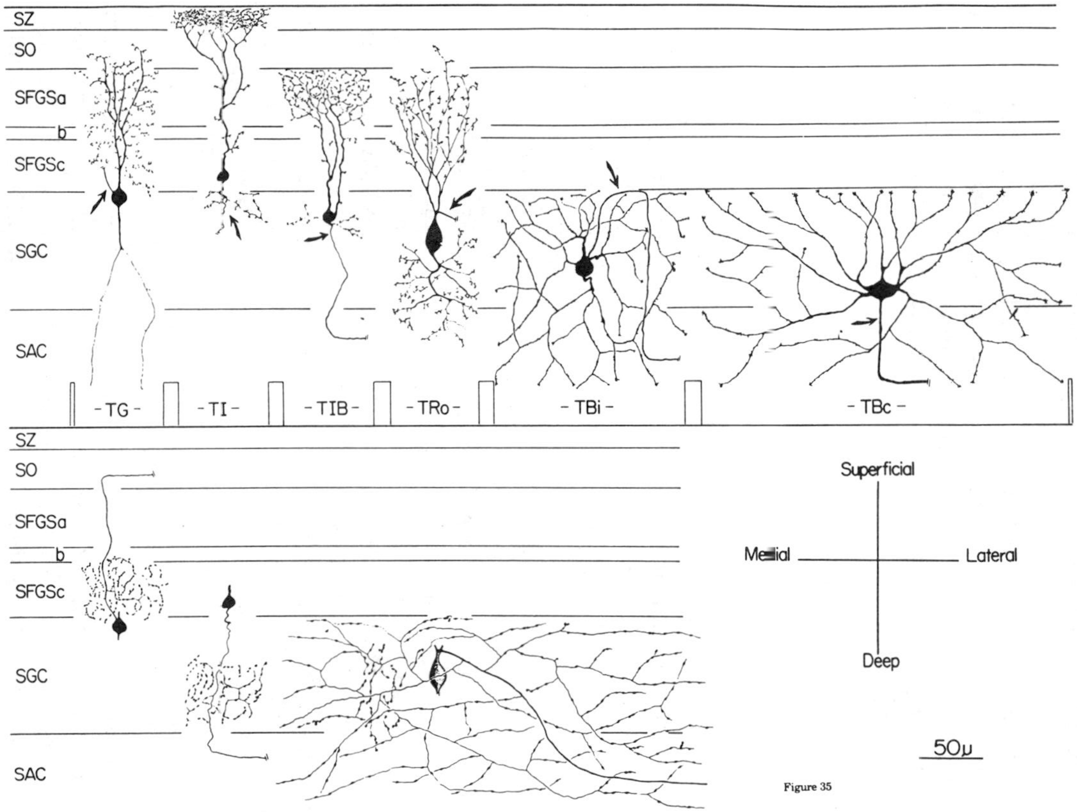

Fig. 4.34. *Thamnophis sirtalis.* Summary of efferent tectal neurons. This figure provides an overall summary of the classes of tectal neurons described in this study. Neurons are all drawn to about the same scale to facilitate comparison. The upper figure shows the morphology of the somata, dendrites, and the principal efferent tectal branch of the axon. Tectogeniculate (TG), tectoisthmi (TI), tectoisthmibulbar (TIB), tectorotundal (TRo), ipsilateral tectobulbar (TBi), and contralateral tectobulbar (TBc) neurons are shown. The layers of the tectum are indicated by horizontal lines. The lower figure shows the somata of the tectogeniculate, tectoisthmi, and the tectorotundal neurons with their intrinsic axon collaterals drawn. The origins of the axons are indicated by arrows. (From Dacey and Ulinski, 1986b.)

the dorsolateral geniculate complex of cats (e.g., Friedlander et al., 1981) that were classically designated as *interneurons* are now known to be efferent neurons. Intrinsic tectal neurons must, therefore, be defined by negative evidence, but several classes of tectal neurons are potential candidates.

B. Horizontal Cells (Type A Cell)

Horizontal cells are neurons with dendrites that extend for considerable distances concentric to the pial surface of the tectum. They have been identified in Golgi preparations of the tectum in lizards (Ramón, 1896; Leghissa, 1962; Butler and Ebbesson, 1975) and turtles (Peterson, 1978b). The most detailed accounts of the morphology are obtained from experiments (Dacey and Ulinski, 1986c) in which horizontal cells were filled with HRP by means of injections in the tectum (Fig. 4.35). The somata of horizontal cells in these preparations are distributed throughout the superficial gray layer but are most frequent just above or below the stratum opticum. They tend to be horizontally fusiform and have thick primary dendrites that taper from the pole of the soma. Each dendrite issues several thinner dendrites that fan out in the horizontal plane for as much as 500 μm, following the curvature of the tectum. They extend further mediolaterally than rostrocaudally so that the overall shape of the dendritic field is elliptical in the horizontal plane. The secondary dendrites bear spines and large varicosities at irregular intervals. The axons of horizontal cells collateralize heavily within their dendritic fields and then extend horizontally in parallel with the dendritic field. Detailed electron microscopic studies have not been conducted on reptiles, but several authors have suggested that the dendritic appendages of horizontal cells in other groups are presynaptic in dendrodendritic synapses (Hayes and Webster, 1975; Angaut and Repérant, 1976) and may use γ-aminobutyric acid (GABA) as a transmitter (Streit et al., 1978). Such synapses raise the possibility that various portions of each horizontal cell function as independent units. Inputs to the distal dendrites will produce relatively large excitatory postsynaptic potentials (EPSPs) due to the large input resistance of the thin dendrites, whereas inputs to the soma will produce relatively small EPSPs. Different regions of the neuron might then differ in their function and participate primarily in local interactions.

C. Small Cells with Descending Axons (Type B Cells)

In contrast with horizontal cells, Type B cells are small, vertically oriented and have an axon that projects to deeper tectal layers. They may correspond to the "pequeños corpusculos de cilindre-eje descendante" of Pedro Ramón (1896). The somata of these neurons are

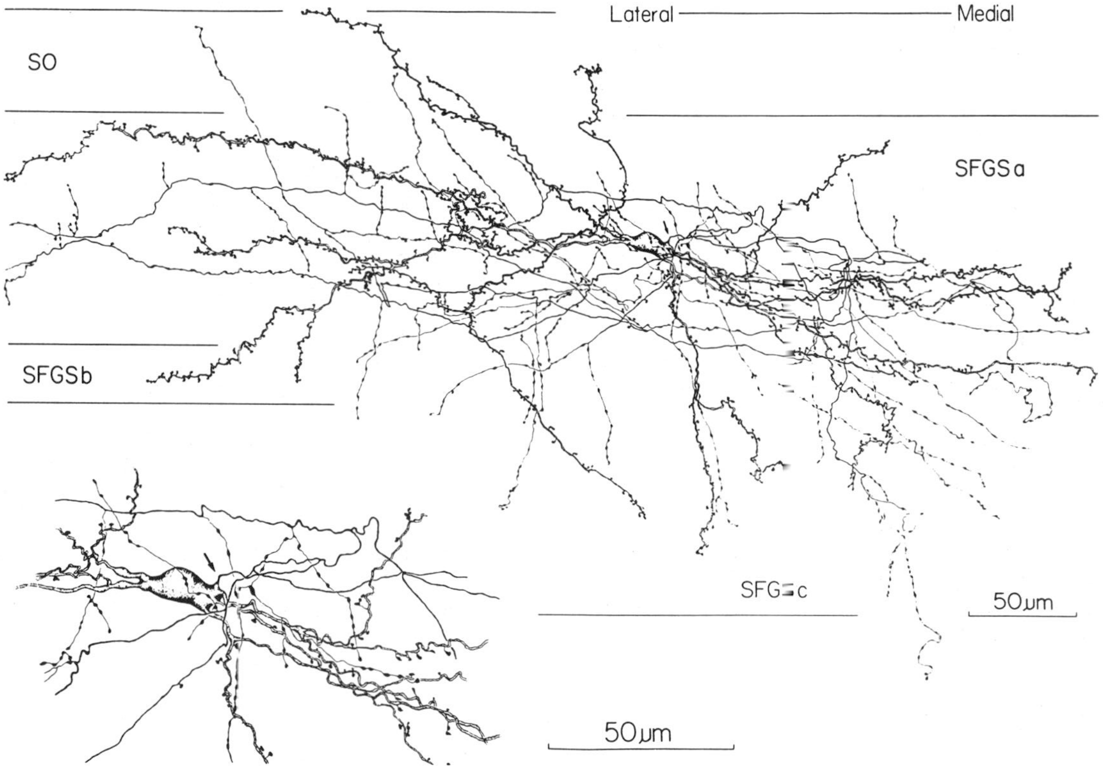

Fig. 4.35. *Thamnophis sirtalis.* Horizontal cell. The dendrites of horizontal cells are extended in the mediolateral axis and fan out from their point of origin. They are thick and varicose and bear numerous appendages in the form of single lollipoplike protrusions or long beaded strands and clusters. The dendritic field of this cell is relatively planar and is restricted to the stratum zonale. The axon is shown in a separate tracing for clarity; the same point in each tracing is indicated by the arrow. (From Dacey and Ulinski, 1986c.)

always positioned within or very close to the cell dense lamina b of the stratum fibrosum et griseum superficiale and seem to participate in clusters of neurons with apposed somata (Fig. 4.36). The dendrites ascend into sublamina a of the stratum fibrosum et griseum superficiale, branching into thin, varicose secondary dendrites. They often recurve at their ends, sometimes forming grapelike clusters or lobulated masses of swellings. The result is a compact bushy arbor that forms a cylindrical field 50 to 80 μm in diameter and extending 40 to 65 μm above the soma. The axons of these neurons originate from the bases of their somata and descend vertically into sublamina c of the stratum fibrosum et griseum superficiale. They recurve and terminate in a spray of collateral branches in sublamina c and bear collaterals that ascend through sublamina c or descend into the upper central gray. The axon system is a vertically aligned field and has dimensions similar to those of the dendritic field, so that the neuron as a whole occupies a cylindrical space 50 to 80 μm in diameter that extends through the superficial gray.

An electron microscopic study of the superficial layers of the asp viper (Repérant et al., 1981) has demonstrated varicose, radially oriented dendrites containing flattened synaptic vesicles that are presynaptic to other dendrites and postsynaptic to retinal ganglion cell terminals. Serial section analysis shows that these profiles arise from small, vertically oriented neurons in the stratum fibrosum et griseum superficiale that may, therefore, correspond to the dendrites of these neurons. It is possible that these neurons are similar to horizontal cells and participate in triadic relations with retinal afferents and the dendrites of other neurons. A study of the optic tectum of the pigeon suggests that these cells accumulate tritiated GABA and may mediate inhibition in the superficial gray (Hunt and Kunzle, 1976).

D. Stellate Cells (Type C Cells)

A third group of intrinsic neurons corresponds to the stellate cells of Golgi preparations but may also be viewed as a form of small radial cells. Similar neurons have been illustrated for the lizard *Pentadactylus* (Quiroga, 1978). Their somata are small and spherical and lie in the middle third of the central gray layer (Fig. 4.37). Their dendritic arbors are spherical or cylindrical with diameters of 60 to 80 μm. One to three primary dendrites arise from each soma and extend radially. The thin and wavy secondary dendrites are laden with complex, varicose appendages. The axons extend laterally in a wavy and often recoiled path, issuing primary collaterals near the dendritic tree of the cell. The primary branches extend vertically through the central gray and form a distinct plexus displaced laterally from the soma. The plexus occupies a cylindrical space 40 to 80 μm in diameter and lies in the central gray.

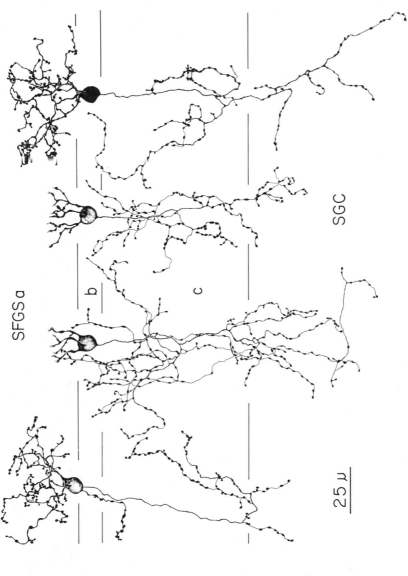

Fig. 4.36. *Thamnophis sirtalis.* Small cells with descending axons. The axons of these cells descend from the base of the soma and arborize in vertical alignment with the cell's dendritic field in sublamina c of the stratum fibrosum et griseum centrale. The axon collaterals are often thick and bear large varicosities (1 to 2 μm in diameter). (From Dacey and Ulinski, 1986c.)

SFGS a

b

c

SGC

25 μ

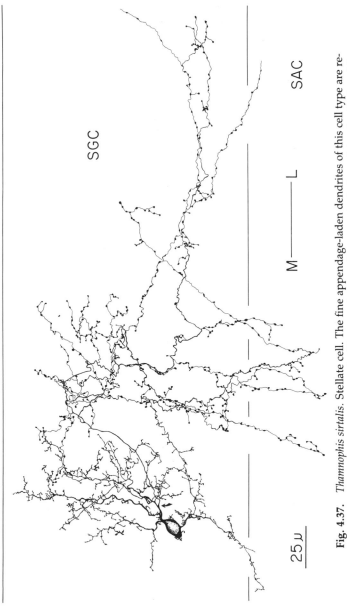

Fig. 4.37. *Thamnophis sirtalis.* Stellate cell. The fine appendage-laden dendrites of this cell type are restricted to the stratum griseum centrale. The axon forms a dense terminal plexus laterally displaced from the cell's dendritic field. Single terminal collaterals may extend laterally for a few hundred micrometers. The origin of the axon from a primary dendrite is indicated by the arrow. M-L, medial-lateral. The soma of this cell is positioned at the border of the central white and periventricular gray. The dendritic field spread is 450 × 260 μm. (From Dacey and Ulinski, 1986c.)

E. Type D Cells

A final group of intrinsic neurons has vertically fusiform somata; most of them lie at the interface of the central and superficial gray layers (Fig. 4.38). The two poles of the somata each issue a single primary dendrite. The ascending dendrites bifurcate within 25 μm of the soma into branches that ascend at slightly oblique angles, forming a conical dendritic field 60 to 80 μm in diameter. The descending dendrites extend to the lower border of the stratum griseum centrale. Both sets of dendrites bear fine terminal branchlets, varicose appendages, and single spinelike protrusions. Axons arise from somata or primary dendrites and extend medially 10 to 50 μm without branching before descending vertically or obliquely into the middle of the central gray, where they issue fine, varicose collaterals. Single collaterals participate in a loose terminal plexus that occupies the central white layer. The cylindrical plexus is cylindrical, 50 to 70 μm in diameter and displaced up to 100 μm medial to the dendritic field of the cell.

F. Summary

Like the efferent neurons, each particular class of putative intrinsic neurons is associated with particular tectal laminae (Fig. 4.39). The horizontal (Type A) neurons and small neurons with descending axons are most superficially located. The radial neurons (types B and D) are located at the border of the superficial and central layers and the stellate cells (type C) lie within the central gray. There is presently no clear evidence for purely intrinsic neurons within the deeper layers. Each set of intrinsic neurons will, therefore, be embedded in its own set of afferents, but all are in potential receipt of retinal input to the superficial layers. The presence of varicosities and appendages on the dendrites of all four types of intrinsic neurons suggests that they are involved in dendrodendritic synapses with other tectal neurons and are likely to participate in triadic relationships with retinal afferents. There is limited direct evidence in reptiles, but studies in frogs (Szekely et al., 1973), pigeons (Hayes and Webster, 1975), and several species of mammals (e.g., Sterling, 1971; Graham and Casagrande, 1980) suggest that intrinsic neurons are involved in relationships of this type. All intrinsic neurons also have axonal systems, but nothing is known about the nature of their synaptic contacts with other neurons.

V. AFFERENTS
A. General

The axons of retinal ganglion cells are the most obvious source of afferents to the optic tectum. They can be traced from the optic nerves, through the optic tracts, and into the substance of the tectum. However, large injections of HRP into the tectum reveal that an over-

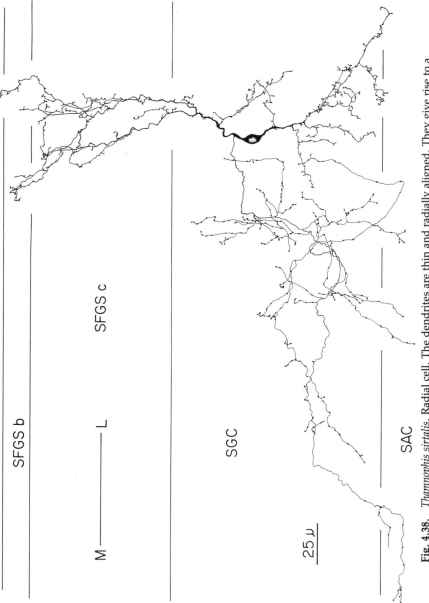

Fig. 4.38. *Thamnophis sirtalis.* Radial cell. The dendrites are thin and radially aligned. They give rise to a moderate density of complex appendages. The appendages may be long and varicose and bear knobby and lobulated protrusions. The dendritic field occupies a narrow cylindrical zone 60 to 80 μm in diameter. (From Dacey and Ulinski, 1986c.)

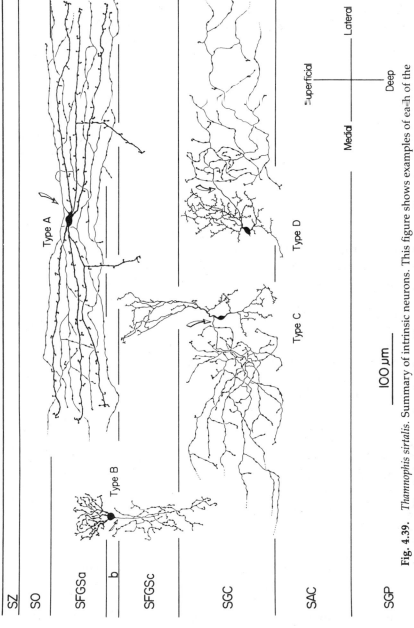

Fig. 4.39. *Thamnophis sirtalis*. Summary of intrinsic neurons. This figure shows examples of each of the four types of intrinsic neurons described here. They are drawn schematically and to the same scale to facilitate comparison. Arrows indicate origins of the axons. (From Dacey and Ulinski, 1986c.)

whelming number of other structures also terminates in the tectum. Fig. 4.40 illustrates the results of such an experiment in *Thamnophis*. No telencephalic afferents to the tectum have been reported in snakes and none are shown, but turtles show projections from the visual cortex to the tectum (Hall et al., 1977). Within the diencephalon of *Thamnophis*, neurons in the geniculate complex, ventral thalamus, and hypothalamus project bilaterally to the tectum. All of the major pretectal structures project bilaterally to the tectum. Several nuclei in the mesencephalic tegmentum and in the torus semicircularis (which is called the posterior colliculi in snakes; Senn, 1969) project to the tectum. The trigeminal nuclei and reticular formation of the rhombencephalon project to the optic tectum. Finally, the spinal cord has a direct projection to the optic tectum.

As far as is possible, afferents to the tectum are here grouped into functional categories for discussion.

B. Visual Afferents

1. VISUAL REPRESENTATION The representation of visual information in the optic tectum can be studied by recording with microelectrodes in anesthetized animals while presenting visual stimuli to one or both eyes. This procedure has been carried out on a wide range of vertebrates (see Kruger, 1969), including snakes (Terashima and Goris, 1975; Hartline et al., 1978), lizards (Stein and Gaither, 1981), turtles (Gusel'nikov et al., 1970), and a crocodilian (Heric and Kruger, 1966). Several general properties have emerged.

There is universally a topological representation of the retinal surface within the optic tectum. The temporal edge of the retina is usually represented rostrally in the tectum with the representation of the nasotemporal axis extending from rostromedial to caudolateral on the tectum. This orientation has been reported for snakes (Terashima and Goris, 1975; Hartline et al., 1978), turtles (Gusel'nikov et al., 1970), and a crocodilian (Heric and Kruger, 1966). A slightly different orientation has been reported in the iguana, *Iguana iguana* (Stein and Gaither, 1981). It is not clear if this difference reflects differences in the shape and orientation of the tectum as a whole or possibly technical difficulties such as the incomplete compensation for the rotation of the eye that often occurs during anesthesia.

Although the complete retinal surface is represented, some regions of the retina may occupy a disproportionate percentage of the total tectal map. Such differences can be quantified by calculating a magnification factor that indicates the number degrees of visual space represented per millimeter of tectal surface. The most accurate estimates of relative magnification have been carried out on *Alligator* (Heric and Kruger, 1966). The magnification factor is fairly constant in

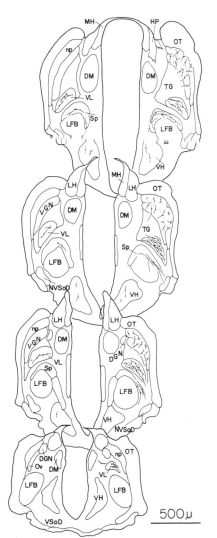

Fig. 4.40. *Thamnophis sirtalis.* Tectal afferents. This figure illustrates the pattern of retrograde cell labeling in the brainstem after a large injection of HRP into the tectum. Diencephalic projections arise from the ventral geniculate nucleus (VGN) ipsilaterally and the ventrolateral nucleus (VL), suprapeduncular nucleus (Sp), and ventral hypothalamus (VH) bilaterally. The lentiform mesencephalic nucleus (LM), pretectal nucleus (PT), geniculate pretectal nucleus (GP), and the lentiform thalamic nucleus (LT) give rise to bilateral projections. An ipsilateral projection arises from the caudal segment of the nucleus of the ventral supraoptic decussation (NVSoD). Projections arise from the nucleus lateralis profundus mesencephali (LPM), the nucleus of the tectothalamic tract (NTTh), and the nucleus isthmi (lst). Labeled neurons are also present within the ventral tectobulbar tract (VTB) bilaterally. The injection site is indicated by the stippling. Projections arise from the posterior colliculus (PC), nucleus of the lateral lemniscus (NLL), and the spinal nucleus of the trigeminal (SpV) bilaterally and a small cluster of neurons associated with the vestibular complex (VeVm) contralaterally. Filled neurons are also embedded in the ventral tectobulbar tract (VTB) ipsilaterally and the predorsal bundle contralaterally. Chartings were traced from 80-μm coronal sections. They are presented as a series of 15 consecutive sections on this and the following three pages. (From Dacey and Ulinski, 1986e.)

this species, and there is only a slight exaggeration of the portion of the retina that sees the central visual fields. Comparable results have been obtained in snakes in a less detailed study. Inferences drawn about the retinal map of *Pseudemys scripta* from such experiments on the red-eared turtle suggest that much of the tectal surface is occupied by a highly magnified representation of the retinal visual streak, which is a linear area of high density of photoreceptors and ganglion cells that extends across the nasotemporal axis of the retina (Brown, 1969; Peterson and Ulinski, 1979).

Experiments in which a microelectrode is systematically advanced into the tectum indicate that most visual units occur in the superficial

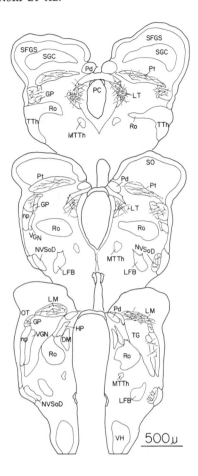

layers of the tectum. The properties of visual units have been studied in turtles (Robbins, 1972; Boyko and Goncharova, 1976; Davydova et al., 1982), snakes (Hartline et al., 1978; Kass et al., 1978), and the iguana, *Iguana iguana* (Stein and Gaither, 1983). Visual units are found throughout the depth of the tectum, but units in the superficial layers typically have smaller receptive fields than those in the deeper layers. Boyko and Goncharova (1976) report receptive field sizes are 5° to 10° in the superficial layers of turtles. In *Iguana*, the receptive field sizes are 0.5° within the central 5° of visual field representation but increase to as large as 40° more peripherally (Stein and Gaither, 1981, 1983). Receptive fields in the deeper layers range from 31° to 110° (Boyko and Goncharova, 1976). Units with small field sizes tend to be elliptical, with the long axis parallel to the horizontal meridian (Boyko and Goncharova, 1976). Tectal units respond with either on, off, or on-off responses to the presentation of visual stimuli (Robbins, 1972; Stein

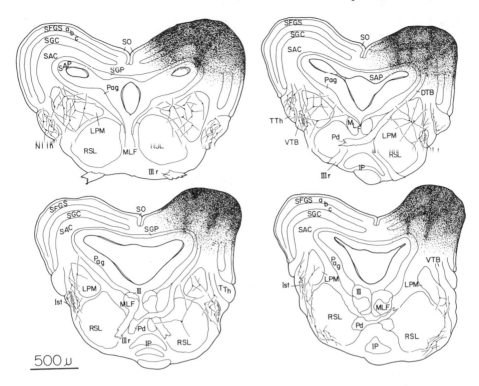

500 μ

and Gaither, 1983). Stimuli of different sizes show that tectal receptive fields in *Iguana* have an internal structure, including a suppressive surround region (Stein and Gaither, 1983). Units in the superficial layers, at least of turtles (Robbins, 1972; Granda and O'Shea, 1972) and *Alligator* (Heric and Kruger, 1966), are color sensitive. Spectral sensitivity curves in turtles vary in shape but usually peak between 600 and 700 μm. Spectral plots for on and off responses indicate antagonistic inputs from spectrally different receptor processes (Robbins, 1972). Tectal units in the superficial layers of *Pseudemys* lack spontaneous activity, whereas those in the stratum griseum periventriculare show high spontaneous rates of discharge (Robbins, 1972). Both directionally selective (Boyko and Goncharova, 1976; Robbins, 1972; Stein and Gaither, 1983) and velocity-selective (Stein and Gaither, 1983) units have been reported. These units tend to be a minority (10% to 33%) of units encountered. Some units will respond to stationary stimuli, but most units respond best to moving ones. Units in the superficial layers respond only to visual stimuli; units in the deeper layers may be multimodal and respond to somatosensory, infrared, or auditory stimuli as well as visual stimuli (Hartline et al., 1978; Stein and Gaither, 1981, 1983).

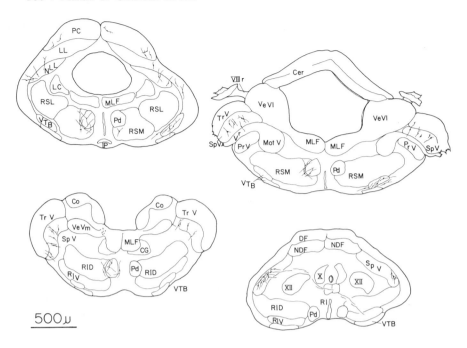

500μ

2. Sources of Visual Information

a. General   Fig. 4.41 summarizes various routes by which visual information can reach the optic tectum. The most familiar involves direct projections from retinal ganglion cells to the optic tectum. However, there are several more complex routes involving one or more relays in other structures. Except for the cortical projection to the tectum, all other routes terminate in the superficial layers. Some of the routes involve topological representations of the retinal surface; others do not.

b. Retinal Afferents
The overall pattern of the retinotectal projection has been studied by degeneration or autoradiographic techniques in turtles (Kosareva, 1967; Knapp and Kang, 1968a, 1968b; Hall and Ebner, 1970a; Belekhova and Kosareva, 1971; Bass and Northcutt, 1981a, 1981b), crocodilians (Burns and Goodman, 1967; Braford, 1973; Repérant, 1975), *Sphenodon* (Northcutt et al., 1974), snakes (Armstrong, 1951; Halpern and Frumin, 1973; Repérant, 1973; Repérant and Rio, 1976; Northcutt and Butler, 1974a), and lizards (Armstrong, 1950; Butler and Northcutt, 1971a; Butler, 1974; Northcutt and Butler, 1974b; Cruce and Cruce, 1975). The pattern is similar in all of the species studied. The

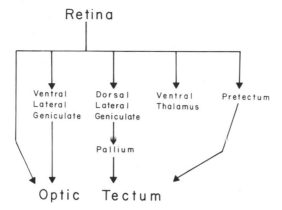

Fig. 4.41. Visual routes to the optic tectum. This diagram summarizes the major routes by which information from the retina can reach the optic tectum. Each route is discussed in detail in the text.

majority of retinal ganglion cell axons cross in the optic chiasm and reach the contralateral tectal lobe through the marginal optic tract, which divides into lateral and medial branches at the rostral pole of the tectal lobe. Some of the axons continue caudally over the tectal surface from the rostral pole, whereas others continue some distance in the two divisions of the marginal optic tract before turning into the tectum, where they terminate.

A minority of retinal ganglion cell axons continue uncrossed to the ipsilateral optic tectum. Evidence for ipsilateral retinotectal projections has been reported in some turtles, crocodilians, lizards, and snakes, but not in *Sphenodon*. The extent of ipsilateral projections is important in understanding the representation of the binocular visual field within the tectum, and we will return to this issue in Section V.B.3. However, it is worth noting here that the ipsilateral projection is fairly difficult to demonstrate with either degeneration or autoradiographic techniques, so that statements about the presence or absence of an ipsilateral projection in any given species should be regarded with caution.

Once the axons of retinal ganglion cells reach the tectum, they can run in the stratum zonale, stratum opticum, or in the stratum fibrosum et griseum superficiale. The stratum opticum contains distinct fascicles of rostrocaudally oriented axons; in transverse sections they appear as vertically elongate columns interposed between the dendritic bundles formed by radial cells. Electron microscopic observations of these fascicles show that they contain very few axon terminals. The overwhelming majority of terminals is present on axons after they have turned out of the fascicles.

The overall distribution of retinal ganglion cell terminal boutons can be studied to a first approximation in Fink-Heimer preparations following retinal lesions or in autoradiographic preparations following intraocular injections of tritiated materials. Such experiments indicate that most of the retinal ganglion cells terminate in the superficial layers of the tectum. Fig. 4.42 is an example of an autoradiogram from the optic tectum of *Pseudemys* following an injection of tritiated proline in the contralateral eye and shows the general pattern of silver grains. Notice that silver grains occur densely over sublayers a and c of the stratum fibrosum et griseum superficiale. In the stratum opticum, regions devoid of grains are bundles of retinal axons. The band of low grain density in sublayer b of the stratum fibrosum et griseum superficiale is a consequence of the densely packed somata. Notice that the density of grains decreases markedly at the upper border of the stratum griseum centrale. This pattern is shown quantitatively in Fig. 4.43, which shows the density of silver grains as a function of tectal depth in a transect through the tectum of *Pseudemys*.

In addition to the retinal projection to the superficial layers, there may be minor retinal projections to the deeper tectal layers. Fig. 4.43, for example, shows grain counts that are above background over the stratum griseum periventriculare. There is some precedent for such projections; thus, Berson and McIlwain (1982) have confirmed direct projections from the Y class of retinal ganglion cells to the deep layers of the superior colliculus of cats. However, sparse projections are difficult to demonstrate convincingly with axonal tracing techniques. Fink-Heimer techniques generate artifact by undersuppression of normal fibers. Autoradiographic techniques can be confounded by the transynaptic transport of labeled material. Thus, the existence of direct retinal projections to the deep layers requires confirmation with electron microscopy or HRP techniques.

It has been traditional to report that the retinal terminal field occupies three layers (layers 14, 12, and 10 of Pedro Ramón) in the superficial layers, which vary in thickness and density—both among species and between tectal areas within the same species. The most careful study is that of Peterson (1981) on the desert iguana, *Dipsosaurus dorsalis*. It utilizes both Fink-Heimer and autoradiographic techniques, with a range of survival times. The autoradiographic experiments show that the density of silver grains over layers 14, 12, and 10 varies for each of the five cytoarchitectonic areas of the tectum (see Section II.A), being quite high in the lateral wall, for example, but low in the dorsolateral area. The Fink-Heimer experiments are consistent with these results but add the additional information that the caliber of degenerating axons varies as a function of tectal locus.

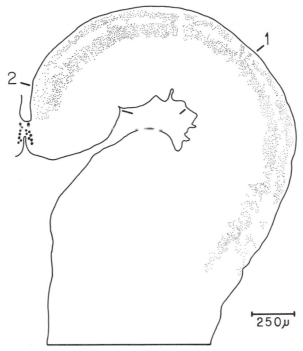

Fig. 4.42. *Pseudemys scripta.* Retinotectal projection. This is a charting made from an autoradiogram of the optic tectum contralateral to the intraocular injection of tritiated proline. The fine dots indicate the distribution of silver grains. The numbered lines indicate the transects used for the grain counts shown in Fig. 4.43.

These experiments give a good impression of the overall structure of the retinotectal projection but do not allow a direct visualization of the terminal arbors of the retinal ganglion cell axons themselves. It is possible to visualize ganglion cell axons in Golgi preparations, but it is often difficult to completely visualize individual terminals because the superficial tectal layers are usually heavily impregnated in Golgi preparations and obfuscate the relations of individual elements. This difficulty can be circumvented by using very small injections of HRP that will only fill small numbers of retinal ganglion cell axons. The labeled arbors can then be carefully studied.

Golgi or HRP preparations show that retinal arbors fall into two groups. The first are flat and form brush-shaped arbors that extend horizontally in the stratum fibrosum et griseum superficiale. These arbors have been described in Golgi preparations of *Pentodactylus schreiberii* (Quiroga, 1978). The second group of retinotectal arbors turn down into the tectum from axons running rostrocaudally in the

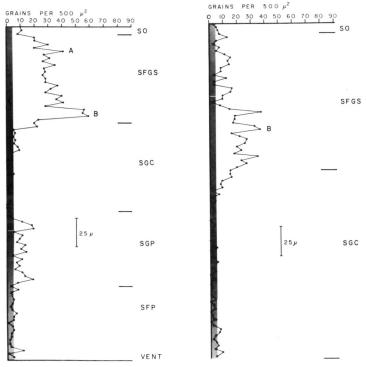

Fig. 4.43. *Pseudemys scripta.* Retinotectal projection. These are grain counts performed on the transects shown in Fig. 4.42. The graph on the left is for transect 1; the graph on the right is for transect 2.

stratum opticum (Fig. 4.44). These form vertically elongate masses of collaterals that bear many varicosities. They have been studied in HRP preparations of *Thamnophis sirtalis* (Dacey and Ulinski, 1986d). Almost all arbors originate from axons that turn down into the stratum fibrosum et griseum superficiale (Fig. 4.44). The parent axons typically branch into several, varicosity-laden collaterals that form a rectangular-shaped mass, about 65 by 65 by 100 μm, regardless of the locus of the terminal in the tectum. Most of the arbors are confined to either sublayer a or sublayer c of the stratum fibrosum et griseum superficiale, although there are a few bilaminar arbors. The arbors can be placed into three groups based on the diameter of their parent axons and the size of their varicosities. Type I arbors have parent axons with diameters of 3.0 to 3.5 μm and varicosities about 3.5 μm long (Fig. 4.45). Type 2 arbors have parent axons 1.5 to 2.5 μm in diameter and varicosities about 2.5 μm long (Fig. 4.46). Type 3 arbors have axons 0.5 to 1.0 μm in diameter and varicosities about 0.5 to 1.0 μm in diameter (Fig. 4.47). All tectal loci contain all three types of arbors, but their relative proportions may vary. The size spectrum histograms

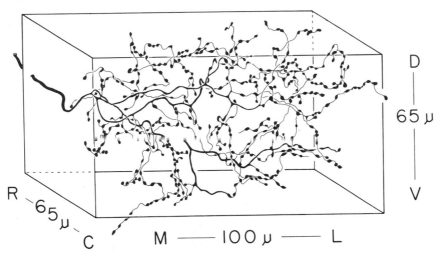

Fig. 4.44. *Thamnophis sirtalis.* Retinotectal terminal arbor. Rapid Golgi impregnation of an axon terminal that was restricted to sublayer a of the SFGS. The primary axon descends into the SFGSa from the overlying optic fiber layer (stratum opticum), splitting into several preterminal branches that in turn issue highly branched, beaded collaterals. The approximate dimensions of the terminal field are shown. Abbreviations: R-C, rostrocaudal; M-L, mediolateral; D-V, dorsoventral. (From Dacey and Ulinski, 1986d.)

of parent axons in the HRP-filled samples may be compared to the histogram of axon diameters taken from electron micrographs of the fiber fascicles in the stratum fibrosum et griseum superficiale (Dacey and Ulinski, 1986d; Repérant et al., 1981). Comparison suggests that the HRP sample is biased in favor of large-caliber terminals. However, the caliber of axons varies as a function of tectal locus, so it appears that the variations in axon caliber seen in degeneration experiments reflect differences in the relative proportions of the several types of arbors seen in different regions of the tectum.

It is likely that many retinal ganglion cell axons branch, sending one collateral to the tectum and a second to other targets such as the lateral geniculate complex. This makes it difficult to directly correlate the diameters of tectal axons in the tectum to those in the optic nerve. However, it seems probable that there is a general correlation, so that the largest-caliber tectal collaterals originate from the largest-caliber axons in the optic nerve and, therefore, from retinal ganglion cells with the largest somata. Data on the size spectra of axons in the optic tract are limited to observations on several species of turtles (Geri et al., 1982; Davydova et al., 1982). These studies report histograms with a distinct peak in the small-diameter range and either a second peak or a tail toward large diameters. These observations are consistent with measurements of conduction velocities of optic nerve axons in

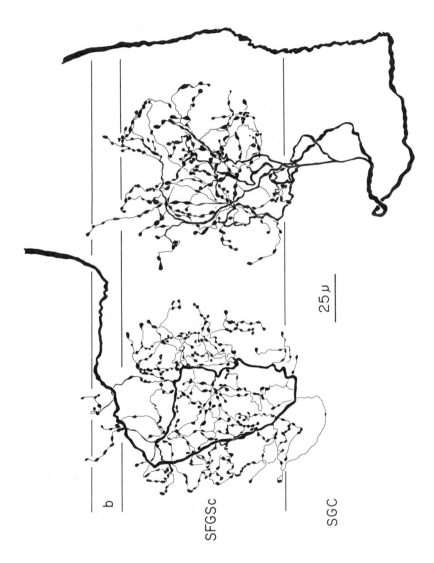

Fig. 4.45. *Thamnophis sirtalis.* Type 1 arbors. Two arbors restricted to the SFGSc are shown in the coronal plane. (From Dacey and Ulinski, 1986d.)

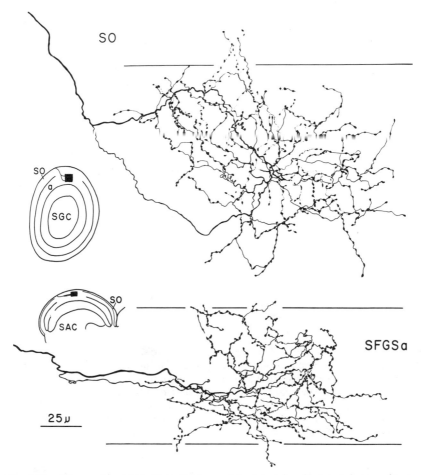

Fig. 4.46. *Thamnophis sirtalis.* Type 2 arbors. A horizontal and a coronal view of two terminals in SFGSa are compared. In the upper arbor, an elongation in the mediolateral axis can be seen. Axon diameter = 1.5 to 2.0 μm in all Type 2 arbors and most terminal boutons were 1.0 to 2.5 μm in diameter. As for the Type 1 arbors, the horizontal dimensions of all medium-caliber terminals = 100 to 110 μm in the mediolateral axis and 60 to 70 μm in the rostrocaudal axis. (From Dacey and Ulinski, 1986d.)

turtles (Davydova et al., 1982) that show the existence of at least two conduction velocity groups.

There are now abundant data from a number of mammalian species showing that conduction velocity groups within the optic nerve reflect the existence of morphologically and physiologically distinct classes of retinal ganglion cells (e.g., Rodieck, 1979; Lennie, 1980). Cells of each class have specific soma sizes, dendritic trees, physiological properties, and patterns of central terminations. There are fewer data on retinal ganglion cells in reptiles. Ramón y Cajal's drawings of

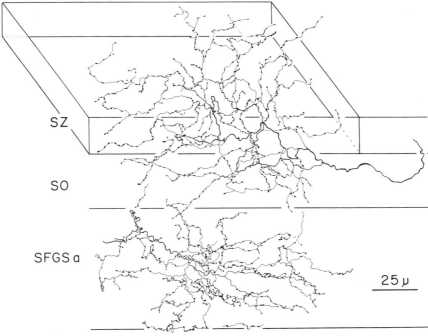

Fig. 4.47. *Thamnophis sirtalis.* Type 3 arbors. This figure shows an arbor with a small-diameter primary axon and small terminal boutons. The upper arbor is viewed in an oblique section. The part seen in the coronal plane extends ventrally into the stratum opticum (SO), and the part seen in the horizontal plane extends caudally in the stratum zonale (SZ). The lower arbor is restricted to the SFGSa and is viewed in the coronal plane; it has precisely the same rostrocaudal extension as the upper arbor. (From Dacey and Ulinski, 1986d.)

the retinal cells of the lizard *Lacerta* show several distinct types of ganglion cells. More recently, observations of ganglion cells in *Pseudemys scripta,* using both HRP-filled and Golgi-impregnated cells (Peterson, 1982; Kolb, 1982), show several morphologically distinct types of cells. At least two groups of ganglion cells can be recognized by physiological criteria (Marchiafava and Weiler, 1980; Marchiafava and Wagner, 1981). Filling of physiologically categorized cells with procion yellow shows that the Type A and Type B ganglion cells are morphologically distinct and have different conduction velocities.

The distribution of ganglion cells in *Pseudemys* (Peterson and Ulinski, 1979, 1982) has been studied with retinal whole mount preparations (Stone, 1981). Ganglion cells are inhomogeneously distributed; they occur in high densities along the visual streak that is aligned along the nasotemporal axis of the retina. The distribution of cell sizes varies in an orderly fashion; cells with small somata are predominant in the visual streak and a particularly large population of cells lies in the temporal-ventral retina. Comparisons to Marchiafava's results suggest that the visual streak consists largely of Type B cells, with Type A cells being spread more evenly over the retinal surface.

Tectal HRP injections have been used to establish the sizes of retinal ganglion cells that project to the tectum (Peterson, 1978a). Comparison of the cell size histograms of labeled and unlabeled ganglion cells indicates that ganglion cells of all sizes project to the tectum and that most cells within each size class project to the tectum. This, of course, does not eliminate the possibility that collaterals of these cells also project to the other targets.

The overall hypothesis, then, is that the nature of the retinotectal terminal zone in the superficial layers reflects retinal organization (Peterson, 1981). Retinal regions containing predominantly ganglion cells with small somata and axons project to tectal regions containing predominantly ganglion cell preterminal axons with small diameters. Conversely, retinal regions containing significant populations of large ganglion cells project to tectal regions of arbors with large diameter. The significance of these differences is presumably that the different regions of the tectum will receive functionally distinct types of information about the visual world.

The nature of the tectal elements that are postsynaptic to retinal arbors has been examined in turtles using electron microscopic preparations following retinal lesions (Davydova and Smirnov, 1973; Davyova et al., 1982) and with somewhat less certainty in normal electron microscopic preparations of the snake *Vipera aspis* (Repérant et al., 1981). Retinal terminals contact predominantly small dendritic profiles. Some of the profiles postsynaptic to retinal terminals contain synaptic vesicles. Some workers have regarded these as axoaxonic synapses, but it is likely that at least some of the vesicle-containing profiles represent presynaptic dendrites. This would be consistent with the presence of varicosities on the dendrites in the superficial layers of several classes of tectal neurons. The studies completed to date do not tell which classes of tectal neurons are postsynaptic to retinal terminals or whether the different types of ganglion cells preferentially contact different classes of neurons. It is now possible to approach these questions by using electron microscopy to examine identified neurons in Golgi or HRP material. However, no such time-consuming studies have yet been undertaken in reptiles.

c. AFFERENTS FROM THE VENTRAL LATERAL GENICULATE NUCLEUS
A second route whereby visual information can reach the optic tectum involves the ventral lateral geniculate nucleus. This is a plate-shaped structure that makes up the caudal part of the geniculate com-

plex. A geniculate occurs in all orders of reptiles and universally comprises a medial cell plate and a lateral neuropil (turtles: Bass and Northcutt, 1981a; snakes: Repérant, 1973; Halpern and Frumin, 1973; lizards: Cruce, 1974; Butler and Northcutt, 1973; crocodilians: Huber and Crosby, 1926).

The geniculotectal projections have been studied only in *Thamnophis*. Neurons in both the neuropil and the cell plate project to the tectum. Fig. 4.48 shows a neuropil cell that was filled by a small injection of HRP in the geniculate. The injection damaged the dendrites and demonstrates the complete morphology. These neurons have spherical somata 12 to 15 μm in diameter and dendritic arbors restricted to the retinorecipient part of the neuropil. The axon arises from the soma, gives off a single collateral in the neuropil, and then proceeds dorsocaudally in the tectogeniculate pathway. Collaterals with restricted arbors terminate en route in the nucleus lentiformis mesencephali of the pretectum. The axon continues in the stratum album centrale and ascends into sublamina a of the stratum fibrosum et griseum superficiale. It ends in a single, sparsely branched arbor with terminal boutons distributed into small clusters separated by bouton-free gaps. The overall spread of the arbor is about 200 to 250 μm in both the horizontal and sagittal planes.

Fig. 4.49 shows a cell plate neuron filled by an injection that impinged on the cell plate. Such neurons have narrow, fusiform dendritic fields that are oriented radial to the optic tract and have a diameter of about 75 μm. The axon (arrow) arises from a primary dendrite and passes dorsocaudally in the tectogeniculate path. Arbors attributable to cell plate neurons are characterized by a single cluster of boutons lying at the border of the superficial and central gray layers (Fig. 4.50). Some arbors occupy a horizontal area as small as 25 μm, whereas others form flattened sheets of boutons that measure 75 to 100 μm by 60 to 80 μm.

These experiments indicate that the ventral lateral geniculate is actually involved in two parallel paths to the optic tectum. The paths involve different populations of geniculate neurons and terminate in different sublayers of the tectum.

d. AFFERENTS FROM THE PALLIUM

A potential third route for transmission of visual information to the optic tectum involves the dorsal lateral geniculate nucleus. The dorsal lateral geniculate complex receives direct retinal projections and projects to the telencephalon. Evidence for this pathway has been assembled for lizards (Bruce, 1982) and turtles (Hall and Ebner, 1970b). Projections from the regions of the telencephalon that receive projec-

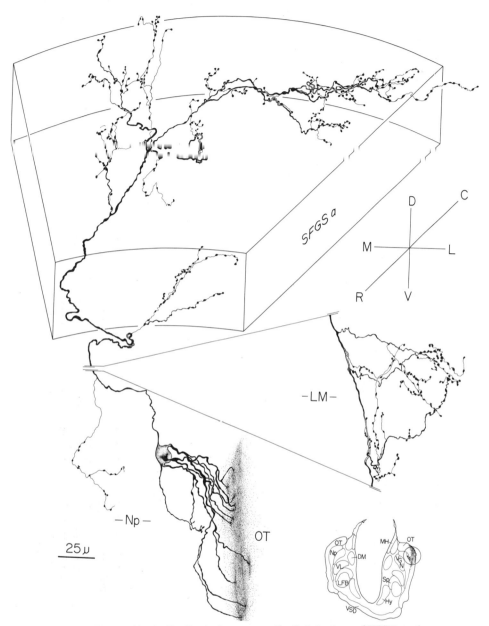

Fig. 4.48. *Thamnophis sirtalis.* Geniculate neuropil cell. Injections of HRP into the optic tract at the level of the ventral geniculate nucleus (circled area in the inset at the lower right) anterogradely filled the axon of this cell. It was traced through nine 80-μm serial sections to its terminus in the tectum. Terminal collaterals arise in the VGN neuropil (np), the lentiform mesencephalic nucleus (LM), and sublamina a of the stratum fibrosum et griseum supeficiale (SFGSa). (From Dacey and Ulinski, 1986e.)

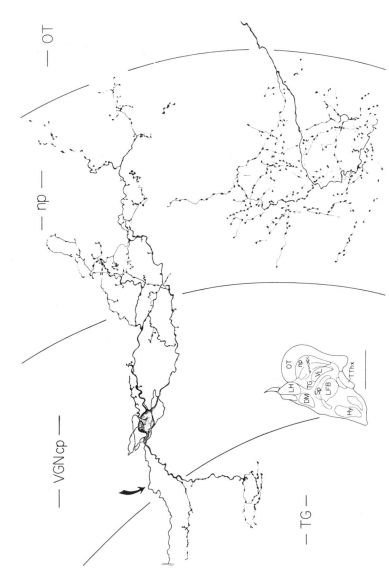

Fig. 4.49. *Thamnophis sirtalis.* Geniculate cell plate neuron. A neuron of the ventral lateral geniculate cell plate retrogradely filled from injections of HRP into the tectum. Laterally directed dendrites occupy a narrow, cylindrical space, approximately 75 μm in diameter in the retinorecipient neuropil. Medially directed dendrites have similarly sized arborizations in the tectorecipient zone. An anterogradely filled optic tract axon and its collateral terminal arbor in the geniculate neuropil is also shown. The diameter of the dimensions of the arbor are closely matched to the dendritic field size of the geniculate cell. (From Dacey and Ulinski, 1986e.)

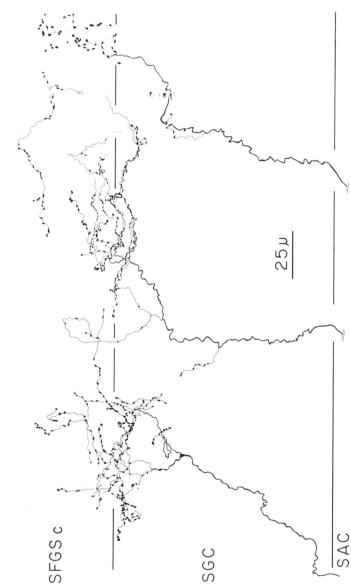

SFGS c

SGC

SAC

25 μ

Fig. 4.50. *Thamnophis sirtalis.* Thalamic arbors. Putative terminal arbors of ventral geniculate cell plate neurons. The parent axon ascends through the central layers of the tectum and arborizes as a single cluster of terminal boutons at the border of the stratum griseum centrale (SGC) and the stratum fibrosum et griseum superficiale (SFGS). (From Dacey and Ulinski, 1986e.)

tions from the dorsal lateral geniculate to the deep layers of the tectum have been reported in *Pseudemys*, but nothing is known of the organization of this projection. Another corticotectal projection has been reported in the lizard *Agama agama* (Elprana et al., 1980), but the projection originates from a region of cortex that probably does not receive visual input from the thalamus. It thus remains to be determined how widespread corticotectal projections are in reptiles and what might be their organization. Projections from the visual Wulst to the optic tectum have been well established by both anatomical and electrophysiological techniques for several species of birds (see Bagnoli and Burkhalter, 1983).

e. Afferents from the Pretectum
Visual information can also reach the optic tectum through the pretectum. This is a complex of nuclei that lies between the rostral pole of the tectum and the diencephalon. No uniform nomenclature has been established, but posterodorsal, mesencephalic lentiform, geniculate pretectal, and pretectal nuclei are usually recognized (e.g., Curwen and Miller, 1939; Cruce, 1974; Butler and Northcutt, 1973; Repérant, 1973). Each of these receives direct and generally bilateral projections from the retina.

The organization of the projection from the pretectum to the tectum has been studied in *Thamnophis* (Dacey and Ulinski, 1986a). Tectal HRP injections indicate that the pretectal, geniculate pretectal, and mesencephalic lentiform nuclei project to the tectum. The mesencephalic lentiform and pretectal nuclei in this species together contain a single, topographically organized projection from both the retina and the tectum (Dacey and Ulinski, 1986d). Soma sizes vary, but neurons in both nuclei issue stout dendrites that branch dichotomously and extend widely within the nuclear complex (Fig. 4.51). The axon arises from a dendrite close to the soma and courses to the contralateral tectum through the posterior commissure. The axon gives rise to a series of terminal branches as it passes caudally. Each branch is heavily studded with boutons and travels parallel to the tectal surface. Such terminal branches often give rise to small, tertiary branchlets that extend dorsoventrally within the stratum fibrosum et griseum superficiale (Fig. 4.52). The projection from a single axon therefore has collaterals that can terminate densely within wide areas of the tectum. Thus, in spite of a topography in the retinopretectal projection, the extensive size of pretectal dendritic fields and the widespread distribution of pretectal axons within the tectum make it unlikely that the retinopretectal-tectal path is carrying topographic information about retinal position.

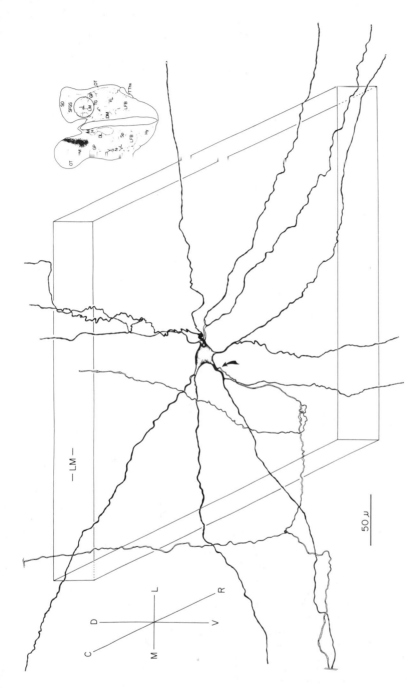

Fig. 4.51. *Thamnophis sirtalis.* Lentiform mesencephalic cell. These cells have long, unbranched dendrites. They are smooth and radiate for at least 250 μm from the soma and tend to be flattened in the horizontal plane. The axon (arrow) gives off primary collaterals that ascend into the superficial gray. The parent axon crossed the midline in the posterior commissure. The inset at the upper right of the figure shows the injection site (stippling) and the position of this cell in the tectum (circled area). (From Dacey and Ulinski, 1986e.)

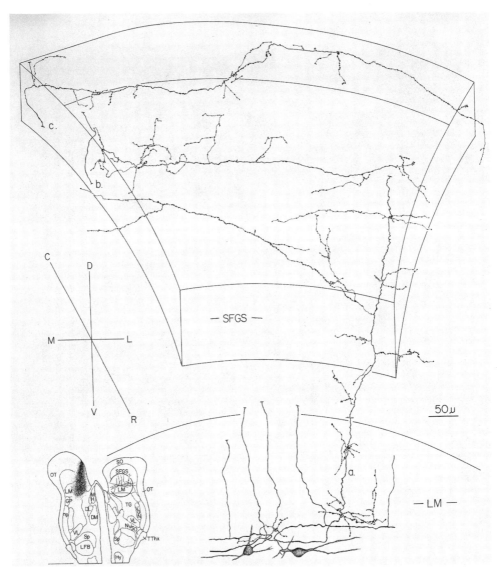

Fig. 4.52. *Thamnophis sirtalis.* Lentiform mesencephalic axons. Anterogradely filled tectal afferent terminals arising from cells of the lentiform mesencephalic nucleus (LM) after injections of HRP into the contralateral pretectal region. Primary collaterals that ascend into the tectum (a, b, and c) were traced for up to 600 μm rostrocaudally within the superficial gray. Collaterals give rise to terminal branches that course parallel to the tectal surface. These branches give rise to short branchlets that extend dorsoventrally within the superficial gray. The inset at the lower left indicates the injection site (stippling) and the position of the labelled cells (circled area). C-R, caudal-rostral; M-L, medial-lateral; D-V, dorsal-ventral. (From Dacey and Ulinski, 1986e.)

3. Binocular Visual Representation

a. General   The discussion has so far considered the routes whereby visual information reaches the optic tectum without regard to the laterality of the projections. However, the issue of laterality is functionally important because certain parts of the visual world are seen by both eyes. It is generally agreed that binocular cues are an important (although not essential) source of information about the distance of objects in the visual world. Most discussions of the binocular visual field in reptiles have been dominated by considerations of the binocular field of humans, which lies principally in front of the individual. Walls (1942), for example, provides measurements of the frontal visual fields that range from 10° to 40° of binocular overlap for several species of reptiles. The general idea has been that the binocular field is relatively restricted in nonmammalian vertebrates and increases as one proceeds up the "phylogenetic scale" to primates. However, it is now clear that many nonmammals have extensive binocular fields that include visual space both above and behind the animal. This point has been demonstrated most clearly for ranid frogs (Grobstein et al., 1983), using both behavioral and electrophysiological mapping techniques to measure the binocular visual field. Many species, such as some arboreal snakes, have extensive frontal binocular fields, but most reptiles probably have significant binocular fields above their heads and behind their bodies.

The rest of this section considers the nature of binocular interactions in the optic tectum. However, two points should be made before becoming embroiled in details. The first is that binocular interactions are not necessary for depth perception. Chameleons (*Chamaeleo jacksoni*), for example, use accommodative cues to make depth discriminations used in catching insects with their prehensile tongues (Harkness, 1977; Collett and Harkness, 1982). The second point is that binocular integration does not necessarily involve the tectum. In birds, both the tectum and visual thalamus seem to receive information only from the contralateral eye, and binocular integration occurs at the telencephalic level, mediated by bilateral thalamotelencephalic projections (Karten et al., 1973; Miceli et al., 1975; Bagnoli and Burkhalter, 1983). Thus the degree of binocular interaction in the tectum is not necessarily a measure of an animal's ability to make depth discriminations.

b. Retinal Afferents

All reptiles have contralateral retinotectal projections (see Section V.B.3). The extent to which ipsilateral retinotectal projections are also present is not entirely certain for technical reasons. However, ipsilateral projections to the tectum from those parts of the two retinas that see the binocular field would be expected a priori. Thus, ipsilateral projections to the rostral medial pole of the tectum (which receives information from the frontal binocular field), to the medial rim of the tectum (dorsal binocular field), and to the caudal pole of the tectum (caudal binocular field) would be expected. Autoradiographic tracing techniques have been used to examine retinotectal projections in the painted turtle, *Chrysemys picta* (Bass and Northcutt, 1981a). Silver grains were distributed in the ipsilateral tectum around the lateral and medial rims of the ipsilateral tectum. The projections to the medial edge of the tectum correspond to the tectal representation of the dorsal and caudal binocular fields. The projection to the lateral edge of the tectum is surprising at first glance because it would bring together information from disjunct points in visual space. However, examination of the retinotectal projection using both Fink-Heimer and the orthograde transport of HRP (Ulinski, unpublished observations) indicates that the fibers running around the lateral edge of the tectum are en route to the caudal pole of the ipsilateral tectum and represent fibers derived from the nasal retina, which sees the caudal binocular field.

c. Nucleus Isthmi

In addition to bilateral retinal projections to the tectum, several brainstem projections potentially involve the tectum in binocular integration. The first of these is nucleus isthmi, which receives afferents from the ipsilateral tectum and projects bilaterally back to the tectum. The contralateral isthmotectal projections thus form a route whereby information from one tectal lobe reaches its opposite member. Comparable structures occur in frogs (Gruberg and Udin, 1978; Gruberg and Lettvin, 1980; Wang et al., 1981; Grobstein et al., 1978), birds (Hunt et al., 1977), and mammals (Sherk, 1979).

Projections from nucleus isthmi to the tectum have been demonstrated in turtles (Foster and Hall, 1975; Sereno, 1983), lizards (Foster and Hall, 1975), and snakes; details of the isthmotectal system have been investigated in *Thamnophis* (Dacey and Ulinski, 1986d) and *Pseudemys* (Sereno and Ulinski, 1987). Small injections of HRP into the tectum disclose topographically organized tectoisthmic and isthmotectal projections. A small patch of anterogradely labeled boutons lies in nucleus isthmi; here it precisely overlaps several retrogradely labeled isthmotectal neurons. The size and position of the label in isthmi varies with the size and position of the tectal injection site, but the details of the maps have not been determined. Isthmotectal neurons have small spherical or slightly larger fusiform somata (Fig. 4.53).

In *Thamnophis*, axons from ipsilaterally projecting neurons arise from a soma or primary dendrite and exit the nucleus dorsally to run

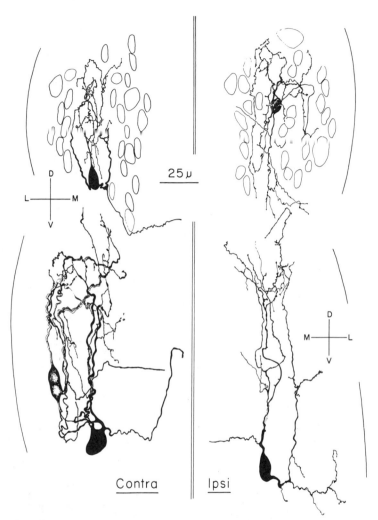

25 µ

Contra   Ipsi

Fig. 4.53. *Thamnophis sirtalis.* Isthmotectal neurons. These cells were retrogradely filled from injections of HRP into the tectum. Neurons both contralateral and ipsilateral to the injection site range from small stellate to larger fusiform types. The dendrites of isthmotectal cells often recurve to form small, spherical dendritic nests around the soma. The axons are indicated by the arrows. The somata of a few counterstained isthmi neurons are included in the top tracings. The curved lines indicate the lateral margin of the brainstem. M-L, medial-lateral; D-V, dorsal-ventral. (From Dacey and Ulinski, 1986e.)

into the stratum album centrale without branching (Fig. 4.54). The axons then ascend through the central and superficial gray and arborize heavily in the stratum zonale and stratum opticum. There is some variation in the morphology of the isthmotectal arbors. Some form conical clusters of terminals that reach the pial surface and spread to

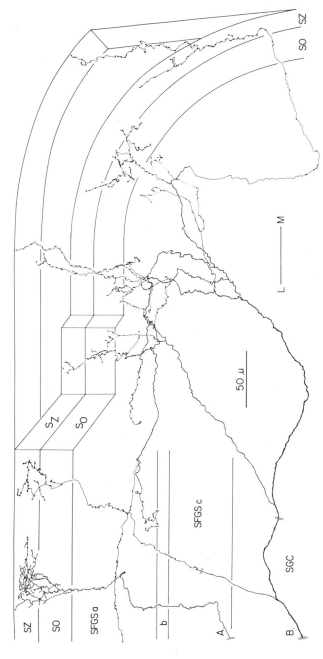

Fig. 4.54. *Thamnophis sirtalis.* Isthmotectal neurons. Tectal afferent axons anterogradely filled after injections of HRP into nucleus isthmi. These fine-caliber axons ascend through the superficial gray and arborize heavily in the stratum opticum (SO) and stratum zonale (SZ). One type of axon (A) forms a single arbor with a maximum diameter of 75 μm. Another type (B) arises from a slightly larger diameter axon that gives rise to multiple, thin collaterals. Terminal arbors issued from these collaterals are patchily distributed over a large area. (From Dacey and Ulinski, 1986e.)

a maximum diameter of 75 μm. Others ascend from the stratum album centrale and travel within the stratum griseum centrale, giving rise to a series of widely spaced, fine-caliber collaterals that eventually ascend to the pial surface, bearing boutons that arborize most heavily in the stratum zonale. The clusters of boutons on different collaterals of single axons are separated by gaps ranging from 20 to 100 μm. The total spread of the terminal field can extend over 500 μm in the mediolateral axis and 250 μm in the rostrocaudal axis. Axons of crossed isthmotectal neurons turn ventrally, join the ventral supraoptic decussation, and ascend dorsocaudally on the surface of the diencephalon just below the optic tract. The axons form a small bundle at the ventrolateral margin of the stratum opticum and course into the superficial layers of the tectum. They are of fine caliber and travel within these layers lateral to medial, parallel to the tectal surface (Fig. 4.55). The axons ascend into the stratum zonale in the interfascicular spaces present in the stratum opticum. They then give off tangentially oriented collaterals of fine caliber that form intertwined clusters of boutons, shaped like small cylinders, about 10 to 20 μm in length and 2 to 3 μm in diameter. The overall size of the arbors ranges from 70 to 100 μm in the mediolateral axis and from 40 to 80 μm in the rostrocaudal axis. Terminals from both the ipsilateral and contralateral nucleus isthmi thus converge in the superficial tectal layers and particularly in the stratum zonale. The topography of the projections has not been examined in reptiles; however, it seems likely that the contralateral isthmotectal projection will supplement the representation of the binocular visual field upon a given tectal lobe with information about corresponding points in the visual world as seen by the ipsilateral retina. It is noteworthy that the terminals of both the ipsi- and contralateral isthmotectal arbors overlap the dendrites of the tectoisthmi cells, which arborize in the stratum zonale.

In *Pseudemys*, nucleus isthmi is a complex of structures (Sereno and Ulinski, 1987). One component of the complex participates in reciprocal, topographically organized projections with the contralateral tectum, as is usually described in a wide range of vertebrates. However, a pair of components participates in connections with the ipsilateral tectum. The caudal nucleus isthmi contains neurons with flattened, bipolar dendritic fields that span only a few percent of the isthmi cell plate. These neurons project topographically to the ipsilateral tectum. The rostral nucleus isthmi contains neurons with large, sparsely branched dendritic fields. These neurons project nontopographically to the tectum by axonal arbors that cover large expanses of the tectum. The rostral nucleus, in turn, receives nontopographically organized projections from the tectum. Conjectures about the functional significance of the components of the nucleus isthmi complex are offered by Sereno and Ulinski (1987).

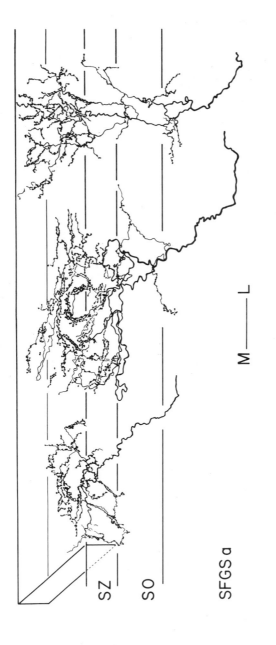

Fig. 4.55. *Thamnophis sirtalis.* Isthmotectal arbors. Tectal afferent axons anterogradely filled after injections of HRP into the ventral supraoptic decussation. These axons are believed to form the crossed isthmotectal projection. Single parent axons ascend through the superficial gray and form a single dense terminal arbor within the stratum zonale (SZ). The dimensions of these arbors in the horizontal plane range from 70 to 100 μm in the mediolateral axis and 10 to 80 μm in the rostrocaudal axis. (From Dacey and Ulinski, 1986e.)

d. PRETECTUM
The pretectum is also a possible route for binocular integration because it projects bilaterally to the tectum. The morphology of the pretectotectal axons in *Thamnophis* has been discussed in Section III.E. In contrast to the isthmotectal projections, pretectal projections to the tectum appear to be nontopologically organized and may therefore serve some function other than the integration of information from homotypic points in the two visual fields

e. COMMISSURAL PROJECTIONS
The two tectal lobes are interconnected by a commissure running in the rostral face of the tectum. Several classic descriptions of tectal anatomy (e.g., Huber and Crosby, 1933) and studies of tectal efferents using degeneration techniques (e.g., Ulinski, 1977; Schroeder, 1981b) describe commissural connections that interconnect the two tectal lobes through the tectal commissure. Such projections could serve as a substrate for binocular interactions; however, these observations are ambiguous because of the potential difficulty posed by axons that originate from outside of the tectum and pass through the tectal commissure to the contralateral tectum. Orthograde tracing studies of reticular-tectal projections with HRP indicate that there are probably several axon systems that follow this trajectory (see Section V.E). Autoradiographic tracing techniques would presumably circumvent this fibers of passage problem, but there is no clear evidence at this time that reptiles have commissural connections. However, commissural interactions occur between the frontal binocular field representations in the superior colliculus of cats (Edwards, 1977).

C. Somatosensory Afferents
1. SOMATOSENSORY REPRESENTATIONS   Two types of somatosensory information are represented in the optic tectum. The first is information derived from mechanoreceptors in the head or postcranial body surface (see von Düring and Miller [1979] for a summary of the mechanoreceptors of reptiles). Tectal responses to somatosensory stimuli have been recorded from the deeper tectal layers in turtles (Robbins, 1972), lizards (Stein and Gaither, 1981, 1983), and snakes (Hartline et al., 1978), but the only detailed studies are mapping studies in the iguana, *Iguana iguana* (Stein and Gaither, 1981, 1983). Units in the optic tectum will be activated by innocuous stimulation of the contralateral of the head or body. Units that respond only to tactile stimulation lie below layer 5 of Pedro Ramón, whereas units that respond to both tactile and visual stimuli occur in layers 5 to 7. Somatic units habituate rapidly and have definable receptive fields that vary systematically in size. The somatic representation is topographically

organized and in register with the overlying visual representation. For instance, the representation of the face is in register with the representation of the central and nasal visual field. Consistent with the orientation of the representations, somatic magnification factors vary, and the face representation occupies approximately 40% of the tectal surface.

Infrared thermosensitive units provide a second type of sensory information that can be represented in the optic tectum. Such units occur on the face and jaws of boid and crotaline snakes, either in specialized pit organs or in labial scales (Barrett, 1970). Because only these two phylogenetically separate lineages of snakes show infrared reception, it is generally assumed that this sensory modality has evolved independently. Details of the infrared system are discussed by Molenaar in Chapter 5 of this volume. It is sufficient for present purposes to note only that the infrared representation resembles general somatosensory information in being restricted to the intermediate and deep layers, being topographically organized, and being in register with the overlying visual representation (Hartline et al., 1978).

2. SOURCES OF SOMATOSENSORY INFORMATION

a. GENERAL   Somatosensory information can reach the optic tectum through the spinal cord, trigeminal nuclei, and thalamus.

b. SPINAL AFFERENTS

Spinotectal projections have been described in the red-eared turtle *Pseudemys scripta* (Ebbesson, 1969), several pleurodiran turtles (Pedersen, 1973), the lizard, *Tupinambis* (Ebbesson, 1967), the boa constrictor, *Boa constrictor* (Ebbesson, 1969), and the caiman, *Caiman* (Ebbesson and Goodman, 1981) using Fink-Heimer techniques following spinal hemisections. These studies all indicate that spinal afferents terminate in a small region of the caudomedial tectum. This is consistent with the representation of the caudal visual field in this part of the tectum and with a relatively small representation of the caudal body surface within the tectum, as suggested by Stein and Gaither (1981) for *Iguana*. The neurons within the spinal gray matter that give rise to this projection are not known.

c. TRIGEMINAL AFFERENTS

Trigeminotectal projections have been reported in the early descriptive literature (Woodburne, 1936) and are the subject of several recent studies of the infrared system in snakes. Ganglion cells that innervate the infrared organs have central processes that terminate in the trigeminal nuclei (Molenaar, 1974; Schroeder and Loop, 1976). Information from these nuclei reaches the contralateral tectum in both boid and crotalid snakes but does so by different routes. In boid snakes, direct projections from the trigeminal complex pass to the deep layers of the optic tectum (Molenaar and Fizaan-Oostveen, 1980; Newman et al., 1980). In crotaline snakes, the trigeminotectal projections involve a relay in reticular nucleus that has been named the nucleus caloris (Gruberg et al., 1979). These projections are discussed more extensively by Molenaar in Chapter 5.

d. THALAMUS

Direct projections from the spinal cord (Ebbesson and Goodman, 1981) and the trigeminal complex (Molenaar and Fizaan-Oostveen, 1980) reach a nucleus situated in the ventral thalamus just dorsal to the forebrain bundle system. It has been variously named the suprapeduncular nucleus or the ventral lateral nucleus, among others, and appears to receive both cranial and postcranial somatosensory information. Tectal HRP injections indicate that this nucleus projects bilaterally to the optic tectum (Dacey and Ulinski, 1986d). The axons of these neurons in the ventrolateral nucleus have been potentially identified in *Thamnophis* following diencephalic injections of HRP (Fig. 4.56). The axons travel caudally in the periventricular gray and white layers and then turn to extend vertically through the stratum album centrale, where they terminate as sparsely branched conical arbors. The axons or arbors branch and issue collaterals within the stratum album centrale and stratum griseum centrale. Single arbors range in diameter from 50 to 100 μm.

D. Auditory Information

Several reports note units in the deep tectal layers that are responsive to auditory stimuli in reptiles (Hartline, 1971; Stein and Gaither, 1981, 1983). HRP injections into the optic tectum retrogradely label a few neurons in the torus semicircularis of the midbrain, thereby indicating a potential route whereby auditory information could reach the optic tectum (Dacey and Ulinski, 1986d).

E. Reticular Afferents

Injections of HRP into the optic tectum indicate that several structures throughout the brainstem reticular formation project to the tectum. Because the tectoreticular neurons project bilaterally and extensively to these same structures, these reticulotectal projections are the return loop of a reciprocal interaction between the tectum and reticular formation.

Only the projections to the tectum from the nucleus lateralis profundus mesencephali of the midbrain reticular formation have been

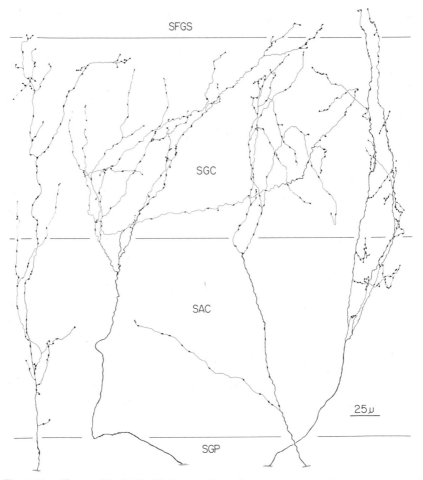

SFGS

SGC

SAC

25μ

SGP

Fig. 4.56. *Thamnophis sirtalis.* Thalamic arbors. Putative terminal arbors of neurons of the ventrolateral nucleus. The parent axons ascend from the periventricular gray (SGP) and form vertically oriented arbors in the central white and gray layers. In comparison with the terminals shown in Figs. 4.8 to 4.10, these arbors have a sparse branching pattern that distributes boutons at a lower density over a more widespread area. (From Dacey and Ulinski, 1986e.)

studied in detail. Tectal HRP injections in *Thamnophis* retrogradely label neurons in the deep mesencephalic nucleus (Fig. 4.57). These neurons have large somata, about 30 μm in diameter, bearing several thick primary dendrites that radiate into the tegmental gray. The dendrites are smooth and bifurcate near the soma. They extend up to 400 μm and form a dendritic field that is relatively flat in the transverse plane. Axons arise from the somata, give off a small collateral within the dendritic field of the cell, and proceed dorsally into the stratum album centrale. The primary branches issue collaterals that terminate

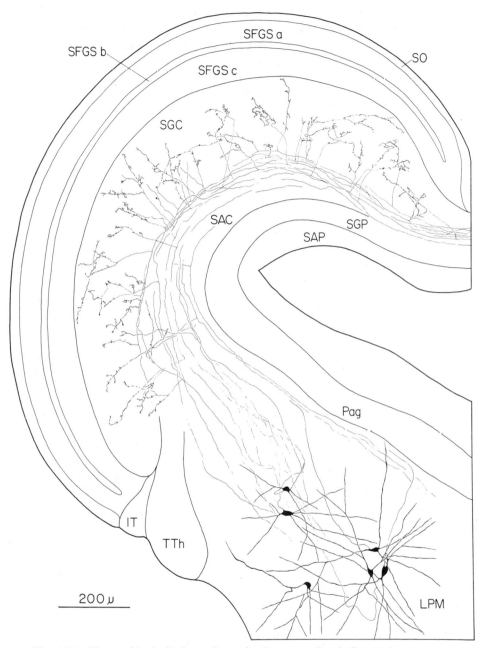

Fig. 4.57. *Thamnophis sirtalis.* Lateralis profundus axons. Tectal afferents from nucleus lateralis profundus mesencephali (LPM) retrogradely filled after an injection of HRP into the opposite tectal hemisphere. Large-caliber axons arising from the multipolar cells of LPM ascend into the stratum album centrale (SAC), course medially, and cross the midline in the tectal commissure. Terminal collaterals are restricted to the stratum griseum centrale (SGC). (From Dacey and Ulinski, 1986e.)

within the stratum griseum centrale, bearing numerous clusters of large boutons that are somewhat patchily distributed. Injections into the midbrain tegmentum anterogradely fill reticulotectal axons and retrogradely fill large tectoreticular neurons. These injections show that the deep mesencephalic terminals appose the somata and primary dendrites of the solid filled neurons, suggesting direct reciprocal relations between tectal and reticular neurons. The reticular axons can be traced through the tectal commissure and into the contralateral tectal lobe, where they terminate in the same fashion. Since the entire axon can often be traced in an individual section, it appears that the trajectories of the reticular axons are restricted to mediolateral planes.

F. Afferents Related to the Striatum

The striatum is a collection of structures in the ventral part of the telencephalon rostral to the anterior commissure. The anatomy of the striatum has been discussed at length (Ulinski, 1983). In brief, it receives afferents from the anterior dorsal ventricular ridge, which lies above it protruding into the lateral ventricle. The anterior dorsal ventricular ridge (ADVR) receives ascending sensory information from visual, auditory, and somatosensory nuclei in the thalamus, so that its projections to striatum presumably carry sensory information to the striatum. The striatum has reciprocal projections to ADVR, but the major striatal efferents run caudally into the brainstem and include two nuclei that project to the optic tectum.

The first of these nuclei lies in the ventrolateral midbrain tegmentum and contains dopaminergic neurons. It is called the substantia nigra in turtles (Parent, 1979) and lizards (Baumgarten and Braak, 1968) and the pedunculopontine nucleus in crocodilians (Brauth and Kitt, 1980). HRP injections into the optic tectum retrogradely label neurons in this nucleus in turtles (*Chrysemys,* Reiner et al., 1980), monitor lizards (*Varanus exanthematicus,* ten Donkelaar and DeBoer-van Huizen, 1981), and crocodilians (*Caiman,* Reiner et al., 1980).

The second link between the striatum and optic tectum lies at the transition between the caudal diencephalon and pretectum. It has been called the dorsal nucleus of the posterior commissure. Like the substantia nigra, it receives descending projections from the striatum (lizards: Hoogland, 1977; Voneida and Sligar, 1979; ten Donkelaar and DeBoer-van Huizen, 1981; Reiner et al., 1980). Nothing is known about the organization of these systems of tectal afferents in reptiles.

G. Monoaminergic Projections

The distribution of neurons and axon systems that contain either serotonin or catecholamines has been studied in turtles and lizards using histofluorescence techniques (reviewed by Parent, 1979). These studies indicate that the patterns of monoaminergic innervation in turtles and lizards are similar. Catecholaminergic terminals are distributed principally in the superficial layers of the tectum. These studies do not establish whether the terminals are noradrenergic or dopaminergic. They presumably derive from neurons with their somata in the locus coeruleus or the substantia nigra, or both, but this point has not been explicitly demonstrated with experimental techniques. Most serotonergic terminals lie in the deeper layers of the tectum and presumably derive from neurons with somata in the raphe nuclei.

H. Summary

The application of modern axonal tracing techniques has greatly increased the number of known afferent systems of the tectum. Investigations with the Golgi technique demonstrated that several morphologically distinct varieties of afferent axons are present in the tectum, but it was generally not possible to identify their sources. The use of the orthograde HRP tracing techniques has now permitted the identification of several of the major sets of axons afferent to the tectum and, when coupled with retrograde tracing techniques, has allowed the determination of the complete morphology of the cell. Orthograde tracing techniques have shown that most of the afferent systems of the tectum terminate preferentially within specific layers or sublayers. There must consequently be specific relationships between particular afferent systems and particular classes of efferent or intrinsic neurons. It has sometimes been possible to suggest relationships, such as the existence of a reciprocal loop between neurons in the reticular formation and the tectoreticular neurons. However, it is for the most part only possible to establish that specific classes of neurons lie within the domains of specific classes of afferents. The general rule seems to be that whenever reciprocal connections occur, the dendritic fields of neurons forming one limb of the loop lie within the terminal domains of the neurons forming the return limb. Like the efferent systems, some of the systems of afferents to the tectum are topologically organized, whereas others are not. The implications of topologically and nontopologically organized projections are considered in the next section.

VI. GENERAL CONCEPTS OF TECTAL ORGANIZATION
A. Overview

The preceding sections have considered the morphology of tectal neurons and afferents to the tectum in some detail. Much of the information reviewed was obtained by the application of contemporary neuroanatomical techniques, particularly those involving the use of HRP. It is inevitable that the accumulation of significant pieces of information on the anatomy of any neural structure will lead to some alteration in the prevailing concepts about its organization. This section will therefore discuss some general aspects of the relation of the optic tectum to the organization of the central nervous system as a whole and of the internal organization of the tectum itself.

B. Position of the Tectum in the Organization of the Nervous System

The concept prevailing up to the 1950s was that the optic tectum receives a variety of sensory inputs and projects to the brainstem reticular formation, thereby participating in the generation of relatively simple orienting movements in "lower" vertebrates. The tectum was held to be relatively unimportant in mammals, in which the cerebral cortex was thought to have taken over most important behavioral tasks through a process called encephalization. This view has been emended in two ways during the past three decades (Fig. 4.58). The first major change in our understanding of the way in which the tectum fits into the overall organization of the brain was the discovery of two circuits that link the optic tectum with the telencephalon in all vertebrates. In reptiles and birds, information from the optic tectum reaches the telencephalon through nucleus rotundus in the dorsal thalamus. Projections from the tectum to the nucleus rotundus were first established in pigeons (Karten, 1965) and then in red-eared turtles *Pseudemys scripta*, (Hall and Ebner, 1970b). They have since been demonstrated in representatives of all of the major groups of reptiles (see Section III.D). As the nucleus rotundus projects to the anterior part of the dorsal ventricular ridge (Section III.D; Ulinski, 1983), the tectorotundal projections are the first link in a pathway that carries information from the optic tectum to the dorsal or pallial component of the telencephalon. For our purposes, the anterior dorsal ventricular ridge has a principal efferent connection to the underlying striatum (Ulinski, 1983), which lies in the ventral or basal component of the telencephalon. The striatum influences the tectum through a pretectal structure (the dorsal nucleus of the posterior commissure) and through a nucleus in the midbrain tegmentum (substantia nigra or pedunculopontine nucleus). These projections thus constitute a feedback loop to the tectum. The sign of the loop—whether it is a negative or positive feedback loop—is not known in reptiles, but physiological studies (e.g., Hikosaka and Wurtz, 1983) on the nigrotectal projections in mammals show they are inhibitory to collicular neurons.

A second loop between the tectum and the telencephalon has been demonstrated in some reptiles (see Section V.B.2.d) and more completely in birds. It involves a direct projection from the pallium to the

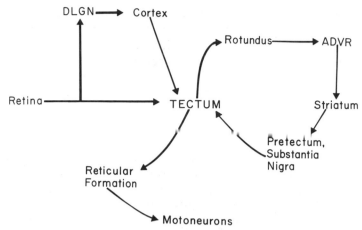

Fig. 4.58. Relation of tectum to other brain structures. This diagram summarizes the relation of some major structures to the optic tectum.

tectum. The regions of the pallium (i.e., pallial thickening or Wulst) that project to the tectum tend to receive visual information from the dorsal part of the geniculate complex (e.g., Hall and Ebner, 1970b; Karten et al., 1973; Miceli et al., 1975; Repérant et al., 1974). As many of the retinal ganglion cells that project to the tectum probably also project to the geniculate complex through collaterals, these pallial projections may represent a feedback to the tectum of part of its visual input. Electrophysiological studies in birds indicate that this loop exerts both an excitatory and an inhibitory effect on tectal units and plays a significant role in the construction of directionally selective units (e.g., Britto, 1978; Leresche et al., 1983).

Consequently, tectal output is a function not only of inputs from the major sensory systems, but also of at least two telencephalic feedback loops. The interaction of direct sensory and telencephalic inputs to modulate tectal output in reptiles is not well understood. There is, however, relevant information from studies on mammals. Both striatal (e.g., Rhoades et al., 1982) and pallial (e.g., Kawamura et al., 1974) projections to the tectum are also present in mammals. Anatomical (e.g., Mize and Sterling, 1976) and electrophysiological evidence indicates that retinal and cortical inputs can converge on the same collicular neuron. As is the case for tectal neurons in reptiles and birds, collicular neurons in mammals frequently show a preference for moving visual stimuli and are often directionally selective (e.g., Goldberg and Robinson, 1975). Experiments in which the visual cortex is either ablated or reversibly inactivated by cooling (e.g., Wickelgren and Sterling, 1969; Rosenquist and Palmer, 1971) suggest that directional selectivity can be conferred upon collicular neurons at least in part by directionally selective neurons in layer V of the visual cortex. The contribution of the telencephalic loops to the behavior of the organism is not clear for either reptiles or mammals.

A second area in which there has been some change in our concept of how the tectum fits into the organization of the brain involves a gradual appreciation of the complexity of the pathways linking the tectum to motoneurons in the spinal cord and brainstem. It has been known for some time that the tectum sends both ipsilateral and contralateral projections from the brainstem reticular formation. Reticular neurons are premotor in the sense that their axons have a relatively direct influence on motoneurons, either in the brainstem or spinal cord. Recent work (e.g, Moschovakis et al., 1988a, 1988b) has shown that the tectoreticular pathways originate from several, morphologically heterogenous groups of neurons and run in several fascicles distinguished by their position in the brainstem and the caliber of their axons (Section III.A). Each axon tends to collateralize rather extensively in the brainstem. Thus, the activation of any particular locus in the tectum will lead to a modulation of the activity of neurons spread throughout the brainstem and at least the cervical levels of the spinal cord.

The nature of the brainstem projections to the spinal cord has been studied in reptiles using both retrograde degeneration (Robinson, 1969; ten Donkelaar, 1976a, 1976b; Cruce and Newman, 1981) and HRP techniques (ten Donkelaar and De Boer-van Huizen, 1978; ten Donkelaar et al., 1980; Newman et al., 1982). These studies show that a large number of nuclei situated throughout the rostrocaudal extent of the neuroaxis project to the spinal cord ipsilaterally, contralaterally, and bilaterally. Thus, the tectum can modulate neuronal activity in the spinal cord through many parallel pathways in the hindbrain and midbrain reticular formations.

A further complexity in the tectal modulation of descending pathways involves the geniculate complex and the pretectum. Both of these structures receive topologically organized projections from the retina, have reciprocal connections with the tectum, and give rise to descending projections to either the spinal cord (Newman, personal communication) or brainstem (Ulinski et al., 1983). The tectogeniculate and tectopretectal projections thus provide the tectum access to additional, parallel routes to motoneurons.

C. Sensorimotor Transformations

From a functional viewpoint, the position of the tectum in the brain is that it modulates orienting movements of the organism through the brainstem reticular formation. In essence, the tectum participates in a transformation of information about the location of objects in extracorporeal space (which must be coded in spatial coordinates) to a set of instructions to motoneurons (which must be coded in the firing frequencies of premotor neurons). Several types of models have been suggested for the mechanisms that underlie such sensorimotor transformations.

The first type of model for sensorimotor transformations—labeled line models—posits a series of direct connections from a locus on a sensory surface, to the tectum, and then through premotor structures (presumably the reticular formation) to those motoneurons that will, when activated, orient the animal toward the stimulus. One could imagine, for example, that stimulation of one region of the retina activates a particular tectal locus that projects, through the reticular formation, to a motoneuron pool, the activity of which causes the animal to orient toward the stimulus. Such connections would be sufficient, at least qualitatively, to account for orienting behaviors. There is evidence in cats, for example, that projections from the region of the tectum that receives a representation of the peripheral visual field (and peripheral auditory field) projects to the vicinity of the facial motoneurons that control pinna movement. These connections could account for orienting movements of the pinnae in response to sudden sounds (Stein and Clamann, 1981).

A drawback to the generality of labeled line models is that orienting behaviors are generally more complex than they imply. Animals do not orient to every stimulus in their environment, and, when they do respond, they have the capability of responding in a graded fashion. Thus, something more than a simple activation of a set of circuitry must be involved. There must be some evaluation of stimulus properties and of the relation between the stimulus and the total environmental context. Similarly, interactions occur when two or more stimuli are presented to the organism, so it is difficult to account for orienting movements in terms of circuitry that activates only restricted tectal loci. The labeled line models also face difficulty when the temporal aspect of orienting movements is considered, the problem being that the amount of time involved in direct activation of motoneurons through the tectum and reticular formation is much shorter than the time course of actual orientation movements.

A second type of model circumvents the first difficulty by suggesting that the efferent projections from the tectum form a distributed system. Distributed systems, in general, are those in which many of the system elements share in a given function (see Anderson and Hinton, 1981, for discussions of the concept of distributed systems). In the case of the tectum, the idea is that each tectoreticular neuron would project to many reticular neurons and influence more than a single pool of motoneurons. Information would be coded in a population of individual tectal neurons. Models of this type were proposed as early as 1947 by Pitts and McCulloch to account for the control of eye movements by the superior colliculus. They suggest that a distributed system involving the superior colliculus and projections to motor pools controls vertical and horizontal eye movements. Data consistent with distributed system models are available in recent neurophysiological studies of the collicular control of eye movements in alert monkeys and cats. These experiments show that neurons throughout a relatively widespread region of the tectum are active before a variety of eye movements occur (e.g., Wurtz and Albano, 1980), making it unlikely that a given cluster of collicular neurons codes particular eye movements. Rather, individual neurons appear to participate in a variety of different movements. Conversely, lesioning or inactivating a particular region of the colliculus appears to have relatively little specific effect on eye movements, suggesting again that the relationship between regions of the colliculus and the particular movements is not one to one (Lee et al., 1988). The little that is known about the anatomy of the tectoreticular projections is also consistent with distributed system models. Grantyn and Grantyn (1982) found that the tectoreticular axons in cats branch extensively throughout the brainstem reticular formation, so that each locus in the colliculus is likely to influence activity in many reticular neurons. Conversely, each region of the reticular formation will receive information from all collicular loci. The same picture emerges from studies in turtles and snakes (see Section III.A), which again emphasize the divergent character of tectoreticular projections.

Distributed system models are attractive because they help to explain how activity in several motoneuron pools is coordinated during a complex movement sequence. However, it is not entirely clear at this point how different movements are produced by the same system. One suggestion that has been explored attempts to explain how the amplitude of horizontal eye movements is coded. It is known that the superior colliculus projects to a center for horizontal gaze, presumably located in the paramedian pontine reticular formation. Neurons in this center drive motoneurons in the abducens nucleus (e.g., Wurtz and Albano, 1980; Robinson, 1981). The force generated by the lateral rectus muscle is a linear function of the firing frequency in abducens motoneurons. The execution of appropriate orienting movements requires the eye to move in the horizontal plane so as to fixate a stimulus located in a specific region of visual space. The problem is how the transformation from position coordinates in visual space to motoneuron firing frequencies is achieved. Orthograde HRP experiments indicate that reticular axons project diffusely within the abducens pool so that there is a divergent-convergent relationship. A simple solution can be achieved if there are differences in the number of neurons that issue from different tectal loci. Thus, the synaptic drive to abducens motoneurons would vary as a function of the tectal locus. This could occur either due to density differences in the number of neurons as a function of tectal locus (see Section VI.D.4), to differences in the magnification factor as a function of position in visual space (McIlwain, 1982), or both.

A further difficulty with models of simple distributed systems is that they do not in themselves explain the temporal features of orientation movements. The relatively long time course of such movements suggests that some sort of computational process is taking place, involving either computations effected entirely within the tectum or feedback loops through the forebrain and brainstem. Behavioral support for more complex processes comes from experiments in which lesions of brainstem or telencephalic structures (e.g., Ewert, 1976; Ingle, 1973) are shown to influence the performance of tectally mediated orienting movements and from other experiments in which interaction effects are demonstrated between two or more simultaneously presented stimuli (Ingle, 1968). Based on their behavioral experiments, Ewert and von Seelen (1974) have produced models of the neural substrates of frog orientation movements that involve telencephalic and brainstem interactions with tectal neurons. Arbib and his colleagues (Lara et al., 1982) have explored a series of computational models involving interactions among tectal modules. They utilize horizontally aligned intratectal connections between neurons positioned throughout the tectum to produce specific tectal outputs. McIlwain (1982) has reported electrophysiological results consistent with extensive intratectal processing. He used microstimulation and recording techniques in cat superior colliculus to show that electrical stimuli that result in orienting movements of the eyes actually activate widespread neurons within the colliculus. The limitations inherent in earlier anatomical techniques led to an underestimation of the spatial extent of tectal axon collaterals. However, HRP techniques have now demonstrated widespread axonal systems within the tectum (see Sections III and IV). The axons of most tectal efferent and intrinsic neurons have extensive vertical components as well as, in many instances, axons that extend horizontally for relatively long distances. Many of these axons are asymmetric, spreading preferentially either laterally or medially. It is clear that such intratectal connections can form the substrate for intratectal processing, but the nature of the processing is not known.

D. Internal Organization of Tectum

1. GENERAL   A major goal in neuroanatomy since the work of Ramón y Cajal has been to understand the patterns of neuronal organization that characterize major regions of the nervous system. Significant advances have been achieved in the case of the cerebellum (Eccles et al., 1967) and cerebral cortex of mammals (Lund, 1981), by using Golgi

and electron microscopic techniques in coordination. In contrast, most of the attention devoted to the optic tectum has focused on the organization of the retinorecipient layers, and we still lack any generally accepted idea of the overall organization of the tectum. However, several aspects of tectal organization merit consideration.

2. Tectal Lamination

The concept that the tectum is a layered or laminated structure stems directly from cytoarchitectual observations that show a number of more or less distinct layers of somata in Nissl preparations. There is an ingrained tendency in neurobiology to assign specific functions to anatomically defined units (e.g., Kaas, 1982), and a subliminal thread that runs through much of the history of ideas of tectal function is an assignment of different functions to different tectal layers. The gradual establishment of relations between the pattern of afferent and efferent connections and that of tectal cytoarchitecture reinforced the concept that tectal layers constitute functional units and, combined with behavioral studies (e.g., Casagrande and Diamond, 1974), led to the general idea that the superficial layers are related through ascending projections to the forebrain with perceptual tasks, whereas the deep layers are related through their descending projections to motor tasks. The behavioral experiments consist of examining the effects of restricted tectal lesions on pattern discrimination versus motor tasks. More recently, studies using electrophysiological recordings from alert, unanesthetized animals have attempted to correlate the firing pattern of tectal neurons with motor activity; these studies found that neurons whose activity correlated with motor activity tend to be located in the intermediate or deep layers.

However, several considerations make it difficult to accept a strict assignment of "perceptual" or "motor" functions to tectal layers. First, it is now clear that auditory, tactile, and infrared information reaches the deeper layers of the tectum, so that no one set of layers can be regarded as "sensory" to the exclusion of others. Rather, there appears to be at least a partial laminar segregation of modalities within the tectum. Second, the idea that the superficial layers are involved in perceptual tasks, such as pattern vision, stems in large part from the demonstration of ascending projections from the superficial layers to the thalamus and ultimately to the telencephalon. It has subsequently been determined that the telencephalon has strong reciprocal projections back to the tectum through both the pallium and striatum. Thus, the ascending projections to the forebrain can also be regarded as one limb in a feedback system to the deeper layers. Third, the idea that the tectal layers are functionally independent also requires some degree of anatomical independence. The apparent absence of interlaminar connections in mammals has been used in support of this notion (e.g., Edwards, 1980; Wurtz and Albano,1980). However, there is good evidence for strong interactions between tectal layers in reptiles, and a good deal of neurophysiological evidence in primates and cats (e.g., Wurtz and Albano, 1980) indicates that there are short latency interactions between neurons in different layers involved in producing eye movements.

A corollate of the idea that tectal layers have functional significance is that the tectal layers should vary during evolution under the force of environmental selection pressures. Thus, several studies have shown that reptiles with large eyes and those that appear visually active have thicker superficial layers with more numerous and more clearly defined sublayers, whereas burrowing or other visually inactive forms have reduced superficial layers (Masai, 1973). Snakes as a group have relatively reduced superficial layers. Senn and Northcutt (1973) and Northcutt (1978) have discussed the tectum of snakes in relation to the theory of Walls (1942) that the snakes are derived from ancestral burrowing forms. (See Chapter 3 for a complete discussion of this theory.) Similarly, snakes with infrared systems tend to show a hypertrophy of those tectal layers that receive trigeminal input (e.g., Auen, 1976). Variations of this sort are clearly present, but the prominence of vertical interactions within the tectum of reptiles complicates their interpretation. For example, the dendrites of neurons with somata in the central and deep layers must constitute a large percentage of the postsynaptic targets in the superficial layers. Thus, hypertrophy of the superficial layers is likely to accompany major changes in neurons through the tectum and in the interaction of central and deep neurons within the superficial layers through interactions through dendritic bundles and axon collaterals.

Regardless of their functional or evolutionary significance, an immediate consequence of the presence of specific sets of connections with the various tectal layers is that each class of tectal efferent and intrinsic neurons is potentially in receipt of a specific pattern of afferent projections. Definitive demonstrations of the relationship between specific afferent systems and specific neuronal classes are for the most part lacking, but such experiments are now possible by combining orthograde tracing techniques with electronmicroscopic observations on individual neurons identified by HRP or Golgi procedures. These experiments are time consuming but are necessary to specify completely the network of connections involved in tectal function.

However, light microscopic information is sufficient to establish a major organizational difference between the tectum in mammals and that of most groups of nonmammalian vertebrates, including reptiles. The receptive surfaces of most types of neurons in mammalian tecta are restricted principally, if not completely, to individual layers or groups of sublayers. Thus, afferent systems terminating in a given layer will make synaptic contacts principally with neurons whose somata are located in that layer. The laminar organization in the superior colliculus is then a relatively clear correlate to its afferent organization. By contrast, the optic tectum in reptiles contains several types of neurons whose dendrites extend across layers, so that an individual neuron can potentially integrate inputs across much, or sometimes the entire thickness, of the tectum. There are again specific relations among afferent systems and the receptive surfaces of the various types of neurons, but the relationship between the position of the soma of a neuron and the afferent organization of the tectum in reptiles is not straightforward.

3. Columns, Modules, and Bands

The radial organization of tectal neurons provides a first and most obvious line of evidence in support of the concept that the tectum has some radial or vertical mode of organization. This mode lies orthogonal to the horizontal organization provided by the tectal laminae and their associated connectional specificities. The dendrites of radial cells form a vertical lattice work that is dramatically obvious in Golgi preparations. Retrograde HRP experiments (see Sections III and IV) reveal that radial neurons, as defined in Golgi preparations, actually include several classes of efferent neurons, each of which is morphologically distinct and has specific connections both within and without the tectum. Observations on light microscopic material suggest that there are interactions between some of these classes of neurons in that the deep and superficial populations of radial neurons have dendrites that come together in dendritic bundles within the superficial layers (Schechter and Ulinski, 1979). Electron microscopic observations indicate that chemical dendrodendritic synapses (Fig. 4.59) are prevalent in these layers (Schechter and Ulinski, 1979), but it is not known which of the various types of radial neurons are interrelated within the bundles.

Radial neurons could in principle serve to integrate inputs across tectal layers. The dendrites of radial neurons in pigeons are known to conduct dendritic spikes (Stone and Freeman, 1971) so that the output from a given radial cell could depend upon synaptic activity in the superficial layers. This activity is conducted to a site of impulse generation on the soma or on the dendritic shaft. The extent to which synaptic activity would be passively conducted to an impulse generation site depends upon the biophysical properties of the neuron. It is possible, for example, that the distal dendrites of radial neurons might serve as a unit that is relatively isolated from the rest of the

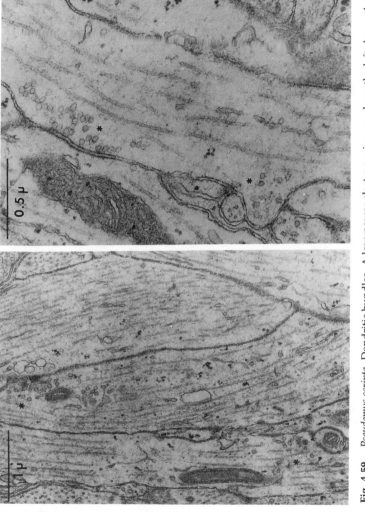

Fig. 4.59. *Pseudemys scripta.* Dendritic bundles. A low-power electron micrograph on the left shows the dendrites of several radial cells within a dendritic bundle. The electron micrograph on the right shows a chemical synapse between two dendrites.

neuron (e.g., Calvin and Graubard, 1979). In this case, the presynaptic dendrites in the superficial layers might serve as one output from a given radial neuron, whereas its axon in the deeper layers of the tectum would serve as a second output. The calculations necessary to determine the extent to which different regions of radial cells are coupled have not been carried out.

A second type of vertical organization within the tectum is provided by radially oriented collaterals of intrinsic and efferent tectal neurons. Some authors (e.g., Edwards, 1980) have denied the existence of interlaminar connections, at least in mammals. However, there is evidence for the existence of such collaterals from Golgi studies (e.g., Ramón, 1896; Ramón y Cajal, 1911), and Sterling (in Ingle and Sprague, 1975) reports radially organized connections within cat superior colliculus using orthograde degeneration techniques. Our HRP studies on tectal neurons in snakes and turtles lead us to believe that statements about the absence of interlaminar connections should be regarded with caution because HRP fills can apparently visualize collateral systems that are difficult to demonstrate with other existing techniques, and several laboratories have now demonstrated interlaminar connections in mammals (Rhoades et al., 1989; Grantyn et al., 1984).

A third type of vertical interaction is provided by the tendency of afferent axons to terminate in arbors that are restricted to domains, usually about 250 μm in diameter. This has been demonstrated directly in reptiles (Section V) with orthograde HRP techniques that explicitly show the morphology of terminal arbors. A similar organization is implied in mammals by autoradiographic tracing experiments that demonstrate several afferent systems terminate in patches or bands in autoradiographs. This was first reported for the retinal projection to the colliculus by Graybiel (1975) and Hubel et al. (1975), who showed that inputs from the two terminate within alternating "patches" and "puffs" within the superficial layers of the binocular zone. Reconstruction of these areas shows that they have the form of rostrocaudally aligned bands. Similar bands were subsequently reported for the nigrotectal (Graybiel, 1978), trigeminotectal (Huerta et al., 1981), and several other systems afferent to the colliculus (Huerta and Harting, 1984). The implication of these findings is that these afferent systems would fit into vertically organized zones within the colliculus.

However, two sets of observations question the degree to which intratectal bands are of functional significance. First Killackey and Erzurumlu (1981) have argued that the banding is an artifact resulting from the way in which axon fascicles are arranged in the tectum. Axons participating in several fiber systems afferent to the tectum tend to run in rostrocaudally oriented fascicles that appear as patches of axons in transverse sections of Fink-Heimer or autoradiographic preparations. Second, Constantine-Paton (e.g., 1981) has shown that intratectal banding of retinal afferents can be experimentally induced in frogs by surgically grafting a supernumerary eye at the tadpole stage. The banding appears to result from competitive effects between the retinotectal axons originating from a normal eye and the supernumerary eye. The banding may reflect developmental interactions that lack functional significance.

4. REGIONAL VARIATION

Compared to considerations of laminar and vertical organization, the concept of regional variation within the tectum has received relatively little attention. An initial clue to the regional organization of the tectum comes from cytoarchitectural observations. Studies of lizards (Peterson, 1981), and birds (Duff et al., 1981) reveal that although the same layers and sublayers occur throughout the tectum, the character and quantitative development of the layers often show significant regional variations at different tectal loci. The same types of neurons occur but appear to vary in the size, number, and packing density. The nature of this variation has been carefully documented in a study of the tectum in the desert iguana, *Dipsosaurus dorsalis* (Peterson, 1981). Tectal laminae are well developed in this species, as is generally the case in diurnal iguanid lizards, and correspond to Pedro Ramón's 14 layers in *Lacerta*. However, the tectum as a whole can be divided into five areas based on regional variation. Regional variation has also been documented in the red-eared turtle, *Pseudemys scripta* (Ulinski, 1978). The periventricular gray layers in this species consist of several well-demarcated sublayers, so it is possible to accurately calculate the density of somata in each sublayer. Using complete sets of serial sections, it was possible to determine the density of somata in the first or outer sublayer as a function of mediolateral position in the section. These data could be used to plot the density of somata onto the tectal surface as if the tectum had been flattened and to construct a series of isodensity lines (Fig. 4.60). The plots show that neurons in the outer sublayer are heterogeneously distributed throughout the tectum. The density of neurons is elevated in the central portion of the tectum and at its rostromedial pole but is low in a rim that extends around most of the tectal margin. There are only a few other studies on regional variation in other species. Regional variations in the synaptic organization of the superficial layers of the tectum in pigeons have been shown by Duff et al. (1981). It seems likely, though, that some degree of regional variation is a general feature of the tectum, at least in non-mammalian vertebrates.

STRATUM   GRISEUM   PERIVENTRICULARE
1st sublayer

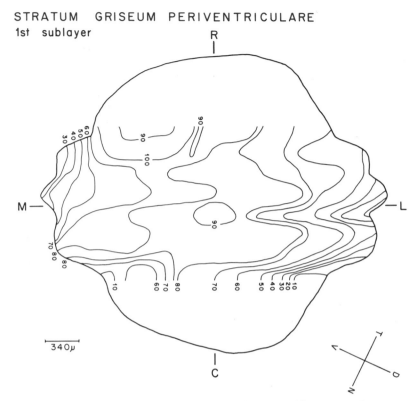

Fig. 4.60. *Pseudemys scripta.* Distribution of tectal neurons. This is an isodensity map for the somata in the outer sublayer of the stratum griseum periventriculare. The rostrocaudal and the medialateral axes are indicated. The inset shows the relation of the dorsoventral and nasotemporal axes of the retina to the tectal surface.

Such cytoarchitectural differences tend to be reflected in the afferent connections of the tectum. The nature of retinotectal input varies in concert with cytoarchitectural variations in lizards (Peterson, 1981), and the same type of variation appears to occur in pigeons (Duff et al. 1981). There also may be regional variations in the nature of somatosensory input (whether from the dorsal column nuclei and spinal cord or from the trigeminal complex). Where ipsilateral retinotectal projections are present, they are restricted to the binocular representation upon the tectal surface. Similarly, there is evidence in mammals that the tectal commissure distributes to one region within the superior colliculus and there appear to be regional variations in the extent of input from the retina and visual cortex, so that much of the representation of central visual fields reaches the colliculus through the visual cortex. It is likely that regional variation also reflects more subtle structures within the afferent projections to the tectum. As different types of ganglion cells are distributed unevenly over the retinal surface in many vertebrates, it follows that the nature of the input to different tectal loci must vary as a reflection of retinal organization. Peterson (1981) suggests that regional variation reflects regional specializations in the retina that include local variations in the number and types of photoreceptors and in the density and soma sizes of retinal ganglion cells. In *Dipsosaurus,* for example, the retina contains a horizontally elongate region of high ganglion cell density—or visual streak, a fovea, and a region in the temporal retina that contains ganglion cells with particularly large somata. It appears that the various areas of the tectum correspond to representations of retinal specializations upon the tectal surface. Similarly, the areas of increased soma density in *Pseudemys* seem to correspond to the representation of the visual streak (Peterson and Ulinski, 1979) and the temporal region of the retina that contains large ganglion cells (Peterson and Ulinski, 1982). Duff et al. (1981) suggest that the regional variation seen in the optic tectum of pigeons reflects the presence of the red and yellow fields of the retina (Galifret, 1968; Blough, 1979).

The extent to which the nature of tectal efferent projections shows regional variations has not been fully evaluated. Studies that place HRP injections in structures that receive input from the tectum and then plot the distribution of retrogradely labeled neurons in the tectum provide some evidence that there are regional variations in the origin of the tectum's efferent systems. Thus, the cells of origin of the tectoreticular (Holcombe and Hall, 1981a), tectopontine (Holcombe and Hall, 1981b), tectogeniculate (e.g., Albano et al., 1979; Harrell et al., 1982), tectotectal (e.g., Magalhaes-Castro et al., 1978), and tectospinal (Murray and Coulter, 1982; Huerta and Harting, 1982) projections each are distributed in particular regions of the superior colliculus.

The existence of such regional variation in cytoarchitecture, afferent connections, and efferent connections is consistent with labeled line models of tectal function in that different regions of the tectum have different relationships to the rest of the nervous system and presumably process information differently. Thus, activation of different tectal loci could possibly result in different behavioral responses. Although a certain degree of regional variation is well established, the ways in which this variation is exploited by the animal is not yet known.

5. MAPS
Each of the afferent and efferent projections of the tectum establishes a spatial relationship between neurons in the tectum and one or more other structures. From a mathematical point of view, the projections

therefore specify a series of maps between the tectum and either the retina or other brain regions. The most familiar of these are point-to-point or topological maps, such as that formed by the projections of the retina upon the tectal surface. Similarly, there are topological maps between the tectum and the geniculate complex, pretectum, nucleus isthmi, and so on. The presumed essential property of these maps is that they preserve information about spatial relationships. Thus, two adjacent loci on the retinal surface are represented by two adjacent loci on the tectal surface and vice versa.

The anatomical substrate for topological maps are axonal arbors with relatively restricted dimensions. For example, the arbors of retinal ganglion cells measure about 65 by 100 μm in *Thamnophis*, regardless of the location in the tectum. Thus, restricted lesions or injections of marker substances into the retina result in degeneration of transported label in a relatively restricted region of the tectum. Conversely, focal injections of HRP in the tectum retrogradely label focal patches of ganglion cells in the retina. The important point is that the size of these arbors is relatively small when compared to the total dimensions of the tectum so that the tectum can embody a relatively detailed representation of the retinal surface.

The existence of topological maps in the visual system is a familiar aspect of nervous system organization and tends to carry with it a certain degree of psychological comfort. However, the functional significance of such maps is not entirely certain. A topological representation of visual space is obviously useful in a variety of behaviorally important computations. It is also possible that topological representations play a role in the generation of complex receptive field properties such as contrast sensitivity and orientation selectivity by forming a substrate for local connections. On the other hand, it is possible that topological maps reflect the underlying developmental properties of the nervous system and have no direct functional significance.

The existence of nontopological maps in the nervous system is less widely appreciated. However, data obtained using the orthograde transport of HRP suggest that many of the tectal afferent and efferent systems consist of axons with widely branching terminals. These axons indicate the existence of maps that do not preserve spatial relationships. Nontopological maps apparently occur in the tectorotundal, tectoreticular, and pretectal-tectal projections. It is interesting to note that in some instances of reciprocal projections, one limb of the loop is topologically organized, whereas the other limb is not. It is possible that nontopological projections, like topological ones, serve to preserve information about spatial relationships. This is somewhat counterintuitive, but a distributed system can code information about spatial relationships. The difference is that spatial information would be coded in the temporal domain, in terms of firing frequencies, in the case of nontopological projections, whereas it is coded in the spatial domain in the case of topological projections. A second possibility is that nontopological projections are involved in the formation of sets of neurons that code for stimulus parameters that are not coded on the receptor sheet. Thus, there is some evidence suggesting that neurons in nucleus rotundus code information about the velocity, or perhaps direction, of a stimulus moving anywhere in the visual field. This would be consistent with the existence of a nontopological projection from the tectum to nucleus rotundus (Dacey and Ulinski, 1983). Finally, nontopological projections can be involved in a transformation from the coding of information in the spatial domain to codes in the temporal domain. This idea has already been mentioned in the case of tectoreticular projections, which might be involved in a sensorimotor transformation (see Section VI.A).

VII. SUMMARY

This chapter has reviewed the anatomical organization of the optic tectum in reptiles. It has emphasized the impact of contemporary neuroanatomical techniques, particularly the use of HRP, upon our data base and other ideas of tectal organization. Until quite recently, information on the tectum was relatively fragmented because the available anatomical techniques only yielded information about parts of the various populations of neurons present in the tectum. However, newer methods, such as the HRP technology, are allowing initial glimpses of the various types of neurons in their entirety. It is often possible with either the intracellular or extracellular application of HRP to completely visualize the soma, dendrites, and axons of individual neurons in the tectum and in those neural structures that project to the tectum. The possession of such a picture serves to highlight areas that deserve future inquiry. The first is what might be called the *microcircuitry* (cf. Sterling, 1983; Gilbert, 1983) of the tectum. Although we now know something about the morphology of the classes of tectal neurons distinguished by their connections, almost nothing is known of the nature of their interactions within the tectum. Thus, we do not know the specific interconnections that obtain between different populations of tectal neurons, whether the synapses are chemical or electrical, excitatory or inhibitory. Many of the interactions that occur between tectal neurons can be determined by the coordinated use of electronmicroscopic techniques on individual neurons that have been studied electrophysiologically and then labeled with HRP. Recently developed neurophysiological techniques that permit the analyses of correlations among firing patterns in neuronal populations should also be useful in determining the physiological significance of neuronal interactions, particularly if distributed representations figure heavily in tectal function.

A related area of investigation has to do with the geometric organization of the tectum. The dendrites of horizontal neurons in rat superior colliculus (Langer and Lund, 1974) and the optic tectum of turtles (Peterson, 1978b) have a local order. Specific patterns in the orientation of the axon systems of tectal intrinsic neurons occur in *Thamnophis* (Dacey and Ulinski, 1986c). It is likely that these are only first views of the patterning of neuronal elements within the tectum. Researchers have actively investigated the geometric arrangement of elements within the striate cortex of cats and monkeys in recent years. The motivation for this line of work is the assumption that the geometric arrangement of axons and dendrites within the striate cortex is important in determining the information-processing properties of the structure. The difficulty in the case of the striate cortex is that we still lack a precise, quantitative concept of what the striate cortex does. The situation is more sanguine in the case of the optic tectum because we have a better idea of the relation of the tectum to orienting movements and are reaching the point at which quantitative descriptions of the sensorimotor transformations seem possible.

Finally, the substantial enrichment of our pictures of tectal neuronal morphology that has resulted from the HRP technique is generating broader and more carefully considered ideas on the nature of topological and nontopological maps in the central nervous system. Ideas of tectal function have been dominated by the idea of a topologically organized retinotectal projection, but it is now apparent that only a minority of the afferent and efferent projections of the tectum are topologically organized. It will be necessary now to carefully rethink our ideas of neuronal information processing and seriously consider models involving distributed representations.

ACKNOWLEDGMENTS

The authors' work on the optic tectum has been supported by PHS Grant NS12518. Maryellen Kurek provided invaluable technical and photographic services. Debra Hawkins, Dorothy Crowder, and Roberta Mims typed the manuscript.

APPENDIX: REPTILIAN SPECIES DISCUSSED

TESTUDINES

Anomalepsis aspinosus
  Senn, 1969
Caretta caretta
  Bass and Northcutt, 1981a

Chelonia mydas
  Granda and O'Shea, 1972
Chelydra serpentina
  Knapp and Kang, 1968a

Chrysemys picta
  Bass and Northcutt, 1981a, 1981b
  Heller and Ulinski, 1987
  Reiner et al., 1980
Emys orbicularis
  Belekhova and Kosareva, 1971
  Boyko, 1976
  Boyko and Davydova, 1975
  Davydova, 1971, 1973
  Davydova et al., 1982
  Davydova and Goncharova, 1970
  Davydova and Mazurskaya, 1973
  Davydova et al., 1976
  Gusel'nikov et al., 1970
  Kosareva, 1967
  Morenkov and Pivovarov, 1975
Pelomedusa subrufa
  Pedersen, 1973

Pelusios subniger
  Pedersen, 1973
Podocnemis unifilis
  Bass et al., 1973
  Knapp and Kang, 1968b
  Pedersen, 1973
Trachemys scripta
  Balaban and Ulinski, 1981a, 1981b
  Brown, 1969
  Curwen and Miller, 1939
  Foster and Hall, 1975
  Granda and Stirling, 1965
  Hall et al., 1977
  Hall and Ebner, 1970a, 1970b
  Heller and Ulinski, 1987
  Kolb, 1982
  Marchiafava and Wagner, 1981
  Schechter and Ulinski, 1979
  Sereno, 1983, 1985

CROCODILIA

Alligator mississippiensis
  Heric and Kruger, 1965, 1966
  Huber and Crosby, 1926
Caiman crocodilus
  Braford, 1972, 1973
  Brauth and Kitt, 1980

Burns and Goodman, 1967
Ebbesson and Goodman, 1981
Reiner et al., 1980
Repérant, 1975
Schapiro and Goodman, 1969

RHYNCHOCEPHALIA

Sphenodon punctatus
  Cairney, 1926

Northcutt, 1978, 1983
Northcutt et al., 1974

SAURIA

Agama agama
  Elprana et al., 1980
Anniella pulchra
  Senn, 1968a
Anolis carolinensis
  Butler and Northcutt, 1971a
Chamaeleo jacksoni
  Collett and Harkness, 1982
  Harkness, 1977
Dipsosaurus dorsalis
  Peterson, 1981
Gekko gecko
  Bruce, 1982
  Butler, 1976, 1978
  Northcutt and Butler, 1974b

Iguana iguana
  Bruce, 1982
  Bruce and Butler, 1979
  Butler and Ebner, 1972
  Butler and Northcutt, 1973
  Butler and Northcutt, 1971a,
    1971b
  Distel, 1978
  Foster and Hall, 1975, 1978
  Stein and Gaither, 1981, 1983
Lacerta viridis
  Baumgarten and Braak, 1968
  Robinson, 1969
Lacerta vivipara
  Armstrong, 1950, 1951

Pentadactylus schreiberii
  Quiroga, 1978
Podarcis muralis
  Baumgarten and Braak, 1968
Podarcis sicula
  Senn, 1968b
Tupinambis nigropunctatus
  Butler and Ebbesson, 1975
  Cruce, 1974
  Cruce and Cruce, 1975, 1978
  Cruce and Newman, 1981
  Ebbesson, 1967, 1981
  Ebbesson and Karten, 1981

Hoogland, 1977
Lohman and van Woerden-
  Verkley, 1978
Voneida and Sligar, 1979
Varanus exanthematicus
  Hoogland, 1982
  ten Donkelaar and De Boer-van
    Huizen, 1981
  ten Donkelaar et al., 1981
Varanus sp.
  Distel and Ebbesson, 1975
Xantusia vigilis
  Butler, 1974

SERPENTES

Boa constrictor
  Ebbeson, 1969
Crotalus atrox
  Molenaar, 1974
Crotalus h. horridus
  Auen, 1976
Crotalus sp.
  Berson and Hartline, 1988
  Hartline, 1971
  Hartline et al., 1978
  Newman and Hartline, 1981
  Terashima and Goris, 1975
Crotalus viridis
  Schroeder, 1981a, 1981b
  Stanford et al., 1981
Liotyphlops albirostris
  Senn, 1969
Natrix natrix
  Senn, 1979
Nerodia sipedon
  Northcutt and Butler, 1974a
  Senn, 1970
  Ulinski, 1977

Python
  Molenaar and Fizaan-Oosteveen,
    1980
  Repérant and Rio, 1976
  Repérant et al., 1981
Thamnophis sirtalis
  Dacey and Ulinski, 1983, 1986a,
    1986b, 1986c, 1986d, 1986e
  Halpern and Frumin, 1973
  Ulinski et al., 1983
  Warner, 1947
Vipera sp.
  Auen, 1978
Vipera aspis
  Repérant, 1973
  Marchiafava and Weiler, 1980
  Peterson, 1978a, 1978b, 1982
  Peterson and Ulinski, 1979, 1982
  Rainey and Ulinski, 1982a, 1982b
  Repérant et al., 1981
  Robbins, 1972
  Sereno, 1985
  Sereno and Ulinski, 1985
  Ulinski, 1978

REFERENCES

Akert, K. (1945). Der visuelle Greifreflex. *Helv. Physiol. Acta* 7, 112–134.
Albano, J. E., Norton, T. T., and Hall, W. C. (1979). Laminar origin of projec-
  tions from the superficial layers of the superior colliculus in the tree shrew,
  Tupaia glis. Brain Res. 173, 1–11.

Altman, J., and Carpenter, M. C. (1961). Fiber projections of the superior colliculus in the cat. *J. Comp. Neurol.* 116, 157–177.

Anderson, J. A., and Hinton, C. E. (1981). Models of information processing in the brain. In *Parallel Models of Associative Memory* (J. A. Anderson and C. E. Hinton, eds.). Erlbaum, Hillsdale, N.J., pp. 9–48.

Angaut, P., and Repérant, J. (1976). Fine structure of the optic fibre termination layers in the pigeon optic tectum: a Golgi and electron microscopic study. *Neuroscience* 1, 93–105.

Ariëns Kappers, C. U., Huber, G. C., and Crosby, E. C. (1936). *The Comparative Anatomy of the Nervous System of Vertebrates, Including Man* (reprinted 1960). Hafner, New York.

Armstrong, J. A. (1950). An experimental study of the visual pathways in a reptile (*Lacerta vivipara*). *J. Anat.* (London) 84, 146–167.

Armstrong, J. A. (1951). An experimental study of the visual pathways in a snake (*Natrix natrix*). *J. Anat.* (London) 85, 275–287.

Auen, E. L. (1976). "A Study of the Optic Tectum of the Snake." Ph.D. dissertation, University of Wisconsin, Madison.

Auen, E. L. (1978). Axonal transport of HRP in descending tectal fibers of the pit viper. *Neurosci. Lett.* 9, 137–140.

Bagnoli, P., and Burkhalter, A. (1983). Organization of the afferent projections to the wulst in the pigeon. *J. Comp. Neurol.* 214, 103–113.

Bagnoli, P., and Magni, F. (1979). Interaction of optic tract and visual wulst impulses on single units of the pigeon's optic tectum. *Brain Behav. Evol.* 16, 19–37.

Balaban, C. D., and Ulinski, P. S. (1981a). Organization of thalamic afferents to anterior dorsal ventricular ridge in turtles. I. Projections of thalamic nuclei. *J. Comp. Neurol.* 200, 95–129.

Balaban, C. D., and Ulinski, P. S. (1981b). Organization of thalamic afferents to anterior dorsal ventricular ridge in turtles. II. Properties of the rotundo-dorsal map. *J. Comp. Neurol.* 200, 131–150.

Barrett, R. (1970). The pit organs of snakes. In *Biology of the Reptilia* (C. Gans and T. S. Parsons, eds.). Academic Press, London, Vol. 2, pp. 277–314.

Bass, A. H., and Northcutt, R. G. (1981a). Primary retinal targets in the Atlantic loggerhead sea turtle, *Caretta caretta. Cell Tissue Res.* 218, 253–264.

Bass, A. H., and Northcutt, R. G. (1981b). Retinal recipient nuclei in the painted turtle, *Chrysemys picta:* an autoradiographic and HRP study. *J. Comp. Neurol.* 199, 97–112.

Bass, A. H., Pritz, M. B., and Northcutt, R. G. (1973). Effects of telencephalic and tectal ablations on visual behavior in the side-necked turtle, *Podocnemis unifilis. Brain Res.* 55, 455–460.

Baumgarten, H. G., and Braak, H. (1968). Catecholamine im Gehirn der Eidechse (*Lacerta viridis* and *Lacerta muralis*). *Z. Zellforsch.* 86, 574–608.

Belekhova, M. G. (1979). Neurophysiology of the forebrain. In *Biology of the Reptilia* (C. Gans, R. G. Northcutt, and P. S. Ulinski, eds.). Academic Press, London, Vol. 10, pp. 287–359.

Belekhova, M. G., and Kosareva, A. A. (1971). Organization of the turtle thalamus: visual, somatic and tectal zones. *Brain Behav. Evol.* 4, 337–375.

Benevento, L. A., and Fallon, J. H. (1975). The ascending projections of the superior colliculus in the rhesus monkey (*Macaca mulatta*). *J. Comp. Neurol.* 160, 339–362.

Benowitz, L. I., and Karten, H. J. (1976). Organization of the tectofugal visual pathway in the pigeon: a retrograde transport study. *J. Comp. Neurol.* 167, 503–520.

Berson, D. M., and Hartline, P. H. (1988). A tecto-rotundal-telencephalic pathway in the rattlesnake: evidence for a forebrain representation of the infrared sense. *J. Neurosci.* 8, 1074–1088.

Berson, D. M., and McIlwain, J. T. (1982). Retinal Y-cell activation of deeplayer cells in superior colliculus of the cat. *J. Neurophysiol.* 47, 700–714.

Blough, P. M. (1979). Functional implications of the pigeon's peculiar retinal structure. In *Neural Mechanisms of Behavior in the Pigeon* (A. M. Granda and J. H. Maxwell, eds.). Plenum Press, New York, pp. 71–88.

Boyko, V. P. (1976). Neuronal receptive fields in mesencephalic visual center of the tortoise *Emys orbicularis*. In *Comparative Neurophysiology and Neurochemistry* (E. M. Kreps, ed.). *Zh. Evol. Biokhim. Fiziol.*, Suppl., pp. 89–96.

Boyko, V. P., and Davydova, T. V. (1975). Morphofunctional changes in the turtle mesencephalic visual centre after enucleation. *Neurofiziologia* (Kiev) 7, 172–177.

Braford, M. R., Jr. (1972). Ascending efferent tectal projections in the South American spectacled caiman. *Anat. Rec.* 172, 275–276.

Braford, M. R., Jr. (1973). Retinal projections in *Caiman crocodilus*. *Am. Zool.* 13, 1345.

Brauth, S. E., and Kitt, C. A. (1980). The paleostriatal system of *Caiman crocodilus*. *J. Comp. Neurol.* 189, 437–465.

Bravo, H., and Pettigrew, J. D. (1981). The distribution of neurons projecting from the retina and visual cortex to the thalamus and tectum opticum of the barn owl, *Tyto alba*, and the burrowing owl, *Speotyto cunicularia*. *J. Comp. Neurol.* 199, 419–441.

Britto, L. R. G. (1978). Hyperstriatal projections to primary visual relays in pigeons: electrophysiological studies. *Brain Res.* 153, 382–386.

Brown, K. T. (1969). A linear area centralis extending across the turtle retina and stabilized to the horizon by non-visual cues. *Vision Res.* 9, 1053–1062.

Bruce, L. L. (1982). "Organization and Evolution of the Reptilian Forebrain: Experimental Studies of Forebrain Connections in Lizards." Ph.D. dissertation, Georgetown University, Washington, D.C.

Bruce, L. L., and Butler, A. B. (1979). Afferent projections to the anterior dorsal ventricular ridge in the lizard *Iguana iguana*. *Neurosci. Abstr.* 5, 140.

Burns, A. H., and Goodman, D. C. (1967). Retinofugal projections of *Caiman sklerops*. *Exp. Neurol.* 18, 105–115.

Butler, A. B. (1974). Retinal projections in the night lizard, *Xantusia vigilis* Baird. *Brain Res.* 80, 116–121.

Butler, A. B. (1976). Telencephalon of the lizard *Gekko gecko* (Linnaeus): some connections of the cortex and dorsal ventricular ridge. *Brain Behav. Evol.* 13, 396–417.

Butler, A. B. (1978). Organization of ascending tectal projections in the lizard *Gekko gecko*: a new pattern of tectorotundal inputs. *Brain Res.* 147, 353–361.

Butler, A. B., and Ebbesson, S. O. E. (1975). A Golgi study of the optic tectum of the tegu lizard, *Tupinambis nigropunctatus*. *J. Morphol.* 146, 215–228.

Butler, A. B., and Ebner, F. F. (1972). Thalamo-telencephalic projections in the lizard *Iguana iguana*. *Anat. Rec.* 172, 282.

Butler, A. B., and Northcutt, R. G. (1971a). Retinal projections in *Iguana iguana* and *Anolis carolinensis*. *Brain Res.* 26, 1–13.

Butler, A. B., and Northcutt, R. G. (1971b). Ascending tectal efferent projections in the lizard *Iguana iguana*. *Brain Res.* 35, 597–601.

Butler, A. B., and Northcutt, R. G. (1970). Architectonic studies of the diencephalon of *Iguana iguana* (Linnaeus). *J. Comp. Neurol.* 149, 439–462.

Butler, A. B., and Northcutt, R. G. (1978). New thalamic visual nuclei in lizards. *Brain Res.* 149, 469–476.

Cairney, J. (1926). A general survey of the forebrain of *Sphenodon punctatum*. *J. Comp. Neurol.* 42, 255–348.

Caldwell, R. B., and Mize, R. R. (1981). Superior colliculus neurons which project to the cat lateral posterior nucleus have varying morphologies. *J. Comp. Neurol.* 203, 53–66.

Calvin, W. H., and Graubard, K. (1979). Styles of neuronal computation. In *The Neurosciences: Fourth Study Program* (F. O. Schmitt and F. G. Worden, eds.). MIT Press, Cambridge, Mass., pp. 513–524.

Casagrande, V. A., and Diamond, I. T. (1974). Ablation study of the superior colliculus in the tree shrew (*Tupaia glis*). *J. Comp. Neurol.* 156, 207–238.

Collett, T. S., and Harkness, L. I. K. (1982). Depth vision in animals. In *Analysis of Visual Behavior* (D. J. Ingle, M. A. Goodale, and R. J. W. Mansfield, eds.). MIT Press, Cambridge, Mass., pp. 111–176.

Constantine-Paton, M. (1981). Induced ocular-dominance zones in tectal cortex. In *The Organization of the Cerebral Cortex* (F. O. Schmitt, F. G. Worden, G. Adelman, and S. G. Dennis, eds.). MIT Press, Cambridge, Mass., pp. 47–68.

Crossland, W. J. (1972). "Receptive Field Characteristics of Some Thalamic Visual Nuclei of the Pigeon (*Columba livia*)." Ph.D. dissertation, University of Illinois, Urbana.

Cruce, J. A. F. (1974). A cytoarchitectonic study of the diencephalon of the tegu lizard, *Tupinambis nigropunctatus*. *J. Comp. Neurol.* 153, 215–238.

Cruce, J. A. F., and Cruce, W. L. R. (1978). Analysis of the visual system in a lizard, *Tupinambis nigropunctatus*. In *Behavior and Neurology of Lizards* (N. Greenberg and P. D. MacLean, eds.). DHEW Publ. No. (ADM) 77-491, Rockville, Md., National Institutes of Health, pp. 79–90.

Cruce, W. L. R., and Cruce, J. A. F. (1975). Projections from the retina to the lateral geniculate nucleus and mesencephalic tectum in a reptile, *Tupinambis nigropunctatus*: a comparison of anterograde transport and anterograde degeneration. *Brain Res.* 85, 221–228.

Cruce, W. L. R., and Newman, D. B. (1981). Brain stem origins of spinal projections in the lizard *Tupinambis nigropunctatus*. *J. Comp. Neurol.* 198, 185–208.

Curwen, A. O., and Miller, R. N. (1939). The pretectal region of the turtle, *Pseudemys scripta troojstii*. *J. Comp. Neurol.* 71, 99–120.

Dacey, D. M., and Ulinski, P. S. (1983). Nucleus rotundus in a snake, *Thamnophis sirtalis:* an analysis of a non-retinotopic projection. *J. Comp. Neurol.* 216, 175–191.

Dacey, D. M., and Ulinski, P. S. (1986a). Optic tectum of the eastern garter snake, *Thamnophis sirtalis.* I. Efferent pathways. *J. Comp. Neurol.* 245, 1–28.

Dacey, D. M., and Ulinski, P. S. (1986b). Optic tectum of the eastern garter snake, *Thamnophis sirtalis.* II. Morphology of efferent neurons. *J. Comp. Neurol.* 245, 198–237.

Dacey, D. M., and Ulinski, P. S. (1986c). Optic tectum of the eastern garter snake, *Thamnophis sirtalis.* III. Morphology of intrinsic neurons. *J. Comp. Neurol.* 245, 283–300.

Dacey, D. M., and Ulinski, P. S. (1986d). Optic tectum of the eastern garter snake, *Thamnophis sirtalis.* IV. Morphology of retinal afferents. *J. Comp. Neurol.* 245, 301–318.

Dacey, D. M., and Ulinski, P. S. (1986e). Optic tectum of the eastern garter snake, *Thamnophis sirtalis.* V. Morphology of brainstem afferents. *J. Comp. Neurol.* 245, 423–453.

Davydova, T. V. (1971). The ultrastructure of nerve terminals in the tectum opticum of the turtle (*Emys orbicularis*). *Cytologia* 13, 433–440.

Davydova, T. V. (1973). The ultrastructure of the midbrain in tectum visual fibers of *Emys orbicularis* L. in the norm and after degeneration. *Cytologia* 15, 150–155.

Davydova, T. V., Boyko, V. P., and Goncharova, N. V. (1982). Correlations between the morpho-functional organization of some parts of visual analyzer and ecology in chelonia: II. Morpho-functional characteristics of the visual nerve and tectum after unilateral enucleation. *J. Hirnforsch.* 23, 433–446.

Davydova, T. V., and Goncharova, N. V. (1970). Cytoarchitectonics and neuronal composition of midbrain tectum in the turtle (*Emys orbicularis*). *Arkh. Anat. Gistol. Embriol.* 59, 53–61.

Davydova, T. V., Goncharova, N. V., and Boyko, V. P. (1976). Retinotectal system of the tortoise, *Testudo horsfieldi,* Gray (Morpho-functional study in the norm and after enucleation). *J. Hirnforsch.* 17, 463–488.

Davydova, T. V., Goncharova, N. V., and Boyko, V. P. (1982). Correlation between the morpho-functional organization of some portions of visual analyzer of chelonia and their ecology: I. Normal morpho-functional characteristics of the optic nerve and the tectum opticum. *J. Hirnforsch.* 23, 271–286.

Davydova, T. V., and Mazurskaya, P. Z. (1973). Ultrastructural features of the nerve terminal degeneration in the turtle tectum opticum after enucleation. *Cytologia* 15, 22–30.

Davydova, T. V., and Smirnov, G. D. (1973). Retinotectal connections in the tortoise: an electron microscope study of degeneration in optic nerve and midbrain tectum. *J. Hirnforsch.* 14, 473–492.

De Britto, L. R. G., Brunelli, M., Francesconi, W., and Magni, F. (1975). Visual response pattern of thalamic neurons in the pigeon. *Brain Res.* 97, 337–343.

Distel, H. (1978). Behavior and electrical brain stimulation in the green iguana, *Iguana iguana.* II. Stimulation effects. *Exp. Brain Res.* 31, 353–367.

Distel, H., and Ebbesson, S. O. E. (1975). Connections of the thalamus in the monitor lizard. *Soc. Neurosci. Abstr.* 1, 559.

Duff, T. A., Scott, G., and Mai, R. (1981). Regional differences in pigeon optic tract, chiasm and retinoreceptive layers of optic tectum. *J. Comp. Neurol.* 198, 231–248.

Ebbesson, S. O. E. (1967). Ascending axon degeneration following hemisection of the spinal cord in the tegu lizard (*Tupinambis nigropunctatus*). *Brain Res.* 5, 178–206.

Ebbesson, S. O. E. (1969). Brain stem afferents from the spinal cord in a sample of reptilian and amphibian species. *Ann. N.Y. Acad. Sci.* 167, 80–101.

Ebbesson, S. O. E. (1978). Somatosensory pathways in lizards: the identification of the medial lemniscus and related structures. In *Behavior and Neurology of Lizards* (N. Greenberg and P. D. MacLean, eds.), DHEW Publ. No. (ADM) 77-491, National Institutes of Health, Rockville, Md. pp. 91–104.

Ebbesson, S. O. E. (1981). Projections of the optic tectum and the mesencephalic nucleus of the trigeminal nerve in the tegu lizard (*Tupinambis nigropunctatus*). *Cell Tiss. Res.* 216, 151–165.

Ebbesson, S. O. E., and Goodman, D. C. (1981). Organization of ascending spinal projections in *Caiman crocodilus*. *Cell Tiss. Res.* 215, 383–396.

Ebbesson, S. O. E., and Karten, H. J. (1981). Terminal distribution of retinal fibers in the tegu lizard (*Tupinambis nigropunctatus*). *Cell Tiss. Res.* 215, 591–606.

Ebbesson, S. O. E., and Vanegas, H. (1976). Projections of the optic tectum in two teleost species. *J. Comp. Neurol.* 165, 161–180.

Eccles, J. C., Ito, M., and Szentagothai, J. (1967). *The Cerebellum as a Neuronal Machine.* Springer-Verlag, New York.

Edwards, S. B. (1977). The commissural projection of the superior colliculus in the cat. *J. Comp. Neurol.* 173, 23–40.

Edwards, S. B. (1980). The deep cell layers of the superior colliculus: their reticular characteristics and structural organization. In *The Reticular Formation Revisited* (J. A. Thompson and M. A. Brazier, eds.). Raven Press, New York, pp. 193–209.

Edwards, S. B., Ginsburgh, C. L., Henkel, C. K., and Stein, B. E. (1979). Sources of subcortical projections to the superior colliculus in the cat. *J. Comp. Neurol.* 184, 309–330.

Elprana, D., Wouterlood, F. G., and Alones, V. (1980). A corticotectal projection in the lizard *Agama agama*. *Neurosci. Lett.* 18, 251–256.

Ewert, J.-P. (1976). The visual system of the toad: behavioral and physiological studies on a pattern recognition system. In *The Amphibian Visual System* (K. V. Fite, ed.). Academic Press, New York, pp. 141–202.

Ewert, J.-P. (1982). Neuronal basis of configurational prey selection in the common toad. In *Analysis of Visual Behavior* (D. J. Ingle, M. A. Goodale, and R. J. W. Mansfield, eds.). MIT Press, Cambridge, Mass., pp. 7–45.

Ewert, J.-P., and von Seelen, W. (1974). Neurobiologie und System-theorie eines visuellen Muster-erkennungs Mechanismus bei Kröten. *Kybernetic* 14, 167–183.

Fish, S. E., Goodman, D. K., Kuo, D. C., Polcer, J. D., and Rhoades, R. W. (1982). The intercollicular pathway in the golden hamster: an anatomical study. *J. Comp. Neurol.* 204, 6–20.

Fite, K. V., and Scalia, F. (1976). Central visual pathways in the frog. In *The*

Amphibian Visual System (K. V. Fite, ed.). Academic Press, New York, pp. 87–118.

Fitzpatrick, D., Carey, R. G., and Diamond, I. T. (1980). The projection of the superior colliculus upon the lateral geniculate body in *Tupaia glis* and *Galago senegalensis*. *Brain Res.* 194, 494–499.

Foster, R. E., and Hall, W. C. (1975). The connections and laminar organization of the optic tectum in a reptile (*Iguana iguana*). *J. Comp. Neurol.* 163, 397–426.

Foster, R. E., and Hall, W. C. (1978). The organization of central auditory pathways in a reptile, *Iguana iguana*. *J. Comp. Neurol.* 178, 783–832.

Friedlander, M. J., Lin, C.-S., Stanford, L. R., and Sherman, S. M. (1981). Morphology of functionally identified neurons in the lateral geniculate nucleus of the cat. *J. Neurophysiol.* 46, 80–129.

Galifret, Y. (1968). Les diverse aires fonctionelles de la retine du pigeon. *Zeit. Zellforsch.* 86, 535–545.

Geri, G. A., Kimsey, R. A., and Dvorak, C. A. (1982). Quantitative electron microscopic analysis of optic nerve of turtle, *Pseudemys*. *J. Comp. Neurol.* 207, 99–103.

Gilbert, C. D. (1983). Microcircuitry of the visual cortex. *Ann. Rev. Neurosci.* 6, 217–248.

Goldberg, M. E., and Robinson, D. L. (1978). Visual system: superior colliculus. In *Handbook of Behavioral Neurobiology* (R. B. Masterson, ed.). Plenum Press, New York, pp. 119–614.

Graham, J. (1977). An autoradiographic study of the efferent connections of the superior colliculus in the cat. *J. Comp. Neurol.* 173, 629–654.

Graham, J., Berman, N., and Murphy, E. H. (1982). Effects of cortical lesions on receptive field properties of single units in the superior colliculus of rabbits. *J. Neurophysiol.* 47, 256–271.

Graham, J., and Casagrande, V. A. (1980). A light microscopic study of the superficial layers of the superior colliculus of the tree shrew (*Tupaia glis*). *J. Comp. Neurol.* 191, 133–151.

Granda, A. M., and O'Shea, P. J. (1972). Spectral sensitivity of the green turtle (*Chelonia mydas mydas*) determined by electrical responses to heterochromatic light. *Brain Behav. Evol.* 5, 143–154.

Granda, A. M., and Stirling, C. E. (1965). Differential spectral sensitivity in the optic tectum and eye of the turtle. *J. Gen. Physiol.* 48, 901–917.

Granda, A. M., and Yazulla, S. (1971). The spectral sensitivity of single units in the nucleus rotundus of pigeon (*Columba livia*). *J. Gen. Physiol.* 57, 363–384.

Grantyn, A., and Grantyn, R. (1982). Axonal patterns and sites of termination of cat superior collicular neurons projecting in the tecto-bulbo-spinal tract. *Exp. Brain Res.* 46, 243–256.

Grantyn, R., Ludwig, R., and Eberhardt, W. (1984). Neurons of the superior tectal gray: an intracellular HRP-study on the kitten superior colliculus *in vitro*. *Exp. Brain Res.* 55, 172–176.

Graybiel, A. M. (1975). Anatomical organization of retinotectal afferents in the cat: an autoradiographic study. *Brain Res.* 96, 1–23.

Graybiel, A. M. (1978). Organization of the nigrotectal connection: an experimental tracer study in the cat. *Brain Res.* 143, 339–348.

Grobstein, P. (1988). Between the retinotectal projection and directed movements: topography of a sensorimotor interface. *Brain Behav. Evol.* 31, 34–48.

Grobstein, P., Comer, C., Hollyday, M., and Archer, S. M. (1978). A crossed isthmotectal projection in *Rana pipiens* and its involvement in the ipsilateral visuotectal projection. *Brain Res.* 156, 117–123.

Grobstein, P., Comer, C., and Kostyk, S. K. (1980). Frog prey capture behavior: between sensory maps and directed motor output. In *Advances in Vertebrate Neuroethology* (J.-P. Ewert, R. R. Capranica, and D. J. Ingle, eds.). Plenum Press, New York, pp. 331–348.

Grover, B. G., and Sharma, S. C. (1981). Organization of extrinsic tectal connections in the goldfish (*Carassius auratus*). *J. Comp. Neurol.* 196, 471–488.

Gruberg, E. R., Kicliter, E., Newman, E. A., Kass, L., and Hartline, P. H. (1979). Connections of the tectum of the rattlesnake *Crotalus viridis:* an HRP study. *J. Comp. Neurol.* 188, 31–42.

Gruberg, E. R., and Lettvin, J. Y. (1980). Anatomy and physiology of a binocular system in the frog *Rana pipiens. Brain Res.* 192, 313–325.

Gruberg, E. R., and Udin, S. (1978). Topographic projections between nucleus isthmi and the tectum of the frog *Rana pipiens. J. Comp. Neurol.* 179, 487–500.

Gusel'nikov, V. I., Morenkov, E. D., and Pivovarov, A. S. (1970). On functional organization of the visual system of the tortoise (*Emys orbicularis*). *Fiziol. Zh. SSSR Sechenova* 56, 1377–1385.

Hall, W. C., and Ebner, F. F. (1970a). Parallels in the visual afferent projections of the thalamus in the hedgehog (*Paraechinus hypomelas*) and the turtle (*Pseudemys scripta*). *Brain Behav. Evol.* 3, 135–154.

Hall, W. C., and Ebner, F. F. (1970b). Thalamotelencephalic projections in the turtle *Pseudemys scripta. J. Comp. Neurol.* 140, 101–122.

Hall, J. A., Foster, R. E., Ebner, F. F., and Hall, W. C. (1977). Visual cortex in a reptile, the turtle (*Pseudemys scripta* and *Chrysemys picta*). *Brain Res.* 130, 197–216.

Halpern, M. (1973a). Retinal projections in blind snakes. *Science* (N.Y.) 182, 390–391.

Halpern, M., and Frumin, N. (1973). Retinal projections in a snake (*Thamnophis sirtalis*). *J. Morphol.* 141, 359–382.

Harkness, L. (1977). Chameleons use accommodation cues to judge distance. *Nature* (London) 267, 346–349.

Harrell, J. V., Caldwell, R. B., and Mize, R. R. (1982). The superior colliculus neurons which project to the dorsal and ventral lateral geniculate nuclei in the cat. *Exp. Brain Res.* 46, 234–242.

Harris, L. R. (1980). The superior colliculus and movements of the head and eyes in cats. *J. Physiol.* (London) 300, 367–391.

Harting, J. K. (1977). Descending pathways from the superior colliculus: an autoradiographic analysis in the rhesus monkey (*Macaca mulatta*). *J. Comp. Neurol.* 173, 583–612.

Harting, J. K., Hall, W. C., Diamond, I. T., and Martin, G. F. (1973). Antero-grade degeneration study of the superior colliculus in *Tupaia glis:* evidence for a subdivision between superficial and deep layers. *J. Comp. Neurol.* 148, 361–386.

Harting, J. K., Huerta, M. F., Frankfurter, A. J., Strominger, N. L., and Royce, G. J. (1980). Ascending pathways from the monkey superior colliculus: an autoradiographic analysis. *J. Comp. Neurol.* 192, 853–882.

Hartline, P. H. (1971). Mid-brain responses of the auditory and somatic vibration systems in snakes. *J. Exp. Biol.* 54, 373–390.

Hartline, P. H. (1984). The optic tectum of reptiles: neurophysiological studies. In *Comparative Neurology of the Optic Tectum* (H. Vanegas, ed.). Plenum Press, New York, pp. 601–618.

Hartline, P. H., Kass, L., and Loop, M. S. (1978). Merging of modalities in the optic tectum: infrared and visual integration in rattlesnakes. *Science* (N.Y.) 199, 1225–1229.

Hayes, B. P., and Webster, K. E. (1975). An electron microscopic study of the retinoreceptive layers of the pigeon optic tectum. *J. Comp. Neurol.* 162, 447–466.

Heller, S. B., and Ulinski, P. S. (1987). Morphology of geniculocortical axons in the turtles, of the genera *Pseudemys* and *Chrysemys. Anat. Embryol.* 175, 505–515.

Heric, T. M., and Kruger, L. (1965). Organization of the visual projection upon the optic tectum of a reptile (*Alligator mississippiensis*). *J. Comp. Neurol.* 124, 101–112.

Heric, T. M., and Kruger, L. (1966). The electrical response evoked in the reptilian optic tectum by afferent stimulation. *Brain Res.* 2, 187–199.

Hess, W. R., Burgi, S., and Bucher, V. (1946). Motorische Function des Tektal and Tegmentalgebietes. *Monatsschr. Psychiatr. Neurol.* 112, 1–52.

Hikosaka, O., and Wurtz, R. H. (1983). Visual and oculomotor functions of monkey substantia nigra pars reticulata. IV. Relation of substantia nigra to superior colliculus. *J. Neurophysiol.* 49, 1285–1301.

Holcombe, V., and Hall, W. C. (1981a). The laminar origin and distribution of the crossed tectoreticular pathways. *J. Neurosci.* 1, 1103–1112.

Holcombe, V., and Hall, W. C. (1981b). Laminar origin of ipsilateral tectopontine pathways. *Neuroscience* 6, 255–260.

Hoogland, P. V. (1977). Efferent connections of the striatum in *Tupinambis nigropunctatus. J. Morphol.* 152, 229–246.

Hoogland, P. V. (1982). Brainstem afferents to the thalamus in a lizard, *Varanus exanthematicus. J. Comp. Neurol.* 210, 152–162.

Hubel, D. H., Le Vay, S., and Wiesel, T. N. (1975). Mode of termination of retinotectal fibers in macaque monkey: an autoradiographic study. *Brain Res.* 96, 25–40.

Huber, G. C., and Crosby, E. C. (1926). On thalamic and tectal nuclei and fiber paths in the brain of the American alligator. *J. Comp. Neurol.* 40, 97–227.

Huber, G. C., and Crosby, E. C. (1933). The reptilian optic tectum. *J. Comp. Neurol.* 57, 57–164.

Huber, G. C., and Crosby, E. C. (1943). A comparison of the mammalian and reptilian tecta. *J. Comp. Neurol.* 78, 133–168.

Huerta, M. F., Frankfurter, A. J., and Harting, J. K. (1981). The trigemino-collicular projection in the cat: patch-like endings within the intermediate gray. *Brain Res.* 211, 1–14.

Huerta, M. F., and Harting, J. K. (1982). Projections of the superior colliculus to the supraspinal nucleus and the cervical spinal cord gray of the cat. *Brain Res.* 242, 326–331.

Huerta, M. F., and Harting, J. K. (1984). The mammalian superior colliculus: studies of its morphology and connections. In *Comparative Neurology of the Optic Tectum* (H. Vanegas, ed.). Plenum Press, New York, pp. 687–774.

Hunt S. P., and Kunzle, H. (1976). Observations on the projections and intrinsic organization of the pigeon optic tectum: an autoradiographic study based on anterograde and retrograde axonal and dendritic flow. *J. Comp. Neurol.* 170, 153–172.

Hunt, S. P., Streit, P., Kunzle, H., and Cuenod, M. (1977). Characterization of the pigeon isthmo-tectal pathway by selective uptake and retrograde movement of radioactive compounds and by Golgi-like horseradish peroxidase labelling. *Brain Res.* 129, 197–212.

Ingle, D. (1968). Visual releasers of prey-catching behavior in frogs and toads. *Brain Behav. Evol.* 1, 500–518.

Ingle, D. (1973). Disinhibition of tectal neurons by pretectal lesions in the frog. *Science* (N.Y.) 180, 422–424.

Ingle, D. (1976). Spatial vision in anurans. In *The Amphibian Visual System* (K. Fite, ed.). Academic Press, New York, pp. 119–140.

Ingle, D. (1980). Some effects of pretectum lesions on the frog's detection of stationary objects. *Behav. Brain Res.* 39–163.

Ingle, D. (1982). Organization of visuomotor behaviors in vertebrates. In *Analysis of Visual Behavior* (D. J. Ingle, M. A. Goodale, and R. J. M. Mansfield, eds.). MIT Press, Cambridge, Mass., pp. 67–110.

Ingle, D., and Sprague, J. M. (1975). Sensorimotor functions of the midbrain tectum. *Neurosci. Res. Prog. Bull.* 13, 173–288.

Ito, H. (1970). Fine structure of the carp tectum opticum. *J. Hirnforsch.* 12, 325–354.

Kaas, J. H. (1982). The segregation of function in the nervous system: why do sensory systems have so many subdivisions? *Contributions Sensory Physiol.* 7, 201–240.

Karten, H. J. (1965). Projections of the optic tectum of the pigeon (*Columba livia*). *Anat. Rec.* 151, 369.

Karten, H. J., Hodos, W., Nauta, W. J. H., and Revzin, A. M. (1973). Neural connections of the "visual Wulst" of the avian telencephalon: experimental studies in the pigeon and owl. *J. Comp. Neurol.* 150, 253–278.

Kass, L., Loop, M. S., and Hartline, P. M. (1978). Anatomical and physiological localization of visual and infrared cell layers in the tectum of pit vipers. *J. Comp. Neurol.* 182, 811–820.

Kawamura, K., Brodal, A., and Hodevik, G. (1974). The projection of the superior colliculus onto the reticular formation of the brainstem: an experimental anatomical study in the cat. *Exp. Brain. Res.* 19, 1–19.

Kawamura, S., Fukushima, N., Hattori, S., and Kudo, M. (1980). Laminar segregation of cells of origin of ascending projections from the superficial layers of the superior colliculus in the cat. *Brain Res.* 184, 486–490.

Kawamura, K., and Hashikawa, T. (1978). Cell bodies of origin of the reticular projection from the superior colliculus in the cat: an experimental study with the use of horseradish peroxidase as a tracer. *J. Comp. Neurol.* 182, 1–16.

Kawamura, S., and Kobayashi, E. (1975). Identification of laminar origin of some tectothalamic fibers in the cat. *Brain Res.* 9, 281–285.

Kawamura, S., Sprague, J. M., and Niimi, K. (1974). Corticofugal projections from the visual cortices to the thalamus, pretectum and superior colliculus in the cat. *J. Comp. Neurol.* 158, 339–362.

Keller, E. L. (1974). Participation of medial pontine reticular formation in eye movement generation in monkey. *J. Neurophysiol.* 37, 316–332.

Killackey, H. P., and Erzurumlu, R. H. (1981). Trigeminal projections to the colliculus of the rat. *J. Comp. Neurol.* 201, 221–242.

Kishida, R., Amemiya, F., Kusunoki, T., and Terashima, S. (1980). A new tectal afferent nucleus of the infrared sensory systems in the medulla oblongata of crotaline snakes. *Brain Res.* 195, 271–279.

Knapp, H., and Kang, D. (1968a). The visual pathways of the snapping turtle (*Chelydra serpentina*). *Brain Behav. Evol.* 1, 19–42.

Knapp, H., and Kang, D. (1968b). The retinal projections of the side-necked turtle (*Podocnemis unifilis*) with some notes on the possible origin of the pars dorsalis of the lateral geniculate body. *Brain Behav. Evol.* 1, 369–404.

Kolb, H. (1982). The morphology of the bipolar cells, amacrine cells and ganglion cells in the retina of the turtle *Pseudemys scripta elegans*. *Phil. Trans. R. Soc. Lond.* B298, 355–393.

Kosareva, A. A. (1967). Projection of optic fibers to visual centers in a turtle (*Emys orbicularis*). *J. Comp. Neurol.* 130, 263–276.

Kosareva, A. A. (1974). Afferent and efferent connections of the nucleus rotundus in the tortoise *Emys orbicularis*. *Zh. Evol. Biokhim. Fiziol.* 10, 395–399.

Kruger, L. (1969). The topography of the visual projection to the mesencephalon: a comparative survey. *Brain Behav. Evol.* 3, 169–177.

Kusunoki, T. (1971). The chemoarchitectonics of the turtle brain. *Yohokama Med. Bull.* 22, 1–29.

Langer, T. P., and Lund, R. D. (1974). The upper layers of the superior colliculus of the rat: a Golgi study. *J. Comp. Neurol.* 158, 405–434.

Lara, R., Arbib, M. A., and Cromarty, A. S. (1982). The role of the tectal column in facilitation of amphibian prey-catching behavior: a neural model. *J. Neurosci.* 2, 521–530.

Lazar, G., and Szekely, G. (1967). Golgi studies on the optic center of the frog. *J. Hirnforsch.* 9, 329–344.

Lee, C., Rohrer, W. H., and Sparks, D. L. (1988). Population coding of saccadic eye movements by neurons in the superior colliculus. *Nature* 332, 357–360.

Leghissa, S. (1962). L'evoluzione del tetto ottico nei bassi vertebrati. *Arch. Ital. Anat. Embryol.* 67, 334–413.

Lennie, P. (1980). Parallel visual pathways: a review. *Vision Res.* 20: 561–594.

Leresche, N., Hardy, O., and Jassik-Gerschenfeld, D. (1983). Receptive field properties of single cells in the pigeon's optic tectum during cooling of the "visual Wulst." *Brain Res.* 267, 225–236.

Lohman, A. H. M., and van Woerden-Verkley, I. (1978). Ascending connections to the forebrain in the tegu lizard. *J. Comp. Neurol.* 182, 555–594.

Luiten, P. G. M. (1981). Afferent and efferent connections of the optic tectum in the carp (*Cyprinus carpis*). *Brain Res.* 220, 51–65.

Lund, J. S. (1981). Intrinsic organization of the primate visual cortex, Area 17, as seen in Golgi preparations. In *The Organization of the Cerebral Cortex* (F. O. Schmitt, F. G. Worden, G. Adelman, and S. G. Dennis, eds.). MIT Press, Cambridge, Mass., pp. 105–124.

Lund, R. D. (1969). Synaptic patterns of the superficial layers of the superior colliculus of the rat. *J. Comp. Neurol.* 135, 179–208.

Lund, R. D. (1972). Synaptic patterns in the superficial layers of the superior colliculus of the monkey (*Macaca mulatta*). *Exp. Brain Res.* 15, 194–211.

Magalhaes-Castro, H. H., Dolabela de Lima, A., Saraiva, P. E. S., and Magalhaes-Castro, B. (1978). Horseradish peroxidase labeling of cat tecto-tectal cells. *Brain Res.* 148, 1–13.

Marchiafava, P. L., and Wagner, H. G. (1981). Interactions leading to colour opponency in ganglion cells of the turtle retina. *Proc. R. Soc. London*, Ser. B 211, 261–267.

Marchiafava, P. L., and Weiler, R. (1980). Intracellular analysis and structural correlates of the organization of inputs to ganglion cells in the retina of the turtle. *Proc. R. Soc. Lond.*, Ser. B 208, 103–113.

Martin, G. F. (1969). Efferent tectal pathways of the opossum, *Didelphis virginiana*. *J. Comp. Neurol.* 125, 209–224.

Masai, H. (1973). Structural patterns of the optic tectum in Japanese snakes of the family Colubridae, in relation to habit. *J. Hirnforsch.* 14, 367–374.

Maxwell, J. H., and Granda, A. M. (1979). Receptive fields of movement-sensitive cells in the pigeon thalamus. In *Neural Mechanisms of Behavior in the Pigeon* (C. A. M. Granda and J. H. Maxwell, eds.). Plenum Press, New York, pp. 177–198.

May, P. J., and Hall, W. C. (1981). A relationship between nigrotectal and crossed tectoreticular pathways in the grey squirrel. *Neurosci. Abst.* 7, 776.

May, P. J., Lin, C.-S., McIlwain, J. T., and Hall, W. C. (1982). Morphology of tecto-pulvinar neurons in the grey squirrel. *Neurosci. Abst.* 8, 406.

McIlwain, J. T. (1976). Large receptive fields and spatial transformations in the visual system. *Int. Rev. Physiol.* 10, 223–248.

McIlwain, J. T. (1982). Lateral spread of neural excitation during microstimulation in intermediate gray layer of cats' superior colliculus. *J. Neurophysiol.* 47, 167–178.

Meek, J. A. (1981a). A Golgi-electron microscopic study of the goldfish optic tectum. I. Description of afferents, cell types, and synapses. *J. Comp. Neurol.* 199, 149–174.

Meek, J. A. (1981b). A Golgi-electron microscopic study of the goldfish optic tectum. II. Quantitative aspects of synaptic organization. *J. Comp. Neurol.* 199, 175–190.

Mendez-Otero, R., Rocha-Miranda, C. E., and Perry, V. H. (1980). The organization of the parabigemino-tectal projections in the opossum. *Brain Res.* 198, 183–189.

Mesulam, M.-M. (1982). *Tracing Neural Connections with Horseradish Peroxidase.* John Wiley, New York.

Miceli, D., Peyrichoux, J., and Repérant, J. (1975). The retino-thalamo-hyperstriatal pathway in the pigeon (*Columba livia*). *Brain Res.* 100, 125–131.

Mize, R. R., Spencer, R. F. S., and Sterling, P. (1981). Neurons and glia in cat superior colliculus accumulate ^{3}H-gamma-aminobutyric acid (GABA). *J. Comp. Neurol.* 202, 300–312.

Mize, R. R., and Sterling, P. (1976). Synaptic organization of the superficial gray layer of cat superior colliculus analyzed by serial section cinematography. *Invest. Ophthalmol. Vis. Sci.* 15, 47.

Molenaar, G. J. (1974). An additional trigeminal system in certain snakes possessing infrared receptors. *Brain Res.* 78, 340–344.

Molenaar, G. J., and Fizaan-Oostveen, J. L. F. P. (1980). Ascending projections from the lateral descending and common sensory trigeminal nuclei in python. *J. Comp. Neurol.* 189, 555–572.

Morenkov, E. D., and Pivavarov, A. S. (1973). Peculiarities of the organization of the visual system in reptiles. In *Functional Organization and Evolution of the Vertebrate Visual System*. Nauka, Leningrad, pp. 95–107.

Morenkov, E. D., and Pivavarov, A. S. (1975). Peculiarities of cell reactions in turtle dorsal and ventral thalamus to visual stimuli. *Zh. Evol. Biokhim. Fiziol.* 11, 70–75.

Moschovakis, A. K., Karabelas, A. B., and Highstein, S. M. (1988a). Structure-function relationships in the primate superior colliculus. I. Morphological classification of efferent neurons. *J. Neurophysiol.* 60, 232–262.

Moschovakis, A. K., Karabelas, A. B., and Highstein, S. M. (1988b). Structure-function relationships in the primate superior colliculus. II. Morphological identity of presaccadic neurons. *J. Neurophysiol.* 60, 263–302.

Murray, E. A., and Coulter, J. D. (1982). Organization of tectospinal neurons in the cat and rat superior colliculus. *Brain Res.* 243, 201–214.

Nagata, T., Magalhaes-Castro, H. H., Saraiva, P. E. S., and Magalhaes-Castro, B. (1980). Absence of tectotectal pathway in the rabbit: an anatomical and electrophysiological study. *Neurosci. Lett.* 17, 125–130.

Newman, D. B., and Cruce, W. L. R. (1982). The organization of the reptilian brainstem reticular formation: a comparative study using Nissl and Golgi techniques. *J. Morphol.* 173, 325–349.

Newman, D. B., Cruce, W. L. R., and Bruce, L. L. (1982). The sources of supraspinal afferents to the spinal cord in a variety of limbed reptiles. I. Reticulospinal systems. *J. Comp. Neurol.* 215, 17–32.

Newman, E. A., Gruberg, E. R., and Hartline, P. H. (1980). Infrared trigemino-tectal pathway in the rattlesnake and in the python. *J. Comp. Neurol.* 191, 465–478.

Newman, E. A., and Hartline, P. H. (1981). Integration of visual and infrared information in bimodal neurons of the rattlesnake optic tectum. *Science* (N.Y.) 213, 789–791.

Niida, A., Oka, H., and Iwata, K. S. (1980). Visual responses of morphologically identified tectal neurons in the carp. *Brain Res.* 201, 361–371.

Northcutt, R. G. (1978). Forebrain and midbrain organization in lizards and its evolutionary significance. In *The Behavior and Neurology of Lizards* (N. Greenberg and P. D. MacLean, eds.). National Institute of Mental Health, Rockville, Md., pp. 11–64.

Northcutt, R. G. (1981). Evolution of the telencephalon in nonmammals. *Ann. Rev. Neurosci.* 4, 301–350.

Northcutt, R. G. (1982). Localization of neurons afferent to the optic tectum in longnose gars. *J. Comp. Neurol.* 204, 325–335.

Northcutt, R. G. (1984). Anatomical organization of the optic tectum in reptiles. In *Comparative Neurology of the Optic Tectum* (H. Vanegas, ed.). Plenum Press, New York, pp. 547–600.

Northcutt, R. G., Braford, M. R., Jr., and Landreth, G. E. (1974). Retinal projections in the turtle, *Cyclemodon punctatus*: an Autoradiographic study. *Anat. Rec.* 178, 428.

Northcutt, R. G., and Butler, A. B. (1974a). Retinal projections in the northern water snake, *Natrix sipedon sipedon. J. Morphol.* 142, 117–136.

Northcutt, R. G., and Butler, A. B. (1974b). Evolution of reptilian visual systems: retinal projections in a nocturnal lizard, *Gekko gecko* (Linnaeus). *J. Comp. Neurol.* 157, 453–465.

Palmer, L. A., and Rosenquist, A. C. (1974). Visual receptive fields of single striate cortical units projecting to the superior colliculus in the cat. *Brain Res.* 67, 27–42.

Parent, A. (1979). Monoaminergic systems in reptile brains. In *Biology of the Reptilia* (C. Gans, R. G. Northcutt, and P. S. Ulinski, eds.). Academic Press, London, Vol. 10, pp. 247–285.

Pedersen, R. (1973). Ascending spinal projections in three species of side-necked turtle: *Podocnemis unifilis, Pelusios subriger,* and *Pelomedusa subrufa. Anat. Rec.* 175, 409.

Peterson, E. H. (1978a). Size classes of ganglion cells which project to the optic tectum in the turtle, *Pseudemys scripta. Anat. Rec.* 190, 509–510.

Peterson, E. H. (1978b). Horizontal cells in the optic tectum of the turtle, *Pseudemys scripta elegans. Am. Zool.* 18, 587.

Peterson, E. H. (1981). Regional specialization in the retinal ganglion cell projection to the optic tectum of *Dipsosaurus dorsalis* (Iguanidae). *J. Comp. Neurol.* 196, 225–252.

Peterson, E. H. (1982). Morphology of retinal ganglion cells in a turtle, *Pseudemys scripta. Neurosci. Abstr.* 8, 48.

Peterson, E. H., and Ulinski, P. S. (1979). Quantitative studies of retinal ganglion cells in a turtle, *Pseudemys scripta elegans.* I. Number and distribution of ganglion cells. *J. Comp. Neurol.* 186, 17–42.

Peterson, E. H., and Ulinski, P. S. (1982). Quantitative studies of retinal ganglion cells in a turtle, *Pseudemys scripta elegans.* II. Size spectrum of ganglion cells and its regional variation. *J. Comp. Neurol.* 208, 157–168.

Pitts, W. H., and McCulloch, W. S. (1947). How we know universals, the perception of auditory and visual forms. *Bull. Math. Biophys.* 9, 127–147.

Pritz, M. B. (1975). Anatomical identification of a telencephalic visual area in crocodiles: ascending connections of nucleus rotundus in *Caiman crocodilus. J. Comp. Neurol.* 164, 323–338.

Quiroga, J. C. (1978). The tectum opticum of *Pantodactylus schreiberii* (Teidae, Lacertilia, Reptilia). *J. Hirnforsch.* 19, 109–131.

Rainey, W. T. (1979). Organization of nucleus rotundus, a tectofugal thalamic nucleus in turtles. I. Nissl and Golgi analyses. *J. Morphol.* 160, 121–142.

Rainey, W. T., and Ulinski, P. S. (1982a). Organization of nucleus rotundus, a tectofugal thalamic nucleus in turtles. II. Ultrastructural analyses. *J. Comp. Neurol.* 209, 187–207.

Rainey, W. T., and Ulinski, P. S. (1982b). Organization of nucleus rotundus, a tectofugal thalamic nucleus in turtles. III. The tectorotundal projection. *J. Comp. Neurol.* 209, 208–223.

Ramón, P. (1896). Estructura de encefalo del camaleon. *Rev. Trim. Micrograf.* 1, 146–182.

Ramón y Cajal, S. (1911). *Histologie du Système Nerveux de l'Homme et des Vértèbres.* Maloine, Paris.

Reiner, A., Brauth, S. E., Kitt, C. A., and Karten, H. J. (1980). Basal ganglionic pathways to the tectum: studies in reptiles. *J. Comp. Neurol.* 193, 565–589.

Repérant, J. (1973). Les voies et les centres optiques primaires chez la vipere (*Vipera aspis*). *Arch. Anat. Micr. Morph. Exp.* 62, 323–352.

Repérant, J. (1975). Nouvelles donnees dur les projections retiniennes chez *Caiman sclerops*. Etude radio-autographique. *C. R. Acad. Sc. Paris,* Serie D. 280, 2881–2884.

Repérant, J., Peyrichoux, J., and Rio, J.-P. (1981). Fine structure of the superficial layers of the viper optic tectum: a Golgi and electron-microscopic study. *J. Comp. Neurol.* 199, 393–417.

Repérant, J., Raffin, J.-P., and Miceli, D. (1974). La voie retino-thalamo-hyperstriatale chez le pouissin (*Callus domesticus*). *C. R. Acad. Sci. Paris* 279, 279–281.

Repérant, J., and Rio, J.-P. (1976). Retinal projections in *Vipera aspis:* a reinvestigation using light radio-autographic and electron microscopic degeneration techniques. *Brain Res.* 107, 603–609.

Repérant, J., Rio, J.-P., Miceli, D., and Lemire, M. (1978). A radioautographic study of retinal projections in Type I and Type II lizards. *Brain Res.* 142, 401–411.

Revzin, A. M. (1970). Some characteristics of wide-field units in the brain of the pigeon. *Brain Behav. Evol.* 3, 195–204.

Revzin, A. M. (1979). Functional localization in the nucleus rotundus. In *Neural Mechanisms of Behavior* (A. M. Granda and J. H. Maxwell, eds.). Plenum Press, New York, pp. 165–176.

Revzin, A. M., and Karten, H. J. (1966). Rostral projections of the optic tectum and nucleus rotundus in the pigeon. *Brain Res.* 3, 264–276.

Rhoades, R. W., Kuo, D. C., Polcer, J. D., Fish, S. E., and Voneida, T. J. (1982). Indirect visual cortical input to the deep layers of the hamster's superior colliculus via the basal ganglia. *J. Comp. Neurol.* 208, 239–254.

Rhoades, R. W., Mooney, R. D., Rhorer, W. H., Nikoletseas, M. M., and Fish, S. (1989). Organization of the projection from the superficial to the deep layers of the hamster's superior colliculus as demonstrated by the anterograde transport of *Phaseolus vulgaris* leucoagglutinin. *J. Comp. Neurol.* 283, 54–70.

Robbins, D. O. (1972). Coding of intensity and wavelength in optic tectal cells of the turtle. *Brain Behav. Evol.* 5, 124–142.

Robinson, D. A. (1981). The use of control systems analysis in the neurophysiology of eye movements. *Ann. Rev. Neurosci.* 4, 463–503.

Robinson, L. R. (1969). Bulbospinal fibres and their nuclei of origin in *Lacerta viridis* demonstrated by axonal degeneration and chromatolysis respectively. *J. Anat.* (London) 105, 59–88.

Robson, J. A., and Hall, W. C. (1976). Projections from the superior colliculus to the dorsal lateral geniculate nucleus of the grey squirrel (*Sciurus carolinensis*). *Brain Res.* 113, 379–385.

Rodieck, R. W. (1979). Visual pathways. *Ann. Rev. Neurosci.* 2, 193–226.

Rosenquist, A. C., and Palmer, L. (1979). Visual receptive fields of cells of the superior colliculus after cortical lesions in the cat. *Exp. Neurol.* 33, 629–652.

Roucoux, A., and Crommelinck, M. (1976). Eye movements evoked by superior colliculus stimulation in the alert cat. *Brain Res.* 106, 349–363.

Rowe, M. H., and Stone, J. (1977). The naming of neurones: the classification and naming of cat retinal ganglion cells. *Brain Behav. Evol.* 14, 185–216.

Rubinson, K. (1968). Projections of the tectum opticum of the frog. *Brain Behav. Evol.* 1, 529–561.

Schapiro, H., and Goodman, D. G. (1969). Motor functions and their anatomical basis in the forebrain and tectum of the alligator. *Exp. Neurol.* 24, 187–195.

Schechter, P. B., and Ulinski, P. S. (1979). Interactions between tectal radial cells in the red-eared turtle, *Pseudemys scripta elegans:* an analysis of tectal modules. *J. Morphol.* 162, 17–36.

Schiller, P. H., and Stryker, M. (1972). Single unit recording and stimulation in superior colliculus of the alert rhesus monkey. *J. Neurophysiol.* 35, 915–924.

Schroeder, D. M. (1981a). Retinal afferents and efferents of an infrared sensitive snake, *Crotalus viridis. J. Morphol.* 170, 29–42.

Schroeder, D. M. (1981b). Tectal projections of an infrared sensitive snake, *Crotalus viridis. J. Comp. Neurol.* 195, 477–500.

Schroeder, D. M., and Ebbesson, S. O. E. (1975). Cytoarchitecture of the optic tectum in the nurse shark. *J. Comp. Neurol.* 160, 443–462.

Schroeder, D. M., and Loop, M. S. (1976). Trigeminal projections in snakes possessing infrared sensitivity. *J. Comp. Neurol.* 169, 1–14.

Senn, D. G. (1966). Über das optische System im Gehirn squamater Reptilien. *Acta Anat,* Suppl. 52, 65, 1–87.

Senn, D. G. (1968a). Über den Bau von Zwischen- und Mittelhirn von *Anniella pulchra* Gray. *Acta Anat.* 69, 239–261.

Senn, D. G. (1968b). Bau und Ontogenese von Zwischen- und Mittelhirn bei *Lacerta sicula* (Rafinesque). *Acta Anat.,* Suppl. 55, 71, 1–150.

Senn, D. G. (1969). Über das Zwischen- und Mittelhirn von zwei typhlopiden Schlangen, *Anomalepis aspinosus* und *Liotyphlops albirostris. Verhandl. Naturf. Ges. Basel.* 80, 32–48.

Senn, D. G. (1970). Zur Ontogenese des Tectum opticum von *Natrix natrix* (L.). *Acta Anat.* 76, 545–563.

Senn, D. G. (1979). Embryonic development of the central nervous system. In *Biology of the Reptilia* (C. Gans, R. G. Northcutt, and P. S. Ulinski, eds.). 9A, 173–244.

Senn, D. G., and Northcutt, R. G. (1973). The forebrain and midbrain of some squamates and their bearing on the origin of snakes. *J. Morphol.* 140, 135–152.

Sereno, M. I. (1985). Tectoreticular pathways in turtles: I. Organization and morphology of tectoreticular axons. *J. Comp. Neurol.* 233, 48–90.

Sereno, M. I., and Ulinski, P. S. (1985). Tectoreticular pathways in turtles: II. Morphology of tectoreticular cells. *J. Comp. Neurol.* 233, 91–114.

Sereno, M. I., and Ulinski, P. S. (1987). Caudal topographic nucleus isthmi and the rostral nontopographic nucleus isthmi in the turtle, *Pseudemys scripta. J. Comp. Neurol.* 261, 319–346.

Sherk, H. (1979). Connections and visual field mapping in cat's parabigeminal circuit. *J. Neurophysiol.* 42, 1656–1688.

Sligar, C. M., and Voneida, T. J. (1976). Tectal efferents in the blind cavefish, *Astynax hubbsi. J. Comp. Neurol.* 165, 107–124.

Smeets, W. J. A. J. (1981). Efferent tectal pathways in two chondrichthyans, the shark, *Scyliorhinus canicula* and the ray, *Raja clavata. J. Comp. Neurol.* 195, 13–24.

Smeets, W. J. A. J. (1982). The afferent connections of the tectum mesencephali in two chondrichthyans, the shark, *Scyliorhinus canicula* and the ray *Raja clavata. J. Comp. Neurol.* 205, 139–152.

Sparks, D. L. (1986). Translation of sensory signals into commands for control of saccadic eye movements: role of primate superior colliculus. *Physiol. Rev.* 66: 118–171.

Sparks, D. L., Holland, R., and Guthrie, B. L. (1976). Size and distribution of movement fields in the monkey superior colliculus. *Brain Res.* 113, 21–34.

Sparks, D. L., and Mays, L. E. (1980). Movement fields of saccade-related burst neurons in the monkey superior colliculus. *Brain Res.* 190, 39–50.

Sparks, D. L., and Nelson, J. S. (1987). Sensory and motor maps in the mammalian superior colliculus. *Trends Neurosci.* 10: 312–317.

Stanford, L. R., Schroeder, D. M., and Hartline, P. H. (1981). The ascending projection of the nucleus of the lateral descending trigeminal tract: a nucleus in the infrared system of the rattlesnake, *Crotalus viridis. J. Comp. Neurol.* 201, 161–173.

Stein, B. E., and Clamann, H. P. (1981). Control of pinna movements and sensorimotor register in cat superior colliculus. *Brain Behav. Evol.* 19, 180–192.

Stein, B. E., and Gaither, N. S. (1981). Sensory representation in reptilian optic tectum: some comparisons with mammals. *J. Comp. Neurol.* 202, 69–87.

Stein, B. E., and Gaither, N. S. (1983). Receptive-field properties in reptilian optic tectum: some comparisons with mammals. *J. Neurophysiol.* 50, 102–124.

Sterling, P. (1971). Receptive field and synaptic organization of the superficial gray layers of the cat superior colliculus. *Vision Res.* 11, Suppl. 3, 309–328.

Sterling, P. (1983). Microcircuitry of the cat retina. *Ann. Rev. Neurosci.* 6, 149–186.

Stone, J. (1981). *The Whole Mount Handbook.* Maitland, Sydney, Australia.

Stone, J., and Freeman, J. A. (1971). Synaptic organization of the pigeon optic tectum: a Golgi and current source-density analysis. *Brain Res.* 27, 203–221.

Streit, P., Knecht, E., Reubi, J.-C., Hunt, S. P., and Cuenod, M. (1978). GABA-specific presynaptic dendrites in pigeon optic tectum: a high resolution autoradiographic study. *Brain Res.* 149, 204–210.

Szekely, G., and Lazar, G. (1976). Cellular and synaptic architecture of the optic tectum. In *Frog Neurobiology* (R. Llinas and W. Precht, eds.). Springer-Verlag, Berlin-Heidelberg, pp. 407–434.

Szekely, G., Octulo, G., and Lazar, C. (1973). Fine structure of the frog's optic tectum: optic fiber termination layers. *J. Hirnforsch.* 14, 189–225.

ten Donkelaar, H. J. (1976a). Descending pathways from the brainstem to the spinal cord in some reptiles. I. Origin. *J. Comp. Neurol.* 167, 421–442.

ten Donkelaar, H. J. (1976b). Descending pathways from the brain stem to the spinal cord in some reptiles. II. Course and site of termination. *J. Comp. Neurol.* 167, 443–464.

ten Donkelaar, H. J., and De Boer-van Huizen, R. (1978). Cells of origin of pathways descending to the spinal cord in a lizard (*Lacerta galloti*). *Neurosci. Lett.* 9, 123–128.

ten Donkelaar, H. J., and De Boer-van Huizen, R. (1981). Basal ganglia projections to the brain stem in the lizard *Varanus exanthematicus* as demonstrated by retrograde transport of horseradish peroxidase. *Neuroscience* 6, 1567–1590.

ten Donkelaar, H. J., Kusuma, A., and De Boer-van Huizen, R. (1980). Cells of origin of pathways descending to the spinal cord in some quadrupedal reptiles. *J. Comp. Neurol.* 192, 827–851.

Terashima, S., and Goris, R. C. (1975). Tectal organization of pit viper infrared reception. *Brain Res.* 83, 490–494.

Tigges, M., Tigges, J., Luttrell, G. L., and Frazier, C. M. (1973). Ultrastructural changes in the superficial layers of the superior colliculus in *Galago crassicaudatus*. *Z. Zellforsch. Mikrosk. Anat.* 140, 291–307.

Tokunaga, A., and Otani, K. (1976). Dendritic patterns of neurons in the rat superior colliculus. *Exp. Neurol.* 52, 189–205.

Torrealba, F., Partlow, G. D., and Guillery, R. W. (1981). Organization of the projection from the superior colliculus to the dorsal lateral geniculate nucleus of the cat. *Neuroscience* 6, 1341–1360.

Ulinski, P. S. (1977). Tectal efferents in the banded water snake (*Natrix sipedon*). *J. Comp. Neurol.* 173, 251–274.

Ulinski, P. S. (1978). Distribution of neurons in the optic tectum of the turtle *Pseudemys scripta*. *Anat. Rec.* 199, 568.

Ulinski, P. S. (1983). *Dorsal Ventricular Ridge: a Treatise on Forebrain Organization in Reptiles and Birds.* John-Wiley Interscience, New York.

Ulinski, P. S., du Lac, S., and Dacey, D. M. (1983). Descending projections of the geniculate complex. *Invest. Ophthalmol. Vis. Sci.* 24, 64.

Valverde, F. (1973). The neuropil in superficial layers of the superior colliculus of the mouse. *Z. Anat. Entwick.* 142, 117–147.

Vanegas, H. (1975). Cytoarchitecture and connexions of the teleostean optic tectum. In *Vision in Fishes* (M. A. Ali, ed.). Plenum Press, New York, pp. 151–158.

von Düring, M., and Miller, M. R., (1979). Sensory nerve endings of the skin and deeper structures of reptiles. In *Biology of the Reptilia* (C. Gans, R. G. Northcutt, and P. S. Ulinski, eds.). New York: Academic Press, 9, 407–441.

Voneida, T. J., and Sligar, C. M. (1979). Efferent projections of the dorsal ventricular ridge and the striatum in the tegu lizard, *Tupinambis nigropunctatus. J. Comp. Neurol.* 186, 43–64.

Walls, G. L. (1940). Ophthalmological implications for the early history of snakes. *Copeia*, 1940, 1–8.

Walls, G. L. (1942). *The Vertebrate Eye.* Cranbrook Institute of Science, Bloomfield Hills, Mich.

Wang, S.-J., Yan, K., and Wang, Y.-T. (1981). Visual field topography in the frog's nucleus isthmi. *Neurosci. Lett.* 23, 37–41.

Wickelgren, B. G., and Sterling, P. (1969). Influence of visual cortex on receptive fields in the superior colliculus of the cat. *J. Neurophysiol.* 32, 16–23.

Wilczinski, W., and Northcutt, R. G. (1977). Afferents to the optic tectum of the leopard frog: an HRP study. *J. Comp. Neurol.* 173, 219–230.

Woodburne, R. T. (1936). A phylogenetic consideration of the primary and secondary centers and connections of the trigeminal complex in a series of vertebrates. *J. Comp. Neurol.* 65, 403–501.

Wurtz, R. H., and Albano, J. E. (1980). Visual-motor functions of the primate superior colliculus. *Ann Rev. Neurosci.* 3, 189–226.

Zagorul'ko, T. M. (1968). Effect of intensity and wavelength of photic stimulus on evoked responses of general cortex and optic tectum in turtles. *Fiziol. Zh. SSSR Sechenova* 54, 436–446.

5

Anatomy and Physiology of Infrared Sensitivity of Snakes

GERARD J. MOLENAAR

Contents

I. INTRODUCTION

Infrared receptors have received much interest over the years. Since an extensive review of earlier studies has been presented in a previous volume of this series (Barrett, 1970), this account gives only a brief survey of the relevant literature.

Infrared receptors, the so-called pit organs, are present in two groups of snakes—the Boidae and the Viperidae. In the latter family, these receptors occur in pit vipers (Crotalinae) but are absent in common vipers (Viperinae) (Barrett, 1970).

The shape of the pit organs in crotaline snakes is different from that in boid snakes (Fig. 5.1). In crotalines, the receptor is contained within a concave membrane that is suspended within a cavity. The cavity expands into a depression of the maxillary bone; as a result the maxillary bone's conformation differs from that of other snakes in order to accommodate the organ (Dullemeijer, 1959, 1961, 1969). The membrane is 15 μm thick and consists of a nervous layer situated between two cornified epidermal layers (Lynn, 1931; Noble, 1934; Noble and Schmidt, 1937; Bullock and Fox, 1957; Cordier, 1964). A rich vascular bed lies directly underneath the nervous layer. The receptors consist of free nerve endings that arise from palmate preterminal swellings (Bullock and Fox, 1957; Bleichmar and De Robertis, 1962). The nerve endings contain densely packed large mitochondria and high concentrations of oxidative enzymes (Meszler and Webster, 1968; Meszler, 1970).

The earliest detailed studies on behavior (Lynn, 1931; Noble and Schmidt, 1937) showed that pit organs allow the snake to avoid objects or to strike toward a warm object. Later electrophysiological studies (Bullock and Cowles, 1952; Bullock and Diecke, 1956; Goris and Nomoto, 1967) revealed these organs to be sensitive radiant heat receptors, which can detect temperature changes as small as 0.003° C of water flowing over the membrane. A generator potential has been recorded from the layer of nerve endings in the membrane (Terashima et al., 1968).

The pit organs of boid snakes are of a simpler construction. They consist of shallow depressions in the labial scales (Lynn, 1931; Ross, 1935; Noble and Schmidt, 1937; Bullock and Barrett, 1968; Warren and Proske, 1968; Barrett, 1970). Their distribution and number vary according to the species (Maderson, 1970). Several boid species have no pit organs but are nevertheless sensitive to radiant heat (Barrett, 1970).

In spite of these morphological differences, the pit organs of both crotaline and boid snakes are analogous and detect infrared radiation. The nature of the transduction processes at the nerve ending membrane is not known, but it seems from several recent studies that the receptor is not only sensitive to radiated but also to conducted heat

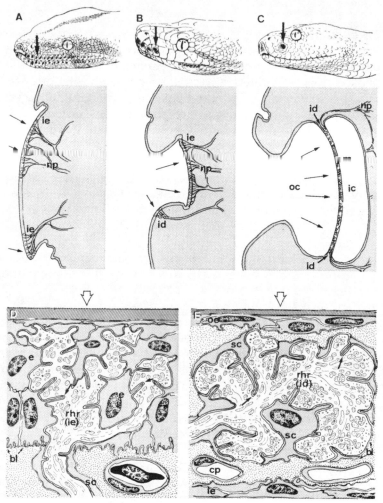

Fig. 5.1. Schematic representation of different types of radiant heat receptors in snakes. (A) *Boa constrictor*. Distribution of the radiant heat receptor in the supra- and infralabials. The intraepidermal radiant heat receptors have been marked in black. No specialization of the scale surface is seen. (B) *Python reticulatus*. Radiant heat receptors. The receptor terminals lie in different grooves of the supra- and infralabials. The grooves are marked in black. (C) *Crotalus durissus*. Radiant heat receptors in the pit organ. Only one pit is present on each side. Large black arrows indicate the scale region, which is shown in schematic representations below. These cross sections show the intraepidermal (ie) and intradermal (id) location of the radiant heat receptors. The nerve fibers form a plexuslike structure (np) within the dermis. In cross section the pit organ shows the pit membrane (pm) dividing the pit into an outer (oc) and inner (ic) cavity. The channel connecting the inner cavity of the scale surface is not shown in the diagram. (D) *Boa constrictor* and *Python reticulatus*. Schematic representation of the intraepidermal radiant heat receptor (rhr; ie). Notice the invaginated crests (arrows) of the epidermal cells (e). (E) *Crotalus*. Schematic drawing of the intradermal radiant heat receptor (rhr; id) in the pit membrane. The invaginated crests (arrow) are formed by the Schwann cell covering (sc). bl, basement lamella; cp, capillary; ie, inner epidermis of the pit membrane; oe, outer epidermis of the pit membrane. (From Von Düring et al., 1979.)

(De Cock Buning et al., 1981a, 1981b) and to other sorts of microwave energy (Harris and Gamow, 1971, 1972). This may indicate that the receptor is an energy detector of a more general nature.

The pit organs play different roles during different phases of feeding behavior (Dullemeijer, 1961) and are most important in triggering and guiding the strike toward the prey. In a biological sense, the receptors detect specific features of objects at a distance and therefore function as telereceptors. Other telereceptors (olfactory organ, eye, ear) are complexly built and are innervated by special nerves that convey their information to specific brain areas. The pit organs are innervated by branches of the trigeminal nerve. This nerve innervates other cutaneous receptors as well. Among these are, at least in mammals, cutaneous thermoreceptors, which send their data to the trigeminal nucleus caudalis.

In reptiles, such common cutaneous thermoreceptors may subserve thermoregulatory behavior by detecting ambient temperature. It may be expected that they are not only present in those species that lack radiant heat sensitivity but also in infrared-sensitive species. If these common thermoreceptors are present in the head, they probably also project to the trigeminal nucleus caudalis. This chapter concerns the extent to which the presence of special radiant heat receptors in some snakes is reflected in the trigeminal sensory system and presents a survey of the structure and function of the infrared system.

II. PERIPHERAL STRUCTURES
A. The Receptor

The infrared receptors in pit vipers are contained within a thin (15 μm) sensory membrane (see Fig. 5.1). The membrane consists of a central layer sandwiched between an outer and inner epithelial layer. The outer epithelial layer contains a few small unmyelinated C fibers, which do not seem to play a part infrared reception (Hirosawa, 1980). The central layer contains the nervous elements involved in infrared reception and a rich capillary bed. In *Trimeresurus flavoviridis* the infrared-sensitive fibers are myelinated and have diameters of 2.5 to 3.5 μm (Terashima et al., 1970) or 2 to 6 μm (Hirosawa, 1980). Before forming the terminal nerve mass, the fibers lose their myelin and gradually taper to 1 μm. Then they expand to form a palmate structure (Bullock and Fox, 1957) of about 4 to 6 μm (Hirosawa, 1980). Secondary and tertiary branches arise from this palmate structure and split further until numerous fine branchlets finally arise; these become closely associated with an aggregation of Schwann cells. The fine terminal branchlets become entwined with fine processes that are given off by these Schwann cells. As a result, a terminal nerve mass is formed in which the Schwann cells are located in the interior and the nerve branchlets on the exterior. The superficial branchlets are partly embedded in the Schwann cells; their bare surfaces are exposed on the side from which the infrared radiation comes.

The size of the terminal nerve mass is 30 to 50 μm in *Crotalus viridis* (Bullock and Fox, 1957) and about 40 μm in *Trimeresurus flavoviridis* (Terashima et al., 1970) or 50 $\times$ 100 μm (same snake, Hirosawa, 1980). This coincides remarkably well with the size of the receptive areas of peripheral units (20 to 50 μm, Terashima and Goris, 1979). The receptive area is defined as the tiny area on the surface of the sensory membrane that is connected to a unit. The unit can be stimulated only if enough energy is applied to that particular part of the sensory membrane. The receptive field is defined as the area in space from which the unit can be stimulated; it forms a cone that extends from the receptive area through the pit opening into space (Fig. 5.2).

Unlike those in pit vipers, the receptors of boid snakes are not contained within a suspended membrane. Instead, they lie at the bottom of the labial pits or in the labial scales if such pits are absent (Von Düring, 1974; Von Düring and Miller, 1979). In either case, the terminal nerve branches arise from preterminal swellings and extend be-

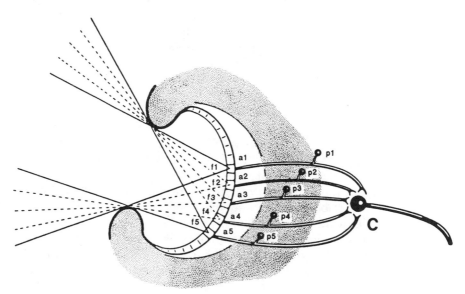

Fig. 5.2. Schematic drawing showing the arrangement of the receptive areas in the pit membrane of a crotaline snake and their corresponding receptive fields. One peripheral neuron (P1–P5) is connected to one receptive area (a1–a5) in the pit membrane and possesses a receptive field (f1–f5), which extends conically into space. The higher-order neuron (C) is connected to a receptive area, which is composed of the receptive areas of the peripheral neurons; likewise its receptive field is composed of the receptive fields of the peripheral neurons.

tween the epithelial cells. The branches lie only 6 to 8 μm below the surface of the scales (Warren and Proske, 1968; Meszler, 1970; Von Düring, 1974). The surface of the nerve terminal is considerably increased in size by means of numerous invaginations. The terminal is closely associated with Schwann cells and is morphologically similar to the terminal nerve mass of pit vipers (Von Düring, 1974). The size of the terminal nerve mass in *Boa* is 15 × 25 μm. The receptor ending exhibits no morphological features that are characteristic for a mechanoreceptor. This observation is in accordance with the electrophysiological finding that the infrared receptors in *Boa* are not responsive to mechanical stimuli (Hensel, 1974, 1975), in contrast to those of *Python*, which are (De Cock Buning et al., 1981b).

The small amount of energy that is sufficient to evoke a response (2 × 10^{-11} cal per receptor area, Bullock and Diecke, 1956) is in itself much less than the energy expended in producing the generator potential. To provide the high amount of energy, the nervous layer of the pit membrane contains very high concentrations of oxidative enzymes, that is, adenosine triphosphatase, succinic dehydrogenase, and lactic acid dehydrogenase (Meszler and Webster, 1968). In addition, the terminal nerve masses are densely packed with large mitochondria, which change their configuration upon infrared stimulation (Meszler, 1970). Vesicles, which are apparently associated with the plasma membrane (Von Düring, 1974; Hirosawa, 1980), are peculiar to this receptor. They are not known from any other receptor; their function is unknown.

Unimodal warm receptors are also present in the oral mucosa of both vipers and pit vipers (Chiszar et al., 1986; Dickman et al., 1987). The receptors are innervated by the superficial maxillary branch of the trigeminal nerve and are activated independently of the pit organs by thermal stimuli. The histological structure of these receptors is remarkably similar to that of receptors in the pit organs (Dickman et al., 1987), but their central projections are not known. Behavioral studies show that both vipers and pit vipers with anesthetized pit organs repeatedly strike toward warm balloons rather than to a cold mouse. This may indicate that these snakes use their intraoral receptors to aim the strike during the final phase, when the mouth is open and the fangs are extended (Chiszar et al., 1986; Dickman et al., 1987).

B. Innervation

The pit organs are innervated by different branches of the trigeminal nerve in boids and crotalines (Fig. 5.3). The pits of the crotaline snakes are innervated by branches of the ophthalmic nerve and by branches of the superficial and deep ramus of the maxillary nerve (Bullock and Fox, 1957; Barrett, 1970; Kishida et al., 1981). The distri-

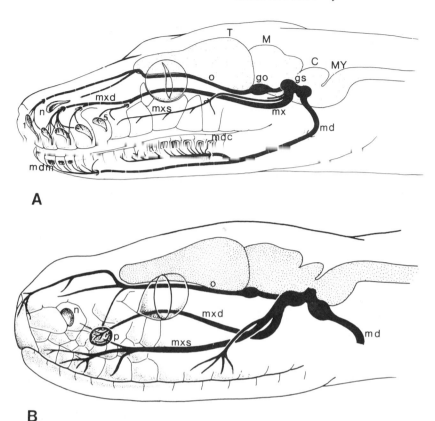

Fig. 5.3. Semischematic drawings of the innervation patterns of the pit organs in (A) a boid snake (*Python reticulatus*) and (B) a pit viper (*Crotalus sp.*). C, cerebellum; go, ganglion of the ophthalmic branch; gs, ganglion of the maxillary and mandibular branches; M, mesencephalon; md, mandibular branch; mdc, mandibular pits; mdm, mental pits; mx, maxillary branch; mxd, deep maxillary branch; mxs, superficial maxillary branch; MY, myelencephalon; n, nasal aperture; o, ophthalmic branch; p, pit organ; T, telencephalon; 1 to 5, maxillary pits.

bution of the infrared receptors in boid snakes varies among species and involves the mandibular region as well. The innervation pattern is more complex when compared to those of the crotaline snakes and varies among the different species. The receptors of the upper labial scales in *Corallus caninus* (Bullock and Barrett, 1968) and *Morelia spilotes* (Warren and Proske, 1968) are innervated by the maxillary branch, and those of the mandibular region are innervated by branches of a single mandibular branch. The innervation pattern in *Python reticulatus* is more complicated (De Cock Buning and Dullemeijer, 1977; Molenaar, 1978c; Molenaar et al., 1979). Distinct groups of infrared receptors are innervated by distinct nerve bundles and seem to converge their projections onto neurons within the medulla

(Molenaar, 1978c; Molenaar et al., 1979) and tectum (Haseltine, 1978). Individual pit organs in the maxillary region may receive nerve branches that diverge from bundles to neighboring pits of the same group (Molenaar, 1978c).

The first pit, located in the rostral scale, and a part of the second pit receive nerve fibers of the ophthalmic nerve; the second to fifth maxillary pits receive fibers from the deep branch of the maxillary nerve, whereas the receptors in the fifth to seventh supralabial scales are innervated by fibers from the superficial branch of the maxillary nerve. The pits in the mandibular region are subdivided into two groups: the rostrally located mental pits, which are innervated by the mental branches of the mandibular nerve, and the caudally located mandibular pits, which are innervated by a superficial ramus of the mandibular nerve.

C. The Ganglion

The three peripheral branches of the trigeminal nerve are represented by two distinct ganglia and three distinct nerve roots in boids and crotalines (Molenaar, 1978a, 1978b; Kishida et al., 1982). The separate ophthalmic ganglion is located near the isthmus cerebri. Its root enters the brainstem rostral and ventral to the roots of the other two trigeminal nerve branches.

The maxillomandibular ganglion is located at the level of the metencephalon, just lateral to the entrance of the trigeminal nerve roots into the brainstem. The cell bodies of the maxillary and mandibular branches from distinct swellings in the ganglion, which are clearly visible macroscopically. The maxillary portion is situated rostral to the mandibular portion. The cell masses of the two divisions are separated by a septum in *Python* (Molenaar, 1978a), except for a group of neurons in the most ventral portion of the ganglion. In *Agkistrodon blomhoffi* such a septum between the maxillary and mandibular division is not present (Kishida et al., 1982).

Descriptions of the internal structure of the trigeminal ganglia are scarce. Kishida et al. (1981, 1982) have studied the cell population and the localization of infrared neurons within the ganglia of the crotaline snake *Agkistrodon blomhoffi*. They concluded from their normal Nissl material that the ganglia contain the same light and dark cells as are reported from avian and mammalian trigeminal ganglia (Hamburger, 1961; Peach, 1972). The light cells exhibit clearly the tendency to be larger than the dark cells. The localization of the infrared neurons within the ganglion was established by Kishida et al. (1981, 1982) after application of horseradish peroxidase (HRP) to the pit membrane. The infrared cells are scattered throughout the ophthalmic ganglion and the maxillary division of the maxillomandibular ganglion. The infrared cells belong to the medium-sized class (20 to 39 μm), but it could not be established whether they are the dark or the light cells. The smallest dark cells, which made up 35.8% of the total population, were not labeled by HRP and are not considered to be involved in infrared perception. Similar small dark cells are considered to be involved in nociception and thermoreception in mammals (Gobel, 1974).

A similar study has been performed in *Python reticulatus* (Tan and Gopalakrishnakone, 1988) after HRP injections in various pits. The labeled cells are scattered throughout the various trigeminal ganglia, but it was not established whether they are the small dark cells. More than two-thirds of the perikarya in the ophthalmic ganglion and 60% of the cells in the maxillary part of the ganglion were positively labeled upon injection of all the maxillary pits. The second maxillary pit is innervated by cells from both the ophthalmic ganglion (3.7%) and the maxillary part of the maxillomandibular ganglion (7%). The mental pits are innervated by 30% and the mandibular pits by 25% of the cells in the ipsilateral mandibular part of the maxillomandibular ganglion.

D. Properties of the Receptor and of the Peripheral Units

A typical peripheral unit exhibits two types of firing patterns: a static and a dynamic response. The static response is evoked by (constant) heat influx from the background, and its frequency is in particular related to the ambient temperature. The static response has been established in all preparations in crotalines and boids (Bullock and Cowles, 1952; Bullock and Diecke, 1956; Bullock and Barrett, 1968; Goris and Nomoto, 1967; Warren and Proske, 1968; Hensel, 1974, 1975; Hensel and Schaefer, 1981; Goris and Terashima, 1976; Terashima and Goris, 1979; De Cock Buning et al., 1981a, 1981b). However, its frequency in *Morelia* and *Python* is less than in crotalines or *Boa,* and it may be absent in some fibers (Warren and Proske, 1968; De Cock Buning et al., 1981b). The static activity shows irregularities in frequency, which appear to be reduced when the receptor is covered with water (Hensel, 1975; De Cock Buning et al., 1981a). These fluctuations have been attributed to small temperature changes at the receptor membrane produced by air turbulence or other factors.

Detailed studies on the static response of peripheral units in *Agkistrodon blomhoffi* and *Boa constrictor* have been performed by application of flowing water at controlled temperatures (Figs. 5.4 and 5.5). These studies showed that this static response is directly related to the temperature of the water and thus to the temperature of the membrane. The static discharges were present only in a limited temperature range, the limits of which varied from one unit to the other. The up-

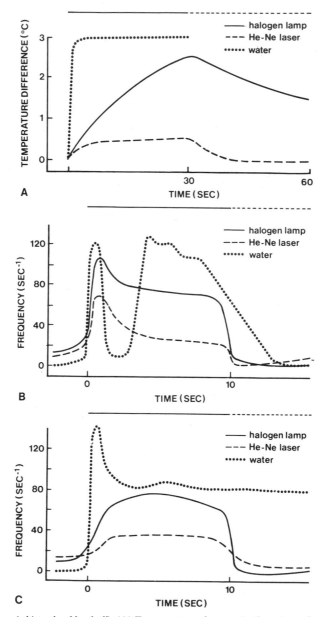

Fig. 5.4. *Agkistrodon blomhoffi*. (A) Temperature changes in the pit produced by three different heat sources: a halogen lamp (solid line) with a time constant for rise of 51 sec and for decline of 71 sec; a He-Ne laser (dash line), with a time constant for both rise and decline of 4 sec.; and flowing water (dotted line), with a time constant for rise of 0.25 sec. (B) Average discharge frequency curves during dynamic stimulation with the same stimuli as in (A) of an untreated pit organ. (C) Average discharge frequency curves during dynamic stimulation with the same stimuli as in (A) of a pit whose heat capacity was increased by filling it with 5 μl of water. The line at the top indicates the duration of the stimulation; the dashed part of the line indicates the continuation of the water stimulation. (After De Cock Buning et al., 1981a.)

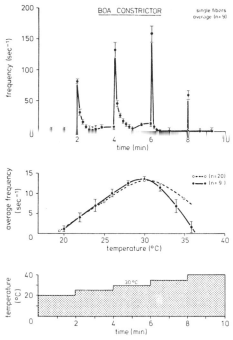

Fig. 5.5. *Boa constrictor.* Dynamic and static responses of peripheral trigeminal nerve fibers innervating the radiant heat receptors upon stepwise 5° C changes in temperature or at constant temperature levels. Top: Averaged responses of nine single fibers upon both dynamic and static stimuli. Middle: Averaged responses of two populations of fibers as a function of constant temperatures. Bottom: Stimulus characteristics of water flowing over the receptor. Bars indicate standard error of mean. (After Hensel, 1975.)

per limit is 29° to 33° C in *Agkistrodon* (De Cock Buning et al., 1981a) and 37° C in *Boa* (Hensel, 1975); the lower limit is 14° to 20° C in *Agkistrodon* and *Boa.* No discharges were observed above and below these limits. The static responses showed a maximum response at a temperature that varied from 23° to 28° C in *Agkistrodon* and around 30° C in *Boa.* The peak frequency of individual units also showed much variation (Fig. 5.5). In *Boa constrictor,* the shape of the static response curve (temperature versus frequency) appears to be related to the average temperature in which the individual snakes had grown (Hensel and Schaefer, 1981). The static responses to high temperatures shown by units in warm-adapted individuals are higher than those of the snakes grown in cold environments.

The dynamic response pattern shows a typical sequence of events, consisting of a latency of 30 to 100 msec at the onset of the stimulus, an initial burst, and a damped oscillation pattern to some adaptive frequency above or below the original static activity. Cessation of the stimulus evokes an afterdischarge, followed by depressed frequency or complete silence and finally a return to the static discharge pattern. Conversely, a cold stimulus evokes a depression of the static activity and a subsequent slow return to the static level. Removal of the cold object evokes a prominent burst of spikes. Each of these phases show irregularities and fluctuations, combined with a considerable variation in duration and intensity, even in the same preparation (Bullock and Cowles, 1952; Bullock and Diecke, 1956; Bullock and Barrett, 1968; Goris and Nomoto, 1967; Terashima et al., 1968; Terashima and Goris, 1979; Warren and Proske, 1968; Hartline, 1974; Hensel, 1974, 1975; Hensel and Schaefer, 1981; Goris and Terashima, 1976; De Cock Buning et al., 1981a, 1981b).

The wide variety of response characteristics has led several authors to assume the primary units lack consistent dynamics or that more than one type of receptor is involved (cf. Hartline, 1974; De Cock Buning et al., 1981a). However, it has been shown by De Cock Buning et al. (1980, 1981a) that these fluctuations can be mimicked in a single unit by varying the parameters of stimulation. They showed that different types of stimuli, as well as different stimulus intensities, evoke different heating characteristics of the pit membrane and thus different response patterns (Fig. 5.4).

A red light laser (He-Ne; 632.8 nm; 2 mW) causes a temperature rise at the membrane of about 0.6° C, which becomes stable in about 4 sec. It evokes a burst of discharges, which reaches a maximum in about 0.5 sec and declines to an adaptation level, slightly above the background level, within 1.0 sec. At offset (after 10 sec), a few afterdischarges were observed, followed by a depression or complete silence of background firing.

A high-intensity radiant source, a halogen lamp (100 V, 650 W) induces a continuous rise in temperature, about 1° C in 10 sec, which does not reach a steady state after 150 sec. It evokes a higher peak of discharges, followed by a fall to an irregular and slowly declining adaptation level, which is well above background level. Offset is followed by a stronger afterdischarge and a longer and more intense silent period than was evoked by the laser. The high and semistatic level of adaptation is attributed to the continuous rise of temperature of the pit membrane.

Flowing water applied to the membrane immediately raises membrane temperature, which becomes stable in 250 msec. Small stepped increases in temperature evoke responses that are comparable to those of the laser. Larger steps evoke the typical pattern of burst, silence, second burst, decline, whereas a steep temperature rise of 10° C evokes a complete silence after an initial burst. Apparently, the receptor is overstimulated. It appears from this study that the variations described in previous studies using similar stimulation methods may be attributed to the stimulus characteristics (De Cock Buning et al., 1980, 1981a).

Response characteristics have been studied in *Boa constrictor* upon successive heating steps of 5° C produced by flowing water within the temperature range of 20° to 40° C. The dynamic response to such stepwise warming consists of a peak in discharges, followed by a depression, after which a new level of static activity is reached (Fig. 5.5). The height of the peak frequency, as well as the level of static activity, appears to be related to the steady temperature. The receptors have their static maximum as well as their highest dynamic differential sensitivity at an optimal temperature. This optimum appears to be related to the temperature in which the snakes have been raised. The maximum dynamic response appears to be at 26° C in boas that were kept at 23° C for 4 years and at 32° C in boas that were raised at 30° C (Hensel and Schaefer, 1981).

De Cock Buning et al. (1981a) have analyzed crotaline pit organs as warm receptors. The authors increased the heat capacity of the pits by filling them with 5 µl of water (see Fig. 5.4). It appeared from their studies that the initial peak in frequency disappeared unless great temperature changes were made. They conclude that the phasic responses on transient stimuli are due to the extremely low heat capacity of the pit membrane.

The fibers innervating the receptors of pit vipers and *Boa* may be considered as specific warm fibers. They have temperature-dependent static discharges, responding with an increase in firing rate to heating and a decrease in firing rate to cooling. They are insensitive to tactile stimuli (Hensel, 1974, 1975; De Cock Buning et al., 1981a). In contrast, the receptors of *Python reticulatus* appear to be nonspecific warm receptors (De Cock Buning et al., 1981b). Their static discharges are much less frequent than in crotalines or may even be absent. Yet their static discharges, if present, show the same characteristics as in *Agkistrodon*. They appear only between 20° and 29° C and have a maximum near 27° C. They respond like warm receptors to an increase in temperature; however, they also respond to tactile stimuli. A response upon stimulation by means of a von Frey hair could only be obtained if the receptive area of the unit under concern was touched; touching of other receptive areas yielded no response. The threshold of the mechanical stimulus is rather high. Thus, the authors suggest these receptors may also be considered as mechanoreceptors that have become specialized to detect changes of temperature (De Cock Buning et al., 1981b). Heat-sensitive, slowly adapting mechanoreceptors with moderately high thresholds are also present in reptilian cutaneous nerves (Kenton et al., 1971).

An infrared unit is driven by a small, restricted area in the pit membrane—the receptive area. The cone that extends from this receptive area through the opening of the pit organ into space, and from which a unit can be stimulated, is called the receptive field (see Fig. 5.2). The size of receptive areas of primary units in the pit viper *Agkistrodon blomhoffi* was established by scanning a 40-μm laser spot over the sensory membrane while making recordings of units within the trigeminal ganglion (Terashima and Goris, 1979). These receptive areas appeared to be more or less rounded, with a diameter of approximately 20 to 50 μm. Other histological studies indicated the sizes of the receptor endings are 30 to 50 μm (Bullock and Fox, 1957), 40 μm (Terashima et al., 1970), and 50 to 100 μm (Hirosawa, 1980). In the last case, the nerve endings appear to overlap. From these findings it can be concluded there is only one terminal nerve ending per ganglion neuron. This contradicts the conclusion reached by De Salvo and Hartline (1978) that more than one receptive area is connected to a fiber. These authors constructed the theoretical response profile that would result if a stimulus were to move through the receptive field of a primary unit innervating a single 50-μm spot on the pit membrane and subsequently to move over the edges of the pit. This theoretical profile was compared with the profile obtained experimentally. The empirical profile appeared to differ in several respects from the theoretical profile, and the authors reached the conclusion that more than one receptive area was connected to a fiber. De Cock Buning (1984) constructed a theoretical profile based on his crotalid geometrical model and using the values that were presented by De Salvo and Hartline. The theoretical profiles of De Cock Buning and the empirical profiles of De Salvo and Hartline now appeared to fit into each other. Apparently, the discrepancy was caused by not taking into account the diameter of the stimulator nor the rounded edges, asymmetry, and concave floor of the pit.

The size of the receptive areas of primary units in *Python reticulatus* appears to be much larger than those of *Agkistrodon rhodostoma* (150 to 250 μm). As with *Agkistrodon*, each single peripheral unit innervates a single receptive area (De Cock Buning et al., 1981b). However, each receptive area has a bimodal sensitivity and responds to tactile stimuli as well.

Threshold sensitivity of the pit organs has been determined in various snakes. It is a function of the distance and temperature of the stimulus and has been measured using different techniques. Threshold sensitivity is defined as the minimum energy necessary to evoke a neural response. Bullock and Diecke (1956) have used recordings of peripheral units to determine threshold sensitivity in *Crotalus*; the threshold is 3.10^{-4} cal/cm².s. In *Boa constrictor*, the threshold is 1.8 ×

10^{-3} cal/cm².s as was determined after recordings of evoked potentials with permanently placed electrodes on the telencephalon (Harris and Gamow, 1972).

Recently, threshold sensitivity has been determined in three species (*Boa constrictor, Python reticulatus, Agkistrodon rhodostoma*) with the simplified formula of Stefan-Boltzman (De Cock Buning, 1983a), which describes the energy flux between two objects of different temperatures:

$$R = \frac{\delta.A.\ (T_2^4 - T_1^4)}{\pi\ D^2}$$

in which R = radiation density (cal/cm².s)
δ = constant of Stefan-Boltzman: 1.35 10 cal/cm².s
A = radiating area
T_2 = temperature (°K) of the heat exchanger
T_1 = temperature (°K) of the heat receiver
D = distance between exchanger and receiver

The radiation density thus depends on the temperature difference and the distance between the sensory membrane and the stimulus. The receptor is predominantly a detector of changes in density of radiation. Thus, the sensor detects the change in heat flux (ΔR). This can be demonstrated experimentally by opening a shutter between an infrared stimulator and the receptor and presenting it with a short flash of radiant heat. Thus:

$$\Delta R = \frac{\delta\ A}{\pi D^2} [(T^4\ \text{stimulus} - T^4\ \text{snake}) - (T^4\ \text{shutter} - T^4\ \text{snake})]$$

$$\Delta R = \frac{\delta\ A}{\pi D^2} (T^4\ \text{stimulus} - T^4\ \text{shutter})$$

The formula implies that the change of radiant heat influx on the sensory membrane is independent of the temperature of the sensors and only results from a change in the intensity of infrared radiation of the environment.

The distance at which only 50% of the stimuli evoked a neural response is denoted as *the critical distance*. The minimal change in density of radiation that can be detected at the level of the membrane was computed by substituting the critical distance in the Stefan-Boltzman formula. It appeared that the most sensitive units responded to changes in heat flux of 42.28×10^{-6} cal/cm².s (*Boa*), 14.28×10^{-6} (*Python*), and 2.57×10^{-6} (*Agkistrodon*).

Once the threshold has been established, this formula can be used to calculate the maximal distance at which a prey animal, such as a mouse or a rat, can be detected, provided the radiating area of the

Table 5.1. The critical radiation density in three species of infrared-sensitive snakes and the maximum distance at which a prey can be detected according to the lowest value of critical radiation density[a]

| Species | Critical Radiation Density (cal/cm².s) | | | Distance of Detection (cm) | | | |
| | Mean | Lowest Value | | Background Temperature | | | |
				25°C	20°C	15°C	10°C
Boa constrictor	48.96×10^{-6}	42.28×10^{-6}	Mouse	11.8	16.4	19.9	22.7
			Rat	28.3	39.6	47.9	54.6
Python reticulatus	19.52×10^{-6}	14.28×10^{-6}	Mouse	20.2	28.3	34.2	39.0
			Rat	48.8	68.1	82.4	93.9
Calloselasma	4.38×10^{-6}	2.57×10^{-6}	Mouse	47.7	66.6	80.6	91.9
rhodostoma			Rat	114.9	160.5	194.2	221.4

Source: After De Cock Buning et al., 1983a.
[a] For this, a mouse is estimated to have a surface temperature of 30° C and an emitting surface of 25 cm²; in the rat these values are 30° C and 145 cm², respectively.

prey and the temperature difference between this and the environment are known.

$$D_{max} \ (cm) \ = \ \sqrt{\frac{\delta \ A \ (T_2^4 \ - \ T_1^4)}{(\Delta R) \ \pi}}$$

Since the radiating area of a mouse of 30 g is estimated to be 25 cm², it can be calculated that a mouse that is 10° C warmer than its surroundings can be detected at a maximum distance of 16.4 cm by a *Boa*, at 28.3 cm by a *Python*, and at 66.6 cm by an *Agkistrodon* (see Table 5.1).

The threshold values, as determined by this method (De Cock Buning, 1983a), agree with those that can be calculated from the parameters presented by Noble and Schmidt (1937) for *Corallus enydris* (30×10^{-6} cal/cm².s). They also agree with those of the common vampire bat (*Desmodus rotundus*): 11.9×10^{-6} cal/cm².s (Kürten and Schmidt, 1982). The values are 50 times lower than those of Bullock and Diecke (1956) (3×10^{-4} cal/cm².s). According to De Cock Buning (1983a) the discrepancy between the earlier results and his may be attributed to differences in the technical procedure.

E. Geometry

Image formation usually requires an optical device such as a lens. Infrared waves, however, are filtered out by the substances that in nature form such optical devices. Therefore image formation in the pit organs is possible only by means of diffraction patterns around edges, as with a pinhole camera. The quality of the image formation in a pinhole camera is determined by the resolving power of the system.

The resolving power can be computed in three different ways using the equations of Fraunhofer or Fresnell or a geometrical model. De Cock Buning (1984) has studied the validity of the three models.

Fraunhofer's diffraction patterns appear inapplicable because of the large projection depth they require (40 cm at a wave length of 10 µm and a pit aperture of 2 mm). Fresnell's diffraction patterns seem equally inapplicable, since the necessary subtle intensity fluctuations of the image projections on the pit floor are lost in the case of natural stimuli.

A better approach is based on the geometrical model (Fig. 5.6), which was initially developed by Otto (1972) to describe the intensity profile that has been projected perpendicularly through a wide aperture onto a flat plane and that has been emitted by a source of given dimensions. This intensity profile consists of a central part having the same shape as the aperture. The central part receives the maximal radiation intensity and is surrounded by a peripheral area in which the intensity decreases radially. The points $R(i)$ indicate the borders of the center and the periphery. They are functions of the direct projection of the aperture edge toward the projection plane (PR), of the projection depth (D), and of the tangent of the angle $\tau(i)$:

$$R(i) = PR - D \tan \tau(i)$$

This basic equation has been adjusted and refined by De Cock Buning (1984), taking into account the size of the stimulus, the angle of incidence, the rounded edges of the pit walls, the asymmetry of the pit organ, and the concave form of the pit floor (Fig. 5.6). As a result, the author obtained different geometrical models for the various pit organs in boids and for the facial pit of crotalines. With these models, the intensity profiles on the pit floor can be computed and used to predict response profiles as a function of the angular position of the stimulator for each unit under study. Thus, such observations as reported in the literature can be better understood, and apparent discrepancies can be explained.

The differences in the shapes between theoretical and empirical response profiles, as obtained by De Salvo and Hartline (1978) for peripheral units (Section II.D), can be eliminated by constructing theoretical response profiles on the basis of the refined crotaline geometrical model and by also taking into account the size of the stimulus. Other aspects of these interesting models will be discussed in the appropriate sections.

F. Concluding Remarks

The radiant heat receptors in crotaline snakes are contained within a thin membrane suspended in a cavity. The receptors of most boid snakes are located at the bottom of shallow grooves in the labial

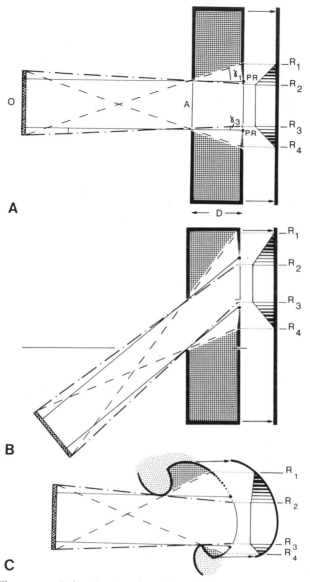

Fig. 5.6. The geometrical projection of an object (O) through the aperture (A) of a pinhole camera with a projection depth (D) gives an intensity profile on the projection plane consisting of a central portion of highest intensity (R2 to R3) and a surrounding halo (R2 to R1 and R3 to R4) of diminishing intensity. The intensity profile is indicated on the scale to the right of each illustration. The points PR indicate the projection of the edges of the aperture upon the floor. (A) Intensity profile of an object that projects perpendicularly upon the flat floor. (B) Intensity profile that projects under a certain angle of incidence upon the flat floor. (C) Intensity profile of an object that projects under a certain angle of incidence upon a concave floor. (After De Cock Buning, 1984.)

scales. In some species these grooves are missing entirely but without loss of infrared sensitivity. The smallest entity of the receptor consists of an aggregation of free nerve endings called the terminal nerve mass. It has a diameter of approximately 40 μm, which is equal to the size of the receptive area. Size and morphology of the terminal nerve masses are the same in boid and crotaline snakes. Special vesicles are associated with the receptor membrane, but their function is unknown.

Each terminal nerve mass is connected to a single peripheral nerve fiber. Other conclusions—that more than one terminal nerve mass should be connected to a single fiber—result from an inappropriate technical approach.

On physiological criteria, two types of infrared receptors can be recognized, depending on the species. The receptors in crotalines and in *Boa* are specific warm receptors, whereas those in *Python* are nonspecific bimodal infrared and tactile receptors. The receptors respond to constant ambient temperature with a static discharge and with a dynamic discharge upon a change in heat flux. Their maximal sensitivity is at a level that depends on the temperature at which the individual snakes have been raised. The disagreement in the literature concerning the nature of the responses is related to the differences in stimulation procedures.

Although the receptors may be regarded as warm receptors, their perikarya in the trigeminal ganglia are not similar to the small dark neurons that, in mammals, are involved in thermoreception. Such small dark neurons are present in the ganglia of infrared-sensitive snakes too, but they are not connected to the pit membrane.

The pit organ functions as a pinhole camera, and its geometry determines its resolving power. Furthermore, the threshold sensitivities of the receptors in *Boa*, *Python*, and *Agkistrodon* are 42.28, 14.28, and 2.57×10^{-6} cal/cm^2.s, respectively. Thus, a mouse that is 10° C warmer than the environment can be detected by the three snakes at distances of 16.4, 28.3, and 66.6 cm, respectively.

The pit organs of crotaline snakes are most specialized and contain specific warm receptors, which are the most sensitive. The pits of *Python* are less refined and contain bimodal nonspecific receptors of intermediate sensitivity. In *Boa* no pits are present; the receptors are specific warm receptors but have the highest threshold.

III. THE SENSORY TRIGEMINAL SYSTEM
A. General

1. INTRODUCTION Upon entering the brainstem, the sensory trigeminal nerve fibers segregate into two distinct fiber bundles that contribute to either of two components of the sensory trigeminal system.

One component is equivalent to the sensory trigeminal system of other vertebrates and is called the common sensory trigeminal system. The second component, the lateral descending system, lies lateral to the former (Fig. 5.7) and consists of a large descending tract that terminates in a huge distinct nucleus. It is the infrared conveying system and is absent from snakes without infrared sensitivity (Molenaar, 1974; Schroeder and Loop, 1976).

The two components of the sensory trigeminal system were first described by Molenaar (1974, 1978a, 1978b) for *Python* and *Crotalus* and were based on normal Nissl-stained material and degeneration studies. The data have been confirmed by degeneration studies for several other species (Schroeder and Loop, 1976) and by use of transganglionic HRP transport for *Agkistrodon* (Kishida et al., 1982). The involvement of the lateral descending system in infrared perception has been demonstrated by Schroeder and Loop (1976), Newman et al. (1980), and Kishida et al. (1982) with transganglionic transport techniques; by Molenaar (1978b, 1978c) and Molenaar et al. (1979) with electrophysiological techniques, and by Auker et al. (1983) and Gruberg et al. (1984) with the deoxyglucose labeling technique.

The lateral descending and the common descending fibers enter the medulla in a mixed order, but they become completely separated as they turn caudally to descend through the medulla. The borderline between the two fiber systems is sharp and can be recognized on morphological criteria. In normal material, the lateral descending tract is distinguished from the common descending tract by the denser packing and finer caliber of its fibers. The difference between the fibers of the two components is even more striking in degenerating Fink-Heimer (1967) stained material (Fig. 5.8). The fibers of the lateral descending tract are more densely packed and much more fragmented and disintegrated than those of the common descending tract. The terminal boutons in the lateral descending system are much coarser than those of the common sensory system. These boutons are also visible after transganglionic HRP transport, when the terminal boutons in the lateral descending system are mostly 4 to 6 μm in diameter and tend to form clusters of 25 to 30 μm, whereas the terminal boutons of the common trigeminal system are only 0.9 to 2.8 μm in diameter (Kishida et al., 1983b).

2. TOPICAL RELATIONS OF THE PRIMARY SENSORY TRIGEMINAL NERVE FIBERS

In *Python*, each of the three trigeminal nerve roots projects into its own territory within the various nuclei of the two trigeminal systems. The ophthalmic territory is most ventral, the mandibular most dorsal, and the maxillary intermediate. The arrangement shifts into a latero-

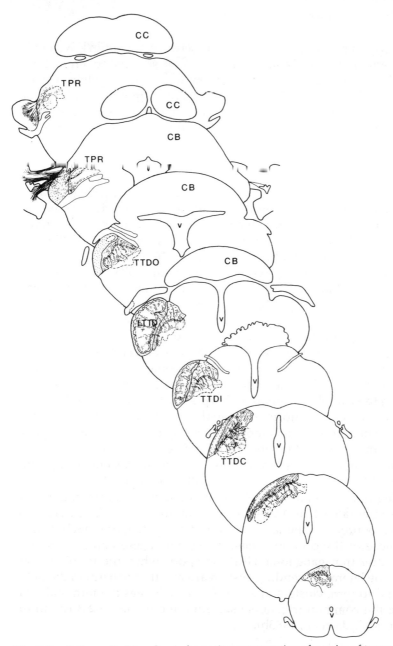

Fig. 5.7. *Python reticulatus.* Semischematic representation of a series of transverse sections through the rhombencephalon and rostral part of the cervical cord to show the topography and arrangement of the two components of the trigeminal sensory system. CB, cerebellum; CC, colliculus caudalis; LTTD, nucleus descendens lateralis n. trigemini; TTDC, nucleus caudalis n. trigemini; TTDI, nucleus interpolaris n. trigemini; TTDO, nucleus oralis n. trigemini; TPR, nucleus sensorius principalis n. trigemini; v, ventricle.

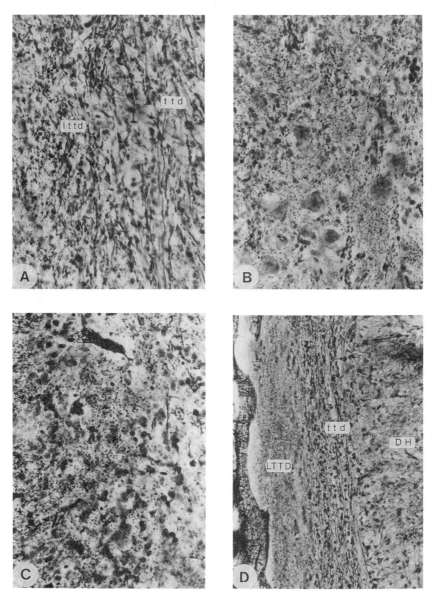

Fig. 5.8. *Python reticulatus.* Photomicrographs showing the appearances of degenerated fibers and terminal boutons in Fink-Heimer (1967) stained sections within the common sensory system and the lateral descending system. Horizontal sections. (A) At the level of the rostral border of the lateral descending nucleus. (B) Corpus portion of the lateral descending nucleus. (C) Transitional area between corpus and cauda of the lateral descending nucleus. (D) Cauda of the lateral descending nucleus and the dorsal horn within the cervical cord. DH, dorsal horn; lttd, lateral descending tract; LTTD, nucleus of the lateral descending tract; ttd, common descending tract. Magnifications: A to C, 400x; D, 250x.

medial one in the cervical cord. A similar arrangement is known from other vertebrates (Van Valkenburg, 1911), various snakes (Schroeder and Loop, 1976), pigeons (Dubbeldam and Karten, 1978), rats (Torvik, 1956; Rustioni et al., 1971), and cats (Kerr, 1963).

The individual territories within the lateral descending nucleus are more circumscribed than those within the common nuclei. Although the fibers of each of the three peripheral branches project into their own territories within the nuclei, a considerable overlap is present between those of the maxillary and the mandibular fibers. This overlap is quite conspicuous in each of the two sensory trigeminal systems. It is, however, most striking in the lateral descending trigeminal system. Here, both the mandibular and maxillary territories are the size of one-half of the total dorsoventral diameter of the lateral descending nucleus but largely overlap one another.

Such overlap has not been described in the literature, although several papers contain indications that such an overlap also exists in mammals (Kruger and Michel, 1962; Darian-Smith et al., 1963; Kerr, 1963; Nord, 1967; Kerr et al., 1968; Woudenberg, 1970). Molenaar (1978b) discusses the phenomenon in more detail.

B. The Common Sensory Trigeminal System

Although the common sensory system is not involved in infrared perception, a brief description of it will permit comparison of the two trigeminal sensory components. A detailed description of the system is presented elsewhere (Molenaar, 1978a, 1978b).

The common sensory system of *Python reticulatus* closely resembles the sensory trigeminal system of other vertebrates (Van Valkenburg, 1911; Woodburne, 1936; Luiten, 1975, carp; Molenaar and Dubbeldam, 1969, mallard; Dubbeldam and Karten, 1978, pigeon; Torvik, 1956, mammals; Clarke and Bowsher, 1962, mammals). Only minor differences are present, which can be attributed to interspecific variations.

The nuclei of the common sensory trigeminal system form a continuous column of gray matter, located at the medial aspect of the common sensory trigeminal fibers. The column can be subdivided into a number of nuclei and subnuclei on the basis of their morphology and cell configuration. The nuclei are distinguished in accordance with the nomenclature proposed by Olszewski (1950) for the mammalian sensory trigeminal nuclei. The column consists of the rostrally located principal sensory trigeminal nucleus and the nucleus of the descending trigeminal tract. The nucleus of the descending tract is further subdivided into the nucleus oralis, which borders the principal sensory nucleus; the nucleus interpolaris; and the nucleus caudalis, which passes into the dorsal horn of the cervical cord. The principal sensory trigeminal nucleus (TPR, see Fig. 5.7) is situated medially and slightly rostrally to the entering trigeminal fibers and extends caudally to the trigeminal motor root. The nucleus contains small and some medium-sized multipolar cells. The larger cells tend to be concentrated in its ventral part.

The nucleus oralis (TTDO, see Fig. 5.7) contains characteristically large neurons, with pale cytoplasm, large Nissl bodies, and a large nucleus with a clear nucleolus. The nucleus contains a dorsal and a ventral part, which can be distinguished on the basis of cell configuration and cell size. The cells in the ventral part are somewhat larger and more densely packed than in the dorsal part. The nucleus oralis passes into the nucleus interpolaris (TTDI, see Fig. 5.7) at the level of the facial motor nucleus. The large cells of the nucleus oralis are absent, but a dorsal and a ventral part are recognizable. In the dorsal part, the small and medium-sized cells are uniformly distributed; the cells in the ventral part are packed closely together. The ventral part extends for a short distance below the nucleus caudalis. The transition between the nucleus interpolaris and the nucleus caudalis (TTDC, see Fig. 5.7) is at the level of the vagal nerve roots. The nucleus caudalis consists of three layers: the subnuclei zonalis, gelatinosus, and magnocellularis.

The subnucleus zonalis is a thin layer of only one or two cells thickness underlying the descending trigeminal tract. It contains small, darkly staining cells; a few are medium sized.

The subnucleus gelatinosus (or substantia gelatinosa layer) is situated more centrally. It contains small and medium-sized multipolar cells with pale cytoplasm.

The subnucleus magnocellularis is situated most centrally. It consists of small and medium-sized cells, which stain somewhat darker and are more closely packed than those of the subnucleus gelatinosus.

The nucleus caudalis passes into the dorsal horn of the cervical cord (DH, see Fig. 5.7). Although essentially the same three layers are present within both nuclei, neither a direct transition nor a clear border between these layers of the two parts could be observed.

The cells within the subnuclei gelatinosus and magnocellularis of both the nucleus caudalis and the dorsal horn are arranged in a peculiar fashion. They are concentrated, like garlic on a straw cord, around bundles of primary sensory trigeminal fibers, which have left the descending trigeminal tract. The arrangement is similar to the trigeminal glomeruli described by Gobel (1974) for the cat.

The nucleus interstitialis of Cajal (1909) is composed of small patches of gray substance scattered between the fibers of the common descending trigeminal tract. These patches consist mainly of neuropil in which neurons are sparse. The neurons are usually small and rounded or ellipsoidal, but a few are larger and multipolar. The patches are most prominent in two places—at the level of the rostral and caudal borders of the nucleus interpolaris. The gray matter of the nucleus interstitialis appears continuous with the substantia gelantinosa layer of the nucleus caudalis. This finding agrees with similar observations of Gobel and Purvis (1972) in the cat.

The fibers that belong to the common sensory system are distributed in much the same way as known from other vertebrates (Torvik, 1956; Kerr, 1961; Clarke and Bowsher, 1962; Rhoton et al., 1966; Molenaar and Dubbeldam, 1969; Fuller and Ebbesson, 1973; Luiten, 1975; Dubbeldam and Karten, 1978). They run to the principal sensory nucleus or descend in the descending tract and terminate within the accompanying nucleus and the nucleus interstitialis. The fibers from the ophthalmic root terminate within the first segment of the cervical cord; the maxillary and mandibular fibers continue caudally until halfway between the roots of the second and third cervical nerves. No fibers were observed to run to the lateral descending nucleus. The terminal debris was confined to the more dorsolateral parts of the nuclei oralis and interpolaris, where the cells are more widely scattered. The ventromedial area, with the densely packed neurons, received few if any direct trigeminal fibers.

Direct projections of primary trigeminal fibers to extratrigeminal regions were only observed to run to the nuclei of the vagosolitarius complex and to the viscerosensory column of the cervical cord. Only a very limited number of fibers is involved in this projection. Direct trigeminal fibers to the cerebellum (Larsell, 1926; Weston, 1936; Woodburne, 1936) or to motor nuclei in the brainstem (Clarke and Bowsher, 1962, rat) were not observed (Schroeder and Loop, 1976; Molenaar, 1978b).

C. The Lateral Descending Trigeminal System

1. THE LATERAL DESCENDING TRACT   The lateral descending trigeminal tract (lttd) is built up by fibers from those sensory branches of the trigeminal nerve that innervate the pit organs. The fibers are densely packed and have a smaller diameter than those of the common descending trigeminal tract. The fibers of the lateral descending tract are either thin myelinated Aδ fibers (1 to 5 μm) or thin unmyelinated C fibers (0.1 to 2.0 μm) (Meszler et al., 1981; Kishida et al., 1982). The C fibers are not primarily trigeminal fibers (Meszler et al., 1981) but primary afferent fibers from the vagus nerve (Kishida et al., 1983a, 1983b, 1984a). A few C fibers have been observed in the pit membrane

of *Trimeresurus flavoviridis*. These fibers were quite scarce and do not seem to be connected to the infrared-sensitive terminal nerve mass (Hirosawa, 1980).

The fibers of the lateral descending tract terminate in the homonymous nucleus. In *Python*, the majority of the fibers enters the nucleus through its rostral pole. Numerous small bundles of fibers traverse the nucleus in the longitudinal direction and run toward more caudal destinations. Many fibers pass over the lateral aspect of the nucleus and enter it from the lateral side. Only a few fibers pass over the other aspects of the lateral descending nucleus to more caudal destinations.

The passage of the primary lateral descending tract fibers over the lateral descending nucleus varies depending on the species. In *Crotalus horridus, C. ruber, Agkistrodon piscivorus* (Meszler et al., 1981), and *Agkistrodon blomhoffi* (Kishida et al., 1982) the fibers pass over the lateral aspect of the nucleus. In the rattlesnake *C. viridis*, on the other hand, the fibers pass over all sides of the nucleus, whereas in *Boa constrictor* and *Corallus enydris* the fibers pass along the medial aspect (Schroeder and Loop, 1976).

2. The Lateral Descending Nucleus

Although the lateral descending nucleus (LTTD) is present in all infrared-sensitive snakes studied hitherto, its localization and cell configuration show interspecific differences. In *Python reticulatus*, the rostral border of the nucleus is at the level of the caudal pole of the nucleus oralis (Figs. 5.7 and 5.9). The nucleus is composed of a huge corpus portion and a narrow cauda portion. The corpus portion protrudes from the lateral surface of the medulla oblongata. Proceeding caudally, the corpus tapers gradually. It passes into the narrow cauda portion at the level of the vagus roots. The cauda extends into the cervical cord. The neuropil of the cauda portion merges with that of the substantia gelatinosa layer of the dorsal horn over the lateral side of the spinal trigeminal tract.

In other boid snakes, the LTTD occupies essentially the same position (Schroeder and Loop, 1976, *Boa constrictor, Corallus enydris;* Molenaar, unpublished, *Python molurus, P. regius*), but in several crotaline snakes it is situated more caudally (Molenaar, 1974, *Crotalus cerastes, C. atrox;* Schroeder and Loop, 1976, *C. viridis;* Kishida et al., 1982, *Agkistrodon blomhoffi*).

One report, however, describes LTTD localization in three crotaline species as similar to that in boid snakes (Auen, 1978, *Crotalus horridus, C. scutulatus, C. ruber*). Remarkably enough, these snakes possess ascending projections from the LTTD that differ from those in other crotalines but that resemble those of *Python*.

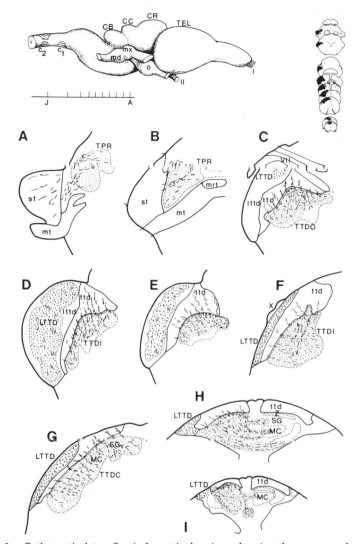

Fig. 5.9. *Python reticulatus.* Semischematic drawings showing the course and termination patterns of the trigeminal primary sensory fibers within the two components of the trigeminal sensory system. The degenerated fibers that leave the tracts are indicated by short lines; the degenerated terminal boutons by dots (fine dots within the nuclei of the common sensory system; coarse dots within the lateral descending nucleus). The inset (upper right) shows the sections on which these drawings are based. The brain with the position of the sections is depicted upper left. c_1, first cervical nerve root; c_2, second cervical nerve root; CB, cerebellum; CC, colliculus caudalis; CR, colliculus rostralis; lttd, tractus descendens lateralis n. trigemini; LTTD, nucleus descendens lateralis n. trigemini; MC, subnucleus magnocellularis; md, ganglion mandibulare; mrt, radix mesencephalicus n. trigemini; mt, radix motoria n. trigemini; mx, ganglion maxillare; o, ganglion ophthalmicus; SG, subnucleus gelatinosa; st, radix sensoria n. trigemini; TEL, telencephalon; ttd, tractus descendens n. trigemini; TTDC, nucleus caudalis; TTDI, nucleus interpolaris; TTDO, nucleus oralis; TPR, nucleus sensorius principalis; Z, subnucleus zonalis.

3. CYTOARCHITECTURE OF THE LATERAL DESCENDING NUCLEUS (FIG. 5.10)

The LTTDs of *Crotalus atrox* and *Python reticulatus* are similar in appearance. Corpus and cauda portions can be distinguished in both snakes, although the cauda in *Python* is more clearly demarcated than in *Crotalus*.

Most neurons in the LTTD are small (5 to 10 μm) to medium sized. They are rounded, and some have a faintly granulated cytoplasm. Large (up to 27 μm) multipolar cells are characteristic for the corpus portion of both snakes (Molenaar, 1974, 1978a, 1978b; Newman et al., 1980; Meszler et al., 1981; Kishida et al., 1982). The cells have several prominent dendritic stems, a clearly granulated cytoplasm, and a large nucleus. These cells are equal in size to the neurons of the vagus motor nucleus. The corpus portion of the LTTD passes into the cauda portion at the level of the vagal roots in *Python*. The large cells are absent from this transitional area, but medium-sized and small cells are still present.

The cauda consists of a narrow dorsal extension that proceeds from this area into the cervical cord. It forms a more or less triangular area in cross-section, which is located at the dorsolateral aspect of the common descending tract. It is present both in *Python* and *Crotalus* and proceeds farther caudally than do the primary trigeminal fibers. This area is called "the ventral trigeminal field" by Schroeder and Loop (1976) in their study of the homology of a similar area in the bullfrog (Fuller and Ebbesson, 1973). In contrast to the cauda portion in *Python*, this area in *Crotalus* is not continuous with the corpus of the LTTD. It is present not only in the infrared-sensitive species used by Schroeder and Loop (1976) (i.e., *Crotalus viridis, Boa constrictor, Corallus enydris*) but also in a non-infrared-sensitive snake (*Coluber constrictor*). It consists of a neuropil in which only a few small cells and fine fibers are present. Although the area receives primary trigeminal fibers, it is apparently not directly connected to the pit membrane, since tracers that were injected in the pit membrane and transported transganglionically were not found in this area in *Crotalus* and *Agkistrodon* (Schroeder and Loop 1976; Kishida et al., 1982).

The cauda portion within the cervical cord of *Python* receives primary trigeminal fibers in a different manner than do the more rostral portions of the LTTD. The degenerated trigeminal fibers reach the area over the dorsal and lateral aspects of the LTTD. Most of the fibers are thinner than the other fibers in the lateral descending tract, whereas only a few are somewhat coarser. The fibers are fragmented in the typical way of the lateral descending tract fibers, and their terminal grains are consistently coarse. However, the dense meshwork of preterminal debris, such as is present in the more rostral parts of

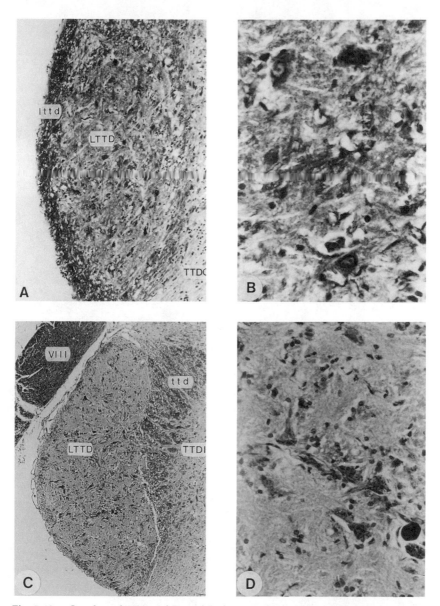

Fig. 5.10. *Crotalus ruber* (A and B) and *Python reticulatus* (C to H). Photomicrographs
showing the cytoarchitecture of the lateral descending trigeminal nucleus. (A and B)
Corpus. (C and D) Corpus portion. (E and F) Transitional area between corpus and
cauda. (G) Cauda within the first cervical segment. (H) Positively labeled large neurons within the corpus portion of LTTD after HRP injection in the tectum. DH, dorsal
horn of the cervical cord; lttd, lateral descending trigeminal tract; LTTD, nucleus of
the lateral descending trigeminal tract; ttd, common descending trigeminal tract;
TTDC, nucleus caudalis; TTDI, nucleus interpolaris; VIII, root of the vestibulocochlear
nerve. Stains: A to G, Klüver-Barrera; H, DAB. Magnifications: (A, C, E, 63x; B, D, F,
G, 400x; H, 250x). See p. 396 for part E–H.

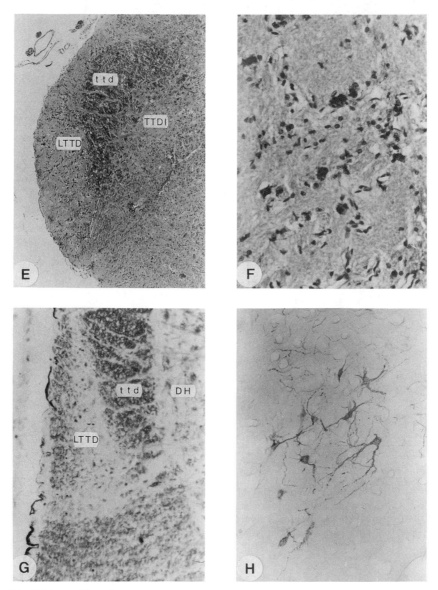

Fig. 5.10 continued

the LTTD, is absent here (see Fig. 5.8). The area projects to the corpus portion of the LTTD through an intranuclear ascending projection (Molenaar et al., 1980).

The transitional area receives primary afferents from the lateral descending trigeminal tract in much the same way as does the corpus portion; distinct bundles of fibers enter this part rostrally or laterally and form a dense preterminal meshwork of fibers. In addition, the

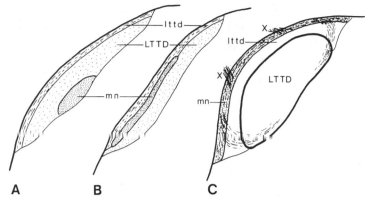

A   B   C

Fig. 5.11. Schematic drawings of transverse sections through the lateral descending nucleus showing the distribution pattern of transganglionically HRP-labeled vagus fibers. (A) *Boa constrictor* transitional zone between corpus and cauda portion. (B) *Python molurus* transitional zone. (C) *Agkistrodon blomhoffi* corpus. lttd, tractus descendens lateralis; LTTD, nucleus descendens lateralis; mn, marginal neuropil; X, radix n. vagi. (After Kishida et al., 1983a and 1983b.)

neuropil of the transitional area in *Python molurus* and *Boa constrictor* receives primary afferent vagus fibers (Fig. 5.11) and thus resembles the marginal neuropil of pit viper LTTD (Kishida et al., 1983a, 1983b, 1984a). The vagus fibers are thin and seem to correspond to C fibers; the trigeminal fibers are coarser, Aδ fibers.

The elucidation of the structure and function of this area awaits further detailed study.

4. SYNAPTIC ORGANIZATION OF LTTD

The synaptic organization of the lateral descending nucleus has been studied in four species of pit vipers (*Agkistrodon piscivorus, Crotalus horridus, C. ruber, C. viridis*) (Meszler et al., 1981) and three species of boids (*Python reticulatus, Corallus enydris, Epicrates striatus*) (Meszler, 1981, 1983) by using conventional, degeneration, Golgi, and electron-microscopical techniques.

The fibers within the lateral descending tract of both pit vipers and boids are either thin, myelinated Aδ fibers, 1 to 5 μm in diameter, or thin, unmyelinated C fibers, 0.2 to 2.0 μm in diameter. The Aδ fibers appeared to be the primary afferent trigeminal fibers (Meszler, 1983), whereas the C fibers are primary afferent vagus fibers (Kishida et al., 1983a, 1983b, 1984a).

The nucleus LTTD is composed of a main neuropil and a marginal neuropil. The fibers of both types terminate in the LTTD; the Aδ fibers in the main neuropil, the C fibers in the marginal neuropil (Meszler, 1981, 1983; Meszler et al., 1981; Kishida et al., 1983a, 1983b, 1984a).

The main neuropil is characterized by the presence of the large multipolar cells, but small and medium-sized cells are also present. The large multipolar cells are absent from the marginal neuropil.

The main neuropil contains the terminals of the primary afferent axons, which terminate on medium or small dendrites in boids (Meszler, 1983) or on the dendrites of the larger cells in pit vipers (Meszler et al., 1981). The synapses are arranged in what appear to be simplified synaptic glomeruli with the afferent fibers in the center. The terminals contain clear spherical vesicles. A single axon may terminate on several dendrites, whereas single dendrites receive terminals of several axons in both pit vipers and boids. Such connections may provide the anatomical basis for the diverging and converging connections that have been postulated on the basis of electrophysiological results concerning the sizes of receptive areas and receptive fields (Terashima and Goris, 1976, 1977, 1979; Stanford and Hartline, 1980). In turn, the primary afferent axons receive presynaptic axon terminals. The terminals come from thin, unmyelinated fibers that make multiple synaptic contacts. These terminals contain pleomorphic vesicles, glycogen, and a few large, dense core vesicles.

Synaptic arrangements of this type suggest an excitatory relay and possible presynaptic modification through the axoaxonic synapses. These axoaxonic synapses make up about one-third (*Python* 27%, *Agkistrodon* 32%, *Crotalus* 30%) of the synapses within the LTTD. Such synapses form only 1% in the substantia gelatinosa of the cat spinal cord (Duncan and Morales, 1978), which suggests that the presynaptic modification within the LTTD is considerable. Such inhibition is shown to play an important role in spatial sharpening of the infrared image for each receptive field. About one-third of the LTTD units in *Crotalus viridis* appear to possess inhibitory characteristics (Stanford and Hartline, 1980).

The origin of the axoaxonic terminals is unknown, although a few indications suggest an intranuclear origin, presumably from the small cells. Such intranuclear projections have been demonstrated in *Python* LTTD after partial lesions within this nucleus (Molenaar and Fizaan-Oostveen, 1980). Additional evidence comes from preliminary data from Golgi-stained pit viper LTTD and from the coincidence of large, dense core vesicles within both the perikaryon of small neurons and some of these terminals (Meszler et al., 1981; Meszler, 1983).

Horseradish peroxidase injections in the target sites of LTTD projections showed that only the large cells were stained (Newman et al., 1980; Welker et al., 1983; Molenaar, 1983a, 1983b). These indications suggest that intranuclear projections may arise from the small cells in LTTD.

The marginal neuropil consists of numerous C fibers that make synaptic contact with small dendritic processes, which seem to come from the marginal neurons. These neurons are small to medium-sized cells. They send their processes into the marginal neuropil as well as into the main neuropil and thus appear to take part in the intranuclear projections. The terminals contain densely packed clear spherical vesicles as well as a few large dense core vesicles. The C fibers are not primary afferent trigeminal fibers (Meszler et al., 1981; Meszler, 1983) but are primary afferent vagal fibers. Such vagus fibers terminate in the marginal neuropil of the LTTD in *Python molurus, Boa constrictor,* and *Agkistrodon* (Kishida et al., 1983a, 1983b). The peripheral innervation of the C fibers is unknown. It is remarkable, though, that C fibers are present in the sensory membrane of pit vipers (Hirosawa, 1980). The C fibers in the pit membrane, however, do not show immunoreactivity to substance P, whereas those in the marginal neuropil are immunopositive (Terashima, 1987).

The large multipolar cells in python LTTD form the efferent projections from this nucleus to the tectum (Newman et al., 1980; Meszler et al., 1981; Meszler, 1983; Molenaar, 1983a, 1983b), whereas in pit vipers, these large tectipetal cells are not localized within the LTTD but form a separate nucleus instead—the nucleus reticularis caloris. In *Python,* these large multipolar cells receive two types of synaptic contacts on both their dendrites and soma. None of these contacts is derived from primary fibers, since they do not degenerate following lesions of the ganglion. The most common terminal contains clear spherical vesicles and has a prominent subsynaptic web at its postsynaptic side. The other type of terminal contains ellipsoid or flattened vesicles. Similar terminals are present on the tectipetal cells of pit viper nucleus reticularis caloris, indicating the close resemblance between the tectipetal cells of boid LTTD and pit viper nucleus reticularis caloris.

5. IMMUNOCYTOARCHITECTURE OF LTTD

The immunocytoarchitecture of the LTTD has been studied in pit vipers only: in *Trimeresurus flavoviridis* (Terashima, 1987) and *Agkistrodon blomhoffi* (Kadota et al., 1984, 1988; Kusunoki et al., 1987). Substance P (SP), serotonin (SER), succinate dehydrogenase (SDH), acetylcholine esterase (AChE), and monoamine oxidase (MAO) are differently distributed over the marginal and main neuropil of LTTD, indicating different functions for the two subdivisions of this nucleus.

The marginal neuropil, which is mainly composed of unmyelinated C fibers from the vagus nerve, stains heavily for SP (Terashima, 1987; Kadota et al., 1988), moderately for AChE and MAO, and weakly for SDH (Kusunoki et al., 1987) or SER (Kadota et al., 1984). The main neuropil, in which the trigeminal fibers from the pit organs terminate, stains heavily for SDH, indicating strong aerobic metabolism, and only weakly for the other enzymes. Thus, it appears that substance P is not used as the primary neurotransmitter of the infrared system, although 30% of the perikarya in the trigeminal ganglia are immuno-positive to SP (Kadota et al., 1988). These perikarya probably project to the common sensory trigeminal system in which the substance P distribution is similar to that in mammals. Substance P is used as a modulator, though, because of the extension of the dendrites from the cells within the main neuropil into the marginal neuropil (Meszler et al., 1981, 1983).

The perikarya of the nucleus reticularis caloris show high SDH activity and weak AChE activity, whereas in the neuropil no significant SDH nor MAO activity occurs (Kusunoki et al., 1987), and SP fibers (Terashima, 1987) are present only occasionally.

6. Properties of LTTD Units

All LTTD units responded to radiant heat stimulation (Terashima and Goris, 1977; Molenaar et al., 1979; Stanford and Hartline, 1980). LTTD units in *Python* seem to respond to tactile stimuli, too. Molenaar (1978c) and Molenaar et al. (1979) mention such a response, although sometimes spikes were recorded before the pit was actually touched. The authors concluded that the response was evoked by a change in heat influx during the mere approach of the pit organ, rather than by tactile stimulus. Later studies in peripheral units demonstrated the bimodal sensitivity of these units to both infrared and moderately high threshold tactile stimuli (De Cock Buning et al., 1981b).

LTTD units show irregularities in static activity, which seem similar to those of peripheral units. The irregularities have been attributed to small temperature changes at the receptor membrane produced by air turbulence or other factors (Hensel, 1975; De Cock Buning et al., 1981a). In the pit vipers *Agkistrodon blomhoffi* (Terashima and Goris, 1977) and *Crotalus viridis* (Stanford and Hartline, 1980) all units are statically active. In *Python reticulatus* the background activity is less than in crotalines and sometimes may even be absent for prolonged periods of time (Molenaar et al., 1979). Analysis of interspike intervals in crotaline snakes under various constant temperature conditions revealed that those of LTTD units do not fit within a normal distribution pattern, whereas those of peripheral units do fit (Terashima and Goris, 1983). This suggests that the interspike intervals that are generated at the spike initiation site of the peripheral units are constant, but that these constant intervals are not used for information processing of steady-state temperatures at the LTTD level. The infrared sensory system functions as a detector of transient temperatures.

The dynamic responses to onset and offset of the stimulus show fluctuations and variations comparable to those of peripheral units. In these units, such variations are affected by the parameters of the stimulus and may be reduced by experimentally increasing the heat capacity of the pit (De Cock Buning et al., 1981a). In the living animal, however, the pit functions by having its own heat capacity, which produces the irregularities in static and dynamic responses. The higher-order neurons reduce these fluctuations. The frequency and irregularities in background activity, the response to a stimulus, and the characteristics of adaptation are intermediate between those of primary and tectal units (Terashima and Goris, 1977). In addition, spatial sharpening (Stanford and Hartline, 1980, 1984) and motion sensitivity—but not directional sensitivity (Terashima and Goris, 1977)—have been observed in LTTD units.

The receptive areas in the pit membrane of *Agkistrodon blomhoffi* are oval; their main axes are not aligned in any particular direction (Terashima and Goris, 1977). Their diameters are 480 to 800 μm (average 640 μm) in the long axis and 360 to 520 μm (average 470 μm) in the short axis. Thus, they are composed of 80 to 400 peripheral receptive areas (Terashima and Goris, 1979).

A theoretical response profile of a LTTD unit can be computed with the crotaline geometrical model of De Cock Buning (1984), taking into account the size of the receptive area, the characteristics of the pit geometry, and the size of stimulus. Such a response profile would be wider at the base and shorter at the top than the response profile of a peripheral unit, because the receptive area of the LTTD unit is larger. However, Stanford and Hartline (1980) found response profiles that were smaller than those expected on the basis of this geometrical model. The authors attributed this phenomenon to the presence of inhibitory interactions between neurons of the LTTD.

Another possible explanation is that the actual transference ratio is less than one; that is, the threshold of LTTD neurons is higher than one spike of a peripheral unit. In that case, only the upper part of the theoretical response profile rises above the threshold value and, consequently, the actual response profile will be smaller (De Cock Buning, 1984).

The combination of a high convergence (ratio K between 80 and 400 for LTTD units; Terashima and Goris, 1979) and a low transference (ratio n between 1/400 and 1/1600, De Cock Buning, 1984) causes a net result smaller than 1: $K \times n = 0.2$. This implies a decrease instead of an increase in sensitivity, in contrast to the situation in other converging neural mechanisms (Van Drongelen et al., 1978). On the other hand, this combination causes a gain in angular resolution due to the smaller top of the intensity profile (De Cock Buning, 1984) and thus enables the snake to locate more precisely the position of its prey.

All receptive areas exhibit excitatory characteristics; no inhibitory receptive areas occur. In contrast, the receptive fields (that is, the cones extending from a receptive area into space, from which stimuli can be presented to a unit) in *Crotalus viridis*, contain both excitatory and inhibitory areas (Stanford and Hartline, 1980, 1984). The excitatory receptive field of LTTD units in these animals is ellipsoidal and has a vertically oriented major axis. The diameter is 16° to 46° (average 30°) in the long axis and 9° to 41° (average 25°) in the short axis. These diameters are smaller than those of primary units (about 65°).

The inhibitory part of the receptive field is not organized in a concentric center-surround fashion, as reported earlier by Hartline (1974), but is usually located on the lateral side of the excitatory field and, occasionally, on the medial side as well. Introduction of an infrared stimulus within the inhibitory field suppresses the static activity. In addition, the excitatory burst evoked by a stimulus within the excitatory field is inhibited by the introduction of a second infrared stimulus into the inhibitory field (Stanford and Hartline, 1980, 1984).

The inhibition is probably the result of inhibitory connections within the LTTD. Inhibitory receptive areas were not present in the pit membrane of *Agkistrodon* (Terashima and Goris, 1977), and inhibitory receptive fields could not be found in primary units of *Crotalus* (De Salvo and Hartline, 1978). The synaptic organization of LTTD in *Python* and *Crotalus* shows the characteristics necessary for inhibitory activity (Meszler et al., 1981; Meszler, 1983).

Such inhibitory activity may account for the small receptive fields of LTTD neurons. These receptive fields are about half the size of those of primary units and approximately the size of those of tectal units. Apparently, a process of temporal and spatial integration takes place at the level of the LTTD (Stanford and Hartline, 1984). Inhibition plays a universal role in spatial analysis of sensory information, not only in infrared perception but also in such senses as vision, electroreception, audition, and touch (Mountcastle and Powell, 1959; Hartline et al., 1961; Darian-Smith, 1965; Zipser and Bennet, 1976; Knudsen and Konishi, 1978).

7. TOPICAL RELATIONS OF LTTD NEURONS
All LTTD units are driven by stimulation of the ipsilateral pit (or pits of boids). The receptive fields are located mainly at the ipsilateral side but may extend contralaterally. No response is obtained upon stimulation of the contralateral pits (Terashima and Goris, 1977; Molenaar et al., 1979; Stanford and Hartline, 1980). *Agkistrodon* shows some topographical relationships, though these have not been studied thoroughly. Units that are located progressively deeper within LTTD are driven by progressively more dorsal receptive areas at the pit

membrane and thus by more ventral parts of space (Terashima and Goris, 1977).

Python has a crude somatotopic projection. The pits from the mandibular region are represented most dorsally in LTTD, the maxillary pits in the intermediate part and the rostral pits in the ventral part of LTTD. Their projection areas show a considerable overlap. These features are in agreement with the anatomical projections of the three trigeminal nerve roots (Molenaar, 1978b).

In *Python*, LTTD units appear to be driven by a group of pits and not by a single pit. In addition, these groups of pits are innervated by distinct fiber bundles of the trigeminal nerve (Molenaar, 1978c) (see Fig. 5.3).

D. Concluding Remarks

The sensory trigeminal system of infrared-sensitive snakes consists of two separate components. One component is the common sensory trigeminal system, which is equivalent to the full complement of the sensory trigeminal system in other vertebrates. The other component is the lateral descending trigeminal system. The lateral descending tract (lttd) is composed of the primary afferent fibers from the pit organs; they terminate in the huge lateral descending nucleus (LTTD). The LTTD consists of a corpus and a cauda portion. The nucleus contains several cell types, of which large multipolar cells are characteristic. These large cells are present in the corpus portion only. The largest cells in boid LTTD—the efferent cells of the nucleus—are probably homologous to those of the nucleus reticularis caloris in pit vipers. The cauda of boid snakes is better developed than in pit vipers. In *Python*, the cauda sends ascending projections to the corpus. It receives vagal as well as trigeminal afferents. The cauda in pit vipers is not continuous with the corpus. It receives primary afferent trigeminal fibers, but apparently not from the pit membrane. It does not receive primary afferent vagus fibers. The primary afferent trigeminal fibers form an elaborate preterminal meshwork in the corpus. Although each of the three peripheral main branches in *Python* projects into its own territory, their preterminal meshwork invades the territories of the others to a considerable extent. A single LTTD unit in *Python* is driven by several neighboring pits. This is probably the result of a converging projection from receptive areas, which are singularly located in neighboring pits, but which share similar receptive fields. Similar converging projections are present on tectal units.

The fibers to the LTTD in pit vipers and boid snakes are either myelinated Aδ fibers or thin, unmyelinated C fibers. The Aδ fibers are primary afferent trigeminal fibers from the pit organs. The C fibers are primary afferent vagal fibers of unknown source. The Aδ fibers terminate on dendrites of neurons within the main neuropil, which are apparently interneurons. They form multitudinal diverging and converging synaptic arrangements. The C fibers terminate on small cells in the marginal neuropil. These small cells are probably interneurons, which may send their axons to form axoaxonic contacts on the Aδ fibers. The large multipolar neurons of *Python* LTTD are the efferent neurons. They receive two types of synaptic contacts, neither of which is a primary afferent. Similar cells with similar synapses form the separate nucleus reticularis caloris in pit vipers.

The receptive areas of LTTD neurons in pit vipers are larger than those of primary neurons and appear to be composed of the terminal nerve masses of 81 to 400 peripheral units. The receptive area of one tectal unit contains 156 to 625 terminal nerve masses. All receptive areas are excitatory.

A low transference ratio may occur between primary afferent and LTTD neurons. It has been calculated that more than 400 spikes of peripheral units are needed to evoke one spike in a LTTD unit. The combination of this low transference ratio and a high convergence ratio reduces sensitivity but increases angular resolution.

The receptive fields of LTTD neurons are smaller than those of peripheral units. They contain inhibitory regions in addition to excitatory regions. Since no inhibitory receptive areas appear to be present, the inhibitory flanks in the receptive fields must be the result of inhibitory processes between LTTD units.

The receptive areas or receptive fields of boid LTTD neurons have not been studied.

IV. EFFERENT PROJECTIONS FROM THE SENSORY TRIGEMINAL NUCLEI
A. Introduction

The efferent projections from the two components of the sensory trigeminal nuclei were studied in detail in *Python reticulatus* both with the anterograde degeneration technique (Molenaar, 1978c, Molenaar and Fizaan-Oostveen, 1980) and the HRP technique (Welker et al., 1983). A comparison between these projections from the two sensory trigeminal components revealed that they differ significantly. The efferent projections of the common sensory trigeminal nuclei appear to resemble those of mammals in that these nuclei have a direct ascending projection to thalamic nuclei (subdivisions of the thalamic ventral nuclear complex) and another, mainly indirect, projection to the central gray layer of the tectum. In contrast, the LTTD sends a massive direct projection to the central gray layer of the tectum, as well as to an apparently unique nuclear complex in the dorsal thalamus and pretectum (the complex of the nuclei rotundus and pararotundus). In pit vipers, an additional nucleus is intercalated (the nucleus reticularis caloris, RC) in the ascending pathway from the LTTD to the tectum (Gruberg et al., 1979; Kishida et al., 1980; Newman et al., 1980; Stanford et al., 1981). At present, accumulating evidence points to a homology of the tectipetal cells in pit viper RC and boid LTTD (see Section IV.F).

After a survey of the relevant thalamic and mesencephalic structures, the efferent projections of the two components of the sensory trigeminal system in *Python* and then of those in pit vipers are described. These descriptions will be followed in time by a comparison of the two infrared pathways. Detailed discussions of the third- and higher-order structures that belong to the infrared system are presented in Chapter V.

B. Normal Morphology of the Tertiary Trigeminal Structures
1. STRUCTURES RELATED TO THE COMMON SENSORY TRIGEMINAL NUCLEI The thalamic ventral nuclear complex is composed of several nuclei, each of which contains multiple cell populations (Fig. 5.12). These nuclei are the nucleus ventrolateralis thalami, the nucleus ventromedialis thalami, and the area triangularis (Senn, 1968, 1974). The ventrolateral nucleus and the triangular area receive trigeminal afferents and will therefore be discussed in more detail.

The nucleus ventrolateralis thalami merges with the ventral part of the lateral geniculate nucleus; demarcation between the two on the basis of cytoarchitectonic criteria alone can only be made with difficulty. Both nuclei consist of a medial cell plate and a lateral neuropil. The borderline between the two nuclei is indistinct in normal preparations. Molenaar (1978c) and Molenaar and Fizaan-Oostveen (1980) provide detailed discussions of their discrimination.

The ventrolateral nucleus contains several populations of cells: magnocellular, parvocellular, and neuropil zones. The two magnocellular cell groups are the most conspicuous; one is dorsomedial to the other. They consist of medium-sized, fusiform cells that stain darkly in Nissl preparations. The cells are rather loosely arranged and have their long axis inclined dorsomedially. The small parvocellular cell group is situated dorsally to the former and consists of small, rounded darkly stained cells. The laterally located neuropil zone merges with that of the lateral geniculate nucleus. It contains a few small, rounded darkly stained cells, in contrast to the neuropil of the lateral geniculate nucleus, which contains characteristically larger rounded cells. Several authors have incorporated these parts within the lateral geniculate nucleus (Warner, 1947; Armstrong, 1951; Halpern and Frumin, 1973; Repérant, 1973; Northcutt and Butler, 1974; Ulinski, 1977; Gruberg et al., 1979; Schroeder, 1981a, 1981b).

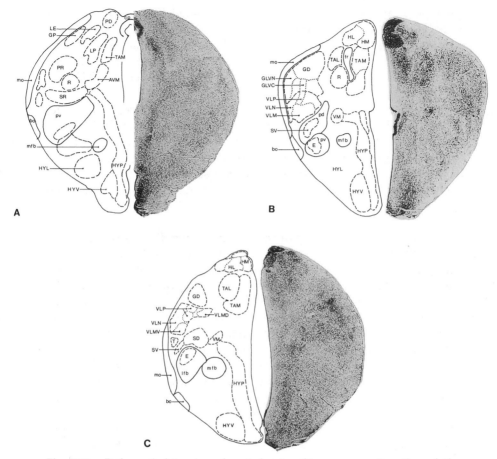

Fig. 5.12. *Python reticulatus.* A caudorostral series of transverse sections through the diencephalon. The distance between sections is 500 μm. Cresyl fast violet stain. AVM, area ventromedialis; bo, tractus opticus basalis; cp, commissura posterior; E, nucleus entopeduncularis; fr, fasciculus retroflexus; GD, nucleus geniculatus lateralis pars dorsalis; GLVC, nucleus geniculatus lateralis pars ventralis, medial cell plate; GLVN, idem, neuropil zone; GP, nucleus geniculatus pretectalis; HL, nucleus habenularis lateralis; HM, nucleus habenularis medialis; HYP, nucleus periventricularis hypothalami; HYL, nucleus lateralis hypothalami, HYV, nucleus ventralis hypothalami; LE, nucleus lentiformis thalami, pars extensa; LP, nucleus lentiformis, pars plicata; mfb, medial forebrain bundle; mo, tractus opticus marginalis; O, nucleus ovalis; PD, nucleus posterodorsalis; pd, pedunculus dorsalis; PR, nucleus pararotundus; pv, pedunculus ventralis; R, nucleus rotundus; SD, nucleus suprapeduncularis pars dorsalis; SR, nucleus subrotundus; SV, nucleus suprapeduncularis pars ventralis; T, area triangularis; TAL, nucleus dorsolateralis thalami; TAM, nucleus ventrolateralis thalami, pars magnocellularis; VLMD, nucleus ventrolateralis thalami, pars magnocellularis dorsalis; VLMV, nucleus ventrolateralis thalami, pars magnocellularis ventralis; VLN, nucleus ventrolateralis thalami, neuropil; VLP, nucleus ventrolateralis thalami, pars ventrolateralis; VM, nucleus ventromedialis thalami.

The area triangularis forms the rostral prolongation of the ventro-lateral nucleus. It can be distinguished from the latter by its slightly smaller cells, which stain more darkly and are more densely packed. Its cells are arranged in clusters, which form the rostral prolongations of the magnocellular and parvocellular cell groups of the ventrolateral nucleus. A neuropil zone, similar to that of the ventrolateral nucleus, is present at the lateral aspect of the area.

2. Structures Related to the Lateral Descending Trigeminal System

The tectum mesencephali of *Python* (Fig. 5.13) resembles that of *Boa* (Senn, 1966). It is composed of three systems of strata: the periventricular strata (layers 1 to 5), the central strata (layers 6 and 7) and the superficial strata (layers 10 to 14). As the central white (layer 6) and gray layers (layer 7) are relevant for the present study, their descriptions follow.

The central white and gray layers are not as sharply separated from each other as they are in other reptiles (Senn, 1966). At their border with the periventricular gray layer, the fibers of the white layer form a rather compact tract, but the more superficial fibers are loosely arranged. The cells of the central gray layer are packed densely where they border the superficial layers; they are more scattered at deeper

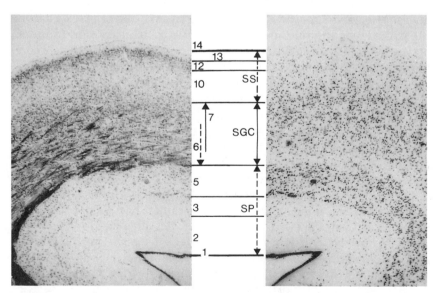

Fig. 5.13. *Python reticulatus.* A transverse section through the colliculus rostralis. Left: Klüver-Barrera stain. Right: Cresyl fast violet stain. 1 to 14, layers of Cajal; SGC, central white and gray layers; SP, periventricular layers; SS, superficial layers.

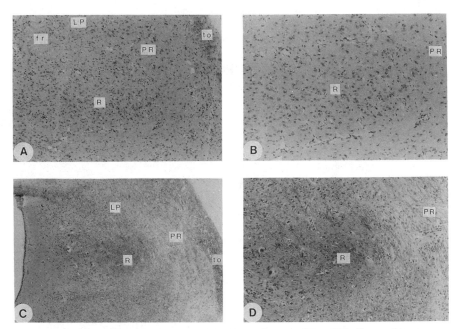

Fig. 5.14. *Python reticulatus* (A and B) and *Crotalus atrox* (C and D). Cytoarchitecture of the nuclei rotundus and pararotundus. fr, fasciculus retroflexus; LP, nucleus lentiformis thalami, pars plicata; PR, nucleus pararotundus; R, nucleus rotundus; to, tractus opticus. (Klüver-Barrera stain. A, 100x; B and D, 200x; C, 50x.)

levels. The most superficial zone of this layer is free of fascicles from the white layer. In this zone, the cells are in general smaller, rounder, and more darkly stained than at deeper levels. The deeper levels contain more medium-sized cells, which stain less darkly. In addition, these levels contain several larger cells with a granulated cytoplasm.

The extensive additional infrared input to the tectum is reflected by the relatively large size of the tectum. In general, the tectal layers are relatively smaller in nocturnal snakes than in diurnal snakes. In the infrared-sensitive snakes (which are nocturnal), the tectum shows the appropriate structural characteristics, but the central gray layer is increased in volume (Goris and Terashima, 1973; Kass et al., 1978).

The nuclei rotundus and pararotundus form a complex that is located within the pretectal area and the caudal half of the thalamus (see Figs. 5.12 and 5.14). The complex is well developed in the boids *Python reticulatus* and *P. regius* and in the pit vipers *Crotalus cerastes* and *C. atrox* but is less developed in the common vipers (Molenaar and Fizaan-Oostveen 1980) and other snakes (Halpern and Frumin, 1973; Repérant, 1973; Northcutt and Butler, 1974; Ulinski, 1977; Dacey and Ulinski, 1983). This indicates its special relation to infrared sensitivity. The complex is easily distinguished from neighboring nuclei by its medium-sized ellipsoidal cells with a light cytoplasm.

The nucleus rotundus is situated medially within the complex; its diencephalic part is situated more medially, and the cells are packed more densely than in its pretectal part.

The nucleus pararotundus contains a similar cell type to that in the nucleus rotundus. Nevertheless, the two nuclei are easily recognized by the different arrangement of their cells. Those of the nucleus pararotundus are packed more loosely and are oriented more tangentially than those of the nucleus rotundus. This part of the complex is well developed in the boid and crotaline species mentioned above, but absent from the common vipers studied and has not been described in other snakes.

C. Ascending Projections of the Common Sensory Trigeminal Nuclei in *Python*

1. SECOND-ORDER PROJECTIONS   The following description is primarily based upon the results of experiments using the anterograde degeneration technique (Molenaar, 1978c, Molenaar and Fizaan-Oostveen, 1980), as well as additional experiments, in which HRP had been injected into the tectum, the tegmentum, the metencephalon, and the lateral aspect of the myelencephalon (Molenaar, 1983a, 1983b; Welker et al., 1983). These experiments confirm and expand the data obtained by the degeneration technique. The second-order projections of the common sensory trigeminal nuclei appear to resemble those of mammals in that these nuclei have a direct ascending projection of thalamic nuclei and another, mainly indirect, projection to the tectum. These projections differ significantly from those of the lateral descending nucleus (Figs. 5.15 to 5.18).

The trigeminal nucleus caudalis (TTDC) sends bilateral projections to the thalamic ventral nuclear complex and the central gray layer of the tectum (Fig. 5.15). It also sends ipsilateral projections to the more rostrally located nuclei of the common sensory trigeminal system, to the reticular formation, and to the cerebellum via the dorsal spinocerebellar tract.

The fibers to the thalamus are of a fine caliber. The contralateral fibers leave the nucleus caudalis at its ventromedial aspect and join the dorsal arcuate fiber system. Here they run ventrally and caudally, decussate, turn rostrally, and form the ventromedial part of the spinal lemniscus. They form a compact tract on their course through the medulla but become more diffusely dispersed within the mesencephalic tegmentum. The majority terminate in the dorsal parts of the magnocellular subnuclei of the thalamic ventrolateral nucleus and triangular

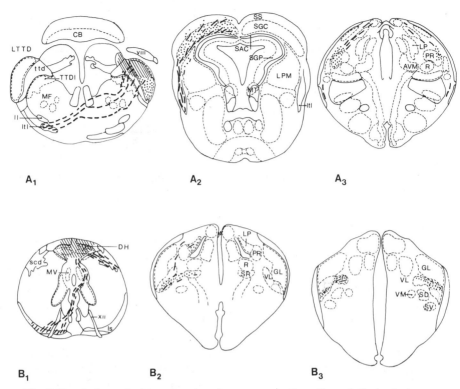

A_1 A_2 A_3

B_1 B_2 B_3

Fig. 5.15. *Python reticulatus.* A series of transverse sections through the brainstem depicting the ascending projections to the mesencephalon and thalamus from the lateral descending nucleus (A) and the trigeminal nucleus caudalis (B) as they appeared after degeneration experiments. A1: Section at the level of the corpus of the lateral descending nucleus, showing the lesion of a particular case. A2: Section through the colliculus rostralis at the level of the trochlear motor nucleus. A3: Section through the caudal part of the thalamus. B1: Section through the transitional area between the myelencephalon and the cervical cord at the level of the motor vagus nucleus showing the extent of the lesion in the nucleus caudalis in this particular case. B2: Section through the caudal part of the thalamus. B3: Section through the rostral part of the thalamus. AVM, area ventromedialis; CB, cerebellum; DH, dorsal horn; GL, nucleus geniculatus lateralis; ll, lemniscus lateralis; LP, nucleus lentiformis thalami, pars plicata; LPM, nucleus lateralis profundus mesencephali; ls, lemniscus spinalis; ltl, lemniscus trigemini lateralis; LTTD, nucleus descendens lateralis n. trigemini; MF, nucleus motorius n. facialis; MT, nucleus motorius n. trochlearis; MV, nucleus motorius n. vagus; PR, nucleus pararotundus; R, nucleus rotundus; SAC, stratum album centrale mesencephali; scd, tractus spinocerebellaris dorsalis; SD, nucleus suprapeduncularis pars dorsalis; SGC, stratum griseum centrale; SGP, stratum griseum periventriculare; SR, nucleus subrotundus; SS, stratum superficiale; SV, nucleus suprapeduncularis pars ventralis; ttd, tractus descendens n. trigemini; VL, nucleus ventrolateralis thalami; VM, nucleus ventromedialis thalami; VIII, nervus vestibulocochlearis; XII, nervus hypoglossus.

area and in the adjacent neuropil. The projection occupies a territory that is dorsal to that of the contralateral projection from the nucleus principalis. The ipsilateral projection is confined to the parvocellular parts of the thalamic ventral nuclear complex and adjacent neuropil. A few fibers from the nucleus caudalis terminate within the contralateral nuclei lateralis profundus mesencephali, lentiformis thalami pars plicata, and subrotundus.

Projections to the tectum could not be demonstrated by the degeneration technique but were established by small injections of HRP into the tectum. These experiments revealed several labeled neurons scattered within the substantia gelatinosa and magnocellular subnuclei of both sides, as well as within the adjacent reticular formation.

The trigeminal nucleus principalis sends bilateral projections to the thalamic ventral nuclear complex. A small contralateral projection to the tectum was identified in the HRP studies only. Other projections could not be studied because of the unexpectedly large extent of the lesions, probably caused by hemorrhages.

The fibers to the thalamus are very thin. The contralateral fibers decussate in the internal arcuate fiber system and ascend through the medial part of the mesencephalic tegmentum. They terminate in the ventral parts of the magnocellular subnuclei of the ventrolateral nu-

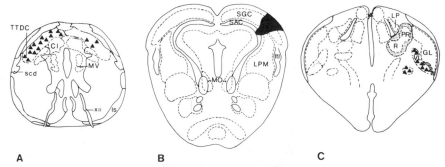

A   B   C

Fig. 5.16. *Python reticulatus.* A caudorostral series of transverse sections through the brainstem, depicting bilateral, retrograde labeling of neurons (black triangles) within the trigeminal nucleus caudalis (A) and anterograde labeling (dots) of the rotundopararotundus complex, as well as both anterograde and retrograde labeling in the thalamic ventrolateral nucleus (C) upon an injection of HRP (black area) within the central gray and white layers of the tectum (B). CI, nucleus commissurae infimae; GL, nucleus geniculatus lateralis; LP, nucleus lentiformis thalami pars plicata; LPM, nucleus lateralis profundus mesencephali; ls, lemniscus spinalis; ltl, lemniscus trigemini lateralis; MO, nucleus motorius n. oculomotorius; MV, nucleus motorius n. vagus; PR, nucleus pararotundus; R, nucleus rotundus; scd, tractus spinocerebellaris dorsalis; SAC, stratum album centrale, SGC, stratum griseum centrale; TTDC, nucleus caudalis n. trigemini; VL, nucleus ventrolateralis thalami.

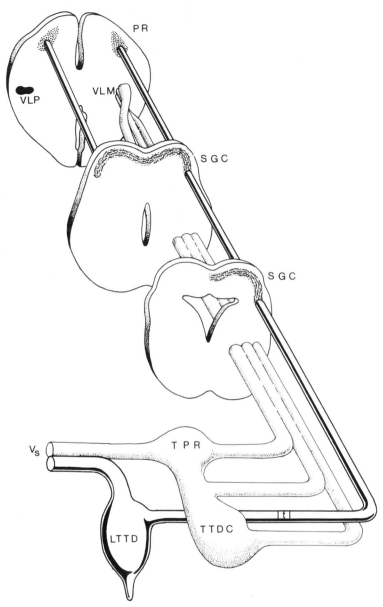

Fig. 5.17. Schematic representation of the two sensory trigeminal systems in *Python* showing their primary afferent tracts, their primary sensory nuclei, and their ascending projections to mesencephalic and diencephalic levels. The projection from the nuclei of the common sensory trigeminal nuclei to the tectum has not been depicted. ltl, lemniscus trigemini lateralis; LTTD, nucleus descendens lateralis n. trigemini; PR, nucleus pararotundus; SGC, stratum griseum centrale; TPR, nucleus sensorius principalis n. trigemini; TTDC, nucleus caudalis n. trigemini; VLM, nucleus ventrolateralis thalami, pars magnocellularis; VLP, nucleus ventrolateralis thalami, pars parvocellularis, which receives ipsilateral projections from the common sensory nuclei; VS, radix sensoria n. trigemini.

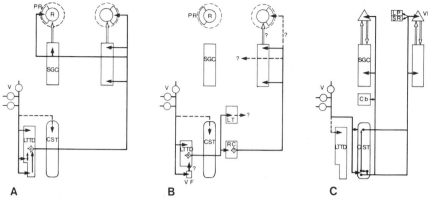

Fig. 5.18. Diagram showing the similarities and dissimilarities of the various sensory trigeminal systems and their ascending projections. (A) The lateral descending trigeminal system in boids. (B) The lateral descending trigeminal system in crotalines. (C) The common sensory trigeminal system in *Python reticulatus*. Cb, cerebellum; CST, common sensory trigeminal system; LP, nucleus lentiformis thalami, pars plicata; LT, nucleus tegmenti lateralis; LTTD, nucleus descendens lateralis n. trigemini; PR, nucleus pararotundus; R, nucleus rotundus; RC, nucleus reticularis caloris; SGC, stratum griseum centrale mesencephali; SR, nucleus subrotundus; VF, ventral trigeminal field; VL, nucleus ventrolateralis thalami.

cleus and triangular area and in the adjacent neuropil. The projection area occupies a more ventral position than that of the nucleus caudalis. The ipsilateral fibers appear to be distributed to the parvocellular subnuclei of the thalamic ventral nuclear complex. Projections to the tectum could not be demonstrated by the degeneration technique, but neurons were labeled when HRP was injected in the contralateral tectum.

Python lacks a quintofrontal tract. According to the classic view, the principal sensory nucleus of reptiles should project through the quintofrontal tract to the striatum (Woodburne, 1936, *Anolis*), as does its avian equivalent (Wallenberg, 1903; Verhaart, 1971; Dubbeldam and Wijsman, 1975).

The trigeminal nuclei interpolaris and oralis project contralaterally to the thalamic ventral nuclear complex and bilaterally to the tectum. The fibers to the thalamus are dispersed through the mesencephalic tegmentum. They terminate within the magnocellular parts of the nucleus ventrolateralis and area triangularis in an area that is intermediate between those of the nuclei caudalis and principalis.

The bilateral projection to the tectum was demonstrated in the HRP experiments. In *Python*, the nuclei oralis and interpolaris contain more neurons that project to the tectum than do the nuclei principalis and caudalis. Furthermore, there are more neurons that project to the contralateral tectum than to the ipsilateral tectum. The projections arise from the medium-sized cells, which lie within the ventral cell concentration of these nuclei, and from neighboring reticular neurons of similar size and shape located within the nucleus reticularis inferior parvocellularis subtrigeminus (Molenaar, 1977). Only occasionally is such a neuron located in the dorsal parts of the nuclei. Although this ventral area remains devoid of degenerating terminals after transsection of the sensory trigeminal nerve roots (Molenaar, 1978b), it seems likely that trigeminal common sensory information is relayed to the tectum through this pathway. Tectal neurons that are responsive to such sensory modalities have not yet been identified in these snakes, although they have been found in the tectum of other vertebrates (Dräger and Hubel, 1975; Stein et al., 1976; Chalupa and Rhoades, 1977).

2. COMPARISON TO OTHER VERTEBRATES

Projections from the common sensory trigeminal nuclei to the tectum have been mentioned for *Python reticulatus* (Welker et al., 1983) and from the contralateral nucleus oralis in *Elaphe quadrivirgata* and *Agkistrodon blomhoffi* (Kishida et al., 1980). No such projections could be demonstrated in *Python reticulatus* (Newman et al., 1980), nor in *Crotalus viridis* (Gruberg et al., 1979; Newman et al., 1980).

Somatosensory projections from medullary or spinal cord origin to the somatosensory ventrocaudal area in the thalamus have been described for other reptiles and amphibians (Neary and Wilczynski, 1977, *Rana*; Vesselkin et al., 1971, *Rana*; Pedersen, 1973, three species of turtles; Belekhova, 1979, turtles, alligator). These projections are consistent with the ascending projections of the common sensory trigeminal nuclei in *Python*. Single units in the thalamic ventrocaudal area of turtles were shown to be extralemniscal or lemniscal, whereas about one-third were multimodal, that is, both somatosensory and visual (Belekhova, 1979).

The trigeminal extralemniscal projection in mammals originates from the nucleus caudalis. The projections to the thalamic ventroposterior nuclei are sometimes reported as bilateral (Stewart and King, 1963; Darian-Smith, 1973) or to originate from the underlying reticular formation (Tiwari and King, 1973). Projections to the tectum have also been mentioned (Stewart and King, 1963; Welker et al., 1983).

The trigeminal lemniscal pathway in mammals originates from the rostral sensory trigeminal nuclei (Torvik, 1957; Michail and Karamanlidis, 1970; Mizuno, 1970; Darian-Smith, 1973; Smith, 1973, 1975; Augustine, 1974; Karamanlidis, 1975); the projection is bilateral in several species (Smith, 1975). Projections to the tectum have also been described (Welker et al., 1983).

The projections frcm the nuclei of the common sensory trigeminal system in *Python* are consistent with these results. The nuclei principalis and caudalis project to neighboring areas in the thalamic ventral nuclear complex, and a tectal projection is also present. It is not known whether the projections from the trigeminal nuclei are actually segregated in lemniscal and extralemniscal pathways, but the structural basis for such segregation is present.

3 THIRD-ORDER PROJECTIONS

Some third-order projections of the common sensory trigeminal system were established by injections of HRP into the tectum (Molenaar, 1983a, 1983b; Welker et al., 1983). Many cell bodies of the thalamic ventral nuclear complex were filled with HRP when relatively small injections were confined to the central gray and white layers of the superior colliculus (Figs. 5.16 and 5.19). The positively labeled neu-

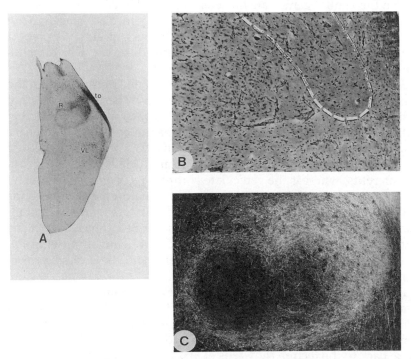

Fig. 5.19. *Python reticulatus.* Distribution of afferent fibers to the nuclei rotundus and pararotundus. (A) Anterograde labeling of the rotundopararotundus complex as well as retrograde cell labeling and anterograde labeling in the neuropil zone of the ventrolateral thalamic nucleus following an injection of HRP in the ipsilateral tectum. (B) The indicated area within the nucleus pararotundus contains the degenerated terminal boutons from the lateral descending trigeminal nucleus. (C) Dark-field photograph of the anterograde HRP labeling of the rotundopararotundus complex after an injection within the ipsilateral tectum. R, nucleus rotundus; to, tractus opticus; VL, nucleus ventrolateralis thalami. (A, 18x; B, 10x; C, 150x.) (Photos courtesy of P. V. Hoogland, Free University Amsterdam.)

rons were located in the magnocellular subnuclei of the ventrolateral nucleus and triangular area. The cell bodies of the parvocellular subnuclei remain free of HRP. In addition, labeled fibers and terminals occur in the lateral neuropil zone. These findings suggest a reciprocal projection between the ventral nuclear complex and the tectum.

Due to inconsistencies in nomenclature, comparison of the results in *Python* with the various results in the literature is restricted to comparison of the illustrations. This, of course, limits the accuracy of such comparisons. Yet it seems that similar projections are present in other snakes, such as *Nerodia* (Ulinski, 1977) and *Crotalus* (Gruberg et al., 1979; Schroeder, 1981a). These studies show extensive connections with pretectal and dorsal and ventral geniculate nuclei. In addition, anterograde degeneration is depicted in an area that seems equal to the ventrolateral thalamic nucleus. This area has been incorporated within the ventral geniculate nucleus (*Nerodia*, Ulinski, 1977; *Crotalus*, Schroeder, 1981a, 1981b) or within the dorsal geniculate nucleus (*Crotalus*, Gruberg, et al., 1979). Moreover, Gruberg et al. (1979) report retrograde HRP labeling of perikarya within this area, which makes the resemblance more striking.

Such incorporation of the ventrolateral thalamic nucleus within the geniculate complex is not uncommon, because these nuclei are not easily discriminated on cytoarchitectonic criteria alone (see Section IV.B.1). They have been recognized as separate entities by Molenaar (1978c) and Molenaar and Fizaan-Oostveen (1980) on the basis of their different afferents. The ventrolateral thalamic nucleus receives trigeminal afferents, whereas the geniculate complex receives visual input.

Direct and indirect (i.e., via the tectum) somatosensory projections to the nuclei within the thalamic somatosensory area have been demonstrated in other reptiles (for references see Belekhova, 1979). The direct projection from the common sensory trigeminal nuclei in *Python* to the thalamic ventral nuclear complex and the indirect projection via the tectum fits within this pattern.

The projection from the common sensory trigeminal nuclei to a thalamic somatosensory area and from there to the tectum resembles the mammalian arrangement in which the trigeminal nuclei project to the zona incerta (Smith, 1973; Erzurumlu and Killackey, 1980) and from there back to the tectum (Edwards et al., 1979; Ricardo, 1981).

D. Projections of the Lateral Descending Nucleus in *Python*

1. INTRANUCLEAR PROJECTIONS   An intranuclear degeneration pattern was observed after partial lesions within the LTTD (Molenaar, 1978c; Molenaar and Fizaan-Oostveen, 1980). This pattern was demonstrated best at survival times that were shorter than those necessary to demonstrate most clearly the extranuclear projections (8 days versus 18 days). The grains of this intranuclear pattern are of a finer caliber than those of the primary fibers. The intranuclear projection

was both ascending and descending and extended over the entire rostrocaudal length within the LTTD. Even in one case having a small lesion only in the cervical part of the cauda of the LTTD, the intranuclear projection ascended to the rostral pole of the nucleus. If a small lesion had been made within the primary projection area of the maxillary root, for instance, not only did the intranuclear projection spread within the maxillary area but also extended into the primary projection areas of the other two roots. Thus, the intranuclear connections seem to give rise to tremendously divergent projections. Yet, sometimes certain areas remained free of intranuclear degenerated boutons following very small lesions.

The intranuclear projection probably originates from the smaller cells within the LTTD, since these cells were not labeled after injections of HRP within the projection areas of the LTTD or within the lateral trigeminal lemniscus (Newman et al., 1980; Molenaar, 1983a, 1983b). The small cells were not even labeled after HRP injections within the nucleus interpolaris, which thereby also involved the efferent fibers of LTTD as soon as these had left the nucleus.

2. EXTRANUCLEAR PROJECTIONS

The efferent fibers of LTTD originate from the large cells within the corpus portion of the nucleus (see Fig. 5.10). This was shown by injections of HRP in the tectum (Newman et al., 1980; Meszler, 1981, 1983), as well as by injections of HRP within the tectum, tegmentum, or the lateral trigeminal lemniscus (Molenaar, 1983a, 1983b).

The efferent fibers leave the LTTD in *Python reticulatus* at the ventromedial side of its corpus portion at the level of the facial motor nucleus (Molenaar and Fizaan-Oostveen, 1980) (Figs. 5.15 and 5.17). They decussate in the internal arcuate fiber system and form a compact tract, which ascends at the ventromedial aspect of the lateral lemniscus. The tract is called the lateral trigeminal lemniscus (ltl). The tract shifts dorsally at the caudal pole of the mesencephalon and resumes its rostral course while it remains just beneath the tectum. Fibers leave the tract throughout its mesencephalic course and turn dorsally, passing through the central white layer of the tectum to terminate within the central gray layer (layer 7 of Senn, 1966). The remainder of the fibers of the ltl proceed rostrally and terminate within the nucleus pararotundus.

The central gray layer of the tectum contains a dense pattern of degenerated terminal boutons, whereas few boutons occur within the superficial layer (layer 10). This is in accordance with the observations that bimodal visual and infrared-sensitive neurons occur within the transitional area of the superficial and central gray layers (Hartline et al., 1978; Haseltine, 1978; Kass et al., 1978; Newman et al., 1981).

Several fibers pass through the central white layer and decussate again at the tectal commissure in the rostral part of the tectum and thus give rise to a bilateral projection. They terminate ipsilaterally within the rostral part of the central gray layer and in the nucleus pararotundus.

An ipsilateral projection to the lateral tegmental nucleus, as described in pit vipers (Stanford et al., 1981), could not be demonstrated in *Python*. Such a nucleus is located at the cerebello-oblongata junction in pit vipers. One injection of HRP in this area in *Python* did not reveal any positive labeling within the LTTD (Molenaar, 1983a). The degeneration material for *Python* was not very useful for studying such a projection, because the surgical procedure led to transsection of the eighth nerve, which evoked an elaborate degeneration pattern in the area.

E. Projections from the LTTD in Pit Vipers

The ascending projections from the LTTD in pit vipers differ from those in *Python* (see Fig. 5.18). *Python* has a direct projection to the central gray layer of the tectum and the nucleus pararotundus, but no such direct projection could be identified in pit vipers. Instead, two additional nuclei are intercalated in the ascending projections.

The lateral tegmental nucleus was identified provisionally by Stanford et al. (1981) after lesioning of the LTTD; its connections are unknown. It is embedded in fibers associated with the cerebellum or caudal colliculus. The nucleus appears not to be homologous to the lateral cerebellar nucleus, the nucleus of the lateral lemniscus, or the nucleus isthmi (Stanford et al., 1981).

The nucleus reticularis caloris (RC) forms the relay in the ascending infrared pathway to the tectum. Its existence has been established and confirmed by various authors using various techniques. Injections of HRP in the tectum revealed that only cells in the RC were labeled, whereas no cells in the LTTD were labeled (Gruberg et al., 1979; Kishida et al., 1980). HRP injections into the RC (Newman et al., 1980) showed retrograde labeling of neurons in the LTTD and anterograde labeling in the central gray layer of the tectum. Other labeled fibers are presumably from fibers of passage, but this assumption needs verification. Lesioning of pit viper LTTD showed that this nucleus projects only to the nucleus reticularis caloris and to the lateral tegmental nucleus (Stanford et al., 1981). The nucleus reticularis caloris is involved exclusively in infrared perception, as was shown by recordings of its units (Kishida et al., 1980; Newman and Hartline, 1981).

The nucleus reticularis caloris is located superficially at the ventrolateral aspect of the medulla (Fig. 5.20). The nucleus is long and narrow and has a rostrocaudal extent from the trigeminal nucleus oralis to the rostral pole of the LTTD (Gruberg et al., 1979; Kishida et al., 1980; Newman et al., 1980; Stanford et al., 1981). It is separated from the trigeminal common sensory and lateral descending structures by the spinocerebellar tracts (Kishida et al., 1980). The nucleus consists of a homogeneous population of medium-sized or large multipolar cells. Cell sizes range from 20 to 30 μm in *Agkistrodon* (Kishida et al., 1980) to 20 or 25 to 45 μm in *Crotalus viridis* (Gruberg et al., 1979; Newman et al., 1980; Stanford et al., 1981). The cells stain heavily in Nissl stains and in stains for acetylcholine esterase (Newman et al., 1980) and succinate dehydrogenase (Kishida et al., 1980) activity. The cells are embedded and completely surrounded by the efferent, large-diameter (5 μm) fibers from the LTTD (Gruberg et al., 1979; Stanford et al., 1981).

Somata and dendrites of these cells receive two types of axon terminals, which are similar to those of the tectipetal neurons of boid LTTD (Meszler, 1983). One type contains clear spherical vesicles and some large granular vesicles, whereas its postsynaptic membrane is associated with a subsynaptic web. The second type contains flat-

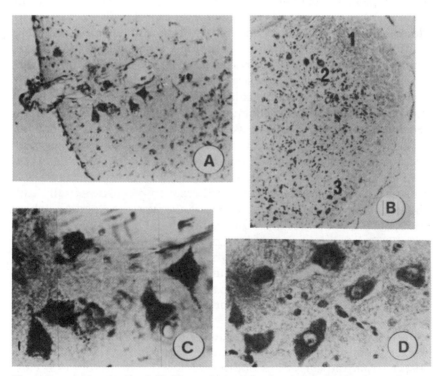

Fig. 5.20. *Agkistrodon blomhoffi*. Photomicrographs showing the localization and cytoarchitecture of the nucleus reticularis caloris. (A and C) HRP-labeled cells after an injection in the contralateral tectum. (B and D) The appearance of this nucleus in the cresyl stain. 1, tractus descendens and tractus descendens lateralis n. trigemini; 2, nucleus oralis of the nucleus descendens n. trigemini; 3, nucleus reticularis caloris. (A, 130x; B, 65x; C and D, 400x.) (From Kishida et al., 1980.)

tened vesicles and symmetrical membranes. The origin of these terminals is unknown, but at least one type has to come from the LTTD.

The efferent fibers of the nucleus reticularis caloris decussate and form a clearly delineated tract in the ventrolateral medulla, in a position similar to that of the lateral trigeminal lemniscus in *Python*. The tract runs rostrally to the caudal tegmentum, where it moves dorsally. Its fibers enter the central gray layer of the tectum and terminate there (Gruberg et al., 1979; Kishida et al., 1980; Newman et al., 1980). A projection to the nucleus pararotundus could not be observed (Newman et al., 1980) upon anterograde transport of HRP.

On the other hand, the report of Auen (1978) is conflicting in that it describes a direct projection from the LTTD to the tectum in three species of rattlesnakes (*Crotalus horridus, C. scutulatus, C. ruber*). These snakes are closely related to those rattlesnakes in which such a direct projection is absent. This direct projection was claimed after HRP injections in the tectum. The LTTD in Auen's paper is situated at the level of the eighth cranial nerve root and thus lies much more rostrally than in other pit vipers. The LTTD in that report is also located more dorsally than the nucleus reticularis caloris in other pit vipers. The situation, as presented in the illustrations of that paper, seems to resemble that in *Python*. Unfortunately, the paper does not provide enough details to draw firm conclusions.

Two reports, based on two different techniques, describe different courses of the efferent LTTD fibers in *Crotalus viridis*. In the one report (Stanford et al., 1981) lesions were placed in LTTD, and its degenerating fibers were followed by the ammoniacal silver impregnation technique of Ebbesson and Heimer (1970). The fibers in this report initially run caudally within the LTTD and leave the medial aspect of this nucleus in its caudal part. The fibers then proceed ventrally, caudally, and medially toward the periventricular gray matter, where they make a sharp turn and run rostrally and laterally to the nucleus reticularis caloris. In the other report, the fibers were retrogradely traced upon HRP injections into the RC (Newman et al., 1980). The fibers in this report leave the lateral aspect of the LTTD and run straight in a rostroventral direction along the lateral surface of the medulla to the nucleus reticularis caloris.

F. Comparison of Secondary Infrared Pathways

Python lacks a nucleus reticularis caloris and instead has a direct projection from the lateral descending nucleus to the tectum (see Figs. 5.17 and 5.18). This projection arises from the large cells. Several criteria support the assumption that the large cells that make up the pit viper nucleus reticularis caloris have become incorporated within the boid LTTD:

1. The LTTD of pit vipers is located at the dorsolateral aspect of the nucleus caudalis; in boids it is relocated more rostrally at the level of the nucleus reticularis caloris (Molenaar, 1974, 1978a; Schroeder and Loop, 1976). The cauda portion of boid LTTD is better developed than that of pit vipers (Schroeder et al., 1976; Molenaar, 1978a) and projects via an elaborate, intranuclearly ascending projection to the rostral pole of the LTTD (Molenaar, 1978c; Molenaar et al., 1980). This might be interpreted as a reflection of the ancient position of boid LTTD.

2. The largest, multipolar cells are the last chain of the brainstem pathway for the system in the two families. Their efferent fibers follow quite a similar course toward the tectum (Newman et al., 1980).

3. The tectipetal cells in the two families stain heavily for acetylcholine esterase and succinate dehydrogenase (Kishida et al., 1980; Newman et al., 1980).

4. The arrangement of the synapses on these cells in the two families is similar in that there appear to be two major types of synaptic input to these cells. At least one of these must come from second-order LTTD neurons. The origin of the other is unknown but is probably not from another extrinsic source (Newman et al., 1980; Meszler, 1983). These synaptic terminals line up along the dendrites and soma of the boid and crotaline tectipetal cells in the same way. The structure of the terminals is quite alike (Meszler, 1983, personal communication).

As a result of the incorporation of the large RC cells within the boid LTTD, the latter nucleus will have obtained a more complicated cell population and internal circuitry than the pit viper LTTD. Moreover, it may be expected that additional interneurons or interconnections are present in the *Python* LTTD related to the presence of several pits with partly overlapping views (Haseltine, 1978). The units within the *Python* LTTD are driven by several pits instead of by a single pit (Molenaar and Fizaan-Oostveen, 1980). It is not known whether this convergent projection is achieved at the level of the LTTD, which would be the case if one peripheral unit would be connected to a single receptive area, as in pit vipers. It is also possible that one peripheral nerve fiber innervates several receptive areas that are separately located within several pits.

It is obvious from the preceding discussion that there is presently insufficient information about the cell populations and internal circuitry of the LTTD or LTTD-RC complex.

G. Concluding Remarks

The second-order projections from the common sensory trigeminal nuclei in *Python* appear to be the full complement of those of the trigeminal nuclei in mammals. The ascending projections consist of direct bilateral projections to nuclei of the thalamic ventral nuclear complex and indirect bilateral projections to the central gray layer of the tectum. The indirect projections run via the ventral cell concentration in the trigeminal nuclei, which does not receive primary trigeminal afferents, or via neurons in the adjacent reticular formation. The central gray layer of the tectum and the thalamic ventral nuclear complex are reciprocally interconnected in what seems to be third-order trigeminal projections.

Within the LTTD, an extensive intranuclear projection is present; it probably originates from the small neurons. The extranuclear projection originates from large cells only in pit vipers and in *Python*. In *Python*, these projections run directly to the central gray layer of the tectum and to the nucleus pararotundus. The tectal projection is mainly contralateral, but the projections to the rostral portion of the tectum and to the nucleus pararotundus are bilateral. In pit vipers, the LTTD projects to the nucleus reticularis caloris. This nucleus is probably incorporated within the LTTD of boid snakes. The nucleus reticularis caloris sends an ascending projection to the contralateral tectum. Bilateral projections to the rostral part of the tectum or to the nucleus pararotundus have not yet been studied properly in pit vipers.

V. HIGHER-ORDER INFRARED PERCEIVING STRUCTURES
A. The Tectum

1. INTRODUCTION The tectum appears to be a major target of the ascending infrared pathways. The fibers are distributed to the central gray layer, whereas a few terminals are observed within the deeper part of the superficial gray layer (Molenaar and Fizaan-Oostveen, 1980). Senn (1966) and Dacey and Ulinski (1986a to 1986e) have presented detailed anatomical descriptions. Electrophysiological studies have shown that the unimodal infrared neurons lie within the central gray layer (Goris and Terashima, 1973; Terashima and Goris, 1975, 1976; Haseltine, 1978; Kass et al., 1978), whereas the visual units lie within the superficial layers (Terashima and Goris, 1975; Haseltine, 1978; Kass et al., 1978; Schroeder et al., 1981a, 1981b). In addition, bimodal neurons, responsive to both infrared and visual stimuli, lie in the more superficial parts of the central gray layer and the deeper parts of the superficial gray layer (Hartline et al., 1978; Haseltine, 1978; Kass et al., 1978; Newman and Hartline, 1981). Occasionally, visual units were encountered within this intermediate zone (Kass et al., 1978). The differential involvement of the superficial and of the central layers in visual and infrared processes, respectively, has also been demonstrated by the deoxyglucose labeling technique (Auker et al., 1983; Gruberg et al., 1984). In other vertebrates, a similar laminated organization occurs with visual cells in the superficial layers, other modalities (such as somatosensory, auditory) within the central layers, and bimodal neurons in the intermediate area (for references see Kass et al., 1978). Reptilian and mammalian tectal cells have comparable receptive field properties, such as on and off areas, habituation, influence of direction, and velocity of movements and corresponding topography of the receptive fields of bimodal and trimodal neurons (Stein and Gaither, 1983). Although *Python* has a projection from the common sensory trigeminal nuclei to the central gray layer, no common trigeminal tactile units have been described in its tectum.

2. PROPERTIES OF TECTAL UNITS

The tectal units show a tendency to respond better to transient stimuli than do peripheral units. The background activity is reduced, latency and the on-burst are shorter, whereas the static discharges to constant temperatures are reduced (Terashima and Goris, 1979). These features fit within the requirements to detect and localize the moving, warm-blooded prey. In case an infrared-sensitive snake encounters a nonmoving warm object, it makes scanning movements with its head (Ross, 1935; Goris and Nomoto, 1967), apparently to create moving shadows on its pit membranes.

In pit vipers, tectal units exhibit static discharges, like the peripheral units, but the frequency of this activity is less than that of peripheral units (Goris and Terashima, 1973). As in peripheral units, the background discharges show considerable fluctuations in frequency.

At high-stimulus intensities, most units respond to onset of the stimulus with a phasic burst, followed by a more tonic response of lower frequency (Goris and Terashima, 1973). In peripheral units, the phasic and tonic responses are related to such stimulus parameters as intensity and the rate of temperature change at the membrane (De Cock Buning et al., 1981a). In tectal units these phases apparently relate as well to the position of the stimulus within the receptive field (Goris and Terashima, 1973). At unnaturally high-stimulus intensities, many units show an afterdischarge that disappears when stimulus intensity is reduced (Goris and Terashima, 1973; Terashima and Goris, 1976). At low intensities, the responses are phasic, adapting within one second (Hartline et al., 1978).

Hartline (1974) and Hartline et al. (1978) found that inhibitory activity—the suppression of background activity—depends on stimulus position. Terashima and Goris (1976) could not find such inhibition.

In boids, the units respond in a similar way to the onset of the stimulus, provided the stimulus falls within the excitatory part of the receptive field. Occasionally a response to offset was encountered but this was restricted to the inhibitory portion of the receptive field (Haseltine, 1978).

In *Agkistrodon blomhoffi*, all units respond to steep fronted, stationary stimuli as well as to moving stimuli (Goris and Terashima, 1973; Terashima and Goris, 1976). They show no difference in response to the direction of the movement but differ in response to the speed of the movement. On the other hand, the tectum of *Crotalus viridis* contains units with directional selectivity (Hartline et al., 1978). In the boids *Python* and *Corallus*, most units respond to steep fronted, stationary stimuli (Haseltine, 1978). These units have not been tested for motion sensitivity, that is, for a more vigorous response to moving stimuli. In addition, some truly motion-selective units respond only to moving stimuli.

3. TOPICAL RELATIONS

Most of the infrared-sensitive units in the central gray layer are driven by the contralateral pit membrane in pit vipers; only a few are driven by the pits of both sides (Goris and Terashima, 1973). In the boid snakes *Python reticulatus* and *Corallus enydris*, the units are driven by a combination of several pits that were usually, but not always, neighboring each other (Haseltine, 1978). A similar finding has been reported from the units within the LTTD (Molenaar et al., 1979).

An infrared tectal unit is driven by the receptive area, a small restricted area on the pit membrane (Figs. 5.2 and 5.21). The location and size of these areas have been studied by focusing a small spot of light with the aid of a concave mirror onto the pit membrane (Terashima and Goris, 1976; Haseltine, 1978). In *Agkistrodon blomhoffi*, the receptive areas have a diameter that varies between 0.5 and 1.0 mm.

The receptive area of a tectal unit (being 500 to 1,000 μm) is larger than that of an LTTD unit (340 to 800 μm) and much larger than that of a peripheral unit (20 to 50 μm). The receptive area of a single peripheral unit coincides with the terminal nerve mass of one peripheral neuron (Terashima and Goris, 1979). Thus, considerable convergence of projections takes place from the sensory membrane to the brain. The receptive area of one LTTD unit thus is composed by the terminal nerve masses of 81 to 400 peripheral units, whereas the receptive area of one tectal unit contains 156 to 625 terminal nerve masses.

In addition, the receptive areas of distinct central units overlap considerably. This can only be explained by the assumption of divergent projections from the ganglion to LTTD neurons as well as from bulbar to tectal neurons.

The geometry of the pit organ determines the receptive field of such an area (see Fig. 5.2). The receptive field is defined as the cone

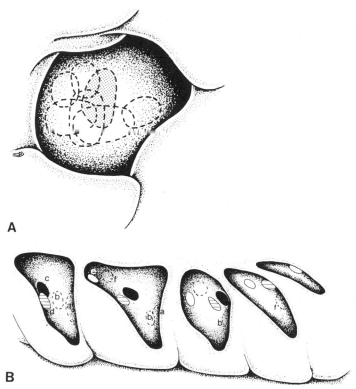

Fig. 5.21. (A) Arrangement of several receptive areas of tectal neurons on the pit membrane of a crotaline snake. One such receptive area has been shaded for clarity. Note its overlapping position with other receptive areas of tectal neurons. (B) Arrangement of receptive areas of several tectal neurons on the pit floor of *Python reticulatus*. Note their increasingly caudal position in the more caudal pits and the change in their relative position. (A, after Terashima and Goris, 1976, 1979; B, after Haseltine, 1978.)

that extends from the receptive area into space and from which a tectal unit can be stimulated. The receptive fields that pertain to these receptive areas have an angle that varies between 25° and 70° (Hartline et al., 1978) (bulbar units 16° to 46°; peripheral units 37° to 85°).

In *Python*, the receptive fields range from 23° to 103°, and in *Corallus*, from 28° to 112°. The receptive areas of *Python* tectal units vary greatly in size and shape (Haseltine, 1978). The convergence from the receptive areas of peripheral units to form the receptive area of tectal units, as is present in *Agkistrodon blomhoffi*, is not present in *Python* (Fig. 5.21). The receptive area of peripheral units is 150 to 250 μm, whereas the area of tectal units is the same or smaller (Haseltine, 1978; De Cock Buning et al., 1981b). On the other hand, peripheral units are driven by only one receptive area (De Cock Buning et al. 1981b), whereas tectal units are driven by a combination of receptive areas with receptive fields that are similar but located separately in neighboring pits (Haseltine, 1978) (Fig. 5.21). This type of convergence is already present at the level of LTTD units (Molenaar, 1978c; Molenaar et al., 1979). Detailed studies concerning the organization of the receptive areas in *Python* have not yet been performed.

The receptive fields of some tectal units appear to be adjoined by an inhibitory region. Introduction of an infrared stimulus within this region depresses the background activity. Removal of the object evokes a postinhibitory rebound. Such inhibitory regions were found in crotalines (Hartline, 1974, *Sistrurus miliarius, Crotalus viridis, Agkistrodon piscivorus*; Hartline et al., 1978, *Crotalus viridis*) and boids (Haseltine, 1978, *Python reticulatus, Corallus enydris*). Such inhibitory regions could not be established by Terashima and Goris (1976) in *A. blomhoffi*. They doubt the existence of such inhibitory surrounds and explain the suppression of a unit's activity by the assumption that the stimulus in Hartline's experiments was too weak (an aluminium plate of 50° C) to stimulate the receptor unless the entire receptive area was covered by the image of the plate. This would only be the case in what Hartline called the excitatory center of the receptive field. Outside of it, the aluminium plate would function as a cool object that cuts off the background radiation, thus producing lower than background activity. The postinhibitory rebound would be evoked by the restoration of background radiation. Later studies (Hartline et al., 1978; Haseltine, 1978) using a hot soldering iron or a tungsten iodide lamp as a stimulus, or using two simultaneously presented stimuli for LTTD units (Stanford and Hartline, 1980), confirm the existence of the inhibitory areas. In addition, inhibitory activity was shown to play an important role in spatial sharpening of units in the LTTD of rattlesnakes (Stanford and Hartline, 1980). Moreover, the synaptic arrangement within the LTTD of pit vipers and *Python* shows the characteristics of inhibitory activity (Meszler et al., 1981; Meszler, 1983). Thus, it seems not unreasonable to assume inhibitory activities to take place in the tectum, too. It becomes the more likely, since complex excitatory and inhibitory interactions take place between visual and infrared bimodal units (Newman and Hartline, 1981).

4. SPATIOTOPY

The projection within the infrared pathway is organized in a way that permits orderly spatiotopic mapping. The mapping is organized in a way that is similar to the spatiotopic maps of the visual system. Thus, tectal units located rostrally in the tectum have rostral receptive fields;

caudal tectal units have caudal receptive fields. Units located in the medial parts of the tectum have dorsal receptive fields, and laterally located units have ventral receptive fields.

Although the ordering of the visual and infrared spatiotopic maps is similar, there are certain differences. In pit vipers, the receptive fields of the tectal infrared units are much larger (60°) than the visual fields (10°) (Terashima and Goris, 1975), and they are more overlapping. The topical relation between tectal neurons and the receptive area on the pit membrane of pit vipers is the exact reverse of that between tectal units and their receptive fields (Terashima and Goris, 1976) (Table 5.2). Thus units within the rostral part of the tectum receive stimuli from the rostral part of space via the caudal part of the pit membrane, whereas caudal units receive stimuli from caudal parts of space via the rostral part of the pit membrane. Units within the medial portion of the tectum receive stimuli from dorsal portions of space via the ventral parts of the membrane, whereas the lateral part of the tectum receives stimuli from the ventral portions of space via the dorsal part of the membrane. Apparently the somatotopic projection from the pit membrane to the tectum has become inverted—in contrast to the somatotopic projection of the common integument— in order to accommodate the crude pinhole camera optics of the pit organs (Otto, 1972) and to bring in register the visual and infrared spatiotopic maps.

There are detailed infrared and visual spatiotopic maps of the rattlesnake tectum (Hartline et al., 1978) (Fig. 5.22), which show the average receptive field centers of tectal multiple units at 200-μm interspaces. Each of these centers is specified by spherical coordinates (latitudes and longitudes; 0° is directly in front of the snake's nose). In the exposed tectal surface, the total infrared field amounts to 30° in the latitudinal direction and 60° in the longitudinal direction as compared to 40° and 110°, respectively, for the total visual field.

Consequently, the infrared system occupies the same tectal area as the visual system, although it covers a smaller part of space. In addi-

Table 5.2. Spatiotopical relationship between the position of a stimulus within the receptive field, the location of its receptive area on the pit membrane, and the location of the neurons within the tectum (crotalines and boids).

Receptive Field	Receptive Area	Tectum
Rostral	Caudal	Rostral
Caudal	Rostral	Caudal
Dorsal	Ventral	Medial
Ventral	Dorsal	Lateral

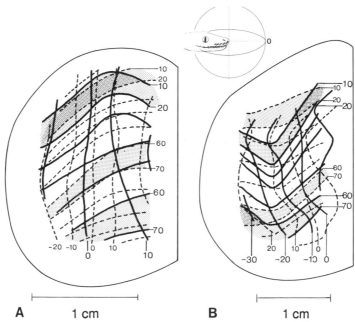

A　　1 cm　　　　　**B**　　1 cm

Fig. 5.22. Spatiotopic maps of visual (broken lines and small numbers) and infrared (heavy lines and large numbers) organization of the tectum in *Crotalus viridis* (A) and *Python reticulatus* (B). The longitudinal and latitudinal lines are drawn on the surface of the left superior colliculus and connect tectal loci, which have receptive field centers with identical spherical coordinates. The latitudinal lines run over the tectum in the rostrocaudal direction. They represent elevation above (positive values) or below (negative values) the horizontal plane (inset). The longitudinal lines run over the tectal surface in the lateromedial direction. They represent spatial positions in the rostrocaudal direction; 0° is directly in front of the snake's nose. For the purpose of clarity, the tectal representations of the longitudinal positions of the receptive field centers between 10° and 20°, as well as those between 60° and 70° of both the infrared and visual systems, have been shaded. Note the increasing disparity of the infrared and visual spatiotopic maps at more caudal positions, particularly in the pit viper. Scale bar indicates 1 cm (A, based upon Hartline et al., 1978; B, based upon Haseltine, 1978.)

tion, the infrared processing layer of the tectum is thicker than the visual layer: 600 μm as compared to 300 μm. Apparently, the tectal volume that is devoted to each particular region of space is larger in the infrared system than in the visual system.

The magnification factor is defined as the distance between two tectal loci whose receptive field centers are separated by 1°. The magnification factors for both the rostrocaudal and mediolateral tectal axes are greater for the infrared system than for the visual system. Consequently, there is no close spatial coincidence between the two maps, although they are arranged in similar axes. Coincidence between the visual and infrared maps is present only in the equatorial latitudinal lines, as well as in the longitudinal lines at about 20°. The more caudally the units are located in the tectum, the more the two maps become out of register. In the caudal tectal region, the centers of the visual fields are located 20° to 40° caudally to the centers of the infrared fields. The spatiotopic properties of bimodal units appear to reflect the disparity between the maps of the unimodal visual and infrared units.

In boids, the spatiotopic organization is similar to that in *Crotalus* (Haseltine, 1978). However, boids possess several pit organs, each with a limited field of view. The projection of the individual pits to the tectum is in register with the spatiotopic organization: the rostral pits, which view the rostral portions of space, are projected onto the rostral part of the tectum; successively more caudally viewing pits project to successively caudal parts of the tectum. Pits with strongly overlapping receptive fields have strongly overlapping projections to the tectum, which remain in register with the overall pattern of spatiotopic projection.

Within this pattern, the receptive areas of the pits project in a strictly reversed way to the tectum, similar to the projection from the receptive areas to the tectum in pit vipers. In addition, different receptive areas, which are located at different positions in neighboring pits but which share the same receptive fields, converge their projections to the same tectal neurons (see Fig. 5.21). The result of these additional projections is a coherent view of infrared space, surrounding the boid's head.

The visual field of one eye in *Python* is 180°; the combined field of view of all unilateral pits is also 180°. Correspondingly, the magnification factors of the infrared and visual systems are similar, and the infrared and visual maps are coincident (Haseltine, 1978).

The tectum plays an important role in the control of orientation movements and the spatial direction of attention in various vertebrates and probably also in the infrared-sensitive snakes (Hartline et al., 1978; Newman and Hartline, 1981). To facilitate orientation functions, several sensory modalities are integrated in the rostral colliculus in similar spatiotopic maps (Dräger et al., 1975; Stein et al., 1976; Stein and Gaither, 1983; Chalupa et al., 1977; Gaither and Stein, 1979; Newman and Hartline, 1981). The closer these spatiotopic maps coincide, the easier the orientation functions may be performed. Remarkably, there is no close coincidence between the infrared and visual maps in *Crotalus*. This seems to reduce the advantage of the similar arrangement of their axes. Yet in that part of the tectum and space in which alignment is most important for the orientation toward prey and aiming the strike (i.e., in front of the mouth), the infrared and visual maps do coincide.

5. PROPERTIES OF BIMODAL UNITS

In *Crotalus*, 103 out of 196 tectal units appear to show some degree of cross-modality interactions (Newman and Hartline, 1981). The units can be classified into six groups on the basis of their responses to either infrared or visual stimulation or to simultaneous bimodal stimulation.

Units have been classified as OR units when they responded well to both infrared or visual unimodal stimulation and gave combined responses to simultaneous visual-infrared stimulation. The combined responses of some units were either greater (cross-modality facilitation) or smaller (cross-modality occlusion) than the linear summation.

Units have been classified as AND units when they responded poorly or not at all to unimodal stimulation but gave reliable responses to simultaneous visual and infrared stimulation.

Bimodal infrared units could be subdivided when their responses were enhanced (visual enhanced infrared units) or depressed (visual depressed infrared units) by simultaneous visual and infrared stimulation. The degree of enhancement or depression varied considerably. In general, the weaker the primary response, the stronger the effect of simultaneous stimulation.

Similarly, the responses of bimodal visual units could be enhanced (infrared enhanced visual units) or depressed (infrared depressed visual units) by simultaneous stimulation.

The different types of neurons are located in the tectal layers as follows (Newman and Hartline, 1981): the unimodal visual cells are located only in the superficial white and gray layers; the infrared enhanced and depressed visual cells are located in the superficial part of the central gray layer and extend into the superficial layers; the visual enhanced and depressed infrared cells are located in the deep part of the central gray layer, whereas the AND and OR cells are located throughout the central gray layer.

The cross-modality interactions are equal in complexity to those in cortical cells of mammals and suggest an integratory activity that may enhance the spatial attention systems of the tectum or increase the sensitivity of feature detectors. Multimodal interactions occur in the tecta of many species, including the mammalian superior colliculus, and may comprise such different modalities as visual, somatic, auditory, and specialized infrared or electroreceptive modalities (cf. Stein and Gaither, 1983).

B. Forebrain Structures

1. THE NUCLEI ROTUNDUS AND PARAROTUNDUS The nuclei rotundus and pararotundus of *Python* receive direct bilateral projections from the lateral descending nucleus as well as indirect projections via the tectum. These tectal fibers originate from the central gray layers and ascend via the ventral tectothalamic tract (Welker et al., 1983). The tectal fibers envelop the complex of the nuclei rotundus and pararotundus and invade also the area that receives the direct bilateral projections from LTTD (Molenaar, 1983b).

The presence of the large complex of the nuclei rotundus and pararotundus in *Python* and *Crotalus* is remarkable (see Figs. 5.16 and 5.19). In the older literature, the nucleus rotundus was described as present in all reptiles except snakes and as forming a large circumscribed mass (Kuhlenbeck, 1977) that forms a link in a visual pathway from the optic tectum to the telencephalon (Hall and Ebner, 1970; Pritz, 1975; Parent, 1976; Lohman and Van Woerden-Verkley, 1978). In snakes, the nucleus rotundus was said to be extremely small or even absent and the tectorotundotelencephalic pathway to be reduced coincidentally with the enlargement of the dorsal part of the geniculate complex and the development of the visual tectodorsal geniculatetelencephalic pathway (Halpern and Frumin, 1973; Repérant, 1973; Northcutt and Butler, 1974; Ulinski, 1977).

The nuclear complex is quite well developed in *Crotalus* and *Python* (see Fig. 5.14) but inconspicuous in the common vipers (*Vipera aspis, V. ammodytes*). A well-demarcated nucleus rotundus is present in *Thammophis sirtalis* and receives a nonspatiotopically organized projection from the central gray layer of the tectum (Dacey and Ulinski, 1983).

The nuclear complex of *Python* receives direct bilateral projections from LTTD (Molenaar and Fizaan-Oostveen, 1980) as well as a unilateral tectal projection (Molenaar, 1983b; Welker et al., 1983). Its efferent projections seem to run to the dorsal ventricular ridge of the telencephalon (observation based on normal material, Molenaar, 1978c; Molenaar et al., 1980; P. V. J. M. Hoogland, 1983, preliminary HRP data, personal communication).

Other tectothalamic connections in *Python* form the presumably visual bilateral pathways, which run via the dorsal tectothalamic tract to the dorsal and ventral geniculate nuclei, and the probably somatosensory, reciprocal connection with the thalamic ventral nuclear complex. These tectothalamic connections do not seem to differ essentially from those in *Nerodia* (Ulinski, 1977) or *Crotalus* (Gruberg et al., 1979; Schroeder, 1981a).

Tectothalamic projections have been studied in *Crotalus* (Gruberg et al., 1979; Schroeder, 1981a; Berson and Hartline, 1988). In the first studies, no mention is made of a tectopararotundal projection resembling that in *Python*. Yet fiber terminations are depicted in a position that might be appropriate. The area under concern is called nucleus lentiformis thalami (Schroeder, 1981a). Other tectal efferent projections form the presumably visual pathways to the dorsal and ventral lateral geniculate nuclei and a pathway to portions of the thalamic ventral nuclear complex, resembling a similar tertiary somatosensory projection in *Python*. They do not seem to differ essentially from similar projections in *Nerodia* (Ulinski, 1977).

Nucleus rotundus and its connections have been identified in *Crotalus viridis* after HRP injections (Berson and Hartline, 1988). Nucleus rotundus receives a bilateral projection from neurons in the central gray and superficial gray and fibrous layers of the tectum via the lateral tectothalamic tract. The nucleus shows weak choline esterase activity.

2. THE ANTERODORSAL VENTRICULAR RIDGE

The nucleus rotundus sends its axons to the rostrolateral part of the ipsilateral part of the ipsilateral anterodorsal ventricular ridge (ADVR) in the telencephalon. The fibers ascend through the lateral forebrain bundle (Berson and Hartline, 1988).

3. PROPERTIES OF ROTUNDUS AND ADVR UNITS

Electrophysiological properties of rotundal and ADVR units are similar (Berson and Hartline, 1988) and will be described together. The units can be excited by visual stimulation, whereas most units also respond to unimodal infrared stimulation, and 20% need simultaneous visual and infrared stimulation. In addition, a few units respond also to vibrational stimulation. Receptive fields are very large, usually extending over the midline. The cells habituate rapidly, but movement of the same stimulus in the receptive field may evoke another brisk response. The cells apparently respond as much to the novelty of a stimulus as to its location or modality.

Neural activity has been recorded from the telencephalon of *Boa constrictor* upon stimulation of their radiant heat receptors (Harris and Gamow, 1972; Gamow and Harris, 1973).

The nuclear complex is quite well developed in *Crotalus* and *Python* (see Fig. 5.14); it is inconspicuous in the common vipers (*Vipera aspis*, *V. ammodytes*) and other snakes. Using cytoarchitectonic criteria, Warner (1947) identified the nuclear complex as the nucleus rotundus in *Crotalus*. A similar area seems to receive tectal afferents (Gruberg et al., 1979) in *Crotalus*, but the figures in this report lack the details to ascertain the exact site of the labeled fibers. Schroeder (1981a) depicts a similar area in *C. viridis* but labeled it nucleus lentiformis thalami. The lateral neuropil of this nucleus in her study receives a bilateral projection from the tectum. This situation closely resembles the situation in *Python* (Molenaar and Fizaan-Oostveen, 1980; Welker et al., 1983), yet Schroeder denies the existence of such a rotundopararotundal complex in *Crotalus* because it would not receive a tectal input.

These data suggest that a tectorotundo, pararotundotelencephalic pathway is present in infrared-sensitive snakes in spite of the presence of the well-developed tectodorsal geniculate-telecenphalic visual pathway.

C. Concluding Remarks

The responses of tectal units show irregularities that are related to stimulus characteristics in a way similar to that in peripheral units. Tectal units show a tendency to respond better to transient stimuli than peripheral units. The receptive fields of tectal units in pit vipers are larger than those of LTTD units and much larger than those of peripheral units (ratio 156–625 to 81–400 to 1). In *Python*, several receptive areas, which possess a similar receptive field and which are located separately in different pits, drive one tectal unit. The size of the receptive area of a tectal unit in *Python* is equal to that of a peripheral unit. The receptive fields of tectal units in *Python* and pit vipers contain an inhibitory area similar to that in LTTD units.

Within the tectum, the visual cells are located superficially, the infrared cells are deep, and the bimodal cells are intermediate. The spatiotopic maps on the tectal surface of visual and infrared-receptive field centers are oriented similarly, although in pit vipers they do not coincide as closely as in *Python*. Cross-modality interactions take place in bimodal infrared and visual neurons. These interactions are equal in complexity to those of tectal cells of mammals.

The tectum is involved in the production of natural orienting movements in response to sensory stimuli (Dacey and Ulinski, 1986e). In infrared-sensitive snakes, this means that the tectum is involved in localizing warm-blooded prey with the aid of infrared cues. The tectum participates in a transformation from topographically organized sensory input to a motor output. Upon a detailed study of cell morphology and the intrinsic and extrinsic connections of the tectum, Dacey and Ulinski (1986e) propose a model of the tectal sensorimotor transformation based upon the role of neuronal geometry in a mechanism for a graded cell population coding. The spatiotopic organization of the tectal part of the infrared system and the organization of the receptive fields as well as other electrophysiological properties of tectal infrared units fulfill the requirements for this function as reviewed by Stein and Gaither (1983) and Hartline (1984).

Two well-developed pathways conveying data from the tectum to the telencephalon are present in *Python* and *Crotalus*. One pathway is via the rotundus-pararotundus complex; the other is via the dorsal geniculate nucleus. The latter pathway is considered to be the lemniscal visual pathway (Berson and Hartline, 1988); the former is involved in infrared perception. A point-to-point relation is absent from the tectorotundal circuit. Unit properties in the rotundotelencephalic part of the infrared system differ from those in the tectal part. The most obvious changes are larger and bilateral receptive fields, sensitivity to multimodal stimuli, and pronounced habituation. Thus spatial and modality selectivity have been reduced, whereas susceptibility to novelty of a stimulus is enhanced. It appears that the rotundotelencephalic circuit is not primarily involved in stimulus localization but in the detection and discrimination of multisensory objects.

VI. BIOLOGICAL ROLE OF THE INFRARED-SENSITIVE SYSTEM
A. Spatial Orientation

To perform a successful strike, a snake needs accurate information on the position of its prey. Since an infrared-sensitive snake is capable of striking accurately on the guidance of the pit organs alone, these receptors are apparently capable of providing the necessary information. To do so, these receptors must accurately assess the direction and distance of the prey.

Because pit organs lack an optical device, they require a mechanism to improve their ability to locate prey precisely. This is provided in pit vipers by the specific combination of large tectal receptive areas and a low transference ratio (see Section III.C.6). This combination causes a loss in sensitivity but a gain in angular resolution (De Cock Buning, 1984). Before striking, the snake makes orientational maneuvers with its head in order to position the prey in front of it. During these movements, the prey shifts from a lateral to a rostral position with respect to the snake's head and, in so doing, it shifts from a lateral position in the receptive field of the pits to a rostral position. The lateral parts of the receptive fields of tectal units contain inhibitory regions (Hartline et al., 1978; Haseltine, 1978). Thus the prey moves out of the inhibitory regions and into the excitatory parts of the receptive fields. As a result, such rostral movements abruptly increase the firing rate of the tectal units; movements in the opposite direction decrease the firing rate. The prey is in exactly the most favorable rostral position when both pits receive an equal-intensity profile.

Distance assessment is basically a sort of trigonometrical surveying of the intensity profiles perceived by both pits. The accuracy of this method depends on the exactness with which the position of the maxima of the intensity profiles is demarcated. The higher-order neurons, with a lateromedial arrangement of inhibitory and excitatory regions within their receptive fields, might be the detectors of that demarcation.

In *Python*, the gain in angular resolution and the accurate assessment of the direction of the prey has to be obtained in a different way because of the fast-adapting character of the dynamic responses and the weak (or absent) static responses of their infrared system. Only an increase of infrared radiation evokes a short activity, whereas both a decrease and a steady state evoke no response in the *Python* system. Yet spatial localization and directional sensitivity may be achieved. This is possible in particular by means of the geometry and compartmentalization of the pit organs. In general, the pit organs have their main axis of view in the rostral direction. Higher-order neurons are driven by several receptive areas located increasingly caudally in adjacent pits. Thus more receptive areas are illuminated at a rostral position of the stimulus than at a lateral position. Accordingly, movement of a stimulus from a lateral to a rostral position constantly increases illumination and thus evokes a constant response, whereas a lateral displacement of the stimulus decreases illumination and thus evokes no response. In this way the pit organs in *Python* may function as excellent detectors of position and direction.

To date, detailed data on response profiles and neuronal connections in boids are lacking.

B. Role in Behavior

The role of infrared sensitivity in prey capture and feeding behavior has been studied in several snakes. Earlier studies showed the importance of the infrared-sensitive system in guiding the strike with an accuracy up to 5° (Noble and Schmidt, 1937; Dullemeijer, 1961; Newman and Hartline, 1982). In recent studies, De Cock Buning et al. (1978, 1981c) and De Cock Buning (1983b) distinguished nine successive phases in this behavior and studied the role of four senses (vision, vibration, chemoreception, and infrared reception) in these phases in snakes with and without infrared sensitivity.

The nine phases may be clustered into three subgroups: prestrike, strike, and poststrike behavior. Prestrike behavior involves alertness reactions, orientational movements ("aiming"), and approaching the prey. Strike follows if the prey is within the appropriate distance. Poststrike behavior involves relocating the prey after it has been killed, spotting its head, and swallowing. Poststrike behavior is guided mainly by chemical or tactile cues in all snakes. Prestrike and strike behavior are guided by the different senses in different ways in various groups of snakes.

De Cock Buning (1983b) distinguished five types of behavior, which he classified according to the role of the various receptors in these phases: colubrid and viperid in snakes without infrared sensitivity and *Crotalus, Python,* and *Boa* types in snakes possessing infrared sensitivity.

In colubrids, prestrike behavior is mainly guided by visual and chemical cues. In the viperid type, low-frequency vibration plays an important role, in addition to visual and chemical cues.

In the crotalid type, infrared sensitivity plays a different role in snakes in which the behavior is guided predominantly by their visual (*Crotalus*) or chemical (*Agkistrodon*) senses. In chemically guided snakes, infrared sensitivity is essential to elicit prestrike behavior; in visually guided snakes infrared sensitivity is additional to visual cues. In both types of snakes, visual and infrared data are supplementary in the strike.

In the *Python* type, visual and infrared data are most important but require the presence of chemical and vibrational cues, although the latter are in themselves insufficient to elicit the strike.

In *Boa,* infrared cues are less important, whereas chemical cues and vibrations are most important. This type resembles that of pitless snakes, which rely on their chemical senses.

The relative importance of infrared perception in the behavior of these snakes is reflected in the threshold sensitivity of their pit organs. The values in threshold sensitivity are in *Agkistrodon* 2.55, in *Python* 14.28, and in *Boa* 48.28 10^{-6} cal/cm^2.s.

De Cock Buning (1983b) concludes that vision and chemoreception play approximately the same role in snakes with or without pit organs. The main difference is found in the role of the thermal and mechanoreception. In those snakes possessing radiant heat perception, this ability overrules the importance of vibration.

The interaction between infrared and visual data in triggering a specific type of behavior is in agreement with the presence of strong cross-modality interactions between visual and infrared unimodal and bimodal units as described in the tectorotundo telencephalic pathway of *Crotalus* (Hartline et al., 1978; Kass et al., 1978; Newman et al., 1981; Berson and Hartline, 1988).

VII. CONCLUSIONS
A. Comparison of Thermosensitive Pathways
1. GENERAL The pit organs are specially constructed sense organs that function by detecting physical characteristics of distant objects (telereceptors). Their presence in the snake's head changes the conformation of the entire craniofacial region (Dullemeijer, 1959, 1961, 1969, 1974). The organs project to a specific nucleus, which in turn

gives rise to a special ascending projection. In these peculiarities the infrared system is similar to other special sense organs, such as nose, eye, and ear. These organs also function as telereceptors, and their presence has consequences for the shape of the head. They are innervated by separate cranial nerves and project to specific brain areas, which in turn give rise to specific secondary projections. The pit organs differ from the other special sense organs in that they are not innervated by a separate cranial nerve.

The system is present in two distantly related families of snakes: the primitive Boidae and in one subfamily (Crotalinae) of the advanced Viperidae. It is absent in other advanced snakes, including the other subfamily (Viperinae) of the Viperidae (Bullock and Barrett, 1968; Barrett, 1970; Maderson, 1970). Thus different snakes may live successfully in the same ecological niches regardless of the presence or absence of these sense organs. In addition, the system is also present in certain mammals (Kürten and Schmidt, 1982; Kishida et al., 1984b). Therefore, one is led to speculate on the ancestry of the infrared system. Did this system evolve especially in these animals, perhaps as an elaboration of some kind of ancestral system common to all vertebrates? Or is it perhaps an elaboration of a common sensory system, in particular the cutaneous thermosensory system? Both ideas are supported by several indications. Section I.A.2 considers arguments supporting the idea that the IR system is an elaboration of a common thermosensory system, and Section V.A.3 considers arguments against such a homology.

2. ARGUMENTS FAVORING HOMOLOGY

a. SPECIFICITY OF THE SENSORS   Behavioral studies of the role of various sense organs in feeding behavior (De Cock Buning, 1983b) have shown that the main difference between snakes with or without pit organs is found in the role of thermal and mechanical (that is, vibrational) cues. In those snakes possessing the ability to perceive radiant heat, this overrules the importance of mechanoreception.

The receptors in *Python* appear to be nonspecific bimodal heat and tactile receptors (De Cock Buning et al., 1981b). Bimodal, heat-sensitive mechanoreceptors are present in other places of the reptilian integument (Kenton et al., 1971). Both mechanoreceptors and infrared receptors consist of blind nerve terminals in the skin.

The infrared receptors of pit vipers can be regarded as an aggregation of unimodal warm receptors. Their extreme sensitivity can be explained by the low heat capacity of the pit membrane (De Cock Buning et al., 1981a), and the infrared spatial properties are a direct consequence of the geometry of the pit organ (De Cock Buning, 1984). It might be postulated from these indications that infrared receptors are an elaboration of common thermoreceptors, which in turn have arisen from mechanoreceptors.

b. SENSORY PATHWAYS

Radiant heat data are mediated by the pit organs via the trigeminal nerve fibers to the lateral descending nucleus (LTTD). In pit vipers, the data are conveyed from the LTTD to the tectum via the nucleus reticularis caloris. Further ascending relay takes place via the nucleus rotundus toward the anterodorsal ventricular ridge in the telencephalon. The latter pathway also forms the "extralemniscal" visual pathway and is, as such, opposed to the strictly visual "lemniscal" retinogeniculotelencephalic system (Berson and Hartline, 1988). In *Python*, a similar projection is present, although with a few interspecies variations: the reticular cells appear to be incorporated within the LTTD and a direct bilateral projection to the nucleus rotundus-pararotundus is present. The existence of a tectorotundotelencephalic pathway is likely but needs experimental verification.

Thermal and crude mechanical data are extralemniscal sensory modalities. They are conveyed via the trigeminal nucleus caudalis. The nucleus caudalis in mammals projects directly to the thalamus and via the reticular formation to the tectum (Tiwari and King, 1973, 1974; Baleydier and Mauguiere, 1978; Burton et al., 1979; Edwards et al., 1979; Killacky and Erzurumlu, 1981; Rhoades, 1981).

One attractive hypothesis would be to consider the infrared pathway as an elaboration of this thermosensitive pathway. In this view, the pathway from the LTTD via the nucleus reticularis caloris to the tectum might be derived from the pathway from the nucleus caudalis via the reticular formation to the tectum. In *Python*, the large cells comprising the nucleus reticularis caloris are incorporated within the LTTD; the infrared pathways in *Python* and pit vipers thus appear to be similar.

Although conclusive evidence is lacking, a few observations support this hypothesis. First, at the ultrastructural level, the LTTD contains a homogeneous population of simplified glomeruli that resemble some of the glomeruli in the nucleus caudalis of mammals (Meszler et al., 1981; Meszler, 1983). Moreover, the efferent LTTD fibers in pit vipers follow a route that is initially similar to that of the efferent fibers of the nucleus caudalis in *Python*, that is, they run toward the dorsal arcuate fiber system. Although the efferent projections of the nucleus caudalis in pit vipers have not been established, it seems justified to assume that they do not differ markedly from those in *Python*. The efferent fibers of the nucleus caudalis decussate via the dorsal arcuate fiber system, but those of the pit viper LTTD turn laterally and run to the nucleus reticularis caloris.

The infrared data are processed within the tectum and forebrain in relation with visual information. Similar relationships exist for the processing of thermosensitive and tactile data.

3. ARGUMENTS AGAINST HOMOLOGY

a. SPECIFICITY OF THE RECEPTORS   Although presently nothing is known of the transductory processes at the receptor membrane, it appears that the sensory nerve endings do have specific peculiarities that distinguish them from common cutaneous receptors.

The nerve endings form a specific terminal nerve mass that is formed in close relation with Schwann cells (Terashima et al., 1970; Hirosawa, 1980). In this terminal nerve mass, vesicles are related with the sensory membrane and may play a role in transductory processes. Such vesicles are peculiar to the infrared receptors and are not present in other cutaneous receptors (Von Düring, 1974; Hirosawa, 1980). In addition, the terminal nerve mass is extremely densely packed with extraordinarily large mitochondria (Meszler, 1970) and contains extremely high concentrations of oxidizing enzymes (Meszler and Webster, 1968; Meszler, 1970).

A common temperature sensitivity in the skin subserving thermoregulatory behavior may be expected in all snakes, including those with radiant heat sensitivity. This type of thermosensitivity is extralemniscal in nature and is probably similar to that in other vertebrates. The bimodal thermosensitive mechanoreceptors that have been located in the reptilian skin (Kenton et al., 1971) are probably related to this function and may also be present in infrared-sensitive snakes.

Behavioral studies have suggested a functional relationship between infrared and certain mechanical stimuli (De Cock Buning, 1983b). These mechanoreceptors are particularly sensitive to vibrations, which may be airborne sound or substrate vibrations. These vibrations are mediated by two systems—the auditory and the somatic system. Neither of the two systems involves the trigeminal nerve.

The auditory system involves the quadrate bone and the columnella auris, which transmit vibrations to the inner ear. The data are conveyed by the eighth cranial nerve. The somatic system involves cutaneous receptors, which are absent from the head (Hartline and Campbell, 1969; Hartline, 1971a, 1971b). In pit vipers and boids, the infrared units in the pit organs are insensitive to these vibrations (Goris and Terashima, 1973; Hartline, 1974; Molenaar, 1978c).

In short, infrared sensitivity is apparently not related to vibration sensitivity, whereas the common cutaneous thermosensory or tactile receptors are not necessarily similar to infrared receptors. To explore the matter further, it is necessary to obtain other than circumstantial evidence. Thus, the occurrence and properties of common thermosensory units in the integument of the head in infrared-sensitive snakes should be studied and compared with those of snakes not sensitive to infrared.

b. Sensory Pathways

The small, darkly staining cells in the trigeminal ganglia, which in mammals are considered to be involved in thermoreception (Gobel, 1974), are not connected to the pit membrane in the crotaline *Agkistrodon blomhoffi* (Kishida et al., 1982).

In *Python*, the full set of the common sensory trigeminal nuclei is present. It includes those parts that in mammals subserve cutaneous thermoreception and it has similar projections as in mammals (Molenaar, 1978a, 1978b; Molenaar and Fizaan-Oostveen, 1980). In addition, a distinct trigeminal structure, the LTTD, is involved in radiant heat perception. A trigeminal structure, which may be regarded as the homologue of the LTTD, has been identified in a variety of vertebrates. The earliest report is from Van Valkenburg (1911) concerning a number of vertebrates. Recent reports also mention the existence of the lateral descending system in the pigeon (Dubbeldam and Karten, 1978) and the vampire bat (Kishida et al., 1984b). In addition, the LTTD in *Python* and *Boa* (Kishida et al., 1983a) and *Agkistrodon* (Kishida et al., 1984a) receive primary afferent vagus C fibers in their marginal neuropil. The LTTD of the pigeon (Dubbeldam and Karten, 1978) receives a similar projection (Dubbeldam, 1984).

The cytoarchitecture of the LTTD is completely different from that of the nucleus caudalis. In particular the large multipolar cells that are so characteristic of the LTTD of both pit vipers and boids are absent from the nucleus caudalis in other snakes and mammals.

Finally, although the ultrastructural organization of the nucleus caudalis consists of several types of glomeruli, it is not known which of these pertains to thermoreception. The presence of a simplified glomerular arrangement within the LTTD thus forms merely circumstantial evidence.

Apart from the considerations discussed above, we emphasize the biological role of the pit organs as specialized telereceptors. Their conformations are special, and their presence requires specific adaptations of the craniofacial region in order to accommodate them. Even though they are innervated by branches of the trigeminal nerve, as are the common cutaneous receptors, they possess special central projections toward special nuclear structures. Their central connections remain completely separated from the common sensory trigeminal connections, at least to the level of the third-order neurons (tectum and thalamus).

These indications support the hypothesis that the facial and labial pit organs have evolved independently of other cutaneous thermoreceptors as specializations that give rise to a separate trigeminal pathway and that may have developed by elaboration of an ancestral system common to all vertebrates. It is obvious that additional evidence to supplement our present fragmentary knowledge of this system is needed.

ACKNOWLEDGMENTS
I wish to thank the following persons for their help in the preparation of this manuscript: Dr. A. A. Macdonald and Dr. D. M. Badoux for criticizing the text; Dr. P. V. J. M. Hoogland and Dr. H. J. ten Donkelaar for kindly lending some of their experimental materials; Mr. H. Halsema and Mr. A. R. Janssen for the drawings; Mr. H. H. Otter for the photographs; Mrs. R. A. Goede for technical assistance; and Mrs. L. J. M. Michielsen for typing.

APPENDIX: REPTILIAN SPECIES DISCUSSED

TESTUDINES

Chrysemys picta
  Parent, 1976
Pelomedusa subrufa
  Pedersen, 1973
Podocnemis unifilis
  Pedersen, 1973

Pelusios subniger
  Pedersen, 1973
Trachemys scripta
  Hall and Ebner, 1970

CROCODILIA

Caiman crocodilus
  Pritz, 1975

SAURIA

Anolis
  Woodburne, 1983
Iguana iguana
  Gaither and Stein, 1979
  Stein and Gaither, 1983

Podarcis sicula
  Senn, 1968
Tupinambis teguixin
  Lohman and Van Woerden-
  Verkley, 1978

SERPENTES

Agkistrodon blomhoffi
  Goris and Terashima, 1976
  Kadota et al., 1988

Kadota et al., 1984
Kishida et al., 1980, 1981, 1982,
  1984a

Agkistrodon piscivorus
    Meszler et al., 1981
    Hartline et al., 1978
Boa (Constrictor) constrictor
    De Cock Buning, 1983a
    De Cock Buning et al., 1981a
    Harris and Gamow, 1971
    Hensel, 1974, 1975
    Hensel and Schaefer, 1981
    Kishida et al., 1983a
    Schroeder and Loop, 1976
    Von Düring, 1974
Calloselasma rhodosfoma
    De Cock Buning, 1983a
    De Cock Buning et al., 1981a,
        1981b
Corallus caninus
    Bullock and Barrett, 1968
    Noble and Schmidt, 1937
Corallus enydris
    Haseltine, 1978
    Meszler, 1981, 1983
    Noble and Schmidt, 1937
    Schroeder and Loop, 1976
Crotalus sp.
    Berson and Hartline, 1988
    Bullock and Diecke, 1956
    De Salvo and Hartline, 1978
    Hartline et al., 1978
    Terashima and Goris, 1976, 1977,
        1979, 1983
    Terashima et al., 1968, 1970
Crotalus adamanteus
    Warner, 1974
Crotalus atrox
    Molenaar, 1974
Crotalus cerastes
    Molenaar, 1974
Crotalus horridus
    Auen, 1978
    Meszler et al., 1981
Crotalus ruber
    Auen, 1978
    Meszler et al., 1981
Crotalus scutulatus
    Auen, 1978
Crotalus viridis
    Bullock and Fox, 1957

Chiszar et al., 1986
Dickman et al., 1978
Gruberg et al., 1979
Gruberg et al., 1984
Hartline et al., 1978
Meszler, 1981
Meszler, 1983
Meszler and Webster, 1968
Newman et al., 1980
Schroeder, 1981a, 1981b
Schroeder and Loop, 1976
Stanford and Hartline, 1980, 1984
Stanford et al., 1981
Elaphe quadrivirgata
    Kishida et al., 1980
Epicrates striatus
    Meszler, 1981, 1983
Morelia spilotes
    Warren and Proske, 1968
Natrix natrix
    Armstrong, 1951
Nerodia sipedon
    Northcutt and Butler, 1974
    Ulinski, 1977
Python sp.
    Haseltine, 1978
    Ross, 1935
Python molurus
    Kishida et al., 1983a
Python regius
    Molenaar, unpublished
Python reticulatus
    De Cock Buning, 1983a
    De Cock Buning and Dullemeijer,
        1977
    De Cock Buning et al., 1978, 1980,
        1981b, 1981c
    Haseltine, 1978
    Molenaar, 1977, 1978a to 1978c
    Molenaar and Fizaan-Oostveen,
        1980
    Molenaar et al., 1979
    Newman and Hartline, 1980
    Tan and Gopalakrishnakone, 1988
    Welker et al., 1983
Sistrurus miliarius
    Hartline, 1974

Thamnophis sirtalis
  Dacey and Ulinski, 1986a to 1986e
  Halpern and Frumin, 1973
Trimeresurus flavoviridis
  Bullock and Fox, 1957
  Goris and Terashima, 1973
  Hirosawa, 1980
  Terashima, 1987
  Terashima and Goris, 1976
  Terashima et al., 1970
Vipera ammodytes
  Repérant, 1973
Vipera aspis
  Repérant, 1973
Pit viper
  Bleichmar and De Robertis, 1962

Bullock and Cowles, 1952
Bullock and Barrett, 1968
Kass et al., 1978
Lynn, 1931
Noble, 1934
Rattlesnake
  Auker et al., 1983
  Gamow and Harris, 1973
  Hartline and Campbell, 1969
  Harris and Gamow, 1971, 1972
  Dullemeijer, 1961
Crotaline snakes
  Goris and Nomoto, 1967
Viperidae
  Dullemeijer, 1959
  Dullemeijer, 1969

REFERENCES

Armstrong, J. A. (1951). An experimental study of the visual pathways in a snake (*Natrix natrix*). *J. Anat.* 85, 275–289.
Auen, E. L. (1978). Axonal transport of horseradish peroxidase in descending tectal fibres of the pit viper. *Neurosci. Lett.* 9, 137–140.
Augustine, J. R. (1974). Certain experimentally demonstrated connections of the chief sensory trigeminal nucleus in the squirrel monkey (*Saimiri sciureus*). *Ala. J. Med. Sci.* 11, 277–301.
Auker, C. R., Meszler, R. M., and Carpenter, D. O. (1983). Apparent discrepancy between single unit activity and ¹⁴C deoxyglucose labeling in optic tectum of the rattlesnake. *J. Neurophysiol.* 49, 1504–1516.
Baleydier, C., and Mauguiere, F. (1978). Projections of the ascending somesthetic pathways to the cat superior colliculus visualized by the horseradish peroxidase technique. *Exp. Brain Res.* 31, 43–50.
Barrett, R. (1970). The pit organs of snakes. In *Biology of the Reptilia*, Vol. 2. *Morphology B* (C. Gans and T. S. Parsons, eds.). Academic Press, London, New York, San Francisco, pp. 277–300.
Belekhova, M. G. (1979). Neurophysiology of the forebrain. In *Biology of the Reptilia*, Vol. 10, *Neurology B* (C. Gans, R. G. Northcutt, and P. Ulinski, eds.). Academic Press, London, New York, San Francisco, pp. 287–359.
Berson, D. M., and Hartline, P. H. (1988). A tecto-rotundo-telencephalic pathway in the rattlesnake: evidence for a forebrain representation of the infrared sense. *J. Neurosci.* 8, 1074–1088.
Bleichmar, H., and De Robertis, E. (1962). Submicroscopic morphology of the infrared receptor of pit vipers. *Z. Zellforsch. mikrosk. Anat.* 56, 748–761.
Bullock, T. H., and Cowles, R. B. (1952). Physiology of an infrared receptor: the facial pit of pit vipers. *Science* (N.Y.) 115, 541–543.

Bullock, T. H., and Diecke, F. P. J. (1956). Properties of an infrared receptor. *J. Physiol.* (London) 134, 47–87.

Bullock, T. H., and Fox, W. (1957). The anatomy of the infrared sense organ in the facial pit of pit vipers. *Q. J. Microsc. Sci.* 98, 219–234.

Bullock, T. H., and Barrett, R. (1968). Radiant heat reception in snakes. *Commun. Behav. Biol.*, Part A, 1, 19–29.

Burton, H., Craig, A. D., Jr., Poulos, D. A., and Molt, J. T. (1979). Efferent projections from temperature sensitive recording loci within the marginal zone of the nucleus caudalis of the spinal trigeminal complex in the cat. *J. Comp. Neurol.* 183, 735–778.

Cajal, S. R. (1909). Histologie du système nerveux de l'homme et des vertebrés, Vol. 1. Maloine, Paris.

Chalupa, L. M., and Rhoades, R. W. (1977). Responses of visual somatosensory and auditory neurons in the golden hamster's superior colliculus. *J. Physiol.* (London) 270, 595–626.

Chiszar, D., Dickman, D., and Colton, J. (1986). Sensitivity to thermal stimulation in prairie rattlesnakes (*Crotalus viridis*) after bilateral anesthetization of the facial pits. *Behav. Neurol. Biol.* 45, 143–149.

Clarke, W. L., and Bowsher, D. (1962). Terminal distribution of primary afferent trigeminal fibres in the rat. *Exp. Neurol.* 6, 372–383.

Cordier, R. (1964). Sensory cells. In *The Cell*, Vol. 6 (J. Bracket and A. Mirsky, eds.). Academic Press, London, New York, pp. 313–386.

Dacey, D. M., and Ulinski, P. S. (1983). Nucleus rotundus in a snake, *Thamnophis sirtalis:* an analysis of a non-retinotopic projection. *J. Comp. Neurol.* 216, 175–191.

Dacey, D. M., and Ulinski, P. S. (1986a). Optic tectum of the eastern garter snake, *Thamnophis sirtalis*. I. Efferent pathways. *J. Comp. Neurol.* 245, 1–28.

Dacey, D. M., and Ulinski, P. S. (1986b). Optic tectum of the eastern garter snake, *Thamnophis sirtalis*. II. Morphology of efferent cells. *J. Comp. Neurol.* 245, 198–237.

Dacey, D. M., and Ulinski, P. S. (1986c). Optic tectum of the eastern garter snake, *Thamnophis sirtalis*. III. Morphology of intrinsic neurons. *J. Comp. Neurol.* 245, 283–300.

Dacey, D. M., and Ulinski, P. S. (1986d). Optic tectum of the eastern garter snake, *Thamnophis sirtalis*. IV. Morphology of afferents from the retina. *J. Comp. Neurol.* 245, 301–318.

Dacey, D. M., and Ulinski, P. S. (1986e). Optic tectum of the eastern garter snake, *Thamnophis sirtalis*. V. Morphology of brainstem afferents and general discussion. *J. Comp. Neurol.* 245, 423–453.

Darian-Smith, I. (1965). Presynaptic component in the afferent inhibition observed within trigeminal brainstem nuclei of the cat. *J. Neurophysiol.* 28, 695–709.

Darian-Smith, I. (1973). The trigeminal system. In *Handbook of Sensory Physiology*, Vol. II. Somatosensory system (A. Iggo, ed.). Springer Verlag, Berlin, Heidelberg, New York, pp. 271–314.

Darian-Smith, I., Proctor, R., and Ryan, R. D. (1963). A single neuron investigation of somatotopic organization within the cat's trigeminal brainstem nuclei. *J. Physiol.* (London) 168, 147–158.

De Cock Buning, T. (1983a). Thresholds of infrared sensitive tectal neurons in *Python reticulatus, Boa constrictor* and *Agkistrodon rhodostoma. J. Comp. Physiol.* A151, 461–468.

De Cock Buning, T. (1983b). Thermal sensitivity as a specialization for prey capture and feeding in snakes. *Am. Zool.* 23, 363–375.

De Cock Buning, T. (1984). A theoretical approach to the heat sensitive pit organs of snakes. *J. Theor. Biol.* 111, 509–530.

De Cock Buning, T., and Dullemeijer, P. (1977). Thermoreceptors in *Python reticulatus. Acta Morphol. Neerl. Scand.* 15, 237–239.

De Cock Buning, T., Poelmann, R. E., and Dullemeijer, P. (1978). Feeding behavior and the morphology of the thermoreceptors in *Python reticulatus. Neth. J. Zool.* 28, 62–93.

De Cock Buning, T., Terashima, S., and Goris, R. C. (1980). Properties of response to various thermal stimuli in pit organs of crotaline snakes. *Neurosci. Lett.* Suppl. 4, S20.

De Cock Buning, T., Terashima, S., and Goris, R. C. (1981a). Crotaline pit organs analyzed as warm receptors. *Cell. Mol. Neurobiol.* 1, 69–85.

De Cock Buning, T., Terashima, S., and Goris, R. C. (1981b). *Python* pit organs analyzed as warm receptors. *Cell. Mol. Neurobiol.* 1, 271–278.

De Cock Buning, T., Goris, R. C., and Terashima, S. (1981c). The role of thermosensitivity in the feeding behavior of the pit viper *Agkistrodon blomhoffi brevicaudus. Jap. J. Herpetol.* 9, 7–27.

De Salvo, J. A., and Hartline, P. H. (1978). Spatial properties of primary infrared sensory neurons in *Crotalidae. Brain Res.* 142, 338–342.

Dickman, J. D., Colton, J. S., Chiszar, D., and Colton, C. A. (1987). Trigeminal responses to thermal stimulation of the oral cavity in rattlesnakes (*Crotalus viridis*) before and after bilateral anesthetization of the facial pit organs. *Brain Res.* 400, 365–370.

Dräger, U. C., and Hubel, D. H. (1975). Responses to visual stimulation and relationship between visual, auditory and somatosensory inputs in mouse superior colliculus. *J. Neurophysiol.* 38, 690–713.

Dubbeldam, J. L. (1984). Afferent connections of N. facialis and N. glossopharyngeus in the pigeon (*Columba livia*) and their role in feeding behavior. *Brain Behav. Evol.* 24, 47–57.

Dubbeldam, J. L., and Wijsman, J. P. M. (1975). The ascending projections of the principal sensory trigeminal nucleus in the mallard (*Anas platyrhynchos*). *Acta Morphol. Neerl. Scand.* 13, 230–231.

Dubbeldam, J. L., and Karten, H. J. (1978). The trigeminal system in the pigeon (*Columba livia*). I. Projection of the Gasserian ganglion. *J. Comp. Neurol.* 180, 661–678.

Dullemeijer, P. (1959). A comparative functional-anatomical study of the heads of some *Viperidae. Morph. Jb.* 99, 881–985.

Dullemeijer, P. (1961). Some remarks on the feeding behavior of rattle snakes. *Proc. Acad. Wetensch. Amsterdam,* Ser. C, 64, 383–396.

Dullemeijer, P. (1969). Growth and size of the eye in viperid snakes. *Neth. J. Zool.* 19, 249–276.

Dullemeijer, P. (1974). *Concepts and Approaches in Animal Morphology.* Van Gorcum, Assen, The Netherlands.

Duncan, D., and Morales, R. (1978). Relative numbers of several types of synaptic connections in the substantia gelatinosa of the cat spinal cord. *J. Comp. Neurol.* 182, 601–610.

Ebbesson, S. O. E., and Heimer, L. (1970). Projection of the olfactory tract fibers in the nurse shark (*Ginglystoma cirratum*). *Brain Res.* 17, 45–57.

Edwards, S. D., Ginsburgh, C. L., Henkel, C. K., and Stein, B. E. (1979). Sources of subcortical projections to the superior colliculus in the cat. *J. Comp. Neurol.* 184, 309–330.

Erzurumlu, R. H., and Killackey, H. P. (1980). Diencephalic projections of the subnucleus interpolaris of the brainstem trigeminal complex in the rat. *Neuroscience* 5, 1891–1902.

Fink, R. P., and Heimer, L. (1967). Two methods for selective silver impregnation of degenerating axons and their synaptic endings in the central nervous system. *Brain Res.* 4, 369–374.

Fuller, P. M., and Ebbesson, S. O. E. (1973). Central projections of the trigeminal nerve in the bull frog. *J. Comp. Neurol.* 152, 193–200.

Gaither, N. S., and Stein, B. E. (1979). Reptiles and mammals use similar sensory organizations in the midbrain. *Science* (N.Y.) 205, 595–597.

Gamow, R. I., and Harris, J. F. (1973). The infrared receptors of snakes. *Sci. Am.* 228, 94–100.

Gobel, S. (1974). Synaptic organization of the substantia gelatinosa glomeruli in the spinal trigeminal nucleus of the adult cat. *J. Neurocytol.* 3, 219–243.

Gobel, S., and Purvis, M. B. (1972). Anatomical studies of the organization of the spinal V nucleus: the deep bundles and the spinal V tract. *Brain Res.* 48, 27–44.

Goris, R. C., and Nomoto, M. (1967). Infrared reception in oriental crotaline snakes. *Comp. Biochem. Physiol.* 23, 879–892.

Goris, R. C., and Terashima, S. (1973). Central response to infrared stimulation of the pit receptors in a crotaline snake (*Trimeresurus flavoviridis*). *J. Exp. Biol.* 58, 59–76.

Goris, R. C., and Terashima, S. (1976). The structure and function of infrared receptors of snakes. In *Progress in Brain Research*, Vol. 43 (A. Iggo and O. B. Ilyinsky, eds.), pp. 159–170. Elsevier, Amsterdam, Oxford, New York.

Gruberg, E. R., Kicliter, E., Newman, E. A., Kass, L., and Hartline, P. H. (1979). Connections of the tectum of the rattle snake *Crotalus viridis*: an HRP study. *J. Comp. Neurol.* 188, 31–42.

Gruberg, E. R., Newman, E. A., and Hartline, P. H. (1984). 2 Deoxyglucose labelling of the infrared sensory system in the rattlesnake, *Crotalus viridis*. *J. Comp. Neurol.* 229, 321–328.

Hall, W. C., and Ebner, F. F. (1970). Parallels in the visual afferent projections of the thalamus in the hedgehog (*Parechinus hypomelas*) and the turtle (*Pseudemys scripta*). *Brain Behav. Evol.* 3, 135–154.

Halpern, M., and Frumin, N. (1973). Retinal projections in a snake (*Thamnophis sirtalis*). *J. Morphol.* 141, 359–382.

Hamburger, V. (1961). Experimental analysis of the dual origin of the trigeminal ganglions in the chick embryo. *J. Exp. Zool.* 148, 91–124.

Harris, J. F., and Gamow, R. I. (1971). Snake infrared receptors: thermal or photochemical mechanism. *Science* (N.Y.) 172, 1252–1253.

Harris, J. F., and Gamow, R. I. (1972). An analysis of heat receptors by means of microwave radiation. *Biomed. Sci. Instrum.* 9, 187–190.

Hartline, P. H. (1971a). Physiological basis for detection of sound and vibration in snakes. *J. Exp. Biol.* 54, 349–371.

Hartline, P. H. (1971b). Midbrain responses of the auditory and somatic vibration systems in snakes. *J. Exp. Biol.* 54, 373–390.

Hartline, P. H. (1974). Thermoreception in snakes. In *Handbook of Sensory Physiology*, Vol. III. Electroreceptors and other specialized receptors in lower vertebrates (A. Fessard, ed.). Springer Verlag, Berlin, Heidelberg, New York, pp. 297–312.

Hartline, P. H. (1984). The optic tectum of reptiles: neurophysiological studies. In *Comparative Neurology of the Optic Tectum* (H. Vanegas, ed.). New York, Plenum Press, pp. 601–618.

Hartline, P. H., and Campbell, H. W. (1969). Auditory and vibratory responses in the midbrains of snakes. *Science* (N.Y.) 163, 1221–1223.

Hartline, H. K., Miller, W. H., and Ratcliff, F. (1961). Inhibitory interactions in the retina and its significance in vision. In *Nervous Inhibition* (E. Florey, ed.). Pergamon Press, London, pp. 241–284.

Hartline, P. H., Kass, L., and Loop, M. S. (1978). Merging of modalities in the optic tectum: infrared and visual integration in rattlesnakes. *Science* (N.Y.) 199, 1225–1229.

Haseltine, E. G. (1978). "Infrared and Visual Organization of the Tectum of Boid Snakes." Ph.D. dissertation, Indiana University, Bloomington.

Hensel, H. (1974). Properties of warm receptors in *Boa constrictor. Naturwissenschaften* 61, 369.

Hensel, H. (1975). Static and dynamic activity of warm receptors in *Boa constrictor. Pflügers Arch.* 353, 191–199.

Hensel, H., and Schaefer, K. (1981). Activity of warm receptors in *Boa constrictor* raised at various temperatures. *Pflügers Arch.* 392, 95–98.

Hirosawa, K. (1980). Electron microscopic observations on the pit organ of a crotaline snake (*Trimeresurus flavoviridis*). *Arch. Histol. Jap.* 43, 65–78.

Kadota, T., Kishida, R., Goris, R. C., Kusunoki, T., Terashima, S., and Yamada, H. (1984). Serotonin-immunoreactive fibers in the trigeminal sensory nuclei of snakes with infrared sensitivity. *Neurosci. Lett.* Suppl. 17, S117.

Kadota, T., Kishida, R., Goris, R. C., and Kusunoki, T. (1988). Substance P like immunoreactivity in the trigeminal sensory nuclei of an infrared sensitive snake, *Agkistrodon blomhoffi. Cell Tissue Res.* 253, 311–317.

Karamanlidis, A. N. (1975). Trigeminothalamic fibre connections in the donkey (*Equus asinus*), studied by means of the retrograde cell degeneration method. *Acta Anat.* 93, 126–134.

Kass, L., Loop, M. S., and Hartline, P. H. (1978). Anatomical and physiological localization of visual and infrared cell layers in the tectum of pit vipers. *J. Comp. Neurol.* 182, 811–820.

Kenton, B., Kruger, L., and Woo, M. (1971). Two classes of slow adapting mechanoreceptor fibres in reptile cutaneous nerve. *J. Physiol.* (London) 212, 21–44.

Kerr, F. W. L. (1961). Structural relation of the trigeminal spinal tract to upper cervical roots and the solitary nucleus in the cat. *Exp. Neurol.* 4, 134–148.

Kerr, F. W. L. (1963). The divisional organization of afferent fibres of the trigeminal nerve. *Brain* 86, 721–732.

Kerr, F. W. L., Kruger, H., Schwassmann, H., and Stein, R. (1968). Somatotopic organization of mechanoreceptor units in the trigeminal nuclear complex of the macaque. *J. Comp. Neurol.* 134, 127–144.

Killackey, H. P., and Erzurumlu, R. H. (1981). Trigeminal projections to the superior colliculus of the rat. *J. Comp. Neurol.* 201, 221–242.

Kishida, R., Amemiya, F., Kusunoki, T., and Terashima, S. (1980). A new tectal afferent nucleus of the infrared sensory system in the medulla oblongata of crotaline snakes. *Brain Res.* 195, 271–280.

Kishida, R., Kusunoki, T., Terashima, S., and Goris, R. C. (1981). Thermal neurons in the trigeminal ganglia of crotaline snakes. *Neurosci. Lett.* Suppl. 6, S91.

Kishida, R., Terashima, S., Goris, R. C., and Kusunoki, T. (1982). Infrared sensory neurons in the trigeminal ganglia of crotaline snakes: transganglionic HRP transport. *Brain Res.* 241, 3–10.

Kishida, R., De Cock Buning, T., and Dubbeldam, J. L. (1983a). Primary vagal nerve projections to the lateral descending trigeminal nucleus in Boidae (*Python molurus* and *Boa constrictor*). *Brain Res.* 263, 132–136.

Kishida, R., Terashima, S., and Kusunoki, T. (1983b). Trigeminal primary afferent terminals in snakes with infrared receptors: differences between the infrared and other sensory systems. *Neurosci. Lett.* Suppl. 13, S126.

Kishida, R., Yoshimoto, M., Kusunoki, T., Goris, R. C., and Terashima, S. (1984a). Vagal afferent C fibers projecting to the lateral descending trigeminal complex of crotaline snakes. *Exp. Brain Res.* 53, 315–319.

Kishida, R., Goris, R. C., Terashima, S., and Dubbeldam, J. L. (1984b). A suspected infrared recipient nucleus in the brainstem of the vampire bat, *Desmodus rotundus*. *Brain Res.* 322, 351–355.

Knudsen, E. I., and Konishi, M. (1978). Center-surround organisation of auditory receptive fields in the owl. *Science* (N.Y.) 202, 778–780.

Kruger, L., and Michel, F. (1962). A morphological and somatotopic analysis of single unit activity in the trigeminal sensory complex of the cat. *Exp. Neurol.* 5, 139–156.

Kuhlenbeck, H. (1977). *The Central Nervous System of Vertebrates*, Vol. 5, Part 1. *Derivatives of the Prosencephalon, Diencephalon and Telencephalon.* S. Karger, Basel.

Kürten, L., and Schmidt, U. (1982). The thermoreception in the common vampire bat (*Desmodus rotundus*). *J. Comp. Physiol.* 146, 223–228.

Kusunoki, T., Kishida, R., Kadota, T., and Goris, R. C. (1987). Chemoarchitectonics of the brainstem in infrared sensitive and nonsensitive snakes. *J. Hirnforsch.* 28, 27–43.

Larsell, O. (1926). The cerebellum of reptiles: lizards and snakes. *J. Comp. Neurol.* 41, 59–94.

Lohman, A. H. M., and Van Woerden-Verkley, I. (1978). Ascending connections to the forebrain in the tegu lizard. *J. Comp. Neurol.* 182, 555–594.

Luiten, P. G. M. (1975). The central projections of the trigeminal, facial and anterior lateral line nerves in the carp (*Cyprinus carpio L.*). *J. Comp. Neurol.* 160, 399–418.

Lynn, W. G. (1931). The structure and function of the facial pit of the pit vipers. *Am. J. Anat.* 49, 97–139.

Maderson, P. F. A. (1970). The distribution of specialized labial scales in the Boidae. In *Biology of the Reptilia*, Vol. 2, *Morphology B* (C. Gans and T. S. Parsons, eds.). Academic Press, London, New York, pp. 301–304.

Meszler, R. M. (1970). Correlation of ultrastructure and function. In *Biology of the Reptilia*, Vol. 2, *Morphology B* (C. Gans and T. S. Parsons, eds.). Academic Press, London, New York, pp. 305–314.

Meszler, R. M. (1981). Fine structure and comparative morphology of the trigeminal thermoreceptor relay in boids and nucleus reticularis caloris of pit vipers (*Crotalus viridis*). *Anat. Rec.* 199, 172A.

Meszler, R. M. (1983). Fine structure and organization of the infrared receptor relays: lateral descending nucleus of V in Boidae and nucleus reticularis caloris in the rattle snake. *J. Comp. Neurol.* 220, 299–309.

Meszler, R. M., and Webster, D. B. (1968). Histochemistry of the rattle snake facial pit. *Copeia* 1968(4), 722–728.

Meszler, R. M., Auker, C. R., and Carpenter, D. O. (1981). Fine structure and organization of the IR receptor relay, the lateral descending nucleus of the trigeminal nerve in pit vipers. *J. Comp. Neurol.* 196, 571–584.

Michail, S., and Karamanlidis, A. N. (1970). Trigeminothalamic fibre connections in the dog and the pig, an experimental study with the retrograde cell degeneration method. *J. Anat.* 107, 557–566.

Mizuno, N. (1970). Projection fibres from the main sensory trigeminal nucleus and the supratrigeminal region. *J. Comp. Neurol.* 139, 457–472.

Molenaar, G. J. (1974). An additional trigeminal system in certain snakes possessing infrared receptors. *Brain Res.* 78, 340–344.

Molenaar, G. J. (1977). The rhombencephalon of *Python reticulatus*, a snake possessing infrared receptors. *Neth. J. Zool.* 27, 133–180.

Molenaar, G. J. (1978a). The sensory trigeminal system of a snake in the possession of infrared receptors. I. The sensory trigeminal nuclei. *J. Comp. Neurol.* 179, 123–136.

Molenaar, G. J. (1978b). The sensory trigeminal system of a snake in the possession of infrared receptors. II. The central projections of the trigeminal nerve. *J. Comp. Neurol.* 179, 137–152.

Molenaar, G. J. (1978c). "Infrared Sensitivity and the Organization of the Sensory Trigeminal System: a Functional Morphological Study in *Python reticulatus Schneid.*" Ph.D. thesis, University of Utrecht.

Molenaar, G. J. (1983a). Unpublished observations in material kindly provided by Dr. H. J. ten Donkelaar, Catholic University of Nijmegen, The Netherlands.

Molenaar, G. J. (1983b). Unpublished observations in material kindly provided by Dr. P. V. Hoogland, Free University of Amsterdam, The Netherlands.

Molenaar, G. J., and Dubbeldam, J. L. (1969). A preliminary study of the sensory trigeminal system in the brainstem of the mallard (Anas platyrhynchos L.). Neth. J. Zool. 19, 637–640.

Molenaar, G. J., and Fizaan-Oostveen, J. L. F. P. (1980). Ascending projections from the lateral descending and common sensory trigeminal nuclei in Python. J. Comp. Neurol. 189, 555–572.

Molenaar, G. J., Fizaan-Oostveen, J. L. F. P., and Van der Zalm, J. M. (1979). Infrared and tactile units in the sensory trigeminal system of Python reticulatus. Brain Res. 170, 372–376.

Mountcastle, W. B., and Powell, T. P. S. (1959). Neural mechanisms subserving cutaneous sensitivity, with special reference to the role of afferent inhibition in sensory perception and discrimination. Bull. Johns Hopkins Hosp. 105, 201–232.

Neary, T. J., and Wilczynski, W. (1977). Ascending thalamic projections from the obex region in ranid frogs. Brain Res. 138, 529–533.

Newman, E. A., Gruberg, E. R., and Hartline, P. H. (1980). The infrared trigeminotectal pathway in the rattle snake (Crotalus viridis) and python (Python reticulatus). J. Comp. Neurol. 191, 465–478.

Newman, E. A, and Hartline, P. H. (1981). Integration of visual and infrared information in bimodal neurons of the rattle snake (Crotalus viridis) optic tectum. Science (N.Y.) 213, 789–791.

Newman, E. A., and Hartline, P. H. (1982). The infrared "vision" of snakes. Sci. Am. 246, 98–107.

Noble, G. K. (1934). The structure of the facial pit of the pit vipers and its probable function. Anat. Rec. 58, Suppl. to No. 2, 4.

Noble, G. K., and Schmidt, A. (1937). The structure and function of the facial and labial pits of snakes. Proc. Am. Philos. Soc. 77, 263–288.

Nord, S. G. (1967). Somatotopic organization in the spinal trigeminal nucleus, the dorsal column nuclei and related structures in the rat. J. Comp. Neurol. 130, 343–355.

Northcutt, R. G., and Butler, A. B. (1974). Retinal projections in the northern watersnake (Natrix sipedon sipedon L.). J. Morphol. 142, 117–136.

Olszewski, J. (1950). On the anatomical and functional organization of the spinal trigeminal nucleus. J. Comp. Neurol. 92, 401–415.

Otto, J. (1972). Das Grubenorgan, ein biologisches System zur Abbildung von Infrarotstrahlern. Kybernetik 10, 103–106.

Peach, R. (1972). Fine structural features of light and dark cells in the trigeminal ganglion of the rat. J. Neurocytol. 1, 151–160.

Parent, A. (1976). Striatal afferent connections in the turtle (Chrysemys picta) as revealed by retrograde axonal transport of horse radish peroxidase. Brain Res. 108, 25–36.

Pedersen, R. (1973). Ascending spinal projections in three species of side-necked turtles: Podocnemis unifilis; Pelusios subniger and Pelomedusa subrufa. Anat. Rec. 175, 409.

Pritz, M. B. (1975). Anatomical identification of a telencephalic visual area in crocodiles: ascending connections of nucleus rotundus in Caiman crocodilus. J. Comp. Neurol. 164, 323–338.

Repérant, J. (1973). Les voiles et les centres optiques primaires chez la vipère (Vipera aspis). Arch. Anat. Microsc. Morphol. Exp. 62, 323–352.

Rhoades, R. (1981). Organization of somatosensory input to the deep collicular laminae in hamster. *Behav. Brain Res.* 3, 201–222.

Rhoton, A. L., O'Leary, J. L., and Ferguson, J. P. (1966). The trigeminal, facial, vagal and glossopharyngeal nerves in the monkey. *Arch. Neurol.* 14, 530–540.

Ricardo, J. A. (1981). Efferent connections of the subthalamic region in the rat: 2. The zona incerta. *Brain Res.* 214, 43–60.

Ross, M. (1935). Die Lippengruben der Pythonen als Temperturorgane. *Jena Z. Naturw.* (N.F.) 70, 1–32.

Rustioni, A., Sanyal, S., and Kuypers, H. G. J. M. (1971). A histochemical study of the distribution of the trigeminal divisions in the substantia gelatinosa of the rat. *Brain Res.* 32, 45–52.

Schroeder, D. M. (1981a). Tectal projections of an infrared sensitive snake (*Crotalus viridis*). *J. Comp. Neurol.* 195, 477–500.

Schroeder, D. M. (1981b). Retinal afferents and efferents of an IR sensitive snake, *Crotalus viridis. J. Morphol.* 170, 29–42.

Schroeder, D. M., and Loop, M. (1976). Trigeminal projections in snakes possessing infrared sensitivity. *J. Comp. Neurol.* 169, 1–14.

Senn, D. G. (1966). Ueber das optische System im Gehirn squamater Reptilien. *Acta Anat.*, Suppl. 52 = 1 ad Vol. 65, 1–87.

Senn, D. G. (1968). Bau und Ontogenese von Zwischen- und Mittelhirn bei *Lacerta sicula* (Rafinesque). *Acta Anat.*, Suppl. 55 = 1 ad Vol. 71, 1–150.

Senn, D. G. (1974). Notes on the amphibian and reptilian thalamus. *Acta Anat.* 87, 555–596.

Smith, R. L. (1973). The ascending fibre projections from the principal sensory trigeminal nucleus in the rat. *J. Comp. Neurol.* 148, 423–446.

Smith, R. L. (1975). Axonal projections and connections of the principal sensory trigeminal nucleus in the monkey. *J. Comp. Neurol.* 163, 347–376.

Stanford, L. R., and Hartline, P. H. (1980). Spatial sharpening by second-order trigeminal neurons in crotaline infrared system. *Brain Res.* 185, 115–123.

Stanford, L. R., and Hartline, P. H. (1984). Spatial and temporal integration in primary trigeminal nucleus of rattle snake infrared system. *J. Neurophysiol.* 51, 1077–1090.

Stanford, L. R., Schroeder, D. M., and Hartline, P. H. (1981). The ascending projection of the nucleus of the lateral descending trigeminal tract: a nucleus in the infrared system of the rattle snake (*Crotalus viridis*). *J. Comp. Neurol.* 201, 161–174.

Stein, B. E., Magelhaes-Castro, N., and Kruger, L. (1976). Relationship between visual and tactile representations in cat superior colliculus. *J. Neurophysiol.* 39, 401–419.

Stein, B. E., and Gaither, N. S. (1983). Receptive field properties in reptilian optic tectum: some comparisons with mammals. *J. Neurophysiol.* 50, 102–124.

Stewart, W. A., and King, R. B. (1963). Fiber projections from nucleus caudalis of the spinal trigeminal nucleus. *J. Comp. Neurol.* 1'21, 271–286.

Tan, C. K., and Gopalakrishnakone, P. (1988). Infrared sensory neurons in the trigeminal ganglia of the python (*Python reticulatus*)—a horse radish peroxidase study. *Neurosci. Lett.* 86, 251–256.

Terashima, S. (1987). Substance P like immunoreactive fibers in the trigeminal sensory nuclei of the pit viper, *Trimeresurus flavoviridis. Neuroscience* 23, 685–691.

Terashima, S., and Goris, R. C. (1975). Tectal organization of pit viper infrared perception. *Brain Res.* 83, 490–494.

Terashima, S., and Goris, R. C. (1976). Receptive area of an infrared tectal unit. *Brain Res.* 101, 155–159.

Terashima, S., and Goris, R. C. (1977). Infrared bulbar units in crotaline snakes. *Proc. Jap. Acad.*, Ser. B., Phys. Biol. Sci. 53, 292–296.

Terashima, S., and Goris, R. C. (1979). Receptive areas of primary infrared afferent neurons in crotaline snakes. *Neuroscience* 4, 1137–1144.

Terashima, S., and Goris, R. C. (1983). Static response of infrared neurons of crotaline snakes—normal distribution of interspike intervals. *Cell. Mol. Neurobiol.* 3, 27–37.

Terashima, S., Goris, R. C., and Katsuki, Y. (1968). Generator potential of crotaline snake infrared receptor. *J. Neurophysiol.* 31, 682–688.

Terashima, S., Goris, R. C., and Katsuki, Y. (1970). Structure of warm fiber terminals in the pit membrane of vipers. *J. Ultrastruct. Res.* 31, 494–506.

Tiwari, R. K., and King, R. B. (1973). Trigeminothalamic projections from nucleus caudalis in primates. *Surg. Forum* 24, 444–446.

Tiwari, R. K., and King, R. B. (1974). Fiber projections from trigeminal nucleus caudalis in primates (squirrel monkey and baboon). *J. Comp. Neurol.* 158, 191–206.

Torvik, A. (1956). Afferent connections to the sensory trigeminal nuclei, the nucleus of the solitary tract and adjacent structures: an experimental study in the rat. *J. Comp. Neurol.* 106, 51–142.

Torvik, A. (1957). The ascending fibres from the main trigeminal sensory nucleus: an experimental study in the cat. *Am. J. Anat.* 100, 1–16.

Ulinski, P. S. (1977). Tectal efferents in the banded watersnake (*Natrix sipedon*). *J. Comp. Neurol.* 173, 251–274.

Van Drongelen, W., Holley, A., and Doving, K. B. (1978). Convergence in the olfactory system: quantitative aspects of odour sensitivity. *J. Theor. Biol.* 71, 39–48.

Van Valkenburg, C. T. (1911). Zur Kenntnis der Radix spinalis nervi trigemini. *Monatsschr. Psychiatr. Neurol.* 29, 407–438.

Verhaart, W. J. (1971). Forebrain bundles and fibre systems in the avian brainstem. *J. Hirnforsch.* 13, 39–64.

Vesselkin, N. P., Agayan, A. L., and Nomonokova, L. M. (1971). A study of thalamotelencephalic afferent systems in frogs. *Brain Behav. Evol.* 4, 295–306.

Von Düring, M. (1974). The radiant heat receptor and other tissue receptors in the scales of the upper jaw of *Boa constrictor. Z. Anat. Entwicklungsgesch.* 145, 299–319.

Von Düring, M., and Miller, M. R. (1979). Sensory nerve endings of the skin and deeper structures. In *Biology of the Reptilia*, Vol. 9, *Neurology* (C. Gans, R. G. Northcutt, and P. S. Ulinski, eds.). Academic Press, London, New York, San Francisco.

Wallenberg, A. (1903). Eine zentrifugal leitende direkte Verbindung der frontalen Vorderhirnbasis mit der Oblongata (+ Rückenmark?) bei der Ente. *Anat. Anz.* 22, 289–292.

Warner, F. J. (1947). The diencephalon and midbrain of the American rattle snake (*Crotalus adamanteus*). *Proc. Zool. Soc. Lond.* 116, 531–550.

Warren, J. W., and Proske, U. (1968). Infrared receptors in the facial pits of the Australian python (*Morelia spilotes*). *Science* (N.Y.) 159, 439–441.

Welker, E., Hoogland, P. V., and Lohman, A. H. M. (1983). Tectal connections in *Python reticulatus. J. Comp. Neurol.* 220, 347–354

Weston, J. K. (1936). The reptilian vestibular and cerebellar gray with fiber connections. *J. Comp. Neurol.* 64, 93–199.

Woodburne, R. Th. (1936). A phylogenetic consideration of the primary and secondary centers and connections of the trigeminal complex in a series of vertebrates. *J. Comp. Neurol.* 65, 403–501.

Woudenberg, R. A. (1970). Projections of mechanoreceptive fields to cuneate-gracile and spinal trigeminal nuclear regions in sheep. *Brain Res.* 17, 417–437.

Zipser, B., and Bennet, M. V. L. (1976). Responses of cells of posterior lateral line lobe to activation of electro-receptors in a mormyrid fish. *J. Neurophysiol.* 39, 693–712.

6

Muscle Spindles, Tendon Organs, and Joint Receptors

ALAN CROWE

CONTENTS

I. INTRODUCTION

This chapter reviews the anatomical and physiological studies of peripheral sense organs that may be important in position sense or the control of the skeletomotor system: the muscle spindles, tendon organs, and joint receptors. It includes only passing references to corresponding receptor organs outside the reptilian system. There are, however, many excellent reviews concerning the functional, comparative, and structural aspects of the subject (e.g., Barker, 1974; Homma, 1976; Kennedy et al., 1981; Matthews, 1972, 1981; Taylor and Prochazka, 1981; Vallbo et al., 1979). Previous reviews have dealt with extrafusal muscle and spindles in reptiles (Proske and Ridge, 1975; Guthe, 1981). Compared to the treatment in mammals, sense organs in reptiles have been sadly neglected, even though references to reptiles appear in the earliest literature on the subject.

Muscle spindles have received more attention than the other muscle and joint receptors, probably because the spindles are more complicated, more numerous, and more easily accessible. Ruffini (1893) summed up the situation: "Apart from the organs of special sense the body possesses no terminal organ that can compare with these in richness of nerve fibers and nerve endings." Muscle spindles have been studied in turtles, crocodilians, lizards, and snakes, but data on their distribution, morphology, and physiology are limited to a relatively few muscles taken from a small number of species. In particular, data on spindles in the crocodilians are extremely sparse. One must therefore be cautioned against too readily accepting the available evidence as generalizations for a particular group. Much work has yet to be done to provide a complete picture of muscle spindles in reptiles.

The study of reptilian joint receptors is due principally to the work of one investigator. Tendon organs have received some attention, but results are mainly limited to the lizard.

II. MUSCLE SPINDLES
A. History and General Structure

Matthews (1972, 1981, 1982) has provided detailed descriptions of the structure and functions of the mammalian spindle, and Barker (1974) has produced a comprehensive comparative account.

Muscle spindles have been found in all the extant vertebrate groups except the fishes. Articles describing muscle spindles, or *Muskelspindeln*, appeared for the first time in the German literature from the 1860s to 1880s (e.g., Kölliker, 1862a, 1862b; Kühne, 1863a, 1863b, 1864; Kerschner, 1888a, 1888b). One of the earliest reviews on the topic appeared at the beginning of the present century (Regaud and Favre, 1904). Muscle spindles are found in skeletal muscles. The essential structural unit is known as an intrafusal muscle fiber (Sherrington, 1894). It is distinguished from the normal, or extrafusal, fiber by having an equatorial region enclosed in a capsular structure and innervated by a sensory nerve. A muscle spindle may consist of just one intrafusal fiber (as in lizards and snakes) or a bundle of several intrafusal fibers with a common sensory region (as in turtles).

Muscle spindles lie parallel to extrafusal fibers (Fig. 6.1), so a stretch applied to a passive muscle also stretches the intrafusal muscle fibers, resulting in excitation of the sensory ending. Since the intrafusal fibers are muscle fibers, they also have contractile properties. In reptiles, the motor innervation needed to produce this contraction is provided by a branch of a motor nerve that also innervates a group of intrafusal fibers. Thus the intrafusal fiber is a component of a motor unit. Stimulation of the motor nerve produces a contraction of the intrafusal fiber, which increases the firing of the sensory nerve. However, extrafusal fibers will simultaneously contract and, since they lie

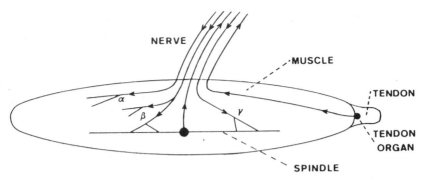

Fig. 6.1. Diagram of muscle containing muscle spindle and tendon organ. A muscle may receive three types of motor innervation: motor fibers that exclusively innervate the extrafusal fibers; motor fibers that innervate both the extrafusal fibers and the intrafusal fibers; and the fibers, which are not present in reptilian muscle, that only contact the intrafusal fiber. The central region of the spindle is enclosed in a capsule that contains the endings of the sensory nerves. Tendon organs receive no motor innervation, but are only innervated by a sensory nerve. (After Naeije, 1977.)

in parallel with the extrafusal fiber, will tend to unload the spindle and reduce the output of the sensory ending.

In summary, the output of the muscle spindle depends upon three parameters: (1) the positive effect of stretch of the muscle, (2) the positive effect of stimulation of the motor nerve to contract the intrafusal fiber, and (3) the negative effect of contraction of the extrafusal fibers.

Note that motor innervation of reptilian spindles differs from that of mammalian spindles in that the motor innervation of the former is always a collateral of a motor nerve supplying extrafusal fibers. This is the equivalent of the β-innervation of the mammalian spindle. No instances of a γ-motor innervation (where certain motor nerves have endings exclusively on the muscle spindles) have been reported in reptiles. Intrafusal fibers are usually smaller in diameter than extrafusal muscle fibers. Since most fibers in a muscle are extrafusal, no instances in which contraction of the intrafusal fibers contributes significantly to the contractile force have been reported.

B. Distribution of Muscle Spindles

Spindles have been observed in most of the muscles in which they have been sought. The turtle muscle most extensively studied is the extensor digitorum brevis I (EDBI) of *Testudo graeca* and *Pseudemys scripta elegans*. Crowe and his group (1967) chose this particular muscle for an extensive series of studies because it, together with its nerve supply, is readily accessible for physiological studies. Detailed observations from serial sections of this muscle reveal that the number of spindles per muscle is not constant (Crowe and Ragab, 1967,

1970a), even when left and right muscles from the same individual are compared. Muscles in four adult *Testudo graeca* specimens contained between four and ten spindles. Huber and DeWitt (1897) studied spindles in the vasti muscles of *Chrysemys picta* and *Emydoidea blandingii* but gave no details on their number and distribution. Perfunctory counts of certain other muscles of *Testudo graeca* revealed 15 spindles in the biceps brachii and 44 in the sartorius (Ragab, 1970). A rather surprising result was the paucity of spindles in the retractor capitis muscle. Ragab (1970) studied four examples. No spindles at all were found in two, and two and three spindles were observed in the other two. Spindles have been observed in the ambiens, ischiofemoralis, and certain flexor muscles of the thigh of the alligator (Hines, 1930). No other references to crocodilian spindles have been found. Viswanathan (1972a, 1972b, 1973, 1976), Viswanathan and Vijayalakshmi (1975), Vijayalakshmi (1976), and Vijayalakshmi and Viswanathan (1976) investigated the distribution of spindles in certain muscles of the garden lizard *Calotes versicolor*. They found 12 spindles in the gastrocnemius/soleus and 20 to 22 in the iliofibularis. From articles on the physiology and innervation of lizard muscle spindles, one can also conclude their presence in iliotibialis, ambiens, ischioflexorius, pectoralis, trapezius, latissimus dorsi (Szepsenwol, 1960), semimembranosus, semitendinosus, and scalenus (Proske and Vaughan, 1968; Proske, 1969a, 1969b, 1973) of about half a dozen species. As in many other vertebrates, no spindles were found in the extraocular muscle of the turtle *Pseudemys scripta elegans*, the lizard *Dipsosaurus dorsalis*, or the snake *Lampropeltis sp.* (Maier et al., 1974).

C. Morphology of Muscle Spindles

1. NUMBER OF INTRAFUSAL FIBERS   Turtle muscle spindles are composed of a bundle of intrafusal muscle fibers (Huber and DeWitt, 1897; Crowe and Ragab, 1967, 1970a, 1970b). Giacomini (1898) reported monofibrillar spindles, but this finding is difficult to evaluate. A monofibrillar spindle in turtle muscle would be difficult to observe using conventional techniques, since turtle spindles are characterized by a capsule containing hardly any fluid space and closely approximated to the intrafusal fibers (Fig. 6.2). Crowe and Ragab took particular care to search for monofibrillar spindles in serial sections or teased preparations. They found none, although they sometimes saw examples of spindles with as few as two to four intrafusal fibers. Huber and DeWitt (1897) observed turtle spindles with from two to ten intrafusal fibers. Later studies indicate a range of 2 to 17 (mean 10.4) intrafusal fibers in the spindles of EDBI muscle (Crowe and Ragab, 1970a), 3 to 17 in the sartorius, 5 to 12 in the biceps brachii, and 2 to 3 in the retractor capitis spindles (Ragab, 1970).

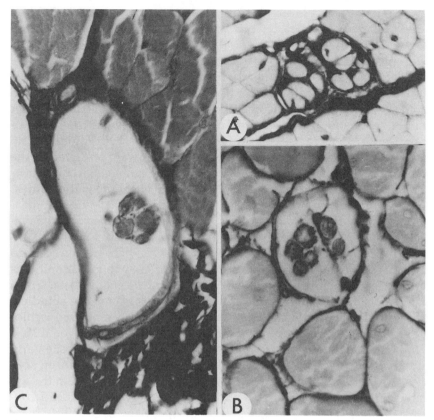

Fig. 6.2. Transverse sections at the capsular region of spindles of (A) *Testudo graeca*, (B) frog, and (C) rat. In each spindle, the intrafusal fibers are thinner than the extra-fusal fibers. Note the absence of the fluid-filled space surrounding the intrafusal fibers of the turtle spindle. (After Crowe and Ragab, 1970a.)

Alligator spindles contain two to five intrafusal fibers (Hines, 1930), but more extensive studies are needed. The spindles of lizards and snakes appear to be unique (Fig. 6.3) in that each is comprised of just one intrafusal fiber (Szepsenwol, 1960).

2. SPINDLE LENGTH AND CAPSULE POSITION

Spindles do not usually exceed 15 mm in length. If the muscle is shorter than this, the spindle may extend the whole length of the muscle from tendon to tendon. In the costocutaneous muscles of the garter snake, the spindles run the whole length of the muscle, which has a resting length of 1 cm (Fukami and Hunt, 1970). A similar result was seen by Crowe and Ragab (1970a) for the EDBI muscle of *Testudo*

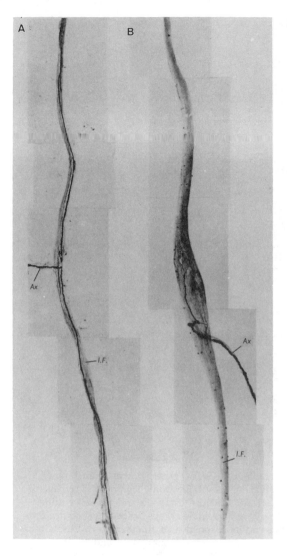

Fig. 6.3. Composite photographs of spindles of the lizard (*Tiliqua nigrolutea*) showing the typical monofibrillar structure common to lizard and snake spindles. The intrafusal fiber (IF) and sensory axon (AX) are marked in each case. Morphological differences between the long-capsule spindle (A) and the short-capsule spindle (B) are clearly seen at the sensory region. (After Proske, 1969a.)

graeca. The mean spindle length was 9.3 mm within the range 6.9 to 11.4 mm. In multifibrillar spindles, intrafusal fibers may be of unequal length and partially overlap so that the longest intrafusal fiber may be shorter than the spindle itself.

The capsule does not extend the whole length of the spindle and is not usually situated precisely at the midpoint of the spindle. The capsule length in the turtle EDBI spindles ranges from 0.6 to 1.3 mm (mean 1.02 mm). Its midpoint is situated on average 4.5 mm from the

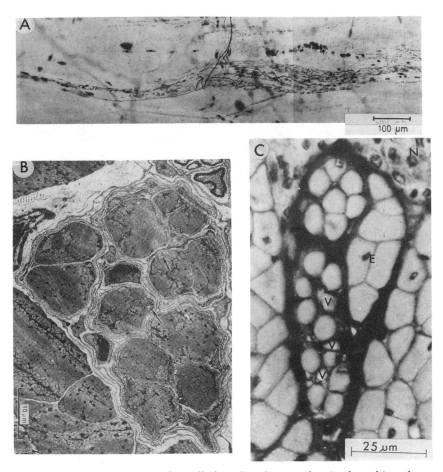

Fig. 6.4. (A) Sensory region of spindle from *Testudo graeca* showing branching of single sensory nerve onto the intrafusal fibers. (B) Electron micrograph of transverse section of turtle spindle showing lamellar structure of capsule and sensory endings (arrows). (C) Transverse section through the spindle at capsular region showing division of intrafusal fibers into groups and inclusion of small number of extrafusal fibers E, capillary blood vessels V, and nerve fibers in cross section N. (A and C, after Crowe and Ragab, 1970a; B, after Crowe and Ragab, 1970b.)

proximal end of the spindle. The capsule in turtle spindles holds the intrafusal fibers together in a tight bundle (Fig. 6.4). However, the intrafusal fibers tend to spread out so that at the polar regions they might form a row or subdivide into groups.

Although lizard and snake spindles have a single intrafusal fiber, they can be easily categorized into two groups on the basis of their capsular structure (see Fig. 6.3). In snakes, the "short capsule" is 100 to 150 μm long, and its diameter, at the widest region, is about 40 μm.

In contrast, the "long capsule" is 250 to 300 μm long with a diameter of 15 to 20 μm (Fukami and Hunt, 1970). The long capsules are distributed along the intrafusal fibers, whereas the short capsules tend to be located near the midpoint (Ichiki, 1973; Ichiki et al., 1976). This is in contrast to the situation in lizards, where, according to Proske (1969a), the short capsules may be situated near the muscle tendon. This is rarely the case for long capsules in lizard spindles. Both short- and long-capsule spindles may be present in the same muscle. Fukami and Hunt (1970) found 28 long-capsule and 18 short-capsule spindles in 16 snake muscles. A predominance of one type of spindle may be found (Proske, 1969b). Only long-capsule spindles were found in semimembranosus, and only short-capsule spindles were seen in scalenus. On the other hand, spindles of both types were found in semitendinosus and iliofibularis. Spindles containing more than one capsule—often referred to as tandem spindles—may be common in certain muscles. Up to one-third of the long-capsule spindles in iliofibularis of lizard may have adjacent capsules supplied by separate sensory endings (Proske, 1969b). Up to 23% of the spindles in the sartorius of the turtle *Testudo graeca* may be tandem (Ragab, 1970). The majority of the intrafusal fibers of such spindles pass through both capsules, but a few may go through just one capsule, either bypassing the other one or being of insufficient length to go through it.

3. Capsular Structure

The capsule in turtle spindles consists of connective tissue layers that either surround the whole spindle or wrap around just one or a group of two or three fibers. Each intrafusal fiber may be separated from the outside of the spindle by five or more layers of connective tissue. The outermost sheaths of the capsular connective tissue often also embrace thicker (presumably extrafusal) fibers (see Fig. 6.4). A unique feature of the turtle spindles is that the capsular tissue is closely approximated to the intrafusal fibers. The bulbous, fluid-filled periaxial space seen in other vertebrate spindles is absent (Figs. 6.4 and 6.5). In contrast, the capsules of lizard and snake spindles have a relatively simple structure (Fig. 6.6). The capsule envelops the sensory region and may extend beyond it (Fukami and Hunt, 1977, garter snake *Thamnophis*). When examined along its length, starting from the polar region and working toward the middle, the space between the outer capsule and intrafusal fiber gradually increases to form the tubular region and reaches maximum diameter in the region of the sensory innervation. The intrafusal fiber is also invested with a so-called inner capsular layer at the tubular and sensory regions. The outer capsule consists of a single layer of flattened cells connected in series. Each

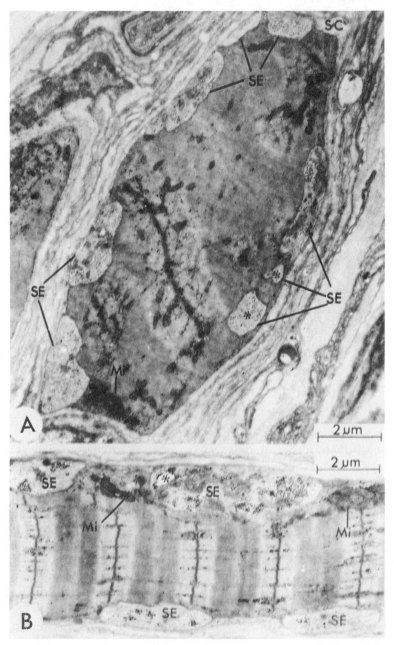

Fig. 6.5. (A) Transverse section of a single intrafusal fiber in a turtle spindle. (B) Longitudinal section through the same fiber to show how sensory terminals (SE) make contact with the muscle fiber. Note the presence of the actin-myosin structure in the sensory region. Also shown in the sections are satellite cells (SC), mitochondria (Mi). In (A) sensory endings completely enclosed by the sarcolemma are indicated by (*). In (B) a fat droplet in the sensory ending is indicated by (*). (After Crowe and Ragab, 1970b.)

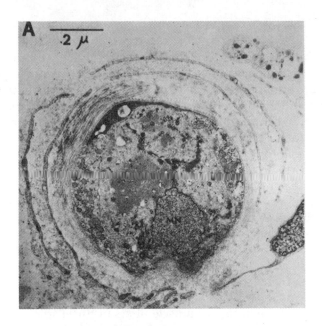

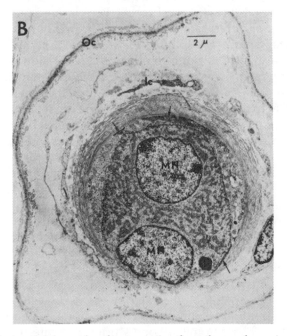

Fig. 6.6. Electron micrographs of cross section through capsular regions of snake spindles. (A) Long capsule showing branches of sensory terminals to the left and right of the muscle nucleus MN. (B) Short capsule showing the inner capsule Ic, the outer capsule Oc, muscle nuclei MN, satellite cells (arrows). Branches of sensory terminals can be seen on the left and right of the section. Note the differences in length scales on the two sections. (After Fukami and Hunt, 1970.)

cell forms a cylinder surrounding the intrafusal fiber (Fukami, 1982). The outer capsule acts as a diffusion barrier between the periaxial space and external medium. Experiments with markers indicate that leakage from the ends of the cylinder occurs to a limited extent (Fukami, 1982). In contrast to the outer capsule cells, which carry numerous pinocytotic vesicles and have a basement membrane on both sides, the cells forming the inner capsule have no basement membrane and form an incomplete covering of the intrafusal fiber. In the short-capsule spindles, the inner capsule may consist of two or three lamellae (Fig. 6.6), but just one layer may be the norm in the long-capsule spindles (Pallot and Ridge, 1972, 1973). Dense layers of collagen lie over the sensory endings within the inner capsule (Fukami and Hunt, 1970).

4. INTRAFUSAL FIBER STRUCTURE

Crowe and Ragab (1970a, 1970b, 1972) have extensively investigated the structure of the intrafusal muscle fibers of turtle spindles. On the basis of length, diameter, nuclear density and distribution, myofibrillar structure, and histochemical tests, there is no evidence of more than one type of intrafusal fiber. The length distribution of 282 intrafusal fibers taken from the spindles of four EDBI muscles is unimodal. The intrafusal fibers are not of uniform diameter along their length. Diameter enlargements are found around the extremities of the capsular region, but at the midcapsular region the fiber diameter is roughly equal to that at the polar regions. The mean intrafusal muscle fiber diameter, measured at a point midway between a pole and the nearest capsule extremity, was found to be 10.6 μm (2.30 μm SD). The surrounding extrafusal muscle fibers have a mean diameter of 27.3 μm (5.71 μm SD). A combined diameter distribution of both extrafusal and intrafusal fibers showed only a small amount of overlap. The intrafusal fibers in the turtle spindle possess no myotube region or regions containing concentrations of nuclei. Thus no counterparts of nuclear bag or nuclear chain structures, as in mammalian spindles, are seen. There is, however, a slight increase in nuclear density at the capsular region. The intrafusal fiber is surrounded by a basement membrane, which also embraces the satellite cells and sensory endings where present. Although elastic tissue is present as strands on the surface of rat and frog intrafusal fibers, none has been located in turtle spindles. Electron microscope studies indicate that the myofibril structure is similar for both intrafusal and extrafusal fibers in the EDBI muscle (Fig. 6.5). The mean sarcomere lengths were 2.60 μm and 2.56 μm, respectively. The T-system and triads are similar in the intrafusal fibers to those described by Page (1968) for turtle extrafusal fibers. The localization of phosphorylase, succinic dehydrogenase, and adenosine activities, together with the demonstration of the presence of lipid by the propylene glycol-Sudan method, have all failed to yield results that could be used to categorize turtle intrafusal fibers into more than one type (Crowe and Ragab, 1972).

The intrafusal fibers of lizard spindles are 10 μm in diameter (Proske, 1969a). Those of snake spindles are 7 to 10 μm at the polar regions (Fukami and Hunt, 1970). In the short-capsule spindles of the snake, the diameter of the intrafusal fiber enlarges at the sensory region, but the myofibrils decrease 75% to 90% (Pallot and Ridge, 1973) or are absent (Fukami and Hunt, 1977). Instead, the fiber is filled with reticulum, nuclei, mitochondria, microtubules, lipid droplets, and electron-dense granules (Fukami and Hunt, 1970). There are no striations in the intrafusal fibers at the sensory region of lizard short-capsule spindles (Proske, 1969b). There is an accumulation of nuclei, and the region is thicker. In the long-capsule spindle of lizards, the intrafusal fibers contain a "slightly larger number of nuclei than in the extra-capsular region and striations are clearly seen throughout the fiber length" (Proske, 1969a). Similarly, a concentration of myofibrils is seen throughout the length of the long-capsule spindle fibers of snakes (Fukami and Hunt, 1970). Ultrastructural studies on snake spindles indicate differences between long-capsule and short-capsule intrafusal fibers. Sarcomeres in the former have a prominent M-line and H zone. No M-lines or well-defined H zones are seen in the short-capsule intrafusal fiber sarcomere (Pallot and Ridge, 1973; Pallot, 1974a, 1974b). The short-capsule intrafusal fibers of snake spindles have a high-density myosin ATPase, succinic dehydrogenase, and phosphorylase. On the other hand, the long-capsule spindles have low-density myosin ATPase and phosphorylase, and intermediate levels of dehydrogenase (Pallot and Taberner, 1974). These fibers have similar histochemical characteristics to tonic extrafusal fibers and twitch extrafusal fibers, respectively.

5. AFFERENT INNERVATION

In turtle spindles, each capsule contains a single sensory nerve. This usually divides into two or three branches before entering the capsule normal to the spindle axis. These branches divide further inside the capsule and ramify to innervate each intrafusal fiber. The fine terminals meander over the surface of the intrafusal fiber, but do not form complete spirals, and appear as varicose threads in methylene blue–stained preparations (Crowe and Ragab, 1969). Contact with the intrafusal fibers is made by the nerve lying in troughlike invaginations on the surface of the intrafusal fibers (Fig. 6.5). The nerve terminations are rich in mitochondria (Crowe and Ragab, 1970b). The membrane of the sensory terminal is separated from the intrafusal muscle fiber membrane by a distance of 20 nm. As is common for the sensory innervation of muscle spindles, a basement membrane does not intervene between the membranes of the sensory terminal and intrafusal fiber membrane. The intrafusal fiber structure at the region of sensory innervation is indistinguishable from that at other regions of the fiber. No light microscopic or ultrastructural evidence seems to be available, which suggests that categorization of the sensory endings into more than one type can be made.

Each capsule of lizard and snake spindles contains a single sensory nerve. Instances have been described of tandem spindles having their sensory regions supplied by separate nerves. Several spindles may be supplied by branches of a common afferent nerve (Szepsenwol, 1960, lizard spindles). Instances have been noted of the sensory regions of long-capsule lizard spindles having independent afferent innervations (Proske, 1969b). In both long- and short-capsule lizard spindles, afferent fiber diameters range from 8 to 13 μm, including the Schwann sheath, as measured between 0.5 and 1.0 mm from the spindle (Proske, 1969a). The diameter distributions show a modal value at 8 μm for short-capsule spindles and modal values of 7 μm and 9 μm for the long-capsule spindles. Thus it is not possible to classify the nerves supplying the two types of spindle on the basis of their diameters. On entering the long-capsule snake spindle, the sensory nerve usually branches into two arms, each extending in opposite directions along the sensory region. Sensory contact is made by way of bulbous enlargements connected by a series of thin links having a length of up to 20 μm (Pallot and Ridge, 1972, *Natrix sp.*), a structure that resembles the sensory terminals of frog muscle spindles as described by Katz (1961). The sensory ending distributes over a region of 500 to 700 μm (Ichiki et al., 1976, rat snake *Elaphe quadrivirgata*). In lizard spindles this region may extend over a distance of 1,500 μm (Proske, 1969b, *Tiliqua nigrolutea*), and the terminations may be varicose in form (Barker, 1974, *Lacerta viridis*) or terminate in several small filaments that lie in close approximation to the intrafusal muscle fiber. The fine terminal filaments may show an enlargement at their contacts (Proske, 1969b, *Tiliqua nigrolutea*). When it enters the short capsule, the sensory nerve immediately divides into several fine branches, each of which comes into close contact with the intrafusal fiber. The sensory region extends 80 to 100 μm along the fiber (Pallot and Ridge, 1973, water snake *Natrix sp.*). The sensory terminals are about 20 μm in length and show no varicosities (Fukami and Hunt, 1970, garter snake). The branching of the sensory nerve forms a complex arborization of sensory bulbs with a smaller link/bulb ratio than in the long-capsule innervation (Ichiki et al., 1976, *Elaphe quadrivir-*

gata). The links between the bulbs are rather short—2 to 3 μm (Pallot and Ridge, 1973, *Natrix sp.*).

The structure of the sensory innervation in spindles of the garden lizard, *Calotes versicolor*, has been studied by Viswanathan and Vijayalakshmi (1975). They describe multiple sensory endings, one of which is designated *primary* and the others *secondary* on the basis of the configurations of the terminations. Both types of spindle have primary and secondary endings. However, these authors state explicitly that the terminals are branches of a single thick sensory axon in the short-capsule spindles. This introduces unnecessary and misleading comparisons with mammalian muscle spindles since, in the latter case, the primary and secondary endings are not supplied by a common nerve fiber and, on the basis of physiological studies, mammalian primary and secondary endings exhibit different patterns of response to muscle stretch. It is difficult to contemplate such physiological differences in lizard spindles, since only one nerve fiber is involved. However, these authors do highlight possible variations in the structure of the spindles they studied as compared to other lizard spindles, and these should be further investigated.

6. EFFERENT INNERVATION

Two types of motor nerve endings (Fig. 6.7) are seen on turtle extrafusal muscle fibers.

a. PLATE ENDINGS

The terminal axons approach the muscle fiber at roughly right angles, forming a T-junction whose arms pass in opposite directions along the surface of the muscle fiber parallel to the axis. Synaptic contacts occur regularly at intervals of approximately two to four sarcomeres.

b. GRAPE ENDINGS

The terminal axons lie across rather than along the muscle fibers. Each synaptic contact usually has its own short terminal axon branch (Crowe and Ragab, 1970a). Plate endings are also seen on the intrafusal muscle fibers, but the arms of the T-junctions tend to be shorter than on the extrafusal fibers. Grape endings similar to those on the extrafusal fibers are also seen on the intrafusal fibers (Fig. 6.7B). Of 21 spindles taken from EDBI muscles of *Testudo graeca*, 18 contained zones of plate endings, two contained distinct zones of grape endings, and 12 contained regions of innervation with both plate and grape endings. Eight spindles contained grape endings only. All the spindles had at least one region of motor innervation on each pole. The question as to whether a single intrafusal fiber may possess both plate and grape endings has apparently not yet been answered.

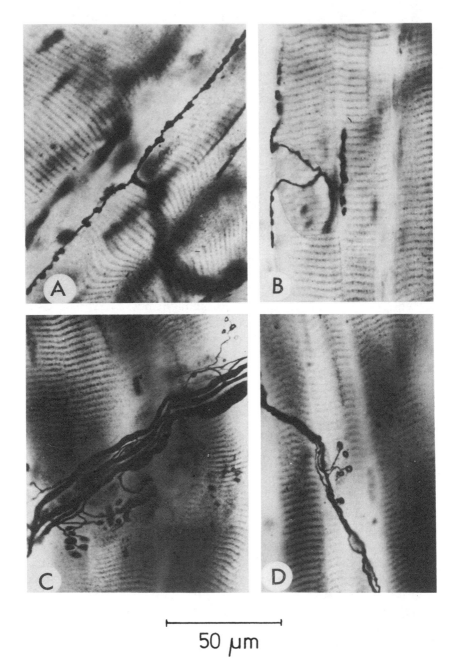

50 μm

Fig. 6.7. Teased silver stained preparations to show the types of motor innervation in turtle muscle. (A) Plate ending on an extrafusal fiber showing the linear array of synaptic contacts along the arms of the T-junction. (B) Plate ending on intrafusal fibers. (C) Grape endings on extrafusal fibers. The synaptic contacts tend to lie across the muscle fiber rather than along the fiber axis. (D) Grape endings on an intrafusal fiber. (After Crowe and Ragab, 1970a.)

In all instances where it has been possible to trace back the nerves, it was seen that the motor nerves to the spindles are collaterals of extrafusal motor nerve fibers. A similar result has been found in the motor innervation to the lizard and snake spindles (Perroncito, 1901; Regaud and Favre, 1904). One cannot absolutely exclude the possibility of an independent motor nerve supply to the reptilian muscle spindles that is analogous to the γ-motor innervation of mammalian spindles. However, if instances of an independent innervation are established, these will probably be exceptions to the rule.

Motor nerve endings to the extrafusal fibers of lizards and snakes are also of the plate and grape type. Szepsenwol (1960) noticed that, in lizard muscles with plate endings, the spindle motor nerves also had plate endings, but in the muscles with both grape and plate endings on the extrafusal fibers, the spindles also had plate and grape endings (Fig. 6.8). Proske (1969b) studied 29 spindles in the semitendinosus—a muscle with predominantly grape endings. Both the long- and short-capsule spindles were found to have grape endings. On the other hand, 12 spindles from the semimembranosus—a muscle having exclusively plate endings and long-capsule spindles—had plate endings (Fig. 6.9). Short-capsule spindles with plate endings have been reported in muscles of the garden lizard, *Calotes versicolor* (Viswanathan and Vijayalakshmi, 1976). Both plate and grape endings have been found on long-capsule spindles of the snake, but only grape endings were found on the short-capsule spindles (Pallot, 1974a, *Natrix sp.*). However, plate endings, albeit rarely, have been found on the short-capsule spindles of *Elaphe quadrivirgata* (Ichiki et al., 1976). These authors reported that the majority of endings on spindles of either type are grape endings. In the long-capsule spindles, the plate endings may be found within the capsular region, as well as at the polar regions.

Giacomini (1898) and Cipollone (1898) first reported polyneural motor innervation of reptilian spindles. Although Fukami and Hunt (1970), using garter snake muscle, reported that most of the spindles examined receive a motor innervation principally or exclusively on one side of the sensory region, this finding does not seem to be generally true for snake muscle spindles. Ichiki et al. (1976) made painstaking measurements of the positions of the motor endings on the short-capsule spindles and found a more or less symmetrical distribution. On the other hand, the extracapsular motor endings on the long-capsule spindles have an asymmetrical distribution about the sensory region, but the positions of the motor endings are by no means exclusively placed on one pole.

Instances of polyneural motor innervation on lizard and snake spindles have been reported. Spindles of the lizard *Anolis cristatellus*

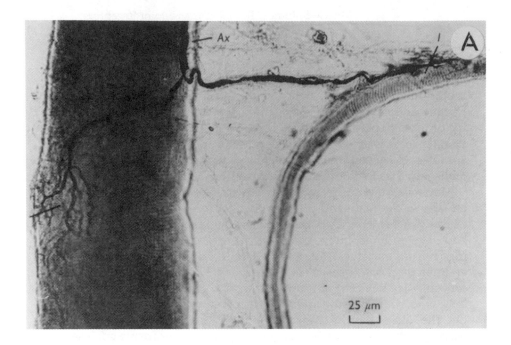

Fig. 6.8. Motor innervation of spindles of lizard *Tiliqua nigrolutea* (A) Axon branching to form plate ending on intrafusal muscle fiber (I) and extrafusal fiber. (B) Axon (Ax) branching to form grape ending on extrafusal fiber (E) and on intrafusal fiber (I). (After Proske, 1969a.)

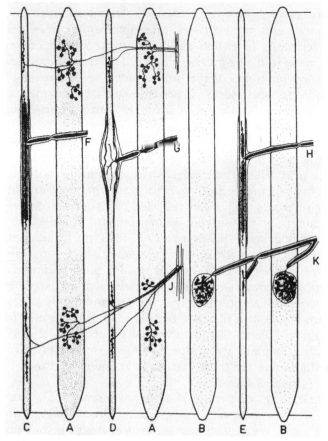

Fig. 6.9. Diagrammatic representation of the structure of lizard muscle spindles and their association with extrafusal fibers. The muscle fibers A represent slow muscle supplied by grape endings J. The fibers B represent twitch muscle fibers receiving plate endings K. The long-capsule spindle E has a sensory ending H and has collaterals of the motor nerve having plate endings K. The short-capsule spindle D has a sensory nerve G and receives grape endings J. (After Proske and Ridge, 1975.)

have several motor endings (Szepsenwol, 1960). In a few cases, the polyneural motor innervation is exclusively distributed along one pole, but the distribution is usually to both poles. Multiple motor innervation is predominantly of one type—either plate or grape. Ichiki et al. (1976) found only 1 out of 20 short-capsule spindles had both plate and grape endings. The other 19 had only grape endings. Three out of 20 long-capsule spindles had both types of ending; 16 had grape endings only; 1 had a single plate ending. The polyneural motor innervation of snake spindles has been studied by using electrophysiological methods (Cliff and Ridge, 1973). Their main finding was that the motor supply to one intrafusal muscle fiber comes from more than one motor unit in most cases. The motor units can be either dissimilar twitch units or twitch and tonic units. Similar observations have been made in lizard muscle (Proske, 1969a). By separately stimulating several filaments of the muscle nerve and recording from a single spindle sensory nerve, it was shown that a spindle could respond to separate stimulation of up to five motor nerve filaments. Further observations have to be made to determine the contraction characteristics of the intrafusal muscle fiber and how these are related to the contraction characteristics of the extrafusal muscle fibers of the same motor unit. This is particularly true for spindles with heterogeneous polyneural innervation. No evidence seems to be available concerning the numbers of spindles supplied by a single motor unit.

The diameters of the efferent axons to lizard spindles can be divided into two groups (Szepsenwol, 1960). One group lies in the range 12 to 15 μm, and the other in the range 3 to 6 μm. The larger-diameter group fibers eventually terminate in plate endings, and the smaller group in grape endings. The measurements were made at some unspecified distance from the nerve terminals. By taking diameter measurements at a distance between 0.5 and 1.0 mm from the spindle, Proske (1969b) found mean values of 5 μm and 3 μm for nerves ending in plate and grape endings, respectively. The corresponding diameters for these nerves near their extrafusal terminations were 7 μm and 4 μm.

The plate endings on the spindle possess a smaller Doyère's eminence than on the extrafusal endings. This is in contrast to the grape endings, which have no Doyère's eminence. Instead, they terminate in minute swellings in groups at short distances from one another along the fiber. From electron microscope studies, the plate endings are seen to be characterized by prominent junctional folds; the grape endings lack them (Fig. 6.10). In both types of ending, the terminals contain many round clear vesicles 40 to 60 nm in diameter and mitochondria (Ichiki et al., 1976). The junctional gap is 60 to 70 nm. The basement membrane runs between the nerve and muscle cells. The contractile characteristics of stimulating the nerves to intrafusal fibers are similar to those of stimulating nerves to extrafusal fibers.

D. Physiology of Muscle Spindles

1. IDENTIFICATION   When a muscle is stretched, it is possible to record afferent activity from the spindles. However, one has to ensure that the observed activity comes from a spindle and not from another type of sensory ending, such as a tendon organ. One of the commonest tests for spindle activity is to study the response of the sensory ending to motor stimulation (Proske, 1969a; Naeije, 1977). The muscle

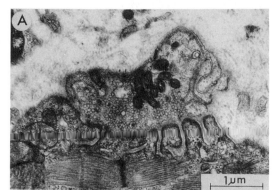

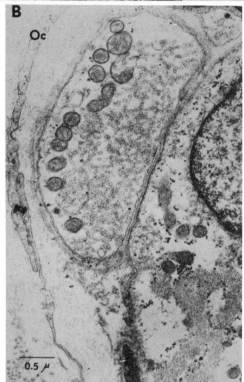

Fig. 6.10. Electron micrographs of motor nerve endings on intrafusal fibers. (A) Turtle spindle intrafusal fiber showing plate ending with junctional folds. (B) Ending on snake spindle with smooth myoneuronal junction characteristic of a grape ending. (A, from Crowe and Ragab, unpublished; B, from Fukami and Hunt, 1970.)

nerve is split into filaments, one of which is used to record the sensory activity. The muscle is held at constant length, and stimulation of one of the other filaments should produce an isometric contraction. If the stimulated filament contains no motor nerves that branch to the muscle spindle under study, the activity of the spindle will only be affected by contraction of the extrafusal muscle fibers lying in parallel. This will have an unloading effect, and the spindle activity should

A

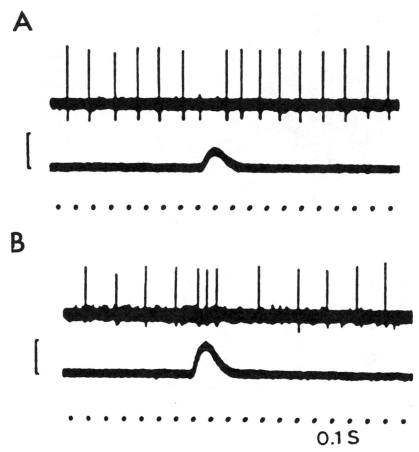

B

0.1 S

Fig. 6.11. Responses of muscle spindle to single shock stimuli applied to two different branches of a split nerve innervating the muscle. In (A) and (B) the upper trace shows the action potentials elicited by the spindle. The middle trace shows the muscle tension, with vertical bar indicating 5 g of force. The bottom trace shows 0.1-sec time pulses. In (A) an "in parallel" behavior is shown; that is, no action potentials are elicited during the rising phase of the tension twitch. This is a typical spindle behavior indicating that only extrafusal fibers contract during the twitch. In (B) an "in series" behavior is shown—extra action potentials are elicited during the rising phase of the twitch, indicating that both extrafusal and intrafusal fibers contract. (After Naeije, 1977.)

decrease or cease altogether. An example is shown in Fig. 6.11. The spindle has a steady discharge before the extrafusal twitch contraction. During the rising phase of the twitch the spindle becomes silent and resumes its activity during the descending phase of the twitch. This is known as the parallel response. Usually the nerve filament that is stimulated will contain several motor nerves, so adjusting the stimulus strength will produce contraction in varying numbers of motor units. By starting with a low stimulus strength one usually excites a motor nerve to give the parallel response. By increasing the stimulus strength to include more motor units, the so-called series response may be obtained if one of the motor nerves has endings on the muscle spindle. The intrafusal fiber will also contract and excite the sensory region of the spindle, producing an increase in firing of the spindle during the twitch (Fig. 6.11B).

Hunt and Wylie (1970) used another method to identify spindles in muscles of the garter snake. The muscle in this case is fairly thin and usually contains two spindles that, under suitable conditions, can be observed with the light microscope. Sensory recordings from the muscle usually show a double unit activity that can be abolished if the spindle structures observed in the muscle are destroyed. This rather drastic method precludes further study on the particular spindle, but it does establish its identity. Another method of studying spindle activity is to dissect out the spindle with its nerves (Hunt and Wylie, 1970; Fukami, 1970b). Such isolated preparations remove any doubt concerning identification of the sense organ and allow one to study the differences between the long-capsule and short-capsule spindles.

2. METHODS OF STUDYING SPINDLE ACTIVITY

A good preparation is one in which the nerve on the recording electrodes contains just one unit, that is, action potentials from only one sensory nerve are seen. These action potentials can be displayed on the oscilloscope as shown in Fig. 6.11. Another common method of display is to use an instantaneous frequency meter. Various models have been in use since the 1960s (e.g., Kay, 1965). Each action potential triggers a bright spot on the oscilloscope. The position of the spot on the horizontal axis is determined by the time of the appearance of the action potential. The vertical position is equal to the reciprocal of the time interval T since the preceding action potential. The function 1/T is the instantaneous firing frequency. If the interval T is of short duration, indicating a high firing frequency, the value of 1/T is large and the spot appears higher on the screen. An example of this type of recording is seen in Fig. 6.12. From a systems engineering point of view, it would perhaps appear that the behavior of the spindle under various conditions of muscle length can be best studied by applying sinusoidal stretches to the muscle and analyzing the corresponding variations in the spindle output. However, a type of stretch pattern favored by many physiologists is the so-called ramp-and-hold stretch. The muscle is held at a constant initial length until the start of the ramp phase. The muscle is then stretched at a constant speed until the desired final length is reached. This signifies the start of the hold phase, during which the muscle is held at its constant final length. This method of muscle stretch enables the dynamic activity of the sense organ to be studied by applying stretches of different speeds

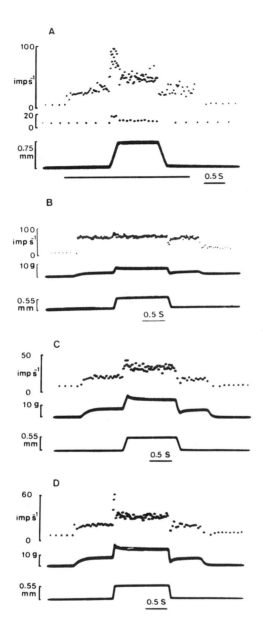

Fig. 6.12. (A) and (B). Two extreme examples of responses to a ramp-and-hold stretch during motor stimulation of turtle spindles. The spindle in (A) shows a relatively large response to the stretch during motor stimulation and the spindle in (B) shows hardly any. (A) The instantaneous frequency responses of the motor-stimulated spindle and passive spindle are shown, as well as the length change of the muscle. The horizontal bar indicates the duration of stimulation. The height of the stretch was 0.75 mm and the speed of stretch was 3.6 mm/sec. Both the static and dynamic sensitivity of the spindle have been increased considerably by the motor stimulation. (B) The instantaneous frequency output of the spindle, the tension of the muscle, and the length of the muscle are shown. Height of the stretch was 0.55 mm, and the speed of stretch was 8.8 mm/sec. The spindle shown in (C) and (D) has an intermediate behavior. It is subjected to stretches of the same height of 0.55 mm but at speeds of 5.5 mm/sec and 11 mm/sec, respectively. The difference in the amount of overshoot is clearly seen at the end of the ramp part of the stretch. For all spindles the stimulation frequency was 30 pulses per second. (After Naeije and Crowe, 1977a.)

during the ramp phase and the static activity at various steady lengths of the muscle.

Muscle spindle activity is often expressed in terms of the *dynamic index* (Jansen and Matthews, 1962), which is defined as the difference between the frequency of firing when it reaches its peak value at the end of the ramp phase of the stretch and the frequency 0.5 sec later during the hold phase of the stretch.

2. Responses to Passive Stretches

Ramp-and-hold stretches have been applied to spindles of the EDBI muscle of *Testudo graeca* (Ottoson, 1972), *Emys orbicularis, Pseudemys scripta elegans,* and *Pseudemys floridana* (Naeije and Crowe, 1974, 1977a; Naeije, 1977), and of the extensor digitorum longus muscle of *Chelodina longicollis* (Proske and Walker, 1975). A passive response is obtained when the spindle is subjected to mechanical stretch only and the motor nerves are not stimulated. A typical passive response to a ramp-and-hold stretch is shown in Fig. 6.12A. Sometimes the spindle has no resting discharge, but usually discharges of 5 to 10 pulses per second are seen. For low velocities of stretch, the frequency of firing increases steadily during the ramp phase to reach a peak value at the point where the hold phase starts. Relatively few action potentials are produced during the ramp phase for high speeds of stretch. The frequency of firing begins to decrease immediately at the start of the hold phase. For faster ramp stretches, one or two spikes of a relatively high frequency are seen at this point, but generally the frequency of firing decays in a fairly exponential manner over the hold phase. If the hold phase is extended in time, the firing rate reaches a fairly steady value, which is higher than the firing rate for the unstretched state. During the release part of the stretch to its original length, the spindle is silent for the higher-velocity releases.

The frequency of firing seems to depend on the preparation used. However, in comparison with spindles of other groups, the turtle spindles have a low firing rate. A frequency of about 30 impulses per second can only be obtained when the spindle is stretched well beyond its in situ length (Proske and Walker, 1975).

Spindle behavior in response to passive stretches is used to distinguish types of spindle ending (e.g., primary and secondary endings in cat spindles), but there appear to be no grounds for subdividing turtle spindles into more than one type. This is in keeping with the morphological studies in which categorization on the basis of structural differences has not been established.

The behavior of the turtle spindles during passive stretch is similar to that of the tonic spindles of the lizard (see below) or the secondary endings of cat muscle spindles (Matthews, 1963). Behavioral differences in response to passive ramp-and-hold stretches are seen when the long-capsule and short-capsule spindles of lizard and snake are compared (Hunt and Wylie, 1970; Fukami, 1972, 1977, 1978; Proske, 1969b). The short-capsule spindles are designated *phasic types*, which are characterized by a high frequency of firing during the ramp phase that abruptly falls to a fairly steady level during the hold phase (Fig. 6.13B). The long-capsule spindles are designated *tonic types*, in which the frequency of firing increases fairly steadily during the ramp phase of the stretch. There is no large abrupt change of frequency at the start of the hold phase, but a more gradual decay in firing frequency to a steady level (Fig. 6.13A). Although differences between long- and short-capsule spindles are observed, caution must be exercised when trying to assess the type of spindle on the basis of its behavior during a single ramp-and-hold stretch. A series of recordings using different speeds and magnitudes of stretch should be made. For instance, long-

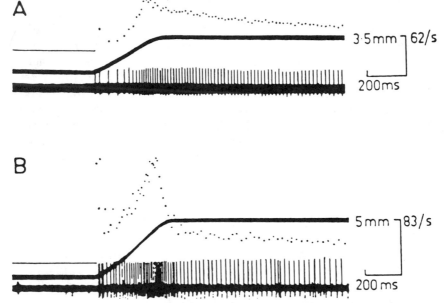

Fig. 6.13. (A) Tonic responses of a muscle spindle from semimembranosus to ramp-and-hold stretching of the muscle. The rate of stretch (mm/sec) is shown in the top left-hand corner of each record. The upper trace shows the frequency of discharge of the ending; each spot represents an action potential and its height above zero (represented by the line at the beginning of the trace) is the reciprocal of the time interval between it and the preceding action potential. The line below the frequency record represents the length change of the muscle while the action potential discharge is shown in the bottom trace. The vertical bar on the right of the figure represents the calibration for both the frequency of the spots and the length of the muscle. The horizontal bar is the time calibration. (B) Phasic responses of a muscle spindle from scalenus to ramp-and-hold stretches. (After Proske and Ridge, 1975.)

capsule spindles in the semimembranosus of lizards have a distinct phasic response at the higher speeds of stretch, but this is reduced or absent at the lower speeds. The differences between the two types of spindle are best displayed by plotting the dynamic index against the speed of stretch. Except for the low speeds of stretch, it is seen that the dynamic index for the phasic spindles is greater than that for the tonic spindles (Fig. 6.14). Here again, caution must be used when

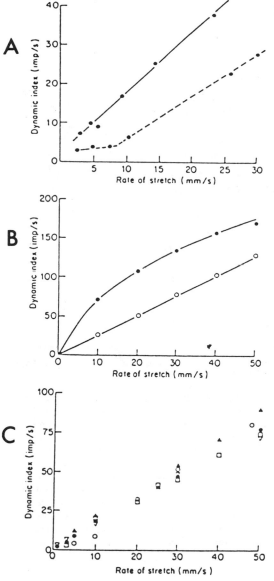

Fig. 6.14. Relation between the rate of stretch and the dynamic index (impulses per second) for various types of spindles. (A) Garter snake. Upper line: Results for phasic (short-capsule) spindle from costocutaneous muscle. Lower line: Results from tonic (long-capsule) spindle from same muscle. (B) Lizard *Tiliqua nigrolutea.* Upper line: Results from phasic spindle from scalenus muscle. Lower line: Results from tonic spindle from semimembranosus. (C) Turtle *Chelodina longicollis* spindle in extensor digitorum longus. The single type of spindle responds like the tonic spindles of squamata. (A, after Fukami, 1970a; B, after Proske, 1969b; C, after Proske and Walker, 1975.)

comparing one spindle with another, particularly if two different muscles are involved. The speed of stretch is usually expressed with respect to the movement of the stretch apparatus and not as a percentage of the spindle length. The actual amount of stretch applied to the spindle will depend upon how much is taken up by the muscle tendon and other factors and the length of the spindle compared to the muscle length. However, in spite of these considerations, it is generally accepted that different physiological responses can be seen when long- and short-capsule spindles are compared. Further research must still be done to explain these differences in terms of structural variations.

4. Fusimotor Stimulation

a. Constant Muscle Length   As already mentioned, motor innervation of reptilian muscle spindles consists of a branch of a motoneuron that also innervates the extrafusal muscle fibers. Fig. 6.15 shows two examples of responses to motor stimulation of turtle spindles held at constant length. In each case, the motor nerve was stimulated at a frequency of 30 per second. In Fig. 6.15A, the frequency changes—compared to changes in extrafusal tension—took place at a rather slow rate; the muscle tension reached a steady value within

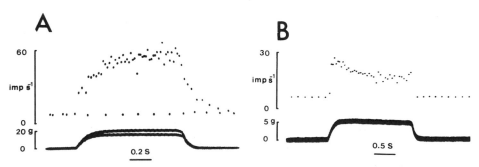

Fig. 6.15.   Two extreme examples of turtle spindle response patterns to motor stimulation. Each figure shows the instantaneous firing frequency output of the spindle and the muscle tension. When the spindle shown in (A) was stimulated, the frequency changes, in comparison to the extrafusal tension changes, took place at a rather slow rate. In (B) the frequency changes of the spindle were more abrupt. For both spindles, the stimulation frequency was 30 pulses per second. Note the different time scales used in (A) and (B). In (A) two superimposed records also show effects of two different stimulus strengths during tetanic stimulation of a motor nerve branch. In the case of the higher stimulus strength, as indicated by the higher muscle tension, the spindle response increased considerably above the resting value. At the lower stimulus strength, the spindle was not activated and an unloading effect, as seen by the slightly reduced firing frequency compared to the resting value, was produced. (After Naeije and Crowe, 1977a.)

about 0.4 sec from the start of stimulation, whereas the spindle firing frequency was still increasing over the whole of the stimulation period. In contrast, the response shown in Fig. 6.15B is more abrupt. At the onset of stimulation, the spindle firing frequency rises immediately to its maximum value and then declines gradually to a steady value at about one second from the start of stimulation. These are two extreme examples of a range of effects seen. It is not possible to divide the responses to fusimotor stimulation into two (or more) categories on the basis of the pattern of firing frequency at constant muscle length. It should be noted, however, that such differences may be seen when several motor fibers supplying the same spindle are studied (Fig. 6.16).

The degree of extrafusal muscle contraction affects the spindle firing frequency. This is shown in Fig. 6.15A, which comprises two superimposed records showing the effects of two different stimulus strengths but identical stimulation frequencies. In the case of the higher stimulus strength, as indicated by the higher muscle tension, the spindle response increased considerably above the resting value. At the lower stimulus strength, no fusimotor stimulation of the

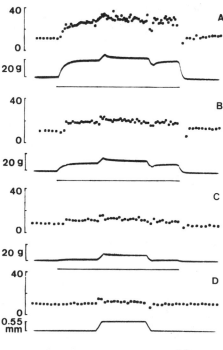

Fig. 6.16. Example of a spindle that was stimulated through three different motor nerve branches. In A to C, responses of the spindle to a ramp-and-hold stretch are shown during successive stimulation of the three nerve branches. In each record, the upper trace shows the instantaneous firing frequency output of the spindle, and the lower trace shows the muscle tension. D shows the response of the nonactivated spindle and the length change of the muscle. Height of the stretch was 0.55 mm, and the speed of stretch was 4.4 mm/sec. (From Naeije and Crowe, 1977a.)

spindle apparently takes place, and an unloading effect, as seen by the slightly reduced firing frequency, was produced.

Clearly, the effects of fusimotor stimulation ought to be studied in the absence of complicating factors due to extrafusal tension unloading. This could be done either by using isolated spindle preparations or by blocking extrafusal contraction by means of the critical curarization technique, as has been successfully used in the studies of other spindles. By using such techniques, it should be possible to resolve the question as to whether there are differing responses to fusimotor stimulation and whether such differences can be related to the morphological differences in the motor endings.

The stimulus frequency of 30 per second was sufficient to produce maximum extrafusal contractile force but not the maximum spindle firing frequency. Higher frequencies of stimulation—up to 70 per second—were needed to produce the maximum spindle firing frequency at the particular muscle length (Fig. 6.17). This suggests that the fusion frequencies for intrafusal and extrafusal fibers are not the same. This is in contrast to the situation for lizard spindles (Proske, 1969a), where the fusion frequency for the extrafusal iliofibularis fibers and the intrafusal (presumably long-capsule) fibers were found to be about 60 per second in both cases (Fig. 6.18).

The techniques of critical curarization and isolated spindle studies have been applied with success to snake spindles. Cliff and Ridge (1973) used the technique to examine the motor innervation of spin-

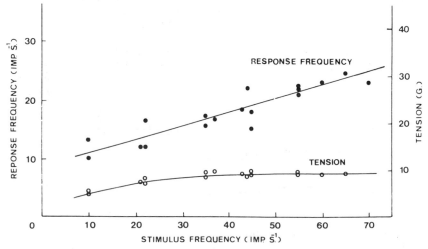

Fig. 6.17. Change in turtle spindle response (measured 0.75 sec after the onset of stimulation) and muscle tension. A stimulation frequency of 35 pulses per second produced the maximum extrafusal contractile force, but clearly this stimulation frequency was not sufficient to elicit the maximum spindle response. (From Naeije and Crowe, 1977a.)

dles by branches of motoneurons that supply twitch and tonic extra-fusal fibers. A single motor axon was stimulated by a microelectrode situated near one of the nerve branches in the muscle. The muscle tension recorded before critical curarization established the twitch or tonic character of the motor unit involved. After critical curarization, the effects upon the intrafusal fibers of (in these experiments) the long-capsule spindles were studied. Although the spindles received both types of motor innervation, their effects upon the spindle re-sponse were similar, a result that was verified in those cases where a single spindle received motor innervation of both types.

The isolated spindle preparations have been used to study the speed of the action potential propagated along the intrafusal fiber fol-lowing motor axon stimulation (Fukami, 1970a). The speed of con-duction is about 0.2 to 0.7 m/sec at a temperature of 25° C, but the speed is increased at the sensory region where the fiber diameter in-creases. The intrafusal fibers have a longer refractory period than the extrafusal fibers.

b. During Muscle Stretch

Fusimotor stimulation applied during a ramp-and-hold stretch to mammalian spindles affects the dynamic index of the spindle ending. For instance, the motor innervation of the cat spindle can be divided into two sorts: the so-called dynamic fibers increase the dynamic in-dex of the primary ending, and the static fibers decrease it (Crowe and Matthews, 1964). Similar experiments have been performed on

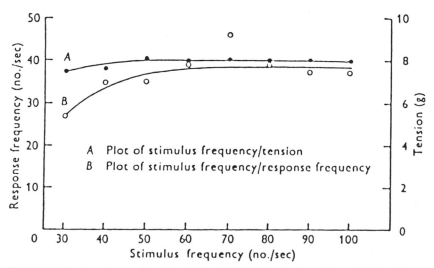

Fig. 6.18. Two graphs representing plots of the tetanic tension of lizard iliofibularis (A) and the response frequency of a spindle during the tetanus (B) at various stimulus frequencies (After Proske 1969b.)

reptilian spindles. Turtle spindles show a wide range of responses to mechanical stretches of the ramp-and-hold type during motor stimulation (Naeije and Crowe, 1977a; Naeije, 1977). Fig. 6.12 shows extreme examples. The motor stimulation has greatly increased the dynamic and static sensitivity in Fig. 6.12A, whereas the response to the stretch is hardly visible in the spindle shown in Fig. 6.12B. The spindle shown in Fig. 6.12C and D has an intermediate behavior. It is subjected to stretches of the same height of 0.5 mm but speeds of 5.5 mm/sec and 11 mm/sec, respectively (the muscle length was 15 mm). The differences in the amount of overshoot are clearly seen at the end of the ramp phase of the stretch. Fig. 6.19A shows a plot of the dynamic index as a function of the ramp speed. The effect of motor stimulation in this experiment is to increase the dynamic index about threefold. The effect of changing the height of the stretch upon the firing frequency during the hold phase compared to the frequency before stretch is shown in Fig. 6.19B. In this case the stimulated spindle is about four times as sensitive to length changes as a passive spindle. The motor stimulation has thus increased the sensitivity to both the amount and speed of stretch. Naeije and Crowe were unable to separate their results into two categories that could indicate different types of motor innervation due to the two types of motor ending. In cases in which it was possible to activate a spindle through two or more motor nerve branches, the effects on the spindle were qualitatively similar. There was no indication that the branches had different effects upon the spindle, despite the fact that in some cases different types of motor ending were stimulated (see Fig. 6.16).

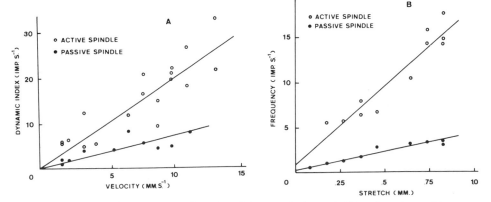

Fig. 6.19. Dynamic index as a function of the speed of stretch (A) and static sensitivity as a function of the height of stretch (B) during motor stimulation (open circles) and without motor stimulation (closed circles). Measurements were made on the spindle shown in Fig. 6.12C and D. The effect of the motor stimulation on this particular spindle was to increase the dynamic sensitivity threefold and the static sensitivity fourfold. (From Naeije and Crowe, 1977a.)

Proske and Walker (1975) performed similar experiments on spindles from the EDL muscle of *Chelodina longicollis*. These authors maintain that two types of effect are seen upon motor stimulation; sometimes a dynamic effect is seen, sometimes a static effect. However, only rarely were the two types of response seen in the same spindle with several motor nerves. Further, Proske and Walker studied 50 spindles, 10 of which had motor filaments exerting a static action and 10 others that had a dynamic action. Two spindles had no response to motor stimulation, and one had a motor innervation of both types. Thus more than 50% of the spindles had motor innervations that could not be categorized. Clearly, more experiments are needed to establish whether there really are two types of effect seen upon motor stimulation of turtle spindles or whether there is in fact just one sort of response for which the range of responses is very wide—one extreme being a static response, and the other a dynamic response.

The effects of motor stimulation upon the response to mechanical stretch have been studied in the spindles of the lizards *Tiliqua nigrolutea* and *Tiliqua scincoides* (Proske, 1973). Spindles in the semimembranosus, a twitch muscle, and semitendinosus, a slow muscle, were used. The firing rate of spindles in the twitch muscle increased before stretch, but there was little further increase in the response during the ramp-and-hold stretch. On the other hand, the spindles in the slow muscle were only moderately excited by the motor stimulation before stretch, but the output was considerably increased during the ramp phase and the hold phase of the stretch. The graphs of Fig. 6.14 show the effects of motor stimulation upon the dynamic index of reptilian spindles.

III. TENDON ORGANS

Rudimentary tendon organs have been reported in *Emydoidea blandingii* (Fig. 6.20B) by Huber and DeWitt (1900), but there seem to be no recent confirmations, either by anatomical studies or physiological investigations. Tendon organs have been reported in python muscle (Kulchitsky, 1924a, 1924b). The nerve fibers are nonmyelinated, and the extensive branches terminate as free endings in the muscular connective tissue. Of reptilian tendon organs, those found in lizards (Fig. 6.20A) have been most extensively studied. The efferent nerve fibers are large and terminate in plaques that are 60 to 100 μm long and 40 to 50 μm wide (Regaud, 1907). The first physiological studies appear to have been carried out by Gregory and Proske (1974, 1975). They used the organs in the tendon of insertion of the caudofemoralis muscle of *Tiliqua scincoides* and studied the afferent discharges to ramp-and-hold tension changes. The organs were silent in the absence of applied tension. There was a high rate of firing during the ramp phase of the stretch, which depended upon the rate of stretch.

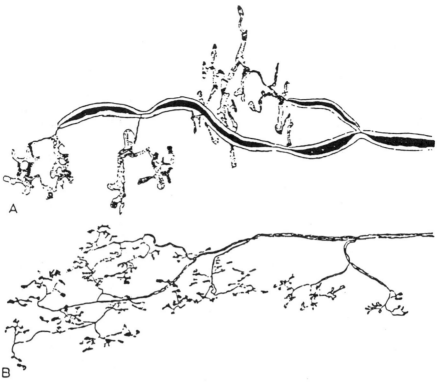

Fig. 6.20. Drawing of sensory plaques innervating the tendons of reptiles. (A), Lizard tendon, gold chloride preparation; from Barker, 1974, after Crevatin, 1901. (B), from turtle (*Emydoidea blandingii*) tendon, methylene blue preparation; from Barker, 1974, after Huber and DeWitt, 1900.)

The firing during the hold phase was considerably less, depending upon the steady (hold) tension levels. The afferent nerves supplying the organs have a large degree of branching at the terminals. The sensory terminations are distributed over a large area. Proske and Gregory (1976) used this anatomical fact to study the mechanism of impulse initiation. A localized mechanical stimulation with a von Frey hair was used, and the nerve branch was severed to see if the total activity in a single afferent nerve during muscle stretch changed as a result of reducing the number of sensory terminations. From these experiments it could be concluded that the tendon organ discharge is a composite response having contributions from several impulse generators.

IV. JOINT RECEPTORS

Joint receptors have been studied in specimens of *Testudo graeca* (Fig. 6.21F and G) and *Emys orbicularis* (Pac, 1968a, 1969, 1976). The organs

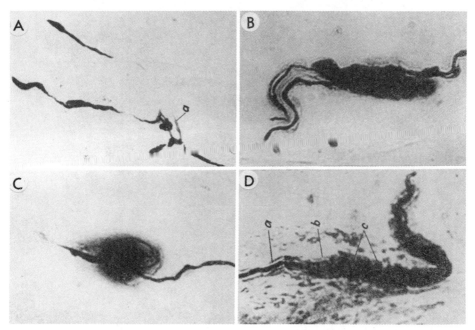

Fig. 6.21. Joint receptors from lizard and turtle (A) Receptor field of encapsulated corpuscles in joint of *Agama stellio* supplying nerve fiber (a). (B) Simple encapsulated corpuscle from hip joint of *Agama stellio*. (C) Nonencapsulated corpuscle from hip joint of *Agama stellio*. (D) Simple encapsulated corpuscle from knee joint of *Agama stellio* showing myelin sheath (a), simple capsule (b), and nuclei of inner core (c). (E) Two simple encapsulated corpuscles from the elbow joint supplying nerve fiber (a). (F) Detail of ramifying encapsulated corpuscle from shoulder joint of turtle *Testudo graeca*. (G) Simple encapsulated corpuscle (b) from the elbow joint of *Agama stellio* showing perineurium passing into capsule at (a). (H) Spraylike ramifying axon from shoulder joint of turtle *Testudo graeca*. (After Pac, 1968b.) See p. 488 for E–H.

are spraylike, ramified encapsulated corpuscles having a primary core and are not unlike the primitive paciniform corpuscles in mammals and birds. Each of the branches of the receptor consists of three components. The terminal axon is surrounded by the inner core cells; the inner core does not have a lamella structure. The capsule consists of one to three layers surrounding the inner core. According to Pac, these joint receptors represent a primitive stage in the phylogenic development of the lamella mechanoreceptors.

Joint receptors in the lizards have been studied by Stefanelli (1934, Gekkonidae) and Pac (1968b, 1972, *Agama stellio* [Fig. 6.21A and B], *Physignathus lesueurii*, and *Lacerta viridis*). Several sorts of sensory nerve endings are found in the joint capsules. Free nerve endings, noncapsulated clawlike structures, noncapsulated spraylike nerve endings, and simple encapsulated corpuscles with a primitive inner core are seen. No reports of physiological studies have been found.

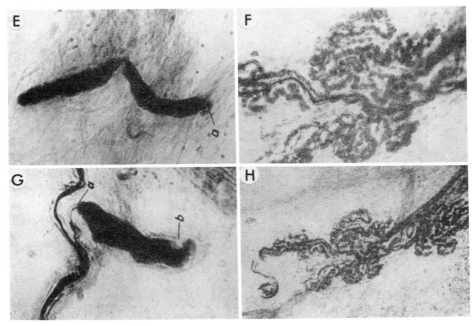

Fig. 6.21 continued

V. CONCLUDING REMARKS

During the past 30 years a great deal of work, much of it inspired by controversy, has been carried out on the mammalian muscle spindle. Resultant findings have led to the realization that it, with its several sorts of motor innervation, different types of intrafusal fiber, and its single primary and multiple secondary endings, is a rather complicated sense organ. Its density and distribution within the skeletal muscles indicate its importance. A similar amount of time and effort should not be devoted to the reptilian spindle, which has already been shown to be a much less complicated sense organ. Further, I doubt that it would be worthwhile to go into its minute morphological details. Discussions over, for instance, the ultrastructure or subtle differences revealed in comparing the spindles of one species with those of another are, at present, of secondary importance. It is now more propitious to study the spindle (and the tendon organs and joint receptors) with other aims in view. More attention should be paid to function and neural connections rather than structural and physiological nuances. The reptiles offer good possibilities to study these sense organs with respect to their role in proprioception and motor control. The spindles are found in their simplest forms in the reptiles. If one accepts that the basic functions of the spindle are common to all the vertebrates in which they are found, then knowledge gained from studying the simpler systems can be used to provide further insight into the more developed systems in the higher vertebrates with their refinements and improvements gained through evolutionary development.

Studies have been done on turtle spindles to find a relationship, known as a transfer function, between the pattern of mechanical input, or stimulation, and the output expressed as the time-dependent frequency of the action potentials recorded in the afferent nerve (Naeije et al., 1976; Naeije and Crowe, 1977a, 1977b). However, the central nervous system receives a train of impulses that it presumably has to interpret. Can this be done given that the incoming signal is a stochastic point process within a limited frequency range? According to Crowe et al. (1980, 1982) a reasonable interpretation of the incoming signals from a turtle spindle might be possible provided the central nervous system somehow has a "knowledge" of the transfer function of the spindle. Whether this method of interpretation is performed in the central nervous system is, of course, open to question, but such studies do indicate the upper limits of interpretation provided certain conditions are met and adequate "computing" facilities are available within the central nervous system.

Such studies, however, beg the question as to the function of the receptor. To be sure, the spindle responds to the length changes, but it is not established that this is what is actually measured, and little is known about where the information is sent, stored, or used within the central nervous system. Electrophysiological techniques in combination with histochemical analyses provide the mechanism for studying the neural pathways. Already, there is evidence to indicate that testudinian spindle afferents make monosynaptic connections with the motoneurons (Rosenberg, 1972; Feenstra, 1983; Ruigrok, 1984; Ruigrok et al. 1984, 1985a, 1985b). Such studies, although very promising, are still in their early stages. The destinations of the spindle afferents to other regions of the central nervous system still need to be traced.

APPENDIX: REPTILIAN SPECIES DISCUSSED

TESTUDINES

Chelodina longicollis
  Proske and Walker, 1975
Chrysemys picta
  Huber and DeWitt, 1897
Emdoidea blandingi
  Huber and DeWitt, 1897
Emys orbicularis
  Pac, 1976

Testudo graeca
  Crowe and Ragab, 1967, 1969,
    1970a, 1970b, 1972
  Pac, 1968a, 1976
  Page, 1968
  Proske and Walker, 1975
  Ragab, 1970
  Rosenberg, 1972

Trachemys floridana
  Naeije, 1977
  Naeije and Crowe, 1974, 1977a,
    1977b
  Naeije et al., 1976

Trachemys scripta elegans
  Feenstra, 1983
  Ruigrok, 1984
  Ruigrok et al., 1984, 1985a, 1985b

CROCODILIA

Alligator
  Hines, 1930

SAURIA

Agama stellio
  Pac, 1972
Anolis cristatellus
  Szepsenwol, 1960
Calotes versicolor
  Vijayalakshmi, 1976
  Vijayalakshmi and Viswanathan,
    1976
  Viswanathan, 1972a, 1972b, 1973,
    1976
  Viswanathan and Vijayalakshmi,
    1975
Dipsosaurus dorsalis
  Maier et al., 1974

Lacerta vividis
  Barker, 1974
  Pac, 1977
Tiliqua nigrolutea
  Proske, 1969a, 1969b, 1973
  Proske and Vaughan, 1968
Tiliqua scincoides
  Proske, 1973
Lizards
  Gregory and Proske, 1974, 1975
  Pac, 1968b

SERPENTES

Elaphe quadrivirgata
  Ichiki, 1973
  Ichiki et al., 1976
Lampropeltis sp.
  Maier et al., 1974
Natrix sp.
  Cliff and Ridge, 1973
  Pallot, 1974a, 1974b

  Pallot and Ridge, 1972, 1973
  Pallot and Taberner, 1974
Python
  Kulchitsky, 1924a, 1924b
Thamnophis
  Fukami, 1970a, 1977, 1978, 1982
  Fukami and Hunt, 1970, 1974,
    1977

REFERENCES

Barker, D. (1974). The morphology of muscle receptors. In *Handbook of Sensory Physiology*, Vol. III/2, *Muscle Receptors* (C. C. Hunt, ed.). Springer, Berlin, 1974, pp. 1–190.

Cipollone, L. T. (1898). Nuove ricerche sul fuso neuro-muscolare. *Richerche. n. lab. anat. norm. r. Univ. Roma* 6, 157–200.

Cliff, G. S., and Ridge, R. M. A. P. (1973). Innervation of extrafusal and intrafusal fibres in snake muscle. *J. Physiol.* 233, 1–18.

Crevatin, F. (1901). Sopra le terminazioni nervose nei tendini dei pipistrelli. *Rend. Sess. Roy. Acad. Sci. Ist Bologna* 5, 31–34.

Crowe, A., and Matthews, P. B. C. (1964). The effects of stimulation of static and dynamic fusimotor fibres on the response to stretching of the primary endings of muscle spindles. *J. Physiol.* 174, 109–131.

Crowe, A., Naeije, M., and Tanke, R. H. J. (1980). A method for the interpretation of signals in afferent nerves from muscle spindles. *Biol. Cybern.* 37, 187–194.

Crowe, A., Naeije, M., and Tanke, R. H. J. (1982). The effects of beta motor stimulation on the interpretation of signals from muscle spindles. *Biol. Cybern.* 44, 9–16.

Crowe, A., and Ragab, A. H. M. F. (1967). Some preliminary studies on spindle-like structures in tortoise muscle. *J. Physiol.* 192, 22–23P.

Crowe, A., and Ragab, A. H. M. F. (1969). The innervation and capsular structure of the tortoise muscle spindle. *J. Physiol.* 201, 5–6P.

Crowe, A., and Ragab, A. H. M. F. (1970a). The structure, distribution and innervation of spindles in the extensor digitorum brevis I muscle of the tortoise *Testudo graeca*. *J. Anat.* 106, 521–538.

Crowe, A., and Ragab, A. H. M. F. (1970b). Studies on the fine structure of the capsular region of tortoise muscle spindles. *J. Anat.* 107, 257–269.

Crowe, A., and Ragab, A. H. M. F. (1972). A histochemical investigation of intrafusal fibers in tortoise muscle spindles. *J. Histochem. Cytochem.* 20, 200–204.

Feenstra, B. W. A. (1983). "Spinal Projections of Hindlimb Afferents in the Terrapin." Doctoral thesis, University of Utrecht.

Fukami, Y. (1970a). Tonic and phasic muscle spindles in snake. *J. Neurophysiol.* 33, 28–35.

Fukami, Y. (1970b). Accommodation in afferent nerve terminals of the snake muscle spindle. *J. Neurophysiol.* 33, 475–489.

Fukami, Y. (1972). Electrical and mechanical factors in the adaptation of reinnervated muscle spindles in snake. In *Research in Muscle Development and the Muscle Spindle* (B. Q. Banker, R. J. Przybylski, J. P. van der Meulen, and M. Victor, eds.). Excerpta Medica, Amsterdam, pp. 379–399.

Fukami, Y. (1977). Snake muscle spindles. *Brain Res.* 128, 527–531.

Fukami, Y. (1978). Receptor potential and spike initiation in 2 varieties of snake muscle spindles. *J. Neurophysiol.* 41, 1546–1556.

Fukami, Y. (1982). Further morphological and electrophysiological studies on snake muscle spindles. *J. Neurophysiol.* 47, 810–826.

Fukami, Y., and Hunt, C. C. (1970). Structure of snake muscle spindles. *J. Neurophysiol.* 33, 9–27.

Fukami, Y., and Hunt, C. C. (1974). Response of the short-capsule spindle in reptiles. *J. Anat.* 119, 192.

Fukami, Y., and Hunt, C. C. (1977). Structures in sensory region of snake spindles and their displacement during stretch. *J. Neurophysiol.* 45, 1121–1131.

Giacomini, E. (1898). Sui fusi neuro-muscolari dei Sauropsidi. *Atti. Accad. Fisiocr. Siena* 9, 215–230.

Gregory, J. E., and Proske, U. (1974). Tendon organs in a lizard. *Proc. Aust. Physiol. Pharmacol. Soc.* 5, 73–74.

Gregory, J. E., and Proske, U. (1975). Responses of tendon organs in a lizard. *J. Physiol.* 248, 519–530.

Guthe, K. F. (1981). Reptilian muscle: Fine structure and physiological parameters. In *Biology of the Reptilia*, Vol. 11, no. 3 *Morphology F* (C. Gans and T. S. Parsons, eds.). Academic Press, London, New York, pp. 267–354.

Hines, M. (1930). The innervation of the muscle spindles. *Proc. Assoc. Res. Nerv. Ment. Dis.* 9, 124–152.

Homma, S., ed. (1976). Understanding the stretch reflex. *Prog. Brain Res.* 44. Elsevier, Amsterdam.

Huber, G. C., and DeWitt, L. M. A. (1897). A contribution on the motor nerve-endings and on the nerve-endings in the muscle-spindles. *J. Comp. Neurol.* 7, 169–230.

Huber, G. C., and DeWitt, L. M. A. (1900). A contribution on the nerve terminations in neuro-tendinous and end-organs. *J. Comp. Neurol.* 10, 159–208.

Hunt, C. C., and Wylie, R. M. (1970). Responses of snake muscle spindles to stretch and intrafusal muscle fiber contraction. *J. Neurophysiol.* 33, 1–8.

Ichiki, M. (1973). Properties of snake muscle spindle fibers. *J. Physiol. Soc. Jap.* 35, 8–9.

Ichiki, M., Nakagaki, I., Konishi, A., and Fukami, Y. (1976). The innervation of muscle spindles in the snake *Elaphe quadrivirgata*. *J. Anat.* 122, 141–168.

Jansen, J. K. S., and Matthews, P. B. C. (1962). The central control of the dynamic response of muscle spindle receptors. *J. Physiol.* 161, 357–378.

Katz, B. (1961). The termination of the afferent nerve fibre in the muscle spindle of the frog. *Proc. Roy. Soc. Lond.* B243, 221–240.

Kay, R. H. (1965). A reciprocal time-interval display using transistor circuits. *Electron. Eng.* 37, 543–545.

Kennedy, W. R., Poppele, R. E., and Quick, D. C. (1981). Mammalian muscle spindles. In *The Physiology of Peripheral Nerve Disease* (A. J. Sumner, ed.). Saunders, Philadelphia, pp. 74–123.

Kerschner, L. (1888a). Beiträge zur Kenntniss der sensiblen Endorgane. *Anat. Anz.* 3, 288–296.

Kerschner, L. (1888b). Bemerkungen über ein besonderes Muskelsystem in willkürlichen Muskeln. *Anat. Anz.* 3, 126–132.

Kölliker, A. (1862a). Untersuchungen über die letzten Endigungen der Nerven. *Z. wiss. Zool.* 12, 149–164.

Kölliker, A. (1862b). On the terminations of nerves in muscles, as observed in the frog: and on the disposition of the nerves in the frog's heart. *Proc. Roy. Soc. (Lond.)* 12, 65–84.

Kühne, W. (1863a). Über die Endigung der Nerven in den Muskeln. *Virchows Arch. path. Anat. Physiol.* 27, 508–523.

Kühne, W. (1863b). Die Muskelspindeln. Ein Beitrag zur Lehre von der Entwickelung der Muskeln und Nervenfasern. *Virchows Arch. path. Anat. Physiol.* 28, 528–538.

Kühne, W. (1864). Neue Untersuchungen über motorische Nervendigung. *Z. Biol.* 23, 1–148.

Kulchitsky, N. (1924a). Nerve endings in muscles. *J. Anat.* 58, 152–164.

Kulchitsky, N. (1924b). Nerve endings in muscles. *J. Anat.* 59, 1–17.

Maier, A., DeSantes, M., and Eldred, E. (1974). The occurrence of muscle spindles in extraocular muscles of various vertebrates. *J. Morphol.* 143, 397–408.

Matthews, P. B. C. (1963). The response of de-efferented muscle spindle receptors to stretching at different velocities. *J. Physiol.* 168, 660–678.

Matthews, P. B. C. (1972). *Mammalian Muscle Receptors and their Central Actions.* Arnold, London.

Matthews, P. B. C. (1981). Evolving views on the internal operation and functional role of the muscle spindle. *J. Physiol.* 320, 1–30.

Matthews, P. B. C. (1982). Where does Sherrington's 'muscular sense' originate? Muscles, joints, corollary discharges. *Rev. Neurosci.* 5, 189–218.

Naeije, M. (1977). "The Chelonian Muscle Spindle." Doctoral thesis, University of Utrecht.

Naeije, M., and Crowe, A. (1974). The response of chelonian muscle spindles to mechanical stimulation. *Life Sci.* 15, 131–136.

Naeije, M., and Crowe, A. (1977a). Responses of the chelonian muscle spindles to mechanical stretch and fusimotor stimulation. *J. Neurophysiol.* 40, 814–821.

Naeije, M., and Crowe, A. (1977b). Model for the beta motor stimulation in chelonian muscle spindles. *Biol. Cybern.* 26, 73–79.

Naeije, M., Crowe, A., and de Klerk, H. (1976). Model of the firing frequency of the chelonian muscle spindle. *Biol. Cybern.* 21, 53–60.

Ottoson, D. (1972). Functional properties of a muscle spindle with no fluid space. *Brain Res.* 41, 471–474.

Pac, L. (1968a). Sensory nerve endings in the joint capsules of the tortoise. *Folia. Morphol. Prague* 16, 43–47.

Pac, L. (1968b). Sensory nerve endings in the joint capsules of some Lacertilia. *Scripta medica. Brno* 41, 155–161.

Pac, L. (1969). Our findings in the phylogenesis of the joint receptors in some vertebrates. Proceedings of the 12th Congress on Morphology, Prague, p. 89.

Pac, L. (1972). Sensory nerve endings in the joint capsules of the green lizard. *Scripta medica. Brno* 45, 115–120.

Pac, L. (1976). Ultrastructure of the joint receptors of the tortoise *Testudo graeca, Emys orbicularis. Z. Mikrosk-Anat. Forsch.* (Leipzig) 89, 1068–1078.

Page, S. G. (1968). Fine structure of tortoise skeletal muscle. *J. Physiol.* 197, 707–715.

Pallot, D. J. (1974a). The structure and innervation of intrafusal fibers in snakes of *Natrix sp. J. Anat.* 118, 281–294.

Pallot, D. J. (1974b). Structural studies on two types of snake spindle. *J. Anat.* 119, 192.

Pallot, D. J., and Ridge, R. M. A. P. (1972). The fine structure of the long capsule spindles of the snake *Natrix sp. J. Anat.* 113, 61–74.

Pallot, D. J., and Ridge, R. M. A. P. (1973). The fine structure of the short capsule muscle spindles in snakes of *Natrix sp. J. Anat.* 114, 13–24.

Pallot, D. J., and Taberner, J. (1974). Histochemistry of muscle spindles in snakes of *Natrix sp. J. Histochem. Cytochem.* 22, 881–886.

Perroncito, A. (1901). Sur la terminaison des nerfs dans les fibres musculaires striées. *Arch. Ital. Biol.* 36, 245–254.

Proske, U. (1969a). The innervation of muscle spindles in the lizard *Tiliqua nigrolutea*. *J. Anat.* 105, 217–230.

Proske, U. (1969b). An electrophysiological analysis of responses from lizard muscle spindles. *J. Physiol.* 205, 289–304.

Proske, U. (1973). The muscle spindles in slow and twitch skeletal muscle of the lizard. *J. Physiol.* 230, 429–448.

Proske, U., and Gregory, J. E. (1976). Multiple sites of impulse initiation in a tendon organ. *Exp. Neurol.* 50, 515–520.

Proske, U., and Ridge, R. M. A. P. (1975). Extrafusal muscle and muscle spindles in the reptiles. In *Progress in Neurobiology* (G. A. Kerkut and J. W. Phillips, eds.). Pergamon Press, Oxford, New York, pp. 1–29.

Proske, U., and Vaughan, P. C. (1968). Histological and electrophysiological investigation of lizard skeletal muscle. *J. Physiol.* 199, 495–509.

Proske, U., and Walker, B. (1975). Responses of muscle spindles in a tortoise. *Brain Res.* 91, 79–88.

Ragab, A. H. M. F. (1970). "The Structure and Innervation of Tortoise Muscle Spindles." Ph.D. thesis, University of Durham, England.

Regaud, C., and Favre, M. (1904). Les terminaisons nerveuses et les organes nerveux sensitifs des muscles striés squelettaux. *Rev. gén. Histol.* 1–140.

Rosenberg, M. E. (1972). Excitation and inhibition of motoneurones in tortoise. *J. Physiol.* 221, 715–730.

Ruffini, A. (1893). Sur la terminaison nerveuse dans les faisceux musculaire et leur signification physiologique. *Arch. ital. Biol.* 18, 106–114.

Ruigrok, T. J. H. (1984). Organization and morphology of motoneurons and primary afferents in the lumbar spinal cord of the turtle *Pseudemys scripta elegans.* Doctoral thesis, University of Utrecht.

Ruigrok, T. J. H., Crowe, A., and ten Donkelaar, H. J. (1985a) Morphology of primary afferents to the spinal cord of the turtle *Pseudemys scripta elegans.* *Anat. Embryol.* 171, 75–81.

Ruigrok, T. J. H., Crowe, A., and ten Donkelaar, H. J. (1985b). Terminations of primary afferents on lumbar motoneurons in the turtle *Pseudemys scripta elegans.* *Brain Res.* 339, 141–145.

Ruigrok, T. J. H., ten Donkelaar, H. J., and Crowe, A. (1984). Anatomy of primary afferents in the spinal cord of the turtle *Pseudemys scripta elegans.* *Neurosci. Lett.,* Suppl., 18, 251.

Sherrington, C. S. (1894). On the anatomical constitution of nerves of skeletal muscles; with remarks on recurrent fibres in the ventral spinal nerve-root. *J. Physiol.* 17, 211–258.

Stefannelli, A. (1934). Su di alcune espansioni nervese nel periosteo e nel periondrio dei Rettili. *Monit. zool. ital.* 55, 115–120.

Szepsenwol, J. (1960). The neuromuscular spindle in the lizard *Anolis cristatellus. Cellule* 61, 21–37.

Taylor, A., and Prochazka, A. (1981). *Muscle Receptors and Movement.* Macmillan, London.

Vallbo, A. B., Hagbarth, K.-E., Torebjörk, H. E., and Wallin, B. G. (1979). Somatosensory, proprioceptive, and sympathetic activity in human peripheral nerves. *Physiol. Rev.* 59, 919–957.

Vijayalakshmi, G. (1976). Nuclear pattern of muscle spindles in garden lizard. *J. Anat. Soc. India* 25, 47–48.

Vijayalakshmi, G., and Viswanathan, C. P. (1976). The muscle spindle pattern in ilio fibularis and gastrocnemius soleus complex in garden lizards. *J. Anat. Soc. India* 25, 47.

Viswanathan, C. P. (1972a). Motor innervation of intrafusal fibres in muscle spindles of the garden lizard *Calotes versicolor*, preliminary report. *J. Anat. Soc. India* 21, 88.

Viswanathan, C. P. (1972b). Structure and sensory innervation of muscle spindles in garden lizard *Calotes versicolor*, preliminary report. *J. Anat. Soc. India* 21, 90.

Viswanathan, C. P. (1973). Further observations on the structure and innervation of muscle spindles in garden lizard *Calotes versicolor*. *J. Anat. Soc. India* 22, 45.

Viswanathan, C. P. (1976). Pattern of sensory axonal distribution to muscle spindles in garden lizard. *J. Anat. Soc. India* 25, 40.

Viswanathan, C. P., and Vijayalakshmi, G. (1975). Structure and innervation of muscle spindles in garden lizard *Calotes versicolor*. *J. Anat. Soc. India* 24, 85–88.

7

The Cerebellum

HANS J. TEN DONKELAAR

AND

GESINEKE C. BANGMA

CONTENTS

I. INTRODUCTION

The cerebellum is present throughout vertebrates as one of the most variable parts of the central nervous system (Larsell, 1967; Nieuwenhuys, 1967). Nevertheless, phylogenetic studies (e.g., Hillman, 1969; Llinás and Hillman, 1969; Llinás, 1981) have demonstrated a similarity between primitive and more advanced cerebellar cortices, which indicates the existence of a basic pattern of neuronal connectivity. This pattern is characterized by two sets of input, the mossy fiber–granule cell–Purkyně cell system and the climbing fiber–Purkyně cell system, and by one output, the Purkyně cell axons (Llinás, 1981).

This similarity in cerebellar organization is matched by a similarity in function. The first adequate experiments to reveal the function of the cerebellum were carried out by Rolando (1809; see Clarke and O'Malley, 1968), whose studies of a wide variety of animals including reptiles showed that by removing portions of the cerebellum, ipsilateral motor activity was impaired. The general idea that the cerebellum functions to maintain posture and muscle tone and to coordinate movements arose from investigations (cerebellar ablations) by Flourens (1824) and Luciani (1891). Cerebellar ablation experiments in reptiles by Leblanc (1923) and Hacker (1931; see also ten Cate, 1937) seem to confirm these functions for the reptilian cerebellum.

Reptiles are particularly interesting objects for research on the cerebellum, as their great variation in form and locomotion is reflected in the central nervous system (see ten Donkelaar, 1976a, 1976b; Kusuma et al., 1979; ten Donkelaar, 1982). The simplest cerebellum is probably found in the limbless lizard *Anniella nigra* (Larsell, 1926); the most highly developed cerebellum, in crocodilians (e.g., de Lange, 1917; Larsell, 1932; Ziehen, 1934). All reptiles have a corpus cerebelli, which forms the main part of the cerebellum, and a small flocculus. Larsell (1926, 1932, 1967) divided the corpus cerebelli into a median zone, the pars interposita, and a lateral zone on either side, the pars lateralis. Larsell noted that the pars interposita of snakes and limbless lizards is relatively thick, whereas the pars lateralis is reduced. In turtles, in which the extremities are the sole organs of locomotion, the pars lateralis is enlarged and the pars interposita is reduced. On the basis of these observations Larsell suggested that the pars lateralis is concerned with the movements of the limbs; the pars interposita, with movements of the axial musculature.

In the present survey, the development, gross anatomy, histology, and fiber connections of the reptilian cerebellum will be discussed with emphasis on evidence gathered with experimental techniques in the red-eared turtle, *Pseudemys scripta elegans*, the savanna monitor lizard, *Varanus exanthematicus*, and a boid snake, *Python regius*. Exper-

imental studies in reptiles with such highly different modes of locomotion can reveal important data on the ways in which the cerebellum modulates the reptilian motor system, especially which parts of
the cerebellum are concerned with movements of the trunk or of the
extremities. The fiber connections can be studied with the already
classic anterograde degeneration techniques (Nauta and Gygax, 1954;
Fink and Heimer, 1967), as has been done for the spinocerebellar
tracts (Ebbesson, 1967, 1969; Jacobs, 1968; Ebbesson and Goodman,
1981), and with the more recent tracer techniques. Thus, retrograde
tracers like horseradish peroxidase (HRP) have been used to study the
cells of origin of cerebellar afferents (Reiner and Karten, 1978;
Schwarz and Schwarz, 1980; Bangma and ten Donkelaar, 1982;
Künzle, 1982, 1983a, 1983b), and anterograde tracers such as ^{3}H-
leucine and ^{35}S-methionine have been used to investigate the projections of the cerebellar nuclei (Bangma, 1983; Bangma et al., 1984; Künzle, 1985b, 1985c). As regards the cerebellar efferents, the organization of the corticonuclear projections (Bangma, 1983; Bangma et al.,
1983; Bangma and ten Donkelaar, 1984)—the projections from the
Purkyně cells in the cerebellar cortex to the cerebellar nuclei and the
vestibular nuclear complex—will be emphasized. Evidence for a longitudinal pattern of organization of the corticonuclear projections will
be presented, showing similarities to the situation in mammals (e.g.,
Jansen and Brodal, 1940, 1942; Voogd, 1964, 1967, 1969; Voogd and
Bigaré, 1980).

In the final sections, we attempt to discuss cerebellar function in
reptiles as inferred from ablation studies, stimulation experiments,
and the available experimental anatomical data. Furthermore, some
comments on the similarities and differences among the cerebellum
of reptiles and that of other terrestrial vertebrates will be made.

II. DEVELOPMENT

As in all vertebrates, the cerebellum in reptiles arises from paired
structures (*Anlagen*) derived from the dorsal part of the alar plate of
the rhombencephalon (Larsell, 1926, 1932; Krabbe, 1939; Senn, 1979).
These *anlagen* are initially connected only by the membranous roof of
the fourth ventricle, but their rostral parts later fuse in the median
plane. At this time, the cerebellum consists of a plate covering the
fourth ventricle (lamina metencephalica; Senn, 1979). The cerebellum
is much slower to develop than other parts of the CNS. This is seen in
^{3}H-thymidine labeling experiments in *Lacerta sicula* (Schwab and Dur-
and, 1974).

The development of the reptilian cerebellum is illustrated for a turtle, *Chrysemys marginata* (Fig. 7.1), and for *Alligator mississippiensis*
(Fig. 7.2). In *Chrysemys marginata* the cerebellum of the 10-mm,
carapace-length stage is represented chiefly by paired, thickened *an-*

lagen of the corpus cerebelli (Larsell, 1932; Fig. 7.1A) connected across the midline by a thin layer. Larsell (1932) noted a small flocculus at the angle between the lateral part of the cerebellar base and the vestibular region. In newly hatched turtles (Fig. 7.1B) the median sagittal furrow persists dorsally, but it eventually disappears. The cerebellum continues to grow in the caudal direction, even in adult turtles (see Krabbe, 1939; Brand and Mugnaini, 1980).

In early stages of development of *Alligator mississippiensis* (staged according to Reese, 1915), the cerebellum consists of two lateral plates (Fig. 7.2A, stage XIX+). These plates grow in the caudal direction (Fig. 7.2B, stage XXI+), still connected across the midline by a thin

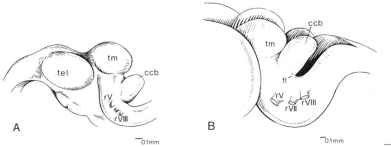

Fig. 7.1. *Chrysemys marginata.* Lateral view of the brain showing the development of the cerebellum. (A) The 10-mm, carapace-length stage. (B) The newly hatched turtle. ccb, Corpus cerebelli; fl, flocculus; tel, telencephalon; tm, tectum mesencephali; rV, rVII, rVIII, roots of the fifth, seventh and eighth cranial nerves. (Redrawn from Larsell, 1932.)

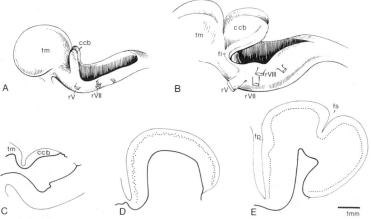

Fig. 7.2. *Alligator mississippiensis.* The development of the cerebellum. (A) Lateral view of stage XIX+; (B) Lateral view of stage XXI+; (C to E) Midsagittal sections of stages XXI+, XXII+, and young adult, respectively. The dots indicate Purkyně cells. ccb, Corpus cerebelli; fl, flocculus; fp, fissura prima; fs, fissura secunda; rV, rVII, rVIII, roots of the fifth, seventh, and eighth cranial nerves; tm, tectum mesencephali. (Redrawn from Larsell, 1932.)

bridge. A clear flocculus can be distinguished, separated from the future corpus cerebelli by a posterolateral fissure. At this stage the crocodilian cerebellum is very similar to that of the newly hatched turtle (Larsell, 1932; see Fig. 7.1B). However, the alligator cerebellum later undergoes extensive transformation, as illustrated in Fig. 7.2. At stage XXII+ (Fig. 7.2D) ascending, horizontal, and descending segments of an irregular dome can be distinguished, the external surface of which has a roughly semicircular outline. Postembryonic growth and differentiation results in the disappearance of the external median fissure by completing the fusion of both halves of the corpus cerebelli. The cerebellum expands rapidly dorsally and dorsocaudally and folds more strongly both anteroposteriorly and from side to side (Larsell, 1932; see Fig 7.2E). Two fissures appear—the fissura prima and the fissura secunda—which divide the crocodilian cerebellum into three lobes: the lobus anterior, medius, and posterior (Fig. 7.2E).

The development of the cerebellar nuclei has been studied in *Chrysemys picta* and *Gongylus ocellatus* (Rüdeberg, 1961) and in *Iguana iguana* (Rüdeberg, 1962). It appeared that the cerebellar nuclei are derived from two migration layers developing from the cerebellar neural epithelium.

III. GROSS ANATOMY

Form and size of the cerebellum differ markedly in the various groups of reptiles. In turtles, the cerebellum forms a caudally directed arch that assumes a helmetlike form (Larsell, 1932; Fig. 7.4B). A bilateral groove is present in its caudal two-thirds (see Fig 7.5A, B, arrows) on the ventricular side of the cerebellum of certain turtles (e.g., *Pseudemys scripta elegans*), roughly separating the median pars interposita and the pars lateralis as distinguished by Larsell (1932). A median sulcus—the sulcus medianus dorsalis (Cruce and Nieuwenhuys, 1974)—separates the two halves of the corpus cerebelli. The softshelled turtle *Trionyx japonica* has a fold of granular layer suspended in the ventricular cavity from the roof (Fuse, 1920); in transverse sections the pars interposita resembles a pendulum. The flocculus of turtles is ill defined, and Larsell (1932) regarded the rostrolateral thinner part of the cerebellum as the flocculus. On the ventricular side a vague groove, the dorsal internal floccular sulcus (see Fig 7.5A), separates this structure from the corpus cerebelli. According to Mugnaini et al. (1974), the flocculus in *Pseudemys scripta elegans* continues caudally as a marginal rim.

In many lizards and in *Sphenodon punctatus* (see Hindenach, 1931), the cerebellar plate is tilted forward or everted. This eversion is especially marked in certain varanids, such as *Varanus salvator* (de Lange,

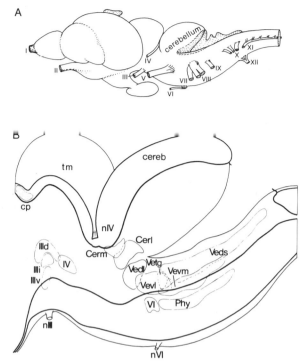

Fig. 7.3. *Pseudemys scripta elegans.* (A) Lateral view of the brain. I–XII, the cranial nerves. (B) Topographical reconstruction of some cell masses in the brain stem as projected upon a sagittal plane. cereb, Cerebellum; Cerl, Cerm, lateral and medial cerebellar nuclei; cp, commissura posterior; nIII, oculomotor nerve; nIV, trochlear nerve; nVI, abducens nerve; Phy, nucleus prepositus hypoglossi; tm, tectum mesencephali; Vedl, Veds, Vetg, Vevl, Vevm, dorsolateral, descending, tangential, ventrolateral, and ventromedial vestibular nuclei; IIId, IIIi, IIIv, dorsal, intermediate, and ventral parts of the oculomotor nucleus; IV, nucleus nervi trochlearis; VI, nucleus nervi abducentis.

1917) and *Varanus exanthematicus* (Fig. 7.6). No distinct sulci appear on the ventricular side of the cerebellum, apart from a dorsal median sulcus (Fig. 7.7A and 7.8A). The flocculus of lizards is small, but distinct, and lies at the transition from the dorsorostral part of the rhombencephalon to the lateral part of the cerebellum (Fig. 7.7B). It is also known as auricle or auricular lobe. The limbless lizard *Anniella nigra* has the smallest and simplest cerebellum found by Larsell among reptiles. In this lizard the cerebellum is almost hidden in a cavity between the midbrain, rostrally, and the eighth nerve roots and nuclei, laterally and caudally.

In snakes, the corpus cerebelli (see Fig. 7.9, *Python regius*) tilts caudally and dorsally from a broad base (see Figs. 7.10, 7.11). The angle of its median axis varies in different species (Larsell, 1926, 1967). In

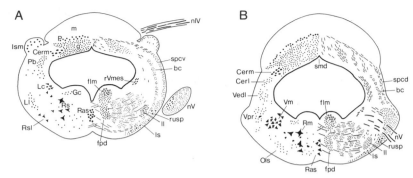

Fig. 7.4 A and B. *Pseudemys scripta elegans.* Transverse sections through the rostral part of the rhombencephalon. At the left the cell picture, based on a Nissl-stained series; at the right the fiber systems based on Häggqvist and Klüver-Barrera preparations. bc, Brachium conjunctivum; Cerl, Cerm, lateral and medial cerebellar nuclei; flm, fasciculus longitudinalis medialis; fpd, fasciculus predorsalis; g, granular layer; Gc, griseum centrale; Ism, nucleus isthmi, magnocellular part; Lc, locus coeruleus; Ll, nucleus of the lateral lemniscus; ll, lateral lemniscus; ls, spinal lemniscus; m, molecular layer; nIV, nervus trochlearis; nV, nervus trigeminus; Ols, oliva superior; P, Purkyně cell layer; Pb, parabrachial region; Ras, nucleus raphes superior; Rm, nucleus reticularis medius; Rs, nucleus reticularis superior; Rsl, nucleus reticularis superior pars lateralis; rusp, rubrospinal tract; rVmes, radix mesencephalicus nervi trigemini; smd, sulcus medianus dorsalis; spcd, dorsal spinocerebellar tract; spcv, ventral spinocerebellar tract; Vedl, dorsolateral vestibular nucleus; Vm, nucleus motorius nervi trigemini; Vpr, nucleus princeps nervi trigemini.

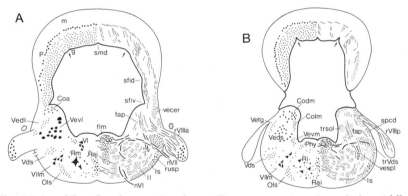

Fig. 7.5 A and B. *Pseudemys scripta elegans.* Transverse sections through the middle part of the rhombencephalon. Codm, Colm, angular, dorsal magnocellular and laminar cochlear nuclei; fap, fibrae arcuatae profundae; flm, fasciculus longitudinalis medialis, g, granular layer; ll, lateral lemniscus; ls, spinal lemniscus; m, molecular layer; nVI, nervus abducens; Ols, oliva superior; P, Purkyně cell layer; Phy, nucleus prepositus hypoglossi; Rai, nucleus raphes inferior; Ri, nucleus reticularis inferior; Rm, nucleus reticularis medius, rusp, rubrospinal tract; rVIIIa, radix anterior nervi octavi; rVIIIp, radix posterior nervi octavi; smd, sulcus medianus dorsalis; trsol, tractus solitarius; trVds, tractus descendens nervi trigemini; vecer, vestibulocerebellar fibers; Vedl, Veds, Vetg, Vevl, Vevm, dorsolateral, descending, tangential, ventrolateral and ventromedial vestibular nuclei; vespl, lateral vestibulospinal tract; Vds, nucleus descendens nervi trigemini; VI, nucleus nervi abducentis; VIIm, nucleus motorius nervi facialis.

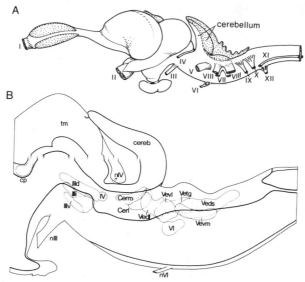

Fig. 7.6. *Varanus exanthematicus.* (A) Lateral view of the brain. I-XII, the cranial nerves. (B) Topographical reconstruction of some cell masses in the brain stem as projected upon a sagittal plane. cereb, Cerebellum; Cerl, Cerm, lateral and medial cerebellar nuclei; cp, commissura posterior; nIII, oculomotor nerve; nIV, trochlear nerve; nVI, abducens nerve; tm, tectum mesencephali; Vedl, Veds, Vetg, Vevl, Vevm, dorsolateral, descending, tangential, ventrolateral, and ventromedial vestibular nuclei; IIId, IIIi, IIIv, dorsal, intermediate, and ventral parts of the oculomotor nucleus; IV, nucleus nervi trochlearis; VI, nucleus nervi abducentis.

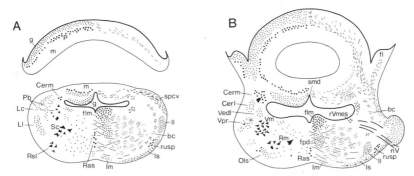

Fig. 7.7 A and B. *Varanus exanthematicus.* Transverse sections through the rostral part of the rhombencephalon. bc, Brachium conjunctivum; Cerl, Cerm, lateral and medial cerebellar nuclei; flm, fasciculus longitudinalis medialis; fpd, fasciculus predorsalis; g, granular layer; Lc, locus coeruleus; Ll, nucleus of the lateral lemniscus; ll, lateral lemniscus; lm, medial lemniscus; ls, spinal lemniscus; m, molecular layer; nV, nervus trigeminus; Ols, oliva superior; P, Purkyně cell layer; Pb, parabrachial region; Ras, nucleus raphes superior; Rm, nucleus reticularis medius; Rsl, nucleus reticularis superior pars lateralis; rusp, rubrospinal tract; rVmes, radix mesencephalicus nervi trigemini; Sc, subcoeruleus area; spcv, ventral spinocerebellar tract; Vedl, dorsolateral vestibular nucleus; Vm, nucleus motorius nervi trigemini; Vpr, nucleus princeps nervi trigemini.

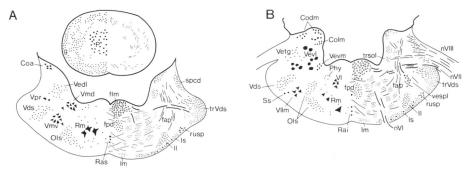

Fig. 7.8 A and B. *Varanus exanthematicus.* Transverse sections through the middle part of the rhombencephalon. Coa, Codm, Colm, angular, dorsal magnocellular and laminar cochlear nuclei; fap, fibrae arcuatae profundae; flm, fasciculus longitudinalis medialis; fpd, fasciculus predorsalis; g, granular layer; ll, lateral lemniscus; lm, medial lemniscus; ls, spinal lemniscus; nVI, nervus abducens; nVII, nervus facialis; nVIII, nervus octavus; Ols, oliva superior; P, Purkyně cell layer; Phy, nucleus prepositus hypoglossi; Rai, nucleus raphes inferior; Ras, nucleus raphes superior; Rm, nucleus reticularis medius; rusp, rubrospinal tract; spcd, dorsal spinocerebellar tract; Ss, nucleus salivatorius superior; trsol, tractus solitarius; trVds, tractus descendens nervi trigemini; Vedl, Vetg, Vevl, Vevm, dorsolateral, tangential, ventrolateral and ventromedial vestibular nuclei; vespl, lateral vestibulospinal tract; Vds, nucleus descendens nervi trigemini; Vmd, Vmv, dorsal and ventral parts, nucleus motorius nervi trigemini; Vpr, nucleus princeps trigemini; VI, nucleus nervi abducentis; VIIm, nucleus motorius nervi facialis.

Eunectes murinus (de Lange, 1917), as in *Python regius* (Fig. 7.9A) and *Python reticulatus* (Fig. 7.9B), the dorsal surface is nearly horizontal; in *Thamnophis sirtalis* it is tilted dorsally. A very small lateral part, which receives a small fascicle of vestibular fibers (Larsell, 1926), represents the flocculus.

The crocodilian cerebellum is the most highly developed cerebellum among reptiles. Two transverse grooves, the sulcus anterior and the sulcus posterior (de Lange, 1917; Ingvar, 1918; see Fig 7.2E), divide the crocodilian cerebellum externally into three lobes: the lobus anterior, medius, and posterior, respectively. Both de Lange (1917) and Ingvar (1918), as well as Larsell (1932, 1967), considered the sulcus anterior to be equivalent to the fissura prima of the mammalian cerebellum. The sulcus posterior is probably comparable to the fissura secunda (Larsell, 1932; Ziehen, 1934). The auricular lobe (flocculus) is well developed in crocodilians (Larsell, 1932; Weston, 1936). It is demarcated from the corpus cerebelli by an external groove. This groove, the homologue of the posterolateral fissure of higher vertebrates, is shallow in the adult alligator but conspicuous in embryonic stages (Larsell, 1932; Fig. 7.2B).

A

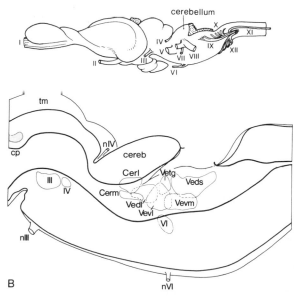

Fig. 7.9. (A) *Python regius*. Lateral view of the brain. I-XII, the cranial nerves. (B) *Python reticulatus*. Topographical reconstruction of some cell masses in the brain stem as projected upon a sagittal plane. cereb, Cerebellum; Cerl, Cerm, lateral and medial cerebellar nuclei; cp, commissura posterior; nIII, nervus oculomotorius; nIV, nervus trochlearis; nVI, nervus abducens; tm, tectum mesencephali; Vedl, Veds, Vetg, Vevl, Vevm, dorsolateral, descending, tangential, ventrolateral and ventromedial vestibular nuclei; III, nucleus nervi oculomotorii; IV, nucleus nervi trochlearis; VI, nucleus nervi abducentis.

IV. HISTOLOGY
A. Topographical Relationships of the Cerebellum

The topographical relationships of the cerebellum are illustrated (Figs. 7.4, 7.5, 7.7, 7.8, 7.10, 7.11) in some selected cross sections of the brain stem based on cresylechtviolet, Klüver and Barrera (1953), and Häggqvist (1936) stained series of the red-eared turtle, *Pseudemys scripta elegans*, the savanna monitor lizard, *Varanus exanthematicus*, and a boid snake, *Python reticulatus*. Each figure schematically shows the cell bodies at the left and the fiber systems at the right. As in previous analyses of the brain stem (ten Donkelaar and Nieuwenhuys, 1979) and spinal cord (Kusuma et al., 1979) for the fiber systems, the general terms small, medium-sized, and coarse are applied to the fibers of the various systems. These terms are not strictly defined, but the small fibers generally range from 0 to 3 μm, the medium-sized ones from 3 to 6 μm, and the coarse fibers from 6 to 12 μm in diameter.

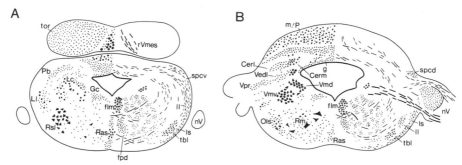

Fig. 7.10 A and B. *Python reticulatus.* Transverse sections through the rostral part of the rhombencephalon. Cerl, Cerm, lateral and medial cerebellar nuclei; flm, fasciculus longitudinalis medialis; fdp, fasciculus predorsalis; g, granular layer; Gc, griseum centrale; Lc, locus coeruleus; Ll, nucleus of the lateral lemniscus; ll, lemniscus lateralis; ls, spinal lemniscus; m/P, molecular-Purkyně cell layer; nV, nervus trigeminus; Ols, oliva superior; Pb, parabrachial region; Ras, nucleus raphes superior; Rm, nucleus reticularis medius; Rsl, nucleus reticularis superior pars lateralis; rVmes, radix mesencephalicus nervi trigemini; spcd, dorsal spinocerebellar tract; spcv, ventral spinocerebellar tract; tbl, lateral tectobulbar tract; tor, torus semicircularis; Vedl, dorsolateral vestibular nucleus; Vmd, Vmv, dorsal and ventral parts, nucleus motorius nervi trigemini; Vpr, nucleus princeps nervi trigemini.

Since the vestibular nuclei are closely related to the cerebellum, both topographically and with regard to their fiber connections, the vestibular nuclear complex will be briefly discussed.

The reptilian *cerebellar cortex*, as first pointed out by Stieda (1875), comprises the three typical layers—molecular, Purkyně cell, and granular—that characterize the cerebellum in vertebrates. The Purkyně cell layer is typically situated between the other two layers (see Figs. 7.4, 7.5, 7.7, 7.12A, B). In snakes, however, Purkyně cells are for the most part found scattered throughout the molecular layer (see Figs. 7.10A, 7.10B, 7.12C). The organization of the Purkyně cell layer—that is, the efferent zone of the cerebellar zone of the cerebellar cortex—will be discussed in the next section.

The cerebellar and vestibular nuclei of reptiles form a more or less continuous complex located in the dorsal part of the rostral rhombencephalon (see Figs. 7.3B, 7.6B, 7.9B). Although Banchi (1903b) and Edinger (1908) described only one cerebellar nucleus, two cerebellar nuclei have been distinguished since the studies of van Hoevell (1916) in *Caiman crocodilus*. The medial cerebellar nucleus (Figs. 7.4A, B, 7.7A, B, 7.11A) is characterized by relatively large, oval, or polygonal cells, with small cells interspersed among the larger ones. The nucleus cerebellaris lateralis (Figs. 7.4B, 7.7B, 7.11A) consists chiefly of small and medium-sized cells. The boundaries between the two cerebellar nuclei are indistinct in most reptiles. In *Sphenodon punctatus*,

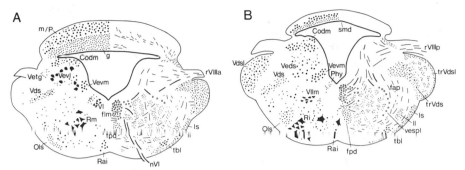

Fig. 7.11 A and B. *Python reticulatus.* Transverse sections through the middle part of the rhombencephalon. Codm, Nucleus cochlearis dorsalis magnocellularis; fap, fibrae arcuatae profundae; flm, fasciculus longitudinalis medialis; fpd, fasciculus predorsalis; g, granular layer; ll, lateral lemniscus; ls, spinal lemniscus; m/P, molecular-Purkyně cell layer; nVI, nervus abducens; Ols, oliva superior; Phy, nucleus prepositus hypoglossi; Rai, nucleus raphes inferior; Ri, nucleus reticularis inferior; Rm, nucleus reticularis medius; rVIIIa, radix anterior nervi octavi; rVIIIp, radix posterior nervi octavi; tbl, lateral tectobulbar tract; trVds, tractus descendens nervi trigemini; trVdsl, tractus descendens lateralis nervi trigemini; Veds, Vetg, Vevl, Vevm, descending, tangential, ventrolateral and ventromedial vestibular nuclei; vespl, lateral vestibulospinal tract; Vdsl, nucleus descendens nervi trigemini; VI, nucleus nervi abducentis; VIIm, nucleus motorius nervi facialis.

Hindenach (1931) reported one continuous cell mass. In Crocodilia, however, both cerebellar nuclei can be clearly separated from each other (van Hoevell, 1916; Weston, 1936; ten Donkelaar and Nieuwenhuys, 1979). As will be shown in Section VI.C the medial cerebellar nucleus has mainly descending projections, and the lateral cerebellar nucleus, predominantly ascending projections via the brachium conjunctivum. A clear brachium conjunctivum could be identified in Klüver-Barrera and Häggqvist material of both *Pseudemys scripta elegans* (Fig. 7.4) and *Varanus exanthematicus* (Fig. 7.7), but not in *Python reticulatus* (see Fig. 7.10A).

Closely related to the cerebellar nuclei is a distinct group of small cells situated above the fourth ventricle and ventrolateral to the medial cerebellar nucleus, surrounding the brachium conjunctivum (Figs. 7.4A, 7.7A, 7.10A). Van Hoevell (1916) described it as the nucleus of the brachium conjunctivum. In more recent studies (see ten Donkelaar and de Boer-van Huizen, 1981b, 1988; ten Donkelaar et al., 1987) this "parabrachial region," which receives the ascending gustatory pathway arising in the nucleus of the solitary tract, was found to project extensively to the forebrain.

As already noted, the vestibular nuclear complex is closely related to the cerebellar nuclei. Particularly the border between the lateral cerebellar nucleus and the dorsolateral vestibular nucleus is ill defined (see Figs. 7.3B, 7.6B, 7.9B). Since Weston (1936), the vestibular nuclear complex of reptiles is divided into at least five nuclei: dorsolateral, ventrolateral, tangential, ventromedial, and descending vestibular. In turtles (see Weston, 1936, *Chrysemys marginata;* Cruce and Nieuwenhuys, 1974, *Testudo hermanni;* Miller and Kasahara, 1979, *Kinosternon leucostomum*), a superior vestibular nucleus also was distinguished. In the present survey, as in previous studies (e.g., Stefanelli, 1944a; Bangma and ten Donkelaar, 1982), this superior nucleus is considered as a rostral extension of the medial part of the dorsolateral vestibular nucleus. The subdivision of the reptilian vestibular nuclear complex into five nuclei as described by Weston (1936; see also Schwab, 1979; ten Donkelaar and Nieuwenhuys, 1979; ten Donkelaar et al., 1987) appears also to be applicable to the reptilian species studied in the present survey.

The nucleus vestibularis dorsolateralis (Figs. 7.4B, 7.5A, 7.7B, 7.8A) has also been called the superior vestibular nucleus by several authors (Beccari, 1911; Larsell, 1926; Papez, 1929; Stefanelli, 1944a). It lies between the ventrolateral vestibular nucleus and the cerebellar nuclei. Its boundaries, particularly with the lateral cerebellar nucleus and principal trigeminal nucleus, are ill defined.

The nucleus vestibularis ventrolateralis (Figs. 7.5A, 7.8B, 7.11B) consists of large cells, among which smaller elements are scattered. The large cell neurons are largest in snakes (Weston, 1936), but relatively less numerous than in other species. Weston suggested that fewer larger neurons, but with more numerous collaterals, are needed to supply the ventral horn neurons related to trunk musculature than are necessary for the functionally more specialized muscles of the extremities possessed by quadrupedal reptiles. The ventrolateral vestibular nucleus is generally considered as the equivalent of the mammalian nucleus of Deiters (see ten Donkelaar, 1976a, 1982).

The nucleus vestibularis tangentialis (Figs. 7.5B, 7.8B, 7.11B) consists of a collection of medium-sized cells, intercalated among the entering fibers of the vestibular root. It lies directly lateral to the ventrolateral vestibular nucleus. Weston (1936), as well as Stefanelli (1944a, 1944b), has suggested that the degree of development of the tangential nucleus is correlated with the relative development of the trunk musculature. This musculature is of great importance, and its activity is well coordinated in the locomotion of snakes and lizards. The absence of trunk musculature in turtles is, according to Weston (1936), correlated with the great reduction of the tangential nucleus in the order Testudines. In keeping with this suggestion is the fact that the tangential vestibular nucleus is only ill defined in *Pseudemys scripta elegans* (Fig. 7.5B).

The nucleus vestibularis ventromedialis (Figs. 7.5B, 7.8B, 7.11A, B) composed of medium-sized cells, is located in the ventrolateral angle of the fourth ventricle. It is found particularly in the caudal one-half of the vestibular nuclear complex.

The nucleus vestibularis descendens (Figs. 7.5B and 7.11B), consisting of small and medium-sized cells, borders rostrally on the ventrolateral vestibular nucleus and extends along the brain stem to the level of the nucleus funiculi dorsalis.

B. Organization of the Purkyně Cell Layer: Longitudinally Oriented Zones of Purkyně Cells

The Purkyně cell layer of *Pseudemys scripta elegans* (see Figs. 7.4 and 7.5) consists rostrally of one to two cell rows; caudally only one cell row appears. No Purkyně cells are present at certain sites (see Fig. 7.5). These places appear to correspond with the position of the sulci on the ventricular side, indicating the boundary between the pars lateralis at each side of the corpus cerebelli.

In lizards such as *Varanus exanthematicus* (Fig. 7.7), the Purkyně cell layer is typically several cells thick. In the boid snakes *Python reticulatus* (see Figs. 7.10, 7.11, 7.12C) and *Python regius* (see Section VI.B), the Purkyně cells are scattered throughout the molecular layer, with a distinct separate Purkyně cell layer only in the lateral, presumably floccular, part (Fig. 7.11) and the caudal part of the cerebellar cortex (Fig. 7.11B). Similar observations were made in *Thamnophis sirtalis* (Weston, 1936) and in several species of Indian and East African snakes (Singh and Chaturvedi, 1979). In the crocodilian cerebellar cortex, the Purkyně cell layer consists of one row of neurons (see Huber and Crosby, 1926; Larsell, 1932; Weston, 1936; ten Donkelaar and Nieuwenhuys, 1979).

The Purkyně cells are the sole output neurons of the cerebellar cortex (see Sections IV.C and VI.A). Myeloarchitectonic studies in mammals (Voogd, 1964, 1967, 1969), based on Häggqvist-stained sections of the cerebellum, revealed that typically six to seven compartments of Purkyně cell axons can be distinguished in the lobular white matter. Each compartment has been associated with a specific target site for its Purkyně cell axons. Although a similar situation was found in birds (*Gallus domesticus*, Feirabend et al., 1976), Häggqvist-stained sections of the reptilian cerebellum did not reveal a subdivision of the white matter into a series of compartments by repeating local differences in axon diameter, as in mammals and birds. However, to find out whether a zonal pattern of organization exists in the cerebellar cortex, a different approach can be taken—a topological analysis of the Purkyně cell layer (for the procedure used see Bangma et al., 1983). Topological analysis of the cerebellum of the turtle *Testudo her-*

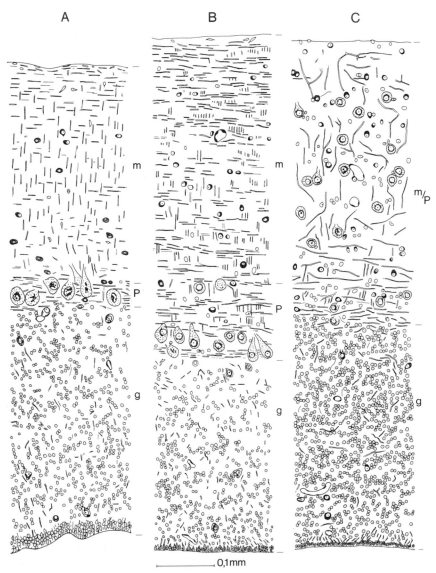

Fig. 7.12. Diagrammatic representations of transverse sections of the corpus cerebelli. (A) *Pseudemys scripta elegans*. (B) *Varanus exanthematicus*. (C) *Python reticulatus*. Bottom: ventricular side. g, Granular layer; m, molecular layer; m/P, molecular-Purkyně cell layer; P, Purkyně cell layer.

manni (Gerrits and Voogd, 1973) showed that each cerebellar half can be divided into a medial and a lateral half. A topological analysis of the Purkyně cell layer of *Pseudemys scripta elegans* (Bangma et al., 1983; Fig. 7.13) revealed a distinct pattern in the Purkyně cell layer. The two halves of the cerebellum are clearly separated by an almost Purkyně cell–free strip along the whole length of the cerebellum. In each half

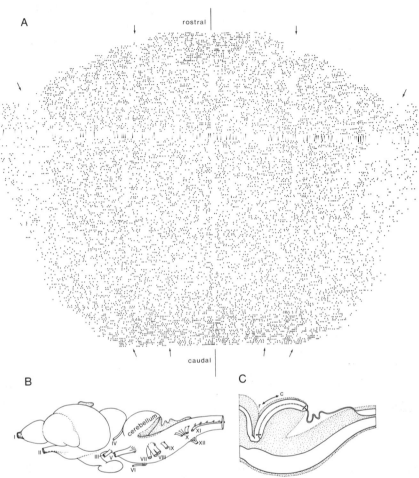

Fig. 7.13. *Pseudemys scripta elegans*. (A) Topographical chart showing the distribution of the Purkyně cells in both halves of the cerebellum. The vertical lines indicate the midline, the arrows the borderlines between the medial, intermediate, and lateral Purkyně cell zone (20 ×). (B) Lateral view of the brain (1.8 ×). (C) Sagittal plane of part of the brain stem showing the analyzed part (dotted line) of the Purkyně cell layer. c, Caudal; r, rostral.

of the cerebellum, three longitudinally oriented zones of Purkyně cells—a *medial*, an *intermediate*, and a *lateral* zone—can be distinguished. The medial zone is separated from the intermediate zone by an almost Purkyně cell–free strip, most clearly visible in the rostral and middle parts of the cerebellum. In the rostral part of the cerebellum, the lateral zone is separated from the intermediate zone by a distinct cell-free area. In the middle part, the lateral zone is characterized by a less dense pattern of Purkyně cells compared to the other zones. This difference diminishes in the most caudal part of the cerebellum. These zones correspond in part to the three different compo-

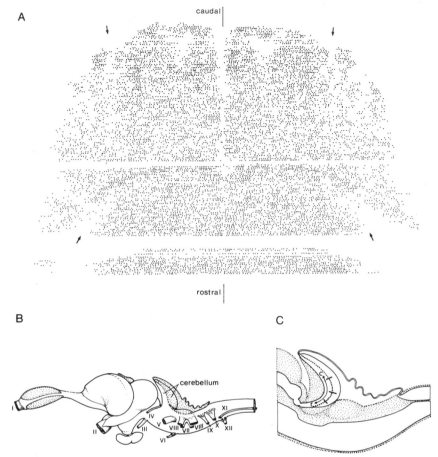

Fig. 7.14. *Varanus exanthematicus.* (A) Topographical chart showing the distribution of the Purkyně cells in both halves of the cerebellum. The vertical lines indicate the midline; the arrows, the borderline between the corpus cerebelli and the flocculus (20 x). (B) Lateral view of the brain (1.6 x). (C) Sagittal plane of part of the brain stem showing the analyzed part (dotted line); the distribution of Purkyně cells has been studied in three parts. c, Caudal; r, rostral.

nents of the cerebellum as distinguished by Larsell (1926, 1932, 1967). The medial zones of the two cerebellar halves together correspond to his pars interposita. Larsell's pars lateralis corresponds to the intermediate zone described in the present survey. The flocculus as distinguished by Larsell corresponds to the rostral part of our lateral zone. The remaining part of the lateral zone of Purkyně cells corresponds to the caudal elongation of the flocculus, the so-called marginal rim (Mugnaini et al., 1974; Brand and Mugnaini, 1980).

In *Varanus exanthematicus* (Bangma and ten Donkelaar, 1984; Fig. 7.14), a less clear subdivision is present than in *Pseudemys scripta ele-*

gans, although at places a similar longitudinal subdivision appeared to exist. A clear lateral, floccular zone can be distinguished.

In *Python reticulatus*, a lateral zone could be distinguished that is particularly distinct in the caudal part of the cerebellum (see also Fig. 7.11). Purkyně cells are less densely packed in certain other parts of the cerebellum, such as the corpus cerebelli (see Fig. 7.10B), indicating a possible subdivision of the corpus cerebelli into two longitudinally oriented parts. So it seems likely that a longitudinal organization of Purkyně cells also exists in pythons.

In short, in the present survey a subdivision of the reptilian cerebellar cortex into various longitudinal zones is advocated, most obviously in *Pseudemys scripta elegans*.

C. Cerebellar Circuitry

As in other vertebrates, the cerebellar cortex of reptiles has two main afferent systems (the mossy and climbing fibers) and one efferent system (the axons of the Purkyně cells). The reptilian cerebellar cortex has been studied with both light microscopic (Golgi) techniques (P. Ramón, 1896; Ramón y Cajal, 1911; Larsell, 1926, 1932; Ochoterena, 1932; Hillman, 1969; Llinás and Hillman, 1969) and electronmicroscopy (EM) (Hillman, 1969; Llinás and Hillman, 1969; Mugnaini et al., 1974). P. Ramón (1896) and Ramón y Cajal (1911) were the first to demonstrate the Purkyně cell dendritic tree, granule cells, and parallel fibers in *Chameleo vulgaris* and *Lacerta stirpium*, respectively.

Climbing and mossy fibers, granule cells, and Purkyně cells form the components of the basic cerebellar circuitry (see Fig. 7.15), which has been changed very little throughout phylogeny (Llinás and Hillman, 1969; Llinás and Nicholson, 1969; Llinás, 1981). Climbing fibers, so named because of the characteristic way each fiber climbs along the dendritic tree of a Purkyně cell, establish direct contacts with Purkyně cell dendrites. Golgi and EM data in the caiman *Caiman crocodilus* (Hillman, 1969; Llinás and Hillman, 1969) showed that (1) climbing fibers consist of an arborizing axon that follows the main Purkyně cell branches as a single fiber: (2) the climbing fiber contacts spines that emerge from the main dendrites of the Purkyně cell; (3) about 150 to 180 synaptic contacts are made between the climbing fiber and the Purkyně cell.

Llinás and Nicholson (1969, 1971) demonstrated the presence of climbing fiber activation of caiman Purkyně cells. They concluded (Llinás and Nicholson, 1969; see also Bantli, 1974) from studying frogs, turtles, caimans, pigeons, and cats that among these forms, climbing fibers show little difference in their mode of action and in their ability to generate bursts of action potentials. As the inferior olive of mammals has been shown (see Llinás, 1981) to be the place of

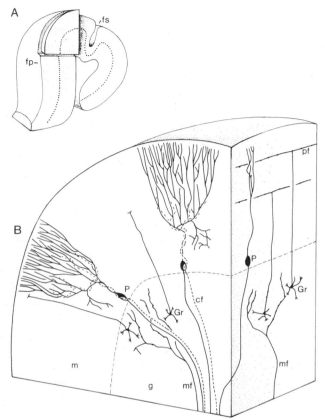

Fig. 7.15. *Caiman crocodilus.* Diagram of the cerebellar cortex. (A) Sagittal section showing the various lobes, separated by fissures. fp, fissura prima; fs, fissura se-cunda. (B) The basic cerebellar circuitry (see text). cf, Climbing fiber; g, granular layer; Gr, granule cell; m, molecular layer; mf, mossy fiber; P, Purkyně cell; pf, parallel fiber. (Based on Hillman, 1969.)

origin for the cerebellar climbing fiber system, the presence of climb-ing fibers in the reptilian cerebellar cortex is indicative for cerebellar afferents arising in an inferior olive (see Section V.C).

Mossy fibers (their terminals appear mosslike) have an indirect in-put to the Purkyně cells, being relayed through the granule cells. Granule cells (Figs. 7.12 and 7.15) are very small cells (6 to 9 μm in diameter in *Pseudemys scripta elegans;* Mugnaini et al., 1974) and fill the granular layer. They give rise to a thin axon that ascends through the granular into the molecular layer. Here it divides (see Fig. 7.15), in a T-shaped fashion, to give rise to two branches (parallel fibers) that run horizontally through the molecular layer (Hillman, 1969). These parallel fibers run perpendicularly through the Purkyně cell dendritic trees. As in other vertebrates, parallel fibers in reptiles have been demonstrated to be excitatory on the dendrites of Purkyně cells (Kitai et al., 1969; Kennedy et al., 1970; Llinás and Nicholson, 1969, 1971). The mossy terminals, as viewed under the electronmicroscope, appear as enlarged swellings, termed rosettes. Tightly grouped around a rosette are numerous terminals, the whole cluster being surrounded by a single layer of glial membranes. The glial membranes demarcate a specific group of terminals (i.e., a synaptic complex), which is termed a glomerulus. Simply organized glomeruli are found in reptiles (see Hillman, 1969; Mugnaini et al., 1974). Glomerular size ranges in *Pseudemys* from 10 to 60 μm and from 7 to 20 μm along the major and minor axes, respectively (Mugnaini et al., 1974). Mossy fiber afferents originate in many extracerebellar sites. Their origin will be discussed in Section V.

The basic cerebellar circuitry (Fig. 7.15), composed of the Purkyně cells and their two afferent systems, is modulated by the action of inhibitory local circuit neurons that operate in both the molecular and granular layer (Hillman, 1969; Kitai et al., 1969; Llinás and Hillman, 1969; Llinás and Nicholson, 1969; Llinás, 1981). Various inhibitory interneurons are present in the cerebellar cortex: (1) stellate cells (10 to 15 μm) which are present in the molecular layer (Fig. 7.12) and synapse on the Purkyně cells (Llinás and Nicholson, 1969; Shimono et al., 1970). P. Rámon (1896, *Chamaeleo vulgaris*) and Ochoterena (1932, *Phrynosoma*) illustrated basketlike formations on Purkyně cell somas and (2) Golgi cells (18 to 25 μm), which can be observed in the granular layer (e.g., Hillman, 1969, and Fig. 7.12).

A second inhibitory system with widespread properties, but lacking the anatomical specificity of the local circuit neurons, is the catecholaminergic system (Llinás, 1981). A catecholaminergic inhibitory system, arising in the locus coeruleus, has been shown in mammals (see Hökfelt and Fuxe, 1969; Hoffer et al., 1971; Olson and Fuxe, 1971) and birds (Mugnaini and Dahl, 1975). In addition, serotonergic afferent fibers have been demonstrated (e.g., Chan-Palay, 1977). In reptiles a monoaminergic innervation of the cerebellum has also been shown (Yamamoto et al., 1977; Parent, 1979; Wolters et al., 1984, 1985) and will be further discussed in Section V.

Purkyně cells are the only source of efferents from the cerebellar cortex. The general morphology of the Purkyně cells has been studied in various reptiles (e.g., *Lacerta stirpium*, Rámon y Cajal, 1911; *Terrapene carolina*, Larsell, 1932; *Caiman crocodilus*, Hillman, 1969; Llinás and Hillman, 1969). The reptilian Purkyně cell, particularly in caimans, is characterized by the highly organized nature of its dendritic tree, which is flattened in one plane. The dendritic tree (Fig. 7.15) stems from a single apical branch situated in the uppermost part of the Purkyně cell soma. The target sites for Purkyně cell output—the cerebellar nuclei and the vestibular nuclear complex—will be discussed in Section VI.

In short, the cerebellar circuitry in reptiles involves two afferent systems: the mossy fiber–granule cell–parallel fiber system and the climbing fiber system, both excitatory to the Purkyně cell. This basic cerebellar circuit is modulated by the action of inhibitory local circuit neurons (stellate and Golgi cells) and probably also by a catecholaminergic inhibitory system.

V. AFFERENT CONNECTIONS
A. General

Classic studies of the reptilian CNS, based mainly on the study of Weigert and pyridine-silver material of various reptiles, described a number of fiber systems afferent to the cerebellum. The major cerebellar afferents suggested are:

1. The spinocerebellar tracts (Edinger, 1908; de Lange, 1917; Ariëns Kappers, 1921; Huber and Crosby, 1926; Larsell, 1926, 1932; Shanklin, 1930; Hindenach, 1931; Weston, 1936; Juh Shen Shyu, 1939; Kawakami, 1954). Most authors recognized dorsal and ventral tracts, both ascending via the lateral funiculus. In Golgi studies, Banchi (1903a, 1903b) noted a nucleus of Clarke in the dorsal horn of a turtle, *Emys orbicularis*, with axons entering the lateral funiculus.

2. The olivocerebellar tract (van Hoevell, 1916; Leblanc, 1923; Larsell, 1926, 1932; Shanklin, 1930), which accompanies the spinocerebellar system. Some authors, however, could not find evidence for the presence of an inferior olive in reptiles (see Kooy, 1917).

3. Primary and secondary vestibulocerebellar fibers (Edinger, 1908; Beccari, 1911; Ingvar, 1918; Ariëns Kappers, 1921; Huber and Crosby, 1926; Larsell, 1926, 1932; Weston, 1936; Kawakami, 1954).

4. Trigeminocerebellar projections (Ariëns Kappers, 1921; Huber and Crosby, 1926, 1933; Larsell, 1926, 1932; Shanklin, 1930; Hindenach, 1931; Weston, 1936).

5. Tectocerebellar projections (Ariëns Kappers, 1921; Huber and Crosby, 1926, 1933; Larsell, 1926, 1932; Shanklin, 1930; Hindenach, 1931; Weston, 1936).

Anterograde degeneration techniques (Nauta and Gygax, 1954; Fink and Heimer, 1967) and the more recently introduced tracer techniques making use of axonal transport phenomena (e.g., the retrograde transport of HRP) have allowed more precise study of the cerebellar afferents. For example, the presence of a distinct olivocerebellar projection has been shown (Bangma and ten Donkelaar, 1982; Künzle and Wiklund, 1982; Künzle, 1983a, 1985a).

B. Spinal Afferents

The classic studies on spinocerebellar projections in reptiles mentioned above distinguished dorsal and ventral spinocerebellar tracts ascending via the lateral funiculus. These spinocerebellar tracts probably relay impulses from cutaneous receptors of the trunk, tail, and extremities and signals from neuromuscular spindles (Larsell, 1967; von Düring and Miller, 1979; see also Chapter 6, this volume) to the cerebellum. The conduction velocities of spinocerebellar fibers vary from 5.1 to 33.3 m/sec (Walsh et al., 1974, *Pseudemys scripta elegans*).

The dorsal and ventral spinocerebellar tracts cannot be delimited as distinct bundles in the spinal cord (see Kusuma et al., 1979) and at caudal sections of the brain stem (see ten Donkelaar and Nieuwenhuys, 1979), where they are still included in the so-called spinal lemniscus. At more rostral rhombencephalic levels, the spinocerebellar tracts apparently arise from the spinal lemniscus (Figs. 7.5 and 7.8A). The ventral tract extends more rostrally (Figs. 7.4A, 7.7A, 7.10A, 7.16, 7.17) than the dorsal one (Figs. 7.4B, 7.8A, 7.10B, 7.16, 7.17) and partly decussates in the anterior medullary velum (see Figs. 7.16C, D, 7.17). Degeneration studies by Jacobs (1968) in *Lacerta viridis* confirmed the subdivision into dorsal and ventral spinocerebellar tracts, both arising from the lateral funiculus. The dorsal spinocerebellar tract is formed by those spinocerebellar fibers situated dorsally in the periphery of the lateral funiculus of the spinal cord (Jacobs, 1968). It sweeps around the rostral edge of the root of the trigeminal nerve. According to Jacobs (1968), the ventral spinocerebellar is formed by those nerve fibers having an intermediate position in the periphery of the lateral funiculus of the spinal cord, ventral to the dorsal spinocerebellar tract. The ventral spinocerebellar tract makes a broad rostrodorsal sweep around both the dorsal spinocerebellar tract and the trigeminal nerve root and attains a position more rostrally in the cerebellar peduncle. Ebbesson (1967, 1969; see also Pedersen, 1973; Ebbesson and Goodman, 1981), however, holds that the dorsal spinocerebellar tract consists of only a few fibers, being a component of the dorsal funicular system, whereas the ventral spinocerebellar tract is considered to consist of all the spinocerebellar fibers passing by way of the lateral funiculus.

The discrepancy between the results of Ebbesson and Jacobs is probably resolved by the degeneration experiment shown in Fig. 7.16, which involved the contralateral dorsal funiculus in the monitor lizard, *Varanus exanthematicus*. Degenerating fibers could be traced in both dorsal funiculi and in the ipsilateral lateral funiculus. The dorsal funicular fibers terminate in the dorsal funicular nuclei (Fig. 7.16J and K) and the descending vestibular nucleus (Fig. 7.16I). At the level of

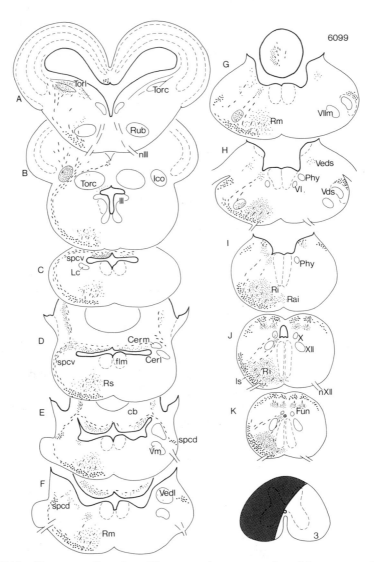

Fig. 7.16. *Varanus exanthematicus.* Diagrammatic representation of the course and distribution of degenerating fibers within the brain stem following a hemisection of the spinal cord (segment 3). Broken (curved) lines indicate degenerating fibers, dots, evidence of preterminal degeneration. cb, Cerebellum; Cerl, Cerm, lateral and medial cerebellar nuclei; fl, flocculus; flm, fasciculus longitudinalis medialis; Fun, nucleus funiculi dorsalis; Ico, nucleus intercollicularis; Lc, locus coeruleus; ls, spinal lemniscus; n IV, nervus trochlearis; n V, nervus trigeminus; n VI, nervus abducens; n XII, nervus hypoglossus; Phg, nucleus prepositus hypoglossi; Ri, nucleus reticularis inferior; Rm, nucleus reticularis medius; Rs, nucleus reticularis superior; Rub, nucleus ruber; spcd, dorsal spinocerebellar tract; spcv, ventral spinocerebellar tract; Torc, Torl, central and laminar nucleus of the torus semicircularis; Vedl, Veds, dorsolateral and descending vestibular nuclei; III, nucleus nervi oculomotorii; Vm, nucleus motorius nervi trigemini; VI, nucleus nervi abducentis; VII m, nucleus motorius nervi facialis; X, nucleus motorius nervi vagi, XII, nucleus nervi hypoglossi.

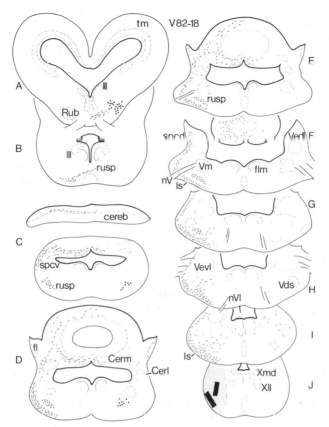

Fig. 7.17. *Varanus exanthematicus.* The distribution of anterogradely labeled fibers and terminal structures, after implantation of HRP slow-release gels in the spinal lemniscus. Broken lines indicate anterogradely labeled fibers (and the retrogradely labeled rubrospinal tract); small dots, labeled terminal structures; large dots, retrogradely labeled cells. cereb, Cerebellum; Cerl, Cerm, lateral and medial cerebellar nuclei; fl, flocculus; flm, fasciculus longitudinalis medialis; ls, lemniscus spinalis, n V, nervus trigeminus; n VI, nervus abducens; Rub, nucleus ruber; rusp, tractus rubrospinalis; spcd, tractus spinocerebellaris dorsalis; spcv, tractus spinocerebellaris ventralis; tm, tectum mesencephali; Vedl, Vevl, dorsolateral and ventrolateral vestibular nuclei; III, nucleus nervi oculomotorii; Vds, nucleus descendens nervi trigemini; Vm, nucleus motorius nervi trigemini; Xmd, nucleus motorius dorsalis nervi vagi; XII, nucleus nervi hypoglossi.

the hypoglossal nuclei, fibers from the dorsal funicular system could be traced to a more ventral position, eventually just dorsal to the descending trigeminal nucleus (e.g., Fig. 7.16G and H). This fiber system, which based on its position in the brain stem should be termed the *dorsal spinocerebellar tract* (Ebbesson, 1967), is joined by fibers that arise from the spinal lemniscus and sweep around the rostral edge of the root of the trigeminal nerve (Fig. 7.16E and F). Both components enter the caudal part of the cerebellar peduncle (Fig. 7.16E). The main part of the spinocerebellar fibers enters the cerebellar peduncle more rostrally (Fig. 7.16C and D). This part will be termed the *ventral spinocerebellar tract.*

Thus it seems likely that at least two spinocerebellar fiber systems are found in reptiles, namely, a dorsal tract composed of dorsal funicular fibers, but also of lateral funicular fibers (probably from the dorsal part of the lateral funiculus), and a ventral spinocerebellar tract. The dorsal funicular projection to the cerebellum in reptiles seems not so well developed as in amphibians. After dorsal root transections in amphibians, an extensive primary spinal afferent projection to the cerebellum was shown with anterograde degeneration techniques (Joseph and Whitlock, 1968a) and with cobalt labeling techniques (Antal et al., 1980; Székely et al., 1980). Dorsal root transections in various reptiles (see Joseph and Whitlock, 1968b; Jacobs and Sis, 1980; Kusuma and ten Donkelaar, 1980) did not show primary spinal afferents extending beyond the nucleus vestibularis descendens, suggesting that the dorsal funicular component to the cerebellum as mentioned above is composed of nonprimary afferents. However, the use of modern tracer techniques has resolved the apparent discrepancy between the various degeneration studies outlined above. Künzle (1982, *Pseudemys scripta elegans*) showed, after injection of ^{35}S-methionine into cervical and lumbar dorsal root ganglia, a primary afferent projection onto the cerebellum. Few, but consistent, patches of silver grains were found within the granular layer, suggesting mossy fiber terminations. This projection was mainly ipsilateral and most intense in rostral cerebellar regions. After HRP injections into the cerebellum of various reptiles (Bangma and ten Donkelaar, 1982), retrogradely labeled fibers could be traced into the dorsal funiculus. Preliminary results with the more sensitive tracer wheat germ agglutinin WGA-HRP (Bangma and ten Donkelaar, unpublished observations) showed that these spinocerebellar fibers passing via the dorsal funiculus originate in dorsal root ganglion cells (Fig. 7.18A). After HRP application to peripheral (hindlimb) nerves in *Pseudemys scripta elegans* a cerebellar projection was also observed (Ruigrok, 1984). Spinocerebellar fibers terminate largely ipsilateral to the side of the lesion, predominantly in the rostral parts of the cerebellar cortex, with a moderate amount of terminal degeneration in the cerebellar nuclei (Ebbesson, 1967, 1969; Jacobs, 1968; Ebbesson and Goodman, 1981; see also Figs. 7.16 and 7.17). In *Pseudemys scripta elegans*, Künzle (1983b) showed that the terminal field of spinocerebellar fibers was not homogeneous but consisted of zones of mossy fiber terminations of varying sites and intensities.

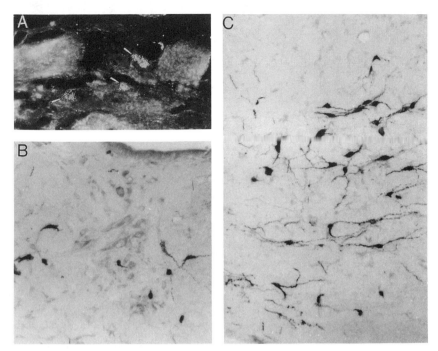

Fig. 7.18. *Pseudemys scripta elegans.* (A) After injections of WGA-HRP into the cerebellum; labeled dorsal root ganglion cells (113 ×). (B) The nucleus prepositus hypoglossi after HRP injection (123 ×). (C) The inferior olive after HRP injection (127 ×).

The cells of origin of the reptilian spinocerebellar tracts are, besides, in the dorsal root ganglia found throughout the spinal gray matter. HRP studies in reptiles (see ten Donkelaar and de Boer-van Huizen, 1978b, *Lacerta galloti;* Bangma and ten Donkelaar, 1982, various reptiles) have probably shown only part of the cells of origin of spinocerebellar fibers. HRP injections into the cerebellum of *Pseudemys scripta elegans* (Fig. 7.19) showed that cells of origin of spinocerebellar fibers are present throughout the spinal cord but occur mainly in the brachial and lumbar enlargements, predominantly in ventral areas of the spinal gray. In the cervical intumescence, the greatest number of retrogradely labeled neurons was found in area VII-VIII (subdivision of the spinal gray after Kusuma et al., 1979), more or less evenly distributed between the ipsilateral and contralateral sides of the spinal cord. The area of motoneurons (IX) also contained a few labeled cells, in keeping with observations in *Lacerta galloti* (ten Donkelaar and de Boer-van Huizen, 1978b). In thoracic segments (e.g., segments 11 and 12, Fig. 7.19), only a few labeled cells were found, most probably due to the lack of trunk musculature. In the lumbar

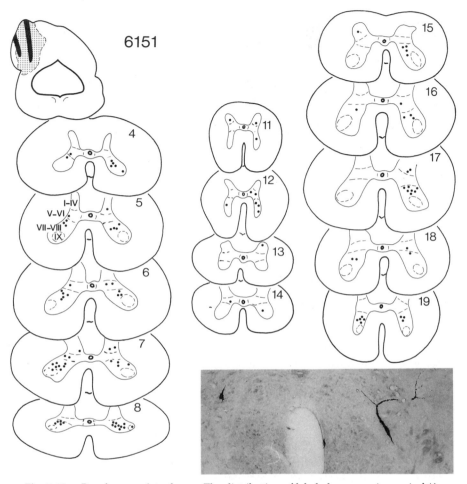

Fig. 7.19. *Pseudemys scripta elegans.* The distribution of labeled neurons in cervical (4 to 8), thoracic (11 to 14), and lumbar (15 to 19) segments of the spinal cord after HRP injections into the cerebellum. Each level represents the composite of the plots of ten consecutive sections. (After Bangma and ten Donkelaar, 1982.) Insert: Retrogradely labeled neurons in the 16th segment of the spinal cord, TMB reaction, 66 ×. I–X, subdivision of the spinal gray into ten areas. (After Kasuma et al., 1979.)

intumescence of *Pseudemys scripta elegans*, spinocerebellar tract neurons were found mostly in area VII-VIII, especially contralateral to the side of the HRP injections (see also Künzle, 1983b).

In *Varanus exanthematicus* (Figs. 7.17 and 7.20), the pattern of labeled cells in the spinal cord after placing an HRP slow-release gel (modified after Griffin et al., 1979) into the spinal lemniscus in the caudal brain stem is roughly comparable to that found after HRP injections into the cerebellum (e.g., Fig. 7.19, *Pseudemys scripta elegans*). In this experiment, the labeled cells in the spinal cord could result partly

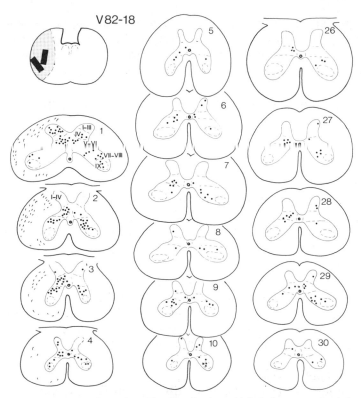

Fig. 7.20. *Varanus exanthematicus.* The distribution of labeled neurons (mainly spino-cerebellar tract cells) in the spinal cord after implantation of HRP slow-release gels in the spinal lemniscus. Each level represents the composite of the plots of ten consecutive sections. I-X, subdivision of the spinal gray into ten areas. (After Kusuma et al., 1979.)

from damage to spinoreticular and spinothalamic fibers. However, as shown in Fig. 7.17, the spinocerebellar tracts (both arising from the spinal lemniscus) are extensively labeled anterogradely. Therefore, it seems likely that a large part of the labeled cells in the spinal cord are cells of origin of spinocerebellar fibers. Such spinocerebellar tracts were found, as in *Pseudemys*, mainly in area VII-VIII of the spinal gray. In addition, however, distinct groups of labeled cells were present ipsilaterally in area V-VI (segments 1 to 4, 26 to 29). In mammals (see Matsushita and Hosoya, 1979; Matsushita et al., 1979; Grant et al., 1982), spinocerebellar tract neurons with uncrossed axons were found at comparable places. In the experiment in *Varanus exanthematicus* shown in Fig. 7.17, spinocerebellar fibers could be traced into the cerebellar peduncle, the dorsal tract entering the caudal part of the cerebellar peduncle (Fig. 7.17E), the ventral tract passing through its rostral part at the level of the isthmus (Fig. 7.17C). In the cerebellar cortex, the spinocerebellar fibers could be traced, mainly ipsilaterally, over about two-thirds of the total length of the cerebellum. Comparable data were found in a recent study in *Pseudemys scripta elegans* (Künzle and Woodson, 1982), in which HRP as used as an anterograde tracer.

In short, although the aforementioned data on the cells of origin, course, and site of termination of spinocerebellar tracts in reptiles are still somewhat fragmentary, it can be concluded that extensive spinocerebellar projections exist in reptiles.

C. Rhombencephalic Afferents

1. GENERAL By far the greatest number of cerebellar afferent sources is found in the rhombencephalon (Bangma and ten Donkelaar, 1982; Künzle, 1983a). Rhombencephalic projections to the cerebellum have been demonstrated from the vestibular nuclear complex, an inferior olive, and from various other structures, including monoamine cell masses. Two HRP experiments, one in *Pseudemys scripta elegans* (Figs. 7.21 and 7.22), the other in *Varanus exanthematicus* (Fig. 7.23), show the retrogradely labeled neurons found in the brain stem and prosencephalon after HRP injections into cerebellum. Similar results were obtained in *Python regius* (see Bangma and ten Donkelaar, 1982).

2. VESTIBULAR AFFERENTS
Both primary and secondary vestibulocerebellar projections were demonstrated in reptiles. Primary vestibulocerebellar fibers—the direct projection from vestibular ganglion cells to the cerebellum—were shown in several degeneration studies (Leake, 1974, *Caiman crocodilus;* Foster and Hall, 1978, *Iguana iguana*) and HRP studies (Bangma and ten Donkelaar, 1982; Barbas-Henry and Lohman, 1988). Eighth nerve fibers travel dorsally and rostrally from their entrance into the brain stem to penetrate the cerebellum via the cerebellar peduncle. Along their course they pass through and ramify within the cerebellar nuclei (Foster and Hall, 1978). After HRP injections into the cerebellum of *Pseudemys scripta elegans*, labeled ganglion cells were found in the ipsilateral vestibular ganglion (Bangma and ten Donkelaar, 1982).

Secondary vestibulocerebellar afferents were demonstrated after lesions including the vestibular nuclear complex (ten Donkelaar, 1976b; Foster and Hall, 1978). HRP studies in various turtles, including *Pseudemys scripta elegans* (Schwarz and Schwarz, 1980; Bangma and ten Donkelaar, 1982; Künzle, 1983a; see Fig. 7.22), *Varanus exanthematicus* (Fig. 7.23), and *Python regius* (Bangma and ten Donkelaar, 1982), showed extensive secondary vestibular projections originating bilaterally in the dorsolateral, ventromedial, and descending vestibular nuclei (Figs. 7.22 and 7.23).

6151

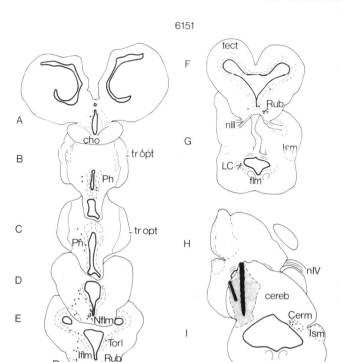

Fig. 7.21. *Pseudemys scripta elegans.* The distribution of labeled neurons in the diencephalon and rostral part of the rhombencephalon after HRP injections into the cerebellum. Bor, Nucleus of the basal optic root; cereb, cerebellum; Cerm, nucleus cerebellaris medialis; cho, chiasma opticum; flm, fasciculus longitudinalis medialis; Iflm, interstitial nucleus of the flm; Ism, nucleus isthmi, pars magnocellularis; LC, locus coeruleus; Nflm, nucleus of the flm; nIII, nervus oculomotorius; nIV, nervus trochlearis; Ph, nucleus periventricularis hypothalami; Rub, nucleus ruber; tect, tectum mesencephali; Torl, laminar nucleus of the torus semicircularis; tr opt, tractus opticus. (From Bangma and ten Donkelaar, 1982.)

3. PRECEREBELLAR NUCLEI

The term *precerebellar nuclei* refers to nuclei that send most of their efferent fibers to the cerebellum (Brodal, 1981). In mammals, these include, among others, the pontine nuclei (probably absent in reptiles), a nucleus funiculi lateralis, and the inferior olive. Although in older studies of the reptilian brain stem the presence of an inferior olive and an olivocerebellar tract was claimed, definite proof for such a projection was obtained only quite recently in tracer studies (Bangma and ten Donkelaar, 1982; Künzle and Wiklund, 1982; Künzle, 1983a, 1985a). After almost all HRP injections into the cerebellum carried out by Bangma and ten Donkelaar (1982), a conspicuously labeled cell mass, contralateral to the injection site, was present in the mediobasal part of the brain stem, adjacent to the nucleus raphes in-

6151

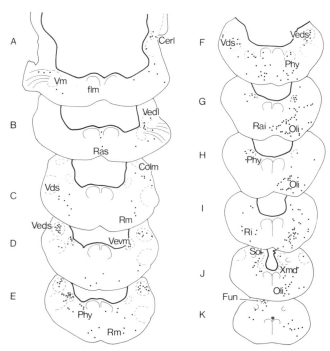

Fig. 7.22. *Pseudemys scripta elegans.* The distribution of labeled neurons in the caudal part of the rhombencephalon after HRP injections into the cerebellum. Cerl, nucleus cerebellaris lateralis; Colm, nucleus cochlearis laminaris; flm, fasciculus longitudinalis medialis; Fun, nucleus funiculi dorsalis; Oli, oliva inferior; Phy, nucleus prepositus hypoglossi; Rai, nucleus raphes inferior; Ras, nucleus raphes superior; Ri, nucleus reticularis inferior; Rm, nucleus reticularis medius; Sol, nucleus tractus solitarii, Vedl, Veds, Vevm, dorsolateral, descending and ventromedial vestibular nuclei; Vds, nucleus descendens nervi trigemini; Vm, nucleus motorius nervi trigemini; Xmd, nucleus motorius dorsalis nervi vagi. (From Bangma and ten Donkelaar, 1982.)

ferior (Figs. 7.18, 7.22G to J, 7.23J to L). Some labeled fibers could also be traced coursing in the ipsilateral spinal lemniscus and crossing the midline to their apparent site of origin. This location and the constant contralateral labeling after both small (mainly molecular layer) and large cerebellar HRP injections point to the conclusion that this mass most probably represents the reptilian homologue of the inferior olive of other vertebrates. This cell mass was also labeled following injection of radioactive D-aspartate into the cerebellar cortex of *Pseudemys scripta elegans* (Künzle and Wiklund, 1982). In the rat, similar D-aspartate injections have been shown to selectively label the climbing fibers and their cells of origin within the inferior olive, whereas mossy fibers or monoaminergic cerebellar afferents remained unlabeled (Wiklund et al., 1982).

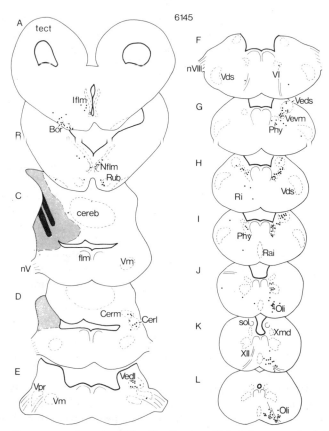

Fig. 7.23. *Varanus exanthematicus.* The distribution of labeled neurons in the brain stem after HRP injections into the cerebellum. Bor, nucleus of the basal optic root; cereb, cerebellum; Cerl, Cerm, lateral and medial cerebellar nuclei; flm, fasciculus longitudinalis medialis; Iflm, interstitial nucleus of the flm; Nflm, nucleus of the flm; nV, nervus trigeminus; nVIII, nervus octavus; Oli, oliva inferior; Phy, nucleus prepositus hypoglossi; Rai, nucleus raphes inferior; Ri, nucleus reticularis inferior; Rub, nucleus ruber; sol, tractus solitarius; tect, tectum mesencephali; Vedl, Veds, Vevm, dorsolateral, descending and ventromedial vestibular nuclei; Vds, nucleus descendens nervi trigemini; Vm, nucleus motorius nervi trigemini; Vpr, nucleus princeps nervi trigemini; VI, nucleus nervi abducentis; Xmd, nucleus motorius dorsalis nervi vagi; XII, nucleus nervi hypoglossi. (From Bangma and ten Donkelaar, 1982.)

The following experimental data provide additional evidence for the presence of an inferior olive in the caudal brain stem of reptiles:

1. Rather massive terminations of spinal afferents in the caudal brain stem of reptiles (Ebbesson, 1967, 1969; Jacobs, 1968; Ebbesson and Goodman, 1981), indicating spino-olivary projections (see also Fig. 7.16).
2. Ipsilateral descending projections from the red nucleus to the caudal brain stem in *Varanus exanthematicus* (see ten Donkelaar and de Boer-van Huizen, 1981a; ten Donkelaar, 1988), indicating a rubro-olivary connection as present in mammals (see Edwards, 1972).

3. Ipsilateral connections of the nucleus of the basal optic root to the caudal brain stem in *Varanus exanthematicus* (see ten Donkelaar and de Boer-van Huizen, 1981a), comparable to projections of the nucleus of the basal optic root to the inferior olive in the pigeon (Clarke, 1977; Brecha et al., 1980).

4. Anterograde tracer (^{3}H-leucine) studies in *Varanus exanthematicus* revealing contralaterally terminating fibers in the molecular layer of the cerebellar cortex after injections into the caudal brain stem (Wolters, Dederen, and ten Donkelaar, unpublished observations, 1984).

5. After injections of WGA-HRP into the vicinity of the inferior olive in *Varanus exanthematicus* a distinct pattern of termination in the contralateral molecular layer of the cerebellar cortex was observed, arranged as longitudinally oriented zones (Fig. 7.24).

6. In *Pseudemys scripta elegans*, Künzle (1985a) demonstrated this climbing fiber projection to the cerebellum in more detail. In animals in which injections of various radioactive-labeled amino acids encompassed the entire area previously identified as the possible source of cerebellar climbing fibers, the basal portion of the molecular layer was labeled almost throughout the contralateral half of the cerebellum. In cases of restricted injections, labeled climbing fibers terminated in quite distinct longitudinally oriented zones.

These observations indicate the existence of an inferior olive in the caudal brain stem of reptiles. Schwarz and Schwarz (1980), however, suggested that a cell mass in the rostral rhombencephalon (their parvocellular isthmic complex) represents a source of climbing fibers in reptiles, possibly corresponding to the inferior olive of mammals. This suggestion conflicts with the phylogenetically constant appearance of the inferior olive in the caudal brain stem as described in all vertebrate classes (see Bangma and ten Donkelaar, 1982, for discussion). Schwarz and Schwarz' (1980) parvocellular isthmic complex might correspond to the parabrachial nucleus distinguished in the present survey (Fig. 7.4A), which contained retrogradely labeled cells in several cases of HRP injections into the cerebellum (Bangma and ten Donkelaar, 1982). This parabrachial region might represent a relay station of telencephalic information to the cerebellum. Reiner (1979) demonstrated that the paleostriatum augmentatum in *Chrysemys picta* projects to the parabrachial isthmic region.

The HRP studies carried out in reptiles (Schwarz and Schwarz, 1980; Bangma and ten Donkelaar, 1982; Künzle, 1983a) do not provide conclusive evidence for the presence of other precerebellar nuclei. A lateral funicular nucleus has been distinguished in Nissl-stained sections (e.g., ten Donkelaar and Nieuwenhuys, 1979). A few labeled neurons observed ipsilaterally at the lateral border of the caudal brain stem (Bangma and ten Donkelaar, 1982—see their Fig. 7.3L; see also Künzle, 1983a) after HRP injections into the cerebellum might represent the lateral funicular nucleus.

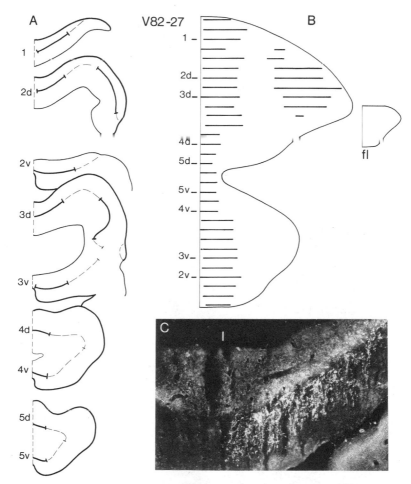

Fig. 7.24. *Varanus exanthematicus.* The distribution of anterogradely labeled fibers and terminal structures after WGA-HRP injections into the caudal brain stem including the inferior olive. (A) Transverse sections of the cerebellum; the bold lines indicate the area of termination of olivocerebellar fibers. (B) A chart in which the termination fields of olivocerebellar fibers are indicated. (C) A dark-field photomicrograph, showing the labeled fibers and terminal structures in the molecular layer; the arrow indicates the midline (section 1 of A) (80 ×).

4. Other Rhombencephalic Afferents

Apart from cerebellar afferents arising in the vestibular nuclear complex and inferior olive, various other rhombencephalic nuclei—the nucleus prepositus hypoglossi, somatosensory nuclei, the nucleus of the solitary tract, the reticular formation, and the locus coeruleus—were found to send fibers to the cerebellum (Schwarz and Schwarz, 1980; Bangma and ten Donkelaar, 1982; Künzle, 1983a).

After HRP injections into the cerebellum of various reptiles, a conspicuous group of bilaterally labeled neurons was observed (Bangma and ten Donkelaar, 1982; see also Künzle, 1983a) in a cell mass directly adjacent to the fasciculus longitudinalis medialis (flm) between the levels of the abducent and hypoglossal nuclei (Figs. 7.22E to J, 7.23G to J). In the tegu lizard (*Tupinambis nigropunctatus*) and in a snake (*Boa constrictor*) this cell mass was termed the nucleus parvocellularis medialis by Ebbesson (1967, 1969). It resembles in its location and spinal afferents (Ebbesson, 1967, 1969; see also Fig. 7.16) the perihypoglossal nuclei of mammals. The fact that the cell mass in question projects extensively to the cerebellum is also in keeping with experimental data in mammals (Brodal, 1952; Kotchabhakdi and Walberg, 1978; Batini et al., 1978). Therefore, this important site of origin of cerebellar afferents has been termed the *nucleus prepositus hypoglossi* (first described as the perihypoglossal nuclear complex; Bangma and ten Donkelaar, 1982). The localization of the main part of this structure between the nucleus nervi abducentis and the nucleus nervi hypoglossi, its spinal afferents, cerebellar projections, and projections to the nuclei that innervate the external eye muscles (see Bangma and ten Donkelaar, 1983; ten Donkelaar et al., 1985) strongly suggest that this cell mass is comparable to the mammalian nucleus prepositus hypoglossi (for references see McCrea and Baker, 1985). In the cat, three perihypoglossal nuclei that project to the vermis can be distinguished. Based on their afferent and efferent connections, these nuclei seem to be involved in the control of eye movements and possibly head movements in relation to the cerebellum and the vestibular system (Kotchabhakdi and Walberg, 1978). Divergent axon collaterals of neurons in the main component of the perihypoglossal nuclei—the nucleus prepositus hypoglossi—were shown in cats to the oculomotor nucleus and the cerebellum (Yingcharoen and Rinvik, 1982).

Somatosensory projections from brain stem nuclei to the cerebellum arise in the descending trigeminal nucleus (Figs. 7.22D, E, F, and J, and 23E) in all reptiles studied so far; in *Pseudemys scripta elegans* these occur also in the dorsal column nucleus (Fig. 7.22K). Künzle (1983a) also reported a small projection to the cerebellum from the principal trigeminal nucleus. A projection from the mesencephalic trigeminal nucleus, as suggested on the basis of normal material (Weston, 1936; Larsell, 1967), could not be confirmed in HRP studies. In mammals (e.g., cats; see Brown-Gould, 1980; Matsushita et al., 1982; Ito, 1984) extensive projections to the cerebellum were shown from the spinal trigeminal nucleus and part of the principal nucleus, but not from the mesencephalic nucleus. In *Python regius*, cerebellar afferents arise in the descending trigeminal nucleus (Bangma and ten Donkelaar, 1982), but not in the lateral descending trigeminal nucleus involved in infrared perception (see Chapter 5).

Projections of the nucleus of the solitary tract to the cerebellum were demonstrated in all reptilian species studied (Schwarz and Schwarz, 1980; Bangma and ten Donkelaar, 1982).

Cells of origin of *reticulocerebellar* fibers were found scattered throughout the nucleus reticularis inferior and medius. Also the superior and inferior raphe nuclei contained labeled neurons in some experiments, but compared with the rather extensive projections of the raphe nuclei demonstrated in the cat (Taber Pierce et al., 1977), this connection seems to be rather limited in reptiles. At least part of the raphecerebellar projection is *serotonergic*. Serotonergic neurons in the brain of turtles are found mostly scattered within the raphe region of the entire rhombencephalon, but mainly in the nucleus raphes inferior (Parent and Poirier, 1971; Parent, 1973, 1979; Yamamoto et al., 1977; Ueda et al., 1983). In *Clemmys japonica*, Ueda et al. (1983) observed a small number of serotonin-immunoreactive fibers in the granular layer of the cerebellar cortex. Immunohistochemical studies (Wolters et al., 1985) in *Varanus exanthematicus* demonstrated a distinct serotonin-immunoreactive fiber projection present in the rostral and middle parts of the cerebellum (see Fig. 7.25). Serotonergic cell bodies were found in several cell masses in the brain stem reticular formation, but particularly in the raphe nuclei (Wolters et al., 1985).

The *locus coeruleus* in mammals is known to give rise to a catecholaminergic inhibitory system (see Section IV.C), most likely *noradrenergic*. A comparable cell mass is known to contain catecholamines in turtles (*Chrysemys picta*, Parent and Poirier, 1971; Parent, 1973, 1979; *Geoclemys reevesii*, Yamamoto et al., 1977), specifically, according to Yamamoto et al. (1977), noradrenaline. Parent (1979) described a sparse catecholaminergic innervation entering the cerebellum via the anterior medullary velum. The presence of catecholamines in the locus coeruleus in *Varanus exanthematicus* was determined by an indirect technique—the demonstration of tyrosine hydroxylase (an enzyme present in catecholamine-containing neurons and fibers). Surprisingly, however, practically no TH immunoreactivity was found in the cerebellum. HRP studies in turtles (Schwarz and Schwarz, 1980; Bangma and ten Donkelaar, 1982) showed a cerebellar projection of the locus coeruleus (e.g., in *Pseudemys scripta elegans*, Fig. 7.21G).

In addition to this monoaminergic innervation of the cerebellum, *peptidergic* projections are found in reptiles. With immunohistochemical techniques a population of apparently mossy fibers and terminals with substance P–like immunoreactivity was observed in the cerebellum of *Pseudemys scripta elegans* (Korte et al., 1980), and *Varanus exanthematicus* (Wolters, ten Donkelaar, and Verhofstad, 1986). Substance P–immunoreactive fibers could be traced into the granular layer of the cerebellum from a distinct bundle of axons in the dorsolateral tegmentum (Korte et al., 1980). Their terminals are most numerous in

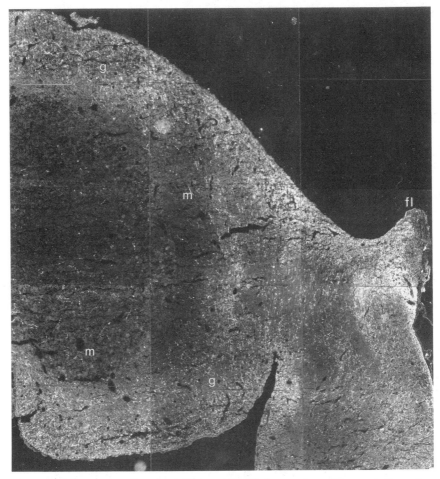

Fig. 7.25. *Varanus exanthematicus.* The serotonergic innervation of the cerebellum. fl, Flocculus; g, granular layer; m, molecular layer.

the rostrolateral part of the cerebellum (Korte et al., 1980; Wolters, ten Donkelaar, and Verhofstad, 1986). A similar peptidergic innervation of the cerebellum was found in the rat (Inakagi et al., 1982). The source of the substance P innervation of the cerebellum is unknown. Possible sites of origin might be the descending trigeminal nucleus, the raphe nuclei, or the nucleus prepositus hypoglossi, all known to send axons to the cerebellum. Immunohistochemical studies in the rat (Cuello and Kanazawa, 1978; Ljungdahl et al., 1978) showed substance P–positive cells in these areas. In *Varanus exanthematicus* a sparse leu- and met-enkephalin-immunoreactive innervation of the cerebellum was found (Wolters et al., 1986).

D. Mesencephalic Afferents

In the mesencephalon, in various turtles (see Reiner and Karten, 1978; Schwarz and Schwarz, 1980; Bangma and ten Donkelaar, 1982; Künzle, 1983a; see Fig. 7.21D and E) and in a lizard, *Varanus exanthematicus* (Bangma and ten Donkelaar, 1982; Fig. 7.23A and B), the most prominent afferent projection to the cerebellum was found to arise in the retinorecipient nucleus of the basal optic root (or nucleus opticus tegmenti), thereby providing a bisynaptic retinal input to the cerebellum. In the turtles *Chrysemys picta picta* and *Pseudemys scripta* (Reiner, 1981), the fibers of the basal optic root originate in large ganglion cells in the ganglionic layer of the retina (see also Chapter 1 above). Lesions of the nucleus of the basal optic root in *Chrysemys picta* (Fite et al., 1979) disrupted optokinetic nystagmus, indicating an important role of this nucleus in oculomotor functions. In the pigeon, *Columba livia*, extensive studies by Brecha et al. (1980) have shown that displaced ganglion cells of the retina give rise to a distinct projection to the nucleus of the basal optic root. The nucleus of the basal optic root in turn projects upon the oculomotor nuclear complex, the nucleus interstitialis of the flm and the vestibulocerebellum, regions that have all been implicated in oculomotor function. These findings strongly suggest that the displaced ganglion cells and the accessory optic system have a major influence upon oculomotor reflexes involving eye and head movements (Brecha et al., 1980). Projections of the nucleus of the basal optic root to the oculomotor nucleus have been reported by Shanklin (1933) and to the ventral part of the interstitial nucleus of the flm by various authors (Beccari, 1923; Huber and Crosby, 1926; Shanklin, 1930). The fact that the nucleus of the basal optic root is very large—forming a large external bulge—in chameleons, with their highly movable eyes (Shanklin, 1930, 1933), also indicates a role in oculomotor function.

In *Caiman crocodilus*, two rostral mesencephalic nuclei were found to project to the cerebellum (Brauth et al., 1978), that is, the nucleus circularis and the interstitial nucleus of the posterior commissure; in other reptiles no comparable cell masses were found to project to the cerebellum (Reiner and Karten, 1978; Schwarz and Schwarz, 1980; Bangma and ten Donkelaar, 1982), although in *Pseudemys scripta elegans* some cells in the vicinity of the posterior commissure also were found to project to the cerebellum (Künzle, 1983a).

Other mesencephalic cell groups found to project to the cerebellum are scattered cells medial to the nucleus of the basal optic root (group A of Reiner and Karten, 1978; Figs. 7.21E and 7.23A), the interstitial nucleus of the flm (Figs. 7.21D and E, 7.23A and B), and the red nucleus. The nucleus ruber has been found to project to the nucleus cerebellaris lateralis with both anterograde degeneration (ten Donkelaar, 1976b) and anterograde tracer (^{3}H-leucine, ten Donkelaar and Dederen, unpublished observations of *Varanus exanthematicus*, 1982) techniques. Therefore, the labeled neurons in the red nucleus after HRP injections into the cerebellum (Figs. 7.21E and F and 7.23B) are probably due to spread of HRP to the lateral cerebellar nucleus. No tectocerebellar pathways, reported by Weston (1936) in normal material in various reptiles, could be demonstrated in degeneration studies (Foster and Hall, 1975; ten Donkelaar, 1976b; Ulinski, 1977) or HRP studies (Reiner and Karten, 1978; Schwarz and Schwarz, 1980; Bangma and ten Donkelaar, 1982; Künzle, 1983a). In this respect, it should be noted that the tectocerebellar tract as described in birds (see Larsell, 1967, for references) in fact arises in the medial spiriform nucleus (Verhaart, 1974; Clarke, 1977).

E. Prosencephalic Afferents

Cerebellar afferents from the prosencephalon in reptiles seem to be rather limited. In the diencephalon of turtles (*Chrysemys picta picta*, Reiner and Karten, 1978; *Pseudemys scripta elegans*, Bangma and ten Donkelaar, 1982; Künzle, 1983a) and *Varanus exanthematicus* (Bangma and ten Donkelaar, 1982), cells of origin of cerebellar afferents were found in the ipsilateral nucleus geniculatus pretectalis. After large cerebellar HRP injections in the red-eared turtle, *Pseudemys scripta elegans*, labeled neurons were also found in hypothalamic nuclei (Bangma and ten Donkelaar, 1982; Künzle, 1983a). It has been suggested (Bangma and ten Donkelaar, 1982) that this hypothalamic labeling may be related to spread of HRP to the locus coeruleus. However, in various mammals, direct connections between the hypothalamus and the cerebellar cortex have been demonstrated (Dietrichs, 1984; Haines et al., 1982), suggesting autonomic input to the cerebellum.

HRP studies (Schwarz and Schwarz, 1980; Bangma and ten Donkelaar, 1982; Künzle, 1983a) revealed no cells of origin of cerebellar afferents in the telencephalon. A striocerebellar projection has been suggested by Ariëns Kappers (1947). Such a direct striocerebellar pathway to the ipsilateral nucleus cerebellaris lateralis has been reported in the tegu lizard, *Tupinambis nigropunctatus*, after large lesions of the striatum (Hoogland, 1977) as well as with the autoradiographic tracing technique (Voneida and Sligar, 1979). This projection could not be confirmed in HRP experiments with spread of the enzyme to the lateral cerebellar nucleus (see Bangma and ten Donkelaar, 1982). It should be noted that the lateral cerebellar nucleus is rather difficult to delimit from the parabrachial nucleus. Therefore, the projection found to the lateral cerebellar nucleus in *Tupinambis nigropunctatus* might in fact be comparable to the projection to the parabrachial isthmic region shown by Reiner (1979, *Chrysemys*). In a turtle, *Chrysemys picta*, Reiner (1979) showed a projection from the paleostriatum augmentatum to parabrachial isthmic regions.

The ways by which the reptilian telencephalon might influence the cerebellum are still unclear. Possible relay stations are present in the rostral mesencephalon, in the parabrachial region, and in the caudal brain stem. In *Caiman crocodilus* two nuclei in the vicinity of the posterior commissure, the nucleus circularis and the interstitial nucleus of the posterior commissure, were found to project to the cerebellum (Brauth et al., 1978). These authors noted the similarity of the nucleus circularis to the avian nucleus spiriformis medialis (see Karten and Finger, 1976) in that both project to the cerebellum. In teleosts (see Karten and Finger, 1976; Ito et al., 1982) there is also an indirect telencephalocerebellar pathway, with a relay in the nucleus paracommissuralis.

In *Varanus exanthematicus*, HRP injections into the caudal brain stem (including the inferior olive) showed retrogradely labeled neurons in the striatum (ten Donkelaar and de Boer-van Huizen, 1981a), which might indicate a projection from the telencephalon to the inferior olive.

F. Summary
The reptilian cerebellum is dominated by vestibular afferents and connections from the nucleus of the basal optic root and from the nucleus prepositus hypoglossi (see Fig. 7.26). In addition, rather extensive

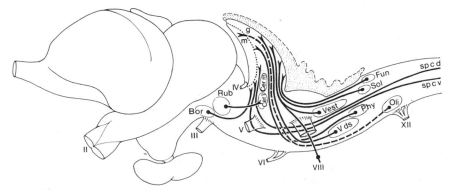

Fig. 7.26. *Varanus exanthematicus*. Diagram summarizing the cerebellar afferent connections. Bor, nucleus of the basal optic root; Cerl, Cerm, lateral and medial cerebellar nuclei; Fun, nucleus funiculi dorsalis; G, granular layer; M, molecular layer; Oli, oliva inferior; Phy, nucleus prepositus hypoglossi; Rub, nucleus ruber; Sol, nucleus tractus solitarii; Vest, vestibular nuclear complex; II, nervus opticus; III, nervus oculomotorius; V, nervus trigeminus; Vds, nucleus descendens nervi trigemini; VI, nervus abducens; VIII, nervus octavus; XII, nervus hypoglossus.

spinal projections have been found, as well as a distinct projection from the inferior olive. The inferior olive probably represents an important relay station to the cerebellum since it receives afferents from the spinal cord, the red nucleus, the nucleus of the basal optic root, and possibly even from the striatum—the main output center of the telencephalon (see Northcutt, 1978; ten Donkelaar and de Boer-van Huizen, 1981a; ten Donkelaar et al., 1987).

VI. EFFERENT CONNECTIONS
A. General

The cerebellar cortex projects to the cerebellar nuclei and the vestibular nuclear complex (Weston, 1936; Senn and Goodman, 1969). These *corticonuclear projections* have been studied with the HRP technique in various reptiles (see Bangma et al., 1983; Bangma and ten Donkelaar, 1984). Projections of the *cerebellar nuclei* were described in the classic studies on the reptilian cerebellum. Weston (1936) regarded the cerebellar efferents, except those to the vestibular nuclei (and a cerebellotectal tract), as forming a continuous cerebellotegmental fiber system composed of a tractus cerebellomotorius et tegmentalis bulbi and a tractus cerebellomotorius et tegmentalis mesencephali. The latter tract includes fibers to the red nucleus, described as brachium conjunctivum by other students of the reptilian CNS (e.g., Ariëns Kappers, 1921; Larsell, 1926, 1933; Papez, 1929; Shanklin, 1930; Hindenach, 1931; Frederikse, 1931). The brachium conjunctivum takes its origin from the lateral cerebellar nucleus (Shanklin, 1930; Larsell, 1932). No cerebellothalamic connection has been observed in these classic studies.

A tractus cerebellovestibularis et cerebellospinalis, arising from the medial cerebellar nuclei on both sides, but also receiving Purkyně cell axons, has been described by Weston (1936). The crossed component presumably represents the reptilian homologue of the uncinate fascicle (or hook bundle of Russell) as found in mammals.

The cerebellar efferents are aimed at motor centers in the brain stem such as the red nucleus, vestibular nuclei, and reticular formation. The role of these cerebellar target sites in motor control will be briefly discussed in Section VII.

B. Corticonuclear Projections

Lesion experiments in *Caiman crocodilus* by Senn and Goodman (1969) support the concept that the reptilian cerebellar cortex is organized into longitudinally oriented zones, each with its own specific projection site. Corticonuclear projections to the cerebellar nuclei originate mainly in the so-called vermal zone (see Goodman and Simpson, 1960; Goodman, 1969)—the medial two-thirds of the corpus cerebelli—but also in the paravermal zone (the lateral one-third of the corpus cerebelli). Corticovestibular projections appeared to be derived from Purkyně cells located in both the vermal and paravermal zones, especially of the posterior lobe (no lesions were made in the flocculus).

The organization of the corticonuclear projections of the cerebellum has been studied in various reptiles with the HRP technique (Bangma et al., 1983; Bangma and ten Donkelaar, 1984). The localization of Purkyně cells projecting to the cerebellar nuclei and vestibular nuclear complex, respectively, is shown in Figs. 7.27 to 7.32. These experiments reveal a longitudinal pattern of organization within the corticonuclear projections. In the experiment in *Pseudemys scripta elegans* shown in Fig. 7.27 an HRP slow-release gel was implanted into the caudal part of the ventrolateral vestibular nucleus. Most of the labeled

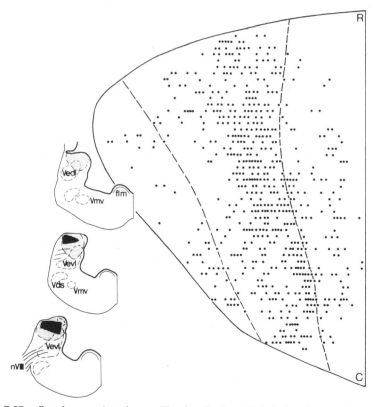

Fig. 7.27. *Pseudemys scripta elegans.* The distribution of labeled Purkyně cells in the cerebellum following HRP slow-release gel implantation into the ventrolateral and dorsolateral vestibular nuclei. C, caudal; flm, fasciculus longitudinalis medialis; nVIII, nervus octavus; R, rostral; Vedl, Vevl, dorsolateral and ventrolateral vestibular nuclei; Vds, nucleus descendens nervi trigemini; Vmv, nucleus motorius nervi trigemini pars ventralis.

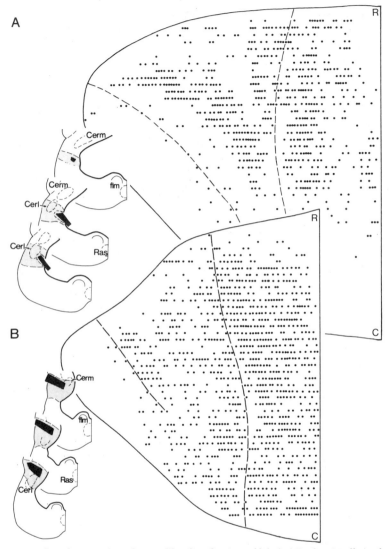

Fig. 7.28. *Pseudemys scripta elegans.* The distribution of labeled Purkyně cells in the cerebellum following HRP slow-release gel implantations aimed at (A) the lateral cerebellar nucleus and (B) the medial cerebellar nucleus. C, caudal; Cerl, Cerm, lateral and medial cerebellar nuclei; flm, fasciculus longitudinalis medialis; R, rostral; Ras, nucleus raphes superior. (From Bangma et al., 1983.)

corticonuclear neurons were found in a distinct rostrocaudally oriented strip in the intermediate zone. The lateral zone contained a few labeled neurons in its rostral, floccular part, as well as at more caudal levels, the so-called marginal rim (Mugnaini et al., 1974). In the medial zone a distinct rostrocaudally oriented strip of labeled Purkyně cells bordering on the intermediate zone was present. As in other ex-

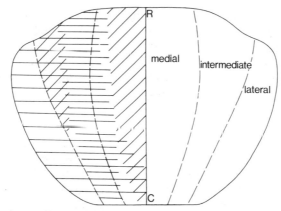

Fig. 7.29. *Pseudemys scripta elegans*. On the left the organization of the corticonuclear projections to the cerebellar nuclei and the vestibular nuclear complex is schematically summarized. Oblique lines correspond to the Purkyně cell areas that project to the medial and lateral cerebellar nuclei, respectively. Horizontal lines correspond to the Purkyně cell areas that project to the vestibular nuclear complex. On the right the three longitudinal zones of Purkyně cells as revealed by the topological analysis of the cerebellum are shown (see Fig. 7.14). C, caudal; R, rostral.

periments in which HRP was applied to the vestibular nuclear complex, the medial part of the medial zone contained no labeled Purkyně cells. It should also be noted that no Purkyně cells were labeled in the rostrolateral part of the intermediate zone. Labeled Purkyně cells in the lateral zone were observed particularly after HRP application to the lateral part of the dorsolateral vestibular nucleus and to the descending and ventromedial vestibular nuclei (Bangma et al., 1983). HRP slow-release gel implantations aimed at the cerebellar nuclei revealed two areas of labeled Purkyně cells not found after the vestibular cases. Apart from labeled *corticovestibular* neurons (due to damage to passing corticovestibular fibers; damaged axons are known to take up and transport HRP), labeled Purkyně cells were observed in the medial part of the medial zone and in the rostrolateral part of the intermediate zone (Fig. 7.28A and B). It seems likely (see Bangma et al., 1983; Fig. 7.28) that in *Pseudemys scripta elegans* the Purkyně cells in the medial part of the medial zone project to the medial cerebellar nucleus. Purkyně cells in the rostrolateral part of the intermediate zone project to the lateral cerebellar nucleus.

In summary, cerebellar corticonuclear projections in the red-eared turtle, *Pseudemys scripta elegans*, are organized as longitudinally oriented zones. Four such zones of Purkyně cells can be distinguished, each with a different target site (Fig. 7.29): (1) the medial part of the medial zone projecting to the medial cerebellar nucleus; (2) the lateral part of the medial zone and the main part of the intermediate zone projecting to the vestibular nuclear complex, particularly the dorsolateral and ventrolateral vestibular nuclei; (3) the rostrolateral part of the intermediate zone projecting to the lateral cerebellar nucleus; and (4) the lateral zone projecting to the vestibular nuclear complex, presumably to the dorsolateral, descending, and ventromedial nuclei. The lateral zone, which includes both Larsell's flocculus and its caudal extension, the marginal rim (Mugnaini et al., 1974; Brand and Mugnaini, 1980), probably constitutes the floccular part of the cerebellum.

Recent experiments in a lizard, *Varanus exanthematicus*, and a snake, *Python regius*, indicate that in these species a comparable pattern of organization of corticonuclear projections is present (Bangma, 1983; Bangma and ten Donkelaar, 1984). A longitudinal pattern of these projections is clearly shown in two experiments in *Varanus exanthematicus*, in which an HRP slow-release gel was implanted into the ventrolateral vestibular nucleus (Fig. 7.31) and the medial cerebellar nucleus (Fig. 7.30), respectively. In the vestibular case, the most medial part of the cerebellar cortex contained only a few labeled Purkyně cells. Lateral to this strip of Purkyně cells a distinct broad zone of labeled cells was found. The most lateral part of the corpus cerebelli contained only a few labeled Purkyně cells. The flocculus was not found to project extensively to the vestibular nuclear complex in this particular experiment. In other experiments (see Section VI.D) in which HRP was applied to the vestibular nuclear complex, the flocculus was distinctly labeled. It seems likely that in *Varanus exanthematicus* the most medial part of the cerebellar cortex and the most lateral part of the corpus cerebelli project to the cerebellar nuclei. Fig. 7.30 shows an experiment in which an HRP slow-release gel was implanted into the medial cerebellar nucleus (with a limited spread of HRP to the lateral cerebellar nucleus). In this case in the medial part of the cerebellar cortex almost all Purkyně cells were found labeled. A few more laterally situated labeled neurons might be due to the spread of HRP to the lateral cerebellar nucleus. The zonal pattern of cerebellar corticonuclear projections found in *Varanus exanthematicus* is summarized in Fig. 7.31C.

Four longitudinally oriented zones of Purkyně cells are present apparently also in *Python regius* (Fig. 7.32). The HRP experiment shown involves the main part of the vestibular nuclear complex. Labeled Purkyně cells were present in two more or less distinct strips. A distinct lateral strip of labeled neurons is bordered by a zone in which almost no labeled cells were observed (see Sections 2 to 6, Fig. 7.32A). More medially again, a distinct zone of labeled Purkyně cells can be observed, whereas the most medial part of the cerebellar half contained only few labeled Purkyně cells (e.g., in sections 3, 6, and 9, Fig. 7.32A).

Thus a zonal pattern in the corticonuclear projections has been found in *Pseudemys scripta elegans*, *Varanus exanthematicus*, and *Python*

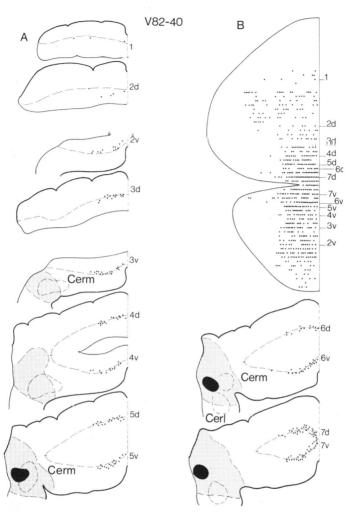

Fig. 7.30. *Varanus exanthematicus.* The distribution of labeled Purkyně cells following HRP slow-release gel implantation into medial, and partly the lateral, cerebellar nuclei. (A) Some transverse sections of the cerebellum with retrogradely labeled Purkyně cells and the implantation site. (B) A chart of the cerebellum in which the distribution of the labeled neurons is indicated. Cerl, Cerm, lateral and medial cerebellar nuclei.

regius. It involves four longitudinal strips of Purkyně cells, each with a different target site (from the midline laterally: the medial cerebellar nucleus, the vestibular nuclear complex, the lateral cerebellar nucleus, and the vestibular nuclear complex). A parallel can be broadly drawn between the present data in reptiles and experimental findings in mammals (see Voogd, 1964, 1967; Voogd and Bigaré, 1980; Haines et al., 1982). The most medial zone of the corpus cerebelli (projecting to the medial cerebellar nucleus) and the large, broad strip of Purkyně

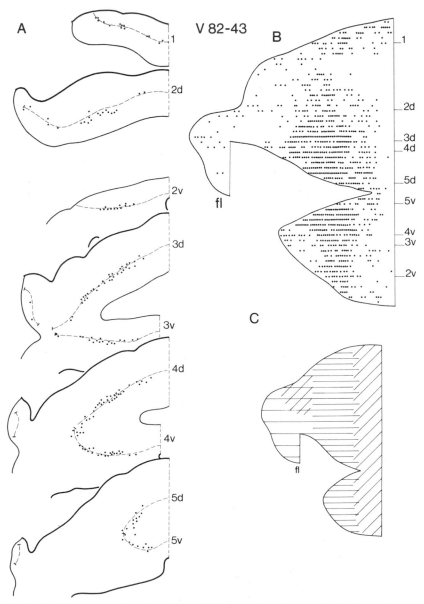

Fig. 7.31. *Varanus exanthematicus.* The distribution of labeled Purkyně cells in the cerebellum following HRP slow-release gel implantation into the middle part of the vestibular nuclear complex. (A) Some transverse sections of the cerebellum with retrogradely labeled Purkyně cells. (B) A chart of the cerebellum in which the distribution of the labeled neurons is indicated. fl, flocculus. (C) Diagram summarizing the corticonuclear projections to the cerebellar nuclei and the vestibular nuclear complex in *Varanus exanthematicus. Oblique lines* correspond to the Purkyně cell zones that project to the medial and lateral cerebellar nuclei, respectively. *Horizontal lines* correspond to the areas that project to the vestibular nuclear complex. (After Bangma and ten Donkelaar, 1984.)

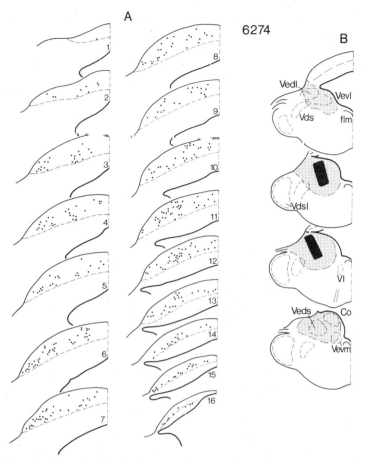

Fig. 7.32. *Python regius.* The distribution of labeled Purkyně cells in the cerebellum after implantation of an HRP slow-release gel into the vestibular nuclear complex. (A) Series of transverse sections of the cerebellum with the labeled Purkyně cells. (B) Site of implantation of the HRP slow-release gel. Co, cochlear nuclear complex; flm, fasciculus longitudinalis medialis; Vedl, Veds, Vevl, Vevm, dorsolateral, descending, ventrolateral, and ventromedial vestibular nuclei; Vds, nucleus descendens nervi trigemini; Vdsl, nucleus descendens lateralis nervi trigemini; VI, nucleus nervi abducentis.

cells with efferents to the vestibular nuclei show clear similarities with the A and B zones in mammals, respectively. The C zones of mammals, projecting to the interposed nuclei, seem to be represented in *Pseudemys scripta elegans* in the rostrolateral part of the intermediate zone and in the lateral part of the corpus cerebelli in *Varanus exanthematicus*. The lateral zone of Purkyně cells as distinguished in *Pseudemys scripta elegans* and *Python regius*, projecting to the vestibular nuclear complex, represents the flocculus in these species. In *Varanus exanthematicus* a clear flocculus is present.

C. Projections of the Cerebellar Nuclei

The projections of the cerebellar nuclei can be studied with various experimental anatomical techniques. Various targets of projections of these nuclei have been demonstrated in a number of HRP studies. The lateral cerebellar nucleus appeared to have mainly contralateral *ascending* connections; the medial cerebellar nucleus, mainly contralateral *descending* projections. In the tegu lizard, *Tupinambis nigropunctatus*, a small direct contralateral projection from the lateral cerebellar nucleus to the lateral part of the dorsal ventricular ridge was found (Lohman and van Woerden-Verkley, 1978). Comparable observations were made in *Varanus exanthematicus* (ten Donkelaar and de Boer-van Huizen, 1981b). In *Caiman crocodilus*, the lateral cerebellar nucleus was found to project to the contralateral thalamus (ventral lateral and ventral medial thalamic areas); ventral lateral area neurons project upon a portion of the rostral telencephalon external to the ventrolateral area (Brauth and Kitt, 1980). These data suggest a bisynaptic cerebellotelencephalic projection with a thalamic relay in *Caiman crocodilus*. In *Varanus exanthematicus*, ascending projections from the lateral cerebellar nucleus were also found to the contralateral diencephalon (ten Donkelaar and de Boer-van Huizen, 1981b). However, spinal projections arising in the lateral cerebellar nucleus were also found (ten Donkelaar and de Boer-van Huizen, 1978a; ten Donkelaar et al., 1980; Wolters, de Boer-van Huizen, ten Donkelaar, and Leenen, 1986).

The medial cerebellar nucleus has only sparse ascending projections in *Varanus exanthematicus* (ten Donkelaar and de Boer-van Huizen, 1981b). Rather extensive, mainly contralateral, spinal projections were found in various reptiles (see ten Donkelaar and de Boer-van Huizen, 1978a; ten Donkelaar et al., 1980; Wolters et al., 1982; Woodson and Künzle, 1982). In *Varanus exanthematicus*, such projections reach as far caudally as the lumbar enlargement; with the multiple retrograde fluorescent tracer technique (Kuypers et al., 1980; Kuypers and Huisman, 1984) it has been shown that such cerebellospinal axons to the lumbar enlargement give off collaterals in the cervical intumescence (Wolters, de Boer-van Huizen, ten Donkelaar, and Leenen, 1986).

A rough idea of the projections of the cerebellar nuclei can be observed in degeneration studies. In Fig. 7.33, a large lesion interrupted the cerebellar peduncle in the savanna monitor lizard, *Varanus exanthematicus*. Both corticonuclear (including corticovestibular) fibers and pathways arising in the cerebellar nuclei were damaged. Since projections from the cerebellar cortex are strictly ipsilateral and restricted to the cerebellar nuclei and the vestibular nuclear complex (Senn and Goodman, 1969; Bangma et al., 1983; Bangma and ten Donkelaar, 1984), degenerating fibers in the ipsilateral vestibular nuclear com-

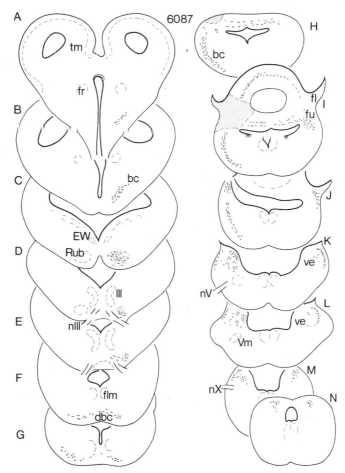

Fig. 7.33. *Varanus exanthematicus.* Diagrammatic representation of the course and distribution of degenerating fibers within the brainstem following lesion of the cerebellar peduncle. Broken lines indicate degenerating fibers, dots, evidence of preterminal degeneration. bc, Brachium conjunctivum; dbc, decussation of the brachium conjunctivum; EW, nucleus of Edinger-Westphal; fl, flocculus; flm, fasciculus longitudinalis medialis; fr, fasciculus retroflexus; fu, fasciculus uncinatus; nIII, nervus oculomotorius; nV, nervus trigeminus; nX, nervus vagus; Rub, nucleus ruber; tm, tectum mesencephali; ve, vestibular nuclear complex; III, nucleus nervi oculomotorii; Xm, nucleus motorius nervi trigemini.

plex (Fig. 7.33K to N) are at least partly due to damage to corticovestibular fibers. However, two bundles of degenerating fibers were found, one ascending and one descending, both arising in the cerebellar nuclei.

The ascending pathway—the brachium conjunctivum—bends ventrally along the lateral wall of the brain stem (Fig. 7.33H to J), decussates in the most caudal part of the midbrain (Fig. 7.33F and G), further ascends in the ventromedial part of the midbrain tegmentum, terminates extensively in the red nucleus (Fig. 7.33D and E), and continues rostrally at least as far as the mesodiencephalic junction. A descending path decussates in the cerebellar commissure and more caudally in the cerebellar base (Fig. 7.33H and I), passes dorsal to as well as through the medial cerebellar nucleus, and hooks around the cerebellar nuclei and the brachium conjunctivum and reaches the vestibular nuclear complex. This bundle represents the reptilian homologue of the mammalian hook bundle or fasciculus uncinatus. The experiment shown in Fig. 7.33 shows that fibers pass from this hook bundle to the vestibular nuclei, but no evidence for preterminal degeneration was found.

Bangma (unpublished observations, 1982) found comparable data in *Pseudemys scripta elegans*, as well as evidence for preterminal degeneration: extensive projections throughout the vestibular nuclear complex and even a sparse projection to various parts of the reticular formation.

With anterograde tracer techniques (^{3}H-leucine and HRP) these two main efferent pathways from the cerebellar nuclei have been further studied. Fig. 7.34 shows an experiment in *Varanus exanthematicus* (Bangma et al., 1984) in which ^{3}H-leucine was injected into the cerebellar peduncle. No ipsilateral descending projection could be shown in this experiment; however, a well-developed contralateral descending projection hooking around the cerebellar nucleus—the fasciculus uncinatus (Fig. 7.34F, G, and H) was found. In various HRP experiments (Bangma et al., 1984) in which HRP was applied to the cerebellar nuclei, extensive contralateral projections to the vestibular nuclei were labeled anterogradely. These included sparse projections to various parts of the reticular formation. The ascending projection, the brachium conjunctivum, was also extensively labeled (Fig. 7.34B to F) in anterograde tracer experiments and found to terminate in the red nucleus (Fig. 7.34B, C, D) and in the interstitial nucleus of the flm (Fig. 7.34A). As in degeneration studies, no labeled fibers could be traced beyond the mesodiencephalic border.

Thus both anterograde degeneration and anterograde tracer studies show two distinct contralateral pathways arising in the cerebellar nuclei: an ascending tract, the brachium conjunctivum, projecting to the red nucleus, and a descending pathway, the fasciculus uncinatus (or hook bundle), to the vestibular nuclear complex and more sparsely to the reticular formation. There is no contralateral descending limb of the brachium conjunctivum, as occurs in mammals (e.g., Brodal and Szikla, 1972; Martin et al., 1974; Chan-Palay, 1977; Faull, 1978; Ito, 1984).

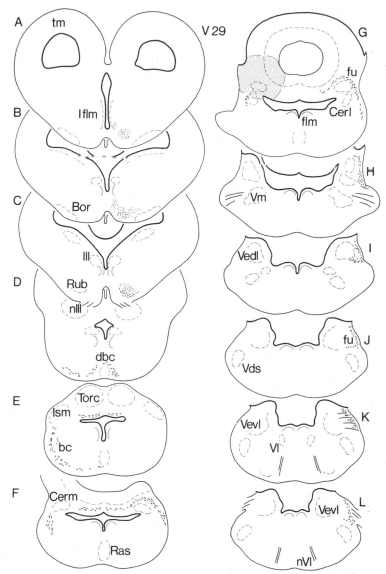

Fig. 7.34. *Varanus exanthematicus.* Diagrammatic representation of the course and distribution of anterogradely labeled fibers in the brainstem after injection of ³H-leucine into the cerebellar nuclei. Broken lines indicate labeled fibers; dots, evidence of labeled terminal structures. bc, Brachium conjunctivum; Bor, nucleus of the basal optic root; Cerl, Cerm, lateral and medial cerebellar nuclei; dbc, decussation of the brachium conjunctivum; flm, fasciculus longitudinalis medialis; fu, fasciculus uncinatus; Iflm, nucleus interstitialis of the flm; Ism, nucleus isthmi, pars magnocellularis; nIII, nervus oculomotorius; nVI, nervus abducens; Ras, nucleus raphes superior; Rub, nucleus ruber; tm, tectum mesencephali; Torc, torus semicircularis, nucleus centralis; Vedl, Vevl, dorsolateral and ventrolateral vestibular nuclei; III, nucleus nervi oculomotorii; Vds, nucleus descendens nervi trigemini; Vm, nucleus motorius nervi trigemini; VI, nucleus nervi abducentis.

In the red-eared turtle, *Pseudemys scripta elegans,* Künzle (1985b, 1985c) studied the projections from cerebellar nuclei with the powerful ^{35}S-methionine anterograde tracer technique. Fibers arising from the cerebellar nuclei were traced via the cerebellar commissure to the contralateral vestibular nuclear complex (particularly to the inferior and ventrolateral vestibular nuclei) and caudal rhombencephalic tegmentum. Ascending projections crossing the midline in the caudal part of the mesencephalic tegmentum terminated in the contralateral red nucleus and nuclei of the flm. Also a minor projection to the diencephalon (suprapeduncular nucleus and dorsal part of the hypothalamus) was found, in keeping with anterograde HRP data in another turtle (*Emys orbicularis;* M. G. Belekhova, personal communication, 1984). Künzle (1985c) also presented intriguing data indicating a projection to a lateral division of the anterior dorsal ventricular ridge (ADVR) arising from the region of the cerebellar and vestibular nuclear complexes. HRP injections into the ADVR in *Pseudemys scripta elegans,* however, revealed only a few retrogradely labeled neurons in the dorsolateral vestibular nucleus (Balaban and Ulinski, 1981). A "Fast Blue" study in *Varanus exanthematicus* (ten Donkelaar and de Boer-van Huizen, 1988) did not reveal ascending vestibular projections to the ADVR, but instead a major projection from the adjacent parabrachial region.

As already mentioned, it seems likely that the lateral cerebellar nucleus is the main site of origin of the brachium conjunctivum. HRP data are supported by recent findings in *Varanus exanthematicus* with the retrograde fluorescent tracer "Fast Blue" (Bangma et al., 1984). Fast Blue injections into the rostral mesencephalon and thalamus led to numerous retrogradely labeled cells in the contralateral lateral cerebellar nucleus, but only relatively few in the medial cerebellar nucleus. These results are in keeping with data in mammals, indicating a small ascending projection of the medial or fastigial nucleus (see Nakano et al., 1980; Bharos et al., 1981; Bentivoglio and Kuypers, 1982).

The medial cerebellar nucleus is probably the main site of origin of the fasciculus uncinatus. In various experiments in which an HRP slow-release gel was implanted into the vestibular nuclear complex of *Pseudemys scripta elegans* (Bangma and ten Donkelaar, 1983), *Varanus exanthematicus* (see Section VI.D), and *Python regius* (Bangma and ten Donkelaar, unpublished observations, 1983), a large number of retrogradely labeled cells was observed in the contralateral medial cerebellar nucleus. However, the lateral cerebellar nucleus also projects to the vestibular nuclear complex (see Figs. 7.37H and I, 7.35B). Such projections might also pass by way of the hook bundle, since this pathway is the only contralateral descending projection to the vestib-

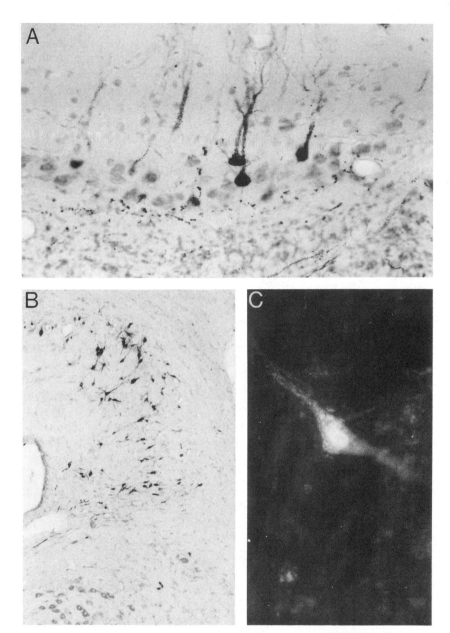

Fig. 7.35. *Varanus exanthematicus.* Retrogradely labeled cells. (A) After HRP slow-release gel implantation into the cerebellar nuclei (see Fig. 7.32): labeled Purkyně cells (300 ×). (B) After HRP slow-release gel implantation into the vestibular nuclear complex (see Fig. 7.33); labeled neurons in the contralateral medial and lateral cerebellar nuclei (75 ×). (C) After injection of the fluorescent tracer FB in the rostral mesencephalon and NY in the cervical spinal cord: double labeled neuron in the contralateral medial cerebellar nucleus (350 ×).

ular nuclear complex demonstrated in the anterograde degeneration and tracer studies (see Figs. 7.33 and 7.34).

The descending projections of the cerebellar nuclei are mainly aimed at the vestibular nuclear complex, although after HRP injections in *Varanus exanthematicus* restricted to the reticular formation, labeled cells also were observed in the contralateral medial cerebellar nucleus and even some in the lateral nucleus (ten Donkelaar and de Boer-van Huizen, unpublished observations, 1983). These findings are in keeping with previously mentioned data on projections of the cerebellar nuclei to the reticular formation in both *Pseudemys scripta elegans* (see also Künzle, 1985b) and *Varanus exanthematicus*.

HRP studies in *Python regius* (Bangma, 1983) showed only sparse projections from the lateral cerebellar nucleus to the contralateral tegmentum mesencephali. The medial cerebellar nucleus projects, predominantly contralaterally, to the vestibular nuclear complex. Also a distinct, bilateral projection of the lateral cerebellar nucleus was found to the vestibular nuclei. Both nuclei also project to the contralateral reticular formation. The medial cerebellar nucleus, as well as some large cells directly below this cerebellar nucleus (see Fig. 7.10B), project to the contralateral spinal cord (ten Donkelaar, 1982; ten Donkelaar et al., 1983). These large cells apparently also belong to the medial cerebellar nucleus.

Since, at least in *Pseudemys scripta elegans* and *Varanus exanthematicus*, both the medial and lateral cerebellar nuclei have ascending as well as descending projections, it is also of interest to establish whether single cerebellar neurons that give rise to fibers ascending to mesencephalon and beyond also give rise to descending collaterals to the medulla oblongata and spinal cord. The multiple retrograde fluorescent tracer technique (see Kuypers et al., 1980; Kuypers and Huisman, 1984) permits study of axonal branching in the cerebellar efferent fibers. In *Varanus exanthematicus* (Bangma et al., 1984) after injections of Fast Blue into thalamus and rostral mesencephalon and Nuclear Yellow into the rostral cervical cord, the pattern of cell labeling was in keeping with the HRP data discussed. However, considerably more neurons were labeled than in HRP studies. The lateral cerebellar nucleus appeared to have predominantly contralateral, ascending projections; fewer neurons with ascending projections were found in the medial nucleus. As regards the spinal projections, the pattern was the reverse. Double-labeled cells (Fast Blue in the cytoplasm; Nuclear Yellow in the nucleus) were observed in both cerebellar nuclei, but only contralaterally and mainly in the medial cerebellar nucleus (see Fig. 7.35C). These experiments indicate that at least part of the neurons in the cerebellar nuclei have both ascending and descending axon collaterals reaching the rostral tegmentum mes-

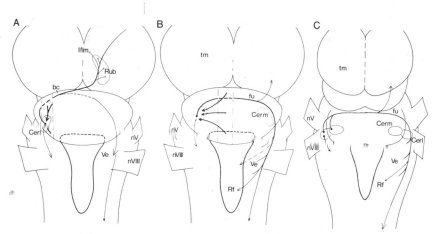

Fig. 7.36. *Varanus exanthematicus.* Diagrams summarizing the cerebellar corticonuclear projections to and the efferent connections of the lateral (A) and medial (B) cerebellar nuclei. (C) Summary of efferent connections of the cerebellar nuclei in *Python regius.* Shaded areas represent the zones of Purkyně cells projecting to the cerebellar nuclei. bc, Brachium conjunctivum; Cerl, Cerm, lateral and medial cerebellar nuclei; Rf, formatio reticularis; fu, fasciculus uncinatus; Iflm, interstitial nucleus of the flm; nV, nervus trigeminus; nVIII, nervus octavus; Rub, nucleus ruber; tm, tectum mesencephali; Ve, vestibular nuclear complex. (C, after Bangma, 1983.)

encephali and rostral spinal cord, respectively. These data are in keeping with recent findings in rats (Bentivoglio, 1982; Bentivoglio and Kuypers, 1982) and cats (Bharos et al., 1981).

In Fig. 7.36 the presently available data on the fiber connections of the reptilian cerebellar nuclei have been summarized for *Varanus exanthematicus* (A and B also apply to *Pseudemys scripta elegans*) and *Python regius* (Fig. 7.36C). In *Varanus exanthematicus* the lateral cerebellar nucleus (Fig. 7.36A), which receives afferents from a lateral strip of Purkyně cells, projects mainly to the contralateral red nucleus via the brachium conjunctivum. HRP studies also indicate projections to the contralateral diencephalon and telencephalon. In addition a well-developed contralateral descending projection arises in the lateral cerebellar nucleus, presumably passing via the hook bundle. This projection is aimed at the vestibular nuclear complex but probably also at the reticular formation. Even a small contralateral spinal projection was found comparable to the interpositospinal projection in mammals (e.g., Fukushima et al., 1977; Matsushita and Hosoya, 1978).

It is striking that apart from the well-developed ascending projection to the red nucleus, only sparse ascending projections beyond the mesodiencephalic junction were found (see, however, Künzle, 1985c). In contrast, the main part of the brachium conjunctivum in mammals, which arises in the lateral (or dentate) and interposed nuclei, proceeds farther rostrally to the thalamus. Fibers are mainly distributed to the ventrolateral and intralaminar nuclei (see Martin et al., 1974; Rinvik and Grofová, 1974; Chan-Palay, 1977; Faull and Carman, 1978; Stanton, 1980; Sugimoto et al., 1981; Ito, 1984).

Another rather striking difference from the mammalian projection pattern is the apparent absence of a contralateral descending limb of the brachium conjunctivum in reptiles. In mammals, the brachium conjunctivum gives off a small contralateral descending limb immediately rostral to its decussation. It is distributed predominantly to the pontine and medullary reticular formation (see Achenbach and Goodman, 1968; Martin et al., 1974; Chan-Palay, 1977; Faull, 1978). An ipsilateral descending limb of the brachium conjunctivum (see Faull, 1978) proceeds ventrally from the brachium conjuctivum before its decussation and terminates mainly in the lateral parvocellular reticular formation of the lower brain stem.

The efferent connections of the medial cerebellar nucleus in reptiles (Fig. 7.36B and C) show striking resemblances with the projections of the mammalian medial (or fastigial) nucleus. The reptilian medial cerebellar nucleus, which receives afferents from the most medial part of the cerebellar cortex, has mainly contralateral descending projections via the hook bundle to the vestibular nuclear complex, relatively sparse projections to the reticular formation, and a well-developed spinal projection. The crossed efferents of the mammalian fastigial nucleus form the uncinate tract, which hooks around the brachium conjunctivum and gives off an ascending and a descending branch (see Thomas et al., 1956; Cohen et al., 1958; Angaut and Bowsher, 1970; Batton et al., 1977). The ascending branch courses dorsal to the brachium conjunctivum. In the midbrain this branch does not distribute fibers to the red nucleus, but some of its fibers terminate in the mesencephalic reticular formation and in the deep layers of the superior colliculus (Angaut and Bowsher, 1970; Kievit and Kuypers, 1972; Batton et al., 1977). The ascending branch of the uncinate tract proceeds farther rostrally and terminates in the ventral thalamus (see also Martin et al., 1974; Haroian et al., 1981; Sugimoto et al., 1981).

The descending branch of the uncinate tract terminates in the vestibular nuclei and the medial pontine and medullary reticular formation (see Thomas et al., 1956; Cohen et al., 1958; Walberg et al., 1962a, 1962b; Martin et al., 1974; Batton et al., 1977). Some of the crossed descending efferents of the medial nucleus reach the spinal cord (e.g., Fukushima et al., 1977; Matsushita and Hosoya, 1978). The medial nucleus has in addition descending uncrossed efferents that are distributed primarily to the ipsilateral vestibular nuclei (Thomas et al., 1956; Walberg et al., 1962a; Batton et al., 1977).

In *Python regius* (Fig. 7.36C) only a small ascending projection of the lateral cerebellar nucleus was found, but no evidence for an ascending projection of the medial cerebellar nucleus.

D. Projections of Cerebellar Target Sites

The reptilian cerebellum sends its main output to the vestibular nuclear complex—directly (corticovestibular projection) as well as via the cerebellar nuclei. The direct corticovestibular projection is strictly ipsilateral; the projection from the cerebellar nuclei, mainly contralateral. The vestibular nuclear complex plays an important role in the control of spinal motor mechanisms such as posture (see ten Donkelaar, 1982), as well as in the control of eye movements (see ten Donkelaar, 1976b; Bangma and ten Donkelaar, 1983). HRP studies in the red-eared turtle, *Pseudemys scripta elegans,* showed distinct vestibulo-oculomotor projections (Bangma and ten Donkelaar, 1983): the dorsolateral vestibular nucleus projects ipsilaterally to the nuclei of the extrinsic eye muscles (III, IV, and VI); the ventromedial vestibular nucleus, mainly contralaterally. The ascending fibers to the third and fourth nerve nuclei pass via the fasciculus longitudinalis medialis (flm). Künzle's (1985b) ^{35}S-methionine data confirm these findings. Vestibulospinal projections in reptiles (see Robinson, 1969; ten Donkelaar, 1976a, 1976b, 1982; ten Donkelaar et al., 1980; Cruce and Newman, 1981; Woodson and Künzle, 1982; ten Donkelaar and de Boer-van Huizen, 1984) are organized as two pathways: the ipsilateral tractus vestibulospinalis lateralis arising in the large-celled nucleus vestibularis lateralis and the tractus vestibulospinalis medialis, which arises in the descending and ventromedial vestibular nuclei, mainly contralaterally.

Recent experiments in *Varanus exanthematicus* (ten Donkelaar et al., 1985) also emphasize the importance of vestibular control for eye movements and spinal motor mechanisms. Figs. 7.37 and 7.38 show results of an experiment in which an HRP slow-release gel was implanted into the ventrolateral vestibular nucleus; spread of the enzyme also took place to the dorsolateral, descending, and ventromedial vestibular nuclei, even some to the sixth nerve nucleus. The vestibulo-oculomotor projections are shown in Fig. 7.37; vestibulo-spinal projections, in Fig. 7.38.

Ascending projections were demonstrated to the ipsi- and contralateral nuclei of the trochlear nerve (Fig. 7.37G), to various parts of the oculomotor nucleus (different subdivisions of this nucleus receive ipsilateral or contralateral vestibulo-oculomotor projections, respectively; see Fig. 7.37D to F), and bilaterally to the interstitial nucleus of the flm. There are also extensive projections to the contralateral sixth nerve nucleus (Figs. 7.37L and 7.38A and B). In several cases in which the fluorescent marker Fast Blue was injected into the rostral tegmen-

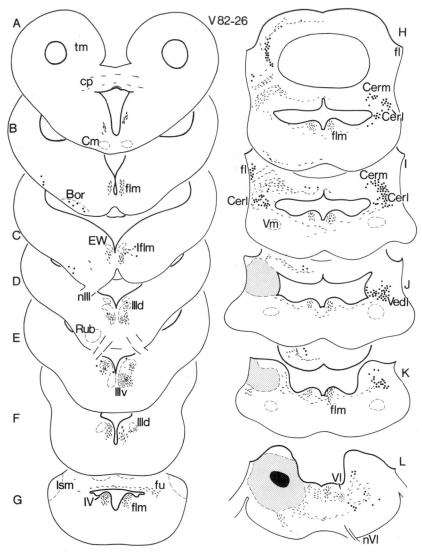

Fig. 7.37. *Varanus exanthematicus.* The distribution of labeled neurons and the course of anterogradely and retrogradely labeled fibers in the rostral and middle part of the brainstem following HRP slow-release gel implantation into the vestibular nuclear complex. Broken lines indicate labeled fibers; small dots indicate evidence of terminal structures. Bor, Nucleus of the basal optic root; Cerl, Cerm, lateral and medial cerebellar nuclei; Cm, corpus mamillare; cp, commissura posterior; EW, nucleus of Edinger-Westphal; fl, flocculus; flm, fasciculus longitudinalis medialis; fu, fasciculus uncinatus; Iflm, nucleus interstitialis of the flm; Ism, nucleus isthmi, pars magnocellularis; nIII, nervus oculomotorius; nVI, nervus abducens; Rub, nucleus ruber; tm, tectum mesencephali; Vedl, dorsolateral vestibular nucleus; IIId, IIIV, dorsal and ventral parts of the oculomotor nucleus; IV, nucleus nervi trochlearis; Vm, nucleus motorius nervi trigemini; VI, nucleus nervi abducentis.

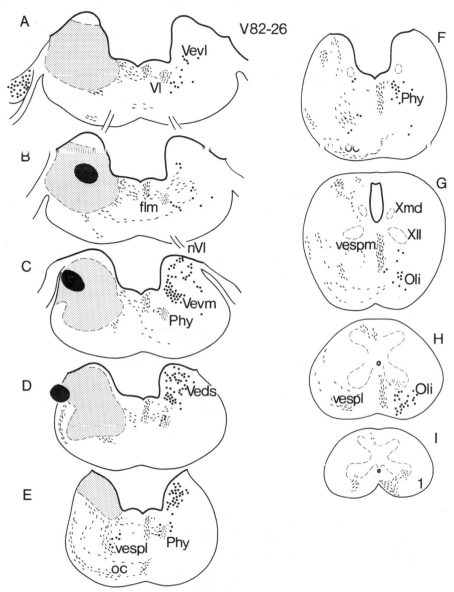

Fig. 7.38. *Varanus exanthematicus.* The distribution of labeled neurons and the course of anterogradely and retrogradely labeled fibers in the caudal part of the brainstem and the first segment of the spinal cord following HRP slow-release gel implantation into the vestibular nuclear complex. Broken lines indicate labeled fibers; small dots indicate evidence of terminal structures. flm, Fasciculus longitudinalis medialis; nVI, nervus abducens; oc, olivocerebellar fibers; Oli, oliva inferior; Phy, nucleus prepositus hypoglossi; Veds, Vevl, Vevm, descending, ventrolateral and ventromedial vestibular nuclei; vespl, vespm, lateral and medial vestibulospinal tracts; VI, nucleus nervi abducentis; Xmd, nucleus motorius dorsalis nervi vagi; XII nucleus nervi hypoglossi.

tum mesencephali (ten Donkelaar et al., 1985, 1987), labeled neurons were preferentially found in the ipsilateral dorsolateral vestibular nucleus and in the contralateral ventromedial vestibular nucleus. In addition, ascending projections of the descending and ventrolateral nuclei were found. These data suggest that, as in *Pseudemys scripta elegans* (Bangma and ten Donkelaar, 1983; Künzle, 1985b), ipsilateral ascending vestibulo-oculomotor fibers arise mainly from the dorsolateral vestibular nucleus. Both projections pass via the flm (Fig. 7.37).

Both vestibulospinal pathways—the ipsilateral lateral vestibulospinal and the contralateral medial vestibulospinal tract—were anterogradely labeled, although only as far as the rostral cervical cord (Fig. 7.38). The contralateral pathway particularly was extensively labeled, with fibers terminating directly on neck motoneurons (Fig. 7.38H and I). Such direct vestibulocollic projections have also been demonstrated in *Pseudemys scripta elegans* (Bangma and ten Donkelaar, 1983; Künzle, 1985c) and *Python regius* (ten Donkelaar et al., 1983). Direct projections of the medial vestibulospinal tract on neck motoneurons also occur in pigeons (Eden and Correia, 1982) and in mammals (see Wilson and Melvill Jones, 1979; Boyle and Pompeiano, 1981).

In anterograde degeneration and retrograde tracer studies (Robinson, 1969; Cruce, 1975, 1979; ten Donkelaar, 1976b; ten Donkelaar et al., 1980; ten Donkelaar, 1982) the ventrolateral vestibular nucleus was found to project ipsilaterally throughout the spinal cord to the ventromedial part of area VII–VIII. However, in *Tupinambis nigropunctatus* it also projects to the medial column of motoneurons innervating axial musculature (ten Donkelaar, 1976b). The descending and ventromedial vestibular nuclei project extensively to the contralateral cervical and thoracic cord, but there is also a distinct lumbar projection (ten Donkelaar et al., 1980; Wolters, de Boer-van Huizen, and ten Donkelaar, unpublished observations, 1982, of *Varanus exanthematicus*). Collateral branching in the spinal cord has been demonstrated for all these projections in *Varanus exanthematicus* with the multiple retrograde fluorescent tracer technique (Wolters, de Boer-van Huizen, ten Donkelaar, and Leenen, 1986).

Fig. 7.39 summarizes the main vestibular connections for *Varanus exanthematicus:* two parts of the cerebellar cortex, that is, a large, intermediate part of the corpus cerebelli and the flocculus project to the vestibular nuclear complex. Experiments in *Varanus exanthematicus* (Bangma and ten Donkelaar, 1984) suggest a corticovestibular organization as in *Pseudemys scripta elegans* (Bangma et al., 1983). The medial corticovestibular projection zone, with efferents mainly to the ventrolateral vestibular nucleus, generally exerts its major influence on the spinal cord via the lateral vestibulospinal tract but presumably also via the medial vestibulospinal tract. The lateral corticovestibular pro-

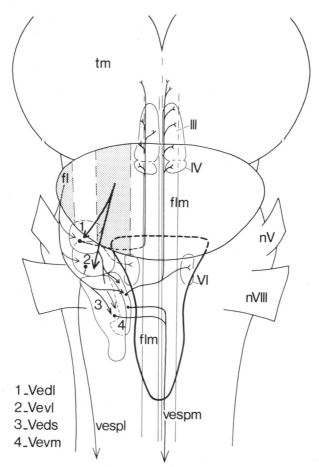

Fig. 7.39. *Varanus exanthematicus.* Diagrammatic representation summarizing the cerebellar corticovestibular projections and the efferent connections of various parts of the vestibular nuclear complex. Shaded areas represent the zones of Purkyně cells projecting to the vestibular nuclear complex. fl, Flocculus; flm, fasciculus longitudinalis medialis; nV, nervus trigeminus; nVIII, nervus octavus; tm, tectum mesencephali; Vedl, Veds, Vevl, Vevm, dorsolateral, descending, ventrolateral and ventromedial vestibular nuclei; vespl, vespm, lateral and medial vestibulospinal tracts; III, nucleus nervi oculomotorii; IV, nucleus nervi trochlearis; VI, nucleus nervi abducentis.

jection zone, the flocculus, supplies mainly other vestibular nuclei that are related to the vestibular influences on oculomotor mechanisms.

As indicated in Fig. 7.39, projections of the vestibular nuclei are mainly to the nuclei of the eye muscles (III, IV, and VI) and to the spinal cord. The dorsolateral vestibular nucleus projects ipsilaterally to these nuclei and to the ventromedial vestibular nucleus contralaterally. These ascending projections from the vestibular nuclei found in reptiles are in keeping with data in birds (*Gallus domesticus*, Wold, 1978) and mammals (see McMasters et al., 1966; Tarlov, 1970; Gacek, 1971).

Vestibulospinal projections are organized in two pathways: an ipsilateral tractus vestibulospinalis lateralis arising from the ventrolateral vestibular nucleus and a mainly contralateral tractus vestibulospinalis medialis originating in the descending and ventromedial vestibular nuclei passing via the flm. The vestibulospinal organization in reptiles is basically similar to that in higher vertebrates (see ten Donkelaar, 1976b, 1982; ten Donkelaar et al., 1980, 1987).

A distinct projection from the lateral cerebellar nucleus is to the contralateral red nucleus (Fig. 7.36B). A well-developed rubrospinal tract is found in quadrupedal reptiles (Robinson, 1969; ten Donkelaar, 1976a, 1976b, 1988; ten Donkelaar et al., 1980; Cruce and Newman, 1981; Woodson and Künzle, 1982) but apparently absent in limbless ones as *Python reticulatus* (ten Donkelaar, 1976a, 1976b) and *Python regius* (ten Donkelaar, 1982). In the experiment shown in Fig. 7.40, an electrolytic lesion in the tegu lizard, *Tupinambis nigropunctatus*, destroyed almost the entire red nucleus (ten Donkelaar, 1976b). A discrete bundle of degenerating descending fibers could be traced, decussating almost immediately (Fig. 7.40B), shifting dorsally along the periphery of the mesencephalic tegmentum, and finally taking a position just ventromedial to the descending tract of the trigeminal nerve (Fig. 7.40E). At caudal levels of the brain stem the rubrospinal tract shifts dorsally from its lateral position and finally takes a position in the most dorsal part of the lateral funiculus of the spinal cord. In its course through the brain stem, degenerating fibers leave the rubrospinal tract and can be traced to the lateral cerebellar nucleus (Fig. 7.40D) and to various nuclei belonging to the trigeminal nerve. Profuse preterminal degeneration has also been noted in the ventral part of the nucleus motorius nervi facialis and less densely in the parvocellular lateral part of the reticular formation. The rubrospinal tract terminates in the lateral part of area V–VI, most extensively within the cervical and lumbar enlargements. Comparable observations were made in *Varanus exanthematicus* after ^{3}H-leucine injections including the red nucleus (ten Donkelaar and Dederen, unpublished observations, 1982). In *Varanus exanthematicus* ipsilateral descending projections also arise from the red nucleus and course to the caudal brain stem (ten Donkelaar and de Boer-van Huizen, 1981a), suggesting a rubro-olivary projection. The red nucleus cannot be subdivided into magno- and parvocellular parts. In mammals, the magnocellular part projects to the spinal cord, the parvocellular part to the inferior olive (see Caughell and Flumerfelt, 1977). Some neurons in the red nucleus of *Varanus exanthematicus* branch to supply the cervical and

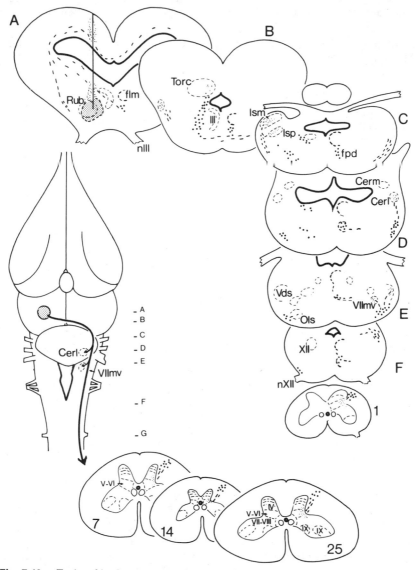

Fig. 7.40. *Tupinambis nigropunctatus.* A series of transverse sections through the brain stem and through representative levels of the spinal cord (segments 7, 14, 25), illustrating the ensuing anterograde degeneration following a lesion within the red nucleus. Coarse dots and broken lines indicate, respectively, transversally and longitudinally cut degenerating fibers, small dots evidence of preterminal degeneration. Cerl, Cerm, lateral and medial cerebellar nuclei; flm, fasciculus longitudinalis medialis; fpd, fasciculus predorsalis; Ism, Isp, magnocellular and parvocellular parts of the nucleus isthmi; nIII, nervus oculomotorius; nXII, nervus hypoglossus; Ols, oliva superior; Rub, nucleus ruber; Torc, torus semicircularis, nucleus centralis; Vds, nucleus descendens nervi trigemini; VIImv, nucleus motorius nervi facialis pars ventralis; III, nucleus nervi oculomotorii; IV, V-VI, VII-VIII, IX, subdivision of the spinal gray into areas according to Kusuma et al. (1979); XII, nucleus nervi hypoglossi. (After ten Donkelaar, 1976b.)

lumbar intumescence (Wolters, de Boer-van Huizen, ten Donkelaar, and Leenen, 1986).

In *Python regius,* no rubrospinal tract could be demonstrated in HRP studies (ten Donkelaar, 1982; ten Donkelaar et al., 1983). However, a distinct number of neurons was found labeled in the contralateral red nucleus after HRP gel implantations involving the descending trigeminal nucleus and nucleus motorius n. facialis (ten Donkelaar and Bangma, 1983), suggesting at least a contralateral rubrobulbar projection. These data suggest that rubral projections can be restricted to the brain stem—without extending into the spinal cord—and stress the importance of rubral control over brain stem mechanisms, particularly the facial motor nucleus. In reptiles, the facial motor nucleus is mainly involved in the control of jaw muscles such as the m. depressor mandibulae and several constrictors (Haas, 1973), which play an important role in the process of mastication (see Gorniak et al., 1982; Smith, 1982; Bramble and Wake, 1985).

It should be noted, however, that recently in a colubrid snake (*Natrix* [*Nerodia*] *sipedon*) after HRP injections into the spinal cord a tight contralateral cluster of small unipolar cells was observed in the tegmentum mesencephali (Cruce et al., 1983). Its size is smaller than that seen in many limbed reptiles; particularly striking is the small size of its neurons.

E. Summary

The cerebellar corticonuclear projections in various reptiles are organized as longitudinally oriented zones. Four longitudinal strips of Purkyně cells, each with a different target site, are found: from the midline lateralward corticonuclear projection zones to the medial cerebellar nucleus, the vestibular nuclear complex, the lateral cerebellar nucleus, and the vestibular nuclear complex. The lateral cerebellar nucleus projects mainly to the contralateral red nucleus via the brachium conjunctivum. In addition, a well-developed contralateral descending projection arises in the lateral cerebellar nucleus. The efferent connections of the medial cerebellar nucleus are mainly descending to the vestibular nuclear complex and relatively sparse to the reticular formation. These projections pass via the hook bundle. Also there is a small contralateral ascending projection. A distinct contralateral spinal projection of the medial cerebellar nucleus is also found in reptiles.

The influence of the cerebellum on spinal motor mechanisms is exerted via the rubrospinal, vestibulospinal, cerebellospinal, and, probably, the reticulospinal pathways.

VII. CEREBELLAR FUNCTION

The first adequate experiments to reveal the function of the cerebellum were carried out by Rolando (1809), who showed that removing parts or the whole cerebellum impaired ipsilateral motor activity. After ablation of the cerebellum in turtles, the animals remained completely paralyzed and lived for 10 to 12 days without making the slightest movement. Rolando erroneously assumed that the cerebellum was responsible for motor activity rather than a regulatory mechanism.

Later, Fano (1884) and Steiner (1886) studied the effects of cerebellar ablations in a turtle (*Emys orbicularis*) and a lizard (*Lacerta viridis*), respectively, but noted no effects on the behavior of these species. Leblanc (1923) studied the effects of cerebellar ablations, in *Uromastyx acanthinurus*, *Varanus griseus*, and *Chameleo vulgaris*, and observed, except in *Varanus griseus*, difficulties in gait and posture, described as ataxia, lack of coordination, and posture problems. He also observed limb tremors. The extent of the lesions was not verified histologically.

Hacker (1931) made cerebellar ablations in two lizard species, the limbless lizard *Ophisaurus apodus* and the quadrupedal *Lacerta viridis*. In *Ophisaurus*, he observed disturbances in locomotion, particularly a lack of coordination in serpentine movement. After two weeks, however, the normal pattern of movement returned. Hypotonia of the muscles was also observed.

In *Lacerta viridis*, the following symptoms were observed: hypotonia alternated with hypertonia, inability to keep the head up, ataxia, lack of coordination, tremor of the limbs. Furthermore, there was dysmetria: the fore- and hindlimbs stepped on each other during locomotion, and during catching of mealworms the animal moved too fast and thus reached beside or above the "target."

Unilateral cerebellar ablations in *Caiman crocodilus* have yielded much information (Goodman, 1969). These ablations involve either the functional vermal or paravermal zone. Lesions in the vermal zone (medial two-thirds of the corpus cerebelli) result in postural effects: ipsilateral limb extension and retraction and contralateral limb flexion and protraction. Sometimes the head and body are also affected posturally: a concavity occurred toward the ipsilateral side when the lesion was in the middle lobe and to the contralateral side when the posterior lobe was involved. The effect on the limbs of lesions in the paravermal zone (lateral one-third of corpus cerebelli) are a mirror image of those in the vermal zone. The head and body sometimes are concave to the ipsilateral side. Both forelimbs appear to be more affected by lesions in the paravermal zone than are the hindlimbs.

Lesions in the vermal zone cause a general hyporeflexia. Proprioceptive placing, pinch reflex, and muscle tone are generally reduced compared to the preoperative condition. The ipsilateral side of the body is more affected than the contralateral side. Lesions in the paravermal zone cause minimal or no detectable effects on the animal's reflexes. Dysmetria follows destruction of the cerebellar cortex in either functional zone but is more severe in animals with vermal zone lesions. The fore- and hindlimbs step on each other during locomotion, and normal coordination is generally disrupted in the stepping associated with forward locomotion.

Stimulation of the cerebellar cortex in the unanesthetized and unrestrained *Caiman crocodilus* elicits three definable postural patterns (Goodman and Simpson, 1960; Goodman, 1964, 1969):

1. Stimulation of the medial two-thirds of the middle lobe and the medial half of the posterior lobe (vermal zone pattern) results in ipsilateral forelimb flexion, adduction, and protraction; ipsilateral hindlimb flexion, adduction, and retraction; contralateral forelimb extension, abduction, and retraction; contralateral hindlimb extension, abduction, and protraction; ipsilateral turning of the head and concavity of the body and tail toward the ipsilateral side.

2. Stimulation of the flocculus results in a floccular zone pattern that is the mirror image of the vermal zone pattern; however, the head was rotated with the contralateral occiput down instead of laterally.

3. Stimulation of the lateral half of the anterior lobe, the lateral one-third of the middle lobe, and the lateral half of the posterior lobe results in a paravermal zone pattern. This consists of body and limb movements similar to those observed for the floccular zone pattern except that the head turns contralaterally.

An additional type of postural adjustment is concerned with the axial musculature. In this, rotation of the head, body, and tail about the longitudinal axis is superimposed upon all three postural patterns mentioned above: stimulation of the vermal zone is associated with clockwise axial rotation; the paravermal zone, with counterclockwise rotation; the floccular zone, with a "corkscrew" axial posture. These rotation movements along the longitudinal body axis caused by cerebellar stimulation may be related to movements peculiar to feeding and aggressive behavior of crocodilians. After a relatively large object has been seized and held firmly by the jaws, these rotate the entire body and head in an attempt to tear away a small piece (Goodman, 1969).

The above observations on the effects of cerebellar ablations and stimulation in various reptiles indicate that the reptilian cerebellum is responsible for the regulation of muscle tone and the coordination of movement and permits a rather precise control of both posture and locomotion. Major signs of cerebellar dysfunction in reptiles are, as in higher vertebrates (see Dow and Moruzzi, 1958; Gilman et al., 1981), loss of coordination—as expressed in ataxia, gait disturbances, and postural effects—and dysmetria. However, compensation occurs. Flourens (1824) showed recovery from motor disturbances produced by surgical lesions of the cerebellum. Negative findings after cerebellar ablations (Fano, 1884; Steiner, 1886; Leblanc, 1923) might be due to incomplete lesions and the ability of other parts of the cerebellum to substitute for the functions destroyed by the lesions.

However, careful behavioral studies after cerebellar ablation in various reptiles would yield a better understanding of cerebellar function in reptiles. It is hoped that in the near future such experimental approaches as reversible cooling of cerebellar nuclei (see Brooks, 1975, 1983) will be applied to reptiles. Furthermore, observations (with subsequent histological verification) of animals with obvious cerebellar symptoms could add to our present knowledge of the role of the cerebellum in motor control in reptiles. Van Valkenburg (1926) reported on an ependymal tumor that involved a large part of the cerebellum unilaterally and also the ipsilateral vestibular region.

As shown in Section VI the influence of the cerebellum on spinal motor mechanisms is exerted through vestibulospinal, rubrospinal, and probably reticulospinal tracts, but also directly via cerebellospinal pathways arising mainly in the medial cerebellar nucleus. The latter, contralateral, projection is presumably mainly for control of neck motoneurons. In cats (see Wilson et al., 1978) direct fastigiospinal connections to neck and shoulder motoneurons have been demonstrated.

The most medial zone of the cerebellar cortex, with projections to the medial cerebellar nucleus, mainly influences spinal motor mechanisms via vestibulospinal and reticulospinal pathways. The lateral part of the corpus cerebelli influences spinal motor mechanisms by way of its projections to the lateral cerebellar nucleus via the rubrospinal tract. Studies on the descending pathways to the spinal cord in various reptiles (Cruce, 1975; ten Donkelaar, 1976a, 1976b, 1982) show them to be classified into medial and lateral systems as advocated by Kuypers (Kuypers, 1964, 1973, 1981; Lawrence and Kuypers, 1968a, 1968b). The medial system terminates in the mediodorsal parts of the ventral horn and the adjacent parts of the intermediate zone. This system (interstitiospinal, reticulospinal, and vestibulospinal pathways) is functionally related to postural activities and progression and constitutes a basic system by which the brain exerts control over movements. The lateral system of brain stem pathways—the rubrospinal tract—terminates in lateral and dorsal parts of the intermediate zone. The latter system, at least in regard to the extremities, super-imposes upon the general motor control by the medial system the capacity for the independent use of the extremities, particularly of the hand. The classification of descending pathways into medial and lateral systems renders it likely that in snakes and limbless lizards the absence of a rubrospinal tract reflects the absence of limbs (see ten Donkelaar, 1982, 1988).

So it appears that a medial strip of Purkyně cells via the medial cerebellar nucleus—but presumably also via the direct corticovestibular projections arising in an intermediate strip of Purkyně cells—influences spinal motor mechanisms via the medial system of brain stem pathways. A lateral strip of Purkyně cells projects to the lateral cerebellar nucleus via the lateral system of brain stem pathways, at least in quadrupedal reptiles. Such an organization within the reptilian cerebellum is in keeping with the concept of localization proposed for the mammalian cerebellum (Chambers and Sprague 1955a, 1955b; see also Brooks and Thach, 1981; Ito, 1984). Reptiles also have two longitudinally organized corticonuclear zones such as those seen in cats. The first is the medial, vermal cortex and the fastigial nucleus, which are concerned with postural tonus, equilibrium, and locomotion of the entire body; the second is the intermediate, paravermal cortex and the interposed nuclei, which are involved in a more discrete control of the use of the ipsilateral limbs only, including the management of postural placing reflexes in these limbs (Chambers and Sprague, 1955a, 1955b). Mammals also have a lateral zone with projections via the dentate (or lateral) nucleus, which is involved in skilled and spatially organized movements of the ipsilateral limbs and face. Thus, the grade of organization of the cerebellum appears to be related to the complexity of motor performance the particular terrestrial vertebrates are capable of.

The ascending projections arising from the lateral cerebellar nucleus in *Python regius* might be aimed at the ill-defined red nucleus. Contralateral rubrobulbar projections as found in *Python regius* (see Section VI.D) are possibly related to a refined control of feeding mechanisms.

The flocculus exerts its influence via the vestibular nuclear complex and is closely connected with the vestibulo-ocular reflex (see Ito, 1982, 1984, for review). This is an elementary and phylogenetically old reflex that produces eye movements to compensate for head movements. Electrophysiological investigations in rabbits (Ito et al., 1970) and cats (Baker et al., 1972) revealed a specific connection from Purkyně cells of the flocculus to relay cells of the vestibulo-ocular reflex in vestibular nuclei.

VIII. CONCLUDING REMARKS
A. General

These remarks summarize our present knowledge of the connectivity of the reptilian cerebellum and provide a comparative perspective. Some remarks on basic features of cerebellar connections throughout terrestrial vertebrates are included.

B. Basic Features of Cerebellar Connectivity in Terrestrial Vertebrates

1. CEREBELLAR AFFERENTS   The cerebellum in reptiles is dominated by vestibular afferents and connections from the nucleus of the basal optic root and the nucleus prepositus hypoglossi, all mainly related to the control of eye, head, and neck movements.

A basic pattern of afferent cerebellar projections appears to be common to amphibians, reptiles, birds, and mammals, including retinal (via the nucleus of the basal optic root), vestibular, precerebellar (inferior olive), somatosensory, and spinal afferents. This relative wealth of information passing to the cerebellum indicates the extent to which the coordination of movement by the cerebellum depends on many factors.

Considerable differences appear to exist in the presence of the so-called precerebellar nuclei (nuclei that send most of their efferent fibers to the cerebellum) in amphibians, reptiles, birds, and mammals. Birds (see Clarke, 1977; Brecha et al., 1980) and mammals (see Brodal, 1981, for review) show cerebellar afferent projections of the pontine nuclei, mediating mainly telencephalic efferent impulses to the cerebellum. Reptiles and amphibians appear to lack a primordium of pontine nuclei relaying input from the telencephalon to the cerebellum. The inferior olivary nucleus is the only precerebellar nucleus common to amphibians, reptiles, birds, and mammals. Recently, an olivocerebellar projection has been shown in anurans (Cochran and Hackett, 1977, 1980; Grover and Grüsser-Cornehls, 1984; Gonzalez et al., 1984) and birds (Clarke, 1977; Feirabend et al., 1976; Freedman et al., 1977).

In *Caiman crocodilus* (see Brauth et al., 1978; Brauth and Kitt, 1980) and the pigeon *Columba livia* (Karten and Finger, 1976), pretectal nuclei were found to project to the cerebellum. In *Caiman crocodilus*, the nucleus circularis and the nucleus interstitialis of the posterior commissure project to the cerebellum (in *Columba livia*, to the nucleus spiriformis medialis). Fossil evidence indicates that of all living reptiles, crocodilians are most closely related to the reptilian ancestor from which birds evolved (Romer, 1959; Hodos, 1970; Walker, 1972). Therefore, since striking similarities exist in cerebellar connectivity as well as in the connections of the basal ganglia (see Brauth and Kitt, 1980;

Reiner et al., 1980, 1984), HRP studies in *Caiman crocodilus* might even show the presence of pontine nuclei in crocodilians.

2. CEREBELLAR CIRCUITRY

A basic pattern of neuronal connectivity is characterized by two sets of input: the mossy fiber–granule cell–parallel fiber system and the climbing fiber system, both of which are excitatory to the Purkyně cell. This basic cerebellar circuit is modulated by the action of inhibitory local circuit neurons (stellate, basket, and Golgi cells; see Hillman, 1969; Llinás and Hillman, 1969; Llinás, 1981) and probably also by a catecholaminergic inhibitory system. Efferents from the cerebellar cortex are formed solely by the Purkyně cells and their axons.

3. CORTICONUCLEAR PROJECTIONS

The projections of the Purkyně cells, the corticonuclear projections, are organized as longitudinally oriented zones. Four longitudinal strips of Purkyně cells, each with a different target site, are found: from the midline lateralward, corticonuclear projection zones are found to the medial cerebellar nucleus, to the vestibular nuclear complex (presumably mainly to the ventrolateral vestibular nucleus), to the lateral cerebellar nucleus, and from the flocculus (in *Varanus exanthematicus*) or lateral (floccular) zone (*Pseudemys scripta elegans, Python regius*) to other parts of the vestibular nuclear complex. This zonal pattern may be basic for terrestrial vertebrates. Experimental data are sparse for amphibians. Degeneration studies in *Rana pipiens* have demonstrated strictly ipsilateral projections to "subcerebellar" nuclei, mainly the vestibular region (Stern and Rubinson, 1971). HRP studies in *Rana pipiens* indicate a zonal subdivision of the corticonuclear projection (Grover, 1983). In birds, a longitudinal organization of the corticonuclear projections is suggested in both myeloarchitectonic studies (Feirabend et al., 1976; Feirabend, 1983) and degeneration and HRP experiments on cerebellovestibular projections (Wold, 1981). In mammals (see Voogd, 1964, 1967; Voogd and Bigaré, 1980; Haines et al., 1982), at least seven longitudinal zones can be distinguished. Of these, the most medial zone (zone A, the medial part of the vermis) projects to the nucleus fastigii; the lateral part of the vermis (zone B) projects to the lateral vestibular nucleus (of Deiters); the intermediate zone (C zones) to the interposed nuclei; and the lateral part of the cerebellar hemisphere (D zones) to the dentate nucleus.

Goodman (1964) presented comparative data (Fig. 7.41) on cerebellar stimulation experiments in the bullfrog, *Rana catesbeiana* (Goodman, 1958), *Caiman crocodilus* (Goodman and Simpson, 1960), the duck, *Anas domesticus* (Goodman et al., 1963b), and rats (Goodman

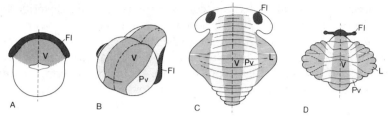

Fig. 7.41. Schematic representation showing the basic subdivisions of the cerebellum in terrestrial vertebrates. (A) The bullfrog *Rana catesbeiana*. (B) *Caiman crocodilus*. (C) The duck *Anas domesticus*. (D) The rat. Fl, floccular zone; L, lateral zone; Pv, paravermal zone; V, vermal zone. (A and B, after Goodman, 1969; C, after Goodman et al., 1963b; D, after Goodman and Simpson, 1961.)

and Simpson, 1961; see also Goodman et al., 1963a). Bullfrogs have only a vermal and a floccular zone pattern. *Caiman crocodilus* shows vermal, paravermal, and floccular patterns, but ducks and particularly rats show an additional lateral zone pattern. The aforementioned anatomical data on corticonuclear projections suggest that the vermal zone, as distinguished by Goodman, comprises both the projection zones to the medial cerebellar nucleus and the (ventro)lateral vestibular nucleus. The paravermal zone has projections to the lateral cerebellar nucleus in reptiles, the intermediate nucleus (and dorsal part of the lateral nucleus) in birds (see Wold, 1981), and the interposed nuclei in mammals. Amphibians have only one cerebellar nucleus (see Larsell, 1923; Nieuwenhuys and Opdam, 1976). The lateral zone, only found in birds and mammals, projects to the dentate (or lateral nucleus) and its avian homologue—the ventrolateral part of the lateral nucleus (see Larsell, 1967; Wold, 1981). A flocculus with its close connection with the vestibulo-occular reflex is found throughout terrestrial vertebrates.

4. Projections of the Cerebellar Nuclei
The present survey shows the reptilian lateral cerebellar nucleus to project mainly rostrally via the brachium conjunctivum and the medial cerebellar nucleus to project mainly caudally via the hook bundle. The lateral cerebellar nucleus influences spinal motor mechanisms predominantly via the lateral system of brain stem pathways and the medial cerebellar nucleus via the medial system.

It seems likely that terrestrial vertebrates show a similar basic projection pattern. Although amphibians show only a single cerebellar nucleus, frogs are capable of fairly sophisticated movements (Ebbesson, 1976). Recently, a rather well-developed rubrospinal projection has been demonstrated (*Xenopus laevis*, ten Donkelaar et al., 1981; see ten Donkelaar, 1982, for further data on rubrospinal projections in amphibians). A small brachium conjunctivum, already described by Larsell (1923), arises in the single cerebellar nucleus (Grover, 1983; Gonzalez et al., 1984; Larson-Prior and Cruce, 1984). In birds, a brachium conjunctivum arises from the lateral cerebellar nucleus (see Shimazono, 1912; Zecha, 1968; Verhaart, 1974). In the pigeon, *Columba livia*, Karten (1964; see also Karten and Revzin, 1966) showed with anterograde degeneration techniques a distinct projection arising in the cerebellar nuclei to the contralateral red nucleus. In mammals, the interposed nuclei project to the magnocellular part of the red nucleus and the dentate (or lateral) nucleus to the parvocellular part of the red nucleus (see Caughell and Flumerfelt, 1977). There are striking differences in the projection pattern of the brachium conjunctivum among reptiles, birds, and mammals. Reptiles have only a sparse projection beyond the mesencephalic border. Birds have a well-developed cerebellothalamic projection (Karten, 1964; Karten and Revzin, 1966), whereas the main part of the brachium conjunctivum of mammals proceeds further rostrally to the thalamus and reaches the precentral motor cortex.

A hook bundle or fasciculus uncinatus probably occurs throughout terrestrial vertebrates. Stern and Rubinson (1971) made cerebellar lesions in *Rana pipiens* and noted decussating fibers within the cerebellum passing to the contralateral vestibular region, suggesting a projection of the amphibian nucleus cerebelli to the contralateral vestibular region. Recent HRP findings in *Rana esculenta* (Grover, 1983) and in *Xenopus laevis* (Gonzalez et al., 1984) confirm this contralateral vestibular projection arising in the cerebellar nucleus. Reptiles (the present survey) have a clear uncinate tract arising in the medial cerebellar nucleus. It remains unclear how contralateral projections of the lateral cerebellar nucleus reach the vestibular region, although various experimental studies suggest that such projections also pass via the hook bundle. Wold (1981) showed bilateral projections of both the medial and intermediate nucleus to the vestibular nuclei in *Gallus domesticus*. In mammals the uncinate tract arises in the contralateral fastigial nucleus.

5. CONCLUSION

Reptiles show a pattern of cerebellar organization that is common to terrestrial vertebrates. This basic pattern in the organization of corticonuclear projections, as well as in pathways arising in the cerebellar nuclei, is superseded in mammals by the development of the lateral zone (the main part of the cerebellar hemisphere), which exerts its influence over spinal motor mechanisms mainly via the dentate nucleus and the corticospinal tract.

ACKNOWLEDGMENTS

The authors would like to thank Mrs. Roelie de Boer-van Huizen and Mr. Jos Dederen for their excellent technical assistance; Mr. Joep de Bekker, Mr. Joop Russon, and Mr. Chris van Huijzen for the drawing; and Mrs. Wanda de Haan, Mrs. Cocky Udo, and Mrs. Marion van de Coevering for typing the manuscript.

APPENDIX: REPTILIAN SPECIES DISCUSSED

TESTUDINES

Chrysemys marginata
  Larsell, 1932
  Weston, 1936
Chrysemys picta
  Parent, 1973
  Parent and Poirier, 1971
  Reiner, 1979
  Reiner and Karten, 1978
Clemmys japonica
  Ueda et al., 1983
Emys orbicularis
  Banchi, 1903a, 1903b
  Fano, 1884
Kinosternon leucostomum
  Miller and Kasahara, 1979
Pelusios subniger
  Pedersen, 1970
Pelomedusa subrufa
  Pedersen, 1970
Podocnemis unifilis
  Pedersen, 1970
Terrapene carolina
  Larsell, 1932
Testudo hermanni
  Cruce and Nieuwenhuys, 1974
  Gerrits and Voogd, 1973
  ten Donkelaar, 1976a, 1976b
Trachemys scripta ex *Pseudemys*
  Balaban and Ulinski, 1981
  Bangma and ten Donkelaar, 1983
  Bangma et al., 1983

Brand and Mugnaini, 1980
Fite et al., 1979
Künzle, 1982, 1983a, 1983b, 1985a,
  1985b, 1985c
Künzle and Wiklund, 1982
Künzle and Woodson, 1982
Kusuma and ten Donkelaar,
  1980a, 1980b
Kusuma et al., 1979
Mugnaini et al., 1974
Mugnaini and Dahl, 1975
Reiner, 1981
Reiner and Karten, 1968
Rüdeberg, 1961
Ruigrok, 1984
Schwarz and Schwarz, 1984
Walsh et al., 1974
Woodson and Künzle, 1982
Trachemys scripta elegans ex *Pseudemys*
  Bangma and ten Donkelaar, 1982
  Korte et al., 1980
  Künzle, 1982, 1983a, 1983b
  Künzle and Wiklund, 1982
  Künzle and Woodson, 1982
  Ruigrok, 1984
  Schwarz and Schwarz, 1980
Trionyx japonica
  Fuse, 1920
Turtle
  Bantli, 1974
  Fuse, 1920

CROCODILIA

Alligator mississippiensis
  Goodman and Simpson, 1960
  Huber and Crosby, 1926
  Llinás and Nicholson, 1969, 1971

Caiman crocodilus
  Brauth and Kitt, 1980
  Brauth et al., 1978
  Ebbesson and Goodman, 1981

Goodman, 1964, 1969
Goodman and Simpson, 1960
Hillman, 1969

Leake, 1974
Llinás and Hillman, 1969
Senn and Goodman, 1969

RHYNCHOCEPHALIA

Sphenodon punctatus
  Gorniak et al., 1982
  Hindendach, 1931

SAURIA

Anniella nigra
  Larsell, 1926
Chamaeleo vulgaris
  Leblanc, 1923
  P. Ramón, 1896
  Shanklin, 1933
Gallotia galloti
  Bangma and ten Donkelaar, 1982
  ten Donkelaar et al., 1978a, 1978b
  ten Donkelaar and de Boer-van
    Huizen, 1978b
Gongylus occellatus
  Rüdeberg, 1961
Iguana iguana
  Foster and Hall, 1978
  Rüdeberg, 1962
Lacerta stirpium
  Ramón y Cajal, 1911
Lacerta viridis
  Hacker, 1923
  Jacobs, 1968
  Kennedy et al., 1970
  Kitai et al., 1969
  Robinson, 1969
  Shimono et al., 1970
Ophisaurus apodus
  Hacker, 1931
Phrynosoma orbiculare
  Ochoterena, 1932
Podarcis muralis
  Beccari, 1911

Podarcis sicula
  Schwab and Durand, 1974
Tupinambis nigropunctatus
  Cruce, 1975
  Cruce and Newman, 1981
  Ebbesson, 1967
  Hoogland, 1977
  Lohman and van Woerden-
    Verkley, 1978
  ten Donkelaar 1976a, 1976b
  Kusuma and ten Donkelaar,
    1980a, 1980b
Uromastyx acanthinurus
  Leblanc, 1923
Varanus exanthematicus
  Bangma and ten Donkelaar, 1984
  Bangma et al., 1984
  Barbas-Henry and Lohman, 1988
  Smith, 1982
  ten Donkelaar et al., 1981a, 1981b,
    1984, 1985, 1987, 1988
  Voneida and Sligar, 1979
  Wolters et al., 1984, 1985, 1986
Varanus griseus
  Leblanc, 1923
Varanus salvator
  de Lange, 1917
Lizard
  Frederikse, 1931

SERPENTES

Boa constrictor
  Ebbesson, 1967, 1968
Eunectes murinus
  de Lange, 1917

Nerodia sipedon
  Cruce et al., 1983
  Ulinski, 1977

Python regius
  Bangma and ten Donkelaar, 1982
  ten Donkelaar, 1983
  ten Donkelaar et al., 1983
Python reticulatus
  ten Donkelaar, 1976a, 1976b

Kusuma and ten Donkelaar, 1980
Thamnophis sirtalis
  Jacobs and Sis, 1980
  Weston, 1936

REFERENCES

Achenbach, K. E., and Goodman, D. C. (1968). Cerebellar projections to pons, medulla and spinal cord in the albino rat. *Brain Behav. Evol.* 1, 43–57.

Angaut, P., and Bowsher, D. (1970). Ascending projections of the medial cerebellar (fastigial) nucleus: an experimental study in the cat. *Brain Res.* 24, 49–68.

Antal, M., Tornai, I., and Székely, G. (1980). Longitudinal extent of dorsal root fibres in the spinal cord and brain stem of the frog. *Neuroscience* 5, 1311–1322.

Ariëns Kappers, C. U. (1921). *Vergleichende Anatomie des Nervensystems.* E. G. Bohn, Haarlem, The Netherlands.

Ariëns Kappers, C. U. (1947). *Anatomie Comparée du Système Nerveux.* De Erven F. Bohn, Haarlem, The Netherlands.

Baker, R., Precht, W., and Llinás, R. (1972). Cerebellar modulatory action on the vestibulo-trochlear pathway in the cat. *Exp. Brain. Res.* 15, 364–385.

Balaban, C. D., and Ulinski, P. S. (1981). Organization of thalamic afferents to anterior dorsal ventricular ridge in turtles. I. Projections of thalamic nuclei. *J. Comp. Neurol.* 200, 95–129.

Banchi, A. (1903a). La minuta struttura della midollo spinale dei Chelonii (*Emys europaea*). *Arch. ital. Anat. Embriol.* 2, 291–307.

Banchi, A. (1903b). Sulle vie di connessione del cerveletto. *Arch. ital. Anat. Embriol.* 2, 426–517.

Bangma, G. C. (1983). "Cerebellar Connections in Some Reptiles." Thesis, University of Nijmegen, The Netherlands.

Bangma, G. C., and ten Donkelaar, H. J. (1982). Afferent connections of the cerebellum in various types of reptiles. *J. Comp. Neurol.* 207, 255–273.

Bangma, G. C., and ten Donkelaar, H. J. (1983). Some afferent and efferent connections of the vestibular nuclear complex in the red-eared turtle, *Pseudemys scripta elegans. J. Comp. Neurol.* 220, 453–464.

Bangma, G. C., and ten Donkelaar, H. J. (1984). Cerebellar afferents in the lizard *Varanus exanthematicus.* I. Corticonuclear projections. *J. Comp. Neurol.* 228, 447–459.

Bangma, G. C., ten Donkelaar, H. J., Dederen, P. J. W., and de Boer-van Huizen, R. (1984). Cerebellar efferents in the lizard *Varanus exanthematicus.* II. Projections of the cerebellar nuclei. *J. Comp. Neurol.* 230, 218–230.

Bangma, G. C., ten Donkelaar, H. J., and Pellegrino, A. (1983). Cerebellar corticonuclear projections in the red-eared turtle, *Pseudemys scripta elegans. J. Comp. Neurol.* 215, 258–274.

Bantli, H. (1974). Analysis of difference between potentials evoked by climbing fibers in cerebellum of cat and turtle. *J. Neurophysiol.* 37, 573–593.

Barbas-Henry, H. A., and Lohman, A. H. M. (1988). The primary projections and efferent cells of the VIIIth cranial nerve in the monitor lizard, *Varanus exanthematicus. J. Comp. Neurol.* 277, 234–249.

Batini, C., Buisseret-Delmas, C., Corvisier, J., Hardy, O., and Jassik-Gerschenfeld, D. (1978). Brain stem nuclei giving fibers to lobulus VI-VII of the cerebellar vermis. *Brain Res.* 153, 241–261.

Batton, R. R., III, Jayaraman, A., Ruggiero, D., and Carpenter, M. B. (1977). Fastigial efferent projections in the monkey: an autoradiographic study. *J. Comp. Neurol.* 174, 281–306.

Beccari, N. (1911). La costituzione, i nuclei terminali, e le vie di connessione del nervo acustico nella *Lacerta muralis*, Merr. *Arch. ital. Anat. Embriol.* 10, 646–698.

Beccari, N. (1923). Il centro tegmentale e interstitiale ed altre formazioni poco note nel mesencefalo e nel diencefalo di un rettile. *Arch. ital. Anat. Embriol.* 20, 560–619.

Bentivoglio, M. (1982). The organization of the direct cerebellospinal projections. In *Progress in Brain Research*, Vol. 57, *Descending Pathways to the Spinal Cord* (H. G. J. M. Kuypers and G. F. Martin, eds.). Elsevier Biomedical Press, Amsterdam, pp. 279–291.

Bentivoglio, M., and Kuypers, H. G. J. M. (1982). Divergent axon collaterals from rat cerebellar nuclei to diencephalon, mesencephalon, medulla oblongata and cervical cord. *Exp. Brain Res.* 46, 339–356.

Bharos, T. B., Kuypers, H. G. J. M., Lemon, R. N., and Muir, R. B. (1981). Divergent collaterals from deep cerebellar neurons to thalamus and tectum, and to medulla oblongata and spinal cord: retrograde fluorescent and electrophysiological studies. *Exp. Brain Res.* 42, 399–410.

Boyle, R., and Pompeiano, O. (1981). Relation between cell size and response characteristics of vestibulospinal neurons to labyrinth and neck inputs. *J. Neurosci.* 1, 1052–1066.

Bramble, D. M., and Wake, D. B. (1985). Feeding mechanisms of lower tetrapods. In *Functional Vertebrate Morphology* (M. Hildebrand, D. M. Bramble, K. F. Liem, and D. B. Wake, eds.). Harvard University Press, Cambridge, Mass., pp. 230–261.

Brand, S., and Mugnaini, E. (1980). Pattern of distribution of acetylcholinesterase in the cerebellar cortex of the pond turtle, with emphasis on parallel fibers. *Anat. Embryol.* 158, 271–287.

Brauth, S. E., and Kitt, C. A. (1980). The paleostriatal system of *Caiman crocodilus. J. Comp. Neurol.* 189, 437–465.

Brauth, S. E., Kitt, C. A., and Ferguson, J. L. (1978). The crocodilian midbrain tegmentum: a key to understanding the avian thalamus. *Soc. Neurosci. Abstr.* 4, 98.

Brecha, N., Karten, H. J., and Hunt, S. P. (1980). Projections of the nucleus of the basal optic root in the pigeon: an autoradiographic and horseradish peroxidase study. *J. Comp. Neurol.* 189, 615–670.

Brodal, A. (1952). Experimental demonstration of cerebellar connections from the perihypoglossal nuclei (nucleus intercalatus, nucleus praepositus hypoglossi and nucleus of Roller) in the cat. *J. Anat.* 86, 110–129.

Brodal, A. (1981). *Neurological Anatomy in Relation to Clinical Medicine.* 3rd ed. Oxford University Press, New York.

Brodal, A., and Szikla, G. (1972). The termination of the brachium conjunctivum descendens in the nucleus reticularis tegmenti pontis: an experimental anatomical study in the cat. *Brain Res.* 39, 337–351.

Brooks, V. B. (1975). Roles of cerebellum and basal ganglia in initiation and control of movements. *Can. J. Neurol. Sci.* 2, 265–277.

Brooks, V. B. (1983). Study of brain function by local reversible cooling. *Rev. Physiol. Biochem. Pharmacol.* 93, 1–109.

Brooks, V. B., and Thach, W. T. (1981). Cerebellar control of posture and movement. In *Handbook of Physiology—the Nervous System,* Sect. 1, Vol. II, *Motor Control* (V. B. Brooks, ed.). American Physiological Society, Bethesda, Md., pp. 877–946.

Brown-Gould, B. (1980). Organization of afferents from the brain stem nuclei to the cerebellar cortex in the cat. *Adv. Anat. Embryol. Cell. Biol.,* Vol. 62, Springer Verlag, Berlin, Heidelberg.

Caughell, K. A., and Flumerfelt, B. A. (1977). The organization of the cerebellorubral projection: an experimental study in the rat. *J. Comp. Neurol.* 176, 295–306.

Chambers, W. W., and Sprague, J. M. (1955a). Functional localization in the cerebellum. I. Organization in longitudinal cortico-nuclear zones and their contribution to the control of posture, both extrapyramidal and pyramidal. *J. Comp. Neurol.* 103, 105–129.

Chambers, W. W., and Sprague, J. M. (1955b). Functional localization in the cerebellum. II. Somatotopic organization in cortex and nuclei. *A.M.A. Arch. Neurol. Psychiatr.* 74, 653–680.

Chan-Palay, V. (1977). *Cerebellar Dentate Nucleus: Organization, Cytology and Transmitters.* Springer, Berlin-Heidelberg, New York.

Clarke, E., and O'Malley, C. D. (1968). *The Human Brain and Spinal Cord.* University of California Press, Berkeley.

Clarke, P. G. H. (1977). Some visual and other connections to the cerebellum of the pigeon. *J. Comp. Neurol.* 174, 535–552.

Cochran, S. L., and Hackett, J. T. (1977). The climbing fiber afferent system of the frog. *Brain Res.* 121, 362–367.

Cochran, S. L., and Hackett, J. T. (1980). Phylogenetically consistent features of cerebellar climbing fibers present in the tadpole. *Brain Res.* 192, 543–549.

Cohen, D., Chambers, W. W., and Sprague, J. M. (1958). Experimental study of the efferent projections from the cerebellar nuclei to the brainstem of the cat. *J. Comp. Neurol.* 109, 233–259.

Cruce, W. L. R. (1975). Termination of supraspinal descending pathways in the spinal cord of the tegu lizard (*Tupinambis nigropunctatus*). *Brain Behav. Evol.* 12, 247–269.

Cruce, W. L. R. (1979). Spinal cord in lizards. In *Biology of the Reptilia, Neurology B* (C. Gans, R. G. Northcutt, and P. S. Ulinski, eds.). Academic Press, London, 9, 111–131.

Cruce, W. L. R., Larson-Prior, L., and Newman, D. B. (1983). Rubrospinal pathways in a colubrid snake. *Soc. Neurosci. Abstr.* 9, 1064.

Cruce, W. L. R., and Newman, D. B. (1981). Brain stem origins of spinal projections in the lizard *Tupinambis nigropunctatus. J. Comp. Neurol.* 198, 185–207.

Cruce, W. L. R., and Nieuwenhuys, R. (1974). The cell masses in the brain stem of the turtle *Testudo hermanni:* a topographical and topological analysis. *J. Comp. Neurol.* 156, 277–306.

Cuello, A. C., and Kanazawa, I. (1978). The distribution of substance P immunoreactive fibers in rat central nervous system. *J. Comp. Neurol.* 178, 129–156.

de Lange, S. J. (1917). Das Hinterhirn, das Nachhirn und das Rückenmark der Reptilien. *Folia Neurobiol. (Lpz).* 10, 385–422.

Dietrichs, E. (1984). Cerebellar autonomic function: direct hypothalamo-cerebellar pathway. *Science* (N.Y.) 223, 591–593.

Dow, R. S., and Moruzzi, G. (1958). *The Physiology and Pathology of the Cerebellum.* University of Minnesota Press, Minneapolis.

Ebbesson, S. O. E. (1967). Ascending axon degeneration following hemisection of the spinal cord in the tegu lizard (*Tupinambis nigropunctatus*). *Brain Res.* 5, 178–206.

Ebbesson, S. O. E. (1969). Brain stem afferents from the spinal cord in a sample of reptilian and amphibian species. *Ann. N.Y. Acad. Sci.* 167, 80–101.

Ebbesson, S. O. E. (1976). Morphology of the spinal cord. In *Frog Neurobiology* (R. Llinás and W. Precht, eds.). Springer Verlag, Heidelberg, pp. 679–706.

Ebbesson, S. O. E., and Goodman, D. C. (1981). Organization of ascending spinal projections in *Caiman crocodilus. Cell Tissue Res.* 215, 383–395.

Eden, A. R., and Correia, M. J. (1982). An autoradiographic and HRP study of vestibulocollic pathways in the pigeon. *J. Comp. Neurol.* 211, 432–440.

Edinger, L. (1908). *Vorlesungen über den Bau der nervösen Zentralorganen des Menschen und der Tiere.* Vogel Verl., Leipzig, Germany.

Fano, G. (1884). Saggio sperimentale sul meccanismo dei movimenti volontari nella Testuggine palustre (*Emys europaea*). Pubbl. Ist. Stud. Super. Firenze, Firenze, Italy.

Faull, R. L. M. (1978). The cerebellofugal projections in the brachium conjunctivum of the rat: the ipsilateral and contralateral descending pathways. *J. Comp. Neurol.* 178, 519–536.

Faull, R. L. M., and Carman, J. B. (1978). The cerebellofugal projections in the brachium conjunctivum of the rat. I. The contralateral ascending pathway. *J. Comp. Neurol.* 178, 495–518.

Feirabend, H. K. P. (1983). "Anatomy and Development of Longitudinal Patterns in the Architecture of the Cerebellum of the White Leghorn *Gallus domesticus.*" Thesis, University of Leiden.

Feirabend, H. K. P., Vielvoye, G. J., Freedman, S. L., and Voogd, J. (1976). Longitudinal organization of afferent and efferent connections of the cerebellar cortex of the white leghorn (*Gallus domesticus*). *Exp. Brain. Res.,* Suppl. 1, 72–78.

Fink, R. P., and Heimer, L. (1967). Two methods for selective impregnation of degenerating axons and their synaptic endings in the central nervous system. *Brain Res.* 4, 369–374.

Fite, K. V., Reiner, A., and Hunt, S. P. (1979). Optokinetic nystagmus and the accessory optic system of pigeon and turtle. *Brain Behav. Evol.* 16. 192–202.

Flourens, P. (1824). *Recherches expérimentales sur les propriétés et les fonctions du système nerveux dans les animaux vertébrés.* Crevot, Paris.

Foster, R. E., and Hall, W. C. (1975). The connections and laminar organization of the optic tectum in a reptile (*Iguana iguana*). *J. Comp. Neurol.* 163, 397–426.

Foster, R. E., and Hall, W. C. (1978). The organization of central auditory pathways in a reptile *Iguana iguana. J. Comp. Neurol.* 178, 783–832.

Frederikse, A. (1931). "The Lizard's Brain." Thesis, University of Amsterdam.

Freedman, S. L., Voogd, J., and Vielvoye, G. J. (1977). Experimental evidence for climbing fibers in the avian cerebellum. *J. Comp. Neurol.* 175, 243–252.

Fukushima, K., Peterson, B. W., Uchino, Y., Coulter, J. D., and Wilson, V. J. (1977). Direct fastigiospinal fibers in the cat. *Brain Res.* 126, 538–542.

Fuse, G. (1920). Studien über die Kleinhirnrinde der Wirbeltiere. I. Die japanische Schildkröte. *Arb. Anat. Inst. Sendai* 5, 92–131.

Gacek, R. R. (1971). Anatomical demonstration of the vestibulo-ocular projections in the cat. *Acta Otolaryng. (Stockh.)*, Suppl. 283, 1–63.

Gerrits, N. M., and Voogd, J. (1973). The distribution of the Purkinje cells in the cerebellum of *Testudo hermanni* (turtle). *Acta Morphol. Neerl.-Scand.* 11, 357–358.

Gilman, S., Bloedel, J. R., and Lechtenberg, R. (1981). *Disorders of the Cerebellum.* F. A. Davis Co., Philadelphia.

Gonzalez, A., ten Donkelaar, H. J., and de Boer-van Huizen, R. (1984). Cerebellar connections in *Xenopus laevis:* an HRP study. *Anat. Embryol.* 169, 167–176.

Goodman, D. C. (1958). Cerebellar stimulation in the unanesthetized bullfrog. *J. Comp. Neurol.* 110, 321–336.

Goodman, D. C. (1964). The evolution of cerebellar structure and function. *Am. Zool.* 4, 33–36.

Goodman, D. C. (1969). Behavioral aspects of cerebellar stimulation and ablation in the frog and alligator and their relationship to cerebellar evolution. In *Neurobiology of Cerebellar Evolution and Development* (R. Llinás, ed.). American Medical Association, Chicago, pp. 467–473.

Goodman, D. C., Hallett, R. E., and Welch, R. B. (1963a). Patterns of localization in the cerebellar corticonuclear projections of the albino rat. *J. Comp. Neurol.* 121, 51–67.

Goodman, D. C., Horel, J. A., and Freemon, F. R. (1963b). Functional localization in the cerebellum of the bird and its bearing on the evolution of cerebellar function. *J. Comp. Neurol.* 123, 45–54.

Goodman, D. C., and Simpson, J. T. (1960). Cerebellar stimulation in the unrestrained and unanesthetized alligator. *J. Comp. Neurol.* 114, 127–136.

Goodman, D. C., and Simpson, J. T. (1961). Functional localization in the cerebellum of the albino rat. *Exp. Neurol.* 3, 174–188.

Gorniak, G. C., Rosenberg, H. I., and Gans, C. (1982). Mastication in the tuatara, *Sphenodon punctatus* (Reptilia: Rhynchocephalia): structure and activity of the motor system. *J. Morphol.* 171, 321–353.

Grant, G., Wiksten, B., Berkley, K. J., and Aldskogius, H. (1982). The localization of cerebellar-projecting neurons within the lumbrosacral spinal cord in the cat: an anatomical study with HRP and retrograde chromatolysis. *J. Comp. Neurol.* 204, 336–348.

Griffin, G., Watkins, L. R., and Mayer, D. J. (1979). HRP pellets and slow-release gels: two new techniques for greater localization and sensitivity. *Brain Res.* 168, 595–601.

Grover, B. G. (1983). Topographic organization of cerebellar efferents in the frog (*Rana esculenta*) as revealed by retrograde transport of wheat germ agglutinin-conjugated horseradish peroxidase. *Neurosci. Lett., Suppl.* 14, S146.

Grover, B. G., and Grüsser-Cornehls, U. (1984). Cerebellar afferents in the frogs *Rana esculenta* and *Rana temporaria*. *Cell Tissue Res.* 237, 259–267.

Haas, G. (1973). Muscles of the jaws and associated structures in the Rhynchocephalia and Squamata. In *Biology of the Reptilia, Morphology D* (C. Gans and T. S. Parsons, eds.). Academic Press, London, 4, 285–490.

Hacker, A. (1931). Zur Physiologie des Reptilienkleinhirnes. *Z. vergl. Physiol.* 15, 679–692.

Häggqvist, G. (1936). Analyse der Faserverteilung in einem Rückenmarkquerschnitt (Th. 3). *Z. mikr. -anat. Forsch.* 39, 1–34.

Haines, D. E., Patrick, G. W., and Satrulee, P. (1982). Organization of cerebellar corticonuclear fiber systems. In *The Cerebellum, New Vistas* (S. L. Palay and V. Chan-Palay, eds.). *Exp. Brain Res., Suppl.* 6, 320–371.

Haroian, A. J., Massopust, L. C., and Young, P. A. (1981). Cerebellothalamic projections in the rat: an autoradiographic and degeneration study. *J. Comp. Neurol.* 197, 217–236.

Hillman, D. E. (1969). Neuronal organization of the cerebellar cortex in Amphibia and Reptilia. In *Neurobiology of Cerebellar Evolution and Development* (R. Llinás, ed.). American Medical Association, Chicago, pp. 279–325.

Hindenach, J. C. R. (1931). The cerebellum of *Sphenodon punctatum*. *J. Anat.* (London) 65, 283–318.

Hodos, W. (1970). Evolutionary interpretation of neural and behavioral studies of living vertebrates. In *The Neurosciences*—Second Study Program (F. O. Schmitt, ed.). Rockefeller University Press, New York, pp. 26–39.

Hoffer, B. J., Siggins, G. R., and Bloom, F. E. (1971). Studies on norepinephrine-containing afferents to Purkinje cells of rat cerebellum. II. Sensitivity of Purkinje cells to norepinephrine and related substances administered by microiontophoresis. *Brain Res.* 25, 523–534.

Hökfelt, T., and Fuxe, K. (1969). Cerebellar monoamine nerve terminals, a new type of afferent fiber to the cortex cerebelli. *Exp. Brain Res.* 9, 63–72.

Hoogland, P. V. J. M. (1977). Efferent connections of the striatum in *Tupinambis nigropunctatus*. *J. Morphol.* 152, 229–246.

Huber, G. C., and Crosby, E. C. (1926). On thalamic and tectal nuclei and fiber paths in the brain of the American alligator. *J. Comp. Neurol.* 40, 97–227.

Huber, G. C., and Crosby, E. C. (1933). The reptilian optic tectum. *J. Comp. Neurol.* 57, 57–163.

Inakagi, S., Saknaka, M., Shiosaka, S., Senba, E., Takagi, H., Takatsuki, K., Kawai, Y., Matsuzaki, T., Iida, H., Hara, Y., and Tohyama, M. (1982). Experimental and immunohistochemical studies on the cerebellar substance P of the rat: localization, postnatal ontogeny and ways of entry to cerebellum. *Neuroscience* 7, 639–645.

Ingvar, S. (1918). Zur Phylo- und Ontogenese des Kleinhirns nebst ein Versuch zu einheitlicher Erklärung der zerebellaren Funktion und Lokulisation. *Folia neurobiol.* 11, 205–495.

Ito, H., Muzakami, T., and Morita, Y. (1982). An indirect telencephalo-cerebellar pathway and its relay nucleus in teleosts. *Brain Res.* 249, 1–13.

Ito, M. (1982). Cerebellar control of the vestibulo-ocular reflex-around the flocculus hypothesis. *Ann. Rev. Neurosci.* 5, 275–296.

Ito, M. (1984). *The Cerebellum and Neural Control.* Raven Press, New York.

Ito, M., Highstein, S. M., and Fukuda, J. (1970). Cerebellar inhibition of the vestibulo-ocular reflex in rabbit and cat and its blockage by picrotoxin. *Brain Res.* 17, 524–526.

Jacobs, V. L. (1968). An experimental study of the course and termination of the spinocerebellar systems in a lizard (*Lacerta viridis*). *Brain Res.* 11, 154–176.

Jacobs, V. L., and Sis, R. F. (1980). Ascending projections of the dorsal column in a garter snake (*Thamnophis sirtalis*): a degeneration study. *Anat. Rec.* 196, 37–50.

Jansen, J., and Brodal, A. (1940). Experimental studies on the intrinsic fibers of the cerebellum. II. The corticonuclear projection. *J. Comp. Neurol.* 73, 267–321.

Jansen, J., and Brodal, A. (1942). Experimental studies on the intrinsic fibers of the cerebellum. III. The corticonuclear projection in the rabbit and the monkey. *Skr. Norske Vidensk.-Akad., I. Mat.-nat. Kl.,* No. 3, pp. 1–50.

Joseph, B. S., and Whitlock, D. G. (1968a). Central projections of selected spinal dorsal roots in anuran amphibians. *Anat. Rec.* 160, 279–288.

Joseph, B. S., and Whitlock, D. G. (1968b). Central projections of brachial and lumbar dorsal roots in reptiles. *J. Comp. Neurol.* 132, 469–484.

Juh Shen Shyu (1939). On fiber connections of the reptilian cerebellum. *J. Orient. Med.* 31, 79–103.

Karten, H. J. (1964). Projections of the cerebellar nuclei of the pigeon (*Columba livia*). *Anat. Rec.* 148, 297–298.

Karten, H. J., and Finger, T. (1976). A direct thalamo-cerebellar pathway in pigeon and catfish. *Brain Res.* 102, 335–338.

Karten, H. J., and Revzin, A. M. (1966). The afferent connections of the nucleus rotundus in the pigeon. *Brain Res.* 2, 368–377.

Kawakami, M. (1954). Contributions to the comparative anatomy of the cerebellar fiber connections in reptiles. *Hiroshima J. Med. Sci.* 2, 295–317.

Kennedy, D. T., Shimono, T., and Kitai, S. T. (1970). Parallel fiber and white matter activation of Purkinje cells in a reptilian cerebellum (*Lacerta viridis*). *Brain Res.* 22, 381–385.

Kievit, J., and Kuypers, H. G. J. M. (1972). Fastigial cerebellar projections to the ventrolateral nucleus of the thalamus and the organization of the descending pathways. In *Corticothalamic Projections and Sensorimotor Activities* (T. Frigyesi, E. Rinvik, and M. D. Yahr, eds). Raven Press, New York, pp. 91–114.

Kitai, S. T., Shimono, T., and Kennedy, D. T. (1969). Inhibition in the cerebellar cortex of the lizard *Lacerta viridis*. In *Neurobiology of Cerebellar Evolution and Development* (R. Llinás, ed.). American Medical Association, Chicago, pp. 481–489.

Klüver, H., and Barrera, E. (1953). A method for the combined staining of cells and fibers in the central nervous system. *J. Neuropathol. Exp. Neurol.* 12, 400–403.

Kooy, F. H. (1917). The inferior olive in vertebrates. *Folia neurobiol.* 10, 205–259.

Korte, G. E., Reiner, A., and Karten, H. J. (1980). Substance P-like immunoreactivity in cerebellar mossy fibers and terminals in the red-eared turtle *Chrysemys scripta elegans*. *Neuroscience* 5, 903–914.

Kotchabhakdi, N., and Walberg, F. (1978). Cerebellar afferent projections from the vestibular nuclei in the cat. *Exp. Brain Res.* 31, 591–604.

Krabbe, K. H. (1939). *Studies on the Morphogenesis of the Brain in Reptiles.* Einar Munksgaard, Copenhagen.

Künzle, H. (1982). Dorsal root projections to the cerebellum in turtle. *Exp. Brain Res.* 45, 464–466.

Künzle, H. (1983a). Supraspinal cell populations projecting to the cerebellar cortex in the turtle, *Pseudemys scripta elegans*. *Exp. Brain Res.* 49, 1–12.

Künzle, H. (1983b). Spinocerebellar projections in the turtle: observations on their origin and terminal organization. *Exp. Brain Res.* 53, 129–141.

Künzle, H. (1985a). Climbing fiber projection to the turtle cerebellum: longitudinally oriented terminal zones within the basal third of the molecular layer. *Neuroscience* 14, 159–168.

Künzle, H. (1985b). The cerebellar and vestibular nuclear complexes in the turtle. I. Projections to mesencephalon, rhombencephalon and spinal cord. *J. Comp. Neurol.* 242, 102–121.

Künzle, H. (1985c). The cerebellar and vestibular nuclear complexes in the turtle. II. Projections to the prosencephalon. *J. Comp. Neurol.* 242, 122–133.

Künzle, H., and Wiklund, L. (1982). Identification and distribution of neurons presumed to give rise to cerebellar climbing fibers in turtle: a retrograde axonal flow study using radioactive D-aspartate as a marker. *Brain Res.* 252, 146–150.

Künzle, H., and Woodson, W. (1982). Mesodiencephalic and other target regions of ascending spinal projections in the turtle, *Pseudemys scripta elegans*. *J. Comp. Neurol.* 212, 349–364.

Kusuma, A., and ten Donkelaar, H. J. (1980). Dorsal root projections in various types of reptiles. *Brain Behav. Evol.* 17, 291–307.

Kusuma, A., ten Donkelaar, H. J., and Nieuwenhuys, R. (1979). Intrinsic organization of the spinal cord. In *Biology of the Reptilia, Neurology B* (C. Gans, R. G. Northcutt, and P. S. Ulinski, eds.). Academic Press, London, 9, 59–109.

Kuypers, H. G. J. M. (1964). The descending pathways to the spinal cord, their anatomy and function. In *Progress in Brain Research*, Vol. 11, *Organization of the Spinal Cord* (J. C. Eccles and J. P. Schadé, eds.). Elsevier, Amsterdam, pp. 178–202.

Kuypers, H. G. J. M. (1973). The anatomical organization of the descending pathways and their contributions to motor control especially in primates. In *New Developments in Electromyography and Clinical Neurophysiology*, Vol. 3 (J. E. Desmedt, ed.). Karger, Basel, Switzerland, pp. 38–68.

Kuypers, H. G. J. M. (1981). Anatomy of the descending pathways. In *Handbook of Physiology—the Nervous System*, Sect. 1, Vol. II, Motor Control (V. D. Brooks, ed.). American Physiological Society, Bethesda, Md., pp. 597–666.

Kuypers, H. G. J. M., and Huisman, A. M. (1984). Fluorescent neuronal tracers. In *Labeling Methods Applicable to the Study of Neuronal Pathways* (S. Fedoroff, ed.). Academic Press, New York, pp. 307–340.

Kuypers, H. G. J. M., Bentivoglio, M., Catsman-Berrevoets, C. E., and Bharos, A. T. (1980). Double retrograde neuronal labeling through divergent axon collaterals, using two fluorescent tracers with the same excitation wavelength which label different features of the cell. *Exp. Brain Res.* 40, 383–392.

Larsell, O. (1923). The cerebellum of the frog. *J. Comp. Neurol.* 36, 89–112.

Larsell, O. (1926). The cerebellum of reptiles: lizards and snakes. *J. Comp. Neurol.* 41, 59–94.

Larsell, O. (1932). The cerebellum of reptiles: chelonians and alligator. *J. Comp. Neurol.* 56, 299–345.

Larsell, O. (1967). *The Comparative Anatomy and Histology of the Cerebellum from Myxinoids through Birds*. The University of Minnesota Press, Minneapolis.

Larson-Prior, L., and Cruce, W. L. R. (1984). A reciprocal connection between cerebellum and nucleus ruber in a frog (*Rana pipiens*). *Anat. Rec.* 208, 101A.

Lawrence, D. G., and Kuypers, H. G. J. M. (1968a). The functional organization of the motor system in the monkey. I. The effects of bilateral pyramidal lesions. *Brain* 91, 1–14.

Lawrence, D. G., and Kuypers, H. G. J. M. (1968b). The functional organization of the motor system in the monkey. II. The effects of lesions of the descending brain-stem pathways. *Brain* 91, 15–36.

Leake, P. A. (1974). Central projections of the stato-acoustic nerve in *Caiman crocodilus*. *Brain Behav. Evol.* 10, 170–196.

Leblanc, E. (1923). L'acérébellation expérimentale chez les lézards. *C.r. Acad. Sci.* (Paris), 176, 1182–1184.

Llinás, R. R. (1981). Electrophysiology of the cerebellar networks. In *Handbook of Physiology—The Nervous System*, Sect. 1, Vol. II, Motor Control (V. B. Brooks, ed.). American Physiological Society, Bethesda, Md., pp. 831–876.

Llinás, R., and Hillman, D. E. (1969). Physiological and morphological organization of the cerebellar circuits in various vertebrates. In *Neurobiology of Cerebellar Evolution and Development* (R. Llinás, ed.). American Medical Association, Chicago, pp. 43–73.

Llinás, R., and Nicholson, C. (1969). Electrophysiological analysis of alligator cerebellar cortex: a study on dendritic spikes. In *Neurobiology of Cerebellar Evolution and Development* (R. Llinás, ed.). American Medical Association, Chicago, pp. 431–465.

Llinás, R., and Nicholson, C. (1971). Electrophysiological properties of dendrites and somata in alligator Purkinje cells. *J. Neurophysiol.* 34, 532–551.

Ljungdahl, A., Hökfelt, T., and Nilsson, G. (1978). Distribution of substance P-like immunoreactivity in the central nervous system of the rat. I. Cell bodies and nerve terminals. *Neuroscience* 3, 861–943.

Lohman, A. H. M., and van Woerden-Verkley, I. (1978). Ascending connections to the forebrain in the tegu lizard. *J. Comp. Neurol.* 182, 555–594.

Luciani, L. (1891). *Il Cervelletto: Nuovi studi de fisiologia normale e patologice.* Le Monnier, Firenze, Italy.

Martin, G. F., King, J. S., and Dom, R. (1974). The projections of the deep cerebellar nuclei of the opossum, *Didelphis marsupialis virginiana. J. Hirnforsch.* 15, 545–573.

Matsushita, M., and Hosoya, Y. (1978). The location of spinal projection neurons in the cerebellar nuclei (cerebellospinal tract neurons) of the cat. A study with the horseradish peroxidase technique. *Brain Res.* 142, 237–248.

Matsushita, M., and Hosoya, Y. (1979). Cells of origin of the spinocerebellar tract in the rat, studied with the method of retrograde transport of horseradish peroxidase. *Brain Res.* 173, 185–200.

Matsushita, M., Hosoya, Y., and Ikeda, M. (1979). Anatomical organization of the spinocerebellar system in the cat, as studied by retrograde transport of horseradish peroxidase. *J. Comp. Neurol.* 184, 81–106.

Matsushita, M., Ikeda, M., and Okado, N. (1982). The cells of origin of the trigeminothalamic, trigeminospinal and trigeminocerebellar projections in the cat. *Neuroscience* 7, 1439–1454.

McCrea, R. A., and Baker, R. (1985). The anatomical connections of the nucleus prepositus hypoglossi of the cat. *J. Comp. Neurol.* 237, 377–407.

McMasters, R. E., Weiss, A. H., and Carpenter, M. B. (1966). Vestibular projections to the nuclei of the extraocular muscles. Degeneration resulting from discrete partial lesions of the vestibular nuclei in the monkey. *Am. J. Anat.* 118, 163–194.

Miller, M. R., and Kasahara, M. (1979). The cochlear nuclei of some turtles. *J. Comp. Neurol.* 185, 221–236.

Mugnaini, E., Atluri, R. L., and Houk, J. C. (1974). Fine structure of granular layer in turtle cerebellum with emphasis on large glomeruli. *J. Neurophysiol.* 37, 1–29.

Mugnaini, E., and Dahl, A. L. (1975). Mode of distribution of aminergic fibers in the cerebellar cortex of the chicken. *J. Comp. Neurol.* 162, 417–432.

Nakano, K., Takimoto, T., Kayahara, T., Takeuchi, Y., and Kobayashi, Y. (1980). Distribution of cerebellothalamic neurons projecting to the ventral nuclei of the thalamus: an HRP-study in the cat. *J. Comp. Neurol.* 194, 427–439.

Nauta, W. J. H., and Gygax, P. A. (1954). Silver impregnation of degenerating axons in the central nervous system: a modified technique. *Stain Technol.* 29, 91–93.

Nieuwenhuys, R. (1967). Comparative anatomy of the cerebellum. In *Progress in Brain Research* Vol. 25, *The Cerebellum* (C. A. Fox and R. S. Snider, eds.). Elsevier, Amsterdam, pp. 1–93.

Nieuwenhuys, R., and Opdam, P. (1976). Structure of the brain stem. In *Frog*

Neurobiology (R. Llinás and W. Precht, eds.). Springer Verlag, Heidelberg, pp. 811–855.

Ochoterena, I. (1932). Histologia del cerebelo del Tepayaxin (*Phrynosoma orbiculare*, Wieg). *An. Inst. Biol. (Méx).* 3, 81–94.

Olson, L., and Fuxe, U. (1971). On the projections from the locus coeruleus noradrenaline neurons: the cerebellar innervation. *Brain Res.* 43, 289–295.

Papez, J. K. (1929). *Comparative Neurology.* Thomas Y. Crowell, New York.

Parent, A. (1973). Distribution of monoamine-containing nerve terminals in the brain of the painted turtle, *Chrysemys picta. J. Comp. Neurol.* 148, 157–166.

Parent, A. (1979). Monoaminergic systems of the brain. In *Biology of the Reptilia, Neurology B* (C. Gans, R. G. Northcutt, and P. S. Ulinski, eds.). Academic Press, London, pp. 247–285.

Parent, A., and Poirier, L. J. (1971). Occurrence and distribution of monoamine containing neurons in the brain of the painted turtle, *Chrysemys picta. J. Anat.* 110, 81–90.

Pedersen, R. (1973). Ascending spinal projections in three species of sidenecked turtle: *Podocnemis unifilis, Pelusios subniger,* and *Pelomedusa subrufa. Anat. Rec.* 175, 409.

Ramón, P. (1896). Las células estrelladas de la capa molecular de cerebelo de los reptiles. *Rev. Trim. Micrográf.* (Cited by Ramón y Cajal, 1909–1911).

Ramón y Cajal, S. (1911). *Histologie du Système Nerveux de l'Homme et des Vertébrés,* Vol. II, A. Maloine, Paris.

Reese, A. M. (1915). *The Alligator and Its Allies.* G. P. Putman's Sons, London.

Reiner, A. (1979). The paleostriatal complex in turtles. *Soc. Neurosci. Abstr.* 5, 146.

Reiner, A. (1981). A projection of displaced ganglion cells and giant ganglion cells to the accessory optic nuclei in the turtle. *Brain Res.* 204, 403–409.

Reiner, A., Brauth, S. E., Kitt, C. A., and Karten, H. J. (1980). Basal ganglionic pathways to the tectum: studies in reptiles. *J. Comp. Neurol.* 193, 565–589.

Reiner, A., Brauth, S. E., and Karten, H. J. (1984). Evolution of the amniote basal ganglia. *Trends Neurosci.* 7, 320–325.

Reiner, A., and Karten, H. J. (1978). A bisynaptic retinocerebellar pathway in the turtle. *Brain Res.* 150, 163–196.

Rinvik, E., and Grofová, I. (1974). Cerebellar projections to the nuclei ventralis lateralis and ventralis anterior thalami. Experimental electron microscopical and light microscopical studies in the cat. *Anat. Embryol.* 146, 95–111.

Robinson, L. R. (1969). Bulbospinal fibres and their nuclei of origin in *Lacerta viridis* demonstrated by axonal degeneration and chromatolysis respectively. *J. Anat.* 105, 59–88.

Rolando, L. (1809). *Saggio sopra la vera struttura del cervello dell'uomo e degli animali e sopra le funzioni del sistema nervosa.* Stampeia da S. S. R. M. Privilegiata, Sassari, Italy.

Romer, A. S. (1959). *Vertebrate Paleontology,* 3rd ed. University of Chicago Press, Chicago.

Rüdeberg, S.-I. (1961). "Morphogenetic Studies on the Cerebellar Nuclei and

Their Homologization in Different Vertebrates Including Man." Thesis, University of Lund, Sweden.

Rüdeberg, S.-I. (1962). Formation of the embryonic migration layers in the cerebellar anlage of the reptile *Iguana iguana*. *Acta Anat.* 51, 329–337.

Ruigrok, T. J. H. (1984). "Organization and Morphology of Motoneurons and Primary Afferents in the Lumbar Spinal Cord of the Turtle *Pseudemys scripta elegans*." Thesis, University of Utrecht, The Netherlands.

Schwab, M. E. (1979). Variation in the rhombencephalon. In *Biology of the Reptilia, Neurology B* (C. Gans, R. G. Northcutt, and P. S. Ulinski, eds.). Academic Press, London, pp. 201–246.

Schwab, M. E., and Durand, M. (1974). An autoradiographic study of neuroblast proliferation in the rhombencephalon of a reptile, *Lacerta sicula*. *Z. Anat. Entw.-Gesch.* 145, 29–40.

Schwarz, I. E., and Schwarz, D. W. F. (1980). Afferents to the cerebellar cortex of turtles studied by means of the horseradish peroxidase technique. *Anat. Embryol.* 160, 39–52.

Senn, D. G. (1979). Embryonic development of the central nervous system. In *Biology of the Reptilia, Neurology A* (C. Gans, R. G. Northcutt, and P. S. Ulinski, eds.). Academic Press, London, pp. 173–244.

Senn, D., and Goodman, D. C. (1969). Patterns of localization in the cerebellar corticofugal projections of the alligator (*Caiman sclerops*). In *Neurobiology of Cerebellar Evolution and Development* (R. Llinás, ed.). American Medical Association, Chicago, pp. 475–480.

Shanklin, W. M. (1930). The central nervous system of *Chameleon vulgaris*. *Acta Zool.* 11, 425–491.

Shanklin, W. M. (1933). The comparative neurology of the nucleus opticus tegmenti with special reference to *Chameleon vulgaris*. *Acta Zool.* 14, 163–184.

Shimazono, J. (1912). Das Kleinhirn der Vögel. *Arch. mikr. Anat.* 80, 397–449.

Shimono, T., Kennedy, D. T., and Kitai, S. (1970). Field potential analysis of the inhibitory pattern in a reptilian cerebellar cortex (*Lacerta viridis*). *Brain Res.* 22, 386–391.

Singh, U., and Chaturvedi, R. P. (1979). Purkinje cells of the cerebellum in some species of Indian and East African snakes. *J. Anat.* (London) 129, 203.

Smith, K. K. (1982). An electromyographic study of the function of the jaw adducting muscles in *Varanus exanthematicus* (Varanidae). *J. Morphol.* 173, 137–158.

Stanton, G. B. (1980). Topographical organization of ascending cerebellar projections from the dentate and interposed nuclei in *Macaca mulatta*: an anterograde degeneration study. *J. Comp. Neurol.* 190, 699–731.

Stefanelli, A. (1944a). I centri statici e della coordinazione motoria dei rettili. *Commentat. pontif. Acad. Scient.* 8, 147–293.

Stefanelli, A. (1944b). La fisiologia dei centri statici alla luce delle ricerche di morfologia ecologia dei rettili. *Arch. Fisiol.* 44, 49–77.

Steiner, J. (1886). *Die Funktionen des Zentralnervensystems und ihre Phylogenese. IV. Abt. Reptilien.* Braunschweig, Germany.

Stern, T. A., and Rubinson, K. (1971). Efferent projections of the cerebellar cortex of *Rana pipiens*. *Anat. Rec.* 169, 438.

Stieda, L. (1875). Ueber den Bau des centralen Nervensystems der Amphibien und Reptilien. Z. wiss. Zool. 25, 1–74.

Sugimoto, T., Mizuno, N., and Itoh, K. (1981). An autoradiographic study on the terminal distribution of cerebellothalamic fibers in the cat. Brain Res. 215, 29–47.

Székely, G., Antal, M., and Görcs, T. (1980). Direct dorsal root projection onto the cerebellum in the frog. Neurosci. Lett. 19, 161–165.

Taber Pierce, E., Hoddevik, G. H., and Walberg, F. (1977). The cerebellar projection from the raphe nuclei in the cat as studied with the method of retrograde transport of horseradish peroxidase. Anat. Embryol. 152, 73–87.

Tarlov, E. (1970). Organization of vestibulo-oculomotor projections in the cat. Brain Res. 20, 159–179.

ten Cate, J. (1937). Physiologie des Zentralnervensystems der Reptilien. Ergebn. Biol. 14, 225–279.

ten Donkelaar, H. J. (1976a). Descending pathways from the brain stem to the spinal cord in some reptiles. I. Origin. J. Comp. Neurol. 167, 421–442.

ten Donkelaar, H. J. (1976b). Descending pathways from the brain stem to the spinal cord in some reptiles. II. Course and site of termination. J. Comp. Neurol. 167, 443–463.

ten Donkelaar, H. J. (1982). Organization of descending pathways to the spinal cord in amphibians and reptiles. In Progress in Brain Research, Vol. 57, Descending Pathways to the Spinal Cord (H. G. J. M. Kuypers and G. F. Martin, eds.). Elsevier, Biomedical Press. Amsterdam, pp. 25–67.

ten Donkelaar, H. J. (1988). Evolution of the red nucleus and rubrospinal tract. Behav. Brain Res. 28, 9–20.

ten Donkelaar, H. J., and Bangma, G. C. (1983). A crossed rubrobulbar projection in the snake Python regius. Brain Res. 279, 229–232.

ten Donkelaar, H. J., Bangma, G. C., Barbas-Henry, H. A., de Boer-van Huizen, R., and Wolters, J. G. (1987). The brain stem in a lizard, Varanus exanthematicus. In Advances in Anatomy, Embryology and Cell Biology, Vol. 107. Springer Verlag, Berlin, Heidelberg.

ten Donkelaar, H. J., Bangma, G. C., and de Boer-van Huizen, R. (1983). Reticulospinal and vestibulospinal pathways in the snake, Python regius. Anat. Embryol. 169, 277–289.

ten Donkelaar, H. J., Bangma, G. C., and de Boer-van Huizen, R. (1985). The fasciculus longitudinalis medialis in the lizard Varanus exanthematicus. 2. Vestibular and internuclear components. Anat. Embryol. 172, 205–215.

ten Donkelaar, H. J., and de Boer-van Huizen, R. (1978a). Cells of origin of pathways descending to the spinal cord in a lizard (Lacerta galloti). Neurosci. Lett. 9, 123–128.

ten Donkelaar, H. J., and de Boer-van Huizen, R. (1978b). Cells of origin of propriospinal and ascending supraspinal fibers in a lizard (Lacerta galloti) Neurosci. Lett. 9, 285–290.

ten Donkelaar, H. J., and de Boer-van Huizen, R. (1981a). Basal ganglia projections to the brain stem in the lizard (Varanus exanthematicus) as demonstrated by retrograde transport of horseradish peroxidase. Neuroscience 6, 1567–1590.

ten Donkelaar, H. J., and de Boer-van Huizen, R. (1981b). Ascending projections of the brain stem reticular formation in the nonmammalian vertebrate (the lizard *Varanus exanthematicus*) with notes on the afferent connections of the forebrain. *J. Comp. Neurol.* 200, 501–528.

ten Donkelaar, H. J., and de Boer-van Huizen, R. (1984). The fasciculus longitudinalis medialis in the lizard *Varanus exanthematicus*. 1. Interstitiospinal, reticulospinal and vestibulospinal components. *Anat. Embryol.* 169, 177–184.

ten Donkelaar, H. J., and de Boer-van Huizen, R. (1988). Brain stem afferents to the anterior dorsal ventricular ridge in a lizard (*Varanus exanthematicus*). *Anat. Embryol.* 177, 465–475.

ten Donkelaar, H. J., de Boer-van Huizen, R., Schouten, F. T. M., and Eggen, S. J. H. (1981). Cells of origin of descending pathways to the spinal cord in the clawed toad (*Xenopus laevis*). *Neuroscience* 6, 2297–2312.

ten Donkelaar, H. J., Kusuma, A., and de Boer-van Huizen, R. (1980). Cells of origin of pathways descending to the spinal cord in some quadrupedal reptiles. *J. Comp. Neurol.* 192, 827–851.

ten Donkelaar, H. J., and Nieuwenhuys, R. (1979). The brainstem. In *Biology of the Reptilia, Neurology B* (C. Gans, R. G. Northcutt, and P. S. Ulinski, eds.). Academic Press, London, pp. 133–200.

Thomas, D. M., Kaufman, R. P., Sprague, J. M., and Chambers, W. W. (1956). Experimental studies of the vermal cerebellar projections in the brain stem of the cat (fastigiobulbar tract.) *J. Anat.* 90, 371–385.

Tohyama, M. (1976). Comparative anatomy of cerebellar catecholamine innervation from teleosts to mammals. *J. Hirnforsch.* 17, 43–60.

Ueda, S., Takeushi, Y., and Sano, Y. (1983). Immunohistochemical demonstration of serotonin neurons in the central nervous system of the turtle *Clemmys japonica. Anat. Embryol.* 108, 1–19.

Ulinski, P. S. (1977). Tectal efferents in the banded water snake, *Natrix sipedon. J. Comp. Neurol.* 173, 251–274.

van Hoevell, J. J. L. D. (1916). De kernen der kleine hersenen. *Proc. Acad. Sci. Amst.* 24, 1485–1498.

van Valkenburg, C. T. (1926). Zur Frage nach einer Funktionsteilung im Eidechsenkleinhirn und ihrer Lokalization. *Psychiatr. Bl. (holl.),* 1926, 83–90.

Verhaart, W. J. C. (1974). Identification of fibre systems of the avian midbrain. *J. Hirnforsch.* 15, 379–386.

von Düring, M., and Miller, M. R. (1979). Sensory nerve endings of the skin and deeper structures. In *Biology of the Reptilia, Neurology A* (C. Gans, R. G. Northcutt, and P. S. Ulinski, eds.). Academic Press, London, pp. 407–441.

Voneida, T. J., and Sligar, C. M. (1979). Efferent projections of the dorsal ventricular ridge and the striatum in the tegu lizard, *Tupinambis nigropunctatus. J. Comp. Neurol.* 186, 43–64.

Voogd, J. (1964). "The Cerebellum of the Cat: Structure and Fiber Connections." Thesis, University of Leiden, The Netherlands.

Voogd, J. (1967). Comparative aspects of the structure and fibre connections of the mammalian cerebellum. In *Progress in Brain Research,* Vol. 25, *The Cerebellum* (C. A. Fox and R. S. Snider, eds.). Elsevier, Amsterdam, pp. 94–134.

Voogd, J. (1969). The importance of fiber connections in the comparative anatomy of the mammalian cerebellum. In *Neurobiology of Cerebellar Evolution and Development* (R. Llinás, ed.). American Medical Association, Chicago, pp. 493–514.

Voogd, J., and Bigaré, F. (1980). Topographical distribution of olivary and corticonuclear fibers in the cerebellum: a review. In *The Inferior Olivary Nucleus: Anatomy and Physiology* (J. Courcille, C. de Montigny, and Y. Lamarre, eds.) Raven Press, New York, pp. 207–234.

Walberg, F., Pompeiano, O., Brodal, A., and Jansen, J. (1962a). The fastigio-vestibular projection in the cat: an experimental study with silver impregnation methods. *J. Comp. Neurol.* 118, 49–75.

Walberg, F., Pompeiano, O., Westrum, L. E., and Hauglie-Hanssen, E. (1962b). Fastigioreticular fibers in the cat: an experimental study with silver methods. *J. Comp. Neurol.* 119, 187–199.

Walker, A. D. (1972). New light upon the origin of birds and crocodiles. *Nature* 237, 257–263.

Walsh, J. V., Houk, J. C., and Mugnaini, E. (1974). Identification of unitary potentials in turtle cerebellum and correlations with structures in granular layer. *J. Neurophysiol.* 37, 30–47.

Weston, J. K. (1936). The reptilian vestibular and cerebellar gray with fiber connections. *J. Comp. Neurol.* 65, 93–199.

Wiklund, L., Toggenburger, G., and Cuénod, M. (1982). Aspartate: possible neurotransmitter in cerebellar climbing fibers. *Science* (N.Y.) 216, 78–80.

Wilson, V. J., and Melvill Jones, G. (1979). *Mammalian Vestibular Physiology.* Plenum Press, New York.

Wilson, V. J., Uchino, Y., Maunz, R. A., Susswein, A., and Fukushima, K. (1978). Properties and connections of cat fastigiospinal neurons. *Exp. Brain Res.* 32, 1–17.

Wold, J. E. (1978). The vestibular nuclei in the domestic hen (*Gallus domesticus*). III. Ascending projections to the mesencephalic eye motor nuclei. *J. Comp. Neurol.* 179, 393–406.

Wold, J. E. (1981). The vestibular nuclei in the domestic hen (*Gallus domesticus*). VI. Afferents from the cerebellum. *J. Comp. Neurol.* 201, 319–341.

Wolters, J. G., de Boer-van Huizen, R., and ten Donkelaar, H. J. (1982). Funicular trajectories of descending brain stem pathways in a lizard. In *Progress in Brain Research*, Vol. 57, *Descending Pathways to the Spinal Cord* (H. G. J. M. Kuypers and G. F. Martin, eds.). Elsevier Biomedical Press, Amsterdam, pp. 69–78.

Wolters, J. G., de Boer-van Huizen, R., ten Donkelaar, H. J., and Leenen, L. (1986). Collateralization of descending pathways from the brainstem to the spinal cord in a lizard, *Varanus exanthematicus*. *J. Comp. Neurol.* 251, 317–333.

Wolters, J. G., ten Donkelaar, H. J., and Verhofstad, A. A. J. (1984). Distribution of catecholamines in the brain stem and spinal cord of the lizard *Varanus exanthematicus*: an immunohistochemical study based on the use of antibodies to tyrosine hydroxylase. *Neuroscience* 13, 469–493.

Wolters, J. G., ten Donkelaar, H. J., Steinbusch, H. W. M., and Verhofstad, A. A. J. (1985). Distribution of serotonin in the brain stem and spinal cord of the lizard *Varanus exanthematicus*: an immunohistochemical study. *Neuroscience* 14, 169–193.

Wolters, J. G., ten Donkelaar, H. J., and Verhofstad, A. A. J. (1986). Distribution of some peptides (substance P, Leu-encephalin, Met-enkephalin) in the brain stem and spinal cord of a lizard, *Varanus exanthematicus*. *Neuroscience* 18, 917–946.

Woodson, W., and Künzle, H. (1982). Distribution and structural characterization of neurons giving rise to descending spinal projections in the turtle *Pseudemys scripta elegans*. *J. Comp. Neurol.* 212, 336–348.

Yamamoto, K., Tohyama, M., and Shimizu, N. (1977). Comparative anatomy of the topography of catecholamine containing neuron systems in the brain stem from birds to teleosts. *J. Hirnforsch.* 18, 229–240.

Yingcharoen, K., and Rinvik, E. (1982). Branched projections from the nucleus prepositus hypoglossi to the oculomotor nucleus and the cerebellum: a retrograde fluorescent double labeling study in the cat. *Brain Res.* 246, 133–136.

Zecha, A. (1968). The brachium conjunctivum in pigeons. *Acta Morphol. Neerl.-Scand.* 7, 97.

Ziehen, T. (1934). Mikroskopische Anatomie des Kleinhirns. In *Bardeleben's Handbuch der Anatomie des Menschen* (H.v. Eggeling, ed.). Vol. 4. G. Fisher, Jena, Germany.

8

Neuropeptides in the Nervous System

ANTON J. REINER

CONTENTS

I. INTRODUCTION

Neuropeptides are now widely accepted as a major new class of substances by which neurons communicate with one another. Although most of the research on neuropeptides has focused on the nervous system of various mammalian species, all neuropeptides found in mammals are the products of the evolutionary history of these neuropeptides in the lineage leading to mammals. Consequently, forerunners of many of these neuropeptides must have been present in the diverse extinct vertebrate species making up the lineage leading to mammals. Studies of various extant nonmammalian species support this viewpoint and show that homologues of many mammalian neuropeptides are present in a wide variety of vertebrate and invertebrate species. Since mammals evolved from reptiles, the neuropeptides found in living reptilian species might be expected to bear particular similarity to those found in mammals. The studies reviewed in this article show that this is, in fact, the case. This chapter reviews the extant data on the molecular structure, location, and function of the neuropeptides found in the nervous system in reptiles, all of which have their correspondents in the nervous systems of mammals. Such information is of value for three major reasons. First, this information helps clarify the evolution of neuropeptides in the lineages leading to birds on one hand and to mammals on the other. Second, this information helps us understand more about the structure, function, and variation of the nervous system among reptiles. Finally, since specific neuropeptides tend to be found within specific brain systems, information on the localization and function of neuropeptides in specific systems in reptiles helps us better understand the evolution of those systems among land vertebrates.

For each neuropeptide discussed in this review, I attempt to show the value of the information presented in the above three respects. The data on individual reptilian neuropeptides will also be related to what is known regarding the corresponding peptides in diverse an- amniotic vertebrate species in order to help place the reptilian data on neuropeptides into clearer evolutionary and functional perspective. Further, since little information is available for the function of neuro- peptides in reptiles, data from mammals will often be used to help illuminate their functions. Before actually discussing the findings in reptiles, a brief review of the history of peptides will be presented, which seems warranted since neuropeptides are a relatively new class of neuroactive substances.

A. Neuropeptides: Historical Background

The role of peptides,[1] or short chains of amino acids, as hormonal agents involved in intercellular communication was recognized dur- ing the early part of this century. Beginning with the extraction, puri- fication, and structural characterization of such peptide hormones as secretin and insulin (Sanger, 1959; Mutt and Jorpes, 1967), it was found that a variety of glandular cells or tissues of the body synthe- size, store, and release (when triggered by the appropriate stimulus) peptide hormones that influence the function of both nearby and dis- tant target tissues. Structures such as the pituitary gland, the pan- creas, and the upper gastrointestinal (GI) tract were found to be ca- pable of the release of significant amounts of specific peptide hormones, which in many cases are carried to distant targets by the vascular system. Work on the peripheral nervous system and at the neuromuscular junction (Loewi, 1921; Dale et al., 1936) suggested that cells of the nervous system utilized a different kind of intercellu- lar communication, one in which the interaction occurs at a special- ized region of apposition (termed the synapse) between two neurons (or their processes) or between the process of a neuron and its target organ. This neuronal communication is effected by chemical messen- gers termed neurotransmitters. A number of different neurotransmit- ters were recognized during the first two-thirds of this century, including acetylcholine, the biogenic amine neurotransmitters (nor- epinephrine, dopamine, and serotonin) (Twarog and Page, 1953; Dahlstrom and Fuxe, 1964a, 1964b; Vogt, 1954; Carlsson, 1959) and later several amino acid transmitters (such as GABA, glutamate, and aspartate) (Okamoto, 1951; Wofsey et al., 1971; Roberts and Frankel, 1950).

Over the last 40 years, the realization has gradually developed that peptides too might be used by neurons in intercellular, and more spe- cifically interneuronal, communication. During the 1940s and 1950s, the work of the Scharrers (Scharrer and Scharrer, 1945) and Bargmann

(1949) revealed that neurons of the supraoptic and paraventricular nuclei of the hypothalamus send their axons into the posterior pituitary or neurohypophysis, where they form terminals that can release substances (of unknown identity at that time) into the bloodstream. Du Vigneaud (1956) and his co-workers subsequently characterized these substances and showed that two different nine-amino-acid (or nonapeptide) peptide hormones were released from the neurohypophysis in placental mammals—vasopressin (which was found to be a hypertensive and antidiuretic hormone) and oxytocin (which was found to stimulate lactation and facilitate uterine contractions, thereby aiding birth in mammals). Subsequently, additional peptides related to pituitary function, namely hypophysiotropic factors controlling the release of anterior pituitary hormones, were extracted from hypothalamic tissue in a variety of mammalian species and purified, characterized, and sequenced: (1) luteinizing hormone–releasing hormone (LHRH), which governs the release of luteinizing hormone; (2) thyrotropin-releasing hormone (TRH), which governs the release of thyroid-stimulating hormone; and (3) somatostatin (SS), which inhibits the release of growth hormone. These hypophysiotropic factors were found to be synthesized in hypothalamic neurons that were able to release these substances into the capillaries of the hypothalamohypophyseal portal system from terminals located in the median eminence of the hypothalamus (Schally et al., 1973). Thus, the concept was established that at least some neurons of the CNS synthesize and, under the control of their neural input, secrete peptide hormones or releasing factors into the bloodstream. During the 1950s, however, it was thought that these neurons *only* released their peptides into the circulatory system.

With the development of radioimmunoassay (RIA) (Yalow and Berson, 1960), it became possible to use antibodies against these various peptide hormones and hypophysiotropic factors to measure their abundance in tissue. A surprising finding emerged from such RIA studies. The posterior pituitary hormones and the hypophysiotropic factors were not only present in hypothalamic-pituitary tissue, as expected, but were also present in extrahypothalamic tissue. The presence of these pituitary hormones and hypophysiotropic peptides in brain regions outside of the hypothalamus was initially of uncertain significance, but it was clear that the peptides outside the hypothalamus were not secreted into the bloodstream. During the 1970s, a virtual explosion of research on peptides in the nervous system led to the realization by the end of the decade that the neurohypophyseal hormones and hypophysiotropic factors, as well as a wide variety of additional peptides, played an important role in the nervous system above and beyond any role in the hormonal functions of the hypothalamopituitary axis. For example, in the early 1970s, substance P (SP) was found to be abundant throughout the nervous system in mammals but to have no specific role in hypothalamic endocrine functions. Subsequently, the opioid peptides were also found to be abundant and have a widespread distribution in the nervous system of a variety of mammalian species. The realization that at least some peptides were abundant in diverse parts of the nervous system in a variety of mammalian species gave impetus to the quest to determine whether yet additional peptides were present in the nervous system of mammals. These further studies focused on both newly discovered peptides (such as vasoactive intestinal polypeptide, VIP) and well-known peptide hormones (such as cholecystokinin, CCK), which in many cases were initially discovered in nonneural tissue and were known for their prominent role outside the nervous system. Invariably such studies revealed that these peptides were also abundant in the nervous system.

Thus, by the late 1970s it became established that a large number of peptides (more than 30) were present and widely distributed in the nervous system in mammals (Snyder, 1980). These peptides came to be grouped together as a class of substances known as *neuropeptides.* The surge of research that led to the discovery of these peptides in nervous tissue has been accompanied by intensive efforts to identify the specific neural systems that contain specific neuropeptides and determine the functions of the neuropeptides in those systems. Although much remains to be learned, some generalizations, largely based on research in mammalian species, are valid (Snyder, 1980). Neuropeptides clearly are chemical agents by which neurons communicate with one another and which satisfy many of the conventional characteristics for acceptance as neurotransmitters. Neuropeptides are synthesized by neurons and packaged in vesicles, as is also true of conventional neurotransmitters. It should be noted, however, that although conventional neurotransmitters such as acetylcholine and the biogenic amines are synthesized by means of an enzyme-mediated mechanism, neuropeptide synthesis does not involve this mechanism. The amino acid sequences for individual neuropeptides are encoded in the genome. Thus, the synthesis of neuropeptides involves transcription from the genome by a ribosomal mechanism. Once packaged in vesicles, the neuropeptides are transported to the axon terminals, from which they are released upon depolarization of the terminal. The released neuropeptide has a receptor-mediated action on the neurons or processes in proximity to that terminal. The distance that neuropeptides may diffuse from their terminal is presently unclear. The suggestion has been made that neuropeptides may influence neurons not directly in synaptic contact with the terminals that release the neuropeptide. In general, neuropeptides appear to have less brisk, lower-amplitude, and more sustained effects on neurons than do conventional transmitters. With increasing frequency neuropeptides have been found in neurons that are known to contain a conventional transmitter (Lundberg and Hokfelt, 1983). On this basis, and on the basis of physiological data, the idea has arisen that peptides are released from neuronal endings that also release conventional transmitters (e.g., acetylcholine) and that peptides act to "modulate" the response of the postsynaptic neurons to the conventional transmitters. Thus, the current ideas about the mechanisms of neurotransmission and the role of peptides in body function are considerably different than 40 years ago. The data clearly support the view that peptides play a major role in interneuronal communication, either as a primary agent of neurotransmission or as a neuromodulator.

B. Neuropeptides: Evolutionary Considerations

Given that peptides play a major role in interneuronal communication, one can ask whether this is a function that has been uniquely developed in mammals. Based on the similarities demonstrated among the members of the different vertebrate classes in the presence and distribution of other neurotransmitters, such as the catecholamines (Parent, 1979) and acetylcholine (Parent, 1979; Reiner, Brauth, and Karten, 1984; Mufson et al., 1984), this possibility appeared extremely unlikely even on an *a priori* basis. In their work on neurosecretion, the Scharrers (Scharrer and Scharrer, 1945) observed that secretory (and thus peptide hormone–containing) neurons were widespread among the members of diverse vertebrate and invertebrate taxa. Further, before the advent of immunohistochemical techniques, the Gomori staining method had been used in a variety of vertebrate groups to demonstrate the presence of hypothalamic secretory neurons that send their axons into the posterior pituitary (Oksche et al., 1964; Ananthanarayanan, 1955). More recent radioimmunoassay and immunohistochemical studies have shown that for nearly every peptide found in the nervous system in mammals, an equivalent peptide of similar immunological reactivity (if not identical structure) is present in the nervous systems of all nonmammalian vertebrates and most invertebrates. For example, the C-terminal octapeptide of CCK (i.e., CCK8) is apparently present in mammalian, reptilian, and amphibian nervous systems, whereas a highly similar, yet distinguishable CCK8-like peptide is present in coelenterates (Grimmelikhuijzen et al., 1980), insects (Dockray, Duve, and Thorpe, 1981), and agnathans (Holmquist et al., 1979). Similarly, SP-like peptides that are immunologically similar but structurally distinct from mammalian SP are present in all nonmammalian vertebrates studied, as well as in molluscs and coelenterates (Erspamer, 1981).

Thus, each neuropeptide in mammals is the product of a long evolutionary history in which that peptide has undergone some change in structure and biological function but has recognizably retained a similar biological activity and chemical structure throughout. The term *isopeptide* has been used to collectively refer to the various non-mammalian peptides that are structurally and evolutionarily related to a given mammalian peptide (Furness et al., 1982). Thus, the SP-like peptide present in the turtle nervous system is an isopeptide of mammalian SP. Not surprisingly, the greatest similarities in chemical structure and tissue localization of a particular peptide are present in closely related groups. As noted above, this chapter will review the extant data on the molecular structure, location, and function of the neuropeptides found in the nervous system in reptiles. In general, the neuropeptides found in reptiles bear remarkable similarity to those found in mammals, both in terms of the chemical structure and in terms of their localization within specific neural systems. Nonetheless, neuropeptides in reptiles show many unique features that presumably are a reflection of the evolutionary histories of the various reptilian species. The present paper will attempt to describe both the conservative features and the evolutionarily more variable features of each neuropeptide found in reptiles.

II. POSTERIOR PITUITARY PEPTIDES AND HYPOTHALAMIC RELEASING/INHIBITING FACTORS

A. Vasotocin and Mesotocin

Studies of members of the different vertebrate classes indicate that each vertebrate species typically has an oxytocinlike peptide hormone and a vasopressinlike peptide hormone (Acher, 1980, 1984). Ten different peptides have been characterized among vertebrates, all nonapeptides with the same general molecular structure (Table 8.1). In all placental mammals (except pigs) and also apparently in monotremes, oxytocin and arginine-vasopressin are found (Chauvet et al., 1985b; Rouille et al., 1988). These peptides differ from one another at the 3 and 8 positions. In pigs, lysine-vasopressin and oxytocin are found. Among marsupials there is considerable variation, with either oxytocin (in opossums) or mesotocin (in Australian marsupials) being found and one or more of three vasopressin peptides being found: arginine-vasopressin, lysine-vasopressin, and/or phenypressin (see Table 8.1) (Gorbman et al., 1983; Chauvet et al., 1985a; 1987). In birds, reptiles, amphibians, and lungfish, mesotocin and vasotocin are found. Mesotocin and vasotocin differ from one another only at the 8 position. Vasotocin is a vasopressinlike peptide that differs from

Table 8.1. Amino acid sequences for the various members of the vasopressin and oxytocin families of peptides

	1	2	3	4	5	6	7	8	9
AVP	Cys	Tyr	Phe	Gln	Asn	Cys	Pro	Arg	Gly-NH$_2$
LVP	Cys	Tyr	Phe	Gln	Asn	Cys	Pro	Lys	Gly-NH$_2$
PhyP	Cys	Phe	Phe	Gln	Asn	Cys	Pro	Arg	Gly-NH$_2$
AVT	Cys	Tyr	Ile	Gln	Asn	Cys	Pro	Arg	Gly-NH$_2$
OT	Cys	Tyr	Ile	Gln	Asn	Cys	Pro	Leu	Gly-NH$_2$
MT	Cys	Tyr	Ile	Gln	Asn	Cys	Pro	Ile	Gly-NH$_2$
VT	Cys	Tyr	Ile	Gln	Asn	Cys	Pro	Val	Gly-NH$_2$
IT	Cys	Tyr	Ile	Ser	Asn	Cys	Pro	Ile	Gly-NH$_2$
GT	Cys	Tyr	Ile	Ser	Asn	Cys	Pro	Gln	Gly-NH$_2$
AT	Cys	Tyr	Ile	Asn	Asn	Cys	Pro	Val	Gly-NH$_2$

Note: As conventional for the depiction of peptides or proteins, the N-terminus is to the left and the C-terminus to the right. Standard abbreviations are used for the amino acids. The two cysteines are linked in all of the above peptides by a disulfide bond. Because of this linkage, the two cysteines are sometimes considered a single entity and the above peptides sometimes referred to as octapeptides. Note that the vasopressin-peptides differ only at the 2, 3, and 8 positions, whereas the oxytocin-peptides differ only at the 4 and 8 positions. See the text and Acher (1984) for more details. Abbreviations: AVP, arginine vasopressin; LVP, lysine vasopressin; PhyP, phenypressin; AVT, arginine vasotocin; OT, oxytocin; MT, mesotocin; VT, vasotocin; IT, isotocin; GT, glumitocin; and AT, aspartocin.

arginine-vasopressin at the 3 position but has the same amino acid (i.e., isoleucine) at this position as mesotocin and oxytocin. Ray-finned fishes and cartilaginous fishes also use vasotocin as their vasopressinlike peptide (Acher, 1980), whereas the oxytocinlike peptide varies: isotocin (ray-finned fishes), glumitocin (rays) and aspartocin and valitocin (sharks). In cyclostomes, only a single oxytocin- vasopressinlike hormone is found (vasotocin), and it has therefore been assumed that the two nonapeptide hormones typical of jawed vertebrates arose by the duplication of the gene for the single oxytocinlike or vasopressinlike hormone found in ancestral jawless vertebrates. Further differences between the two hormones are presumed to have arisen by point mutations at the 3, 4, or 8 positions. A number of different schemes have been proposed to account for the evolution of the various oxytocinlike and vasopressinlike hormones (Gorbman et al., 1983).

The available data suggest that in each species, the oxytocinlike and vasopressinlike hormones appear to be synthesized as part of a considerably larger precursor, with each hormone being derived from a different precursor (Saayman et al., 1985; Nojiri et al., 1987; Licht, Pickering, Papkoff, et al., 1984; Chauvet et al., 1986; Chauvet et al., 1988; Brownstein et al., 1980). In amphibians and mammals, the available data show that each of these precursors consists of approximately 200 amino acids, with a large peptide at the N-terminus, the hormone in the middle, and a carrier peptide (termed a *neurophysin*) for the hormone at the C-terminus. The neurophysins from each precursor are different, and each hormone thus is associated with its own neurophysin. In mammals, the neurophysin associated with oxytocin is termed neurophysin I (also termed VLDV-neurophysin) and that with vasopressin, neurophysin II (also termed MSEL-neurophysin). Neurophysin I and neurophysin II (or their isopeptides) have been demonstrated in amphibians, reptiles, birds, and mammals (Saayman et al., 1985; Nojiri et al., 1987; Licht, Pickering, Papkoff, et al., 1984; Chauvet et al., 1986; Chauvet et al., 1988; Brownstein et al., 1980). In mammals, oxytocin and vasopressin are synthesized in separate neuronal populations of the magnocellular secretory neurons of the supraoptic and paraventricular nuclei, each of which sends its axons into the neurohypophysis (De Mey et al., 1975). Similarly, in reptiles, as also in birds (Berk et al., 1982) and amphibians (Vandesande and Dierickx, 1976), the mesotocinergic and vasotocinergic terminals in neurohypophysis are the endings of the hypothalamic magnocellular secretory neurons, which in reptiles are found in the supraoptic and paraventricular nuclei and in a series of accessory clusters lying between these two nuclei (Fig. 8.1). The hormonal actions of vasotocin and mesotocin in reptiles, as in other vertebrate groups, are mediated by the release of these peptides from axonal terminals in the neurohypophysis. In birds and reptiles, vasotocin has the antidiuretic action of increasing water reabsorption by the distal portion of the renal tubule, as is also true of vasopressin in mammals (Turner and Bagnara, 1976; Gorbman et al., 1983). In mammals, oxytocin is a hormone that stimulates lactation and uterine contractions (Turner and Bagnara, 1976; Gorbman et al., 1983). Mesotocin in birds and reptiles appears to promote oviposition, an effect apparently homologous to the effect of oxytocin on the uterus of female mammals (Turner and Bagnara, 1976). Vasotocin, however, is an even more potent stimulant of oviposition in birds and reptiles than is mesotocin (Gorbman et al., 1983). Thus the functional importance of mesotocin in birds and reptiles is not entirely certain.

The hypothalamohypophyseal system of reptiles was initially demonstrated using nonspecific staining methods (e.g., Gomori methods)

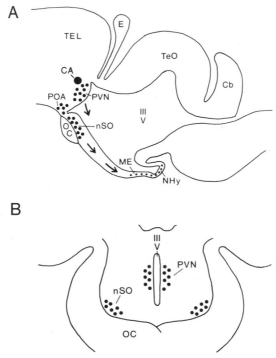

Fig. 8.1. Schematic illustrations (based on Bons, 1973) of a lizard brain in sagittal view (A) and of a lizard hypothalamus in transverse view (B) showing the location of mesotocin- and vasotocin-containing perikarya (large dots). Although mesotocin and vasotocin are not contained in the same neurons, neurons containing these peptides are found in the same cell groups, including the preoptic area (POA), the paraventricular nucleus (PVN), and the supraoptic nucleus (nSO). These neurons give rise to fibers (small dots) that terminate in the median eminence (ME) and the neurohypophysis (NHy). Other abbreviations: TEL, telencephalon; OC, optic chiasm; E, epiphysis; TeO, optic tectum; Cb, cerebellum; III V, third ventricle.

(Ananthanarayanan, 1955; Abdel Messeih and Tawfik, 1963; Gesell and Callard, 1972; Zaloglu, 1973; Prasada Rao and Subhedar, 1977). More recently, this system has been demonstrated using immunohistochemical techniques with mesotocin-specific antisera and vasotocin-specific antisera (Bons, 1983; Goosens et al., 1979; Gabrion et al., 1978; Stoll and Voorn, 1985; Thepen et al., 1987; Fernandez-Llebrez, 1988) in a number of different lizard species (*Ctenosaura pectinata, Basiliscus vittatus, Anolis cybotes, Anolis roquet, Calotes versicolor, Podarcis muralis, Acanthodactylus pardalis, Acanthodactylus boskianus, Tarentola mauritanica; Gekko gecko),* in snakes (*Natrix maura),* and in turtles (*Mauremys caspica)* (Fig. 8.1). Magnocellular secretory neurons have also been observed in the supraoptic and paraventricular nuclei of red-eared turtles, using an antiarginine vasopressin antiserum that cross-reacts with vasotocin and an antioxytocin antiserum that cross-reacts with mesotocin (Reiner, unpublished observation) (Fig. 8.2). As demonstrated in mammals, vasotocin and mesotocin are apparently found in separate neurons in reptiles (Goosens et al., 1979). Vasotocin and mesotocin have also been demonstrated in the neurohypophysis of turtles, lizards, snakes, and crocodilians by bioassay (Sawyer, 1961; Acher et al., 1969, 1972; Holder et al., 1982). Higher levels of vasotocin are present in the turtle neurohypophysis than in those of snakes or lizards (Holder et al., 1982). Within the neurohypophysis, the central portion of the lobe is dominated by vasotocinergic fibers in lizards and turtles (Goosens et al., 1979; Reiner, unpublished observation). The axons of the magnocellular secretory neurons course caudally along the base of the hypothalamus at the lateral edge of the median eminence of the hypothalamus. In snakes, lizards, and turtles, the neurons and axons of the hypothalamo-ME projection system have been shown to contain a substance that cross-reacts immunologically with antibodies against mammalian neurophysins (Gonzalez and Rodriguez, 1980; Wolf, 1976; Fernandez-Llebrez, 1988). This is further evidence that mesotocin and vasotocin derive from neurophysin-containing precursors similar to those in mammals.

Many vasotocinergic terminals terminate in the external zone of the median eminence in lizards, snakes, and turtles (Goosens et al., 1979; Stoll and Voorn, 1985; Thepen et al., 1987; Fernandez-Llebrez, 1988; Reiner, unpublished observations). Mesotocinergic fibers are present in the median eminence but are fewer than the vasotocinergic fibers and do not cluster in the external zone of the median eminence (Goosens et al., 1979; Bons, 1983; Stoll and Voorn, 1985; Thepen et al., 1987; Reiner, unpublished observations). The distribution of vasotocinergic and mesotocinergic terminations in the median eminence has not been studied in snakes and crocodilians. Vasopressin and oxytocin are also found in the mammalian median eminence (Vandesande et al., 1977), from which they are presumably released into the hypothalamohypophyseal portal circulatory system. Although the vasopressin and oxytocin released from the median eminence (ME) in mammals presumably influence the release of adenohypophyseal hormones, the nature of this influence is uncertain. Similarly, vasotocin released from the reptilian ME may also influence the release of anterior pituitary hormones, but the nature of such an influence is uncertain.

Using Gomori's chromalum-haematoxylin staining technique (Gomori, 1941), to specifically stain secretory granules, changes in the magnocellular neurons of the supraoptic and paraventricular nuclei

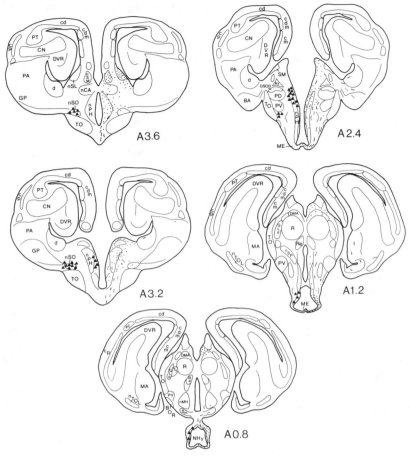

Fig. 8.2. A rostral to caudal series of line drawings of transverse sections through the hypothalamus of *Chrysemys* illustrating the distribution of mesotocin + neurons (triangles) and fibers (dots), as revealed by an antiserum against oxytocin. See list of abbreviations to identify the various brain structures. Numbers to the lower right of each drawing identify the level of the section in terms of the stereotaxic atlas of Powers and Reiner (1980). Abbreviations: AT, area triangularis; BA, basal amygdaloid nucleus; CN, core nucleus of the DVR; CO, optic chiasm; cd, cortex dorsalis; cdm, cortex dorsomedialis; cm, cortex medialis; cp, cortex pyriformis; DLA, nucleus dorsolateralis anterior; DMA, nucleus dorsomedialis anterior; DSOD, nucleus of the dorsal supraoptic decussation; DVR, dorsal ventricular ridge; d, area d; GLd, nucleus geniculatus lateralis, pars dorsalis; GLv, nucleus geniculatus lateralis, pars ventralis; GP, globus pallidus; HC, labenular commissure; HL, lateral habenular nucleus; MA, medial amygdaloid nucleus; ME, median eminence; NHy, neurohypophysis; nCA, nucleus of the anterior commissure; nDB, nucleus of the diagonal band; nMH, nucleus medialis hypothalami; nPH, nucleus periventricularis hypothalami; nSL, nucleus septalis lateralis; nSM, nucleus septalis medialis; nSO, nucleus supraopticus; nSP, nucleus suprapeduncularis; nVH, nucleus ventromedialis hypothalami; nTOL, nucleus of the lateral olfactory tract; PA, paleostriatum augmentatum; PD, dorsal peduncle of the lateral forebrain bundle; PH, primordium hippocampi; PT, pallial thickening; PV, ventral peduncle of the lateral forebrain bundle; R, nucleus rotundus; Re, nucleus reuniens; SM, stria medullaris; Sm, supramammillary region; TT, tractus tectothalamicus; TO, tractus opticus.

have been observed in reptiles under a variety of conditions: during the reproductive cycle (Saint Girons, 1963, 1967; Zaloglu, 1973), with osmotic stress (Pandalai, 1960; Sheela and Pandalai, 1968; Philibert and Kamemoto, 1965), during starvation (Abdel Messeih and Tawfic, 1963), and following injection of male hormone (Pandalai, 1960). Responses of the supraoptic and paraventricular neurons to osmotic stress or sexual stimuli are interpretable in terms of the roles of the hormones secreted by the neurons in these nuclei. Bons (1983) has reported a depletion of mesotocinergic fibers from the external zone of the ME in cold-treated lizards. The role of mesotocin and its influence on the adenohypophysis in the response of lizards to cold stress is unclear. Morphine and chlorpromazine have been found to lead to depletion of secretory granules (presumably containing vasotocin, mesotocin, or both) from the ME and neurohypophysis in lizards (Haider, 1974). The occurrence of this morphine effect suggests that opioid systems may modulate vasotocin and/or mesotocin release in reptiles. As will be described later, numerous opioid peptide-containing fibers and terminals are present in the ME and neurohypophysis of reptiles and may be the basis of the morphine effect on posterior pituitary peptides in reptiles.

In mammals, oxytocin and vasopressin (and their neurophysins) are found in pathways arising from the hypothalamus and projecting to a variety of brain regions outside the hypothalamoneurohypophyseal system, including the intermediolateral column of the spinal cord, the nucleus of the solitary tract, locus coeruleus, the parabrachial region, the medial amygdaloid nucleus, the bed nucleus of the stria terminalis, the central gray, and the septum (Buijs, 1978; Swanson, 1977). This is also true for mesotocinergic fibers and vasotocinergic fibers in reptiles, birds, and amphibians (Jokura and Urano, 1987; Panzica et al., 1988). The hypothalamic neurons giving rise to these projections, at least in mammals, are different than those projecting to the median eminence and the neurohypophysis (Hancock, 1976; Ono et al., 1978). Vasotocin has also been found in the pineal gland of fish, rats, turtles, and snakes (Vivien-Roels et al., 1979; Bowie and Herbert, 1976). In desert lizards, Bons (1983) has found vasotocin-positive fibers coursing from the preoptic area to the olfactory bulbs.

In unpublished studies in turtles (Fig. 8.2), extrahypothalamic mesotocinergic fibers were more extensive than extrahypothalamic vasotocinergic fibers (Reiner, unpublished observations) whereas in published studies on *Gekko gecko* (Stoll and Voorn, 1985; Thepen et al., 1987), the opposite was true. This difference may reflect an authentic difference between vasotocin and mesotocin systems in turtles and lizards, or it may reflect differences in antisera sensitivity between the two sets of investigators. In turtles and lizards, fibers containing mesotocin were observed to be abundant in the medial and lateral septum, in nucleus accumbens septi (termed area c in turtles by Riss et al., 1969), in the diagonal band, and in the preoptic area (Reiner, unpublished observations; Goosens et al., 1979; Stoll and Voorn, 1985; Thepen et al., 1987). Scattered fibers were observed in the amygdala, dorsomedial and dorsolateral thalamic nuclei, the tectum, the central gray, and the parabrachial region. Fibers containing vasotocin were observed in all of the above regions in lizards (Goosens et al., 1979; Stoll and Voorn, 1985; Thepen et al., 1987) but only in the medial and lateral septum and area c in turtles. Of particular interest, Stoll and Voorn (1985) have noted a sex difference in the distribution of vasotocinergic fibers in *Gekko gecko*, with males having much higher levels of these fibers than females in the lateral septum, nucleus sphaericus, and the periaqueductal gray. A similar sex difference has been reported in mammals (de Vries et al., 1981). Fernandez-Llebrez et al. (1988) have reported that neurophysin-containing fibers are present in turtles and snakes in all of the same forebrain and midbrain regions noted above to contain either vasotocin- or mesotocin-containing fibers.

As in birds and mammals (Keyser et al., 1982; Swanson, 1977), vasotocinergic or mesotocinergic fibers have been observed in the spinal cord or the nucleus of the solitary tract in reptiles (Stoll and Voorn, 1985; Thepen et al., 1987). This is consistent with the observation that retrogradely labeled neurons are found in the paraventricular nuclei after spinal cord injections of horseradish peroxidase in reptiles (Cruce and Newman, 1981; ten Donkelaar et al., 1980). The function of oxytocin, vasopressin, and their neurophysins in these various extrahypothalamic regions is uncertain, but a role in synaptic transmission seems likely (Riphagen and Pittman, 1986).

Perikarya containing vasotocin and mesotocin are generally restricted to the paraventricular and supraoptic nuclei, but Bons (1983) has reported that some vasotocinergic and mesotocinergic neurons are present in the tuberal region of the hypothalamus in lizards. This also appears to be the case in turtles and snakes (Fernandez-Llebrez, et al., 1988; Reiner, unpublished observations). In addition, in turtles numerous mesotocinergic perikarya were observed in the neurohypophysis. These neurons were observed to send labeled processes into the lumen of the third ventricle. These neurons may correspond to the neurons of the ventral paraventricular nucleus of Bons (1983). In addition, mesotocinergic perikarya were observed in the central portion of the neurohypophysis in turtles. This has not been reported for other reptiles. In mammals, dynorphin has been reported in vasopressinergic neurons in rats (Watson et al., 1982a) and CCK8 has been reported in oxytocinergic neurons (Vanderhaeghen et al., 1981;

Martin et al., 1983). In turtles, CCK8 is found in supraoptic neurons and is found in high levels in the external zone of the ME and neurohypophysis. These portions of the median eminence and neurohypophysis are, however, low in mesotocinergic fibers in turtles. Thus, although CCK8 may be present in some mesotocinergic fibers in turtles, apparently the bulk of the CCK8-containing fibers in the ME and neurohypophysis do not contain mesotocin. Fibers containing dynorphin A have been observed in the external zone of the median eminence and neurohypophysis of turtles, and cell bodies containing dynorphin A have been observed in the supraoptic and paraventricular nuclei (Reiner, unpublished observations). Thus, a dynorphinlike peptide may be present in some hypothalamic and pituitary vasotocinergic fibers in reptiles.

B. Thyrotropin-Releasing Hormone (TRH)

Thyrotropin-releasing hormone was the first of the hypothalamic releasing factors isolated, and in mammals it has been identified as the tripeptide, pyroglutamylhistidylprolinamide (Burgus et al., 1970; Nair et al., 1970). TRH is also found in the nervous system of the members of a wide variety of nonmammalian groups, and it typically appears indistiguishable in chemical structure from mammalian TRH (Jackson and Reichlin, 1974; Jackson, 1978). In mammals, birds, and amphibians, fibers containing TRH have been reported to be present in the external zone of the median eminence and the neurohypophysis (Seki et al., 1983; Hokfelt et al., 1975; Brownstein et al., 1974; Mimnagh et al., 1987; Peczely and Kiss, 1988; Jozsa et al., 1988). No published data are available on reptiles. Unpublished immunohistochemical studies in red-eared turtles indicate that TRH-positive fibers are present in the central area of the external zone of the caudal median eminence and neurohypophysis (Reiner, unpublished observations). TRH released from the median eminence enters the portal blood vessels and, in mammals, birds, and reptiles (and apparently also in amphibians and teleosts), stimulates the release of thyrotropin (also known as thyroid-stimulating hormone, TSH) from the anterior pituitary (Scanes, 1974; Eales and Himick, 1988; Darras and Kuhn, 1982; Licht and Denver, 1988; Preece and Licht, 1987; Denver and Licht, 1987). TSH, in turn, stimulates the release of thyroxine from the thyroid gland. The function of the TRH found in the posterior pituitary is uncertain.

The hypothalamus in members of all vertebrate classes has been found to be high in TRH (rats, chickens, snakes, frogs, tadpoles, and salmon) (Jackson and Reichlin, 1974; Seki et al., 1983). Immunohistochemical studies indicate that the neurons of origin of the TRH fibers and terminals in the ME in birds, amphibians, and mammals are located along the midline of the rostral hypothalamus and preoptic area (Hokfelt et al., 1975; Brownstein et al., 1974; Seki et al., 1983), which is also the hypothalamic region found by physiological studies to control TSH release from the anterior pituitary (Greer, 1957). It seems likely, although not anatomically demonstrated, that the location of the TRH+ cell bodies projecting to the ME in reptiles is similar. Little was known of the function of TRH in reptiles until recently. Since TRH controls TSH release in both birds and mammals, it seemed plausible a priori that this characteristic function of TRH would have been passed on from ancestral reptiles to modern reptiles, birds, and mammals. Biochemical studies have, in fact, isolated a peptide from pituitary of turtle that is indistinguishable from mammalian TSH (MacKenzie et al., 1981). Nonetheless, one earlier study in turtles failed to find a TSH-mediated effect of TRH on systemic thyroxine levels (Sawin et al., 1981). More recently, however, Licht and his co-workers have shown that TRH does evoke TSH release from turtle and lizard pituitaries in vitro, and TRH injected into anoline lizards leads to an increase in systemic thyroxine (Denver and Licht, 1987; Preece and Licht, 1987; Licht and Denver, 1988). Thus the role of TRH in TSH-mediated regulation of thyroxine release appears to be present in modern reptiles and therefore must have also been present in the stem reptiles. There is also evidence that in reptiles TRH can increase prolactin and growth hormone (somatotropin) release from the anterior pituitary (Hall and Chadwick, 1984; Preece and Licht, 1987; Denver and Licht, 1987; Ball, 1981). In mammals, TRH is known to influence prolactin release, and in birds TRH is known to influence both prolactin and growth hormone release (Fehrer et al., 1985a, 1985b). TRH also appears to affect αMSH release from the pars intermedia of the pituitary in lizards (Licht and Denver, 1988), as also appears to be the case in amphibians (Leroux et al., 1983).

Jackson and Reichlin (1974) have found that substantial amounts of TRH are also present in extrahypothalamic brain tissue in members of all vertebrate groups, including lamprey. For example, RIA studies have revealed that rat hypothalamus contains only about 30% of the brain's total amount of TRH (Oliver et al., 1974; Winokur and Utiger, 1974). Immunohistochemical studies have shown that TRH-containing fibers in rats, birds, and frogs are prominent in extrahypothalamic nuclei such as the nucleus accumbens, the lateral septal nucleus, and many motor cell groups of the brain stem and spinal cord (Hokfelt et al., 1975; Mimnagh et al., 1987; Peczely and Kiss, 1988; Jozsa et al., 1988; Seki et al., 1983). Such extrahypothalamic TRH appears to be synthesized in extrahypothalamic TRH+ neurons (Jackson and Reichlin, 1977; Brownstein, Utiger, Palkovits, and Kizer, 1975). The extrahypothalamic distribution of TRH+ cell bodies and fibers has not been studied extensively in reptiles. In turtles, extrahypothalamic TRH+ fibers have been observed in the lateral septum, the nucleus of the diagonal band, the preoptic area, and the dorsomedial and dorsolateral nuclei of the thalamus (Reiner, unpublished observations). Available evidence suggests that extrahypothalamic TRH may act as a neurotransmitter. For example, spinal motor neurons in frogs and rats have been found to show excitatory responses to TRH (Nicoll et al., 1980a; Oka and Fukuda, 1984; Takahashi, 1985). Based on the observation that lampreys possess TRH (Youngs et al., 1985) but lack TSH, and amphioxus and snails possess TRH (Jackson and Reichlin, 1974; Grim-Jorgenson, 1978), but lack a pituitary, Jackson (1978, 1981) has suggested that TRH may originally, in fact, have been a neurotransmitter and/or a hormone unrelated to pituitary function and only later in evolution acquired its functions in regulating TSH release from the anterior pituitary.

C. Luteinizing Hormone–Releasing Hormone (LHRH)

Luteinizing hormone–releasing hormone is a decapeptide that stimulates the release of both luteinizing hormone and follicle-stimulating hormone from the anterior pituitary in a wide variety of vertebrate species. Because its releasing role is not restricted to luteinizing hormone, LHRH is sometimes called gonadotropin-releasing hormone (GnRH). As in the case of TRH, the effects of LHRH on the anterior pituitary are brought about by the transport of this hormone from nerve endings in the median eminence to the anterior pituitary via the hypothalamohypophyseal portal system. Although LHRH is a decapeptide in all vertebrate species studied, LHRH shows some structural variation among the different vertebrate species. Further, in many species, more than one form of LHRH is present. All variation in the LHRH molecule is limited to amino acid positions 5, 7, and/or 8 of the C-terminus (Sherwood, 1986). The N-terminus of the molecule, which is the portion that is necessary for receptor binding, is evolutionarily stable (Sherwood, 1986). The evolutionary conservatism of this molecule is perhaps not surprising in light of the essential role LHRH plays in reproduction. Nonetheless, variation in the C-terminus portion of the molecule does occur and does affect the binding affinity of LHRH for its receptor. Thus any evolutionary changes in the structure of LHRH needed to be accompanied by concomitant changes in the LHRH receptor if there was to be no loss in biological potency attending the change in LHRH structure.

The structures of four different LHRH molecules (Table 8.2) are known: mammalian LHRH, chicken I LHRH, chicken II LHRH, and salmon I LHRH (Matsuo et al., 1971; Burgus et al., 1972; Tan and Rousseau, 1982; Sherwood et al., 1983; King and Millar, 1982a, 1982b;

Table 8.2. Amino acid sequences for LHRH in several vertebrate groups.

	1	2	3	4	5	6	7	8	9	10
Porcine LHRH	pGlu	His	Trp	Ser	Tyr	Gly	Leu	Arg	Pro	Gly-NH$_2$
Chicken I LHRH	pGlu	His	Trp	Ser	Tyr	Gly	Leu	Gln	Pro	Gly-NH$_2$
Chicken II LHRH	pGlu	His	Trp	Ser	His	Gly	Trp	Tyr	Pro	Gly-NH$_2$
Salmon I LHRH	pGlu	His	Trp	Ser	Tyr	Gly	Trp	Leu	Pro	Gly-NH$_2$

Note: Ovine and porcine LHRH are identical. "pGlu" indicates the presence of a pyr-rolidone carboxylic acid at the N-terminus (occupied by glutamic acid).

Sherwood, 1986; Miyamoto et al., 1984). Chromatographic and im-munological studies suggest that several more LHRH molecules may exist (Matsuo et al., 1971; Burgus et al., 1972; Tan and Rousseau, 1982; Sherwood et al., 1983; King and Millar, 1982a, 1982b; Sherwood, 1986; Miyamoto et al., 1984). Mammalian LHRH is the only form of LHRH that has been reported in the mammalian nervous system. Chicken I LHRH and chicken II have both been found in a variety of birds, in-cluding chickens, ostriches, and starlings (Powell et al., 1987; Sher-wood et al., 1988), whereas sparrows only possess chicken I LHRH in their nervous system (Sherwood et al., 1988). The major form of LHRH found in teleosts is salmon I (Sherwood, 1986). Many teleosts, sturgeons (a chondrostean fish), and cartilaginous fish also possess a form of LHRH termed salmon II LHRH (Sherwood 1986; Yu et al., 1988; Sherwood et al., 1984; Powell et al., 1986b). The structure of salmon II is unknown, and thus the salmon II peptides of different species may not all be alike. There is some evidence, nonetheless, that salmon II of some piscine species and chicken II LHRH may pos-sess the same amino acid sequence, although this has yet to be dem-onstrated directly (Sherwood, 1986; Yu et al., 1988). In amphibians and reptiles, the situation is somewhat more complicated. In amphib-ians, mammalian LHRH is the major form of LHRH found in the brain (King and Millar, 1979a, 1980a; Rivier et al., 1981). Consistent with this finding, mammalian LHRH has marked gonadotropin-releasing activity in amphibians (Ball, 1981). Two other forms of LHRH, however, appear to be found in amphibian nervous system. Some of the LHRH in frog brain and all of the LHRH in frog periph-eral tissues (including sympathetic ganglia) is distinct from mammal-ian LHRH and appears to be highly similar and possibly identical to salmon I LHRH (Eiden et al., 1982; Eiden and Eskay, 1980; Branton et al., 1982). In addition, frog brain appears to also contain a salmon II-like molecule (Sherwood et al., 1986; King and Millar, 1986).

What these results suggest in evolutionary terms is that piscine LHRH (salmon I) may have been the form of LHRH ancestral to that found in land vertebrates and that in amphibians this form of LHRH was retained but at reduced levels. In the amphibians ancestral to mammals, an additional form of LHRH evolved and became predominant in the CNS, namely the form of LHRH retained in the mammalian and amphibian lineage. Since mammals evolved from reptiles, one might expect that mammalian LHRH might have been the pro dominant form of LHRH in the stem reptiles and also in living reptiles. Although this inference for stem reptiles is likely to be true, the inference for living reptiles is clearly not. The results in living reptiles are simple for some groups, complex for others. In turtles and crocodilians, both chicken I and chicken II LHRH are present (Sherwood and Whittier, 1988; Powell, Garcia, Lance, et al., 1986). In snakes, only chicken I has been found (Sherwood and Whittier, 1988). In the skink (*Chalcides ocellatus tiligugu*), chicken II was the predominant form of LHRH found, whereas in the lizard *Cordylus nigra* salmon II LHRH appeared to be the major form of LHRH found. This is consistent with previous suggestions that lizards and snakes might have only a single gonadotropic hormone (Licht, 1979). These results indicate that following the divergence of the mammalian and reptilian lineages, two forms of LHRH, chicken I and II (note that if chicken II is salmon II, then chicken II did not evolve in the reptilian lineage) evolved in reptiles from that found in ancestral reptiles (in whom mammalian LHRH was the predominant form). Chicken I and II appear to have been typically retained in several of the major groups of the sauropsid lineage, including turtles, crocodilians, and birds. The evolution of LHRH among snakes and lizards, however, appears much more complex and in need of further study.

In general, it appears that LHRH stimulates follicle-stimulating hormone and luteinizing hormone release in reptiles (Ball, 1981). As noted above, in reptiles as in all other vertebrates in whom the distribution of LHRH in the hypothalamus has been studied (which includes several reptilian species), the LHRH+ neurons regulating gonadotropin release from the pituitary are located in the preoptic area and rostral portions of the hypothalamus (Witkin et al., 1982; Setalo et al., 1976; King and Gerall, 1976; Kelly and Ronnekleiv, 1981; King et al., 1988; Kah et al., 1986; Goos et al., 1985; Crim, 1985; Jozsa and Mess, 1982; Nozaki et al., 1984). These LHRH+ neurons are the source of LHRH+ fibers in the external layer of the median eminence (Hokfelt et al., 1975; Ibata et al., 1979; King et al., 1988; Kah et al., 1986; Goos et al., 1985; Crim, 1985; Jozsa and Mess, 1982; Nozaki et al., 1984). This has been demonstrated immunohistochemically in a number of different reptilian species, including a variety of snake species (*Elaphe climacophora, E. quadrivirgata, E. conspicillata, Rhabdophis ti-*

grinus), lizard species (*Gekko gecko* and *Eumeces okadae*), and turtle species (*Pseudemys scripta* and *Geoclemys reevesii*) (Nozaki and Kobayashi, 1979; Doerr-Schott and Dubois, 1978; Reiner, unpublished observations). That the hypothalamus is the location of the LHRH-containing neurons regulating gonadotropin release in reptiles is further supported by the fact that the hypothalamus is the site at which sex steroids exert their suppressive effects on LHRH-controlled gonadotropin secretion (Lisk, 1967). Further, lesions of the anterior preoptic area of hypothalamus in lizards have been shown to disrupt gonadotropin secretion, as evidenced by testicular regression in males (Crews, 1979). Immunohistochemical studies show that LHRH+ fibers are also found in the neurohypophysis in reptiles (Nozaki et al., 1984; Doerr-Schott and Dubois, 1978; Reiner, unpublished observations).

It seems very likely that native LHRH in reptiles does effectively promote gonadotropin release, since hypothalamic extracts have been shown to cause gonadotropin release in turtles (Hall et al., 1975) and since LHRH facilitates sexual behavior in female lizards (Alderete et al., 1980). Studies directly examining the effects of LHRH on gonadotropin release in reptiles, however, have yielded conflicting and ambiguous results. For example, injections of mammalian LHRH have been found in a number of studies to be ineffective in promoting gonadotropin release in turtles and snakes (Callard and Lance, 1977; Licht et al., 1982; Licht, Millar, King, et al., 1984; Licht and Porter, 1985a). In contrast, other studies utilizing injections of mammalian LHRH in some species of turtles and alligators have noted increases in LH secretion (Callard and Lance, 1977; Licht, 1980; Lance et al., 1985; Licht and Porter, 1985a). Further, studies involving administration of mammalian LHRH to turtle pituitaries in vitro have shown effective stimulation of LH release (Licht, Millar, King et al., 1984; Licht and Porter, 1985a, 1985b, 1987). Many of the studies yielding negative results may have been complicated by a number of factors. First, earlier studies in reptiles tended to examine gonadotropin release in reptiles in response to mammalian LHRH. Since it is now known that the nervous systems of reptiles do not contain mammalian LHRH but instead primarily contain either or both chicken I and chicken II LHRH, it might be expected that reptilian pituitaries do not respond as well to mammalian LHRH as to chicken I and II LHRH. This may in part account for the inconsistent results obtained in reptiles in studies using mammalian LHRH. One set of in vitro studies has further shown that reproductive status, sex, species, and individual variation all affect the ability of reptile pituitaries to respond to mammalian LHRH (Licht and Porter, 1985a, 1985b).

More recently, Licht and Porter (1987) have compared the efficacy of chicken I LHRH and mammalian LHRH in releasing LH from turtle pituitaries in vitro. Surprisingly, they found no difference in this study. Mammalian LHRH has also consistently been found to be very effective in producing gonadotropin release in alligators (Lance et al., 1985). The result in alligators is not surprising, however, in that similar results are observed in birds (Fehrer et al., 1985; Sherwood, 1986), in whom the pituitary responds nearly equally well to mammalian and chicken I LHRH (Furr et al 1973; Johnson et al., 1984; Sherwood, 1986). In contrast, mammalian pituitaries respond poorly to chicken LHRH (Sherwood, 1986). These various results in mammals, crocodilians, and birds, and more recently in turtles, suggest that (1) the LHRH receptor in mammalian pituitary may differ from that in birds and many reptilian species and (2) the pituitary LHRH receptor may itself have undergone a very conservative evolution in the reptilian-avian lineage. Nonetheless, further studies are required on the comparative efficacies of mammalian and chicken LHRHs in a variety of turtle, lizard, and snake species before any detailed conclusions can be made about the response of the pituitary in reptiles to LHRH. Further, the relative roles of the two forms of LHRH typically found in reptiles are uncertain. Some authors have suggested that the two different LHRHs may separately control the release of the two different gonadotropins (Sherwood and Whittier, 1988), but there is no evidence supporting this possibility.

In nonmammals, as well as in mammals, LHRH is found outside the hypothalamus. For example, in frogs, ostrich, and mammals, 10% to 30% of total brain LHRH is found outside the hypothalamus (Albert et al., 1976; Jackson, 1978; Powell et al., 1987). In mammals, LHRH+ perikarya and fibers are found in numerous extrahypothalamic locations, including the accessory olfactory bulb, the nucleus of the diagonal band, and the septal nuclei (Silverman and Krey, 1978; Witkin et al., 1982). In addition, LHRH-positive fibers are observed in the organum vasculosum of the lamina terminalis (OVLT) and several of the other circumventricular organs. Similar results have been reported in birds and frogs (Crim, 1985; Foster et al., 1988; Jokura and Urano, 1986; Jozsa and Mess, 1982; Sterling and Sharp, 1982; Bons et al., 1978; Nozaki et al., 1984). In reptiles, extrahypothalamic LHRH is present in fibers of the OVLT, the septal nucleus, and the diagonal band (Nozaki et al., 1984; Reiner, unpublished observations). Extrahypothalamic LHRH-positive perikarya have been reported in the septal region and nucleus accumbens region of the telencephalon in reptiles (Nozaki et al., 1984). It is uncertain whether reptiles (or birds) possess an LHRH+ terminal nerve projection system, which has been reported to be present in mammals, cartilaginous fish, and bony fish (Schwanzel-Fukuda and Silverman, 1980; Witkin and Silverman, 1983; Stell, 1984; Munz et al., 1982). In turtles, lizards, and snakes, sex steroid (e.g., estrogen)-concentrating cells have been reported in the tuberal hypothalamus, the hypothalamic preoptic area, and the telencephalic septal area (Kim et al., 1978; Kim et al., 1981; Martinez-Vargas et al., 1978; Halpern et al., 1982; Morrel et al., 1979; Morrel and Pfaff, 1978). Since the sex steroids influence gonadotropin release, these regions may control gonadotropin release by influencing the LHRH-positive cells found in these regions. The LHRH-containing fibers found outside the hypothalamus presumably play a role in synaptic transmission at those extrahypothalamic sites. An LHRH-like peptide is found in preganglionic fibers in bullfrog sympathetic ganglia, where LHRH has been shown to have an excitatory effect on postsynaptic neurons that is slow in onset and long in duration (Jan and Jan, 1983).

D. Somatostatin

Somatostatin (SS), or somatotropin-release inhibiting factor (SRIF), is a 14-amino-acid peptide that inhibits growth hormone (or somatotropin) release from the anterior pituitary. The somatostatin-containing neurons regulating the anterior pituitary are found in the periventricular hypothalamic nucleus in a variety of vertebrate species. These neurons send their terminals into the median eminence, from which SS is released into the hypothalamohypophyseal portal system. The structure of SS in mammals was originally determined by Brazeau et al. (1973) using ovine hypothalamic extracts. More recently, the amino acid sequence of the precursor for SS was determined using rat thyroid tissue (Goodman et al., 1982). As is typically the case with neuropeptides, SS derives from a large precursor molecule, which in this case contains SS at its C-terminus. The final 28 amino acids of the C-terminus of the precursor molecule constitute a high-molecular-weight form of SS (SS-28) that has been found in tissue extracts and is biologically active (Tannenbaum et al., 1982; Vayse et al., 1981). King and Millar (1979b) have found that SS is present in hypothalamus, extrahypothalamic parts of the brain, pancreas, and stomach of not only mammals, but birds, reptiles, amphibians, bony fish, and cartilaginous fish. Similar findings have been reported by Vale et al. (1976). Using chromatographic techniques and somatostatin-14 RIA dose-response curves, King and Millar (1979b) concluded that SS-14 was indistinguishable in structure among mammals (rat), birds (pigeon), reptiles (turtle), bony fish (cichlid), and cartilaginous fish (dogfish shark). In studies of the retina of members of different vertebrate groups (but not including any reptiles), Marshak and Yamada (1984) have also concluded that the structure of SS-14 appears to be remarkably phylogenetically conservative, although the ratio of SS-14 to SS-28 appears to vary among species. This conclusion is further confirmed by the finding that the amino acid sequence of SS-14 in *Torpedo marmorata* (an elasmobranch) and pigeons is identical to that found in mammals (Conlon et al., 1985; Spiess et al., 1979). Finally, the structure of the cDNA encoding somatostatin and its precursor was recently determined for anglerfish and catfish and compared to that found in mammals (Hobart et al., 1980; Taylor et al., 1981; Magazin et al., 1982; Su et al., 1988). Although two different forms of somatostatin cDNA are present in both fish species, in each species one of the forms is highly similar to the cDNA for the somatostatin precursor found in mammals. Further, the SS-14 portion of the cDNA was identical among mammals and the piscine species studied. Thus, the evolution of SS-14 does appear to have been highly conservative, and it is likely that SS-14 in reptiles possesses the same amino acid sequence as SS-14 in mammals.

In turtles and lizards, as in other terrestrial vertebrates (Crowley and Terry, 1980; Hokfelt et al., 1975; Finlay et al., 1981b; Weindl et al., 1984; Vandesande and Dierickx, 1980; Inagaki et al., 1981b; Dierickx et al., 1981; Van Vossel-Daeninck et al., 1981; Blahser et al., 1978; Takatsuki et al., 1981; Shiosaka et al., 1981), SS+ perikarya in the periventricular hypothalamic nucleus project to the external zone of the ME via the hypothalamohypophyseal tract (Bear and Ebner, 1983; Fasolo and Gaudino, 1982; Doerr-Schott and Dubois, 1977, 1978; Weindl et al., 1984; Goosens et al., 1980; Reiner and Oliver, 1987). These SS+ hypothalamic neurons in reptiles are distinct from those containing vasotocin and mesotocin (Goosens et al., 1980; Fasolo and Gaudino, 1982). Somatostatin released from these terminals is thought to inhibit pituitary growth hormone secretion in reptiles, although physiological studies on this issue have not been carried out (Gorbman et al., 1983). Immunohistochemical studies have clearly shown, however, that a growth hormone-like peptide is present in cells of the anterior pituitary in lizards and turtles (Naik et al., 1980; Pearson et al., 1983; Pearson and Licht, 1982) and the amino acid sequence for growth hormone is known for sea turtles (Yasuda et al., 1989). Further, SS has been shown to inhibit growth hormone release in teleosts (Marchant et al., 1987). Thus this function of SS is likely to be evolutionarily conservative and also found in reptiles. In mammals, somatostatin also inhibits TSH and prolactin release from the anterior pituitary (Turner and Bagnara, 1976), but this possibility has also not been explored in reptiles. As is also found in other vertebrate species,

SS+ perikarya and fibers are also abundant in additional hypothalamic regions. For example, neurons containing somatostatin (SS) have also been observed in the ventromedial hypothalamus and arcuate region in turtles, lizards, and caiman, and SS+ fibers have been observed in the neurohypophysis in turtles and lizards (Reiner, unpublished observations; Goosens et al., 1980; Weindl et al., 1984).

Somatostatinergic neurons are also very abundant and widespread outside the hypothalamus in reptiles (see below), as well as in a wide variety of other vertebrate groups, including bony fish (Dubois et al., 1979; Tohyama et al., 1981), amphibians (Vandesande and Dierickx, 1980; Inagaki et al., 1981b; Dierickx et al., 1981; Van Vossel-Daeninck et al., 1981), birds (Blahser et al., 1978; Takatsuki et al., 1981; Shiosaka et al., 1981), and mammals (Finlay et al., 1981b; Krisch, 1979; Bennett-Clark and Romagnano, 1979; Brownstein et al., 1975a; Patel and Reichlin, 1978; Kobayashi et al., 1977; King and Millar, 1979b; Vale et al., 1976). In reptiles, as in the other amniote species that have been studied, extrahypothalamic SS-containing neurons are particularly numerous in the telencephalon. This has been shown for the cortex, dorsal ventricular ridge (DVR), and the basolateral telencephalon (in the regions equivalent to the mammalian striatum and amygdala) (Fig. 8.3) in lizards and turtles (Goosens et al., 1980; Bear and Ebner, 1983; Weindl et al., 1984; Davila et al., 1988; Perez-Clausell and Fredens, 1988). Preliminary observations on the telencephalon of *Caiman crocodilus* suggest that the same may be true in crocodilians (Reiner, unpublished observations). Bear and Ebner (1983) have reported that three distinct somatostatin-containing cell types are present in turtle telencephalic cortex and that the somatostatin-containing cells of the mammalian neocortex include the same three types. In turtles, additional SS-containing perikarya are present in the ventral thalamus, the central gray, the medial tegmentum, and the nucleus of the solitary tract (Weindl et al., 1984; Reiner, unpublished observations). Weindl et al. (1984) also report SS-containing perikarya in the lateral habenular nucleus, the interpeduncular nucleus, the nucleus of the lateral lemniscus, the superior raphe, and the pontine reticular formation of turtles. These results are similar to those in mammals, in which SS-containing cells are prominent in the cerebral cortex, hippocampus, amygdala, striatum, nucleus accumbens, zona incerta, central gray, the nucleus of the solitary tract and dorsal root ganglia (Finlay et al., 1981b). The SS+ perikarya of the telencephalon in turtles also typically contain neuropeptide Y (Reiner and Oliver, 1987), which is also characteristic of SS+ perikarya of the telencephalon in birds and mammals (Vincent, Skirboll, Hokfelt, et al., 1982; Vincent, Johansson, Hokfelt, et al., 1982; Smith and Parent, 1986; Anderson and Reiner, 1990b).

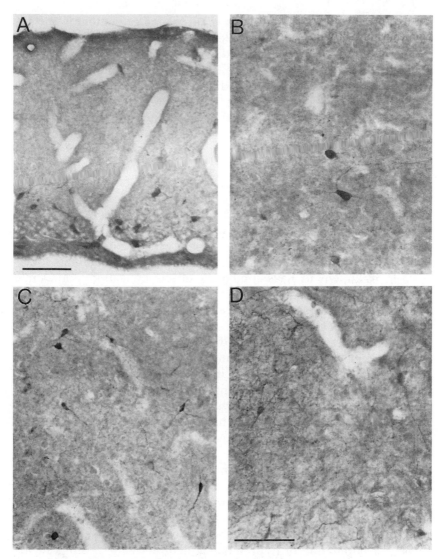

Fig. 8.3. *Pseudemys scripta*. Photomicrographs of sections through the telencephalon labeled for somatostatin (A to C) or neuropeptide Y (NPY)(D). Numerous somatostatin (SS)-containing neurons are present in the cellular and molecular layer of medial cortex (A), the dorsal ventricular ridge (B) and the medial striatum (area d) of the basal ganglia (C). Telencephalic neurons that are somatostatin + are morphologically similar to neurons that are labeled for the presence of NPY, as illustrated by a comparison of the SS + neurons of area d (C) to the NPY + neurons in area d (D). Double-label studies show that SS and NPY are largely contained in the same telencephalic neurons. Medial is toward the top in (A), toward the right in (B), and toward the left in (C) and (D). Scale bars: (A) = 100 μm; D = 100 μm. (B) and (C) are at the same magnification as (D).

Fibers that contain somatostatin in turtles are prominent in the superficial molecular layer of the telencephalic cortex, in area d of the basolateral telencephalon (the medial part of the striatum), in the nucleus of the diagonal band, in the dorsomedial/dorsolateral thalamic nuclei, in the lateral habenula, in the deep layers of the optic tectum, in the interpeduncular nucleus, in the raphe region, in the caudal nucleus of the solitary tract, and in the dorsal and intermediate gray of the spinal cord (Weindl et al., 1984; Reiner, unpublished observations). Preliminary observations in caiman indicate that SS+ fibers are prominent in the medial striatum (termed the small-celled part of the ventrolateral area in crocodilians), in the dorsomedial nucleus of the thalamus, in the habenula, in the deep tectum and interpeduncular nucleus of the midbrain, in the nucleus of the solitary tract, and in the dorsal and intermediate gray of the spinal cord (Reiner, unpublished observations). The function of these various extrahypothalamic SS+ fibers has not been examined in reptiles, but in mammals somatostatin is known to have inhibitory effects on extrahypothalamic neurons (Delfs and Dichter, 1985).

Somatostatin is also found in cells of the stomach, intestine, and pancreas in a variety of vertebrate species, and on this basis is one of several peptides that are referred to as *brain-gut peptides* (Snyder, 1980). Within the GI tract of mammals, somatostatin is known to inhibit the release of glucagon, insulin, and gastrin (Bloom et al., 1974). In alligator, SS+ endocrine cells have been observed in the GI tract mucosa and in the pancreas (Buchan et al., 1982, 1983), whereas in lizards and turtles SS+ cells have been described in the pancreas (Lopez et al., 1988; Gapp et al., 1985; Agulleiro et al., 1985).

E. Growth Hormone–Releasing Factor

A growth hormone–releasing factor, long thought to exist, was finally isolated and characterized in mammalian nervous system during the early 1980s (Guillemin et al., 1982). Immunocytochemical studies subsequently demonstrated growth hormone–releasing factor in mammals in a population of hypothalamic neurons with endings in the median eminence (Bloch et al., 1983). Since turtle hypothalamic extracts are able to release growth hormone from turtle and chicken pituitary in vitro (Hall and Chadwick, 1979), it appears likely that a growth hormone–releasing factor is present in turtle hypothalamus. Immunochemical techniques using antisera directed against mammalian growth hormone–releasing factor have not yet been used to determine if turtle growth hormone–releasing factor is immunologically similar to mammalian growth hormone–releasing factor. Further, the distribution of the putative growth hormone–releasing factor-like substance in turtles or any other reptiles has not been studied with immunohistochemical techniques and is presently unknown.

F. Corticotropin-Releasing Factor

A hypothalamic releasing factor responsible for controlling ACTH release (termed corticotropin-releasing factor, or CRF) has long been recognized (Guillemin and Rosenberg, 1955) but has only recently been characterized and synthesized from ovine hypothalamic extracts (Spiess et al., 1981; Vale et al., 1981). CRF in mammals is a 41-amino-acid peptide that now has been demonstrated immunohistochemically in a set of hypothalamic neurons near the paraventricular nucleus that send their axonal projections to the median eminence (Bloom et al., 1982; Merchenthaler et al., 1982). In turtles, also, secretion of ACTH appears to depend on a hypothalamic CRF (Holmes and Ball, 1974). Hypothalamic lesions and corticosteroid implants into the hypothalamus in reptiles, both of which will inhibit CRF and (consequently) ACTH release if placed in the proper hypothalamic site, indicate that the anterior hypothalamus may be the location of neurons containing a CRF-like substance (Callard and Chester-Jones, 1971; Daugherty and Callard, 1972; Callard and Callard, 1978). Using antisera against ovine CRF, the distribution of CRF+ neurons both in hypothalamus and outside of hypothalamus has recently been described for members of a number of vertebrate groups, including mammals, birds, reptiles, amphibians, and bony fish (Fasolo et al., 1984; Fellmann et al., 1984; Merchenthaler et al., 1982; Olschowka et al., 1982; Gonzalez and Lederis, 1988). In turtles, as in all other vertebrate groups, CRF+ perikarya are present in the paraventricular region of the hypothalamus (Fellman et al., 1984). These neurons apparently project to the median eminence and presumably thereby control ACTH secretion from the anterior pituitary. In turtles, CRF+ perikarya and fibers were also prominent outside of the hypothalamus (Fellman et al., 1984). Extrahypothalamic CRF+ perikarya were particularly prominent in the nucleus accumbens (or rostral area d of Riss et al., 1969), in the medial cortex, in the amygdala, and in the reticular formation of the medulla. Fibers containing CRF were observed in the hypothalamus, the amygdala, the tectum, and the pons. Extrahypothalamic neurons and fibers containing CRF are also widespread in the CNS in birds and mammals and have been reported in the pallium of the telencephalon or its derivatives (in mammals including the cerebral cortex, primarily layers I to III), the septal region, the amygdala, the central gray, the locus coeruleus, the parabrachial region, some vestibular nuclei, and the nucleus of the solitary tract

(Fellman et al., 1984; Merchenthaler et al., 1982; Olschowka et al., 1982; Bons et al., 1985; Knigge and Piekut, 1985). CRF presumably plays some type of neurotransmitter role in these extrahypothalamic circuits.

III. OPIOID PEPTIDES

Over the last several years, the opioid peptides, the brain peptides with opiatelike actions, have been the most intensively studied peptides. The initial studies on opiatelike substances in the nervous system were directed toward elucidating the neurobiological basis of morphine's action on the nervous system. During the early 1970s, Snyder and Pert and their co-workers (Pert and Snyder, 1973a, 1973b) succeeded in demonstrating the existence of receptor sites in the mammalian nervous system that are specific for morphine and other related opiate compounds. Based on the rationale that the presence of opiate receptors in nervous tissue implied that nervous tissue should also contain endogeneous opiatelike ligands (that act at these receptor sites), several groups of investigators sought such substances and successfully found two opioid pentapeptides (five amino acids) in mammalian brain tissue, namely leucine-enkephalin and methionine-enkephalin (Hughes et al., 1975; Simantov and Snyder, 1976). Subsequent studies by many have succeeded in identifying a considerable number of additional, larger opioid peptides, all of which have one of the pentapeptides at their N-terminus or C-terminus. The various opioid peptides were found to be widespread and abundant in the central and peripheral nervous system in mammals, as well as in various peripheral tissues, including endocrine cells and nerve fibers of the gut and the adrenal gland (Johansson et al., 1978; Alumets, et al., 1978b; Weber et al., 1983). During the late 1970s and early 1980s, the interrelationships of the various opioid peptides became clarified by the discovery that each opiate peptide in mammals is derived from one of three different large precursor molecules (Weber et al., 1983), prodynorphin, proenkephalin, and proopiomelanocortin (POMC). As shown in Fig. 8.4, each of these precursors is the parent molecule for a number of related opioid peptides. The members of the three families of opioid peptides appear to be found in different neurons and fibers of the nervous system of mammals (Weber et al., 1983).

Two important generalizations emerged from studies on the functional importance of the opiate peptides in mammals. First, the various opiate peptides clearly have a neurotransmitter-type action (typically inhibitory) in a wide variety of neural systems. Related to this idea, it is clear therefore that the opioid peptides play a role in a wide variety of neural functions and that the well-recognized analgesic effects of the opioid peptides are mediated at only a few of the sites in the central nervous system in which opioid peptides are found (Snyder, 1980). Second, opioid peptides produce their actions by binding at opiate receptor sites. As there are three families of opioid peptides, three classes of opiate receptors have been identified in the nervous system in mammals. These receptor classes have been termed mu, delta, and kappa. One might expect that the members of each family of opioid peptides would be most specific for one of these types of receptors, and this appears to be generally true (Zukin and Zukin, 1984)

The opiate peptides and their receptors have also been studied in a variety of nonmammalian species, including reptilian species. In general, the opiate peptides and their receptors found in mammals are very similar in their major characteristics to those found in other land vertebrates. The findings in reptiles are summarized below.

A. POMC and Beta-endorphin

Fig. 8.4 shows the structure of the precursor for the beta-endorphin and the other peptides derived from this precursor, as it has been determined from various studies in mammals (Akil and Watson, 1983; Herbert et al., 1983; 1985). As can be seen, it is actually a complex

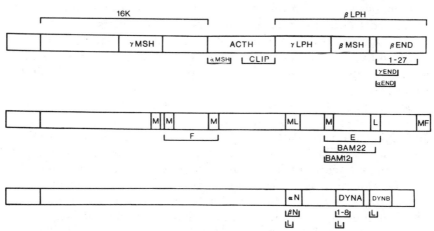

Fig. 8.4. Schematic illustration of the three opioid peptide precursors: POMC (top), proenkephalin (middle), and prodynorphin (bottom). As usual, the N-terminus is to the left and the C-terminus to the right and the relative location within each precursor of its major opioid peptide products is shown. Abbreviations: 16K, 16K fragment; MSH, melanocyte stimulating hormone; ACTH, adrenocorticotropic hormone; CLIP, corticotropinlike intermediate lobe peptide; LPH, lipotropin; END, endorphin; 1-27, beta-endorphin 1-27; M, MENK; F, peptide F; ML, MERGL; L, LENK; E, peptide E; MF, MERF; alpha-N, alpha-neo-endorphin; DYNA, dynorphin A 1-17; 1–8, dynorphin A 1-8; DYNB, dynorphin B.

molecule that serves as the precursor for the anterior pituitary hormone adrenocorticotropic hormone (ACTH) and melanocyte-stimulating hormone (MSH), as well as the opioid peptide beta-endorphin (Akil and Watson, 1983). Note that beta-endorphin contains the methionine-enkephalin (MENK) sequence at its N-terminus. Because of the opioid and nonopioid products of this precursor, the precursor has been termed pro-opiomelanocortin. POMC is found in great abundance in many cells of the anterior lobe of the pituitary and all cells of the intermediate lobe of the pituitary, from which ACTH and MSH are released, respectively (Akil and Watson, 1983). POMC is also found in considerable abundance in the central nervous system, where beta-endorphin appears to be the major active product of POMC. The enzymatic processing of POMC into its different product peptides in mammals has been found to be different in the anterior than in the intermediate lobe of the pituitary. In the anterior lobe, POMC is processed to ACTH and beta-lipotropin (the 91-amino-acid C-terminal peptide of POMC), with approximately one-third of the beta-lipotropin being further processed to beta-endorphin (the 30-amino-acid C-terminal peptide of POMC). In intermediate lobe, ACTH is processed further to corticotropinlike intermediate lobe peptide (CLIP) and alpha-MSH, and all beta-lipotropin is processed to beta-endorphin. Processing of POMC in mammalian brain is thought to be similar to that in the intermediate lobe. Beta-endorphin is not processed to MENK. All the opioid peptide products derived from POMC are present and immunohistochemically detectable in the same individual neurons. The actions of beta-endorphin are mediated via opiate receptors, and beta-endorphin shows equal affinity for the mu and delta sites (Zukin and Zukin, 1984).

Both mu and delta receptor sites are present in the nervous system of the members of all vertebrate classes (Pert et al., 1974; Edley et al., 1982; Nishimura and Pasternak, 1982; Buatti and Pasternak, 1981; Reiner et al., 1989; Albin et al., 1988), implying the presence of endogeneous opioid neuropeptides in all vertebrates. Immunohistochemical and radioimmunoassay techniques have shown that one of these opioid peptides is a peptide similar to mammalian beta-endorphin, which is present in the nervous system and pituitary of fish, frogs, pigeons, and reptiles (Dubois et al., 1979; Jackson et al., 1980; Follenius and Dubois, 1978; Doerr-Schott et al., 1981; Khachaturian et al., 1984; Dores, 1982a; Naik et al., 1980; Nozaki and Gorbman, 1984). Within the nervous system of all vertebrate groups, beta-endorphin is prominent in terminals in the hypothalamus, and beta-endorphin-containing cells are found in the tuberal or arcuate hypothalamus. Within the pituitary of all vertebrates, POMC and beta-endorphin are found in high levels in cells of the anterior and intermediate lobe of the pituitary. The localization, chemical characteristics, and processing of POMC-derived peptides in the brain and pituitary of reptiles have been extensively studied, particularly in *Anolis carolinensis*, by Dores and his co-workers. These comprehensive studies on POMC in reptiles are described in further detail below and compared to the results in other species.

Using immunohistochemical techniques with antisera against various peptides derived from mammalian POMC, Dores (1982a) has demonstrated the presence of cells in the anterior pituitary that contain an ACTH-like peptide, a beta-endorphin-like peptide, and a 16K fragment-like peptide (Fig. 8.4). Cells of the intermediate lobe were found to contain an ACTH-like peptide, an alpha-MSH-like peptide, a beta-endorphin-like peptide, and a 16K fragment-like peptide. Extracts from both lobes, when analyzed by RIA in ACTH and beta-endorphin assays, yielded dose-response curves that were parallel to a standard ACTH dose-response curve and a standard beta-endorphin dose-response curve (Fig. 8.5). Only intermediate lobe extract showed an RIA dose-response curve that was parallel to the standard dose-response curve in an alpha-MSH assay (Fig. 8.5). In contrast, extract from neither lobe showed a parallel dose-response

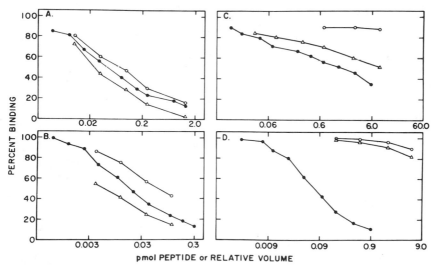

Fig. 8.5. RIA parallelism curves for extracts from a single anterior lobe and a single intermediate lobe of *Anolis carolinensis*. Both tissues showed serial dilution curves parallel to those for ACTH (A) and beta-endorphin (B) in ACTH and beta-endorphin RIAs, respectively. Only intermediate lobe showed parallelism to an alpha-MSH standard in an alpha-MSH RIA (C), and neither showed parallelism to a 16K fragment standard in a 16K fragment RIA (D). Solid circles, peptide standard; triangle, intermediate lobe extract; open circle, anterior lobe extract. (From Dores 1982a).

curve to that for 16K fragment in a 16K fragment assay (Fig. 8.5). Thus, peptides highly similar to beta-endorphin, ACTH, and alpha-MSH are present in the anoline pituitary. In addition, a 16K fragment-like peptide is present in anolis pituitary but is of a somewhat different structure than the 16K fragment peptide derived from mammalian POMC. Since antisera against either ACTH or beta-endorphin were found to detect a high-molecular-weight peptide (although somewhat less in weight than mammalian POMC), it seemed likely that the ACTH-like peptide, the beta-endorphin-like peptide, and the 16K fragment-like peptide all derive from a POMC-like precursor in *Anolis*. In the anterior lobe of *Anolis*, the POMC-like peptide is proteolytically processed to an ACTH-sized peptide and a beta-endorphin-sized peptide (Dores, 1982a; Dores and Surprenant, 1984). In the intermediate lobe, the POMC-like peptide is processed to a beta-endorphin-sized peptide, a CLIP-sized peptide, and an alpha-MSH-sized peptide (Dores, 1982a). The levels of the ACTH-like substance and the beta-endorphin-sized peptide in intermediate lobe are eight times higher than in the anterior lobe. These differences between anterior lobe and intermediate lobe in POMC-peptide content and POMC processing are comparable to those observed between anterior and intermediate lobes in mammals. In mammals, however, anterior lobe beta-lipotropin is less fully processed to beta-endorphin than in lizards.

Partial structural characterization of *Anolis* beta-endorphin indicates that this is highly similar to but distinct from mammalian beta-endorphin (Dores, 1983). *Anolis* beta-endorphin is perhaps more similar in structure to avian beta-endorphin than to mammalian beta-endorphin (Naude and Oelofsen, 1981; Yamashiro et al., 1980). Ostrich beta-endorphin shows approximately 75% sequence similarity in terms of amino acid composition to human beta-endorphin (Naude and Oelofsen, 1981), and ostrich beta-lipotropin shows approximately 85% similarity in amino acid composition to ovine beta-lipotropin (Li et al., 1965). All forms of beta-endorphin that have been sequenced or at least partially characterized, including salmon, share in common the N-terminal sequence Tyr-Gly-Gly-Phe-Met (Dores, 1983; Kawauchi et al., 1980), which is identical to the sequence of methionine-enkephalin (MENK).

Dores (1982b) has also studied the posttranslational steps involved in the enzymatic cleavage of *Anolis* POMC to its final end products in the intermediate lobe. He incorporated radioactive amino acids to radioactively label reptile POMC during its biosynthesis from messenger RNA in cells of the intermediate lobe of *Anolis*. A single type of POMC molecule, 23.3 K^2 in size, was found in *Anolis*. By contrast, in mammals POMC is a 31K molecule (Mains et al., 1977). In *Xenopus*, two types of POMC have been found in the pituitary: one 32K in size and one 29.5K in size (Loh, 1979), whereas in the pituitary of *Rana ridibunda* POMC is reportedly 36K in size (Vaudry et al., 1984). Rodrigues et al. (1983) have reported that the neurointermediate lobe of the rainbow trout also appears to synthesize two distinct POMC molecules. In *Anolis* intermediate lobe, POMC is first cleaved to a structure termed ACTH biosynthetic intermediate 1 by Dores, plus beta-lipotropin. Subsequently, beta-lipotropin is cleaved to produce beta-endorphin, while ACTH biosynthetic intermediate 1 is cleaved to produce an ACTH-sized peptide (termed by Dores ACTH biosynthetic intermediate 2) and the (approximately) 8K N-terminal peptide of POMC. The smaller size of *Anolis* POMC as compared to mammalian POMC appears to be explained by the fact that the N-terminal peptide abutting ACTH in mammalian POMC is 16K in size, while it is approximately 8K in size in *Anolis*. As noted above, immunological techniques confirm that the N-terminal 8K peptide of *Anolis* POMC is similar to mammalian N-terminal 16K fragment peptide but distinguishable from it. In the final step in the processing of POMC in the *Anolis* intermediate lobe, ACTH biosynthetic intermediate 2 is cleaved to form a CLIP-like and an alpha-MSH-like peptide. These steps are shown in Fig. 8.6. Beta-endorphin also can be generated from POMC in *Anolis* by several minor biosynthetic pathways (Dores, 1982b).

The steps involved in the major pathway for the processing of *Anolis* POMC are similar to those observed in mammals, though beta-endorphin seems to be more rapidly derived from POMC than is the case in mammals. In further studies, Dores has shown that, much as in mammals, nearly all beta-lipotropin in *Anolis* intermediate lobe is processed to beta-endorphin (Dores, 1983). Also as in mammals,

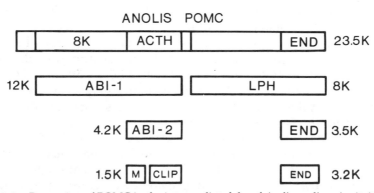

Fig. 8.6. Processing of POMC in the intermediate lobe of *Anolis carolinensis* pituitary, based on the work of Dores and adapted from Dores 1983. See text for further explanation. ABI-1 and ABI-2 stand for ACTH biosynthetic intermediate 1 and ACTH biosynthetic intermediate 2, respectively.

in which intermediate lobe beta-endorphin is further processed to beta-endorphin 1-27 by C-terminal proteolytic cleavage, the beta-endorphin-sized peptide of *Anolis* intermediate lobe is cleaved at the C-terminal to a beta-endorphin 1-27-sized material. Unlike in mammals, however, in which intermediate lobe beta-endorphin is acetylated at the N-terminus and consequently rendered devoid of opiate activity, lizard intermediate lobe beta-endorphin is not N-terminally acetylated and remains a biologically active opioid peptide (Dores, 1983; Dores and Surprenant, 1983). Thus, beta-endorphin may be released from intermediate lobe in lizards and act hormonally on peripheral opiate receptor sites.

In the anterior pituitary in mammals, ACTH is not cleaved to CLIP and alpha-MSH. The absence of CLIP-like labeling in the *Anolis* anterior pituitary with immunohistochemical techniques suggested that the anterior pituitary in *Anolis* also does not process ACTH into CLIP. Dores and Surprenant (1984) have recently directly examined POMC biosynthesis in *Anolis* anterior pituitary and confirmed the earlier observation that the major POMC products are an ACTH-sized material and a beta-endorphin-sized material. Thus, unlike in rat anterior lobe, the processing of the C-terminus of anterior lobe POMC in *Anolis* does not largely stop with the production of a beta-lipotropin-like peptide but continues through to the production of a beta-endorphin-like peptide. In contrast, in mammalian anterior lobe, only approximately one-third of the beta-lipotropin is processed to beta-endorphin. Thus, except for the cleavage of ACTH biosynthetic intermediate 2 to a CLIP-like and an alpha-MSH-like substance in intermediate lobe, the anterior and intermediate lobe appear to process POMC in similar ways. Recent studies in mammals indicate that beta-endorphin can influence temperature regulation, respiration, and cardiovascular functions (Martin, 1984). Beta-endorphin released from the lizard pituitary may influence similar functions.

Immunohistochemical techniques have been used to study the CNS distribution of beta-endorphin in lizards (Khachaturian et al., 1984) (Fig. 8.7). Since antibodies against beta-endorphin, ACTH, and alpha-MSH all appear to label the same neurons and fibers in the *Anolis* nervous system, it appears likely that processing of POMC in *Anolis* brain is similar to that observed in the intermediate lobe of the pituitary (Khachaturian et al., 1984; Dores et al., 1984). POMC cells were observed in the arcuate region of the hypothalamus and in the medial tegmentum. The POMC cells in *Anolis* tegmentum do not correspond to any POMC cell group previously described in mammals. Two populations of beta-endorphin-containing neurons have been identified in the nervous system in mammals: one in the arcuate region of the hypothalamus and the second in the nucleus of the solitary tract (Akil and Watson, 1983; Knigge and Joseph, 1984). The dis-

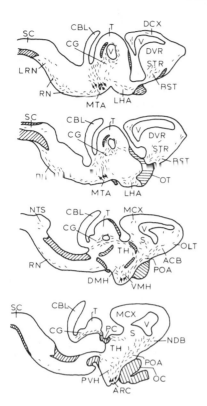

Fig. 8.7. Distribution of POMC-containing neurons and fibers in the brain of *Anolis carolinensis*. The sagittal sections shown represent a lateral to medial (top to bottom) series through the brain and rostral spinal cord. Dashes represent fibers and large dots perikarya. Abbreviations: ACB, nucleus accumbens; ARC, arcuate nucleus; BST, bed nucleus of the stria terminalis; CBL, cerebellum; CG, central gray; DCX, dorsal cortex; DMH, dorsomedial hypothalamic nucleus; DVR, dorsal ventricular ridge; LHA, lateral hypothalamic area; LRN, lateral reticular nucleus; MTA, medial tegmental area; NDB, nucleus of the diagonal band; NTS, nucleus of the solitary tract; OC, optic chiasm; OLT, olfactory tubercle; OT, optic tract; PC, posterior commissure; POA, preoptic area; PVH, periventricular hypothalamus; RN, reticular nucleus; S, septum; SC, spinal cord; STR, striatum; T, tectum; TH, thalamus; V, ventricle; VMH, ventromedial hypothalamic nucleus. (From Khachaturian et al., 1984.)

tribution of beta-endorphin-positive fibers was similar to that in mammals; numerous fibers were observed in the olfactory tubercle, nucleus accumbens, the septal nuclei, the bed nucleus of the stria terminalis, the amygdala, the preoptic region, the periventricular hypothalamus, the deep tectum, and the central gray. Additional fibers were also observed in the ventromedial and posterior hypothalamic nuclei, the dorsomedial thalamus, the ventral mesencephalon, the nucleus of the solitary tract, the raphe nuclei, and the rhombencephalic reticular formation. These various beta-endorphin fibers presumably arise from the tuberal or the tegmental beta-endorphin-containing neurons or both. Although these regions of beta-endorphin-containing fibers are similar to those reported in mammals, beta-endorphin-positive fibers are apparently more sparse in lizard telencephalon than generally observed in the telencephalon in mammals. Two areas containing scattered beta-endorphin-positive fibers were observed in lizard telencephalon that apparently do not contain such fibers in the mammalian telencephalon, namely the striatum and the medial cortex.

Vallarino (1984) has recently described the distribution of alpha-

MSH in the brain of *Podarcis muralis*. Since both alpha-MSH and beta-endorphin derive from POMC, the distribution of alpha-MSH should be comparable to that of beta-endorphin. Nonetheless, Vallarino (1984) reports immunoreactive perikarya in the preoptic region and in the paraventricular and supraoptic nuclei, which contrasts with the findings of Khachaturian et al. (1984). Vallarino's description of the distribution of beta-endorphin-containing perikarya and fibers in extrahypothalamic brain areas of *Podarcis* also differs markedly from that of Khachaturian et al. (1984). Studies in turtles indicate that the distribution of beta-endorphin is fundamentally the same as reported in anoline lizards (Reiner, unpublished observations). Cells containing a beta-endorphin-like peptide, however, have not been observed in the tegmentum of turtles, and beta-endorphin-positive fibers have not been observed in turtle medial cortex (Reiner, unpublished observations). In turtles, beta-endorphin-positive fibers were observed in the median eminence, and beta-endorphin-positive cells and fibers were observed in the neurohypophysis. Using an antiserum against the POMC product alpha-MSH, Vallarino (1984) also observed labeled terminals in the median eminence in *Podarcis*. Analyses of extracts from different regions of lizard brain by gel filtration chromatography and RIA indicate that beta-endorphin and alpha-MSH are the major POMC end products in brain (Dores et al., 1984). Processing of POMC in the lizard hypothalamus was found to be similar to that in the intermediate lobe. In forebrain, more low-molecular-weight forms of beta-endorphin are present than in the hypothalamus (2.5K–3.0K). In the hindbrain, 70% of the beta-endorphin is 2.6K in size, and little ACTH is detectable. Thus in both the tegmental and the hypothalamic POMC neuronal systems, significant processing of POMC end products occurs in terminals or during transport to terminals. Although the postsynaptic influence of beta-endorphin on neurons in reptiles has not been examined, it is likely that beta-endorphin binds to mu and delta receptors and has a neurotransmitter type function.

B. Enkephalin

Proenkephalin in mammals contains four methionine[5]-enkephalin (MENK) molecules, one leucine[5]-enkephalin (LENK) molecule, and two C-terminally extended methionine-enkephalin molecules (methionine[5]-enkephalin-arginine[6]-phenylalanine[7], or MERF, and methionine[5]-enkephalin-arginine[6]-glycine[7]-leucine[8], or MERGL) (see Fig. 8.4). The enkephalins are prominent in a variety of brain regions and in the adrenal medulla, where enkephalin is present in the same storage granules of chromaffin cells as adrenalin (Weber et al., 1983; Viveros et al., 1979). The concentration of these six enkephalin peptides in any tissue will depend on the extent to which proenkephalin is enzymatically cleaved in that tissue. For example, although the smaller enkephalin molecules are the primary products derived from proenkephalin in the brain, larger enkephalin peptides (e.g., peptide E, BAM22P, and BAM12P; see Fig. 8.4) are the more prominent products in the adrenal medulla (Udenfriend and Kilpatrick, 1983). The various enkephalins and proenkephalin are found in numerous neurons of diverse cell groups of the nervous system in mammals (Khachaturian et al., 1983a, 1983b; Finlay et al., 1981a; Wamsley et al., 1980; Haber and Elde, 1982). Individual enkephalinergic neurons have been shown to co-contain the various pro-enkephalin-derived peptides (Khachaturian et al., 1983c). Enkephalin is also prominent in other peripheral tissues, such as in endocrine cells and nerve fibers of the gut (Johansson et al., 1978; Alumets, 1978b) and in the retina, where it has been observed in amacrine cells in a wide variety of species (Brecha et al., 1984).

Enkephalin peptides appear to be present in abundance in the members of all vertebrate groups studied. Edley et al. (1982) have reported the presence of delta receptors (which are the preferred receptors for at least the smaller enkephalin peptides) in the telencephalon of bony fish, frogs, turtles, birds, and mammals, as well as in insects. Immunohistochemical studies indicate that enkephalin peptides are present in birds (La Valley and Ho, 1983; Bayon et al., 1980; De Lanerolle et al., 1981; Blahser and Dubois, 1980; Ryan et al., 1981; Reiner, Karten, and Brecha, 1982; Reiner, Davis, Brecha, and Karten, 1984), reptiles (Naik et al., 1981; Brauth and Reiner, 1982; Reiner, 1983; Brauth, 1984), amphibians (Doerr-Schott et al., 1981; Kuljis and Karten, 1982; Schulman et al., 1981), bony fish (Finger, 1981; Schulman et al., 1981; Reaves and Hayward, 1980; Reiner and Northcutt, 1987, 1988); cartilaginous fish (Northcutt et al., 1983, 1988) and agnathans (Gold and Finger, 1982). Enkephalin bioassay (using opiate receptor–rich membranes from mammalian nervous tissue) has also been used to demonstrate the presence of enkephalin pentapeptides in members of the above vertebrate groups, as well as in the members of such invertebrate groups as planarians, molluscs, arthropods, including insects, and spiders (Simantov et al., 1976). Radioimmunoassay and immunohistochemistry have also been used to demonstrate enkephalin in nervous tissue in further invertebrate groups such as leeches (Zipser, 1980), the urochordate *Ciona* (Georges and Dubois, 1984), and earthworms (Rzasa et al., 1984). Dose-response curves, showing the displacement of radioactive opiates from opiate receptor–rich membranes by enkephalin extract from toad, turtle, hagfish, chick, and mouse, are parallel to one another in enkephalin bioassay (Simantov et al., 1976), thereby suggesting that a considerable similarity exists among the enkephalin pentapeptides present in the nervous systems of the members of these different vertebrate groups. Gel chromatography indicates that the small enkephalin peptides present in shark, turtle, and calf are of a similar molecular weight (Simantov et al., 1976). Using RIA for LENK and MENK, Rzasa et al. (1984) showed that the enkephalinlike substances in earthworm nervous tissue showed dose-response curves that were parallel to those for authentic LENK and MENK. Using the same approach, King and Millar (1980b) have also shown that MENK is present and of a highly similar (if not identical) chemical structure in the nervous systems of rats, toads, teleosts, and snails.

A number of recent studies using more refined biochemical techniques and molecular techniques have clarified the structure of proenkephalin and its derived peptides in members of the different classes of land vertebrates. Martens and Herbert (1984) used cDNA hybridization and cloning techniques to isolate and characterize the gene coding for proenkephalin in *Xenopus laevis*. They found that two proenkephalin genes are present in *Xenopus*, and both contain five copies of MENK, rather than four of MENK and one of LENK. The LENK sequence is entirely absent in both proenkephalin genes (Martens and Herbert, 1984). Further, although MERF is present, MERGL as such is not present in *Xenopus* proenkephalin. Instead, an altered form of MERGL, in which tyrosine has replaced leucine, is present in *Xenopus*. Since fundamentally similar results have been obtained in two additional anuran species (Kilpatrick et al., 1983; Merchenthaler et al., 1987, 1989), it appears likely that the major opioid peptide products of proenkephalin are largely similar in all frogs. Since no comparable data are available for urodeles or any fish species, it is uncertain whether the features of frog proenkephalin are primitive for land vertebrates or are derived in frogs (Lindberg, 1986). Some data suggest that LENK may be absent in cartilaginous fish and bony fish (Lindberg, 1986), thereby suggesting that frog proenkephalin may, in fact, possess the features that are primitive for land vertebrates.

Biochemical and immunohistochemical studies indicate that the structure of proenkephalin and the proenkephalin-derived peptides in reptiles are similar to those in birds but different from those found in frogs or mammals (Reiner, 1987a; Lindberg, 1986; Lindberg and White, 1986; Maderdrut et al., 1986; Kilpatrick et al., 1983; Petrusz et al., 1985; White et al., 1985). In the nervous systems of chickens, turtles (*Pseudemys scripta*), crocodilians (*Alligator mississippiensis*), and lizards (*Anolis carolinensis*), proenkephalin appears to contain four copies of MENK, one of LENK, and one of MERF. MERGL as such is not present, although the form of MERGL with a tyrosine substitution at the C-terminus may be present. The simplest interpretation of these results in terms of the evolution of proenkephalin is that LENK was added at some point during the evolution of the reptiles ancestral to living birds, reptiles, and mammals from their amphibian fore-bears, and that MERGL evolved during the mammalian lineage after its divergence from the lineage leading to modern birds and reptiles (Lindberg, 1986; Lindberg and White, 1986). MERF may have been added during the lineage leading from bony fish to amphibians, since MERF is reportedly absent in bony and cartilaginous fish (Lindberg, 1986).

Thus a precursor for the enkephalin peptides is clearly present in the nervous system of diverse reptilian species, and this precursor (which will be called proenkephalin) has as its major derived peptides LENK, MENK, and MERF. The biochemical studies of Lindberg and White (1986) show, however, that proenkephalin may be processed differently in the brains of different groups of reptiles. In turtles, the MENK:MERF:LENK ratio is nearly 4:1:1, indicating nearly complete processing of all copies of each of these peptides from the precursor. In *Alligator mississippiensis* and *Anolis carolinensis*, however, the MENK:MERF:LENK ratio is approximately 4:0.4:0.7 and 6:1:1, respectively. Thus, LENK and MERF may be processed incompletely from proenkephalin in crocodilians and lizards. Unpublished RIA studies (Table 8.3) (White, Reiner, and McKelvy, unpublished observations) confirm that the MENK:LENK ratio in a number of different brain regions in red-eared turtles is similar to that reported for whole brain by Lindberg and White (1986). These studies in turtles also show that the concentration and ratio of the enkephalin pentapeptides for a number of brain regions is similar to that observed in mammals and birds (Table 8.3) (White, Reiner, and McKelvy, unpublished observations).

Table 8.3. Comparison of the amounts of LENK and MENK measured by RIA in several comparable regions of the brains of turtle, rabbit, and pigeon.

	Turtle		Mammal		Pigeon	
	MENK	MENK LENK	MENK	MENK LENK	MENK	MENK LENK
Basal telencephalon, including the basal ganglia	341	3.2	391	3.0	100	5.7
Hypothalamus	659	2.7	330	5.8	200	5.7
Tectum	145	4.3	—	—	20	5.0
Spinal cord	163	4.0	—	—	—	—
Cortex	77	2.1	24	1.5	8	4.0
Cerebellum	< 1	—	30	—	8	4.0

Note: The amounts of MENK shown are in pg/mg wet weight tissue. The amount of LENK can be calculated from the MENK/LENK ratio for that tissue. Turtle data from unpublished studies by J. D. White, A. Reiner and J. F. McKelvy. Mammalian data from Hughes et al. (1977) and pigeon data from Bayon et al. (1980). The value shown for the avian hypothalamus is actually for the entire diencephalon.

The distribution of the enkephalin peptides has been studied in members of each order of reptiles, including rhynchocephalians. In brief, the distribution of enkephalin appears to be very similar among birds, mammals, and the members of these three orders of reptiles, with the major exception that an enkephalin-positive pretectal cell group that is prominent in birds and reptiles is not evident in mammals. The distribution of enkephalin-positive neurons and fibers has been reported in detail by several authors (Naik et al., 1981; Brauth and Reiner, 1982; Reiner, 1983; Brauth, 1984; Wolters et al., 1986; Reiner, 1987a; Reiner, Brauth, and Karten, 1984; Russchen et al., 1987). Only the major features of this distribution pattern will be summarized here. The original articles contain further detail. Since MENK, LENK, and MERF have been shown by immunohistochemical double-label studies to be in the same neurons and fibers in turtles and since this is likely to be true for all reptiles, the distributions of MENK, LENK, and MERF will not be considered separately in the following discussions.

Several of the published studies on enkephalin distribution in reptiles (except those of Wolters et al., 1986, Russchen et al., 1987, and Reiner, 1987a) have used the same antisera (supplied by K. J. Chang; antisera described in Miller et al., 1978), and the results of these studies are generally comparable to one another and to the several studies in mammals and birds using the same antisera (Finlay et al., 1981a; Sar et al., 1978; Reiner, Karten, and Brecha, 1982; Reiner, Davis, Brecha, and Karten, 1984). Results of preliminary studies (A. Reiner) on the distribution of enkephalin in a sceloporine lizard (*Sceloporus undulatus*), a snake (*Thamnophis sirtalis*), and a crocodilian (*Caiman crocodilus*), using Chang's LENK antiserum, will also be presented below. Unpublished studies on *Sphenodon* have been carried out using commercially available anti-LENK (from ImmunoNuclear Corp.) (Reiner, unpublished observations). Additional studies have been performed in turtles using antisera against several of the larger enkephalin peptides, including Peptide E, BAM-22P, and MERF (Reiner, 1987a). As noted above, double-label studies (Reiner, 1987a) show these peptides are present in the same neurons as LENK and MENK, and the distribution of these peptides in turtles is similar to those observed for LENK and MENK (though regional differences do appear in the relative labeling intensities for these different enkephalin peptides). The similarity in distribution among the various enkephalin peptides in turtles serves to confirm that the distribution of labeling observed with LENK antisera in turtles does reflect the distribution of enkephalinergic perikarya and fibers (and does not reflect cross-reactive staining of dynorphinergic cells and fibers).

Within the telencephalon, enkephalin-positive neurons and fibers are abundant in the striatum of turtles (Reiner, 1983, 1987a), crocodilians (Brauth, 1984), gekkonid, anoline, and sceloporine lizards (Naik et al., 1981; Russchen et al., 1987; Reiner, unpublished observations), snakes (Reiner, unpublished observations), and *Sphenodon* (Reiner, unpublished observations) (Fig. 8.8). Unpublished RIA studies indicate that the levels of the enkephalin pentapeptides found in turtle basal ganglia are similar to those found in mammals (see Table 8 3) (White, Reiner, and McKelvy, unpublished observations). In turtles, lizards, and snakes, a clearly distinct region of enkephalin-positive fibers in the ventrolateral telencephalon appears to correspond to the globus pallidus of mammals (Reiner, 1983, 1987a, unpublished observations; Reiner, Brauth, and Karten, 1984; Russchen et al., 1987). Brauth (1984) has reported that globus pallidus is less identifiable as a discrete cell group in crocodilians. Consistent with this observation, enkephalinergic fibers, though present in the ventrolateral area of crocodilians, are not aggregated into a discrete pallidal cell field. In all reptiles, however, the enkephalinergic fibers in globus pallidus appear to originate from enkephalinergic neurons in the striatum (Reiner, Brauth, and Karten, 1984). The striatum of all reptiles is also rich in fibers that appear to be the processes of the enkephalin-positive striatal neurons (see Fig. 8.8). Recent double-label studies indicate that the enkephalinergic neurons of the striatum are largely a separate population from those that contain substance P, dynorphin, or both (Reiner, 1986a; Anderson and Reiner, 1990a), but the enkephalinergic neurons do typically also contain the neurotransmitter GABA (Reiner, unpublished observations).

Within the telencephalon, additional enkephalin-positive neurons have been observed in the septum and medial cortex of turtles, anoline lizards, and crocodilians (Naik et al., 1981; Reiner, 1983, 1987a; Brauth, 1984). In snakes, numerous, heavily labeled enkephalinergic neurons have been observed in the septum (Reiner, unpublished observations). In the septal region of turtles, snakes, sceloporine and anoline lizards, and caiman, enkephalinergic fibers have been observed to make prominent pericellular terminations on nonenkephalinergic septal neurons. In turtles and crocodilians, enkephalin-positive neurons are more numerous in the caudal portions of the pallial roof of the telencephalon than in the rostral. In snakes, enkephalinergic neurons have been observed in the pyriform cortex. A small number of enkephalin-positive neurons have been observed in the dorsal ventricular ridge (DVR) of turtles, lizards, and crocodilians. In colchicine-treated turtles, enkephalinergic neurons have also been observed in the superficial cell plate of the DVR. In snakes, enkephal-

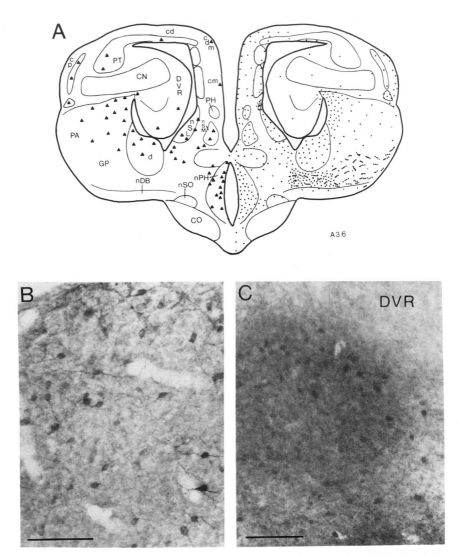

Fig. 8.8. (A) Line drawing of a transverse section through a midtelencephalic level of *Pseudemys* depicting the distribution of enkephalinergic neurons (triangles) and fibers (dots). Each triangle represents 10 neurons. The number to the lower right of the line drawing represents the level of this section in relation to the stereotaxic atlas of the turtle brain of Powers and Reiner (1980). See list of abbreviations below to identify the various brain structures. (B) Photomicrograph of enkephalinergic neurons in medial striatum (area d) of *Pseudemys* (medial to the left). The neurons have been labeled according to the PAP procedure using an antiserum against peptide E. (C) Photomicrograph of enkephalinergic neurons in the striatum of *Sceloporus undulatus* (medial to the right) labeled according to the PAP technique using Chang's anti-LENK antiserum. The scale bars for both (B) and (C) equal 100 μm. Abbreviations: CN, core nucleus of the DVR; CO, optic chiasm; cd, cortex dorsalis; cdm, cortex dorsomedialis; cm, cortex medialis; cp, cortex pyriformis; DVR, dorsal ventricular ridge; d, area d; GP, globus pallidus; nDB, nucleus of the diagonal band; nPH, nucleus periventricularis hypothalami; nSL, nucleus septalis lateralis; nSM, nucleus septalis medialis; nSO, nucleus supraopticus; PA, paleostriatum augmentatum; PH, primordium hippocampi; PT, pallial thickening.

inergic neurons were observed in the amygdalar region. In lizards and snakes, enkephalinergic neurons were observed in nucleus sphaericus. These neurons were particularly well labeled in snakes. In all reptiles studied, prominent telencephalic enkephalin-positive fibers (in addition to those in the basal ganglia), were found in the medial cortical regions, the septum, and the ventral paleostriatum (in anoline lizards comprising the so-called stria terminalis and ventral nucleus accumbens of Naik ct al. (1981; Reiner, 1987a, unpublished observations; Brauth, 1981, Russ lion et al., 1987).

Within the diencephalon, enkephalin-positive neurons were numerous in the rostral two-thirds of the hypothalamus and mammillary region and largely absent from the thalamus and habenula in turtles, anoline lizards, and crocodilians (Naik et al., 1981; Reiner, 1983, 1987a, unpublished observations). In turtles, enkephalinergic neurons in the preoptic region send processes through the ependyma that appear to be in contact with the third ventricle. In snakes, preoptic and paraventricular enkephalinergic neurons apparently contain large amounts of enkephalin (since they are very heavily labeled). In both snakes and sceloporine lizards, a few enkephalinergic neurons were observed in the median eminence and the neurohypophysis. In sceloporine lizards, enkephalinergic neurons were observed in the medial habenular nucleus. In snakes, cells and fibers of the habenula were extremely heavily labeled for LENK. Enkephalin-positive fibers were densely present in the hypothalmus and neurohypophysis and prominent in the dorsomedial and dorsolateral thalamic nuclei and the medial habenular nucleus of all reptiles examined. Enkephalinergic fibers tend to be conspicuously absent from the thalamic cell groups that project to the telencephalon and process visual or auditory information (Reiner, 1987a; Hall and Ebner, 1970; Reiner and Powers, 1978, 1980, 1983; Ulinski, 1983, 1988). Enkephalinergic fibers have been observed in the ventromedial hypothalamic region and median eminence of lizards, turtles, and snakes. In snakes, median eminence fibers were extremely heavily labeled for LENK. It is uncertain if the heavy labeling of the ME represents authentic labeling of LENK-containing processes or cross-reactive staining of dynorphin-positive processes. Since antisera against the larger enkephalin peptides that do not cross-react with any dynorphin peptides (such as BAM-22P and peptide E) also label fibers of the median eminence in turtles, it does seem likely that enkephalin is present in fibers of the median eminence of turtles, lizards, and snakes.

Within the midbrain, enkephalin-positive neurons and fibers are prominent in the optic tectum of members of all orders of reptiles studied (Reiner, 1983, 1987a; Brauth and Reiner, 1982; Naik et al., 1981; Reiner, unpublished observations). The enkephalin-positive neurons in the tectum give rise to enkephalin-positive processes that ascend radially and ramify in the major tectal retinorecipient layer

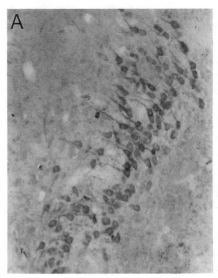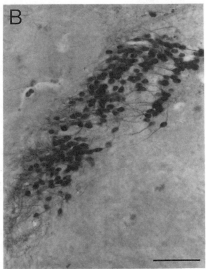

(Brauth and Reiner, 1982; Reiner, 1987, unpublished observations) (Fig. 8.9). In turtles, these neurons are located in the deep periventricular layer of the tectum, whereas in anoline and sceloporine lizards and snakes they are located in the intermediate tectal layers. These neurons are presumably homologous to one another, despite the differences in laminar location, which may reflect the tendency of tectal cells to vary among different reptilian groups in the extent to which they migrate from the ventricle during development (Senn, 1971). In sceloporine lizards and snakes, enkephalinergic neurons have also been observed in the superficial layers of the tectum. In birds, enkephalin-positive neurons in intermediate layers of the tectum also appear to give rise to enkephalin-positive processes that ascend radially and ramify in the main retinorecipient layer (Reiner et al., 1982a, 1982b). Enkephalin-positive fibers are also present in the deeper layers of the avian and reptilian tectum (Reiner et al., 1982b; Reiner, 1983, 1987a, unpublished observations; Naik et al., 1981; Brauth and Reiner, 1982).

In birds, the enkephalin-positive fibers in the deeper layers of the tectum are the axons and terminals of the enkephalin-positive neurons of the pretectal cell group termed the nucleus spiriformis lateralis (SpL). All neurons of this cell group contain enkephalin in birds. A homologous cell group (termed the dorsal nucleus of the posterior commissure in reptiles, nDCP), all neurons of which contain enkephalin, is clearly present in the pretectum of turtles, lizards, and crocodilians (Brauth and Reiner, 1982; Reiner, 1983, 1987a, unpublished

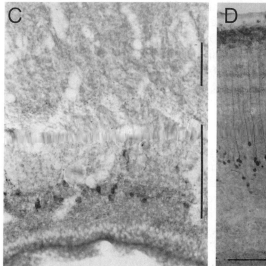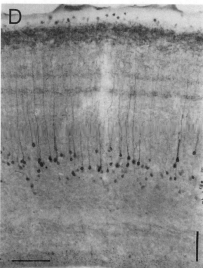

Fig. 8.9. Enkephalin-positive labeling in neurons of nDCP of *Pseudemys* (A), neurons of nDCP of *Sceloporus undulatus* (B), neurons and fibers of the deep tectum of turtle (C), and neurons and fibers in the tectum of *Sceloporus undulatus* (D). The turtle tissue was processed with an antiserum against peptide E, whereas the lizard tissue was processed with Chang's anti-LENK antiserum. Seemingly all neurons of nDCP in turtles and lizards are positive for enkephalin. This cell group projects to the tectum and gives rise to enkephalinergic fibers in the deeper layers of the tectum. Neurons of nDCP in turtles are more heavily labeled by LENK antisera than by antisera against the large enkephalin peptides (such as peptide E and BAM-22P). Medial is to the right in (A) and (B). Tectal neurons in turtles and lizards give rise to radially directed processes that ramify in the major retinorecipient layer, though the processes of the neurons in turtle are not evident in the photomicrograph shown. In turtle these enkephalinergic neurons are present in the deep, periventricular layers (its extent is shown by the solid bar to the lower right in C). In contrast, the apparently homologous enkephalinergic neurons in lizards are found in the intermediate tectal layers, not the deep layers (indicated by the bar in D). A few rare enkephalinergic neurons have, however, been observed in the intermediate layers of the turtle tectum superficial to the layer of large multipolar neurons (which is often called the stratum griseum centrale and is indicated by the upper bar in C). The scale bars for (B) and (D) both equal 100 μm. Photomicrographs (A) and (C) are at the same magnification as (B).

observations) (Fig. 8.9). This cell group also appears to be present in snakes, though the presence of large amounts of enkephalin in the neurons of this cell group has been less clearly evident in preliminary studies in snakes (Reiner, unpublished observations). In reptiles (as in birds), the enkephalin-positive neurons of nDCP appear to be the source of enkephalinergic fibers in the deep layers of the tectum. Pathway tracing studies have shown that nDCP projects massively to the reptilian optic tectum (Reiner et al., 1980; Brauth and Kitt, 1980;

ten Donkelaar and de Boer-van Huizen, 1981; Dacey, 1982). In immunohistochemical studies, enkephalin-positive fibers can be traced
from nDCP into the deep layers of the reptilian optic tectum. This
enkephalin-positive pretectal-tectal pathway is prominent in birds
and all reptiles in which it has been studied extensively. Although this
pathway does appear to be present in mammals and amphibians, it is
not as prominent as in birds and reptiles (Reiner, Karten, and Brecha,
1982; Reiner, Brauth, and Karten, 1984). This pathway, however, may
also be enkephalinergic in amphibians (Merchenthaler et al., 1989;
Lazar et al., 1990). Thus the enkephalinergic nature of the SpL/nDCP
projection to the tectum may be unique to nonmammalian land vertebrates (Reiner, Brauth, and Karten, 1984). SpL and nDCP have been
shown to receive massive input from the basal ganglia in birds and
reptiles (Reiner et al., 1980; Brauth and Kitt, 1980; Reiner, Brecha, and
Karten, 1982) and to project to the tectal layers that give rise to the
output pathways of the tectum (Reiner and Karten, 1982; Reiner,
Brauth, and Karten, 1984). The frog homologue of nDCP appears to
possess similar connectivity (Reiner, Brauth, and Karten, 1984; Wilczynski and Northcutt, 1983). Since basal ganglia outputs to the thalamus are sparse and since no telencephalic motor cortex is clearly
evident in amphibians, birds, and reptiles, the basal ganglia of amphibians, birds, and reptiles does not appear to be able to mediate its
role in motor functions by a pathway to motor cortex via the thalamus, as is the case in mammals. Accordingly, the pathway from the
basal ganglia to tectum via SpL/nDCP may be the major pathway by
which the avian and reptilian basal ganglia influence motor functions
(Reiner et al., 1980; Reiner, Brecha, and Karten, 1982; Reiner, Karten,
and Brecha, 1982; Reiner, Brauth, and Karten, 1984). The behavioral
data of Bugbee (1979) in birds is consistent with this hypothesis. This
circuit may be the ancestral circuit by which the basal ganglia influence motor functions.

Enkephalin-positive neurons and fibers were evident in the medial
tegmentum of turtles, and enkephalin-positive fibers and a few enkephalinergic neurons were evident in the central and periventricular
gray and area pretectalis of lizards and turtles (Naik et al., 1981; Reiner, 1983, 1987a Reiner, unpublished observations). Fibers labeled
with Chang's LENK antiserum were more prominent in the ventral
tegmental area and the substantia nigra area in snakes than in other
reptiles. In turtles, snakes, and sceloporine lizards, enkephalinergic
neurons were prominent in the laminar nucleus of the torus semicircularis. Weiler (1985) has recently reported that enkephalinergic neurons within the isthmic region (Reiner, 1987a), located between locus
coeruleus and the nucleus isthmi pars magnocellularis, are the source
of centrifugal projections to the eye in red-eared turtles.

Within the hindbrain and spinal cord, enkephalin-positive fibers and neurons were prominent in the dorsal horn of the spinal cord and the caudal nucleus of the descending trigeminus in lizards and snake (Reiner, unpublished observations; Naik et al., 1981) and in the tract of Lissauer of the spinal cord and the caudal descending tract of the trigeminal nerve in turtles (Reiner, 1983, 1987a). Since opiates are known for their analgesic properties, it has been proposed that the opiates may promote analgesia by preventing the transmission of pain input from the primary sensory fibers to the descending tract of the trigeminal nerve and the dorsal horn of the spinal cord (Nicoll et al., 1980b; Snyder, 1980). In light of the presence of enkephalin-positive neurons in dorsal horn regions of the reptilian spinal cord and the presence of enkephalin-positive fibers in the central gray of reptiles (which is another region implicated in pain gating), some central enkephalin-positive systems in reptiles may be involved in suppression of transmission of nociceptive information in reptiles. This hypothesis for reptiles is made further plausible by the finding that opioid peptides are involved in analgesia in frogs (Pezalla and Dicig, 1984). In anoline and varanid lizards and turtles, enkephalinergic neurons have been observed in the nucleus of the solitary tract (Naik et al., 1981; Wolters et al., 1986; Reiner, 1987a). In turtles, enkephalinergic neurons are present in the pontine region near the motor nucleus of the seventh cranial nerve (Reiner, 1987a), which is reported to give rise to olivocochlear projections (Strutz, 1982). In mammals, such neurons are known to contain acetylcholine and enkephalin (Altschuler et al., 1984). In both turtles and varanid lizards, enkephalin-positive neurons have been observed in the descending vestibular nucleus (Reiner, 1987a; Wolters et al., 1986). In all reptiles studied, enkephalin-positive fibers have been noted to be present in abundance in the raphe region, the ventrolateral pons and medulla, locus coeruleus, the parabrachial region, the nucleus of the solitary tract, and the dorsal motor nucleus of the vagus nerve (Naik et al., 1981; Reiner, 1983, 1987a; Wolters et al., 1986; Reiner, unpublished observations). A few enkephalin-positive fibers were observed in the cerebellum of turtles (Reiner, 1987a) and sceloporine lizards (Reiner, unpublished observations).

The function of enkephalin in the turtle nervous system has not been extensively studied, but it seems reasonable to assume that enkephalin acts as an inhibitory neurotransmitter or neuromodulator, as has been found in mammals (Nicoll, 1982). Electrophysiological studies in turtles indicate that enkephalin may act as an inhibitory substance in turtle olfactory bulb (Nicoll, Alger and Jahr, 1980). It seems likely that the opioid peptides in the nervous system are acting at endogeneous opiate receptor sites. Mu and delta sites have been reported in reptiles (as noted above), and since kappa receptors are present in birds (Reiner et al., 1989), it is likely they are also present in reptiles. LENK and MENK show a high specificity for the delta site, whereas the longer enkephalin peptides (e.g., BAM22P, BAM12P, and peptide E) show considerable kappa potency (Zukin and Zukin, 1984; Magnen et al., 1982; Quirion and Weiss, 1983). Thus, depending on how proenkephalin is processed, the derived peptides released from an individual enkephalinergic terminal can bind preferentially to either delta or kappa receptors.

Finally, as noted above, the opioid peptides are also present in various peripheral tissues, for example, the adrenal gland, gut, and retina (Johansson et al., 1978; Epstein et al., 1981; Alumets et al., 1978b; Brecha et al., 1984). Although few data are available on the localization of enkephalin in peripheral tissues in reptiles, enkephalin has been observed in amacrine cells in turtles and lizards (Brecha et al., 1984; Eldred and Karten, 1985). It is likely that enkephalin is present in additional peripheral tissues in reptiles, such as gut and adrenal gland, since it is found in such tissues in other groups of nonmammals (LeBoulenger et al., 1983). In the periphery, opioid peptides typically function to inhibit the contraction of smooth muscle (Snyder, 1980).

C. Dynorphin

Prodynorphin (the last of the three opioid precursors discovered) contains three distinct opioid peptides of the dynorphin family, dynorphin A, dynorphin B, and alpha-neo-endorphin, each of which contains LENK at its N-terminus. Despite the presence of the LENK sequence at the N-terminus of the three major dynorphin peptides, the dynorphin peptides are, in general, not thought to be processed into LENK. Dynorphin is found in less abundance in the nervous system of mammals than is enkephalin (Giraud et al., 1983), and the dynorphin peptides, which are colocalized in the same neurons, are found in different neurons than are the proenkephalin peptides (Weber et al., 1982; Watson et al., 1983). Nonetheless, dynorphin distribution in the nervous system of mammals is highly similar to, yet distinct from, enkephalin distribution (Watson et al., 1982b). Dynorphin peptides are present in their highest concentration in the posterior pituitary and median eminence of the hypothalamus, from which they were first extracted and characterized (Goldstein et al., 1979). Based on the overall structural similarities between proenkephalin and prodynorphin, particularly with respect to the location of the LENK/MENK sequences, several investigators have suggested that the proenkephalin and prodynorphin gene arose by the duplication and subsequent divergent evolution of a single gene present in some early ancestral vertebrate or invertebrate species (Lewis and Erickson, 1986; Herbert et al., 1983).

Since the dynorphin family of peptides was only recently discovered (Goldstein et al., 1979), however, little information is actually available on the biochemistry or localization of dynorphin peptides among the various vertebrate species. Cone and Goldstein (1982) reported the presence of a dynorphin A-like peptide in bullfrog brain and posterior pituitary, based on RIA and chromatographic studi s, whereas Sei et al. (1989) have found that dynorphin A, as well as alpha-neo-endorphin, are present in *Xenopus* brain. Using either of three anti-dynorphin A antisera (courtesy of A. Goldstein, L. Terenius, and S. J. Watson) or two dynorphin B (also known as rimorphin) antisera (courtesy of A. Goldstein and S. J. Watson), dynorphinlike substances have been detected in turtle brain and spinal cord using immunohistochemical techniques (Reiner, 1983, 1986a; Anderson and Reiner, 1990a). For example, as in mammals (Vincent et al., 1982; Anderson and Reiner, 1988), a dynorphin-containing projection from neurons of the striatum to the substantia nigra is clearly present in turtles (Reiner, 1983, 1986a; Anderson and Reiner, 1990a). Staining of this and other dynorphinergic systems in turtle brain with a dynorphin A 1-17 antiserum (from S. J. Watson) was not affected when the antiserum was blocked with 50 μM synthetic LENK but was abolished when the antiserum was blocked with 50 μM synthetic dynorphin A 1-17 (Fig. 8.10). Similar results have been obtained in pigeons (Reiner, 1986a; Anderson and Reiner, 1990a; Reiner, Brauth, and Karten, 1984; Reiner, Davis, Brecha, and Karten, 1984). In addition, immunohistochemical evidence was obtained in turtles for the presence of a peptide resembling the bridge peptide portion of the mammalian prodynorphin molecule (spanning between alpha-neo-endorphin and dynorphin A) (see Fig. 8.4). The results of the biochemical studies in frogs, combined with the immunohistochemical studies in turtles and pigeons using various antidynorphin antisera, suggest that a prodynorphin peptide is present in reptiles and contains sequences closely resembling alpha-neo-endorphin, dynorphin A, and dynorphin B.

Few published data are available, however, on dynorphin localization in reptiles. In the telencephalon of turtles, dynorphin-positive neurons are found in the striatum and appear to give rise to prominent dynorphin-positive projections to the globus pallidus of the basal ganglia and the substantia nigra, as is also true in birds and mammals (Reiner, 1986d; Reiner et al., 1984a; Reiner, 1986a; Anderson and Reiner, 1988, 1990a; Vincent et al., 1982) (Fig. 8.10). Double-label immunohistochemical studies have been carried out and have shown that nearly all of these dynorphinergic striatal neurons and their fibers in the pallidum and nigra also contain the neuropeptide

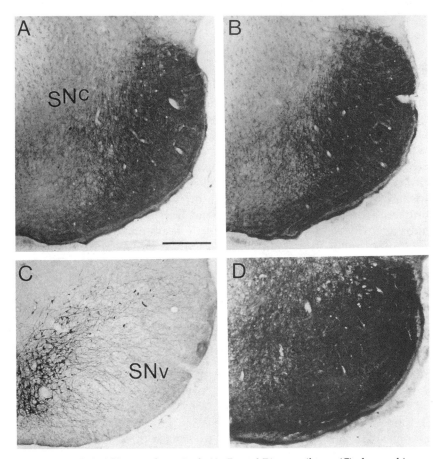

Fig. 8.10. Labeled fibers and terminals (A, B, and D) or perikarya (C) observed in transverse sections of the ventrolateral tegmentum of turtle in sections processed with an antiserum against dynorphin A 1–17 (from S. J. Watson and H. Akil) (A), with the same antiserum blocked with 100 micromolar LENK (B), with an antiserum against tyrosine hydroxylase (courtesy, T. Joh) (C), and with an antiserum against SP (D). Medial is to the left in all four photomicrographs and the magnification is the same in all four. The labeling is equally intense in both A and B, demonstrating that the antiserum is labeling a dynorphin A–like substance and is not cross-reacting with LENK. The dynorphin + fibers are present in a region termed the substantia nigra, pars ventralis (SNv), which also contains a dense accumulation of SP-containing fibers. The SP + fibers are more intensely labeled than the dynorphinergic fibers but have the same distribution. Double-label studies show that the SP and dynorphin co-occur in the same fibers. The relation of the SP-containing and the dynorphinergic fibers and terminals to the dopaminergic neurons of the substantia nigra, pars compacta (SNc) can be seen by comparing A, B, and D to C. The fibers appear to principally terminate in relation to the dendrites of the SNc neurons, as is also true of SP + and dynorphin + striatonigral fibers in mammals. All four photomicrographs are at the same magnification and the scale bar in (A) equals 250 μm.

substance P, as well as the neurotransmitter GABA (Reiner, 1986a; Anderson and Reiner, 1987, 1990a; Reiner and Anderson, 1987). Although similar studies have not been carried out in other reptiles, it is likely that dynorphinergic striatal projection neurons are also abundant in other reptilian species and that these also contain substance P and GABA. The basis of this conclusion is that dynorphinergic striatal projection neurons are abundant also in pigeons and rats and typically contain substance P and GABA also in these animals (Reiner, 1986a; Anderson and Reiner, 1988, 1990a; Reiner and Anderson, 1987). Thus an abundance of striatonigral and striatopallidal projection neurons containing dynorphin, substance P, and GABA appears to be common to the striatum in all amniotes in which this issue has been studied, and it is therefore likely to be characteristic of the striatum in all reptiles.

No other published information is available on the distribution of dynorphin in other parts of the CNS in reptiles. Unpublished studies (Reiner, unpublished observations) indicate that dynorphinergic cell bodies and fibers are widespread and numerous throughout the CNS of red-eared turtles. Dynorphinergic fibers are present in the dorsomedial and dorsolateral thalamic nuclei, in the hypothalamus, in the median eminence, in the neurohypophysis, in the deep tectum, in the ventrolateral rhombencephalon, and in the dorsal horn region of the spinal cord. Although dynorphinergic perikaya are present throughout the diencephalon, midbrain, and hindbrain, prominent accumulations of dynorphinergic perikarya are present in the supraoptic and paraventricular nuclei of the hypothalamus. These cell bodies are presumably the source of the dynorphinergic fibers in the median eminence and neurohypophysis. Since vasotocinergic fibers are also present in the median eminence and neurohypophysis of turtles, it is possible that the dynorphin and vasopressin are colocalized in the same fibers, as has been reported to be true for vasopressin and dynorphin in the median eminence and neurohypophysis in mammals (Watson et al., 1982a). Although it seems likely that dynorphin peptides are widespread in the nervous system of other reptiles as well, this possibility has not yet been explored.

The function of dynorphin in the turtle central nervous system is presumably similar to that in the mammalian CNS, namely that of a peptidergic neurotransmitter/neuromodulator. Dynorphin is known to have inhibitory effects on neurons (Nicoll, 1982), as is also true of enkephalin, although enkephalin and dynorphin appear to effect this inhibition via different ionic mechanisms (Werz and McDonald, 1983, 1984). Dynorphin A and dynorphin B show a high specificity and affinity for the kappa site, whereas the shorter dynorphin peptides (e.g., dynorphin A 1-8) show considerable delta potency (Zukin and Zukin, 1984; Magnen et al., 1982; Quirion and Weiss, 1983). Thus, depending on how the dynorphin peptides are processed, the opioid peptides released from an individual dynorphinergic terminal can have delta or kappa effects or both. As noted above, it is likely that both delta and kappa opiate receptors are present in the nervous system of reptiles.

IV. SUBSTANCE P AND NEUROTENSIN
In the early 1970s, Leeman and co-workers purified, characterized, and sequenced a bovine hypothalamic peptide that stimulated salivation (Chang and Leeman, 1970). They found that their peptide was identical to a hypotensive compound termed substance P (SP), whose extraction from horse brain and intestine had been reported in 1931 (von Euler and Gaddum, 1931). While working on substance P, Leeman and co-workers also isolated, characterized, and sequenced a second peptide that was present in great abundance in bovine hypothalamus (Carraway and Leeman, 1973). Both substance P and the second neuropeptide, termed neurotensin for its localization in nervous system and for its hypotensive properties, were subsequently shown to be widespread and abundant in the nervous system and the GI tract. These two peptides have been grouped together in the present section, largely for the above noted historical reason.

A. Substance P
Substance P is an 11-amino-acid neuropeptide (an undecapeptide) that is now known to belong to a larger class of substances known as tachykinins because of their quick actions on smooth muscle. The tachykinins all are similar in the last five amino acids of their C-terminus (Table 8.4). Many tachykinins, such as eledoisin, physalaemin, and kassinin, are present in various nonneural tissues in nonmammals, for example, frog skin and octopus salivary glands (Erspamer, 1981; Nakajima, 1981). Such tachykinins have not, however, been observed in neural tissues in mammals or nonmammals. SP-like peptides have, however, been identified by RIA and immunohistochemistry in the nervous systems of diverse nonmammal taxa, including pigeons (Reubi and Jessell, 1979; Reiner et al., 1983; Erichsen et al., 1982a, 1982b; Gamlin et al., 1982), lizards (Engbretson et al., 1982; Gaudino and Fasolo, 1984), turtles (Korte et al., 1980; Brauth et al., 1983; Reiner, Krause, Keyser, et al., 1984), crocodilians (Brauth et al., 1983), frogs (Lorez and Kemali, 1981; Inagaki et al., 1981a; Kuljis and Karten, 1982; Eskay et al., 1981), newts (Taban and Catheini, 1983), fish (Eskay et al., 1981; Brecha et al., 1984; Northcutt et al.,

Table 8.4. Amino acid sequences for several members of the tachykinin family of peptides (the family of peptides to which SP belongs)

	1	2	3	4	5	6	7	8	9	10	11
SP		Arg	Pro	Lys	Pro	Gln	Gln	Phe	Phe	Gly	Leu	Met-NH$_2$
PHYS		Pyr	Ala	Asp	Pro	Asn	Lys	Phe	Tyr	Gly	Leu	Met-NH$_2$
ELED		Pyr	Pro	Ser	Lys	Asp	Ala	Phe	Ile	Gly	Leu	Met-NH$_2$
KASS	Asp	Val	Pro	Lys	Ser	Asp	Gln	Phe	Val	Gly	Leu	Met-NH$_2$
SK		His	Lys	Thr	Asp	Ser	Phe	Val	Gly	Leu	Met-NH$_2$	
NK		Asp	Met	His	Asn	Phe	Phe	Val	Gly	Leu	Met-NH$_2$	

Abbreviations. PHYS, physalaemin; ELED, eledoisin; KASS, kassinin; SK, substance K; NK, beta neurokinin.

1988; Reiner and Northcutt, 1987, 1988), and the coelenterate *Hydra* (Taban and Catheini, 1978; Grimmelikhuijzen et al., 1981).

Studies in mammals have characterized the gene encoding the precursor for SP and have found that a single gene gives rise to two different SP precursors (termed alpha- and beta-preprotachykinin). Both of the precursors contain the sequence for substance P, and, in addition, one of the two precursors (beta-preprotachykinin) contains the sequence for a 10-amino-acid tachykinin peptide that has been termed substance K (SK; now more typically known as neurokinin A) due to its similarities at the N-terminus to the tachykinin, kassinin (Nawa et al., 1983, 1984; Krause et al., 1987) (see Table 8.4). The two precursors appear to be generated from the single gene in a tissue-specific fashion by differential RNA splicing mechanisms (Nawa et al., 1984). Substantial levels of SK have been demonstrated in the nervous system in mammals (Maggio et al., 1983; Maggio and Hunter, 1984). The relative distributions of SP and SK in mammalian nervous system, not surprisingly, are highly similar (Valentino et al., 1986). In addition, a third SP-like tachykinin, termed neurokinin B, has also been found in mammalian nervous system (specifically, the spinal cord) (Harmar, 1984).

The structure of the SP has not been determined in any nonmammalian species. Nonetheless, considerable data are available on the biochemical characteristics of the SP in diverse nonmammalian species. In an extensive study on "representative" species from two invertebrate phyla (anemone, Coelenterata; snail, Mollusca) and from several nonmammalian vertebrates (skate, mackerel, frogs, lizards, and pigeons), Creagh et al. (1980) used RIA to determine the amounts of the SP-like material present in the nervous system and peripheral tissues. Moderate to large amounts of the SP-like material were found in the tissue examined in all animal species. Similar results have been

Table 8.5. Amounts of SP, in pg/mg wet weight tissue, in a number of different regions of the central and peripheral nervous system of *Pseudemys scripta*

	pg substance P/mg tissue
Olfactory bulb	87 ± 39
Cortex	202 ± 69
Dorsal ventricular ridge	107 ± 20
Basal telencephalon, including basal ganglia	368 ± 82
Thalamus and epithalamus	416 ± 112
Hypothalamus and neurohypophysis	479 ± 116
Tectum	208 ± 41
Tegmentum	607 ± 169
Cerebellum	24 ± 3
Rhombencephalon	291 ± 59
Cervical spinal cord	231 ± 68
Dorsal root ganglion	32 ± 3
Retina	12 ± 3

Note: Data from Reiner, Krause, Keyser, et al., 1984.

obtained by Lembeck et al. (1985), who also found that a peptide closely resembling SP is present in abundance in the nervous system of a wide variety of vertebrates, including hagfish, rainbow trout, frogs (*Xenopus laevis* and *Bufo bufo*), reptiles (*Testudo kleinmanni* and *Lacerta sp.*). The work of Creagh et al. (1980) further shows that the SP-like material in the nervous system of frogs, lizards (*Podarcis muralis*), and pigeons is highly similar to mammalian SP, since in a dose-response study using RIA, the nervous tissue extracts gave results parallel to those for undecapeptide SP. In contrast, the SP-like peptide from the nervous systems of mackerel, skate, snails, and anemones did not show parallelism to SP in the RIA dose-response study (Creagh et al., 1980). Thus a peptide closely resembling mammalian SP appears to be present in considerable amounts in the nervous system in all vertebrate species and many invertebrate species, although this resemblance is greater among land vertebrates.

The chemical characteristics of SP and its distribution in the nervous system of red-eared turtles have been studied in detail (Reiner, Krause, Keyser, et al.; 1984; Reiner et al., 1985). Using RIA with an antiserum specific for the C-terminus of SP, the amounts of SP in various regions of turtle central nervous system and several peripheral tissues were assayed (Table 8.5). Large amounts of an SP-like substance were present in all regions of turtle brain and spinal cord, retina, and dorsal root ganglia (DRGs). The presence of high levels of SP in turtles is consistent with the work of Lembeck et al. (1985). In turtles, particularly high levels of SP were found in the hypothalamus, the basal telencephalon (including the basal ganglia), the tegmentum, and the rhombencephalon. The high levels of SP in the teg-

Table 8.6. Amount of SP in a number of comparable neural regions in turtle, rat, and pigeon (in pg/mg wet weight tissue)

	Turtle	Mammal	Pigeon
Basal telencephalon,			
including the basal ganglia	368	303	410
Hypothalamus	479	539	234
Tectum	208	—	52
Spinal cord	231	—	
Cortex	202	23	116
Cerebellum	24	11	2
Tegmentum,			
including the substantia nigra	607	830	594

Turtle data from Reiner, Krause, Keyser, et al., 1984; rat data from Kanazawa and Jessell, 1976; pigeon data from Reubi and Jessell, 1979. Dashes indicate that data for that structure are not presented in the cited study.

mentum were shown by immunohistochemical studies to be due to a dense SP-positive terminal field in the substantia nigra, as has also been found in birds and mammals (Reiner et al., 1983; Jessell et al., 1978). The overall concentration of SP in the turtle nervous system was similar to that in mammals and birds, and it was found that comparable regions of avian, mammalian, and reptilian central nervous systems contained comparable concentrations of SP (Table 8.6). Thus the tegmentum (which contains the substantia nigra) was found to contain the highest concentration of SP in the brain in turtles, pigeons, and rats. As noted above, dose-response curves for the SP-like material for all neural regions in turtles (except the olfactory bulb) were parallel in SP-RIA to the dose-response curve for synthetic mammalian undecapeptide SP (Fig. 8.11). In contrast, such tachykinins as eledoisin and physalaemin (which are similar but not identical to undecapeptide SP in their C-terminus), did not show parallelism to undecapeptide SP in this RIA.

Thus, the RIA studies show that an SP-like peptide is present in the nervous systems of a wide variety of vertebrate species, including reptilian species. Although these studies also show that the SP-like peptides in these species typically are highly similar to SP, conclusions from these studies as to the precise similarities in SP sequences among mammalian species and various nonmammalian species must take into consideration the specificity of the antisera used in the RIA studies. Since the antisera used typically were specific for the C-terminus of SP, the results of these studies only show that the SP-like peptides in various nonmammals (including turtles) are likely to be identical to mammalian SP in the C-terminus of the molecule, but not necessarily in the whole molecule.

Using a more sensitive method of comparison, namely gradient

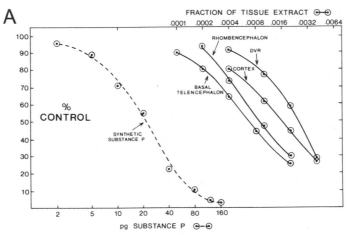

Fig. 8.11. (A) Graph showing that serial dilutions of tissue extract from four different regions of turtle brain (telencephalic cortex, DVR, basal telencephalon, and rhombencephalon) show a dose-related effect on the binding of radioactive substance P (in substance P RIA), an effect that is parallel to that of synthetic substance P. One hundred percent binding is defined as the difference between specific and nonspecific binding in the absence of nonradioactive standard or extract. The parallelism between the standard curve and the tissue extracts suggests that the SP-like material in these tissues in turtle brain is highly similar if not identical to SP, at least through the portion of the molecule recognized by the anti-SP antiserum used (the C-terminus). All tissues from turtle shown in Table 8.5 (except the olfactory bulb) showed parallelism to synthetic SP in this RIA system. (From Reiner, Krause, Keyser, et al., 1984). (B) Elution profile for turtle (*Pseudemys scripta*) basal telencephalon extract, using a Videc C-4 column and a 1% to 30% exponential gradient of acetonitrile (J. E. Krause and A. Reiner, unpublished observations). The amounts of substance P-like material in each fraction were measured by RIA. Nearly all of the substance P-like material in the basal telencephalon eluted as a single peak that was distinct from those for other known tachykinins (see Table 8.4) but close to the peaks for two tachykinins that are structurally similar to one another: substance K (SK) and kassinin (KASS). The small amount of substance P-like material eluting after the major peak does not coincide with the peak for substance P and was not observed in other HPLC studies examining the turtle basal telencephalon or telencephalic cortex (Reiner et al., 1985). The results shown in (A) and (B) suggest that nearly all of the substance P-like material in turtle basal telencephalon (and possible other neural regions as well) is a single molecular species that is highly similar to substance P in its C-terminus and possibly similar to SK or kassinin in its N-terminus. Other abbreviations: LHRH, luteinizing hormone releasing hormone; PHYS, physalaemin; ELED, eledoisin; NK, beta-neurokinin.

elution high-performance liquid chromatography (HPLC), differences have, in fact, been found between SP in mammals and the members of some nonmammalian groups. For example, using this approach, Eskay et al. (1981) has found that all of the SP-like material in frog and carp retina differed from mammalian SP, but most of the SP-like material in chicken retina was indistinguishable from mammalian SP. White et al. (1985) and Davis et al. (1988) have also both

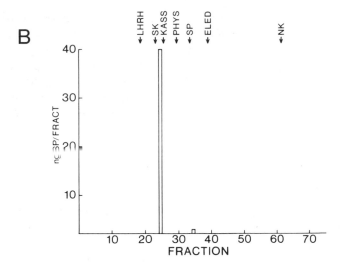

reported, based on the use of gradient elution HPLC, that the SP-like peptide in the pigeon nervous system is indistinguishable from mammalian undecapeptide SP. Unlike studies in birds, detailed studies in red-eared turtles (as discussed below), using the same HPLC system as used in the White et al. (1985) and Davis et al. (1988) studies in pigeons, have shown that SP in turtles does differ from mammalian SP (Reiner et al., 1985). The turtle SP-like material eluted much earlier than SP (Fig. 8.11) and consistently showed a retention time similar to (but distinct from) SK and kassinin (J. E. Krause and A. Reiner, unpublished observations). The HPLC results for turtles are identical for the two neural regions thus far examined—the telencephalic cortex and the basal telencephalon. The HPLC results suggest that turtle SP may show some similarity to kassinin and SK, presumably in its N-terminus. Further studies are required, however, to precisely determine the structure of the SP-like substance present in turtle nervous system. Although the HPLC data suggest that a single SP-like peptide predominates in turtle nervous system, studies using antisera against SK (K. D. Anderson and A. Reiner, unpublished observations) show that SK is also present in turtles.

Thus, these overall results suggest that SP has been very conservative in its evolution. Nonetheless, SP in fish, amphibians, and at least turtles, if not other reptiles as well, differs from that found in mammals and birds, in which SP appears to be identical. At face value, these results in birds suggest that the chemical structure of SP characteristic of birds and mammals is as found in their common ancestor (which is presumably the same common ancestor as for birds, mammals, and turtles) and that the structure of SP as found in turtles has changed from that of this common ancestor. It is uncertain, however, whether birds and mammals have inherited their SP sequences from that in the common reptilian ancestor, or whether the similarity is due to convergent evolution. Data on SP in various other groups of reptiles would be useful to determine the structure of SP that is primitive for reptiles. Such information would help to clarify the evolution of SP among land vertebrates. Further studies are also needed on the structure of the SP precursor in nonmammals. No data are currently available on this issue, although immunohistochemical studies using antisera specific for SK have shown that an SK-like neuropeptide (presumably derived from preprotachykinin) is present in the nervous systems in pigeons and turtles and possesses a distribution largely similar to that of SP (K. D. Anderson and A. Reiner, unpublished observations).

The localization and distribution of SP (or more correctly of the reptilian isopeptide or isopeptides of mammalian SP) in the nervous system have been studied in detail in turtles (Reiner, Krause, Keyser, et al., 1984) and varanid lizards (Wolters et al., 1986) using immunohistochemical techniques. Several specific SP-containing pathways and brain regions have been described in crocodiles (Brauth et al., 1983), lizards (Engbretson et al., 1982; Gaudino and Fasolo, 1984; Russchen et al., 1987), and snakes (Terashima, 1987; Kadota et al., 1988). The distribution of the SP-like peptide in turtles, as revealed by immunohistochemistry, is consistent with that indicated by RIA (Reiner, Krause, Keyser, et al., 1984). Most prominently, numerous SP-positive perikarya and fibers were observed in the striatal subdivision of the basal ganglia (Fig. 8.12), the hypothalamus, and the tectum. The tegmentum contained a very dense field of SP-containing terminals and fibers among and in the region ventrolateral to the dopaminergic neurons of the substantia nigra, as also true in birds and mammals (see Fig. 8.10) (Reiner, Brauth, and Karten, 1984; Reiner, 1986d). In a separate study, the SP-containing terminals in the substantia nigra were found to arise from the SP-containing neurons within the striatum (Brauth et al., 1983). Preliminary results (Reiner, unpublished observations) using electron microscopic immunohistochemical techniques indicate that the ultrastructural localization of SP-like immunoreactivity in the substantia nigra in turtles is very similar to that in mammals (DiFiglia et al., 1982a, 1982b). In both vertebrate groups, SP-like immunoreactivity is present within the nigra in large dense core vesicles and in the cytoplasm of many terminals that contact long dendritic profiles, which are presumably the processes of dopaminergic neurons. In addition to a projection to the substantia nigra, striatal SP-containing neurons in turtles project to the pallidal

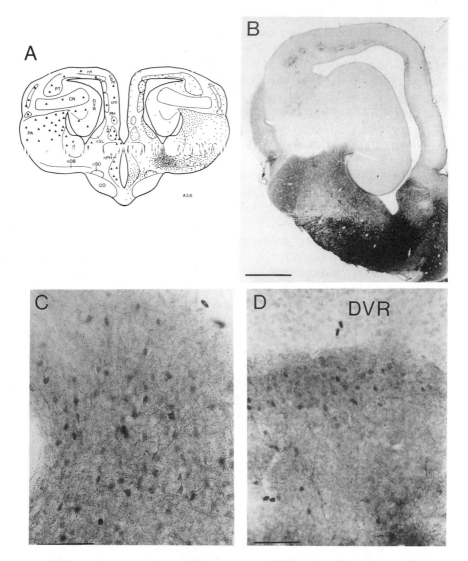

Fig. 8.12. (A) Line drawing of a transverse section through a midtelencephalic level of the turtle brain showing the distribution of SP+ neurons (triangles) and fibers (dots). (B) Low-power photomicrograph (scale bar equals 1 mm) of a section of turtle brain, labeled for the presence of SP according to immunohistochemical techniques, at the same level as shown in (A). The various regions of the telencephalon can be identified by referring to (A). (C) SP+ neurons in the striatum (PA) of a red-eared turtle and (D) SP+ neurons in the striatum of *Sceloporus undulatus*. Medial is to the left in (C) and to the right in (B) and (D). The scale bars in (C) and (D) equal 100 μm. Abbreviations: CN, core nucleus of the DVR; CO, optic chiasm; cd, cortex dorsalis; cdm, cortex dorsomedialis; cm, cortex medialis; cp, cortex pyriformis; DVR, dorsal ventricular ridge; d, area d; GP, globus pallidus; nDB, nucleus of the diagonal band; nPH, nucleus periventricularis hypothalami; nSL, nucleus septalis lateralis; nSM, nucleus septalis medialis; nSO, nucleus supraopticus; PA, paleostriatum augmentatum; PH, primordium hippocampi; PT, pallial thickening.

portion of the basal ganglia (Figs. 8.10 and 8.12). Double-label immunohistochemical studies have shown that the SP+ striatonigral and
striatopallidal projection neurons in turtles and their terminals in the
nigra and pallidum typically also contain dynorphin and GABA (Reiner, 1986a; Reiner and Anderson, 1987; Anderson and Reiner, 1990a).

In crocodilians, also, SP-containing neurons in the striatum appear
to have an SP-containing projection to the region of the midbrain
equivalent to the substantia nigra (Brauth et al., 1983). In crocodilians, striatal neurons also are presumably the source of coarse, thick,
SP-containing fibers in the ventrolateral portion of the basal ganglia
(the region termed the crocodilian basal ganglia in the present study
corresponds to the region termed the ventrolateral area by Crosby,
1917); this portion of the crocodilian basal ganglia may be comparable
to the region termed the globus pallidus in turtles. As noted in the
section on enkephalin, in crocodilians the pallidal portion of the basal
ganglia is cytoarchitectonically less clearly identifiable as a discrete
cell field than in turtles and birds. The coarse SP+ fibers in the pallidal field of the crocodilan basal ganglia do not form a distinct field as
is the case in turtles, and as in birds and mammals (Reiner et al., 1983,
Reiner, Brauth, and Karten, 1984; Reiner, Krause, Keyser, et al., 1984;
Brauth et al., 1983). SP-containing neurons are also present in the
striatum of lizards (Russchen et al., 1987; Reiner, unpublished observations) (Fig. 8.12). These are presumably the source of the numerous
SP-containing terminals in the nigral portion of the tegmentum and
in the pallidal portion of the basal ganglia. Although SP-positive fibers were evident in the striatum of snakes, SP-containing neurons
were not observed in the striatum and SP-positive terminals characteristic of the pallidal cell field were not evident (Reiner, unpublished
observations). These results suggest that SP-containing neurons must
be present in the snake striatum but contain lower levels of SP than in
other reptiles. In birds, turtles, crocodilians, and lizards, the SP-
containing fibers from the striatum that terminate in the substantia
nigra course via the medial forebrain bundle through the ventromedial telencephalon. This region of the telencephalon, termed the ventral paleostriatum in birds, turtles, and crocodilians (Reiner et al.,
1983; Brauth et al., 1983), contains high levels of SP and is comparable
to the ventral pallidum of mammals. In garter snakes, however, the
SP+ fiber labeling in the medial forebrain bundle and the ventral paleostriatum region is much lighter than in members of other reptilian
groups (Reiner, unpublished observations). The distribution of SP in
the basal ganglia and its projection targets in *Sphenodon* is largely similar to that in turtles (Reiner, unpublished observations). As discussed
in the section on dynorphin, since SP+ striatal projection neurons in
species as diverse as turtles, pigeons, and rats also typically contain

dynorphin and GABA (Reiner, 1986a; Reiner and Anderson, 1987; Anderson and Reiner, 1988, 1990a), it is very likely that this is a conservative feature of basal ganglia organization among amniotes. Thus, it is likely that SP+ striatal neurons in all reptiles also typically contain dynorphin and GABA.

The SP-containing processes of the SP+ neurons of the striatum are abundant in the turtle and crocodilian striatum (Brauth et al., 1983; Reiner, Krause, Keyser, et al., 1984). The abundance of SP in the striatum makes it readily distinguishable from the overlying DVR in immunohistochemically stained sections (see Fig. 8.12). The DVR contains a conspicuously lower immunohistochemically evident amount of SP than do the basal ganglia in turtles, crocodilians, and lizards. This is also revealed in the RIA studies in turtles by comparing the levels of SP in the basal telencephalon to those in the DVR (Reiner, Krause, Keyser, et al., 1984). In crocodilians, however, immunohistochemical data indicate that the medial part of the portion of the DVR termed the intermediolateral area by Crosby (1917) does contain somewhat greater amounts of SP than other portions of the DVR. The basal ganglia in garter snakes are also much richer in SP than the DVR, although the dorsomedial portion of the rostral DVR does contain a prominent collection of SP-positive fibers along the ventricular surface (Reiner, unpublished observations). In birds, also, the overlying DVR (including the neostriatum, ectostriatum, and hyperstriatum ventrale) are much lower in their SP content than the basal ganglia (Reubi and Jessell, 1979; Reiner et al., 1983). These observations on the SP content of the basal ganglia in birds and reptiles are consistent with the observations on the SP content of telencephalic regions in mammals. In mammals, the basal ganglia contain much higher levels of SP than does the cortex. The observations in birds and reptiles support previous suggestions that the sauropsid basal ganglia are highly similar to the mammalian basal ganglia and that the sauropsid DVR is comparable to mammalian neocortex (Nauta and Karten, 1970; Reiner, Brauth, and Karten, 1984).

A collection of SP-containing fibers and terminals is, however, present throughout the cortex in turtles (see Fig. 8.17). This system, which is confined to the cellular layer of cortex, is described in greater detail in the sections on CCK8 and VIP because double-labeling studies have shown that these fibers in the perikaryal layer contain SP, CCK8, and VIP (Reiner et al., 1985). These SP+ fibers of the cortical cellular layer arise from lateral hypothalamic and supramammillary regions (Reiner et al., 1985). A similar projection to telencephalic cortex has been described in mammals (Gall and Selawski, 1984; Ino et al., 1988; Seroogy et al., 1988). Additional SP+ fibers, which do not contain CCK8, are present in the superficial fourth of the molecular layer of medial and dorsomedial cortex in turtles (Reiner, Krause, Keyser, et al., 1984; Reiner et al., 1985). Fibers containing SP are also present in the cortex of crocodilians (Brauth et al., 1983), *Sphenodon*, and snakes (Reiner, unpublished observations). Although SP-containing perikarya and fibers are evident in the septal region in turtles and caiman (Brauth et al., 1983; Reiner, Krause, Keyser, et al., 1984), in snakes the septum is noteworthy for the strikingly high levels of SP observed in cells and fibers in immunohistochemically processed sections (Reiner, unpublished observations). In snakes and lizards, nucleus sphaericus was devoid of SP+ fibers and neurons (Reiner, unpublished observations).

Within the diencephalon, SP+ fibers are present in the dorsomedial and dorsolateral thalamic nuclei of turtles, crocodilians, lizards, and snakes and conspicuously absent from thalamotelencephalic projection nuclei such as nucleus rotundus and nucleus reuniens, which process visual or auditory information (Hall and Ebner, 1970; Reiner and Powers, 1978, 1980, 1983; Ulinski, 1983, 1988; Reiner, Krause, Keyser, et al., 1984; Reiner, unpublished observations). In general, SP-containing cell bodies are not prominent in the thalamus of turtles, crocodilians, and lizards, and only a few were noted in the habenula in turtles. In sceloporine lizards, a few SP-containing neurons were observed in the dorsomedial nucleus. The thalamus of garter snakes was also largely devoid of SP+ neurons, but numerous, extremely SP-rich perikarya were observed in the habenula of non-colchicine-treated snakes (Reiner, unpublished observations). The inability to find large numbers of SP-containing cells in the habenula of turtles, crocodilians, and lizards, even in colchicine-treated lizards and turtles (Reiner, Krause, Keyser, et al., 1984; Engbretson et al., 1982) and crocodilians (Brauth and Reiner, unpublished observations), is surprising, since an SP-positive habenulointerpeduncular pathway has been described in mammals and numerous SP-containing neurons are present in the medial habenula of birds and frogs (Engbretson et al., 1981; Reiner, Krause, Keyser, et al., 1984). In pigeons, fibers of the habenulointerpeduncular tract have been observed to be SP+ (Reiner, unpublished observations). In all reptiles, however, the medial habenular nucleus was rich in SP-containing fibers. In lizards with a parietal eye, Engbretson et al. (1981) have found that the parietal eye projects only to the left habenula and terminates in a discrete bipartite field within the dorsal portion of the medial habenular nucleus. It was further found that the habenula receiving parietal eye input was considerably larger than the habenula on the other side of the brain. In several species of lizards with parietal eyes (*Uta stansburiana*, *Sceloporus undulatus*, and *S. occidentalis*), the parietal eye–recipient habenula but not the non-parietal eye–

recipient habenula was found to contain a field of SP-positive terminals in the medial habenular nucleus. These SP-positive terminals did not arise from the parietal eye and did not overlap with terminals from the parietal eye (Engbretson et al., 1982). In a subsequent study of habenular asymmetry in lizards, it was found that in nearly all lizard species with a parietal eye, one habenula (presumably the parietal eye–recipient left habenula) was larger than the other (Engbretson and Reiner, unpublished observations). In all cases of habenular asymmetry, the asymmetry stemmed from a hypertrophy of a portion of the medial habenular nucleus, the pars dorsolateralis. The source of the SP-containing terminals in the habenula of reptiles is presently uncertain.

Of tissues processed with immunohistochemical techniques, the hypothalamus in reptiles appears to contain higher levels of SP than the thalamus. In turtles, however, RIA indicated comparable amounts of SP in the epithalamus/thalamus as in the hypothalamus/hypophysis. This discrepancy may be the consequence of the course of the ventral peduncle of the lateral forebrain bundle through the ventral thalamus. The ventral peduncles were presumably largely included with the thalamus during tissue dissections for RIA in turtles (Reiner, Krause, Keyser, et al., 1984). The ventral peduncle contains the SP-rich efferent fibers of the SP+ neurons of the striatum. Within the hypothalamus of turtles, SP+ neurons are prominent dorsal to and around the OVLT, in the rostral periventricular hypothalamus, and in a more lateral field extending from the lateral hypothalamus to the supramammillary nucleus (Fig. 8.17). Cells in the lateral hypothalamic-supramammillary field are also rich in CCK8 and VIP, as will be detailed below (Fig. 8.17). Double-label studies have shown that SP and CCK8 co-occur in neurons in this region (Reiner et al., 1985), but studies have not been performed to determine whether VIP is also found in the SP/CCK8-containing neurons. Fibers containing SP are abundant in the hypothalamus, particularly in the more medial hypothalamus. Fibers rich in SP were also observed in the external zone of the median eminence. The source or functional significance of these fibers is uncertain. Although hypothalamic SP+ neurons and fibers are also abundant in lizards, snakes, and caiman, SP+ fibers have been reported to be absent from the median eminence in at least one species of lizard (Gaudino and Fasolo, 1984). The diencephalic zone between the hypothalamus and the thalamus (namely the ventral thalamus) contains the ventral peduncle, which (as noted) contains the SP-rich striatonigral axons, which are en route to the tegmentum. The SP-containing fibers of the ventral peduncle are particularly prominent in turtles, crocodilians, and lizards (Reiner, Krause, Keyser, et al., 1984; Brauth et al., 1983; Reiner, unpublished observations). In garter snakes, both the SP-positive fibers in the ventral peduncle and the SP-positive fibers in the tegmentum were observed to be less prominent than in other reptiles (Reiner, unpublished observations). In caiman, (sceloporine) lizards, and garter snakes, SP+ neurons were observed in the same hypothalamic regions as in turtles and SP-containing fibers were particularly prominent in the preoptic region, paraventricular region, and median eminence of snakes.

Within the midbrain, SP-positive neurons and fibers were observed in tectum of turtles, crocodilians, snakes, and lizards (Reiner, Krause, Keyser, et al., 1984; Brauth et al., 1983; Reiner, unpublished observations). The SP-containing neurons give rise to radially ascending processes that ramify in the superficial retinorecipient layers of the tectum. The ramifications are not, however, confined to a cytoarchitectonically defined layer. The SP-positive neurons are found in the intermediate layers in lizards and snakes, in the deep and intermediate layers in crocodilians, and almost exclusively in the deep layers in turtles. These neurons are presumably homologous to one another, despite the differences in laminar location, which may reflect the tendency of tectal cells to vary among different reptilian groups in the extent to which they migrate from the ventricle during development (Senn, 1971). In lizards, snakes, and crocodilians, many SP-containing neurons are also located in the superficial tectal layers. These neurons may also be present in the tectum of turtles and snakes but contain levels of SP that are immunohistochemically less readily detected. Fibers containing SP are observed in deep and intermediate tectal layers as well, but the source of these fibers is uncertain. The organization of SP-containing neurons and fibers typically observed in the tectum in reptiles is in many respects similar to that observed in the tectum in birds and frogs (Karten et al., 1982; Kuljis and Karten, 1982).

The tegmentum, as noted, is dominated by the SP-containing fibers and terminals of the striatonigral projection in turtles, crocodilians, and lizards (Brauth et al., 1983; Reiner, Krause, Karten, et al., 1984; Wolters et al., 1986; Reiner, unpublished observations). In all reptiles studied, prominent SP+ fibers were observed along the midline of the interpeduncular nucleus. Similar SP+ fibers have been observed in the interpeduncular nucleus of various mammalian species, in which they have been shown to be the terminals of a projection from the medial habenula to the interpeduncular nucleus via the fasciculus retroflexus (Nicoll et al., 1980b). In snakes, the interpeduncular nucleus was also observed to contain a particularly striking number of SP-rich neurons (Reiner, unpublished observations). Numerous SP-containing fibers and a number of SP-containing neurons were observed in the central gray and the laminar nucleus of the torus semi-circularis in turtles.

Within the rhombencephalon in reptiles, SP-containing fibers are prominent in much the same regions as are enkephalinergic fibers: the locus coeruleus, the parabrachial region, the raphe region, the ventrolateral rhombencephalon, the nucleus of the solitary tract, the dorsal motor nucleus of the vagus, the lateral margin of the descending tract of the trigeminus, the dorsal root terminal zone in the spinal cord, and much of the dorsal and ventral horns of the spinal cord (Reiner, Krause, Keyser, et al., 1984; Wolters et al., 1986; Terashima, 1987; Kadota et al., 1988). In snakes, SP-positive fibers are very prominent in the nucleus of the solitary tract and the dorsal motor nucleus of the vagus. In snakes, SP has also been demonstrated in perikarya of the trigeminal ganglion (Terashima, 1987; Kadota et al., 1988). It seems likely that these neurons are the source of the SP+ fibers along the descending tract of the trigeminus in snakes. In turtles, SP-containing neurons have been demonstrated in the dorsal root ganglia, which presumably give rise to at least some of the SP-containing terminals in the tract of Lissauer and dorsal horn region of the spinal cord in turtles (Reiner, Krause, Keyser, et al., 1984). Since SP-rich fibers are present in the dorsal horn region of all reptiles, it seems likely that a SP+ projection from the dorsal root ganglia terminates in the dorsal horn in all reptiles. Studies in mammals have shown that SP+ fibers projecting to the dorsal horn and descending tract of the trigeminus play a role in transmitting nociceptive input from the periphery to the CNS (Otsuka and Yanagisawa, 1987). SP may be a transmitter of nociception in reptiles, as well.

In turtles, caiman, and sceloporine lizards, SP-containing neurons are observed in the superior raphe region (Reiner, Krause, Keyser, et al., 1984; Reiner, unpublished observations) and may be a source of some of the SP-positive terminals in the gray matter of the spinal cord, as reported in mammals (Helke et al., 1982). Finally numerous SP-positive mossy fibers were observed throughout the granule cell layer of the cerebellum of turtles, caiman, lizards and garter snakes, (Korte et al., 1980; Wolters et al., 1986; Reiner, unpublished observations). The source of these SP+ mossy fibers is uncertain. The only cell group afferent to the cerebellum with neurons known to contain SP are the dorsal root ganglia. Kunzle (1982) has shown that primary afferent fibers of the dorsal root ganglia in turtles do project directly to the cerebellum. The extent of the input to the cerebellum from the DRGs is, however, presently uncertain. The alternative possibility exists that brainstem neurons that project to the cerebellum are the source of the SP-positive mossy fibers and that colchicine treatment is needed to enhance the visualization of the SP in these neurons. In unpublished immunohistochemical studies, SP has also been observed in cerebellar mossy fibers in frogs, rats, and pigeons (Reiner, unpublished observations).

The function of SP in turtle nervous system has not been studied. The observation that SP in reptiles is likely to be highly similar in structure to mammalian SP, is found in many of the same neural systems, and is present in the same overall concentration strongly argues that the neural function of SP is fundamentally similar in mammals and reptiles. Receptor-binding studies have shown the existence of multiple tachykinin receptor sites in the nervous systems of mammals. SP appears to bind preferentially to one of these sites, whereas SK binds preferentially to a different site (Hunter and Maggio, 1984; Buck and Burcher, 1986; Maggio, 1988). No data are available on the presence or distribution of such receptor sites in reptiles. In mammals, SP has been shown to act as an excitatory substance in a variety of neural systems (Otsuka and Konishi, 1976; Nicoll et al., 1980a, 1980b; Otsuka et al., 1984). Since SP has been shown to excite spinal motoneurons and sympathetic ganglia neurons in frogs, it is likely that SP has similar actions on neurons in reptiles (Nicoll et al., 1980a, 1980b; Adams et al., 1983). The action of SP in the nervous system (slow, sustained, and often below threshold for action potential initiation) is, however, more characteristic of a modulatory compound than a primary transmitter. SP has also been proposed to be a neuromodulator on the basis of its colocalization with other transmitter substances or neuropeptides, and it has now been shown to be colocalized with other neuroactive substances in a variety of brain regions and in a variety of species (Chan-Palay et al., 1978; Skirboll et al., 1982; Dalsgaard et al., 1982; Erichsen et al., 1982a, 1982b). As described above, SP has been colocalized with other neuroactive substances in several regions of turtle brain, thus supporting the view that SP in turtles may also typically be a modulatory substance.

Little is known of the distribution of SP in peripheral tissues in reptiles. Based on the demonstrated presence of SP in sensory neurons in turtles, birds, and mammals, it appears likely that SP is present in sensory fibers in the periphery (Hayashi et al., 1983). As shown to be the case in mammals (Holzer, 1988), it is likely these sensory nerve endings in the periphery can release SP, particularly during injury, and thereby act to dilate local blood vessels. Based on the presence of SP in neurons, nerves, and endocrine cells of the gut in birds and mammals (Brodin et al., 1981), it appears likely that the same is true in reptiles. SP-positive fibers and neurons have been observed in alligator GI tract (Buchan et al., 1983). SP in the gut appears to play a role in GI tract motility (Holzer, 1988). Amacrine cells that label lightly for SP with immunohistochemical techniques have been reported in turtle retina (Reiner, Krause, Keyser, et al., 1984) and are also present in the retinas of *Lacerta* (Osborne et al., 1982), *Uta stansburiana*, *Gekko gecko*, and *Anolis carolinensis* (Brecha et al., 1984; Engbretson and Reiner, unpublished observations).

B. Neurotensin

In the course of purifying substance P from bovine hypothalamus, a fraction of the hypothalamic extract distinct from that containing SP was found to produce vasodilation in test rats. The vasoactive substance in this non-SP-containing fraction was subsequently purified, sequenced, identified as a tridecapeptide, and named neurotensin (Carraway and Leeman, 1973). Neurotensin (NT) was found to be abundant in the nervous system of mammals (Goedert, 1984), as well as in such peripheral tissues as the gut and adrenal medulla (Lundberg, Rokaeus, Hokfelt, et al., 1982, Goedert and Emson, 1983). Carraway et al. (1982b), using immunochemical and chromatographic techniques, has found that NT, or its isopeptides, are present in a wide variety of vertebrate (including turtles, crocodilians, and snakes) and invertebrate species. Immunochemical characterization of NT among these diverse species shows that the C-terminus has been evolutionarily conservative but not the N-terminus. In further support of this conclusion, partially purified NT-like peptides from chicken, turtle, dogfish shark, and lobster displayed biological activity similar to mammalian NT in several different bioassay systems, indicating that the biologically active portion of NT, the C-terminus, has been conserved during evolution. Three major groupings of vertebrates were noted in terms of the chemical characteristics of NT found in the members of the various vertebrate taxa. First, in mammals, authentic NT is present. Second, NT in reptiles is very similar or identical to that in birds, but NT in both is distinct from that in mammals. Later studies in chickens using RIA and three antisera (Carraway and Bhatnagar, 1980a), each specific for a different portion of bovine NT, confirmed the presence of large amounts of an NT-like peptide in chicken brain and gut. This peptide (Table 8.7), which will be called chicken NT, was subsequently found to be a tridecapeptide identical to mammalian tridecapeptide NT (except for amino acid substitutions at the 3, 4, and 7 positions) with biological activity largely identical to mammalian NT (Carraway and Bhatnagar, 1980b). Subsequent studies revealed that NT in turtles shows identical chromatographic properties in HPLC to chicken NT, thus suggesting that turtle NT is identical to chicken NT. This may be the case in all or most reptilian species. Finally, in amphibians, cartilaginous fish, and agnathans, NT-like peptides distinct from mammalian NT and chicken NT are present in the nervous system (Carraway et al., 1982b). The NT-like peptide in am-

Table 8.7. Amino acid sequences for mammalian neurotensin (NT) and chicken neurotensin and several other neurotensin-related peptides.

	1	2	3	4	5	6	7
Human-bovine NT	pGlu	Leu	Tyr	Glu	Asn	Lys	Pro
Chicken NT	pGlu	Leu	His	Val	Asn	Lys	Ala
LANT6							
Neuromedin N							
Xenopsin						pGlu	Gly

	8	9	10	11	12	13
Human-bovine NT	Arg	Arg	Pro	Tyr	Ile	Leu-OH
Chicken NT	Arg	Arg	Pro	Tyr	Ile	Leu-OH
LANT6	Lys	Asn	Pro	Tyr	Ile	Leu-OH
Neuromedin N	Lys	Ile	Pro	Tyr	Ile	Leu-OH
Xenopsin	Lys	Arg	Pro	Trp	Ile	Leu-OH

Note: See text for further details.

phibians is clearly distinct from xenopsin (Table 8.7), which is a neu-rotensin-related peptide originally found in *Xenopus laevis* skin (Carraway et al., 1982a).

A number of peptides related in structure to NT, but distinct from it, also appear to be commonly present in the nervous systems of a wide variety of vertebrate species. For example, Carraway and co-workers (Carraway and Bhatnagar, 1980a, 1980b; Carraway et al., 1983) have recently discovered a neurotensin-related hexapeptide (six amino acids) that is present in gut and brain tissue. This peptide consists of the last four amino acids of the C-terminal peptide of NT, with lysine-asparagine present as an N-terminus extension (see Table 8.7). The peptide has been termed LANT6 (Carraway et al., 1983) and was discovered originally in chicken gut. Carraway et al. (1983), using a highly specific radioimmunoassay for LANT6, showed that LANT6 is present in abundance in a variety of GI tract tissues, thymus, and brain. Within chicken brain, LANT6 was most abundant in the telen-cephalon, hypothalamus, and midbrain. Fractionation of chicken brain demonstrated that LANT6 was most highly concentrated in the synaptic terminals, thereby suggesting a role in neurotransmission. Although structurally related to NT, LANT6 peripherally produces hypertension (rather than hypotension as in the case of NT) (Carra-way and Ferris, 1983) and competes poorly with NT for receptor sites. Thus, NT and LANT6 may act on different receptor sites. Subsequent studies revealed that a LANT6-like peptide is also present in the ner-vous systems in turtles, rats, and hamsters, although LANT6 in these species appears to be typically contained within higher-molecular-weight molecules, whereas in pigeons and chickens LANT6 appears to be present as a hexapeptide (Reiner and Carraway, 1987). Nonetheless, LANT6 can be cleaved from these larger molecules in turtles and hamsters by pepsin treatment of the larger molecules (Reiner and Carraway, 1987). The LANT6-like peptide derived in this fashion in turtles appears to be highly similar to that in birds, whereas that in hamsters does appear to differ from avian LANT6 (Reiner and Carraway, 1987). The chromatographic studies of Carraway et al. (1982b) suggest that a LANT6-like peptide (as well as NT) may also be present in snakes, amphibians, cartilaginous fish, and agnathans. Yet a third NT-related peptide is also present in mammals and turtles, but seemingly not in birds. This peptide is termed neuromedin N (NMN) and it is identical in structure to LANT6, except that isoleucine is substituted for asparagine as the second residue from the N-terminus (Minamino et al., 1984) (Table 8.7). HPLC analysis suggests that NMN in turtles differs in structure from that found in mammals (Reiner and Carraway, 1987). Since NMN has been localized to synaptosomes and binds to NT receptors, it is likely that NMN is a neuroactive substance (Carraway and Mitra, 1987; Checler et al., 1986)

Recent studies using cloned cDNAs derived from canine enteric mucosa cells have revealed the structure of the NT precursor and somewhat clarified the relationships of these diverse NT and NT-related peptides (Dobner et al., 1987). Both NT and NMN were found to derive from the same 170-amino-acid precursor. This finding is consistent with the observation that NT and NMN show highly similar distributions in the nervous system in mammals (Carraway and Mitra, 1987; Reiner and Carraway, unpublished observations). Since immunohistochemical studies reveal that NMN and NT in turtles also show highly similar distributions in the nervous system (Reiner and Carraway, unpublished observations), it appears likely that NT and NMN also derive from the same precursor in turtles. Whether NT and NMN are both present in lizards, snakes, and crocodilians is uncertain. The precursor for LANT6 is also uncertain. The NT/NMN precursor identified in mammals does not contain the LANT6 sequence, although it does contain an additional hexapeptide sequence (other than NMN) that bears some resemblance to LANT6 (Dobner et al., 1987). It is uncertain, however, whether this sequence accounts for the LANT6-like immunoreactivity found in the nervous system of mammals in previous studies (Carraway et al., 1983; Reiner and Carraway, 1987). To some extent, it appears unlikely that LANT6 derives from the same sequence as NT and NMN, since the distribution of LANT6 in the nervous system and gut in birds, turtles, and mammals differs from that of NT and NMN (Carraway et al., 1983; Reinecke, 1985; Reiner and Carraway, 1985, 1987; unpublished observations).

Few immunohistochemical data are available on the neural distribution of NT in reptiles. Eldred and Karten (1983) have demonstrated the presence of NT+ amacrine cells in turtles, as found in members of other vertebrate classes (Brecha et al., 1984). In turtle brain, using a C-terminally specific NT antiserum that does not cross-react with LANT6 (i.e., HC-8, courtesy of R. E. Carraway), numerous NT-positive neurons and fibers were observed. Within hypothalamus, NT-positive neurons and fibers were abundant, and some NT-positive fibers were observed in the median eminence, as also true in mammals (Ibata et al., 1983). In mammals, these NT+ fibers in the median eminence arise from neurons of the arcuate and periventricular regions that contain dopamine as well as NT (Ibata et al., 1983). Since both dopaminergic neurons and NT+ neurons are found in similar locations in turtles (Smeets et al., 1987; Reiner, unpublished observations), the same may be true in reptiles. Neurotensin released from the ME may regulate anterior pituitary hormone release, and it has been shown that NT affects the release of prolactin and TSH in mammals (Rivier et al., 1977; Goedert, 1984). Numerous NT+ fibers and some neurons were also observed in the striatal portion of the basal ganglia of turtles. Similar results have been reported in mammals (Uhl et al., 1977, 1979).

One of the most striking features of NT distribution in turtles is that it is found in many neurons of the dorsal thalamus, including the dorsomedial and dorsolateral thalamic nuclei, and the dorsal geniculate nucleus (Fig. 8.13). All or nearly all neurons of the dorsal geniculate contain NT (or the turtle isopeptide of NT). Further, the terminal zones of these thalamic nuclei in the telencephalic cortex also contain NT, implying that these neurons have NT+ thalamocortical projections. Terminals containing NT are particularly dense in the superficialmost portion of the molecular layer of the dorsal cortex and lateral to the pallial thickening (the cortical projection targets of the dorsal geniculate nucleus) (Ulinski, 1988). In dorsomedial cortex, NT+ terminals are also observed in the superficialmost molecular layer but are more lightly labeled. In medial cortex, lightly labeled NT+ terminals are present in a band 50 to 100 μm deep to the cortical surface. Although the distribution of NT has been studied in the thalamus and cortex of mammals, there has been no evidence presented of NT+ thalamocortical projections in mammals. As with other neuropeptides, all available lines of evidence (based largely on work in mammals) support the suggestion that NT acts as a neurotransmitter/neuromodulator in brain: (1) it is found in neurons and their terminals (Uhl et al., 1977, 1979; Uhl and Snyder, 1980); (2) it is released from depolarized nerve terminals (Iversen et al., 1978); (3) it affects neuronal firing activity (Andrade and Aghajanian, 1981); and (4) the distribution of neurotensin-specific receptor sites in the brain largely

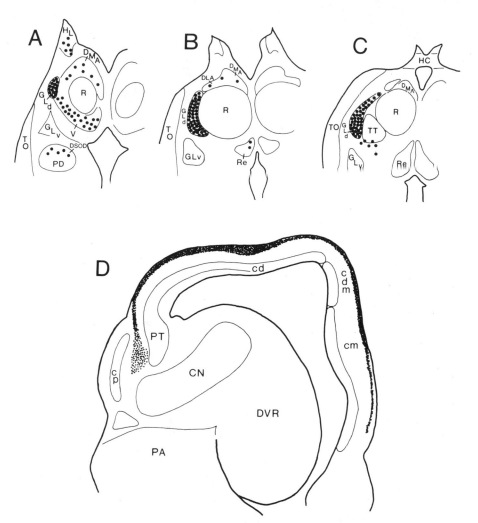

Fig. 8.13. Line drawings of the distribution of LANT6/neurotensin+ neurons in a rostrocaudal series of transverse sections in the thalamus of red-eared turtle (A to C) and of LANT6/neurotensin+ fibers in the thalamorecipient zones (Hall and Ebner, 1970) of the dorsal cortex and pallial thickening of the telencephalon (as seen in a transverse section) of a red-eared turtle. LANT6/neurotensin+ neurons are present in several dorsal thalamic nuclei that project to the telencephalon and are particularly abundant in the dorsal geniculate nucleus, all of whose neurons appear to be LANT6/neurotensin+. In the telencephalon, LANT6/neurotensin+ fibers are particularly prominent in the lateral half of the dorsal cortex and the pallial thickening, which are the targets of the GLd. Abbreviations: CN, core nucleus of the DVR; cd, cortex dorsalis; cdm, cortex dorsomedialis; cm, cortex medialis; cp, cortex pyriformis; DLA, nucleus dorsolateralis anterior; DMA, nucleus dorsomedialis anterior; DSOD, nucleus of the dorsal supraoptic decussation; DVR, dorsal ventricular ridge; d, area d; GLd, nucleus geniculatus lateralis, pars dorsalis; GLv, nucleus geniculatus lateralis, pars ventralis; HC, habenular commissure; HL, lateral habenular nucleus; PA, paleostriatum augmentatum; PD, dorsal peduncle of the lateral forebrain bundle; PT, pallial thickening; R, nucleus rotundus; Re, nucleus reuniens; TT, tractus tectothalamicus; TO, tractus opticus; V, nucleus ventralis.

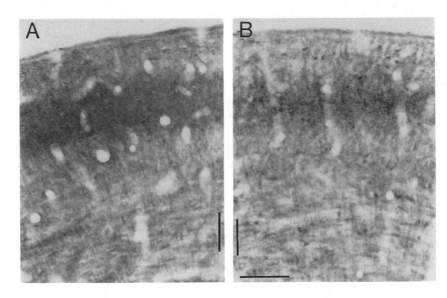

matches the distribution of NT (Quirion et al., 1982; Brauth et al., 1986).

In gut, NT is found in endocrine cells and nerve fibers in a wide variety of vertebrate species, including reptilian species (Gapp et al., 1985; Helmstaedter et al., 1977; Reinecke, 1985; Goedert, 1984; Carraway and Reinecke, 1984; Buchan et al., 1983). Neurotensin acts as a potent relaxant of contraction of the circular muscular layer of gut. Release of neurotensin from endocrine cells of the gut may play a role in the demonstrated ability of NT to inhibit gastric acid secretion and in the inhibition of gastric motility (Goedert, 1984).

The distribution of LANT6 has also been studied in turtle brain and shows some similarities to the distribution of NT, being present in neurons and fibers of the hypothalamus and in neurons of the same thalamic cell groups (and their cortical projection fields) as NT (Eldred et al., 1988). In the case of the dorsal lateral geniculate nucleus, all or nearly all neurons were labeled for LANT6, thus implying the coexistence of NT and LANT6 in many of the same neurons. In turtle retina, also, LANT6 appears to be present in the same amacrine cells, or at least the same amacrine cell types, as NT (Eldred et al., 1987). In addition, LANT6 is present in numerous retinal ganglion cells of diverse types in turtles (Eldred et al., 1988). Further, it has been shown that LANT6+ fibers are present in all retinorecipient areas of the turtle brain, and these fibers disappear unilaterally after monocular enucleation of the contralateral eye (Eldred et al., 1988) (Fig. 8.14). Similar results have also been obtained in pigeons (Reiner,

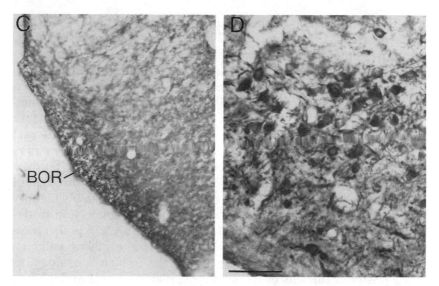

Fig. 8.14. (A) LANT6+ labeling in the superficial layers of the tectum in a *Pseudemys scripta*. The band of heavy LANT6+ labeling observed is present in the major retino-recipient layer of the turtle tectum and most of the LANT6+ fibers contributing to this band arise from LANT6+ retinal ganglion cells. (B) The superficial layers of the contralateral tectum in the same turtle as shown in (A). Retinal input to this tectum had been removed by contralateral monocular enucleation two months before immunohistochemical processing of the brain tissue (retinal input to the tectum is almost entirely crossed in turtles: Bass and Northcutt, 1981). Comparison of (A) and (B) shows that LANT6+ fiber labeling in the retinorecipient tectum is considerably reduced by the elimination of retinal input, thus indicating the existence of a LANT6+ input from the retina. The source of residual LANT6+ fibers in the retinorecipient layers of the tectum after enucleation is uncertain. Dark vertical bars indicate the extent and location of the stratum griseum centrale of the tectum. (C) Dark punctate profiles in the BOR are the LANT6+ axons of the BOR in a red-eared turtle. (D) LANT6+ neurons of the globus pallidus in a red-eared turtle. Medial is to the right in both (C) and (D). The scale bars in (B) and (D) both equal 100 μm. The magnification in (A) and (C) is the same as in (B).

1986c). LANT6+ retinal ganglion cells have also been observed in goldfish, frogs, chickens, and mammals (Eldred et al., 1987). Some previous studies have provided suggestive evidence of the existence of peptidergic retinal ganglion cells (Reiner, Karten, and Brecha, 1982; Kuljis and Karten, 1982, 1983; Kuljis and Karten, 1984, 1988), but peptides have rarely been unequivocally observed directly in large numbers of retinal ganglion cells. LANT6 thus was the first peptide clearly visualized in large numbers of retinal ganglion cells and their central projections.

LANT6 is also present in neurons of numerous cell groups of the central nervous system. It is present in neurons of the globus pallidus and the ventral paleostriatum in turtles and crocodilians (Reiner and Carraway, 1985, 1987) (Fig. 8.14). This has also been observed to be true in pigeons, hamsters, monkeys, frogs, lungfish, and sharks (Reiner and Carraway, 1985, 1987; Reiner, 1987b; Reiner and Northcutt, 1987). In turtles and pigeons, these LANT6+ neurons have been shown to project to the nigral cell groups of the midbrain (Reiner, 1986d; Reiner and Carraway, 1987). In light of the reported excitatory effects of NT on dopaminergic neurons in mammals (Andrade and Aghajanian, 1981), these findings suggest that pallidal neurons may use LANT6 to influence tegmental dopaminergic neurons. Further, the apparent conservation of LANT6 in pallidal neurons throughout phylogeny is striking and affords a means by which to readily identify pallidal neurons in diverse vertebrate groups. In pigeons and turtles both (Reiner and Carraway, 1987; Reiner, 1986b), LANT6 is also present in neurons of several projection targets of the globus pallidus, such as the entopeduncular nucleus of turtles (which is comparable to the posterior nucleus of the ansa lenticularis in birds and is often homologized to the subthalamic nucleus in mammals) and the nucleus profundus mesencephali of turtles (which is comparable to the dorsolateral subdivision of the nucleus tegmentipedunculopontinus in birds and is often homologized to the substantia nigra, pars reticulata of mammals) (Baker-Cohen, 1968).

In turtles, LANT6+ neurons were also observed in the DVR and cortex (Reiner and Carraway, 1987; Reiner, 1991). In turtles, as well as in pigeons, LANT6 was observed in neurons of the rhombencephalic cochlear nuclei and in neurons in the central and laminar nuclei of the torus semicircularis (comparable to nucleus mesencephali lateralis, pars dorsalis, and the nucleus intercollicularis in birds). Thus, in contrast to other peptides, LANT6 is present in neurons of both visual and auditory relay cell groups of the brain and in projection neurons of the retina. As noted by Carraway et al. (1983), LANT6 appears to be associated with synaptic vesicles in birds and thus may have a transmitter-like function. In light of the presence of LANT6 in many projection systems of the brain in reptiles and birds, further studies on the functional role of LANT6 in the nervous system are of interest.

V. ADDITIONAL BRAIN-GUT PEPTIDES

Several neuropeptides present in both brain and gut have been described above. A number of additional peptides that are prominent in gut or were first discovered in gut are also present in the nervous system. Included among these are such well-known GI tract hormones as cholecystokinin (CCK), vasoactive intestinal polypeptide (VIP), and the pancreatic polypeptide family of peptides.

A. Cholecystokinin

Cholecystokinin (CCK) is a 33-amino-acid peptide (Mutt and Jorpes, 1967, 1968) that was first discovered in the GI tract in mammals, where it has two distinct actions: (1) contraction of the gallbladder (hence its name cholecystokinin), thereby increasing bile flow, and (2) stimulation of pancreatic enzyme secretion (Turner and Bagnara, 1976; Gorbman et al., 1983). These effects are produced by CCK released from endocrine cells of the upper intestinal tract. A related peptide hormone that has the same pentapeptide sequence at the C-terminus as CCK is found in endocrine cells of the stomach. This hormone, gastrin, stimulates gastric acid secretion. In an initial study of brain extracts using RIA, Vanderhaeghen et al. (1975) used antigastrin antibodies and reported that large amounts of a gastrinlike substance are present in the brain of a variety of vertebrates, including mammals. Subsequently, several investigators showed that the gastrinlike peptide detected by Vanderhaeghen et al. (1975) was actually a CCK-like peptide. Eventually it was shown that authentic CCK (identical in structure to gut CCK) was present in mammalian brain but that most of the CCK-like material in brain was the C-terminal octapeptide of CCK (termed CCK8). It is now clear that CCK8 is also abundant in gut (Straus et al., 1981). Central receptor sites for CCK have been found in the brain of mammals, but interestingly, the brain CCK receptor appears to differ from the GI tract CCK receptor (Vigna, 1985). Studies in mammals have shown that CCK8 demonstrates such further characteristics of a neurotransmitter as localization within synaptic terminals, release upon depolarization from synaptic terminals, and the ability to affect neuronal membrane potentials (Dockray, 1982).

Cholecystokinin is among the more extensively studied neuropeptides from a comparative viewpoint. Cholecystokininlike peptides have been found in the brain and gut of invertebrates such as Hydra (Grimmelhuijzen et al., 1980), annelid worms (Engelhardt et al., 1982), blowfly (Dockray, Duve, and Thorpe, 1981; Duve and Thorpe, 1981), dungeness crab (Larsen and Vigna, 1983), molluscs (Straus et al., 1975), as well as protochordates (Pestarino, 1985; Thorndyke and Dockray, 1986), and cephalochordates (Salin, 1988). The CCK-like materials in the brain and gut of invertebrates, however, appear to be distinct from CCK8 and CCK of mammalian brain and gut and do not resemble CCK any more than they resemble gastrin.

Vigna (1979), based on biological assays of extracts of stomach and gut in the members of a variety of vertebrate groups, has suggested that a distinct CCK-like peptide made its evolutionary appearance in cells of the upper intestinal tract and brain of vertebrates at the very beginning of the vertebrate lineage. Immunohistochemical studies support this suggestion, since CCK+ labeling has been observed in the GI tract and brain of lamprey, various species of bony and cartilaginous fish, and amphibians (Buchan et al., 1983; Holmquist et al., 1979; Rajjo et al., 1988). A distinct gastrinlike peptide, however, does not appear to be present in the brain or GI tract of anamniotic vertebrates (Vigna, 1985, 1986; Rajjo et al., 1988; Larsson and Rehfeld, 1977; Holmgren et al., 1982; Noaillac-Depeyre and Hollande, 1981). In reptiles, birds, and mammals, immunohistochemical and biochemical studies have shown that a CCK-like peptide is present in the intestine, a gastrinlike peptide is present in the stomach, and authentic CCK8 (or a peptide nearly identical to CCK8) is present in the brain (Larsson and Rehfeld, 1977; Dimaline et al., 1982; Dimaline, 1983; Beinfeld, Trubatch and Brownstein, 1983; Reiner et al., 1985; Reiner and Beinfeld, 1985; Vigna, 1986). These results suggest that gastrin may have evolved by the duplication and subsequent differentiation of the gene coding for CCK during the amphibian-reptile transition (Vigna, 1985, 1986). It is of interest to note that the chemical structure of CCK8 in brain appears to be the same (or highly similar) in frogs, reptiles (at least turtles), birds (at least chickens), and mammals (Beinfeld, Trubatch, and Brownstein, 1983; Reiner et al., 1985; Reiner and Beinfeld, 1985; Fan et al., 1987). Thus, CCK8 appears to be more highly conservative in structure than most neuropeptides, with the exception of such neuropeptides as somatostatin and the enkephalin pentapeptides. Interestingly, the evolution of the CCK receptor appears to have been less conservative. In mammals and birds, the nervous system CCK receptor differs from that in the periphery, whereas such a difference is not evident in fish, amphibians, and reptiles (garter snakes) (Vigna, 1985; Vigna et al., 1986). Based on these data, Vigna (1985) has suggested that only a single CCK receptor was present in ancestral reptiles and this receptor was ancestral to the distinct neural and GI tract CCK receptors observed in birds and mammals.

The distribution of CCK8 in the nervous system has been studied in some detail in reptiles, chiefly turtles. In the periphery, CCK8+ labeling of amacrine cells has been reported in the retinas of lacertid lizards (Osborne et al., 1982), and low but measurable amounts of CCK8 have been reported in the retinas of lizards, *Anolis*, and turtles, *Chrysemys* (Eskay and Beinfeld, 1982). The distribution of CCK8 in turtle brain has been studied in detail by Reiner and Beinfeld (1985). As alluded to above, CCK8 in turtle brain has been chemically characterized as being indistinguishable from mammalian CCK8-sulfate, based on HPLC (Fig. 8.15). The concentration of CCK8 in turtle CNS is, in general, comparable to that in mammal CNS, except much lower concentrations of CCK8 are present in turtle basal telencephalon and telencephalic cortex than in mammalian basal telencephalon and telencephalic cortex (Table 8.8). In turtle brain (as in frog brain), the highest concentration of CCK8 is found in the hypothalamus and neurohypophysis (Reiner and Beinfeld, 1985; Beinfeld, Trubatch, and Brownstein, 1983). The concentration of CCK8 in several other brain regions is also high but considerably less than in the hypothalamus and neurohypophysis. These other regions include the thalamus, tegmentum, rhombencephalon, and spinal cord.

Immunohistochemical studies confirm that CCK8-containing neurons and fibers are widespread in the CNS of red-eared turtles (Reiner and Beinfeld, 1985; Reiner et al, 1985). In the spinal cord, CCK8-containing fibers are present in the dorsal and ventral horns, the dorsolateral funiculus, and the tract of Lissauer. The presence of CCK8 + fibers in the tract of Lissauer, which contains numerous SP + terminals (Fig. 8.16) that presumably are the terminals of dorsal root gan-

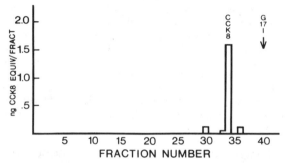

Fig. 8.15. HPLC of turtle brain tissue extract, using a linear gradient of acetonitrile on an Altech column. The bulk of the CCK8-like material in turtle brain (87%) coelutes with CCK8 sulfate, the same CCK8 peptide found in mammalian brain. The minor peaks before and after CCK8 may correspond to a high-molecular-weight CCK-like peptide and CCK8 desulfate, respectively. (From Reiner and Beinfeld, 1985.)

Table 8.8. Amounts of CCK8 in various neural regions in *Pseudemys scripta* and rat (in pg/mg wet weight tissue)

	Turtle	Mammal
Basal telencephalon, including the basal ganglia	10	382
Hypothalamus	212	194
Tectum	21	—
Spinal cord	58	—
Cortex	10	528
Cerebellum	5	1
Rhombencephalon	51	27

Turtle data from Reiner and Beinfeld (1985) and rat data from Beinfeld et al. 1981. Note that although the brain stem content of CCK8 is comparable, CCK8 content in the cerebral hemispheres is much greater in mammals than in reptiles.

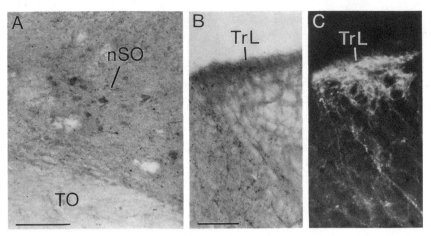

Fig. 8.16. (A) CCK8+ neurons in the supraoptic nucleus (seen in transverse view) of *Pseudemys scripta*. Medial is to the right. (B) CCK8+ fibers in Lissauer's zone or tract (TrL) in the spinal cord of a red-eared turtle. Medial is to the right. (C) SP+ fibers, labeled according to the immunofluorescence procedure in the same portion of the turtle spinal cord as shown in (B). (D) CCK8+ fibers in the median eminence of a red-eared turtle. Medial is to the right. (E) Mesotocin+ fibers (labeled with an antioxytocin antiserum) in the median eminence of a red-eared turtle. Medial is to the right. (F) Vasotocin+ fibers (labeled with an antivasopressin antiserum) in the median eminence of a painted turtle. Medial is to the right. Scale bars in (A), (B), and (D) equal 100 μm. (C) is at the same magnification as (B). The scale bar in (E) equals 250 μm and (F) is at the same magnification as (E).

glion neurons, suggests that SP and CCK8 may be present in the same primary afferent terminals, as reportedly the case in mammals (Dalsgaard et al., 1982). In the rhombencephalon, CCK8+ fibers are present in the reticular formation, the raphe nuclei, the nucleus of the solitary tract, the dorsal motor nucleus of the vagus, the locus coeruleus, the parabrachial region, and the descending tract of the trigeminal nerve. The most prominent sites of CCK8+ cell bodies were the nucleus of the solitary tract, the dorsal motor nucleus of the vagus, and the region of nucleus ambiguus. Within the midbrain, prominent CCK8+ fibers were observed in the deep and intermediate tectal layers and in the medial tegmentum. In the medial midbrain tegmentum of colchicine-treated turtles, a number of CCK8+ cells were observed in rostral and caudal portions of the ventral tegmental area (as defined in Brauth et al., 1983). In mammals, numerous CCK8+ cells are also present in the ventral tegmental area, and many of these cells also contain dopamine (Vanderhaeghen, 1985).

Within the diencephalon, CCK8+ neurons and fibers are numerous and heavily labeled in the hypothalamus. A large field of CCK8+ neurons is present in the lateral hypothalamic/supramammillary regions of the hypothalamus (Fig. 8.17). These neurons also contain SP

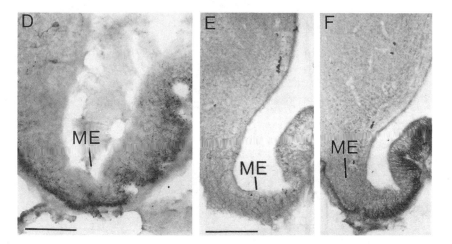

and apparently VIP in turtles (Reiner et al., 1985) (Fig. 8.17). In mammals, neurons of this field are known to project to the cerebral cortex, as also is seemingly true in reptiles (Seroogy et al., 1988; Reiner et al., 1985). There is also evidence of a CCK8+ projection from the magnocellular hypothalamic secretory cell groups to the ME and neurohypophysis, since CCK8+ labeling is present in supraoptic and paraventricular neurons and in terminals in the external zone of the ME and neurohypophysis (Fig. 8.16). As noted previously, this projection system also contains vasotocin and mesotocin in reptiles. Since CCK8+ terminals cluster heavily in the external zone of the ME and neurohypophysis, and mesotocinergic terminals do not, CCK8 and mesotocin may not be present in the same terminals or they may be colocalized in only a small number of terminals (Fig. 16). Vasotocin is present in the same regions of the hypothalamopituitary pathways as CCK8 and these two peptides may thus be colocalized with one another (Fig. 8.16). In mammals, oxytocin and CCK8 are reportedly colocalized in the same neurons and their terminals in the ME (Vanderhaeghen et al., 1981; Martin et al., 1983). Within both mammalian and turtle hypothalamus, CCK8+ fibers are present in abundance in the ventromedial nucleus. In the remainder of the diencephalon, CCK8+ terminals are sparse. Some lightly labeled CCK8+ terminals are observed in the dorsomedial/dorsolateral thalamic nucleus and in the lateral portion of the medial habenular nucleus.

In the telencephalon of *Pseudemys*, CCK8+ fibers and neurons are few (Reiner and Beinfeld, 1985; Reiner et al., 1985). Fibers containing CCK8+ are present in the septal region, the nucleus of the diagonal band/ventral paleostriatum region, and in the telencephalic cortex. CCK8+ neurons were also observed in the rostral ventromedial te-

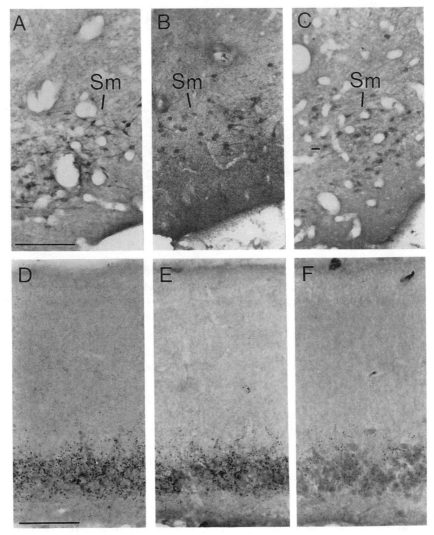

Fig. 8.17. Neurons within the supramammillary region of the turtle hypothalamus (in transverse view) labeled immunohistochemically for SP (A), CCK8 (B), and VIP (C) and fibers within the cellular layer of the dorsal cortex of turtle labeled for SP (D), CCK8 (E), and VIP (F). The magnification for (A), (B) and (C) is the same and the scale bar in (A) equals 150 μm. Medial is to the left in all three. The magnification is the same in (D), (E), and (F) and the scale bar in (D) equals 100 μm. Double-label procedures have shown that SP +, CCK8 + and VIP + labeling is present in the same fibers of the cellular layer of cortex in turtle. Double-label procedures have also shown that SP + and CCK8 + labeling both are present in the same neurons of the supra-mammillary region, which is the probable source of the cortical SP + /CCK8 + /VIP + fibers. The colocalization of VIP with SP/CCK8 in Sm neurons has not yet been studied, though it seems likely that such colocalization does occur in Sm neurons.

lencephalon and in the medial striatum (area d). The CCK8+ cortical fiber system has been studied in greater detail by Reiner et al. (1985). The CCK8+ fibers of the cortex are restricted to the cellular layer of the cortex, and all CCK8+ fibers of the cortex also contain an SP-like peptide. These fibers appear to originate from SP/CCK8-containing neurons of the lateral hypothalamic/supramammillary region, as noted above. HPLC studies of cortex indicate that authentic CCK8-sulfate and an SP-like material highly similar to mammalian SP are present in turtle cortex. The amounts of turtle SP and CCK8 in these SP/CCK8+ cortical fibers is comparable, though SP appears to be present in a somewhat higher concentration (.31 pg CCK8/μg cortical protein). The terminals containing SP/CCK8 make asymmetrical synapses onto cell bodies and their primary dendrites. In terminals containing SP/CCK8, the SP and CCK8 appear to be present in the same large dense core vesicles (Reiner et al., 1985). In mammals, also, SP and CCK8 are found within large dense core vesicles (Rehfeld et al., 1979; DiFiglia et al., 1982a, 1982b), but no data are available on the ultrastructural localization of these peptides in neurons or fibers where they co-occur. Terminals containing SP and CCK8 in turtle cortex also contain numerous small vesicles that do not contain either SP or CCK8 and a number of large dense core vesicles that do not contain either SP or CCK. The small vesicles are clustered near synaptic release sites and presumably contain the primary transmitter used by these terminals. The SP/CCK8-containing large dense core vesicles are not clustered near synaptic release sites but presumably do release their contents into the extracellular space, where they in all likelihood act as modulatory agents affecting the postsynaptic target cell. SP as well as CCK8 has been shown to excite cortical neurons in mammals (Phillis and Limacher, 1974; Phillis and Kirkpatrick, 1980; Lamour et al., 1983; Rogawski, 1982). It should be noted, as will be discussed in the section on VIP, that the CCK8/SP-containing fibers of turtle cortex also contain VIP (Reiner et al., 1985).

In immunohistochemical studies in mammals, it has been found that the most striking aspect of the distribution of CCK8 in the brain is its presence in large numbers of cortical neurons. RIA studies in mammals have confirmed that CCK8 is present in its highest brain concentration in the cortex (Beinfeld et al., 1981; Beinfeld, Lewis, Eiden, et al., 1983). As noted, the CCK8-RIA data indicate that turtle cortex has only approximately 1/20 to 1/40 the CCK8 concentration of mammalian cortex (Beinfeld, Lewis, Eiden, et al., 1983). The DVR in turtles contains even less CCK8 than the cortex. Since many investigators have noted that both DVR and cortex in turtles appear comparable to the cortex of mammals (Northcutt, 1978, 1981; Ulinski, 1983;

Nauta and Karten, 1970), it might be expected that cortical and DVR CCK8 levels be more nearly comparable in turtles. Since they are not, this raises the question as to why turtle cortex and DVR differ in their CCK8 content from one another if both are thought to be comparable to mammalian neocortex and why CCK8 levels are so low in both turtle cortex and DVR (as compared to mammalian cortex). In mammals, the bulk of the cortical CCK8 is present in intrinsic cortical neurons of layers II to III (Peters et al., 1983; McDonald et al., 1982c). In turtles, intrinsic CCK8+ cortical neurons and CCK8+ DVR neurons are extremely rare and very lightly labeled when observed. The SP/CCK8-containing fibers of turtle cortex (which contain nearly all of the cortical CCK8 in turtles) appear to arise from the SP/CCK8-containing neurons of the lateral hypothalamic/supramammillary region. Pathway tracing data confirm that this cell group projects to the cortex in turtle (Desan, 1984). Thus there would appear to be little intrinsic CCK8 in either DVR or cortex in turtles (except for the rare cortical and DVR CCK8+ neurons).

Thus regardless of the degree to which CCK8 in turtle cortex is intrinsic in origin, CCK8 is clearly a much more prominent cortical peptide in mammals than in turtles. CCK8 may not have become a prominent intrinsic cortical peptide until the mammalian lineage in evolution. Ebner (1976) has previously suggested that turtle cortex may be largely equivalent to layers IV to VI of mammalian cortex based on pathway tracing data. The results in turtles clearly suggest that CCK8+ neurons of cortical layers II to III (and possibly other neurons of these layers as well) have proliferated greatly during the transition from reptiles to mammals. Beinfeld, Trubatch, and Brownstein (1983) have noted that CCK8 is also not abundant in frog cortex (or cerebrum). Preliminary observations indicate that the same is true in crocodilians and *Sphenodon* (Reiner, unpublished observations). The functional implications of the increase in abundance in CCK8 interneurons in cortex in mammals is uncertain, and understanding of this issue requires further clarification of the role of CCK8+ neurons in cortical function.

B. Vasoactive Intestinal Polypeptide

Vasoactive intestinal polypeptide (VIP) is a potent vasodilator that was first isolated from porcine small intestine in 1970 by Said and his colleagues (Said and Mutt, 1970, 1972). Following its sequencing as a 28-amino-acid peptide (or octacosapeptide), immunohistochemical studies demonstrated that VIP is not found in endocrine cells of the mammalian gut but rather is found in neurons of the intramural neural plexi of the gut and in nerve fibers throughout all layers of the gut and in fibers in association with blood vessels and glandular epithelium of the gut (Costa and Furness, 1982; Said, 1984). VIP was later revealed to be present in neurons of many autonomic ganglia throughout the body and in the postganglionic fibers of ganglia innervating blood vessels, the respiratory and urogenital tracts, exocrine glands, and the thyroid gland. In these various target organs, VIP appears to relax smooth muscle, thereby promoting such peripheral effects as vasodilation, enhanced secretion, and sphincter opening. In sympathetic and parasympathetic ganglia that innervate sweat glands (sympathetic ganglia) or innervate blood vessels (pterygopalatine and submandibular parasympathetic ganglia), VIP is known to be present in cholinergic neurons and their postganglionic fibers. In these systems of peripheral nerves, VIP and acetylcholine have been shown to be coreleased from the same terminals and have complementary actions on their target structures (Said, 1984; Lundberg, Anggard, and Fahrenkrug, 1982; Lundberg et al., 1979). Following its discovery, VIP was also found to be a prominent CNS peptide in mammals. The chemical structure of the VIP in nerves of the mammalian GI tract is apparently the same as (or highly similar to) that of VIP in the mammalian CNS (Costa and Furness, 1982).

VIP-like peptides appear to be widespread among both vertebrates and invertebrates. A VIP-like substance has been observed by immunohistochemical techniques in the nervous system of snails (Schot et al., 1981) and in intestinal endocrine cells of protochordates (Reinecke, 1981). In cartilaginous fish, bony fish, amphibians, and reptiles, VIP+ intramural nerve cells and fibers and VIP+ endocrine cells have been reported in the intestinal tract (Reinecke et al., 1980; Van Noorden and Falkmer, 1980; Falkmer et al., 1980; Buchan et al., 1981; El-Salhy, 1984; Tagliafierro et al., 1988). Several other studies in nonmammalian species (cartilaginous and bony fish, reptiles, and birds), however, have observed VIP-containing nerves and fibers in the GI tract wall but have failed to find VIP-containing endocrine cells in the gut (Holmgren and Nilsson, 1983; Holmgren et al., 1982; Buchan et al., 1983; Vaillant et al., 1980; Rawdon and Andrew, 1981; Sundler et al., 1979; Fontaine-Perus, 1984; Saffrey et al., 1982; Hayashi et al., 1983). This discrepancy may have arisen because of the presence of a VIP-like peptide, distinct from that found in neurons in gut endocrine cells in fish and reptiles. In support of this possibility, a VIP-like peptide that appears to differ from the VIP in the nervous system has been demonstrated in gut endocrine cells in mammals (Dimaline et al., 1980; Furness et al., 1982). This may also be the case in nonmammals, with the endocrine cell VIP possibly less differentiated from neural VIP than in mammals, thereby leading to a greater tendency to be labeled by the same antiserum. Nerves and fibers containing VIP may be as widespread in bodily tissues in nonmammals as in mammals since VIP+ nerves are present in diverse peripheral tissues in a variety of nonmammals (Gibbins, 1983; LeBoulenger et al., 1983; Reiner, 1987c; Chipkin et al., 1988).

VIP is a member of a group of gut peptides that are related in structure. This group includes secretin, glucagon, gastrin-inhibitory peptide (GIP), PHI (porcine histidine-isoleucine amide-containing peptide), as well as VIP (Fahrenkrug and Emson, 1982; Dockray, 1979). The structure of these is shown in Table 8.9. Secretin, glucagon, and GIP will be discussed in greater detail in a subsequent section. PHI is found in the same nerves and fibers as VIP and is derived from the same large precursor molecule as VIP (Tatemoto, 1984; Yanaihara et al., 1983). The similarities among VIP/PHI, secretin, glucagon, and GIP suggest that all are derived from an ancestral molecule (Dockray, 1977, 1979). Distinct glucagon, VIP and secretin peptides have been isolated from the chicken intestine and pancreas (Dockray, 1979). Thus it seems likely that reptiles also possess their own versions of these peptides. The structure for chicken VIP, which is also an octacosapeptide, is given in Table 8.9. As can be seen, only 4 of the 28 amino acids differ between chicken and porcine VIP. Among mammals, VIP in humans, dogs, goats, cows, and rats is identical to porcine VIP (Said, 1984; Eng et al., 1986). This striking conservatism seems to have arisen because, unlike many other peptides in which only one end is necessary for biological activity, nearly the entire VIP molecule is needed for full activity (Dockray, 1979). As the structure of reptilian peptides is generally similar to that of avian and mammalian ones, it is likely that this similarity pertains to VIP as well.

Little is published on the distribution of VIP in reptiles. Buchan et al. (1983) have found prominent VIP+ nerve fibers underlying the surface epithelium and in the submucosa of the entire GI tract of *Alligator mississippiensis*. Fibers containing VIP were particularly prominent in the stomach. Neurons containing VIP were observed in the submucosal plexus but not in the myenteric plexus. Nerve fibers containing VIP have also been found to end on endocrine cells of the alligator pancreas (Buchan et al., 1982) and may play a role in the release of pancreatic hormones. The peripheral distribution of VIP outside the gut has not been studied in reptiles.

Unpublished observations (Reiner) indicate that VIP is widespread in the CNS of red-eared turtles. Fibers containing VIP are present in the hypothalamus, the external zone of the median eminence, the deep layers of the tectum, the cellular layer of the telencephalic cortex, and the tract of Lissauer of the spinal cord. In mammals also, VIP+ fibers are found in the external zone of the median eminence and are presumably of hypothalamic origin (Loren et al., 1979b; Fahrenkrug and Emson, 1982). Since VIP has been found to increase pro-

Table 8.9. Amino acid sequences for various members of the VIP family of peptides.

	1	2	3	4	5	6	7	8	9	10	11	12	13	14	15	16	17	18	19	20
Porcine VIP	His	Ser	Asp	Ala	Val	Phe	Thr	Asp	Asn	Tyr	Thr	Arg	Leu	Arg	Lys	Gln	Met	Ala	Val	Lys
Chicken VIP	His	Ser	Asp	Ala	Val	Phe	Thr	Asp	Asn	Tyr	Ser	Arg	Phe	Arg	Lys	Gln	Met	Ala	Val	Lys
PHI	His	Ala	Asp	Gly	Val	Phe	Thr	Ser	Asp	Phe	Ser	Arg	Leu	Leu	Gly	Gln	Leu	Ser	Ala	Lys
Porcine secretin	His	Ser	Asp	Gly	Thr	Phe	Thr	Ser	Glu	Leu	Ser	Arg	Leu	Arg	Asp	Ser	Ala	Arg	Leu	Gln
Porcine glucagon	His	Ser	Gln	Gly	Thr	Phe	Thr	Ser	Asp	Tyr	Ser	Lys	Tyr	Leu	Asp	Ser	Arg	Arg	Ala	Gln
Chicken glucagon	His	Ser	Gln	Gly	Thr	Phe	Thr	Ser	Asp	Tyr	Ser	Lys	Tyr	Leu	Asp	Ser	Arg	Arg	Ala	Gln
Duck/alligator glucagon	His	Ser	Gln	Gly	Thr	Phe	Thr	Ser	Asp	Tyr	Ser	Lys	Tyr	Leu	Asp	Thr	Arg	Arg	Ala	Gln
Porcine GIP	Tyr	Ala	Glu	Gly	Thr	Phe	Ile	Ser	Asp	Tyr	Ser	Ile	Ala	Met	Asp	Lys	Ile	Arg	Gln	Gln

	21	22	23	24	25	26	27	28	29	30	31	32	33	34	35	36	37	38	39	40
Porcine VIP	Lys	Tyr	Leu	Asn	Ser	Ile	Leu	Asn-NH$_2$												
Chicken VIP	Lys	Tyr	Leu	Asn	Ser	Val	Leu	Thr-NH$_2$												
PHI	Lys	Tyr	Leu	Glu	Ser	Leu	Ile-NH$_2$													
Porcine secretin	Arg	Leu	Leu	Gln	Gly	Leu	Val-NH$_2$													
Porcine glucagon	Asp	Phe	Val	Gln	Trp	Leu	Met	Asn	Thr											
Chicken glucagon	Asp	Phe	Val	Gln	Trp	Leu	Met	Ser	Thr											
Duck/alligator glucagon	Asp	Phe	Val	Gln	Trp	Leu	Met	Ser	Thr											
Porcine GIP	Asp	Phe	Val	Asn	Trp	Leu	Leu	Ala	Gln	Gln	Lys	Gly	Lys	Lys	Ser	Asp	Trp	Lys	His	Asn

	41	42	43
Porcine GIP	Ile	Thr	Gln

Data from Dockray, 1977, 1979; Markussen et al., 1972; Sundby et al., 1972; Pollock and Kimmel, 1975; Lance et al., 1984 lactin release in birds and mammals, it is possible that VIP + fibers in the ME in both turtles and mammals may mediate this function in vivo (Said, 1984; MacNamee et al., 1986). Large neurons containing VIP are also present in the raphe nuclei in turtles at intermediate levels of the rhombencephalon. VIP + neurons are also present in turtles in the nucleus of the solitary tract and in the lateral hypothalamic/ supramammillary cell field (Fig. 8.17). The presence of VIP + fibers in the tract of Lissauer, the terminal zone of the primary afferent fibers, suggests that VIP may be present in DRG neurons and in primary afferent fibers, as reported in mammals (Hokfelt et al., 1980). The fibers containing VIP in the cellular layer of the telencephalic cortex also contain SP and CCK8 (Reiner et al., 1985) (Fig. 8.17). Neurons in the lateral hypothalamic/supramammillary region contain CCK8 and SP, as well as VIP, although it has not yet been directly demonstrated that all three are present in the same individual neurons (Fig. 8.17). Since telencephalic cortical fibers show colocalization for SP, CCK8, and VIP, and since the lateral hypothalamic/supramammillary region is the only region of the turtle brain where CCK8, VIP, and SP have all been observed in neurons, it seems very likely that all three co-occur in these neurons and that these neurons are the source of the SP/CCK8/VIP + cortical fibers (Reiner et al., 1985). Cortex in mammals is also known to receive CCK8/VIP-containing inputs from the supramammillary region (Seroogy et al., 1988). It should be emphasized that, unlike in mammals, in which VIP + cortical neurons are prominent and abundant, chiefly in layers II to III (Snyder, 1980; Said, 1984; McDonald et al., 1982a; Loren et al., 1979b; Obata-Tsoto et al., 1983), VIP + neurons have not been observed in the telencephalic cortex of turtles (Reiner, 1991). In fact, VIP (along with CCK) is one of the few widely studied peptides known to have its highest concentration in mammals in cerebral cortex (Beinfeld, Lewis, Eiden, et al., 1983). The seeming absence of VIP + cells from turtle cortex, together with the similar absence of CCK8 + cells, lends credence to the notion that these cell types were added to the pallium during the evolution of mammalian neocortex (Reiner, 1991).

Electrophysiological studies have shown that VIP excites neurons in various brain areas and in the gut in mammals (Fahrenkrug and Emson, 1982; North and Egan, 1982). Further consistent with a role in neurotransmission, VIP has been localized to vesicles in synaptic terminals in mammals. In addition, VIP has been shown to be released from nerve terminals by depolarization, and VIP-specific receptor binding sites have been demonstrated in the mammalian nervous system (Said, 1984; Fahrenkrug and Emson, 1982). It is likely that VIP has similar neural properties in reptiles.

C. The Pancreatic Polypeptide Family

A large number of peptides belong to the pancreatic polypeptide family. Avian pancreatic polypeptide (APP), which inhibits gastric acid secretion and glycogenolysis (Hazelwood et al., 1973), was the first member of the family isolated, purified, and sequenced using extracts from chicken pancreas (Kimmel et al., 1968a). Subsequently, clearly homologous peptides have been found in the pancreas of all mammalian species examined (Lin and Chance, 1974). In all cases, the pancreatic polypeptide (PP) was a 36-amino-acid peptide. PP was also found to be present in GI tract endocrine cells (El-Salhy, Wilander, Juntii Berggren, and Grimelius, 1983). Encouraged by the success of investigators who had localized other GI tract peptides to the CNS (e.g., CCK and VIP), several investigators used anti-APP antisera or antibovine PP antisera to study the localization of PP in the nervous system of mammals. A PP-like substance or substances were found to have a remarkably widespread distribution, occurring in numerous neurons from cerebral cortex and basal ganglia to the spinal cord and in sympathetic adrenergic neurons and fibers (Emson and DeQuidt, 1984). However, RIA studies of the levels of PP in the CNS revealed only low levels of PP (DiMaggio et al., 1985). Subsequent studies by Tatemoto and Mutt (1980), Tatemoto, Carlquist, and Mutt (1982), and Tatemoto, Lundberg, Terenius, and Mutt (1982) revealed the existence of two further members of the PP family (Table 8.10): (1) peptide YY (or PYY, termed such because of the presence of tyrosine at both N- and C-termini), which is found in great abundance in the gut, and (2) neuropeptide Y (or NPY), which is highly similar to PYY and is found in great abundance in the nervous system (Emson and DeQuidt, 1984). In subsequent immunohistochemical studies, PYY and PP were shown to be differentially distributed in endocrine cells of the intestinal tract (El-Salhy, Grimelius, Lundberg, et al., 1982; El-Salhy, Wilander, Grimelius, et al., 1982). Studies of the nervous system in mammals have shown that the distribution of NPY in the nervous system is identical to that observed with anti-PP antisera, implying that the previously used anti-PP antisera cross-react with NPY and label NPY-containing neurons and fibers (DiMaggio et al., 1985). It is now generally assumed that earlier studies purportedly localizing a PP-like substance in the nervous system were, in fact, labeling NPY-containing neurons and fibers.

In the pancreas, pancreatic polypeptide appears to be largely restricted to endocrine cells of the pancreatic islet tissue in all jawed vertebrates (El-Salhy, 1984; Johnson et al., 1982; Emson and DeQuidt, 1984; Lopez et al., 1988; Garcia-Ayala et al., 1987; Gapp et al., 1985; Abad et al., 1988; Falkmer et al., 1985). A PP-like peptide has also

Table 8.10. Amino acid sequences for various members of the pancreatic polypeptide family of peptides

	1	2	3	4	5	6	7	8	9	10	11	12	13	14	15	16	17	18
NPY	Tyr	Pro	Ser	Lys	Pro	Asp	Asn	Pro	Gly	Glu	Asp	Ala	Pro	Ala	Glu	Asp	Leu	Ala
PYY	Tyr	Pro	Ala	Lys	Pro	Glu	Ala	Pro	Gly	Glx	Asx	Ala	Ser	Pro	Glx	Glx	Leu	Ser
BPP	Ala	Pro	Leu	Glu	Pro	Gln	Tyr	Pro	Gly	Asp	Asp	Ala	Thr	Pro	Glu	Gln	Met	Ala
PPP	Ala	Pro	Leu	Glu	Pro	Val	Tyr	Pro	Gly	Asp	Asn	Ala	Thr	Pro	Glu	Gln	Met	Ala
APP	Gly	Pro	Ser	Gln	Pro	Thr	Tyr	Pro	Gly	Asp	Asp	Ala	Pro	Val	Glu	Asp	Leu	Ile
AMPP	Thr	Pro	Leu	Gln	Pro	Lys	Tyr	Pro	Gly	Asp	Gly	Ala	Pro	Val	Glu	Asp	Leu	Ile

	19	20	21	22	23	24	25	26	27	28	29	30	31	32	33	34	35	36
NPY	Arg	Tyr	Tyr	Ser	Ala	Leu	Arg	His	Tyr	Ile	Asn	Leu	Ile	Thr	Arg	Gln	Arg	Tyr-NH$_2$
PYY	Arg	Tyr	Tyr	Ala	Ser	Leu	Arg	His	Tyr	Leu	Asn	Leu	Val	Thr	Arg	Gln	Arg	Tyr-NH$_2$
BPP	Gln	Tyr	Ala	Ala	Glu	Leu	Arg	Arg	Tyr	Ile	Asn	Met	Leu	Thr	Arg	Pro	Arg	Tyr-NH$_2$
PPP	Gln	Tyr	Ala	Ala	Glu	Leu	Arg	Arg	Tyr	Ile	Asn	Met	Leu	Thr	Arg	Pro	Arg	Tyr-NH$_2$
APP	Arg	Phe	Tyr	Asn	Asp	Leu	Gln	Gln	Tyr	Leu	Asn	Val	Val	Thr	Arg	Pro	Arg	Tyr-NH$_2$
AMPP	Gln	Phe	Tyr	Asn	Asp	Leu	Gln	Gln	Tyr	Leu	Asn	Val	Val	Thr	Arg	Pro	Arg	Phe-NH$_2$

Note: Peptides include avian pancreatic polypeptide (APP), bovine pancreatic polypeptide (BPP), porcine pancreatic polypeptide (PPP), *Alligator mississippiensis* pancreatic polypeptide (AMPP), neuropeptide Y (NPY), and peptide YY (PYY). The sequences presented are from Emson and DeQuidt (1984) and Lance et al. (1984). Glx means that it is uncertain if the amino acid is Glu (glutamic acid) or Gln (glutamine). Asx means that it is uncertain if the amino acid is aspartic acid or asparagine.

been found in endocrine cells of the GI tract in members of all vertebrate classes, as well as in GI tract endocrine cells of *Branchiostoma* (Reinecke, 1981; Noaillac-Depeyre and Hollande, 1981; El-Salhy and Grimelius, 1981; El-Salhy, 1984; Alumets et al., 1978a; El-Salhy, Grimelius, Lundberg, et al., 1982; El-Salhy, Wilander, Grimelius, et al., 1982; El-Salhy et al., 1981; Falkmer et al., 1985). Although PP+ immunoreactivity has been detected in the nervous system of a variety of nonmammalian and mammalian species using immuno histochemical techniques (Loren et al., 1979a, DiMaggio et al., 1985; Emson and DeQuidt, 1984; Karten et al., 1982; Kuljis and Karten, 1982), the available data indicate that PP is not present in nervous system and that immunohistochemical labeling in nervous system with anti-PP antisera reflects cross-reactive labeling of NPY, which is similar in structure to PP and is abundant in nervous tissue in a variety of vertebrate species (Vallarino et al., 1988; Pontet et al., 1989; Noe et al., 1986; Danger et al., 1985; DiMaggio et al., 1985; Chronwall et al., 1985; Isayama and Eldred, 1988; Verstappen et al., 1986; Falkmer et al., 1985).

Using an antiserum against the C-terminal hexapeptide of PP, Greeley et al. (1984) have confirmed the presence of large amounts of PP in the pancreas of mammals, birds, reptiles (*Chrysemys picta* and *Anolis carolinensis*), and amphibians (*Rana pipiens* and *Necturus maculosus*). Dose-response curves for the PP-like peptide in the pancreas were parallel to a standard bovine PP curve in RIA for mammals, birds, reptiles, and amphibians. No PP-like peptide was detected in bony fish pancreas with the antiserum used in this assay. Several immunohistochemical studies have, however, reported the presence of a PP-like peptide in the pancreatic endocrine cells in both cartilaginous and bony fish (Stephan et al., 1978; Van Noorden and Paten, 1978; Langer et al., 1979; Klein and Van Noorden, 1980; Johnson et al., 1982; El-Salhy, 1984). The absence of PP-like immunoreactivity in bony fish pancreas in the Greeley et al. (1984) study presumably indicates that bony fish PP differs from PP of terrestrial vertebrates in at least the last six amino acids of the C-terminus. Lance et al. (1984) have recently characterized the structure of pancreatic PP in *Alligator mississippiensis* (Table 8.10) (where it is contained in endocrine cells; Buchan et al., 1982) and found that it is of the same length as chicken and mammalian PP. They noted the existence of 7 substitutions in the amino acid sequence compared to chicken PP and 18 compared to bovine PP, but these substitutions are of a conservative nature since they do not alter the conformational characteristics of the molecule. By contrast bullfrog PP (which is also 36 amino acids in length) shows 61% sequence similarity to bovine PP (Pollock, Hamilton, Rouse, et al., 1988), whereas even among mammalian species there is variation in the structure of PP, with human, dog, pig, cow, and sheep PP differing among one another by one to three positions and rat PP differing from these by eight to nine positions (Larhammar et al., 1987). Thus, the evolution of PP, although conservative in terms of the length and conformation of the molecule, appears to be marked by more amino acid substitutions than is typically the case for neuropeptides.

As noted above, PP is but one in a family of peptides. Another member of the PP family, PYY, has also been demonstrated immunohistochemically in the gut of mammals, chickens, and cartilaginous fish (El-Salhy, Wilander, Grimelius, et al., 1982; El-Salhy, Grimelius, Wilander, et al., 1983; El-Salhy, Wilander, Juntii-Berggren, et al., 1983; El-Salhy, 1984). The PYY+ cells of the gut are endocrine cells and have a different distribution in the gut than the PP+ cells of the gut. PYY is also present in endocrine cells of the gut in some amphibian and reptilian species, and these PYY+ cells differ in their distribution from PP+ endocrine cells (El-Salhy, Grimelius, Lundberg, et al., 1982; Bottcher et al., 1985). Cells containing PYY have also been reported in the pancreas of mammals and reptiles (Falkmer et al., 1985). PYY is scarce in the nervous system of the vertebrate species studied, which includes one reptilian species (*Lacerta vivipara*) (Bottcher et al., 1985). It seems likely that PYY differs somewhat in chemical structure among the members of the different vertebrate groups, though few directly relevant data are available on this point (Bottcher et al., 1985).

NPY has been shown to be abundant in the central and peripheral nervous systems of all vertebrate species studied and in some invertebrate species (Schoofs et al., 1988; Vallarino et al., 1988; Pontet et al., 1989; Noe et al., 1986; Danger et al., 1985; DiMaggio et al., 1985; Chronwall et al., 1985; Isayama and Eldred, 1988; Verstappen et al., 1986; Falkmer et al., 1985). In mammals, NPY has also been found in neurons of both the submucosal and myenteric plexi of the gut wall (Furness et al., 1983; Sundler et al., 1983). As noted above, early studies reporting the distribution and abundance of PP in the nervous system of various vertebrate species appear to actually have been detecting NPY, which is similar in structure to PP (approximately 50% amino acid similarity) and would therefore bind anti-PP antisera. Little is known of the variation in the structure of NPY among the various vertebrate species. In mammals, NPY is 36 amino acids in length and much more conservative in structure than PP. For example, rat and human NPY are identical, and porcine NPY differs by only one amino acid (Larhammar et al., 1987). The structure of NPY in bony fish and frogs is highly similar to that in mammals, since the behavior of NPY from the members of these groups parallels the behavior of mammalian NPY in RIA (Pontet et al., 1989; Danger et al., 1985). Kimmel et al. (1986) have reported that NPY extracted from pancreas (where it is found in nerve fibers) of Coho salmon is a 36-amino-acid peptide that shows 83% sequence similarity to mammalian NPY. These various data make it reasonable to assume that reptilian NPY is also a 36-amino-acid neuropeptide that is highly similar to mammalian NPY in chemical structure.

In reptiles, NPY is present in numerous neurons and fibers of the central and peripheral nervous systems. In the periphery, NPY has been observed in autonomic nerves in several reptilian species (Falkmer et al., 1985) and in amacrine cells of the turtle retina (Isayama and Eldred, 1988). Similar observations have been made in other vertebrate species (Falkmer et al., 1985; Verstappen et al., 1986). In mammals, NPY has been found in adrenergic autonomic fibers, including those innervating blood vessels. Studies of the effect of NPY on blood vessels in mammals show NPY to be a potent vasoconstrictor (Emson and DeQuidt, 1984; Lundberg and Tatemoto, 1982). This may also be true for the NPY of reptiles.

Few published data are available on the central distribution of NPY in reptiles. In my own unpublished studies in turtles and caiman, NPY-like immunoreactivity-containing cells and fibers were observed to be widespread and numerous (A. Reiner, unpublished observations). Observations in caiman brain are based on the use of an anti-APP antiserum (courtesy of J. Kimmel), whereas those in turtles are based on the use of both anti-NPY and anti-APP (Reiner and Oliver, 1987). The PP+ labeling is assumed to represent cross-reactive labeling of NPY and is referred to as NPY+ labeling. Neurons containing NPY immunoreactivity are prominent in the telencephalon of turtle and caimans, being present in cortex and DVR (which together correspond to the neocortex in mammals) and basal ganglia (Figs. 8.3 and 8.18). NPY neurons are also abundant in the neocortex of mammals (where they are found in all cortical layers) in the DVR and Wulst of birds, and in the basal ganglia in mammals and birds (McDonald et al., 1982b; Vincent, Skirboll, Hokfelt, et al., 1982; Vincent, Johansson, Hokfelt, et al., 1982; Smith and Parent, 1986; Anderson and Reiner, 1990b). The distribution, numbers, and morphology of the telencephalic NPY+ cells in reptiles are identical to those of the SS-containing telencephalic cells, implying that SS and NPY may be present in the same telencephalic neurons. This has been directly confirmed for turtles using double-labeling methods (Reiner and Oliver, 1987). The NPY+ neurons of the telencephalon in mammals and birds have also been shown to contain SS (Vincent, Hokfelt,

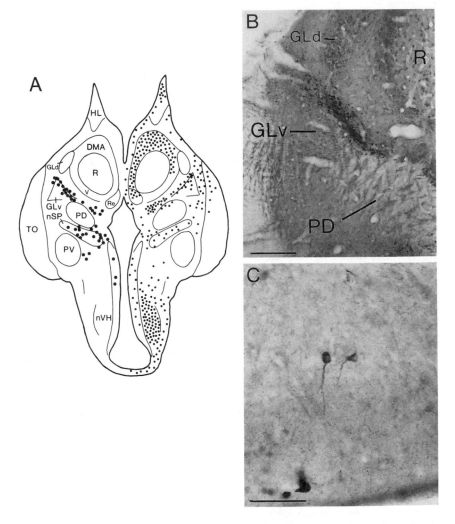

Fig. 8.18. (A) A line drawing of a transverse section through a rostral thalamic level of the red-eared turtle diencephalon illustrating the distribution of NPY+ neurons (large filled circles) and fibers (dots) in the intergeniculate region. Each large, filled circle corresponds to five labeled neurons. (B) A photomicrograph of the NPY+ neurons in the region between the two geniculate nuclei. Medial is to the right and the scale bar equals 250 μm. (C) NPY+ neurons in the superficial cell plate of the ventral rostral DVR. Medial is to the right and the scale bar equals 100 μm. Abbreviations: DMA, nucleus dorsomedialis anterior; GLd, nucleus geniculatus lateralis, pars dorsalis; GLv, nucleus geniculatus lateralis, pars ventralis; HL, lateral habenular nucleus; nSP, nucleus suprapeduncularis; nVH, nucleus ventromedialis of the hypothalamus; PD, dorsal peduncle of the lateral forebrain bundle; PV, ventral peduncle of the lateral forebrain bundle; R, nucleus rotundus; Re, nucleus reuniens; TO, tractus opticus; V, nucleus ventralis.

Christensson, and Terenius, 1982; Smith and Parent, 1986; Anderson and Reiner, 1990b). In turtle and caiman, NPY+ fibers are found in abundance in many telencephalic regions known to receive adrenergic input (e.g., medial portions of the striatum and the molecular layer of telencephalic cortex) in turtles and caiman (Parent, 1979). This observation is consistent with the possible colocalization of NPY and noradrenalin in the same fibers in the reptilian central nervous system. In mammals, NPY has been found in adrenergic neurons of the CNS (Hokfelt et al., 1983; Hunt et al., 1981; Lundberg, Terenius, Hokfelt, et al., 1982, 1983; Jacobowitz and Olschowka, 1982). NPY+ cell bodies have not been observed, however, in any of the adrenergic cell groups of the hindbrain in reptiles.

In turtles, additional NPY+ neurons were observed in the ventral portions of the dorsal thalamus, in the periventricular and periarcuate regions of the hypothalamus, in the neurohypophysis, and in the dorsal pretectal nucleus (Fig. 8.18). In *Caiman*, the latter cell group (termed the spiriform nucleus, Reiner et al., 1980) also contains NPY+ neurons. NPY+ neurons of the dorsal pretectal nucleus appear to be homologous to the NPY+ neurons of the nucleus pretectalis of birds (Karten et al., 1982). Nucleus pretectalis of birds has a bilateral NPY+ projection to the major retinorecipient layer of the tectum (Karten et al., 1982). The dorsal pretectal nucleus of turtles and caiman also appears to have an NPY+ projection to the major tectal retinorecipient layer. In turtles, Kunzle and Schnyder (1983) have shown that the dorsal thalamic cell field that includes area triangularis and rostral portions of the ventral geniculate nucleus receives both spinal cord and retinal input. Based on these observations they proposed that this region was comparable to the intergeniculate leaflet of mammals, which is similar in connectivity. The suggestion of Kunzle and Schnyder (1983) is strengthened by the fact that neurons of the intergeniculate leaflet in mammals (Harrington et al., 1984) and the dorsal edge of the area triangularis/ventral geniculate complex in turtles both contain NPY+ neurons (Reiner, unpublished observations) (Fig. 8.18). In turtles, fibers containing NPY+ are prominent in the molecular layer of the telencephalic cortex, in the medial striatum, in the dorsomedial and dorsolateral thalamic nuclei, in the ventromedial hypothalamus, in the external zone of the median eminence and the neurohypophysis, in the tectum, lateral to the caudal descending tract of the trigeminus, and in Lissauer's tract of the spinal cord. As with the other neuropeptides, all available lines of evidence (based largely on studies in mammals) favor the view that NPY plays the role of a neuromodulator/neurotransmitter in the nervous system of reptiles (Martel et al., 1986).

VI. OTHER GUT PEPTIDES

A number of peptides in addition to CCK8, VIP, and the PP peptides have been found in mammalian gut. A number of these have also been found in the mammalian CNS, albeit in lower levels than the aforementioned gut peptides. Although few data are available in reptiles, these peptides may also be present in low levels in the nervous system of reptiles.

A. Gastrin

As noted above, early studies of gut peptides in brain identified the CCK/gastrin-like peptide in brain as a gastrinlike peptide. It was later recognized, however, that the greater bulk of the CCK/gastrin-like material in the CNS was CCK8 (Rehfeld et al., 1984). Nonetheless, several investigators have demonstrated low levels of gastrin in the CNS in mammals, notably in the hypothalamohypophyseal system. Although low levels of a gastrinlike peptide may be present in the nervous system of the members of nonmammalian groups as well, no evidence for this was obtained in frogs (Beinfeld, Trubatch, and Brownstein, 1983) or turtles (Reiner et al., 1985; Reiner and Beinfeld, 1985). Gastrin+ endocrine cells have been observed in the stomach of alligators (Buchan et al., 1983; Dimaline et al., 1982).

B. Bombesin

Bombesin is a 14-amino-acid peptide originally isolated from frog skin (Anastasi et al., 1971). Bombesin was subsequently found to influence the mammalian GI tract, increasing gastrin secretion (and hence gastric acid secretion) and altering intestinal motility (Nemeroff et al., 1983; Snyder, 1980). Immunohistochemical studies subsequently indicated the presence of a bombesinlike peptide in GI tract endocrine cells and in intramural nerve fibers and neurons of the stomach and small intestines in a variety of mammalian and nonmammalian species, including reptilian species (Dockray et al., 1979; Polak et al., 1976; Nemeroff et al., 1983; Langer et al., 1979; Van Noorden and Falkmer, 1980; El-Salhy, 1984; Holmgren and Nilsson, 1983; Walsh et al., 1982; Lechago et al., 1978; Erpsamer et al., 1979; Buchan et al., 1983). A bombesinlike peptide is also present in the brain in various vertebrate species (Nemeroff et al., 1983; Snyder, 1980; Walsh et al., 1982; Kuljis and Karten, 1982, 1983; Moody et al., 1979, 1981). Although the precise chemical structure of the bombesinlike substance found in brain and GI tract was initially uncertain, this issue was soon resolved by the work of McDonald et al. (1979), who isolated a 27-amino-acid peptide (called gastrin-releasing peptide, GRP) from porcine gut that has the same sequence for the last ten amino acids of its C-terminus as bombesin, except for the presence of glutamine instead of histidine as residue 7 of bombesin. GRP therefore appears to be the bombesin-like (BOMLI) substance detected by RIA and immunohistochemistry in nervous tissues and gut (Brown et al., 1980; Yanaihara et al., 1981; Greeley et al., 1986). Within the nervous system, the overall evidence, such as in terms of localization within synaptic terminals and calcium-dependent depolarization-evoked release of the bombesin-like/GRP peptide, supports the idea that the bombesinlike/GRP peptide is a neuroactive CNS peptide (Moody et al., 1980; Nemeroff et al 1983). In addition receptor sites specific for a bombesinlike peptide have been found in the mammalian CNS (Moody et al., 1978; Wolf and Moody, 1985).

Little information is available on the neural distribution of bombesin/GRP in reptiles. Using antibombesin antiserum (courtesy of M. Brown), detectable levels of a bombesin/GRP-like peptide were observed in the red-eared turtle CNS (A. Reiner, unpublished observations). Labeling was prominent in fibers in the preoptic hypothalamus and in neurons of the substantia nigra (Fig. 8.19). The neurons of the turtle substantia nigra containing the bombesin/GRP-like peptide appear to be those containing dopamine. Although double-label studies have not been performed, this similarity is evident based on

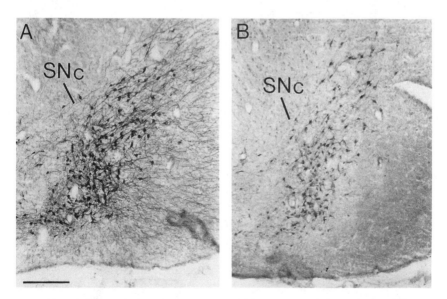

Fig. 8.19. (A) Dopaminergic neurons of the substantia nigra, pars compacta in red-eared turtle labeled with an antiserum against tyrosine hydroxylase and (B) seemingly nearly all neurons of the same cell group labeled with an antiserum against bombesin. These results suggest that a bombesinlike substance and dopamine may co-occur in SNc neurons in turtles. Medial is to the left in both photomicrographs; both are at the same magnification and the scale bar in (A) equals 100 μm.

the location, numbers, and morphology of the neurons labeled for dopamine (by immunohistochemistry with antityrosine hydroxylase antibodies) and for bombesin/GRP (using the antibombesin antiserum). Consistent with a possible bombesin/GRP-like projection to the striatum from the substantia nigra, numerous fibers and terminals containing a bombesin/GRP-like substance were observed in the striatum. Along this line, it is of interest to note that high levels of receptor sites specific for a bombesin/GRP-like substance have been observed in the mammalian striatum (Moody et al., 1978; Wolf and Moody, 1985). In turtles, fibers containing a bombesin/GRP-like substance were also observed in the septum, the nucleus of the solitary tract, and the tract of Lissauer in the spinal cord. High levels of the bombesinlike substance are also present in mammals in the substantia gelatinosa of the spinal cord and the nucleus of the solitary tract (Moody et al., 1979).

C. VIP-Related Peptides: Secretin, Glucagon, and Gastric-Inhibitory Polypeptide

1. Secretin and Gastric-Inhibitory Polypeptide   A group of peptides related in structure to VIP has also been found in mammalian gut, as noted above. These peptides include secretin, glucagon, and gastric-inhibitory polypeptide (GIP) (see Table 8.9), as well as VIP and PHI. All are large peptides (approximately 30 or more amino acids), and the similarities are most marked through the C-terminus. Secretin and GIP are peptides of intestinal endocrine cells, whereas glucagon is found in endocrine cells of pancreatic islets. A GIP peptide has only been reported thus far in mammals. A secretinlike peptide has been isolated from bony fish, but this peptide is apparently more similar to VIP than secretin (Dockray, 1974, 1977, 1979). It has been proposed that a VIP-like peptide was the ancestral peptide of the VIP/PHI-GIP-secretin-glucagon group and that the other members of the group arose by gene duplication (Van Noorden and Polak, 1979). Along these lines, it is of interest to note that although a secretinlike peptide is present in birds, avian VIP rather than avian secretin reportedly regulates the release of nonenzymatic digestive fluids into the intestines from the exocrine pancreas (Dockray, 1977, 1979). In mammals, secretin regulates this function. Although no evidence is available that secretin is present in the mammalian, avian, or reptilian CNS, the structural similarity of secretin to neuropeptides such as VIP should be kept in mind during studies using antisera against secretin. Secretin+ endocrine cells have been observed in alligator GI tract (Buchan et al., 1983).

2. Glucagon

Glucagon, a 29-amino-acid peptide found in pancreatic endocrine cells, regulates glycogenolysis. Glucagon has had a highly conservative evolutionary history (Van Noorden and Polak, 1979; Lopez et al., 1984). Among the mammalian species studied (except for guinea pig), the structure of glucagon has been fully conserved (Conlon and Thim, 1985). The structure for pancreatic glucagon in chickens and turkeys differs from porcine glucagon by only one amino acid substitution (see Table 8.9) (Dockray, 1977, 1979; Pollack and Kimmel, 1975), whereas duck glucagon differs by two positions (Table 8.9) (Markussen et al., 1972; Sundby et al., 1972). Lance et al. (1984) have recently determined the structure of pancreatic glucagon in *Alligator mississippiensis* and found it to be identical to duck glucagon. Further underlining the degree of conservatism in the evolution of glucagon, even in species such as *Torpedo marmorata* (a cartilaginous fish) and *Lepisosteus spatula* (a holostean bony fish), glucagon differs from porcine glucagon by only three amino acid positions and glucagon in bullfrogs differs by only one amino acid from porcine glucagon (Conlon and Thim, 1985; Pollack, Hamilton, Rouse, et al., 1988; Pollack, Kimmel, Ebner, et al., 1988). In contrast, glucagon in teleosts and in ratfish (a primitive cartilaginous fish) does show a greater difference from that found in mammals, differing by between seven to nine amino acid positions, depending on the fish species (Andrews and Ronner, 1985; Plisetskaya et al., 1986; Andrews et al., 1986; Lund et al., 1981; Conlon et al., 1989). In addition to its location within the pancreas, glucagon appears to be found in neural tissue in the members of a wide variety of vertebrate groups. For example, amacrine cells containing glucagon have been observed in the retina of red-eared and painted turtles (Eldred and Karten, 1983; Brecha et al., 1984), as has also been observed in avian, amphibian, and fish species (Brecha et al., 1982, 1984; Kuwayama et al., 1982). Unpublished immunohistochemical observations indicate the presence of glucagon + fibers in the region of the hypothalamohypophyseal tract of turtles (A. Reiner, unpublished observations) (glucagon antiserum courtesy of T. Yamada). Similarly in mammals, glucagon + labeling has been observed in the mammalian CNS, particularly in the hypothalamus (Snyder, 1980; Tominaga et al., 1981). Among reptiles, glucagon + endocrine cells have been reported in the GI tract and pancreas of alligators, turtles, and lizards (Buchan et al., 1982, 1983; Lopez et al., 1988; Garcia-Ayala et al., 1987).

D. Insulin

Insulin is a 51-amino-acid pancreatic peptide whose hormonal action is generally the opposite of glucagon (Turner and Bagnara, 1976; Gorbman et al., 1983). Whereas glucagon increases glycogen breakdown, insulin promotes uptake and utilization of glucose and prevents the breakdown of glycogen. An insulin peptide is present in the pancreas of all vertebrates (Turner and Bagnara, 1976; Gorbman et al.,

1983). Structurally, insulin shows remarkable conservatism, with only 14 amino acids differing even between lampreys and mammals (Plisetskaya et al., 1988; Dockray, 1979). Lance et al. (1984) have determined the structure of alligator pancreatic insulin and found it to be more similar to avian insulin (Kimmel et al., 1968b; Evans et al., 1988) than to either snake insulin (Kimmel et al., 1976; Zhang et al., 1981) or mammalian insulin. Insulin+ pancreatic cells have been observed in all vertebrate species studied, including such reptilian groups as alligators, turtles, and lizards (Buchan et al., 1982; Lopez et al., 1988; Garcia-Ayala et al., 1987). Snyder (1980) has suggested that insulin may be a peptide of the CNS in mammals, and an insulinlike peptide has been found in mammalian brain by RIA (Havrankova et al., 1978). No data are available regarding the possible presence of insulin in the reptilian CNS.

VII. OTHER PEPTIDES

A few additional peptides have been found in the nervous system of vertebrates and proposed to play a role in the chemical transmission of information between neurons. Although the presence of some of these peptides in nervous system in reptiles has not been investigated, the overall similarities between reptilian neuropeptides and those in other vertebrates suggest that all of the peptides discussed below may eventually also be shown to be present in the nervous system in reptiles.

A. Bradykinin

Bradykinin was discovered as a biologically active factor that is released from alpha-globulin fractions of mammalian blood upon incubation with trypsin. The peptide was called bradykinin because of its ability to elicit slow contractions in a guinea pig ileum assay. Peripherally bradykinin appears to be involved in the response to tissue injury, helping produce such effects as pain, cardiovascular shock, and inflammation (Snyder, 1980). Bradykinin+ neurons have been reported in the hypothalamus, septum, and central gray of mammalian brain (Correa et al., 1979). There is no evidence that it is present in the nervous system of reptiles.

B. Angiotensin II

Angiotensin II is an octapeptide that is cleaved by the enzymatic actions of renin (which is released from the kidney) and by angiotensin-converting enzyme from a large plasma peptide called angiotensinogen (Wilson, 1984a, 1984b). Angiotensin II is a potent vasoconstrictor and promotes renal sodium retention by stimulating release of steroids from the adrenal cortex (Snyder, 1980; Wilson, 1984a, 1984b).

The structure, function, and generation of angiotensin II has been evolutionarily conservative (Wilson, 1984a, 1984b). For example, angiotensin II shows little or no structural variation among mammals, and angiotensin II in teleosts, bullfrogs, snakes, and chickens differs from the most common form of angiotensin II in mammals by only one or two amino acid positions (Wilson, 1984a, 1984b). Immunohistochemical studies in mammals have shown angiotensin II+ staining in several regions of the nervous system, including in the paraventricular neurons of the hypothalamus and in fibers of the substantia gelatinosa of the spinal cord, the spinal trigeminal nucleus, and the periventricular gray (Snyder, 1980). Since dorsal root lesions reduce angiotensin II in the dorsal horn, angiotensin II may be present in primary sensory neurons of the dorsal root ganglia. The angiotensin II-like peptide in the CNS may differ in structure from peripheral angiotensin II (Snyder, 1980). There is no evidence, however, that angiotensin II or a similar peptide is present in the nervous system of non-mammals, including reptiles.

C. Carnosine

Carnosine is a dipeptide (beta-alanylhistidine) that is highly concentrated in the pathway from the olfactory epithelium to the olfactory bulb. Unlike other peptides, which are manufactured by a ribosomal mechanism, carnosine is manufactured in a single enzymatic step. Carnosine appears to be almost exclusively a peptide of the olfactory bulbs (Snyder, 1980). Nicoll et al. (1980a) have reported that carnosine excites neurons of the olfactory bulb in turtles.

D. FMRFamide

Molluscan cardioacceleratory peptide, also known as FMRFamide because of its amino acid composition—Phe-Met-Arg-Phe-Amide (Price and Greenberg, 1977)—was isolated from the nervous system of clams and shown to have potent actions on many molluscan tissues. Since FMRFamide was subsequently shown to excite neurons of rat brain stem (Gayton, 1982), interest was raised in the possible presence of a FMRFamide-like peptide in the nervous system of vertebrates. Immunohistochemical techniques, in fact, revealed the presence of a FMRFamide-like substance in the gut and nervous system of fish, amphibia, birds, and mammals, as well as in tissues of several invertebrate species (Boer et al., 1980; Dockray, Vaillant, and Williams, 1981; Dockray, Vaillant, Williams, et al., 1981; Dockray and Williams, 1981; Grimmelhuijzen et al., 1982; Weber et al., 1981). Since FMRFamide has the same last four amino acids as the opioid peptide MERF (although MERF has a C-terminus that is devoid of the amide characteristic of the C-terminus of FMRFamide), immunohistochemical studies using anti-FMRFamide antisera must, however, be careful to use antisera that do not cross-react with MERF. The converse caution must be exercised in studies using antisera against MERF. Further, at least some anti-FMRFamide antisera reportedly cross-react slightly with members of the PP family of peptides (Lundberg et al., 1984; Verhaert et al., 1985; Triepel and Grimmelikhuijzen, 1984). Despite these cross-reactivity problems, recent studies have shown that mammals and birds possess a novel peptide or peptides resembling FMRFamide (Dockray and Williams, 1983; Dockray et al., 1983, 1986; O'Donohue et al., 1984). Although it therefore appears likely that a FMRFamide-like peptide is present in the nervous system of reptiles, no data are currently available on its chemical identity or its distribution in the nervous system in reptiles.

E. Calcitonin Gene-Related Peptide

Calcitonin gene-related peptide (CGRP) is a 37-amino-acid peptide derived from the same gene as calcitonin. Calcitonin and CGRP are expressed in a tissue-specific fashion by alternative RNA processing, with calcitonin expressed in thyroid cells and CGRP expressed in neural cells (Fischer and Born, 1985; Goodman and Iversen, 1986). Although little is known about the presence of CGRP in nonmammals, immunohistochemical studies provide evidence for its occurrence in the nervous system of birds and reptiles (Reiner, 1987c; Reiner, unpublished observations). Within the nervous system, CGRP appears to have a neuroactive role (Henke et al., 1985; Twery and Moss, 1985). Since the forms of CGRP that predominate in nervous system in humans and rats differ by only four amino acids (Fischer and Born, 1985), it is likely that CGRP in reptiles is highly similar in structure to that in mammals. CGRP has been found to have a widespread distribution in the nervous system of mammals (Kawai et al., 1985; Skofitsch and Jacobowitz, 1985). One of the more interesting aspects of the localization of CGRP is its colocalization with substance P in peripheral sensory fibers and their central endings. This has been demonstrated in both birds and mammals (Goodman and Iversen, 1986; Reiner, 1987c), in whom some of these fibers innervate blood vessels and have a CGRP-mediated vasodilatory effect. In light of this and in light of the immunohistochemical localization of CGRP in fibers in the tract of Lissauer in turtles (Reiner, unpublished observations), it is likely that the same is true in reptiles. Although few further details on the localization of CGRP in the nervous systems of reptiles are available, Reiner and Brauth (1989) have recently found that cells of nucleus reuniens and the surrounding region in turtles contain CGRP, as does the telencephalic target of nucleus reuniens within the DVR in both turtles and caiman. Thus, this auditory thalamic cell group appears to have a CGRP+ projection to the DVR in reptiles. This result is of interest because the comparable cell groups in pigeons and rats (nucleus ovoidalis and the medial geniculate, respectively) and their surrounding regions also appear to have a CGRP+ projection to their telencephalic targets (Field L and auditory cortex, respectively) (Reiner and Brauth, 1989). These results suggest that nucleus reuniens, nucleus ovoidalis, and the medial geniculate are homologous cell groups (as previously suggested by Nauta and Karten, 1970) and also that this may be true of the telencephalic targets of these thalamic cell groups.

VIII. SUMMARY AND CONCLUSIONS
A. Evolutionary Conservatism of Neuropeptides

It is clear that peptides have played a prominent role in nervous system function throughout vertebrate evolution. Further, for each neuropeptide identified in mammals, a similar or identical neuropeptide has generally been identified in the members of the other vertebrate groups. The various nonmammalian peptides structurally related to an individual mammalian peptide have been termed isopeptides of that mammalian peptide (Furness et al., 1982). Individual peptides of the mammalian nervous system and their nonmammalian isopeptides are characteristically not only similar in amino acid sequence (typically differing only by a few amino acid positions) but typically identical in the total number of amino acids composing the molecule. Further, just as individual mammalian peptides are derived from large peptide precursor molecules, nonmammalian isopeptides also appear to derive from large precursor molecules. In the case of several peptides (for example, the enkephalin family of peptides), the mammalian peptides and their nonmammalian isopeptides derive from similar precursor molecules. Further, mammalian peptides and their nonmammalian isopeptides are generally found within the same bodily tissues and generally serve similar functions. Those reptiles that are phylogenetically closest to mammals have individual isopeptides with a chemical structure closer to that of the mammals than to the isopeptides of other vertebrates. These isopeptides generally occur in the same parts of the nervous system as their mammalian counterparts. In the case of some mammalian neuropeptides, the identical neuropeptides appear to be present in both reptiles (or at least some groups of reptiles) and mammals. In terms of the CNS localization of peptides, the general rule appears to be that if a region of the nervous system appears to be neuroanatomically similar in reptiles and mammals (in terms of cytoarchitecture and connections), then similar or identical neuropeptides will be found in that region.

Although the entire evolutionary history of many peptides is uncertain, for several peptides considerable data are available. For example, the structure of neuropeptides such as CCK8, somatostatin, the enkephalin pentapeptides, and MERF has apparently not changed since at least the evolutionary appearance of land vertebrates. In contrast, other neuropeptides, such as SP, neurotensin, beta-endorphin, oxytocin, and arginine-vasopressin, did not achieve the chemical structures characteristically observed in mammals till at some point in the synapsid reptile-mammalian lineage. The evolution of the structure of LHRH among land vertebrates has apparently been conservative in the lineage leading from amphibians to mammals and nonconservative in the lineage leading from amphibians to reptiles to living birds and reptiles. The mammalian structure of LHRH was apparently achieved in an extinct common ancestor of frogs and mammals (presumably an early amphibian) and retained in the lineages leading to modern frogs and mammals. This conclusion implies that mammalian LHRH was widespread among amphibians and also present in synapsid reptiles. Apparently, in the extinct reptiles ancestral to living birds and reptiles, the structure of LHRH must have evolved to the forms typical of living birds and reptiles. The findings on the comparative structure of LHRH suggest that the line of reptiles leading to mammals must have diverged from the common reptilian stock before LHRH had acquired the forms typical in living birds and reptiles. This conclusion—that mammallike reptiles diverged from stem reptiles before the lineages leading to turtles, lizards, snakes, and crocodilians—has also been reached on paleontological grounds by other researchers (Hopson, 1979). Thus, data on neuropeptide evolution confirm conclusions about vertebrate evolution.

The structural conservatism of neuropeptides is presumably in large part a function of structure-function requirements for exercising biological function. To produce its biological actions, a neuropeptide must bind to its receptor. Since the binding affinity of a neuropeptide for its receptor is affected by its structure, changes in neuropeptide structure that affect its binding efficacy will deleteriously affect the ability of the neuropeptide to produce its biological effects. Although evolutionary changes in peptide receptors are possible, simultaneous and compatible mutations in the genes for a peptide and its receptor (such that binding efficacy is retained unaltered) are seemingly low probability events. Nonetheless, simultaneous evolution of a neuropeptide and its receptor does occur. For example, fish LHRH stimulates release of gonadotropins from fish pituitary, but mammalian LHRH has no such effect on fish pituitary (Ball, 1981). Nonetheless, in the case of neuropeptides in which nearly the entire molecule is required for the full biological activity of the peptide, almost all random point mutations of the gene coding for the peptide would typically be evolutionarily disadvantageous and would tend to be weeded out of the gene pool. Thus, a large peptide such as VIP (28 amino acids), nearly all of which is required for full biological activity, should have a conservative evolutionary history. This appears to be true, since VIP differs in only four amino acid positions between birds and mammals. The phylogenetic conservatism of some peptides (e.g., SS, CCK8, and several of the shorter enkephalin peptides) would suggest, therefore, that their biological activity resides in the entire molecule. In the case of peptides such as SP and neurotensin, in contrast, the C-terminus appears to be the biologically active region and (presumably consequently) both peptides show phylogenetic conservatism in the C-terminus and phylogenetic variation in the N-terminus.

Finally, although point mutations along the gene coding for an individual peptide have apparently occurred during peptide evolution, the number of amino acids making up a neuropeptide is strongly conserved during the evolution of that peptide. Thus, amino acid insertions or duplications appear, as a general rule, to be rare events during neuropeptide evolution. However, it seems likely that the entire genes or portions of the genes coding for a number of specific peptides have, in many cases, undergone duplication during evolution. Such duplication has been proposed to be the basis of the presence of related peptides such as CCK and the gastrin in members of individual vertebrate groups. In the case of CCK and gastrin, it has been proposed that the gene coding for CCK became duplicated during the evolution of land vertebrates, with one of the duplicates eventually evolving into land vertebrate CCK and the other into gastrin (Vigna, 1985, 1986). A similar proposal has been advanced to account for the existence of the three families of opioid peptides (Herbert et al., 1983).

B. Similarities and Differences among Reptiles and Other Vertebrates in Neuropeptide Localization

As noted, the localization of individual neuropeptides within the CNS also tends to be similar among vertebrates, particularly in the case of neural systems that show great interphyletic resemblance. One striking instance of this are the basal ganglia and their projection systems (Reiner, Brauth, and Karten, 1984). For example, anatomical studies have demonstrated the existence of a projection from the striatum to the substantia nigra in turtles, lizards, and crocodilians, as well as in birds and mammals. Immunohistochemical studies indicate that SP and dynorphin are colocalized in this pathway in reptiles, birds and mammals. Similarly, peptides found in the median eminence of reptiles also are found in the mammalian median eminence, including SP, LHRH, DYN, TRH, SS, CCK8, vasopressin/vasotocin, and mesotocin/oxytocin. Further, CCK8, SP, the enkephalin peptides, VIP, and a bombesin/GRP-like peptide all are present in the mammalian substantia gelatinosa of the spinal cord and equivalent or identical peptides are present in the comparable spinal cord regions of reptiles. On the other hand, CCK8 and VIP appear to be largely absent (particularly from neurons) in the telencephalic cortex in reptiles, in contrast to the situation in mammalian cortex, where they are abundant. This reptile-mammal difference in cortical peptide localization suggests that an extensive proliferation of neuron types may have occurred during cortical evolution in the mammalian lineage, types that are absent or rare in telencephalic cortex in living reptiles. The neuron types that are few or absent in living reptiles appear to include the correspondents of neurons that are present in layers II to III of mammalian cortex. Peptidergic neurons whose localization within mammalian cortex includes layers V to VI appear to have correspondents present in reptilian cortex. These peptidergic neurons include those that contain SP, SS, NPY, LANT6, and the enkephalin pentapeptides (Reiner, 1991). Ebner (1976) has previously suggested, based on pathway tracing data, that all three layers of dorsal cortex in turtles might be largely equivalent only to layers V to VI of mammalian neocortex cortex. Although the cortex of turtles may contain undiscovered peptidergic neurons or nonpeptidergic neurons (such as intracortical projection neurons) that are typical of neurons in layers II to III of mammals, current data on peptidergic neurons in cortex are consistent with Ebner's suggestion (Fig. 8.20) (Reiner, 1991).

Peptide localization in the pretectum and tectum also appears to differ between reptiles and mammals. Peptide localization in this part of reptile brain is more similar to that seen in frogs and birds. In nonmammalian amniotes, individual layers of the tectum are much more clearly defined than is the case with the superior colliculus of mammals. For many peptides, prominent layer-specific bands of fibers can be observed in the tectum of nonmammalian vertebrates, particularly for such peptides as SP, the enkephalins, LANT6, and NPY. In contrast, in mammals, fibers containing these peptides are less distinctly localized to specific layers or sublayers of the superior colliculus. Concomitant with the elaboration of the sauropsid tectum, individual pretectal cell groups related to the tectum are more readily identifiable in sauropsids than in mammals. In some cases, the sources of specific peptidergic fibers in the sauropsid tectum have been found to be some of these pretectal cell groups. For example, enkephalinergic fibers in deep tectal layers apparently arise from enkephalinergic neurons of a prominent pretectal cell group termed nucleus spiriformis lateralis (SpL) in birds and the dorsal nucleus of the posterior commissure (nDCP) in reptiles. The prominence of enkephalin in SpL/nDCP may be a uniquely avian/reptilian brain trait among amniotes, since

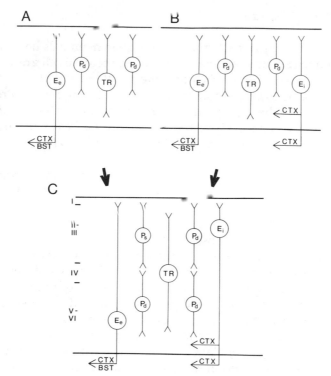

Fig. 8.20. In mammalian neocortex (C), several neurotransmitter/neuropeptide-specific and connectionally specific cell types are present: (1) thalamorecipient neurons (TR), which consist of two major types, layer IV granular neurons receiving direct axosomatic thalamic input and those outside layer IV that receive primarily axodendritic thalamic input; (2) neurons, infragranular in location, with efferent projections that are largely extrinsic to the cortex (E_e), i.e., to the brain stem and basal ganglia (BST); (3) neurons, supragranular in location, with projections that are intrinsic to the cortex (E_i), i.e., ramifying within the immediate area of the cortical column or projecting to other ipsilateral or contralateral cortical areas; (4) intrinsic peptidergic neurons that are abundant in the deep, or infragranular, layers but are also present to varying extents in superficial layers (P_d); and (5) intrinsic peptidergic neurons that are found almost exclusively in the supragranular layers (P_s). Immunohistochemical studies in turtles indicate that P_s neurons observed in mammalian cortex are absent or extremely scarce in turtle telencephalic cortex. Immunohistochemical and pathway tracing studies have shown that TR (of the type that receive axodendritic input), E_e and P_d cell types are present in turtle cortex. It is uncertain whether E_i type neurons are present in turtle cortex. These results suggest that a major event that may have occurred during the evolution of neocortex was the development and proliferation of cell types characteristic of layers II and III of mammalian neocortex, specifically of P_s neurons, P_d neurons of the supragranular layers, and possibly E_i neurons, and also of layer IV granular neurons. This suggestion presupposes that turtle dorsal cortex possesses the morphological features characteristic of the cortical fields present in the early reptiles ancestral to mammals. It would be of interest to determine whether or not E_i neurons are present in the cortex of living reptiles in order to gain clues as to whether the common reptilian ancestors of mammals and living reptiles possessed a cortex more similar to that in (A) or (B). If (A), then neuron types found in layers II and III of mammalian cortex would be largely unique to the mammalian lineage. Ebner (1976) has previously offered a proposal on the evolution of the neocortex similar to the present one.

the neurons of the homologous cell group in mammals have not been observed to contain enkephalin. A second pretectal cell group termed *nucleus pretectalis (PT)* in birds and the *dorsal pretectal nucleus (NPd)* in turtles appears to be the source of NPY+ fibers in the main retinorecipient layer of the sauropsid tectum. Prominent LANT6+ fibers are present in the retinorecipient superficial tectal layers of birds, reptiles, and frogs and appear to originate from LANT6+ retinal ganglion cells.

These examples of the presence of the same individual peptide (or of isopeptides) in homologous neural systems in the members of different vertebrate groups indicate that peptide localization may be an aid in studying brain evolution. If, for example, structure "a" in species A is tentatively proposed to be homologous to structure "b" in species B, then the presence of the same peptide (or of isopeptides) in structures a and b would help to more firmly establish the proposed homology. Thus, by using the localization of brain peptides to help identify homologous cell groups, it should be possible to increase our knowledge of the changes that have occurred during brain evolution. The presence or absence of peptides in individual cell groups or neuronal populations within cell groups should not, however, be the sole criterion on which conclusions of homology are based. Other characteristics, such as the connectivity and embryology of the cell group as well as the characteristics of the cell group in other lineages, should also be considered (Campbell and Hodos, 1970; Northcutt, 1984).

C. Neuropeptide Function

Few data are available on the functions of neuropeptides in reptiles. It seems reasonable to presume that individual reptilian neuropeptides play the same types of neuroactive roles in the reptilian nervous system, both at the cellular level and at the neural circuit level, as do the equivalent peptides in the mammalian nervous system. In the case of some systems that are clearly common to both reptiles and mammals, it might prove profitable to further elucidate the functions of particular peptides in particular systems using a reptilian model system, for example, the role of co-occurrent SP and dynorphin in the striatonigral system. In the case of some other neural systems that appear to be common to mammals and reptiles, the function of the homologous peptides appears to have undergone a change during evolution. For example, in reptiles, mesotocin occurs in parts of the nervous system homologous to those containing oxytocin in mammals. Although the central synaptic actions of oxytocin and mesotocin may be very similar in reptiles and mammals, the peripheral hormonal actions presumably differ since mesotocin in reptiles clearly cannot play a lactogenic role, a role that is played by oxytocin in mammals. Although mesotocin may play a role in oviposition (a role homologous to that of oxytocin in parturition), vasotocin is much more potent than mesotocin in promoting oviposition (Gorbman et al., 1983). What then is the function of mesotocin in reptiles? It will be of interest to determine how the function of mesotocin/oxytocin has changed during evolution as part of the overall issue of how the functions of homologous entities may change during evolution.

Regarding the evolution of peptide function, it would also be of interest to study the evolution of the affinity or specificity of individual peptides for their receptors. Is evolutionary change in the structure of a peptide accompanied by change in receptor specificity? Does, for example, reptilian neurotensin act more effectively on neurons of reptilian brain than it does on neurons of mammalian brain? What are the effects, for example, of mammalian neurotensin on neurons of the reptilian brain? In other words, does reptilian neurotensin have its optimal effects on reptilian neurons and mammalian neurotensin on mammalian neurons? Or does a neuropeptide evolve so as to be increasingly effective in binding to a receptor of evolutionarily unchanging structure? Finally, it will be of interest to determine if homologous structures process homologous precursor peptides in different ways, thereby arriving at different end products (or different proportions of the same end products). Dores (as discussed previously) has shown that this appears to be the case in the processing of POMC to beta-endorphin in both the anterior and the intermediate lobe of the pituitary in lizards as compared to mammals. Such differences in posttranslational processing of peptide precursors may provide suggestions as to how organisms adapt particular structures or molecules to their own specific ecological needs.

D. Conclusion

In conclusion, studies of neuropeptides in the reptilian CNS further our understanding of reptile brain organization and function. These studies also provide model systems by which to clarify the functions of particular peptides in specific neural systems common to other vertebrates. Further, study of peptides in the nervous system of reptiles can advance our understanding of nervous system evolution by helping clarify the changes the brain has undergone during amniote brain evolution and suggest some of the principles or mechanisms underlying those changes.

ACKNOWLEDGMENTS

I would like to thank Dr. J. E. Krause of the Department of Anatomy and Neurobiology at Washington University, St. Louis, and Dr. J. D. White of the Department of Endocrinology at SUNY at Stony Brook for allowing me to present some of our previously unpublished biochemical data in the present review. I would also like to thank Patricia A. Lindaman, Gary Henderson, and Debora Romeo for technical assistance in collecting the previously unpublished immunohistochemical data presented in this review and Ann Haight and Ellen Karle for secretarial and editorial assistance. Drs. R. M. Dores of the Department of Biological Sciences at the University of Denver and H. Khachaturian of the National Institute of Mental Health have kindly allowed me to reproduce figures from some of their published papers. I also thank Dr. Dores for his helpful comments on the manuscript. The research was supported by NS-19620.

APPENDIX: REPTILIAN SPECIES DISCUSSED

TESTUDINES

Caretta caretta
  Pearson et al., 1983
Chelonia mydas
  Licht, 1980
  Licht et al., 1984b
  MacKenzie et al., 1981
  MacNamee et al., 1986
  Yasuda et al., 1989
Chelydra serpentina
  Senn, 1971
Chrysemys picta
  Bass and Northcutt, 1981
  Bear and Ebner, 1983
  Gapp et al., 1985
  Greeley et al., 1984
  Hall et al., 1975
  Licht and Porter, 1985a, 1985b
  Powers and Reiner, 1980
  Reiner and Powers, 1978, 1980, 1983
  Sawin et al., 1981
Geoclemys reevesii
Lepidochelys olivacea
  Licht et al., 1982
Mauremys caspica
  Fernandez-Llebrez et al., 1988
Mauremys japonica
  Ueda et al., 1983
Terrapene ornata
  Strutz, 1982

Testudo graeca
  Garcia-Ayala et al., 1987
Testudo hermanni
  Weindl et al., 1984
Testudo kleinmanni
  Lembeck et al., 1985
Trachemys scripta elegans
  Agulleiro et al., 1985
  Desan, 1984
  Eldred and Karten, 1983, 1985
  Eldred et al., 1988
  Hall and Ebner, 1970
  Isayama and Eldred, 1988
  Korte et al., 1980
  Kunzle, 1982
  Kunzle and Schnyder, 1983
  Mufson et al., 1984
  Reiner, 1987a
  Reiner and Beinfeld, 1985
  Reiner and Carraway, 1987
  Reiner and Oliver, 1987
  Reiner et al., 1984c
  Reiner et al., 1985
  Sherwood and Whittier, 1988
  Smeets et al., 1987
  Ulinski, 1988
Turtle
  Preece and Licht, 1987

CROCODILIA

Alligator mississippiensis
Buchan et al., 1982, 1983
Crosby, 1917
Emson and DeQuidt, 1984
Lance et al., 1984
Lance et al., 1985

Caiman crocodilus
Brauth, 1984
Brauth and Kitt, 1980
Brauth and Reiner, 1982
Crocodylus niloticus
Dimaline et al., 1982

RHYNCHOCEPHALIA

Sphenodon punctatus
Reiner, pers. obs.

SAURIA

Acanthodactylus boskianus
Bons, 1983
Acanthodactylus pardalis
Bons, 1983
Agama bibroni
Saint-Girons, 1967
Anolis sp.
Osborne et al., 1982
Anolis carolinensis
Alderete et al., 1980
Dores, 1982a, 1982b
Dores and Surprenant, 1984
Dores et al., 1984
Greeley et al., 1984
Khachaturian et al., 1984
Licht and Denver, 1988
Morrel et al., 1979
Naik et al., 1980, 1981
Anolis cybotes
Goosens et al., 1979
Anolis roquet
Goosens et al., 1979
Basiliscus vittatus
Goosens et al., 1979
Calotes versicolor
Goosens et al., 1979
Pandalai, 1960
Prasada Rao and Subhedar, 1977
Sheela Pandalai, 1968
Chalcides ocelatus tiligugu
Sherwood and Whittier, 1988
Cordylus nigra
Sherwood and Whittier, 1988
Ctenosaura pectinata
Goosens et al., 1979, 1980

Dipsosaurus dorsalis
Gesell and Callard, 1972
Lisk, 1967
Eumeces okadae
Nozaki and Kobayashi, 1979
Gekko gecko
Russchen et al., 1987
Stoll and Voorn, 1985
Thepen et al., 1987
Hemidactylus flavoviridis
Haider, 1974
Iguana iguana
Acher et al., 1972
Lacerta sp.
Osborne et al., 1982
Lacerta vivipara
Bottcher et al., 1985
Mabuya quinquetaeniata
El-Salhy and Grimelius, 1981
Ophisops elegans
Zaloglu, 1973
Podarcis hispanica
Lopez et al., 1988
Perez-Clausell and Fredens, 1988
Podarcis muralis
Bons, 1983
Doerr-Schott and Dubois, 1977
Khachaturian et al., 1984
Vallarino, 1984, 1986
Podarcis sicula
Gaudino and Fasolo, 1984
Lopez et al., 1988
Sceloporus cyanogenys
Callard and Chester-Jones, 1971
Daugherty and Callard, 1972

Sceloporus occidentalis
  Engbretson et al., 1982
Sceloporus undulatus
  Engbretson et al., 1982
Tarentola mauritanica
  Bons, 1983
Tupinambis nigropunctatus
  Cruce and Newman, 1981
Uta stansburiana
  Engbretson et al., 1981, 1982

Varanus exanthematicus
  ten Donkelaar and de Boer-van
    Huizen, 1981
  Wolters et al., 1986
Varanus griseus
  Abdel-Messeih and Tawfik, 1963
Lizard
  Bottcher et al., 1985

SERPENTES

Agkistrodon blomhoffi
  Kadota et al., 1988
Crotalus atrox
  Kimmel et al., 1976
Diadophis punctatus
  Philibert and Kamemoto, 1965
Elaphe climacophora
  Nozaki and Kobayashi, 1979
Elaphe conspicillata
  Nozaki and Kobayashi, 1979
Elaphe quadrivirgata
  Nozaki and Kobayashi, 1979
Natrix maura
  Fernandez-Llebrez et al., 1988

Rhabdophis tigrinus
  Nozaki and Kobayashi, 1979
Thamnophis
  Halpern et al., 1982
Thamnophis sirtalis
  Dacey, 1982
  Sherwood and Whittier, 1988
Trimeresurus flavoviridis
  Terashima, 1987
Zaocys dhumnades
  Zhang et al., 1981

NOTES

1. By convention molecules consisting of fewer than 100 amino acids are referred to as peptides, not proteins.

2. The weight of an individual molecule or element is numerically the same as the weight of Avogadro's number of that molecule or element (i.e., the molecular weight). The common unit of measure for an individual molecule or element is the dalton; 1,000 daltons or a kilodalton is abbreviated as K.

REFERENCES

Abad, M. E., Taverne-Thiele, J. J., and Romboult, J. H. W. M. (1988). Immunocytochemical and ultrastructural characterization of co-existence of pancreatic polypeptide and glucagon-like immunoreactivity in the pancreatic endocrine cells of *Sparus auratus* L. (Teleostei). *Gen. Comp. Endocrinol.* 70, 9–19.

Abdel-Messeih, G., and Tawfik, J. (1963). Effect of starvation on the activity of the hypothalamo-hypophyseal system of *Varanus griseus* Daud. *Z. Zellforsch.* 59, 395–404.

Acher, R. (1980). Molecular evolution of biologically active polypeptides. *Proc. Roy. Soc. Lond.* B210, 21–43.

Acher, R. (1984). Evolution of neurohormonal peptides: from genetic machinery to functional tailoring. In *Evolution and Tumour Pathology of the Neuroendocrine System.* (S. Falkmer, R. Hakanson, and F. Sundler, eds.). Elsevier, Amsterdam, pp. 181–201.

Acher, R., Chauvet, J., and Chauvet, M. T. (1969). The neurohypophyseal hormones of reptiles: comparison of the viper, cobra and *Elaphe* active principles. *Gen. Comp. Endocrinol.* 13, 357–360.

Acher, R., Chauvet, J. and Chauvet, M. T. (1972). Reptilian neurohypophyseal hormones: the active peptides of a saurian *Iguana iguana. Gen. Comp. Endocrinol.* 19, 345–348.

Adams, P. R., Brown, D. A., and Jones, S. W. (1983). Substance P inhibits the M-current in bullfrog sympathetic neurones. *Br. J. Pharmacol.* 79, 330–333.

Agulleiro, B., Garcia Ayala, A., and Abad, M. E. (1985). An immunocytochemical and ultrastructural study of the endocrine pancreas of *Pseudemys scripta elegans* (Chelonia). *Gen. Comp. Endocrinol.* 60, 95–103.

Akil, H., and Watson, S. J. (1983). Beta-endorphin and biosynthetically related peptides in the central nervous system. *Handbook Psychopharmacol.* 16, 209–253.

Albert, L. C., Brauer, J. R., Jackson, I. M. D., and Reichlin, S. (1976). Localization of LH-RH in neurons in frog brain *Rana pipiens* and *Rana catesbeiana. Endocrinology* 98, 910–921.

Albin, R. L., Richfield, E. K., Reiner, A., Young, A. B., and Penney, J. B., Jr. (1988). The distribution of glutamate, GABA, benzodiazepine, muscarinic and opiate receptors in the avian tectum. *Soc. Neurosci. Abs.* 14, 56.

Alderete, M. R., Tokarz, R. R., and Crews, D. (1980). Luteinizing hormone-releasing hormone and thyrotropin-releasing hormone induction of female sexual receptivity in the lizard, *Anolis carolinensis. Neuroendocrinology* 30, . 200–205.

Altschuler, R. A., Fex, J., Parakkal, M. H., and Eckenstein, F. (1984). Co-localization of enkephalin-like and choline acetyltransferase-like immunoreactivities in olivocochlear neurons of the guinea pig. *J. Histochem. Cytochem.* 32, 839–843.

Alumets, J., Hakanson, R., and Sundler, F. (1978a). Distribution, ontogeny and ultrastructure of pancreatic polypeptide (PP) cells in the pancreas and gut of the chicken. *Cell Tiss. Res.* 194, 377–386.

Alumets, J., Hakanson, R., Sundler, F., and Chang, K. J. (1978b). Leu-enkephalin-like material in nerves and enterochromaffin cells in the gut. *Histochemistry* 56, 187–196.

Ananthanarayanan, V. (1955). Nature and distribution of neurosecretory cells of the reptilian brain. *Z. Zellforsch.* 43, 8–16.

Anastasi, A., Erspamer, V., and Bucci, M. (1971). Isolation and structure of bombesin and alytesin, two analagous active peptides from the skin of the European amphibians *Bombina* and *Alytes. Experientia* 27, 166–167.

Anderson, K. D., and Reiner, A. (1987). Striatonigral projection neurons: a retrograde labeling study of the relative numbers that contain substance P or enkephalin. *Soc. Neurosci. Abs.* 13, 1574.

Anderson, K. D., and Reiner, A. (1988). Extensive co-occurrence of substance

P and dynorphin in striatal projection neurons in rats. *Soc. Neurosci. Abs.* 14, 76.

Anderson, K. D., and Reiner, A. (1990a). The extensive co-occurrence of substance P and dynorphin in striatal projection neurons: an evolutionarily conserved feature of basal ganglia organization. *J. Comp. Neurol.* 295, 339–369.

Anderson, K. D., and Reiner, A. (1990b). Distribution and relative abundance of neurons in the pigeon telencephalon containing somatostatin, neuropeptide Y, or both. *J. Comp. Neurol.* 299, 261–282.

Andrade, R., and Aghajanian, G. K. (1981). Neurotensin selectively activates dopaminergic neurons in the substantia nigra. *Soc. Neurosci. Abs.* 7, 573.

Andrews, P. C., Hawke, D. H., Lee, T. D., Legesse, K., Noe, B. D., and Shively, J. E. (1986). Isolation and structure of the principal products of preproglucagon processing, including an amidated glucagon-like peptide. *J. Biol. Chem.* 261, 8128–8133.

Andrews, P. C., and Ronner, P. (1985). Isolation and structures of glucagon and glucagon-like peptide from catfish pancreas. *J. Biol. Chem.* 260, 3910–3914.

Baker-Cohen, K. F. (1968). Comparative enzyme histochemical observations on submammalian brains. *Adv. Anat. Embryol.* 40, 1–70.

Ball, J. N. (1981). Hypothalamic control of the pars distalis in fish, amphibians and reptiles. *Gen. Comp. Endocrinol.* 44, 135–170.

Bargmann, W. (1949). Über die neurosekretische Verknüpfung von Hypothalamus and Neurohypophyse. *Z. Zellforsch.* 34, 610–634.

Bass, A. H., and Northcutt, R. G. (1981). Retinal recipient nuclei in the painted turtle, *Chrysemys picta*: an autoradiographic study. *J. Comp. Neurol.* 199, 97–112.

Bayon, A., Koda, L., Battenberg, E., Azad, R., and Bloom, F. E. (1980). Regional distribution of endorphin, met[5]-enkephalin and leu[5]-enkephalin in the pigeon brain. *Neurosci. Lett.* 16, 75–80.

Bear, M. F., and Ebner, F. F. (1983). Somatostatin-like immunoreactivity in the forebrain of *Pseudemys* turtles. *Neuroscience* 9, 297–309.

Beinfeld, M. C., Lewis, M. E., Eiden, L. E., Nilaver, G., Pert, C. B., and Pert, A. (1983). The distribution of cholecystokinin and vasoactive intestinal peptide in rhesus monkey brain as determined by radioimmunoassay. *Neuropeptides* 3, 337–344.

Beinfeld, M. C., Meyer, K. K., Eskay, R. L., Jensen, R. T., and Brownstein, M. J. (1981). The distribution of cholecystokinin immunoreactivity in the central nervous system of the rat as determined by radioimmunoassay. *Brain Res.* 212, 51–57.

Beinfeld, M. C., Trubatch, J. R., and Brownstein, M. J. (1983). Cholecystokinin peptides in the brain and pituitary of the bullfrog *Rana catesbeiana*: Distribution and characterization. *Brain Res.* 268, 192–196.

Bennet-Clarke, C., and Romagnano, M. A. (1979). Distribution of somatostatin in the rat brain: telencephalon and diencephalon. *Anat. Rec.* 193, 481–482.

Berk, M. L., Reaves, T. A., Jr., Hayward, J. N., and Finkelstein, J. A. (1982).

The localization of vasotocin and neurophysin neurons in the diencephalon of the pigeon, *Columba livia. J. Comp. Neurol.* 204, 382–406.

Blahser, S., and Dubois, M. P. (1980). Immunocytochemical demonstration of met-enkephalin in the central nervous system of the domestic fowl. *Cell Tiss. Res.* 213, 53–68.

Blahser, S., Fellman, D., and Bugnon, C. (1978). Immunocytochemical demonstration of somatostatin-containing neurons in the hypothalamus of the domestic mallard. *Cell Tiss. Res.* 195, 183–187.

Bloch, B., Brazeau, P., Ling, N., Bohlen, P., Esch, F., Wehrenberg, W. B., Benoit, R., Bloom, F., and Guillemin, R. (1983). Immunohistochemical detection of growth hormone-releasing factor in brain. *Nature* (London) 301, 607–608

Bloom, F. E., Battenberg, E. L. F., Rivier, J., and Vale, W. (1982). Corticotropin releasing factor (CRF): immunoreactive neurons and fibers in rat hypothalamus. *Regulatory Peptides* 4, 43–48.

Bloom, S. R., Besser, G. M., Roy, V. M., Russell, R. C. G., and Schally, A. V. (1974). Inhibition of gastrin and gastric acid secretion by growth hormone release-inhibiting hormone. *Lancet* 2, Pt. 2, 1106–1109.

Boer, H. H., Schot, L. P. C., Veenstra, J. H., and Riechelt, D. (1980). Immunocytochemical identification of neural elements in the central nervous systems of a snail, some insects, a fish, and a mammal with an antiserum to the molluscan cardio-excitatory tetrapeptide, FMRFamide. *Cell Tiss. Res.* 213, 21–27.

Bons, N. (1983). Immunocytochemical identification of the mesotocin- and vasotocin-producing systems in the brain of temperate and desert lizard species and their modifications by cold exposure. *Gen. Comp. Endocrinol.* 52, 56–66.

Bons, N., Bouille, C., Vaudry, H., and Guillaume, V. (1985). Localization of corticotropin-releasing factor producing neurons in the pigeon brain: an immunofluorescence study. *C. R. Acad. Sc. Paris,* 300 [3](2), 49–52.

Bons, N., Kerdelhume, B., and Assenmacher, I. (1978). Immunohistochemical identification of an LH-RH producing system originating in the preoptic nucleus of the duck. *Cell Tiss. Res.* 188, 99–106.

Bottcher, G., Skagerberg, G., Ekman, R., Hakanson, R., and Sundler, F. (1985). PYY-like peptides in the central and peripheral nervous system of a frog and a lizard. *Peptides* 6, 215–221.

Bowie, E. P., and Herbert, D. C. (1976). Immunocytochemical evidence for the presence of arginine-vasotocin in the rat pineal gland. *Nature* (London) 26, 66–68.

Branton, W. D., Jan, L. Y., and Jan, Y. N. (1982). Nonmammalian luteinizing hormone-releasing factor (LRF) in tadpole and frog brain. *Soc. Neurosci. Abs.* 8, 14.

Brauth, S. E. (1984). Enkephalin-like immunoreactivity within the telencephalon of the reptile *Caiman crocodilus. Neuroscience* 11, 345–358.

Brauth, S. E., and Kitt, C. A. (1980). The paleostriatal system of *Caiman crocodilus. J. Comp. Neurol.* 189, 437–465.

Brauth, S. E., Kitt, C. A., Reiner, A., and Quirion, R. (1986). Neurotensin receptors in the forebrain and midbrain of the pigeon. *J. Comp. Neurol.* 253, 358–373.

Brauth, S. E., and Reiner, A. (1982). A pretectal-tectal enkephalin connection: immunohistochemical studies of homologous systems in reptiles. *Soc. Neurosci. Abs.* 8, 116.

Brauth, S. E., Reiner, A., Kitt, C. A., and Karten, H. J. (1983). The substance P–containing striato-tegmental path in reptiles: an immunohistochemical study. *J. Comp. Neurol.* 219, 305–327.

Brazeau, P., Vale, W., Burgus, R., Ling, N., Butcher, M., Rivier, J., and Guillemin, R. (1973). Hypothalamic polypeptide that inhibits the secretion of immunoreactive pituitary growth hormone. *Science* (N.Y.) 179, 77–79.

Brecha, N., Ciluffo, M., and Yamada, T. (1982). Localization and characterization of glucagon-like immunoreactivity in the retina. *Soc. Neurosci. Abs.* 8, 586.

Brecha, N. C., Eldred, W. D., Kuljis, R. O., and Karten, H. J. (1984). Identification and localization of biologically active peptides in the vertebrate retina. In *Progress in Retinal Research*, Vol. 3 (N. Osborne and J. Chader, eds.). Pergamon Press, New York, pp. 185–226.

Brodin, E., Alumets, J., Hakanson, R., Leander, S., and Kundler, F. (1981). Immunoreactive substance P in the chicken gut: distribution, development and possible functional significance. *Cell Tiss. Res.* 216, 455–469.

Brown, M., Mark, M., and Rivier, J. (1980). Is gastrin-releasing peptide mammalian bombesin? *Life Sci.* 27, 125–128.

Brownstein, M. J., Arimura, A., Sato, H., Schally, A. V., and Kizer, J. S. (1975). The regional distribution of somatostatin in the rat brain. *Endocrinology* 96, 1456–1461.

Brownstein, M. J., Palkovits, M., Saavedra, J. M., Bassiri, R. M., and Utiger, R. D. (1974). Thyrotropin-releasing hormone in specific nuclei of rat brain. *Science* (N.Y.) 185, 267–269.

Brownstein, M. J., Russel, J. T., and Gainer, H. (1980). Synthesis, transport and release of posterior pituitary hormones. *Science* (N.Y.) 207, 373–378.

Brownstein, M. J., Utiger, R. F., Palkovits, M., and Kizer, J. S. (1975). Effect of hypothalamic deafferentation on thyrotropin-releasing hormone levels in rat brain. *Proc. Natl. Acad. Sci. USA* 72, 4177–4179.

Buatti, M. C., and Pasternak, G. W. (1981). Multiple opiate receptors: phylogenetic differences. *Brain Res.* 218, 400–405.

Buchan, A. M. J., Lance, V., and Polak, J. M. (1982). The endocrine pancreas of *Alligator mississippiensis*—an immunocytochemical study. *Cell Tiss. Res.* 224, 117–128.

Buchan, A. M. J., Lance, V., and Polak, J. M. (1983). Regulatory peptides in the gastrointestinal tract of *Alligator mississippiensis*. *Cell Tiss. Res.* 321, 439–449.

Buchan, A. M. J., Polak, J. M., Bryant, M. G., Bloom, S. R., and Pearse, A. G. E. (1981). Vasoactive intestinal polypeptide (VIP)-like immunoreactivity in anuran intestine. *Cell Tiss. Res.* 216, 413–422.

Buck, S. H., and Burcher, E. (1986). The tachykinins: a family of peptides with a brood of receptors. *Trends Pharmacol. Sci.* 7, 65–68.

Bugbee, N. (1979). "The Effect of Bilateral Lesions in the Nucleus Spiriformis Lateralis on Visually Guided Behavior in the Pigeon." Ph.D. thesis, University of Maryland, Md.

Buijs, R. M. (1978). Intra- and extra-hypothalamic vasopressin and oxytocin pathways in the rat. Pathways to the limbic system, medulla oblongata and spinal cord. *Cell Tiss. Res.* 192, 423–435.

Burgus, R., Butcher, M., Amoss, M., Ling, N., Monahan, M., Rivier, J., Blackwell, R., Vale, W., and Guillemin, R. (1972). Primary structure of the ovine hypothalamic luteinizing hormone-releasing factor (LRF). *Proc. Natl Acad. Sci. USA* 69, 278–282.

Burgus, R., Dunn, T. F., Desiderio, D., Ward, D. N., Vale, W., and Guillemin, R. (1970). Characterization of ovine hypothalamic hypophysiotropic TSH-releasing factor. *Nature* (London) 226, 321–325.

Callard, I. P., and Callard, G. V. (1978). The adrenal gland in Reptilia. Part 2. Physiology. In *General Comparative and Clinical Endocrinology of the Adrenal Cortex*, Vol. 2 (I. Chester-Jones and I. W. Henderson, eds.). Academic Press, New York, pp. 370–418.

Callard, I. P., and Chester-Jones, I. (1971). The effect of hypothalamic lesions and hypophysectomy on adrenal weight in *Sceloporus cyanogenys*. *Gen. Comp. Endocrinol.* 17, 194–202.

Callard, I. P., and Lance, V. (1977). The control of reptilian follicular cycles. In *Reproduction and Evolution* (J. H. Calaby and C. H. Tyndale-Biscoe, eds.) Australian Academy of Science, Canberra City, pp. 199–209.

Campbell, C. B. G., and Hodos, W. (1970). The concept of homology and the evolution of the nervous system. *Brain Behav. Evol.* 3, 353–367.

Carlsson, A. (1959). Occurrence, distribution and physiological role of catecholamines in the nervous system. *Pharmacol. Rev.* 11, 300–304.

Carraway, R. E., and Bhatnagar, Y. M. (1980a). Immunochemical characterization of neurotensin-like peptides in chicken. *Peptides* 1, 159–165.

Carraway, R. E., and Bhatnagar, Y. M. (1980b). Isolation, structure and biological activity of chicken intestinal neurotensin. *Peptides* 1, 167–174.

Carraway, R. E., and Ferris, C. F. (1983). Isolation, biological and chemical characterization and synthesis of neurotensin-related hexapeptide from chicken intestine. *J. Biol. Chem.* 258, 2475–2479.

Carraway, R. E., and Leeman, S. E. (1973). Isolation of a new hypotensive peptide, neurotensin, from bovine hypothalamus. *J. Biol. Chem.* 248, 6854–6861.

Carraway, R. E., and Mitra, S. P. (1987). The use of radioimmunoassay to compare the tissue and subcellular distributions of neurotensin and neuromedin N in the cat. *Endocrinology* 120, 2092–2100.

Carraway, R. E., and Reinecke, M. (1984). Neurotensin-like peptides and a novel model of the evolution of signaling systems. In *Evolution and Tumor Pathology of the Neuroendocrine System* (S. Falkmer, R. Hakanson, and F. Sundler, eds.). Elsevier, Amsterdam, pp. 245–283.

Carraway, R., Ruane, S. E., Fuerle, G. E., and Taylor, S. (1982a). Amphibian neurotensin (NT) is not xenopsin (XP): dual presence of NT-like and XP-like peptides in various amphibia. *Endocrinology* 110, 1094–1101.

Carraway, R. E., Ruane, S. E., and Kim, H. R. (1982b). Distribution and immunochemical character of neurotensin-like material in representative vertebrates and invertebrates: apparent conservation of the COOH-terminal region during evolution. *Peptides* 1, 115–123.

Carraway, R. E., Ruane, S. E., and Ritsema, R. S. (1983). Radioimmunoassay for Lys^8, Asn^9 Neurotensin$^{8-13}$: tissue and subcellular distribution of immunoreactivity in chickens. *Peptides* 4, 111–116.

Chang, M. M., and Leeman, S. E. (1970). Isolation of a sialogogic peptide from bovine hypothalamic tissue and its characterization as substance P. *J. Biol. Chem.* 245, 4784–4790.

Chan-Palay, V., Jonsson, V., and Palay, S. L. (1978). Serotonin and substance P co-exist in neurons of the rat's central nervous system. *Proc. Natl. Acad. Sci. USA* 75, 1582–1586.

Chauvet, J., Hurpet, D., Colne, T., Michel, G., Chauvet, M. T., and Acher, R. (1985a). Neurohypophyseal hormones as evolutionary tracers: identification of oxytocin, vasopressin and arginine vasopressin in two South American opossums (*Didelphis marsupialis* and *Philander opossum*). *Gen. Comp. Endocrinol.* 57, 320–328.

Chauvet, J., Hurpet, D., Michel, G., Chauvet, M. T., Carrick, F. N., and Acher, R. (1985b). The neurohypophyseal hormones of the egg-laying mammals: identification of arginine vasopressin in the platypus (*Ornithorhynchus anatinus*). *Biochem. Biophys. Comm.* 127, 277–282.

Chauvet, J., Michel, G., Chauvet, M. T., and Acher, R. (1988). An amphibian two-domain 'big' neurophysin: conformational homology with mammalian MSEL-neurophysin/copeptin intermediate precursor shown by trypsin-Sepharose proteolysis. *FEBS Letters* 230, 77–80.

Chauvet, J., Rouille, Y., Chauvet, M. T., and Acher, R. (1987). Evolution of marsupials traced by their neurohypophyseal hormones: microidentification of mesotocin and arginine vasopressin in two Australian families, Dasyuridae and Phascolarctidae. *Gen. Comp. Endocrinol.* 67, 399–408.

Chauvet, M. T., Levy, B., Michel, G., Chauvet, J., and Acher, R. (1986). Precursors of mesotocin and vasotocin in birds: identification of VLDV- and MSEL-neurophysins in chicken, goose and ostrich. *Biosci. Rep.* 6, 381–385.

Checler, F., Vincent, J. P., and Kitabgi, P. (1986). Neuromedin N: high affinity interaction with brain neurotensin receptors and rapid inactivation by brain synaptic peptidases. *Eur. J. Pharmacol.* 126, 239–244.

Chipkin, S. R., Stoff, J. S., and Aronin, N. (1988). Immunohistochemical evidence for neural mediation of VIP activity in the dogfish rectal gland. *Peptides* 9, 119–124.

Chronwall, B. M., DiMaggio, D. A., Massari, V. J., Pickel, V. M., Ruggiero, D. A., and O'Donohue, T. L. (1985). The anatomy of neuropeptide Y-containing neurons in rat brain. *Neuroscience* 15, 1159–1181.

Cone, R. I., and Goldstein, A. (1982). A dynorphin-like opioid in the central nervous system of an amphibian. *Proc. Natl. Acad. Sci. USA* 79, 3345–3349.

Conlon, J. M., Agoston, D. V., and Thim, L. (1985). An elasmobranch somatostatin: primary structure and tissue distribution in *Torpedo marmorata*. *Gen. Comp. Endocrinol.* 60, 406–413.

Conlon, J. M., Goke, R., Andrews, P. C., and Thim, L. (1989). Multiple molecular forms of insulin and glucagon-like peptide from the pacific ratfish (*Hydrolagus colliei*). *Gen. Comp. Endocrinol.* 73, 136–146.

Conlon, J. M., and Thim, L. (1985). Primary structure of glucagon from an elasmobranch fish, *Torpedo marmorata*. *Gen. Comp. Endocrinol.* 60, 398–405.

Correa, F. M. A., Innis, R. B., Uhl, G. R., and Snyder, S. H. (1979). Bradykinin-like immunoreactive neuronal systems localized histochemically in rat brain. *Proc. Natl. Acad. Sci. USA* 76, 1489–1493.

Costa, M., and Furness, J. B. (1982). Neuronal peptides in the intestine. *Br. Med. Bull.* 38, 247–252.

Creagh, T., Skrabanek, P., Cannon, D., Balfe, A., and Powell, D. (1980). Phylogeny of substance P *Gen. Comp. Endocrinol.* 40, 503–506.

Crews, D. (1979). Neuroendocrinology of lizard reproduction. *Biol. Reprod.* 20, 51–73.

Crim, J. W. (1985). Immunocytochemistry of luteinizing hormone-releasing hormone and sexual maturation of the frog brain: comparisons of juvenile and adult bullfrogs (*Rana catesbeiana*). *Gen. Comp. Endocrinol.* 59, 424–433.

Crosby, E. C. (1917). The forebrain of *Alligator mississippiensis*. *J. Comp. Neurol.* 27, 325–403.

Crowley, W. R., and Terry, L. C. (1980). Biochemical mapping of somatostatinergic systems in rat brain: effects of periventricular hypothalamic and medial basal amygdaloid lesions on somatostatin-like immunoreactivity in discrete brain nuclei. *Brain Res.* 200, 283–291.

Cruce, W. L. R., and Newman, D. B. (1981). Brainstem origins of spinal projections in the lizard, *Tupinambis nigropunctatus*. *J. Comp. Neurol.* 198, 185–207.

Dacey, D. (1982). "The Optic Tectum in the Snake *Thamnophis sirtalis*." Ph.D. thesis, University of Chicago.

Dahlstrom, A., and Fuxe, K. (1964a). Evidence for the existence of monoamine-containing neurons in the central nervous system. I. Demonstration of monoamines in the cell bodies of brain stem neurons. *Acta Physiol.* 62, 1–55.

Dahlstrom, A., and Fuxe, K. (1964b). A method for the demonstration of monoamine-containing fibers in the central nervous system. *Acta Physiol. Scand.* 60, 293–295.

Dale, H. H., Feldberg, W., and Vogt, M. (1936). Release of acetylcholine at voluntary motor nerve endings. *J. Physiol.* 86, 353–380.

Dalsgaard, C. J., Vincent, S. R., Hokfelt, T., Lundberg, J. M., Dahlstrom, A., Schultzberg, M., Dockray, G. J., and Cuello, A. C. (1982). Co-existence of cholecystokinin- and substance P-like peptides in neurons of the dorsal root ganglia of the rat. *Neurosci. Lett.* 33, 159–163.

Danger, J. M., Guy, J., Benyamina, M., Jegou, S., LeBoulenger, F., Cote, J., Tonon, M. C., Pelletier, G., and Vaudry, H. (1985). Localization and identification of neuropeptide Y (NPY)-like immunoreactivity in the frog brain. *Peptides* 6, 1225–1236.

Darras, V. M., and Kuhn, E. R. (1982). Increased plasma levels of thyroid hormones in a frog *Rana ridibunda* following intravenous administration of TRH. *Gen. Comp. Endocrinol.* 48, 469–475.

Daugherty, D. R., and I. P. Callard (1972). Plasma corticosterone levels in the male iguanid lizard *Sceloporus cyanogenys* under various physiological conditions. *Gen. Comp. Endocrinol.* 19, 69–72.

Davila, J. C., Guirado, S., and De la Calle, A. (1988). Immunocytochemical localization of somatostatin in the cerebral cortex of lizards. *Brain Res.* 447, 52–59.

Davis, B. M., Krause, J. E., Bogan, N., and Cabot, J. B. (1988). Intraspinal substance P-containing pathways to sympathetic preganglionic neuropil in pigeon, *Columba livia*: high-performance liquid chromatography, radioimmunoassay and electron microscopic evidence. *Neuroscience* 26, 655–668.

De Lanerolle, N. C., Elde, R. P., Sparber, S. B., and Frick, M. (1981). Distribution of methionine-enkephalin immunoreactivity in the chick brain: an immunohistochemical study. *J. Comp. Neurol.* 199, 513–533.

Delfs, J. R., and Dichter, M. A. (1985). Somatostatin. In *Neurotransmitter Actions in the Vertebrate Nervous System* (A. Rogawski and J. L. Barker, eds.) Plenum Publishing, New York, pp. 411–437.

De Mey, J., Dierickx, K., and Vandesande, F. (1975). Immunohistochemical demonstration of neurophysin I and neurophysin II-containing nerve fibers in the external region of the bovine median eminence. *Cell Tiss. Res.* 157, 517–519.

Denver, R. J., and Licht, P. (1987). Thyroid hormones act at the level of the pituitary to regulate basal and TRH-mediated secretion in the turtle. *Am. Zool.* 27, 78A.

Desan, P. (1984). "The Organization of the Cerebral Cortex in the Pond Turtle, *Pseudemys scripta elegans*." Ph.D. thesis, Harvard University, Cambridge, Mass.

de Vries, G. J., Buijs, R. M., and Swaab, D. F. (1981). Ontogeny of the vasopressinergic neurons of the suprachiasmatic nucleus and their extrahypothalamic projections in the rat brain—presence of a sex difference in the lateral septum. *Brain Res.* 218, 67–78.

Dierickx, K., Goosens, N., and Vandesande, F. (1981). The origin of somatostatin fibers in the median eminence and neural lobe of *Rana temporaria*. *Cell Tiss. Res.* 215, 41–45.

DiFiglia, M., Aronin, N., and Leeman, S. E. (1982a). Light and ultrastructural localization of immunoreactive substance P in the dorsal horn of monkey spinal cord. *Neuroscience* 7, 1127–1139.

DiFiglia, M., Aronin, N., and Leeman, S. E. (1982b). Immunoreactive substance P in the substantia nigra of the monkey: light and electron microscopic localization. *Brain Res.* 233, 381–388.

DiMaggio, D. A., Chronwall, B. M., Buchanan, K., and O'Donohue, T. L. (1985). Pancreatic polypeptide immunoreactivity in rat brain is actually neuropeptide Y. *Neuroscience* 15, 1149–1157.

Dimaline, R. (1983). Is caerulein amphibian CCK? *Peptides* 4, 457–462.

Dimaline, R., Rawdon, B. B., Brando, S., Andrew, A., and Loveridge, J. P. (1982). Biologically active gastrin/CCK-related peptides in the stomach of a reptile, *Crocodylus niloticus*: identified and characterized by immunochemical methods. *Peptides* 3, 977–984.

Dimaline, R., Vaillant, C., and Dockray, G. (1980). The use of region specific antibodies in the characterization and localization of vasoactive intestinal polypeptide-like substances in the rat gastrointestinal tract. *Regulatory Peptides* 1, 1–16.

Dobner, P. R., Barber, D. L., Villa-Komaroff, L., and McKiernan, C. (1987). Cloning and sequence analysis of cDNA for the canine neurotensin/neuromedin N precursor. *Proc. Natl. Acad. Sci. USA* 84, 3516–3520.

Dockray, G. J. (1974). Extraction of a secretin-like factor from the intestines of the pike (*Esox lucius*). *Gen. Comp. Endocrinol.* 23, 340–347.

Dockray, G. J. (1977). Progress in gasteroenterology. Molecular evolution of gut hormones: application of comparative studies on the regulation of digestion. *Gastroenterology* 72, 344–358.

Dockray, G. J. (1979). Evolutionary relationships of the gut hormones. *Fed. Proc.* 38, 2295–2301.

Dockray, G. J. (1982). The physiology of cholecystokinin in brain and gut. *Br. Med. Bull.* 38, 253–258.

Dockray, G. J., Duve, H., and Thorpe, A. (1981). Immunochemical characterization of gastrin/cholecystokinin-like peptides in the brain of the blowfly, *Calliphora vomitoria*. *Gen. Comp. Endocrinol.* 45, 491–496.

Dockray, G. J., Reeve, J. R., Jr., Shively, J., Gayton, R. J., and Banard, C. S. (1983). A novel active pentapeptide from chicken brain identified by antibodies to FMRFamide. *Nature* (London) 305, 328–330.

Dockray, G. J., Sault, C., and Holmes, S. (1986). Antibodies to FMRFamide, and the related pentapeptide LPLRFamide, reveal two groups of immunoreactive peptides in chicken brain. *Regulatory Peptides* 16, 27–37.

Dockray, G. J., Vaillant, C., and Walsh, J. H. (1979). The neuronal origin of bombesin-like immunoreactivity in the rat gastrointestinal tract. *Neuroscience* 4, 1561–1568.

Dockray, G. J., Vaillant, C., and Williams, R. G. (1981). New vertebrate brain-gut peptide related to a molluscan neuropeptide and an opioid peptide. *Nature* (London) 293, 656–657.

Dockray, G. J., Vaillant, C., Williams, R. G., Gayton, R. J., and Osborne, N. N. (1981). Vertebrate brain-gut peptides related to FMRFamide and Met-enkephalin Arg⁶Phe⁷. *Peptides*, Suppl. 2, 25–30.

Dockray, G. J., and Williams, R. G. (1981). FMRFamide immunoreactivity in rat brain: a new mammalian neuropeptide. *Neurosci. Lett.*, Suppl. 7, S452.

Dockray, G. J., and Williams, R. G. (1983). FMRF-amide-like immunoreactivity in rat brain: development of a radioimmunoassay and its application in studies of distribution and chromatographic properties. *Brain Res.* 266, 295–303.

Doerr-Schott, J., and Dubois, M. P. (1977). Immunohistochemical demonstration of an SRIF-like system in the brain of the reptile: *Lacerta muralis* Laur. *Experientia* 33, 947–948.

Doerr-Schott, J., and Dubois, M. P. (1978). Immunohistochemical localization of different peptidergic substances in the brain of amphibians and reptiles. In *Comparative Endocrinology* (P. J. Gaillard and H. H. Boer, eds.). Elsevier, Amsterdam, pp. 367–370.

Doerr-Schott, J., Dubois, M. P., and Lichte, C. (1981). Immunohistochemical localization of substances reactive to antisera against alpha- and beta-endorphin and met-enkephalin in the brain of *Rana temporaria* L. *Cell Tiss. Res.* 217, 79–92.

Dores, R. M. (1982a). Localization of multiple forms of ACTH- and beta-endorphin-related substances in the pituitary of the reptile, *Anolis carolinensis*. *Peptides* 3, 913–924.

Dores, R. M. (1982b). Evidence for a common precursor for alpha-MSH and beta-endorphin-related in the intermediate lobe of the pituitary of the reptile *Anolis carolinensis*. *Peptides* 3, 925–935.

Dores, R. M. (1983). Further characterization of the major forms of reptile beta-endorphin. *Peptides* 4, 897–905.

Dores, R. M., Khachaturian, H., Watson, S. J., and Akil, H. (1984). Localization of neurons containing pro-opiomelanocortin-related peptides in the hypothalamus and midbrain of the lizard, *Anolis carolinensis*: evidence for region specific processing of beta-endorphin. *Brain Res.* 324, 384–389.

Dores, R. M., and Surprenant, A. (1983). Biosynthesis of multiple forms of beta-endorphin in the reptile intermediate pituitary. *Peptides* 4, 889–896.

Dores, R. M., and Surprenant, A. (1984). *In vitro* synthesis of ACTH- and B-endorphin-related substances in the pars distalis of *Anolis carolinensis*. *Gen. Comp. Endocrinol.* 56, 90–99.

Dubois, M. P., Billard, R., Breton, B., and Peter, R. E. (1979). Comparative distribution of somatatostatin, LH-RH, neurophysin and alpha-endorphin in the rainbow trout: an immunocytochemical study. *Gen. Comp. Endocrinol.* 37, 220–232.

Duve, H., and Thorpe, A. (1980). Localization of pancreatic polypeptide (PP)-like immunoreactive material in neurons of the brain of the blowfly, *Calliphora erythrocephala* (Diptera). *Cell Tiss. Res.* 210, 1–109.

Duve, H., and Thorpe, A. (1981). Gastrin/cholecystokinin (CCK)-like immunoreactive neurons in the brain of the blowfly, *Calliphora erythrocephala* (Diptera). *Gen. Comp. Endocrinol.* 43, 381–391.

du Vigneaud, V. (1956). Hormones of the posterior pituitary gland: oxytocin and vasopressin. *Harvey Lecture* 50, 1–26.

Eales, J. G., and Himick, B. A. (1988). The effects of TRH on plasma thyroid hormone levels of rainbow trout (*Salmo gairdneri*) and Arctic Charr (*Salvelinus alpinus*). *Gen. Comp. Endocrinol.* 72, 333–339.

Ebner, F. F. (1976). The forebrain of reptiles and mammals. In *Evolution of Brain and Behavior in Vertebrates* (R. B. Masterton, M. E. Bitterman, C. B. G. Campbell, and N. Hotton, eds.). John Wiley and Sons, New York, pp. 147–167.

Edley, S. M., Hall, L., Herkenham, M., and Pert, C. B. (1982). Evolution of striatal opiate receptors. *Brain Res.* 249, 184–188.

Eiden, L. E., and Eskay, R. L. (1980). Characterization of LRF-like immunoreactivity in the frog sympathetic ganglia: non-identity with LRF decapeptide. *Neuropeptides* 1, 29–37.

Eiden, L. E., Loumaye, E., Sherwood, N., and Eskay, R. L. (1982). Two chemically distinct forms of luteinizing hormone-releasing hormone are differentially expressed in frog neural tissues. *Peptides* 3, 323–327.

Eldred, W. D., Isayama, T., Reiner, A., and Carraway, R. E. (1988). Ganglion cells in the turtle retina contain the neuropeptide LANT6. *J. Neurosci.* 8, 119–132.

Eldred, W. D., and Karten, H. J. (1983). Characterization and quantification of peptidergic amacrine cells in the turtle retina: enkephalin, neurotensin and glucagon. *J. Comp. Neurol.* 221, 371–381.

Eldred, W. D., and Karten, H. J. (1985). Ultrastructure and synaptic contacts of enkephalinergic amacrine cells in the retina of turtle (*Pseudemys scripta*). *J. Comp. Neurol.* 232, 36–42.

Eldred, W. D., Li, H. B., Carraway, R. E., and Dowling, J. E. (1987). Immunocytochemical localization of LANT6-like immunoreactivity within neurons in the inner nuclear and ganglion cell layers in vertebrate retinas. *Brain Res.* 424, 361–370.

El-Salhy, M. (1984). Immunocytochemical investigation of gastro-enteropancreatic (GEP) neurohormonal peptides in the pancreas and gastrointestinal tract of the dogfish *Squalus acanthias*. *Histochemistry* 80, 193–205.

El-Salhy, M., and Grimelius, L. (1981). The endocrine cells of the gastrointestinal mucosa of a squamate reptile, the grass lizard (*Mabuya quinquetaeniata*): a histological and immunohistochemical study. *Biomed. Res.* 2, 639–658.

El-Salhy, M., Grimelius, L., Lundberg, J. M., Tatemoto, K., and Terenius, L. (1982). Immunocytochemical evidence for the occurrence of PYY, a newly isolated gut polypeptide in endocrine cells in the gut of amphibians and reptiles. *Biomed. Res.* 3, 303–306.

El-Salhy, M., Grimelius, L., Wilander, E., Abu-Sinna, G., and Lundquist, G. (1981). Histological and immunohistochemical studies of the endocrine cells of the gastrointestinal mucosa of the toad (*Bufo regularis*). *Histochemistry* 71, 53–65.

El-Salhy, M., Grimelius, L., Wilander, E., Ryberg, B., Terenius, L., Lundberg, J. M., and Tatemoto, K. (1983). Immunocytochemical identification of polypeptide YY (PYY) cells in the human gastrointestinal tract. *Histochemistry* 77, 15–23.

El-Salhy, M., Wilander, E., Grimelius, L., Terenius, L., Lundberg, J. M., and Tatemoto, K. (1982). The distribution of polypeptide YY (PYY) and pancreatic polypeptide (PP)-immunoreactive cells in the domestic fowl. *Histochemistry* 75, 25–30.

El-Salhy, M., Wilander, E., Juntii-Berggren, L., and Grimelius, L. (1983). The distribution and ontogeny of polypeptide YY (PYY)- and pancreatic polypeptide (PP)-immunoreactive cells in the gastrointestinal tract of rat. *Histochemistry* 78, 53–60.

Emson, P. C., and DeQuidt, M. E. (1984). NPY-a new member of the pancreatic polypeptide family. *Trends Neurosci.* 7, 31–35.

Eng, J., Du, B. H., Raufman, J. P., and Yalow, R. S. (1986). Purification and amino acid sequences of dog, goat and guinea pig VIPs. *Peptides* 7, 17–20.

Engbretson, G. A., Brecha, N., and Reiner, A. (1982). Substance P-like immunoreactivity in the parietal eye visual system of the lizard *Uta stansburiana*. *Cell Tiss. Res.* 227, 543–554.

Engbretson, G. A., Reiner, A., and Brecha, N. C. (1981). Habenular asymmetry and the central connections of the parietal eye of the lizard. *J. Comp. Neurol.* 198, 155–165.

Engelhardt, R. P., Dhainaut-Courtois, N., and Tramu, G. (1982). Immuno-histochemical demonstration of a CCK-like peptide in the nervous system of a marine annelid worm, *Nereis diversicolor* O. F. Muller. *Cell Tiss. Res.* 227, 401–411.

Epstein, M. L., Lindberg, I., and Dahl, J. L. (1981). Development of enke-phalinergic neurons in the gut of the chick. *Peptides* 2, 271–276.

Erichsen, J. T., Karten, H. J., Brecha, N. C., and Eldred, W. D. (1982a). Local-ization of substance P-like and enkephalin-like immunoreactivity within preganglionic terminals of the avian ciliary ganglion: light and electron mi-croscopy. *J. Neurosci.* 2, 994–1003.

Erichsen, J. T., Reiner, A., and Karten, H. J. (1982b). Co-occurrence of sub-stance P-like and leu-enkephalin-like immunoreactivities in neurons and fibers of the avian nervous system. *Nature* (London) 295, 407–410.

Erspamer, V. (1981). The tachykinin peptide family. *Trends Neurosci.* 44, 267–269.

Erspamer, V., Falconeri-Erspamer, G., Melchiorri, P., and Negri, L. (1979). Occurrence and polymorphism of bombesin-like peptides in the gastroin-testinal tract of birds and mammals. *Gut* 20, 2047–2056.

Eskay, R. L., and Beinfeld, M. C. (1982). HPLC and RIA of cholecystokinin peptides in the vertebrate neural retina. *Brain Res.* 246, 315–318.

Eskay, R. L., Furness, J. B., and Long, R. T. (1981). Substance P activity in the bullfrog retina: localization and identification in several vertebrate species. *Science* (N.Y.) 212, 1049–1051.

Evans, T. K., Litthauer, D., and Oelofsen, W. (1988). Purification and primary structure of ostrich insulin. *Int. J. Peptide Protein Res.* 31, 454–462.

Fahrenkrug, J., and Emson, P. C. (1982). Vasoactive intestinal polypeptide: functional aspects. *Br. Med. Bull.* 38, 265–270.

Falkmer, S., Dafgard, E., El-Salhy, M., Engstrom, W., Grimelius, L., and Zet-terberg, A. (1985). Phylogenetic aspects of islet hormone families: a mini-review with particular reference to insulin as a growth factor and to the phylogeny of PYY and NPY immunoreactive cells and nerves in the endo-crine and exocrine pancreas. *Peptides* 6, 315–320.

Falkmer, S., Fahrenkrug, J., Alumets, F., Hakanson, R., and Sundler, F. (1980). Vasoactive intestinal polypeptide (VIP) in epithelial cells of the gut mucosa of an elasmobranchian cartilaginous fish, the ray. *Endocrinol. Jap.* 1, 401–405.

Fan, Z.-W., Eng, J., Miedel, M., Hulmes, J. D., Pan, Y. C. E., and Yalow, R. S. (1987). Cholecystokinin octapeptides purified from chinchilla and chicken brains. *Brain Res. Bull.* 18, 757–760.

Fasolo, A., Andreone, C., and Vandesande, F. (1984). Immunohistochemical localization of corticotropin-releasing factor (CRF)-like immunoreactivity in the hypothalamus of the newt, *Triturus cristatus. Neurosci. Lett.* 49, 135–142.

Fasolo, A., and Gaudino, G. (1982). Immunohistochemical localization of so-matostatin-like immunoreactivity in the hypothalamus of the lizard, *Lacerta sicula. Gen. Comp. Endocrinol.* 48, 205–212.

Fehrer, S. C., Silby, J. L., Behnke, E. J., and El Halawani, M. E. (1985a). The influence of thyrotropin releasing hormone on *in vivo* prolactin release and *in vitro* prolactin, luteinizing hormone and growth hormone release from dispersed pituitary cells of the young turkey (*Meleagris gallopavo*). *Gen. Comp. Endocrinol.* 59, 64–72.

Fehrer, S. C., Silby, J. L., Behnke, E. J., and El Halawani, M. E. (1985b). Hypothalamic and serum factors influence on prolactin and luteinizing hormone release by the pituitary gland of the young turkey (*Meleagris gallopavo*). *Gen. Comp. Endocrinol.* 59, 73–81.

Fellmann, D., Bugnon, C., Bresson, J. L., Gouget, A., Cardot, J., Clavequin, M. C., and Hadjiyiassemis, M. (1984). The CRF neuron: immunocytochemical study. *Peptides* 5, 19–33.

Fernandez-Llebrez, P., Perez, J., Nadales, A. E., Cifuentes, M., Grondona, J. M., Mancera, J. M., and Rodriguez, E. M. (1988). Immunocytochemical study of the hypothalamic magnocellular neurosecretory nuclei of the snake *Natrix maura* and the turtle *Mauremys caspica*. *Cell Tiss. Res.* 253, 435–445.

Finger, T. E. (1981). Enkephalin-like immunoreactivity in the gustatory lobes and visceral nuclei in the brains of goldfish and catfish. *Neuroscience* 6, 2747–2758.

Finlay, J. C. W., Maderdrut, J. L., and Petrusz, P. (1981a). The immunocytochemical localization of enkephalin in the central nervous system of the rat. *J. Comp. Neurol.* 198, 541–565.

Finlay, J. C. W., Maderdrut, J. L., Roger, L. J., and Petrusz, P. (1981b). The immunocytochemical localization of somatostatinergic neurons in the rat central nervous system. *Neuroscience* 6, 2173–2192.

Fischer, J. A., and Born, W. (1985). Novel peptides from the calcitonin gene: expression, receptors and biological function. *Peptides* 6, 265–271.

Follenius, E., and Dubois, M. P. (1978). Immunocytochemical detection and localization of a peptide reacting with an alpha-endorphin antiserum in the corticotropic and melanotropic cells of the trout (*Salmo irideus* Gibb). *Cell Tiss. Res.* 188, 273–283.

Fontaine-Perus, J. (1984). Development of VIP in the peripheral nervous system of avian embryo. *Peptides* 5, 195–200.

Foster, R. G., Panzica, G. C., Parry, D. M., and Viglietti-Panzica, C. (1988). Immunocytochemical studies on the LHRH system of the Japanese quail: influence of photoperiod and aspects of sexual differentiation. *Cell Tiss. Res.* 253, 327–335.

Furness, J. B., Costa, M., Emson, P. C., Hakanson, R., Moghinzadeh, E., Sundler, F., Taylor, I. L., and Chance, R. E. (1983). Distribution, pathways and reactions to drug treatment of nerves with neuropeptide Y- and pancreatic polypeptide-like immunoreactivity in the guinea pig digestive tract. *Cell Tiss. Res.* 234, 71–92.

Furness, J. B., Costa, M., Murphy, R., Beardsley, A. M., Oliver, J. R., Llewellyn-Smith, I. J., Eskay, R. L., Shulkes, H. A., Moody, T. W., and Meyer, D. K. (1982). Detection and characterization of neurotransmitters, particularly peptides, in the gastrointestinal tract. *Scand. J. Gastroenterol.* 17, Suppl. 71, 61–70.

Furr, M. J. A., Onura, G. I., Bonney, R. C., and Cunningham, F. J. (1973). The effect of synthetic hypothalamic releasing factors on plasma levels of luteinizing hormone in the cockerel. *J. Endocrinol.* 59, 495–502.

Gabrion, J., Kerdelhume, B., Alonso, G., Bosler, O., Assenmacher, I., and Calas, A. (1978). Vastoocin immunoreactive neurons in the hypothalamus of nonmammalian vertebrates. In *Neurosecretion and Neuroendocrine Activity: Evolution, Structure and Function* (W. Bargmann, A. Oksche, A. Polenov, and B. Scharrer, eds.). Springer, New York, pp. 186–189.

Gall, C., and Selawski, L. (1984). Supramammillary afferents to guinea pig hippocampus contain substance P-like immunoreactivity. *Neurosci. Lett.* 51, 171–176.

Gamlin, P. D. R., Reiner, A., and Karten, H. J. (1982). Substance P-containing neurons of the avian suprachiasmatic nucleus project directly to the nucleus of Edinger-Westphal. *Proc. Natl. Acad. Sci. USA* 79, 3891–3895.

Gapp, D. A., Kenny, M. P., and Polak, J. M. (1985). The gastro-entero-pancreatic system of the turtle, *Chrysemys picta*. *Peptides* 6, 347–352.

Garcia-Ayala, A., Lozano, M. T., and Agulleiro, B. (1987). Endocrine pancreas of *Testudo graeca* L. (Chelonia) in summer and winter: an immunocytochemical and ultrastructural study. *Gen. Comp. Endocrinol.* 68, 235–248.

Gaudino, G., and Fasolo, A. (1984). Immunohistochemical localization of substance P-like immunoreactivity in the hypothalamus of the lizard, *Podarcis sicula*, R. *Gen. Comp. Endocrinol.* 56, 32–40.

Gayton, R. J. (1982). Mammalian neuronal actions of FMRF amide and the structurally related opioid met-enkephalin-Arg6-Phe7. *Nature* (London) 298, 275–276.

Georges, D., and Dubois, M. P. (1984). Methionine-enkephalin-like immunoreactivity in the nervous ganglion and ovary of a protochordate, *Ciona intestinalis*. *Cell Tiss. Res.* 236, 165–170.

Gesell, M. S., and Callard, I. P. (1972). The hypothalamic-hypophyseal neurosecretory system in the iguanid lizard, *Dipsosaurus dorsalis:* a qualitative and quantitative study. *Gen. Comp. Endocrinol.* 184, 1–13.

Gibbins, I. L. (1983). Peptide-containing nerves in the urinary bladder of the toad, *Bufo marinus*. *Cell Tiss. Res.* 229, 137–144.

Giraud, P., Castanas, E., Patey, G., and Rossier, J. (1983). Regional distribution of methionine-enkephalin-Arg⁶-Phe⁷ in the rat brain: comparative study with the distribution of other opioid peptides. *J. Neurochem.* 41, 154–160.

Goedert, M. (1984). Neurotensin—a status report. *Trends Neurosci.* 7, 3–5.

Goedert, M., and Emson, P. C. (1983). The regional distribution of neurotensin-like immunoreactivity in central and peripheral tissues of the cat. *Brain Res.* 272, 291–297.

Gold, M. R., and Finger, T. E. (1982). Localization of enkephalin-like immunoreactivity in the brain of the lamprey. *Soc. Neurosci. Abs.* 7, 85.

Goldstein, A., Tachibana, S., Lowney, L. I., Hunkapillar, M., and Hood, L. (1979). Dynorphin-(1-13), an extraordinarily potent opioid peptide. *Proc. Natl. Acad. Sci. USA* 77, 6207–6210.

Gomori, G. (1941). Observations with differential stains on human islets of Langerhans. *Am. J. Pathol.* 17, 395–406.

Gonzalez, C. B., and Rodriguez, E. M. (1980). Ultrastructure and immunocytochemistry of neurons in the supraoptic and paraventricular nuclei of the lizard *Liolaemus cyanogaster*. *Cell Tiss. Res.* 207, 463–477.

Gonzalez, G. C., and Lederis, K. (1988). Sauvagine-like and corticotropin-releasing factor-like immunoreactivity in the brain of the bullfrog (*Rana catesbeiana*). *Cell Tiss. Res.* 253, 29–37.

Goodman, E. C., and Iversen, L. L. (1986). Calcitonin gene-related peptide: Novel neuropeptide. *Life Sci.* 38, 2169–2178.

Goodman, R. H., Jacobs, J. H., Dee, P. C., and Haber, J. F. (1982). Somatostatin-28 encoded in a cloned cDNA obtained from a rat medullary thyroid carcinoma. *J. Biol. Chem.* 257, 1156–1159.

Goos, H. J. Th., de Leeuw, R., De Zoeten-Kamp, C., Peute, J., and Blahser, S. (1985). Gonadotropin releasing hormone-immunoreactive structures in the brain and pituitary of the African catfish, *Clarias gariepinus* (Burchell). *Cell Tiss. Res.* 241, 593–596.

Goosens, N., Dierickx, K., and Vandesande, F. (1979). Immunocytochemical localization of vasotocin and mesotocin in the hypothalamus of lacertilian reptiles. *Cell Tiss. Res.* 200, 223–227.

Goosens, N., Dierickx, K., and Vandesande, F. (1980). Immunocytochemical localization of somatostatin in the brain of the lizard, *Ctenosaura pectinata*. *Cell Tiss. Res.* 208, 499–505.

Gorbman, A., Dickhoff, W. W., Vigna, S. R., Clark, N. B., and Ralph, C. L. (1983). *Comparative Endocrinology*. John Wiley and Sons, New York.

Greeley, G. H., Jr., Partin, M., Spannagel, A., Dinh, T., Hill, F. L. C., Trowbridge, J., Salter, M., Chuo, H. F., and Thompson, J. C. (1986). Distribution of bombesin-like peptides in the alimentary canal of several vertebrate species. *Regulatory Peptides* 16, 169–181.

Greeley, G. H., Jr., Trowbridge, J., Burdett, J., Hill, F. L. C., Spannagel, A., and Thompson, J. C. (1984). Radioimmunoassay of pancreatic polypeptide in mammalian and submammalian vertebrates using a carboxyl-terminal hexapeptide antiserum. *Regulatory Peptides* 8, 177–187.

Greer, M. A. (1957). Studies on the influence of the central nervous system on physiological above: these peptides include was most anterior pituitary function. *Recent Progr. Hormone Res.* 13, 67–104.

Grim-Jorgenson, Y. (1978). Immunoreactive thyrotropin-releasing factor in a gastropod: distribution in the central nervous system and hemolymph of *Lymnaea stagnalis*. *Gen. Comp. Endocrinol.* 35, 387–390.

Grimmelikhuijzen, C. J. P., Balfre, A., Emson, P. C., Powell, D., and Sundler, F. (1981). Substance P-like immunoreactivity in the nervous system of hydra. *Histochemistry* 71, 325–333.

Grimmelikhuijzen, C. J. P., Dockray, G. J., and Schot, L. P. C. (1982). FMRFamide-like immunoreactivity in the nervous system of hydra. *Histochemistry* 74, 499–508.

Grimmelikhuijzen, C. J. P., Sundler, F. S., and Rehfeld, J. F. (1980). Gastrin-CCK-like immunoreactivity in the nervous system of coelenterates. *Histochemistry* 69, 61–68.

Guillemin, R., Brazeau, P., Bohlen, P., Esch, F., Ling, N., and Wehrenberg, W. (1982). Growth hormone-releasing factor from a human pancreatic tumor caused acromegaly. *Science* (N.Y.) 218, 585–587.

Guillemin, R., and Rosenberg, B. (1955). Humoral hypothalamic control of anterior pituitary: a study with combined tissue cultures. *Endocrinology* 57, 599–607.

Haber, S., and Elde, R. P. (1982). The distribution of enkephalin immunoreactive fibers and terminals in the monkey central nervous system: an immunohistochemical study. *Neuroscience* 7, 1049–1095.

Haider, S. (1974). The effect of chlorpromazine and morphine on the neurosecretory system of the Indian Wall Lizard *Hemidactylus flaviuridis* (Ruppell). *Z. mikrosk.-anat. Forsch.* 88, 449–454.

Hall, T. R., and Chadwick, A. (1979). Hypothalamic control of prolactin and growth hormone secretion in different vertebrate species. *Gen. Comp. Endocrinol.* 38, 333–342.

Hall, T. R., and Chadwick, A. (1984). Effects of synthetic mammalian thyrotropin releasing hormone, somatostatin and dopamine on secretion of prolactin and growth hormone from amphibian and reptilian pituitary glands incubated *in vitro. J. Endocrinol.* 102, 175–180.

Hall, T. R., Chadwick, A., and Callard, I. P. (1975). Control of prolactin secretion in the terrapin (*Chrysemys picta*). *J. Endocrinol.* 67, 52–54.

Hall, W. C., and Ebner, F. F. (1970). Thalamotelencephalic projections in the turtle (*Pseudemys scripta*). *J. Comp. Neurol.* 140, 101–122.

Halpern, M., Morrell, J. I., and Pfaff, D. W. (1982). Cellular [^{3}H]-estradiol and [^{3}H]-testosterone localization in the brain of garter snakes: an autoradiographic study. *Gen. Comp. Endocrinol.* 46, 211–224.

Hancock, M. B. (1976). Cells of origin of hypothalamo-spinal projections in the rat. *Neurosci. Lett.* 3, 179–184.

Harmar, A. J. (1984). Three tachykinins in mammalian brain. *Trends Neurosci.* 7, 57–60.

Harrington, M. E., Nance, O. M., and Rusak, B. (1984). NPY-like immunoreactivity in the geniculo-suprachiasmatic tract. *Soc. Neurosci. Abs.* 10, 502.

Havrankova, J., Schmechel, D., Roth, J., and Brownstein, M. (1978). Identification of insulin in rat brain. *Proc. Natl. Acad. Sci. USA* 75, 5737–5741.

Hayashi, M., Edgar, D., and Thoenen, H. (1983). The development of substance P, somatostatin and vasoactive intestinal polypeptide in sympathetic and spinal sensory ganglia of the chick embryo. *Neuroscience* 10, 31–39.

Hazelwood, R. L., Turner, S. D., Kimmel, J. R., and Pollock, H. G. (1973). Spectrum effects of a new polypeptide (third hormone) isolated from chicken pancreas. *Gen. Comp. Endocrinol.* 21, 485–496.

Helke, C. J., Neil, J. J., Massari, V. J., and Loewy, A. D. (1982). Substance P neurons project from the ventral medulla to the intermediolateral cell column and ventral horn in the rat. *Brain Res.* 243, 147–152.

Helmstaedter, V., Taugher, C. H., Feurle, G. E., and Forssman, W. G. (1977). Localization of neurotensin-immunoreactive cells in the small intestine of man and various mammals. *Histochemistry* 53, 35–44.

Henke, H. Tschopp, F. A., and Fischer, J. A. (1985). Distinct binding sites for calcitonin gene-related peptide and salmon calcitonin in rat central nervous system. *Brain Res.* 360, 165–171.

Herbert, E., Civelli, O., Douglass, J., Martens, G., and Rosen, H. (1985). Generation of diversity of opioid peptides. In *Biochemical Actions of Hormones,* Vol. XII, Academic Press, New York, pp. 1–36.

Herbert, E., Oates, E., Martens, G., Comb, M., Rosen, H., and Uhler, M. (1983). Generation of diversity and evolution of opioid peptides. *Cold Spring Harbor Symposium on Quantitative Biology* 48, 375–384.

Hobart, P., Crawford, R., Shen, L. P., Pictet, R., and Rutter, W. J. (1980). Cloning and sequence analysis of cDNAs encoding two distinct somatostatin precursors found in the endocrine pancreas of anglerfish. *Nature* (London) 288, 137–141.

Hokfelt, T., Fuxe, K., Johansson, O., Jeffcoate, S., and White, N. (1975). Distribution of thyrotropin-releasing hormone (TRH) in the central nervous system as revealed with immunohistochemistry. *Eur. J. Pharmacol.* 34, 389–392.

Hokfelt, T., Johansson, O., Ljungdahl, A., Lundberg, J. M., and Schultzberg, M. (1980). Peptidergic neurons. *Nature* (London) 284, 515–521.

Hokfelt, T., Lundberg, J. M., Tatemoto, K., Mutt, V., Terenius, L., Polak, J., Bloom, S., Sasek, O., Elde, R., and Goldstein, M. (1983). Neuropeptide Y (NPY)- and FMRFamide neuropeptide-like immunoreactivities in catecholamine neurons of the rat medulla oblongata. *Acta Physiol. Scand.* 117, 315–318.

Holder, F. C., Schroeder, D. M., Pollatz, M., Guerne, J. M., Vivien-Roels, B., Pevet, P., Buijs, R. M., Dogterom, J., and Meiniel, A. (1982). A specific and sensitive bioassay for arginine-vasotocin: description and some applications in lower and higher vertebrates. *Gen. Comp. Endocrinol.* 47, 483–491.

Holmes, R. L., and Ball, J. N. (1974). *The Pituitary Gland: a Comparative Account.* Cambridge University Press, London.

Holmgren, S., and Nilsson, S. (1983). Bombesin-, gastrin/ CCK-, 5-hydroxytryptamine-, neurotensin-, somatostatin-, and VIP-like immunoreactivity and catecholamine fluorescence in the gut of the elasmobranch, *Squalus acanthias. Cell Tiss. Res.* 234, 595–618.

Holmgren, S., Vaillant, C., and Dimaline, R. (1982). VIP-, substance P-, gastrin/CCK-, bombesin-, somatostatin-, and glucagon-like immunoreactivities in the gut of the rainbow trout, *Salmo gairdneri. Cell Tiss. Res.* 223, 141–153.

Holmquist, A. L., Dockray, G. J., Rosenquist, G. L., and Walsh, J. H. (1979). Immunochemical characterization of cholecystokinin-like peptides in lamprey gut and brain. *Gen. Comp. Endocrinol.* 37, 474–481.

Holzer, P. (1988). Local effector functions of capsaicin-sensitive sensory nerve endings: involvement of tachykinins, calcitonin gene-related peptide and other neuropeptides. *Neuroscience* 24, 739–768.

Hopson, J. A. (1979). Paleoneurology. In *Biology of the Reptilia. Neurology A,* Vol. 9 (C. Gans, R. G. Northcutt, and P. Ulinski, eds.). Academic Press, New York, pp. 39–146.

Hughes, J., Kosterlitz, H. W., and Smith, T. W. (1977). The distribution of methionine-enkephalin and leucine-enkephalin in the brain and peripheral tissues. *Br. J. Pharmacol.* 61, 639–647.

Hughes, J., Smith, T. W., Kosterlitz, H. W., Forthergill, L. A., Morgan, B. A., and Morris, H. R. (1975). Identification of two related pentapeptides from the brain with potent opiate agonist activity. *Nature* (London) 258, 577–579.

Hunt, S. P., Emson, P. C., Gilbert, R., Goldstein, M., and Kimmel, J. R. (1981). Presence of avian pancreatic polypeptide-like immunoreactivity in catecholamine- and methionine-enkephalin-containing neurons within the central nervous system. *Neurosci. Lett.* 21, 125–130.

Hunter, J. C., and Maggio, J. E. (1984). A pharmacological study with substance K: evidence for multiple types of tachykinin receptors. *Eur. J. Pharmacol.* 195, 149–153.

Ibata, Y., Fukui, K., Okamura, H., Kawakami, T., Tanaka, M., Obata, H. L., Tsuto, T., Terubayashi, H., Yanaihara, C., and Yanaihara, N. (1983). Coexistence of dopamine and neurotensin in hypothalamic arcuate and periventricular neurons. *Brain Res.* 269, 177–179.

Ibata, Y., Watanabe, K., Kinoshita, H., Kibo, S., and Sano, Y. (1979). The location of LHRH neurons in the rat hypothalamus and their pathways to the median eminence. *Cell Tiss. Res.* 198, 381–395.

Inagaki, S., Senba, E., Shiosaka, S., Takagi, H., Kawai, Y., Takatsuki, K., Sanaka, M., Matsuzaki, T., and Tohyama, M. (1981a). Regional distribution of substance P-like immunoreactivity in the frog brain and spinal cord: immunohistochemical analysis. *J. Comp. Neurol.* 201, 243–254.

Inagaki, S., Shiosaka, S., Takatsuki, K., Sakanaka, M., Takagi, H., Senba, E., Matsuzaki, T., and Tohyama, M. (1981b). Distribution of somatostatin in the frog brain, *Rana catesbeiana*, in relation to location of catecholamine-containing neuron system. *J. Comp. Neurol.* 202, 89–101.

Ino, T., Itoh, K., Sugimoto, T., Kaneko, T., Kamiya, H. and Mizuno, N. (1988). The supramammillary region of the cat sends substance P-like immunoreactive axons to the hippocampal formation and the entorhinal cortex. *Neurosci. Lett.* 90, 259–264.

Isayama, T., and Eldred, W. D. (1988). Neuropeptide Y-immunoreactive amacrine cells in the retina of the turtle *Pseudemys scripta elegans. J. Comp. Neurol.* 271, 56–66.

Iversen, L. L., Iversen, S. D., Bloom, F. E., Douglas, C., Brown, M., and Vale, W. (1978). Calcium-dependent release of somatostatin and neurotensin from the brain *in vitro. Nature* (London) 273, 161–163.

Jackson, I. M. D. (1978). Extrahypothalamic and phylogenetic distribution of hypothalamic peptides. In *The Hypothalamus* (S. Reichlin, R. J. Baldessarini, and J. B. Martin, eds.). Raven Press, New York, pp. 217–231.

Jackson, I. M. D. (1981). Evolutionary significance of the phylogenetic distribution of the mammalian hypothalamic releasing hormones. *Fed. Proc.* 40, 2545–2552.

Jackson, I. M. D., Bolaffi, J. L., and Guillemin, R. (1980). Presence of immunoreactive beta-endorphin and enkephalin-like material in the retina and other tissues of the frog, *Rana pipiens. Gen. Comp. Endocrinol.* 42, 505–508.

Jackson, I. M. D., and Reichlin, S. (1974). Thyrotropin-releasing hormone (TRH): distribution in hypothalamic and extrahypothalamic brain tissues of mammalian and submammalian chordates. *Endocrinology* 95, 854–862.

Jackson, I. M. D., and Reichlin, S. (1977). Brain thyrotropin-releasing hormone is independent of the hypothalamus. *Nature* (London) 267, 853–854.

Jacobowitz, D. M., and Olschowka, J. A. (1982). Bovine pancreatic polypeptide-like immunoreactivity in brain and peripheral nervous system: coexistence with catecholaminergic nerves. *Peptides* 3, 569–590.

Jan, Y. N., and Jan, L. Y. (1983). Co-existence and co-release of cholinergic and peptidergic transmitters in frog sympathetic ganglia. *Fed. Proc.* 42, 2929–2933.

Jessell, T. M., Emson, P. C., Paxinos, G., and Cuello, A. C. (1978). Topographic projections of substance P and GABA pathways in the striato-pallido-nigral system: a biochemical and immunohistochemical study. *Brain Res.* 152, 487–498.

Johansson, O., Hokfelt, T., Elde, R. P., Schultberg, M., and Terenius, L. (1978). Immunohistochemical distribution of enkephalin neurons. *Adv. Biochem. Psychopharmacol.* 18, 51–70.

Johnson, A. L., Dickerson, R. W., and Advis, J. P. (1984). Comparative ability of chicken and mammalian LHRH to release LH from rooster pituitary cells *in vitro*. *Life Sci.* 34, 1847–1851.

Johnson, D. E., Noe, B. D., and Bauer, G. E. (1982). Pancreatic polypeptide (PP)-like immunoreactivity in the pancreatic islet of the anglerfish (*Lophius americanus*) and the channel catfish (*Ictalurus punctatus*). *Anat. Rec.* 204, 61–67.

Jokura, Y., and Urano, A. (1986). Extrahypothalamic projections of luteinizing hormone-releasing hormone fibers in the brain of the toad, *Bufo japonicus*. *Gen. Comp. Endocrinol.* 62, 80–88.

Jokura, Y., and Urano, A. (1987). Extrahypothalamic projection of immunoreactive vasotocin fibers in the brain of the toad, *Bufo japonicus*. *Zool. Sci.* 4, 675–681.

Jozsa, R., Korf, H.-W., Csernus, V., and Mess, B. (1988). Thyrotropin-releasing hormone (TRH)-immunoreactive structures in the brain of the domestic mallard. *Cell Tiss. Res.* 251, 441–449.

Jozsa, R., and Mess, B. (1982). Immunohistochemical localization of the luteinizing hormone-releasing hormone (LHRH)-containing structures in the central nervous system of the domestic fowl. *Cell Tiss. Res.* 227, 451–458.

Kadota, T., Kishida, R., Goris, R. C., and Kusunoki, T. (1988). Substance P-like immunoreactivity in the trigeminal sensory nuclei of an infrared-sensitive snake, *Agkistrodon blomhoffi*. *Cell Tiss. Res.* 253, 311–317.

Kah, O., Breton, B., Dulka, J. G., Nunez-Rodriguez, J., Peter, R. E., Corrigan, A., Rivier, J. E., and Vale, W. (1986). A reinvestigation of the Gn-RH (gonadotropin-releasing hormone) systems in the goldfish brain using antibodies to salmon Gn-RH. *Cell Tiss. Res.* 244, 327–337.

Kanazawa, I., and Jessell, T. M. (1976). Postmortem changes and the regional distribution of substance P in the rat and mouse nervous system. *Brain Res.* 117, 362–367.

Karten, H. J., Reiner, A., and Brecha, N. C. (1982). Laminar organization and origins of neuropeptides in the avian retina and optic tectum. In *Cytochemical Methods in Neuroanatomy* (S. L. Palay and V. C. Palay, eds.). Alan R. Liss, New York, pp. 189–204.

Kawai, Y., Takami, K., Shiosaka, S., Emson, P. C., Hillyar, C. J., Girgis, S., MacIntyre, I., and Tohyama, M. (1985). Topographic localization of calcitonin gene-related peptide in the rat brain: an immunohistochemical study. *Neuroscience* 15, 747–763.

Kawauchi, H., Tsubokawa, M., Kanezawa, A., and Kitagawa, H. (1980). Occurrence of two different endorphins in the salmon pituitary. *Biochem. Biophys. Res. Commun.* 92, 1278–1288.

Kelly, M. J., and Ronnekleiv, O. K. (1981). Identification of luteinizing hormone-releasing hormone (LHRH) neurons in the arcuate nucleus of hypothalamic slices. *Soc. Neurosci. Abs.* 7, 20.

Keyser, K. T., Karten, H. J., and Cabot, J. B. (1982). Descending hypothalamic and oxytocin/mesotocin projections end preferentially on pulmonary and baroreceptor regions of the nucleus solitarius. *Soc. Neurosci. Abs.* 8, 112.

Khachaturian, H., Dores, R. M., Watson, S. J., and Akil, H. (1984). Beta-endorphin/ACTH immunocytochemistry in the CNS of the lizard, *Anolis carolinensis:* evidence for a major mesencephalic cell group. *J. Comp. Neurol.* 229, 576–584.

Khachaturian, H., Lewis, M. E., Hollt, V., and Watson, S. J. (1983a). Telencephalic enkephalinergic systems in the rat brain. *J. Neurosci.* 3, 844–855.

Khachaturian, H., Lewis, M. E., and Watson, S. J. (1983b). Enkephalin systems in the diencephalon and brain stem of the rat. *J. Comp. Neurol.* 220, 310–320.

Khachaturian, H., Lewis, M. E., and Watson, S. J. (1983c). Co-localization of proenkephalin peptides in rat brain neurons. *Brain Res.* 279, 369–373.

Kilpatrick, D. L., Howells, R. D., Lahn, H. W., and Udenfriend, S. (1983). Evidence for a proenkephalin-like precursor in amphibian brain. *Proc. Natl. Acad. Sci. USA* 80, 5772–5775.

Kim, Y. S., Stumpf, W. E., and Sar, M. (1981). Anatomical distribution of estrogen target neurons in turtle brain. *Brain Res.* 230, 195–204.

Kim, Y. S., Stumpf, W. E., Sar, M., and Martinez-Vargas, M. C. (1978). Estrogen and androgen target cells in the brain of fishes, reptiles and birds: ontogeny and phylogeny. *Am. Zool.* 18, 425–433.

Kimmel, J. R., Maher, M. J., Pollock, H. G., and Vensel, W. H. (1976). Isolation and characterization of reptilian insulin: partial characterization of reptilian insulin; partial amino acid sequence of rattlesnake (*Crotalus atrox*) insulin. *Gen. Comp. Endocrinol.* 28, 320–333.

Kimmel, J. R., Plisetskaya, E. M., Pollock, H. G., Hamilton, J. W., Rouse, J. B., Ebner, K. E., and Rawitch, A. B. (1986). Structure of a peptide from Coho salmon endocrine pancreas with homology to neuropeptide Y. *Biochem. Biophys. Res. Comm.* 141, 1084–1091.

Kimmel, J. R., Pollock, H. G., and Hazelwood, R. C. (1968a). Isolation and characterization of a new pancreatic polypeptide hormone. *J. Biol. Chem.* 250, 9369–9376.

Kimmel, J. R., Pollock, H. G., and Hazelwood, R. L. (1968b). Isolation and characterization of chicken insulin. *Endocrinology* 83, 1323–1330.

King, J. A., and Millar, R. P. (1979a). Heterogeneity of vertebrate luteinizing hormone-releasing hormone. *Science* (N.Y.) 206, 67–69.

King, J. A., and Millar, R. P. (1979b). Phylogenetic and anatomical distribution of somatostatin in vertebrates. *Endocrinology* 105, 1322–1329.

King, J. A., and Millar, R. P. (1980a). Comparative aspects of luteinizing hormone releasing hormone: structure and function in vertebrate phylogeny. *Endocrinology* 106, 707–717.

King, J. A., and Millar, R. P. (1980b). Radioimmunoassay of methionine[5]-enkephalin sulphoxide: phylogenetic and anatomical distribution. *Peptides* 1, 211–216.

King, J. A., and Millar, R. P. (1982a). Structure of chicken hypothalamic luteinizing hormone-releasing hormone. I. Structural determination of partially purified material. *J. Biol. Chem.* 257, 10722–10728.

King, J. A., and Millar, R. P. (1982b). Structure of chicken hypothalamic luteinizing hormone-releasing hormone. II. Isolation and characterization. *J. Biol. Chem.* 257, 10729–10732.

King, J. A., and Millar, R. P. (1986). Identification of His[5], Trp[7], Tyr[8]-GnRH (Chicken GhRH II) in amphibian brain. *Peptides* 7, 827–834.

King, J. C., and Gerall, A. A. (1976). Localization of luteinizing hormone-releasing hormone. *J. Histochem. Cytochem.* 24, 829–845.

King, J. C., Sower, S. A., and Anthony, E. L. P. (1988). Neuronal systems immunoreactive with antiserum to lamprey gonadotropin-releasing hormone in the brain of *Petromyzon marinus. Cell Tiss. Res.* 253, 1–8.

Klein, C., and Van Noorden, S. (1980). Pancreatic polypeptide (PP) and glucagon cells in the pancreatic islet of *Xiphophorus hetleri* H. (Teleostei). *Cell Tiss. Res.* 205, 187–198.

Knigge, K. M., and Joseph, S. A. (1984). Anatomy of opioid-systems of the brain. *Can. J. Neurol. Sci.* 11, 14–23.

Knigge, K. M., and Piekut, D. T., (1985). Distribution of CRF- and tyrosine hydroxylase-immunoreactive neurons in the brainstem of the domestic fowl (*Gallus domesticus*). *Peptides* 6, 97–101.

Kobayashi, R. M., Brown, M., and Vale, W. (1977). Regional distribution of neurotensin and somatostatin in the nonlesioned rat. *Brain Res.* 126, 167–176.

Korte, G. E., Reiner, A., and Karten, H. J. (1980). Substance P-like immunoreactivity in cerebellar mossy fibers and terminals in the red-eared turtle, *Chrysemys scripta elegans. Neuroscience* 5, 903–914.

Krause, J. E., Chirgwin, J. M., Carter, M. S., Xu, Z. S. and Hershey, A. D. (1987). Three rat preprotachykinin mRNAs encode the neuropeptides substance P and neurokinin A. *Proc. Natl. Acad. Sci. USA* 84, 881–885.

Krisch, B. (1979). Hypothalamic and extrahypothalamic distribution of somatostatin-immunoreactive elements in the rat brain. *Cell Tiss. Res.* 195, 499–513.

Kuljis, R. O., and Karten, H. J. (1982). Laminar organization of peptide-like immunoreactivity in the anuran optic tectum. *J. Comp. Neurol.* 212, 188–201.

Kuljis, R. O., and Karten, H. J. (1983). Modifications in the laminar organization of peptide-like immunoreactivity in the anuran optic tectum following retinal deafferentation. *J. Comp. Neurol.* 217, 239–251.

Kuljis, R. O., and Karten, H. J. (1984). Substance P and leucine-enkephalin in ganglion cell axons in the anuran retina. *Soc. Neurosci. Abs.* 10, 458.

Kuljis, R. O., and Karten, H. J. (1988). Neuroactive peptides as markers of retinal ganglion cell populations that differ in anatomical organization and function. *Visual Neurosci.* 1, 73–81.

Kunzle, H. (1982). Dorsal root projections to the cerebellum in turtle. *Exp. Brain Res.* 45, 464–466.

Kunzle, H., and Schnyder, H. (1983). Do retinal and spinal projections overlap within the turtle thalamus? *Neuroscience* 10, 161–168.

Kuwayama, Y. Ishimoto, I., Fukuda, M., Shiosaka, Y. S. S., Inagaki, S.,

Senba, E., Sakanaka, M., Takagi, H., Takatsuki, K., Hara, Y., Kawai, Y., and Tohyama, M. (1982). Over all distribution of glucagon-like immunoreactivity: an immunohistochemical study with flat-mounts. *Invest. Ophthalmol. Vis. Sci.* 22, 681–686.

Lamour, Y., Dutar, P., and Jobert, A. (1983). Effects of neuropeptides on rat cortical neurons: laminar distribution and interaction with effect of acetylcholine. *Neuroscience* 10, 107–117.

Lance, V. A., Hamilton, J. W., Rouse, J. B., Kimmel, J. R., and Pollock, H. G. (1984). Isolation and characterization of reptilian insulin, glucagon and pancreatic polypeptide: complete amino acid sequence of Alligator (*Alligator mississippiensis*) insulin and pancreatic polypeptide. *Gen. Comp. Endocrinol.* 55, 112–124.

Lance, V. A., Vliet, K. A., and Bolaffi, J. L. (1985). Effect of mammalian luteinizing hormone-releasing hormone on plasma testosterone in male alligators, with observations on the nature of alligator hypothalamic gonadotropin-releasing hormone. *Gen. Comp. Endocrinol.* 60, 138–143.

Langer, M., Van Noorden, S., Polak, J. M., and Pearse, A. G. E. (1979). Peptide hormone-like immunoreactivity in the gastrointestinal tract and endocrine pancreas of eleven teleost species. *Cell Tiss. Res.* 199, 493–508.

Larhammar, D., Ericsson, A., and Persson, H. (1987). Structure and expression of rat neuropeptide Y gene. *Proc. Natl. Acad. Sci. USA* 84, 2068–2072.

Larsen, B. A., and Vigna, S. R. (1983). Gastrin/cholecystokinin-like immunoreactive peptides in the Dungeness crab, *Cancer magister* (Dana): immunochemical and biological characterization. *Regulatory Peptides* 7, 155–170.

Larsson, L. I., and Rehfeld, J. F. (1977). Evidence for a common evolutionary origin of gastrin and cholecystokinin. *Nature* (London) 269, 335–338.

LaValley, A. L., and Ho, R. H. (1983). Substance P, somatostatin and methionine-enkephalin immunoreactive elements in the spinal cord of the domestic fowl, *Gallus domesticus. J. Comp. Neurol.* 213, 406–413.

Lazar, G., Merchenthaler, I., and Maderdrut, J. L. (1990). Some enkephalinergic pathways in the brain of *Rana esculenta:* an experimental analysis. *Brain Res.* 521, 238–246.

LeBoulenger, F., Leroux, P., Tonon, M. C., Coy, D. H., Vaudry, H., and Pelletier, G. (1983). Coexistence of vasoactive intestinal peptide and enkephalins in the adrenal chromaffin granules of the frog. *Neurosci. Lett.* 37, 221–225.

Lechago, J., Holmquist, A. L., Rosenquist, G. L., and Walsh, J. H. (1978). Localization of bombesin-like peptides in frog gastric mucosa. *Gen. Comp. Endocrinol.* 36, 553–558.

Lembeck, F., Bernatzky, G., Gamse, R., and Saria, A. (1985). Characterization of substance P-like immunoreactivity in submammalian species by high performance liquid chromatography. *Peptides* 6, 231–236.

Leroux, P., Tonon, M. C., Stoeckel, M. E., Jegou, S., Leboulenger, F., Delarue, C., Perroteau, I., Netchitailo, P. P., Kupryszewski, C., and Vaudry, H. (1983). Role of TRH in the control of melatropin release in amphibians. In *Thyrotropin-Releasing Hormones* (E. C. Griffiths and G. W. Bennett, eds.). Raven Press, New York, pp. 229–240.

Lewis, R. V., and Erickson, B. W. (1986). Evolution of proenkephalin and pro-dynorphin. *Am. Zool.* 26, 1027–1032.

Li, C. H., Barnafi, L., Chretien, M., and Chung, D. (1965). Isolation and structure of beta-lipotropin from sheep pituitary glands. *Excerpta Med.* 3, 111–112.

Licht, P. (1979). Reproductive endocrinology of reptiles and amphibians: gonadotropins. *Ann. Rev. Physiol.* 41, 337–351.

Licht, P. (1980). Evolutionary and functional aspects of pituitary gonadotropins in the green turtle, *Chelonia midas. Am. Zool.* 20, 565–574.

Licht, P., and Denver, R. J. (1988). Effects of TRH on hormone release from pituitaries of the lizard, *Anolis carolinensis. Gen. Comp. Endocrinol.* 70, 355–362.

Licht, P., Millar, R., King, J. A., McCreery, B. R., Mendonca, M. T., Bona-Gallo, A., and Lofts, B. (1984). Effects of chicken and mammalian gonadotropin-releasing hormones (GnRH) on *in vivo* pituitary gonadotropin release in amphibians and reptiles. *Gen. Comp. Endocrinol.* 54, 89–96.

Licht, P., Owens, D. W., Clifton, K., and Penaflores, C. (1982). Changes in LH and progesterone associated with the nesting cycle and ovulation in olive Ridley sea turtle, *Lepidochelys olivacea. Gen. Comp. Endocrinol.* 48, 247–253.

Licht, P., Pickering, B. T., Papkoff, H., Pearson, A., and Bona-Gallo, A. (1984). Presence of a neurophysin-like precursor in the green turtle (*Chelonia mydas*). *J. Endocrinol.* 103, 97–106.

Licht, P., and Porter, D. A. (1985a). LH secretion in response to gonadotropin releasing hormone (GnRH) by superfised pituitaries from two species of turtles. *Gen. Comp. Endocrinol.* 59, 442–448.

Licht, P., and Porter, D. A. (1985b). *In vivo* and *in vitro* responses to gonadotropin releasing hormone in the turtle, *Chrysemys picta,* in relation to sex and reproductive stage. *Gen. Comp. Endocrinol.* 60, 75–85.

Licht, P., and Porter, D. A. (1987). Specificity of amphibian and reptilian pituitaries for various species of gonadotropin releasing hormones *in vitro. Gen. Comp. Endocrinol.* 66, 248–255.

Lin, T. M., and Chance, R. E. (1974). Bovine pancreatic polypeptide (BPP) and avian pancreatic polypeptide (APP). *Gastroenterology* 67, 737–738.

Lindberg, I. (1986). On the evolution of proenkephalin. *Trends Pharmacol. Sci.* 7, 216–217.

Lindberg, I., and White, L. (1986). Reptilian enkephalins: implications for the evolution of proenkephalin. *Archives Biochem. Biophys.* 245, 1–7.

Lisk, R. D. (1967). Neural control of gonad size by hormone feedback in the desert iguana *Dipsosaurus dorsalis dorsalis. Gen. Comp. Endocrinol.* 8, 258–266.

Loewi, O. (1921). Über humorale Übertragbarkeit der Herzennervenwirkung. I. Mitteilung. *Pfluger's Arch. Ges. Physiol.* 189, 239–242.

Loh, Y. P. (1979). Immunological evidence for two common precursors to corticotropins, endorphins and melanotropins in the neurointermediate lobe of the toad pituitary. *Proc. Natl. Acad. Sci. USA* 76, 796–800.

Lopez, J., Echevarria, M., and Vazquez, J. J. (1988). Histological and immunocytochemical study of the endocrine pancreas of the lizard *Podarcis hispanica* Steindachner, 1870 (Lacertidae). *Gen. Comp. Endocrinol.* 71, 212–228.

Lopez, L. C., Li, W. H., Frazier, M. L., Luo, C. C., and Saunders, G. F. (1984). Evolution of glucagon genes. *Mol. Biol. Evol.* 1, 335–344.

Loren, I., Alumets, J., Hakanson, R., and Sundler, F. (1979a). Immunoreactive pancreatic polypeptide (PP) occurs in the central and peripheral nervous system: preliminary immunocytochemical observations. *Cell Tiss. Res.* 200, 179–186.

Loren, I., Emson, P. C., Fahrenkrug, J., Bjorkland, A., Alumets, J., Hakanson, R., and Sundler, F. (1979b). Distribution of vasoactive intestinal polypeptide in the rat and mouse brain. *Neuroscience* 4, 1953–1976.

Lorez, H. P., and Kemali, M. (1981). Substance P-, met-enkephalin and somatostatin-like immunoreactivity distribution in the frog spinal cord. *Neurosci. Lett.* 26, 119–124.

Lund, P. K., Goodman, R. H., and Habener, J. F. (1981). Pancreatic preproglucagons are encoded by two separate mRNAs. *J. Biol. Chem.* 256, 6515–6518.

Lundberg, J. M., Anggard, A., and Fahrenkrug, J. (1982). Complementary role of vasoactive intestinal polypeptide (VIP) and acetylcholine for cat submandibular gland blood flow and secretion. *Acta Physiol. Scand.* 114, 329–337.

Lundberg, J. M., and Hokfelt, T. (1983). Coexistence of peptides and classical neurotransmitters. *Trends Neurosci.* 6, 325–333.

Lundberg, J. M., Hokfelt, T., Schultzberg, M., Uvnas-Wallenstein, K., Kohler, C., and Said, S. I. (1979). Occurrence of vasoactive intestinal polypeptide (VIP)-like immunoreactivity in certain cholinergic neurons of the cat: evidence from combined immunohistochemistry and acetylcholinesterase staining. *Neuroscience* 4, 1539–1561.

Lundberg, J. M., Rokaeus, A., Hokfelt, T., Rosell, S., Brown, M., and Goldstein, M. (1982). Neurotensin-like immunoreactivity in the preganglionic nerves and in the adrenal medulla of the cat. *Acta Physiol. Scand.* 114, 153–155.

Lundberg, J. M., and Tatemoto, K. (1982). Pancreatic polypeptide family (APP, BPP, NPY and PYY) in relation to sympathetic vasoconstriction resistant to alpha-adrenergic blockade. *Acta Physiol. Scand.* 116, 393–402.

Lundberg, J. M., Terenius, L., Hokfelt, T., and Goldstein, M. (1983). High levels of neuropeptide Y in peripheral noradrenergic neurons in various mammals including man. *Neurosci. Lett.* 42, 167–172.

Lundberg, J. M., Terenius, L., Hokfelt, T., Martling, C. R., Tatemoto, K., Mutt, V., Polak, J., Bloom, S. R., and Goldstein, M. (1982). Neuropeptide Y (NPY)-like immunoreactivity in peripheral noradrenergic neurons and effects of NPY on sympathetic function. *Acta Physiol. Scand.* 116, 447–480.

Lundberg, J. M., Terenius, L., Hokfelt, T., and Tatemoto, K. (1984). Comparative immunohistochemical and biochemical analysis of pancreatic polypeptide-like peptides with special reference to presence of neuropeptide Y in central and peripheral neurons. *J. Neurosci.* 4, 2376–2386.

MacKenzie, D. S., Licht, P., and Papkott, H. (1981). Purification of thyrotropin from the pituitaries of two turtles: the green sea turtle and the snapping turtle. *Gen. Comp. Endocrinol.* 45, 39–48.

MacNamee, M. C., Sharp, P. J., Lea, R. W., Sterling, R. J., and Harvey, S. (1986). Evidence that vasoactive intestinal polypeptide is a physiological prolactin-releasing factor in the bantam hen. *Gen. Comp. Endocrinol.* 62, 470–478.

Maderdrut, J. L., Merchenthaler, I., Sundberg, D. K., Okado, N., and Oppenheim, R. W. (1986). Distribution and development of proenkephalin-like immunoreactivity in the lumbar spinal cord of the chicken. *Brain Res.* 377, 29–40.

Magazin, M., Minth, C. D., Funckes, C. L., Deschenes, R., Tavianini, M. A., and Dixon, J. E. (1982). Sequence of a cDNA encoding pancreatic presomatostatin-22. *Proc. Natl. Acad. Sci. USA* 79, 5152–5156.

Maggio, J. E. (1988). Tachykinins. *Ann. Rev. Neurosci.* 11, 13–28.

Maggio, J. E., and Hunter, J. C. (1984). Regional distribution of kassinin-like immunoreactivity in rat central and peripheral tissues and the effect of capsaicin. *Brain Res.* 307, 370–373.

Maggio, J. E., Hunter, J. C., Sandberg, B. E. B., Iversen, L. L., and Hanley, M. R. (1983). Substance K: a novel tachykinin in mammalian CNS. *Soc. Neurosci. Abs.* 9, 17.

Magnen, J., Paterson, S. J., and Kosterlitz, H. (1982). The interaction of [Met^5]enkephalin and [Leu^5]enkephalin sequences, extended at the C-terminus, with mu, delta and kappa-binding sites in the guinea pig brain. *Life Sci.* 31, 1359–1361.

Mains, R. E., Eipper, B. A., and Ling, N. (1977). Common precursor to corticotropins and endorphins. *Proc. Natl. Acad. Sci. USA* 74, 3014–3018.

Marchant, T. A., Fraser, R. A., Andrews, P. C., and Peter, R. E. (1987). The influence of mammalian and teleost somatostatins on the secretion of growth hormone from goldfish (*Carrasius auratus* L.) pituitary fragments *in vitro. Regulatory Peptides* 17, 41–52.

Markussen, J., Frandson, E., Heding, L. G., and Sundby, F. (1972). Turkey glucagon: crystallization, amino acid composition and immunology. *Horm. Metab. Res.* 4, 360–363.

Marshak, D., and Yamada, T. (1984). Characterization of somatostatin-like immunoreactivity in vertebrate retinas. *Invest. Ophthalmol. Vis. Sci.* 25, 112–115.

Martel, J. C., St.-Pierre, S., and Quirion, R. (1986). Neuropeptide Y receptors in rat brain: autoradiographic localization. *Peptides* 7, 55–60.

Martens, G. J. M., and Herbert, E. (1984). Polymorphism and absence of leu-enkephalin sequences in pro-enkephalin genes in *Xenopus laevis. Nature* (London) 310, 251–254.

Martin, R., Geis, R., Holl, R., Schafer, M., and Voight, K. H. (1983). Co-existence of unrelated peptides in oxytocin and vasopressin terminals of rat neurohypophysis: immunoreactive methionine^5-enkephalin-, leucine^5-enkephalin- and cholecystokinin-like substances. *Neuroscience* 8, 213–227.

Martin, W. R. (1984). Pharmacology of opioids. *Pharmacological Rev.* 35, 283–323.

Martinez-Vargas, M. C., Keefer, D. A., and Stumpf, W. E. (1978). Estrogen localization in the brain of the lizard. *J. Exp. Zool.* 205, 141–147.

Matsuo, H., Baba, Y., Nair, R. M. G., Arimura, A., and Schally, A. V. (1971). Structure of the porcine LH- and FSH-releasing hormone. I. The proposed amino acid sequence. *Biochem. Biophys. Res. Commun.* 43, 1334–1339.

McDonald, J. K., Parnevelas, J. G., Karamanlidis, A. N., and Brecha, N. (1982a). The morphology and distribution of peptide-containing neurons in the adult and developing visual cortex of the rat. II. Vasoactive intestinal polypeptide. *J. Neurocytol.* 11, 825–837.

McDonald, J. K., Parnevelas, J. G., Karamanlidis, A. N., and Brecha, N. (1982b). The morphology and distribution of peptide-containing neurons in the adult and developing visual cortex of the rat. IV. Avian pancreatic polypeptide. *J. Neurocytol.* 11, 985–995.

McDonald, J. K., Parnevelas, J. G., Karamanlidis, A. N., Rosenquist, G., and Brecha, N. (1982c). The morphology and distribution of peptide-containing neurons in the adult and developing visual cortex of the rat. III. Cholecystokinin. *J. Neurocytol.* 11, 881–895.

McDonald, T. J., Jornvall, H., Nilsson, G., Vagne, M., Ghatei, M. A., Bloom, S. R., and Mutt, V. (1979). Characterization of a gastrin releasing peptide from porcine I non-antral gastric tissue. *Biochem. Biophys. Res. Commun.* 90, 227–233.

Merchenthaler, I., Lazar, G., and Maderdrut, J. L. (1989). Distribution of proenkephalin-derived peptides in the brain of *Rana esculenta*. *J. Comp. Neurol.* 281, 23–39.

Merchenthaler, I., Maderdrut, J. L., Lazar, G., Gulyas, J., and Petrusz, P. (1987). Immunocytochemical analysis of proenkephalin-derived peptides in the amphibian hypothalamus and optic tectum. *Brain Res.* 416, 219–227.

Merchenthaler, I., Vigh, S., Petrusz, P., and Schally, A. V. (1982). Immunocytochemical localization of corticotropin-releasing factor (CRF) in the rat brain. *Am. J. Anat.* 165, 385–396.

Miller, R. J., Chang, K. J., Cooper, B., and Cuatrecasas, P. (1978). Radioimmunoassay and characterization of enkephalins in rat tissue. *J. Biol. Chem.* 253, 531–538.

Mimnagh, K., Bolaffi, J. L., Montgomery, N. M., and Kaltenbach, J. C. (1987). Thyrotropin-releasing hormone (TRH): immunohistochemical distribution in tadpole and frog brain. *Gen. Comp. Endocrinol.* 66, 394–404.

Minamino, N., Kangawa, K., and Matsuo, H. (1984). Neuromedin N: a novel neurotensin-like peptide identified in porcine spinal cord. *Biochem. Biophys. Res. Comm.* 122, 542–549.

Miyamoto, K., Hasegawa, Y., Nomura, M., Igarashi, M., Kangawa, K., and Matsuo, H. (1984). Identification of the second gonadotropin-releasing hormone in chicken hypothalamus: evidence that gonadotropin secretion is probably controlled by two distinct gonadotropin-releasing hormones in avian species. *Proc. Natl. Acad. Sci. USA* 81, 3874–3878.

Moody, T. W., O'Donohue, T. L., and Jacobowitz, D. M. (1981). Biochemical localization and characterization of bombesin-like peptides in discrete regions of rat brain. *Peptides* 2, 75–79.

Moody, T. W., Pert, C. B., and Jacobowitz, D. M. (1979). Bombesin-like peptides: regional distribution in rat brain. In *Peptides—Structure and Biological*

Function. Proceedings of the 6th American Peptide Symposium (E. Gross and J. Meienhofer, eds.). Pierce Chemical Co., Rockford, Ill., pp. 865–868.

Moody, T. W., Pert, C. B., Rivier, J., and Brown, M. R. (1978). Bombesin: specific binding to rat brain membranes. *Proc. Natl. Acad. Sci. USA* 75, 5372–5376.

Moody, T. W., Thoa, N. B., O'Donohue, T. L., and Pert, C. B. (1980). Bombesin-like peptides in rat brain: localization in synaptosomes and release from hypothalamic slices. *Life Sci.* 26, 1707–1712.

Morrel, J. I., Crews, D., Ballin, A., Morgentaler, A., and Pfaff, D. (1979). 3H Estradiol, 3H-testosterone and 3H-dihydrotestosterone localization in the brain of the lizard *Anolis carolinensis:* an autoradiographic study. *J. Comp. Neurol.* 188, 201–224.

Morrel, J. I., and Pfaff, D. W. (1978). A neuroendocrine approach to brain function: localization of sex steroid concentrating cells in vertebrate brains. *Am. Zool.* 18, 447–460.

Mufson, E. J., Desan, P. H., Mesulam, M. M., Wainer, B. H., and Levey, A. I. (1984). Choline acetyltransferase-like immunoreactivity in the forebrain of the red-eared pond turtle (*Pseudemys scripta elegans*). *Brain Res.* 323, 103–108.

Munz, H., Claas, B., Stumf, W. E., and Jennes, L. (1982). Centrifugal innervation of the retina by luteinizing hormone releasing hormone (LHRH)-immunoreactive telencephalic neurons in teleostean fishes. *Cell Tiss. Res.* 222, 313–323.

Mutt, V., and Jorpes, J. E. (1967). Contemporary developments in the biochemistry of the gastrointestinal hormones. *Rec. Progr. Hormone Res.* 23, 483–503.

Mutt, V., and Jorpes, J. E. (1968). Structure of porcine cholecystokinin-pancreozymin. *Eur. J. Biochem.* 6, 156–162.

Naik, D. R., Sar, M., and Stumpf, W. E. (1980). Immunohistochemical identification of cells in pars distalis of the pituitary of the lizard *Anolis carolinensis. Histochemistry* 69, 19–26.

Naik, D. R., Sar, M., and Stumpf, W. E. (1981). Immunohistochemical localization of enkephalin in the central nervous system and pituitary of the lizard, *Anolis carolinensis. J. Comp. Neurol.* 198, 583–601.

Nair, R. M. G., Barret, J. F., Bowers, C. Y., and Schally, A. V. (1970). Structure of porcine thyrotropin releasing hormone. *Biochemistry* 9, 1103–1106.

Nakajima, T. (1981). Active peptides in amphibian skin. *Trends Pharmacol. Sci.* 2, 202–205.

Naude, R. J., and Oelofsen, O. (1981). Isolation and characterization of beta-lipotropin from the pituitary gland of the ostrich, *Struthio camelus. Int. J. Pept. Protein Res.* 18, 135–137.

Nauta, W. J. H., and Karten, H. J. (1970). A general profile of the vertebrate brain with sidelights on the ancestry of the cerebral cortex. In *The Neurosciences, Second Study Program* (G. C. Quarton, T. Melnechek, and F. O. Schmitt, eds.). Rockefeller University Press, New York, pp. 7–26.

Nawa, H., Hiroshi, T., Takashima, H., Inayama, S., and Nakanishi, S. (1983). Nucleotide sequences of cloned cDNAs for two types of bovine brain substance P precursor. *Nature* (London) 306, 32–36.

Nawa, H., Kotani, H., and Nakanishi, S. (1984). Tissue-specific generation of two preprotachykinin mRNAs from one gene by alternative RNA splicing. *Nature* (London) 312, 729–734.

Nemeroff, C. B., Luttinger, D., and Prange, A. J., Jr. (1983). Neurotensin and bombesin. In *Handbook of Psychopharmacology*, Vol. 16 (L. Iversen, S. D. Iversen, and S. H. Snyder, eds.). Plenum Press, New York, pp. 363–466.

Nicoll, R. A. (1982). Response of central neurons to opiates and opioid peptides. In *Regulatory Peptides: from Molecular Biology to Function* (E. Costa and M. Trabucchi, eds.). Raven Press, New York, pp. 337–345.

Nicoll, R. A., Alger, B. E., and Jahr, C. E. (1980). Peptides as putative excitatory neurotransmitters: carnosine, enkephalin, substance P and TRH. *Proc. Roy. Soc. London* 210, 133–149.

Nicoll, R. A., Schenker, C., and Leeman, S. (1980). Substance P as a transmitter-candidate. *Ann. Rev. Neurosci.* 3, 227–268.

Nishimura, S., and Pasternak, G. W. (1982). Opiate and opioid peptide binding in rat and goldfish: further evidence for opiate receptor heterogeneity. *Brain Res.* 248, 192–195.

Noaillac-Depeyre, J., and Hollande, E. (1981). Evidence for somatostatin, gastrin and pancreatic polypeptide-like substances in the mucosa cells of the gut in fishes with and without stomach. *Cell Tiss. Res.* 216, 193–203.

Noe, B. D., McDonald, J. K., Greiner, F., and Wood, J. G. (1986). Anglerfish islets contain NPY immunoreactive nerves and produce the NPY analog aPY. *Peptides* 7, 147–154.

Nojiri, H., Ishida, I., Miyashita, E., Sato, M., Urano, A., and Deguchi, T. (1987). Cloning and sequence analysis of cDNAs for neurohypophysial hormones vasotocin and mesotocin for the hypothalamus of toad, *Bufo japonicus*. *Proc. Natl. Acad. Sci. USA* 84, 3043–3046.

North, R. A., and Egan, T. M. (1982). Electrophysiology of peptides in the peripheral nervous system. *Br. Med. Bull.* 38, 291–296.

Northcutt, R. G. (1978). Forebrain and midbrain organization in lizards and its phylogenetic significance. In *Behavior and Neurology of Lizards* (N. Greenberg and P. D. MacLean, eds.). National Institutes for Mental Health, Bethesda, Md., pp. 11–64.

Northcutt, R. G. (1981). Evolution of the telencephalon in nonmammals. *Ann. Rev. Neurosci.* 4, 301–350.

Northcutt, R. G. (1984). Evolution of the vertebrate central nervous system: patterns and processes. *Am. Zool.* 24, 701–716.

Northcutt, R. G., Reiner, A., and Karten, H. J. (1983). The basal ganglia of spiny dogfish: an immunohistochemical study. *Anat. Rec.* 208, 128A.

Northcutt, R. G., Reiner, A., and Karten, H. J. (1988). An immunohistochemical study of the telencephalon of the spiny dogfish, *Squalus acanthias*. *J. Comp. Neurol.* 277, 250–267.

Nozaki, M., and A. Gorbman (1984). Distribution of immunoreactive sites for several components of pro-opiocortin in the pituitary and brain of adult lamprey *Petromyzon marinus* and *Entosphenus tridentatus*. *Gen. Comp. Endocrinol.* 53, 335–352.

Nozaki, M., and Kobayashi, H. (1979). Distribution of LHRH-like substances in the vertebrate brain as revealed by immunohistochemistry. *Arch. Histol. Jap.* 42, 201–219.

Nozaki, M., Tsukahara, T., and Kobayashi, H. (1984). Neuronal systems producing LHRH in vertebrates. In *Endocrine Correlates of Reproduction* (K. Ochai, Y. Arai, T. Shioda, and M. Takahashi, eds.). Japan Scientific Society Press, Tokyo/Springer-Verlag, Berlin, pp. 3–27.

Obata-Tsoto, H. L., Okamura, H., Tsuto, T., Terubayashi, H., Fukui, F., Yanaihara, N., and Ibata, Y. (1983). Distribution of the VIP-like immunoreactive neurons in the cat central nervous system. *Br. Res. Bull.* 10, 653–660.

O'Donohue, T. L., Bishop, J. F., Chronwall, B. M., Groome, J., and Watson, W. I I., III. (1901). Characterization and distribution of FMRFamide immunoreactivity in the rat central nervous system. *Peptides* 5, 563–568.

Oka, J.-I., and Fukuda, H. (1984). Properties of the depolarization induced by TRH in the isolated frog spinal cord. *Neurosci. Lett.* 46, 167–172.

Okamoto, S. (1951). Epileptogenic action of glutamate directly applied into brain of animals and inhibitory effects of proteins and tissue emulsions on its actions. *J. Physiol. Soc. Jap.* 13, 555–562.

Oksche, A., Wilson, W. O., and Farmer, D. S. (1964). The hypothalamic neurosecretory system of *Coturnix coturnix japonica. Z. Zellforsch.* 61, 688–709.

Oliver, C., Eskay, R. L., Ben-Jonathan, N., and Porter, J. C. (1974). Distribution and concentration of TRH in rat brain. *Endocrinology* 95, 540–546.

Olschowka, J. A., O'Donohue, T. L., Mueller, G. P., and Jacobowitz, D. M. (1982). The distribution of corticotropin releasing factor-like immunoreactive neurons in cat brain. *Peptides* 3, 995–1015.

Ono, T., Nishino, H., Sasaka, K., Muramoto, K., Yano, I., and Simpson, A. (1978). Paraventricular nucleus connections to spinal cord and pituitary. *Neurosci. Lett.* 10, 141–146.

Osborne, N. N., Nichols, D. A., Dockray, G. J., and Cuello, A. C. (1982). Cholecystokinin and substance P immunoreactivity in retinas of rats, frogs, lizards and chicks. *Exp. Eye Res.* 34, 639–649.

Otsuka, M., and Konishi, S. (1976). Substance P as an excitatory transmitter of primary sensory neurons. *Cold Spring Harbor Symposium Quant. Biol.* 40, 135–143.

Otsuka, M., Konishi, S., Yanagisawa, M., and Akagi, H. (1984). Role of substance P as a neurotransmitter in a slow spinal reflex. *Biomed. Res.*, Suppl. 5, 85–90.

Otsuka, M., and Yanagisawa, M. (1987). Does substance P act as a pain transmitter? *Trends Pharmacol. Sci.* 8, 506–510.

Pandalai, K. R., (1960). Reactions of neurosecretory cells of *Calotes versicolor* to dehydration, stress, variations in temperature, blinding and injection of sex hormones. *J. Anat. Soc. India* 9, 88–96.

Panzica, G. C., Calcagni, M., Ramieri, G., and Viglietti-Panzica, C. (1988). Extrahypothalamic distribution of vasotocin-immunoreactive fibers and perikarya in the avian central nervous system. *Bas. Appl. Histochem.* 32, 89–94.

Parent, A. (1979). Monaminergic systems of the brain. In *Biology of the Reptilia. Neurobiology B*, Vol. 10 (C. Gans, R. G. Northcutt, and P. Ulinski, eds.). Academic Press, New York, pp. 247–285.

Patel, Y. C., and Reichlin, S. (1978). Somatostatin in hypothalamic, extrahypothalamic and peripheral tissue of the rat. *Endocrinology* 102, 523–530.

Pearson, A. K., and Licht, P. (1982). Morphology and immunocytochemistry of the turtle pituitary gland with special reference to the pars tuberalis. *Cell Tiss. Res.* 222, 81–100.

Pearson, A. K., Wurst, G. Z., and Cadle, J. E. (1983). Ontogeny and immunocytochemical differentiation of the pituitary gland in a sea turtle, *Caretta caretta. Anat. Embryol.* 167, 13–37.

Peczely, P., and Kiss, J. Z. (1988). Immunoreactivity to vasoactive intestinal polypeptide (VIP) and thyreotropin (sic)-releasing hormone (TRH) in hypothalamic neurons of the domesticated pigeon (*Columba livia*): alterations following lactation and exposure to cold. *Cell Tiss. Res.* 251, 485–494.

Perez-Clausell, J., and Fredens, K. (1988). Chemoarchitectonics in the telencephalon of the lizard *Podarcis hispanica*: distribution of somatostatin, enkephalin, cholecystokinin and zinc. In *The Forebrain of Reptiles: Current Concepts of Structure and Function* (W. K. Schwerdtfeger and W. J. A. J. Smeets, eds.) Karger, Basel, Switzerland, pp. 85–96.

Pert, C. B., Aposhian, D., and Snyder, S. H. (1974). Phylogenetic distribution of opiate receptor binding. *Brain Res.* 75, 356–361.

Pert, C. B., and Snyder, S. H. (1973a). Opiate receptors: demonstration in nervous tissue. *Science* (N.Y.) 179, 1011–1014.

Pert, C. B., and Snyder, S. H. (1973b). Properties of opiate receptor binding in rat brain. *Proc. Natl. Acad. Sci. USA* 70, 2243–2247.

Pestarino, M. (1985). CCK-like peptides in the neural complex of a protochordate. *Peptides* 6, 389–392.

Peters, A., Miller, M., and Kimerer, L. M. (1983). Cholecystokinin-like immunoreactive neurons in rat cerebral cortex. *Neuroscience* 8, 431–448.

Petrusz, P., Merchenthaler, I., and Maderdrut, J. L. (1985). Distribution of enkephalin-containing neurons in the central nervous system. In *Handbook of Chemical Neuroanatomy*, Vol. 3, (A. Bjorklund and T. Hokfelt, eds.). Elsevier, Amsterdam, pp. 273–334.

Pezalla, P. D., and Dicig, D. (1984). Stress-induced analgesia in frogs: evidence for involvement of an opioid system. *Brain Res.* 296, 356–360.

Philibert, R. L., and Kamemoto, F. I. (1965). The hypothalamo-hypophyseal neurosecretory system of the ring-necked snake, *Diadophis punctatus. Gen. Comp. Endocrinol.* 5, 326–335.

Phillis, J. W., and Kirkpatrick, J. R. (1980) The actions of motilin, luteinizing hormone-releasing hormone, cholecystokinin, somatostatin, vasoactive intestinal peptide and other peptides on rat cerebral cortical neurons. *Can. J. Physiol. Pharmacol.* 58, 612–623.

Phillis, J. W., and Limacher, J. J. (1974). Excitation of cerebral cortical neurons by various polypeptides. *Exp. Neurol.* 43, 413–423.

Plisetskaya, E. M., Pollock, H. G., Elliot, W. M., Youson, J. H., and Andrews, P. C. (1988). Isolation and structure of lamprey (*Petromyzon marinus*) insulin. *Gen. Comp. Endocrinol.* 69, 46–55.

Plisetskaya, E., Pollock, H. G., Rouse, J. B., Hamilton, J. W., Kimmel, J. R., and Gorbman, A. (1986). Isolation and structures of coho salmon (*Oncorhynchus kisutch*) glucagon and glucagon-like peptide. *Regulatory Peptides* 14, 57–67.

Polak, J. M., Bloom, S. R., Hobbs, S., Solcia, E., and Pearse, A. G. E. (1976).

Distribution of bombesin-like peptide in human gastrointestinal tract. *Lancet* 1109–1110.

Pollock, H. G., Hamilton, J. W., Rouse, J. B., Ebner, K. E., and Rawitch, A. B. (1988). Isolation of peptide hormones from the pancreas of the bullfrog (*Rana catesbeiana*): amino acid sequences of pancreatic polypeptide, oxyntomodulin and two glucagon-like peptides. *J. Biol. Chem.* 263, 9746–9751.

Pollock, H. G., and Kimmel, J. R. (1975). Chicken glucagon: isolation and amino acid sequence studies. *J. Biol. Chem.* 250, 9377–9380.

Pollock, H. G., Kimmel, J. R., Ebner, K. E. Hamilton, J. W., Rouse, J. B., Lance, V., and Rawitch, A. B. (1988). Isolation of alligator gar (*Lepisosteus spatula*) glucagon, oxyntomodulin and glucagon-like peptide: amino acid sequences of oxyntomodulin and glucagon-like peptide. *Gen. Comp. Endocrinol.* 69, 133–140.

Pontet, A., Danger, J. M., Dubourg, P., Pelletier, G., Vaudry, H., Calas, A., and Kah, O. (1989). Distribution and characterization of neuropeptide Y-like immunoreactivity in the brain and pituitary of the goldfish. *Cell Tiss. Res.* 255, 529–538.

Powell, R. C., Garcia, G., Lance, V., Millar, R. P., and King, J. A. (1986). Identification of diverse molecular forms of GnRH in reptile brain. *Peptides* 7, 1101–1108.

Powell, R. C., Jach, H., Millar, R. P., and King, J. A. (1987). Identification of Gln[8]-GnRh and His[5], Trp[7], Tyr[8]-GnRH in the hypothalamus and extrahypothalamic brain of the ostrich (*Struthio camelus*) *Peptides* 8, 185–190.

Powell, R. C., King, J. A., and Millar, R. P. (1985). [Trp[7], Leu[8]]LH-RH in reptilian brain. *Peptides* 6, 223–227.

Powell, R. C., Millar, R. P., and King. J. A. (1986). Diverse forms of gonadotropin-releasing hormone in an elasmobranch and a teleost fish. *Gen. Comp. Endocrinol.* 63, 77–85.

Powers, A. S., and Reiner, A. (1980). A stereotaxic atlas of the forebrain and midbrain of the eastern painted turtle (*Chrysemys picta picta*). *J. Hirnforsch.* 21, 125–159.

Prasada Rao, P. D., and Subhedar, N. (1977). A cytoarchitectonic study of the hypothalamus of the lizard, *Calotes versicolor*. *Cell Tiss. Res.* 180, 63–85.

Preece, H., and Licht, P. (1987). Effects of thyrotropin-releasing hormone *in vitro* on thyrotropin and prolactin release from the turtle pituitary. *Gen. Comp. Endocrinol.* 67, 247–255.

Price, D. A., and Greenberg, M. J. (1977). Structure of a molluscan cardioexcitatory neuropeptide. *Science* (N.Y.) 197, 670–671.

Quirion, R., Gaudreau, P., St.-Pierre, S., Rioux, F., and Pert, C. B. (1982). Autoradiographic distribution of [3H]-neurotensin receptors in rat brain: Visualization by tritium-sensitive film. *Peptides* 3, 757–763.

Quirion, R., and Weiss, A. S. (1983). Peptide E and other proenkephalin-derived peptides are potent kappa opiate receptor agonists. *Peptides* 4, 445–449.

Rajjo, I. M., Vigna, S. R., and Crim, J. W. (1988). Cholecystokinin immunoreactivity in the digestive tract of bowfin (*Amia calva*), bluegill (*Lepomis macrochirus*), and bullfrog (*Rana catesbeiana*). *Gen. Comp. Endocrinol.* 70, 133–144.

Rawdon, B. B., and Andrew, A. (1981). An immunocytochemical survey of endocrine cells in the gastrointestinal tract of chicks at hatching. *Cell Tiss. Res.* 220, 279–292.

Reaves, T. A., Jr., and Hayward, J. N. (1980). Functional and morphological studies of peptide-containing neuroendocrine cells in goldfish hypothalamus. *J. Comp. Neurol.* 193, 777–788.

Rehfeld, J. F., Gotterman, N., Larsson, L. I., Emson, P. C., and Lee, C. M. (1979). Gastrin and cholecystokinin in the central and peripheral neurons. *Fed. Proc.* 38, 2325–2329.

Rehfeld, J. F., Hansen, H. F., Larsson, L. I., Stengaard-Pedersen, K., and Thorn, N. A. (1984). Gastrin and cholecystokinin in pituitary neurons. *Proc. Natl. Acad. Sci. USA* 81, 1902–1905.

Reinecke, M. (1981). Immunohistochemical localization of polypeptide hormones in endocrine cells of the digestive tract of *Branchiostoma lanceolatum*. *Cell Tiss. Res.* 219, 445–456.

Reinecke, M. (1985). Neurotensin: immunohistochemical localization in central and peripheral nervous system and in endocrine cells and its functional role as a neurotransmitter and endocrine hormone. *Progr. Histochem. Cytochem.* 16, 1–175.

Reinecke, M., Schulter, P., Yanaihara, N., and Forssman, W. G. (1980). VIP-like immunoreactive cells in the gut of vertebrates. *11th Congress on Anatomy*, Mexico City.

Reiner, A. (1983). Comparative studies of opioid peptides: enkephalin distribution in turtle central nervous system. *Soc. Neurosci. Abs.* 9, 439.

Reiner, A. (1986a). The co-occurrence of substance P-like immunoreactivity and dynorphin-like immunoreactivity in striatopallidal and striatonigral projection neurons in birds and reptiles. *Brain Res.* 371, 155–161.

Reiner, A., (1986b). The extensive co-occurrence of GABA and the neurotensin-related hexapeptide LANT6 in the avian brain. *Anat. Rec.* 214, 106A.

Reiner, A. (1986c). Evidence for the presence of the neurotensin-related hexapeptide LANT6 in ganglion cells of the pigeon retina. *Invest. Ophthalmol. Vis. Sci.*, Suppl. 27, 185.

Reiner, A. (1986d). Transmitter-specific projections from the basal ganglia to the tegmentum in pigeons. *Soc. Neurosci. Abs.* 12, 873.

Reiner, A. (1987a). The distribution of proenkephalin-derived peptides in the central nervous system of turtle. *J. Comp. Neurol.* 259, 65–91.

Reiner, A. (1987b). A LANT6-like peptide that is distinct from Neuromedin N is present in striatal and pallidal neurons in the monkey basal ganglia. *Brain Res.* 422, 186–191.

Reiner, A. (1987c). The presence of substance P/CGRP-containing fibers, VIP-containing fibers and numerous cholinergic fibers on blood vessel of the avian choroid. *Invest. Ophthalmol. Vis. Sci.*, Suppl., 28, 81.

Reiner, A. (1991). A comparison of the neurotransmitter-specific and neuropeptide-specific neuronal cell types present in turtle cortex to those present in mammalian isocortex: implications for the evolution of isocortex. *Brain Behav. Evol.* (In press.)

Reiner, A., and Anderson, K. D. (1987). The co-occurrence of substance P and GABA in striatal projection neurons of the basal ganglia. *Soc. Neurosci. Abs.* 13, 1574.

Reiner, A., and Beinfeld, M. C. (1985). The regional content and distribution of cholecystokinin-8 in the turtle central nervous system: an immunohisto-chemical and biochemical study. *Brain Res. Bull.* 15, 167–81.

Reiner, A., and Brauth, S. E. (1989). CGRP is an evolutionarily conserved marker of thalamic auditory relay neurons. *Soc. Neurosci. Abs.* 15, 376.

Reiner, A., Brauth, S. E., and Karten, H. J. (1984). Evolution of the amniote basal ganglia. *Trends Neurosci.* 7, 320–325.

Reiner, A., Brauth, S. E., Kitt, C. A., and Karten, H. J. (1980). Basal gangli-onic pathways to the tectum: studies in reptiles. *J. Comp. Neurol.* 193, 565–589.

Reiner, A., Brauth, S. E., Kitt, C. A., and Quirion, R. (1989). The distribution of mu, delta and kappa opiate receptors in the pigeon forebrain and mid-brain. *J. Comp. Neurol.* 280, 359–382.

Reiner, A., Brecha, N. C., and Karten, H. J. (1982). Basal ganglia pathways to the tectum: the afferent and efferent connections of the lateral spiriform nucleus of pigeons. *J. Comp. Neurol.* 208, 16–36.

Reiner, A., and Carraway, R. E. (1985). The presence and phylogenetic con-servation of a neurotensin-related hexapeptide in neurons of globus palli-dus. *Brain Res.* 341, 365–371.

Reiner, A., and Carraway, R. E. (1987). Immunohistochemical and biochemi-cal studies on Lys[8]-Asn[9]-Neurotensin[8-13] (LANT6)-related peptides in the basal ganglia of pigeons, turtles and hamsters. *J. Comp. Neurol.* 257, 453–476.

Reiner, A., Davis, B. M., Brecha, N. C., and Karten, H. J. (1984). The distri-bution of enkephalin-like immunoreactivity in the telencephalon of the adult and developing domestic chicken. *J. Comp. Neurol.* 228, 245–262.

Reiner, A., Eldred, W. D., Beinfeld, M. C., and Krause, J. E. (1985). The co-occurrence of a substance P-like peptide and cholecystokinin-8 in a fiber system of turtle cortex. *J. Neurosci.* 5, 1522–1536.

Reiner, A., and Karten, H. J. (1982). Laminar distribution of the cells of origin of the descending tectofugal pathways in the pigeon (*Columba livia*). *J. Comp. Neurol.* 204, 165–187.

Reiner, A., Karten, H. J., and Brecha, N. C. (1982). Enkephalin-mediated basal ganglia influences over the optic tectum: immunohistochemistry of the tectum and the lateral spiriform nucleus in pigeon. *J. Comp. Neurol.* 208, 37–53.

Reiner, A., Karten, H. J., and Solina, A. (1983). Substance P: localization within paleostriatal-tegmental pathways in the pigeon. *Neuroscience* 9, 61–85.

Reiner, A., Krause, J. E., Keyser, K. T., Eldred, W. D., and McKelvy, J. F. (1984). The distribution of substance P in turtle nervous system: a radioim-munoassay and immunohistochemical study. *J. Comp. Neurol.* 226, 50–75.

Reiner, A., and Northcutt, R. G. (1987). An immunohistochemical study of the telencephalon of the African lungfish. *J. Comp. Neurol.* 256, 463–481.

Reiner, A., and Northcutt, R. G. (1988). An immunohistochemical study of the telencephalon of *Polypterus*. *Soc. Neurosci. Abs.* 14, 53.

Reiner, A., and Oliver, J. R. (1987). Somatostatin and neuropeptide Y are al-most exclusively found in the same neurons in the telencephalon of turtles. *Brain Res.* 426, 149–156.

Reiner, A., and Powers, A. S. (1978). Intensity and pattern discrimination in turtles following lesions of nucleus rotundus. *J. Comp. Physiol. Psychol.* 92, 1156–1168.

Reiner, A., and Powers, A. S. (1980). The effects of extensive forebrain lesions on visual discriminative performance in turtles (*Chrysemys picta picta*). *Brain Res.* 192, 327–338.

Reiner, A., and Powers, A. S. (1983). The effects of lesions of telencephalic visual structures on visual discriminative performance in turtles (*Chrysemys picta picta*). *J. Comp. Neurol.* 218, 1–24.

Reubi, J. C., and Jessell, T. M. (1979). Distribution of substance P in the pigeon brain. *J. Neurochem.* 31, 359–361.

Riphagen, C. L., and Pittman, Q. J. (1986). Arginine vasopressin as a central neurotransmitter. *Fed. Proc.* 45, 2318–2322.

Riss, W., Halpern, M., and Scalia, F. (1969). The quest for clues to forebrain evolution—the study of reptiles. *Brain Behav. Evol.* 2, 1–50.

Rivier, J. E., Brown, M. R., and Vale, W. (1977). Effects of neurotensin, substance P and morphine sulfate on the secretion of prolactin and growth hormone in the rat. *Endocrinology* 100, 751–754.

Rivier, J., Rivier, C., Brandon, D., Millar, R., Spiess, J., and Vale, W. (1981). In *Peptides: Synthesis-Structure-Function* (D. H. Rich and E. Gross, eds.). Proceedings of the Seventh American Peptide Symposium, Pierce Chemical Co., Rockford, Ill., pp. 771–776.

Roberts, E., and Frankel, S. (1950). Gamma-aminobutyric acid in brain: its formation from glutamic acid. *J. Biol. Chem.* 187, 55–63.

Rodrigues, K. T., Jenks, B. G., and Sumpter, J. P. (1983). Biosynthesis of proopiomelanocortin-related peptides in the neurointermediate lobe of the pituitary gland of the rainbow trout (*Salmo gairdneri*). *J. Endocrinol.* 98, 271–282.

Rogawski, M. A. (1982). Cholecystokinin octapeptide: Effects on the excitability of cultured spinal neurons. *Peptides* 3, 545–551.

Rouille, Y., Chauvet, M.-T., Chauvet, J., Acher, R., and Hadley, M. E. (1988). The distribution of lysine vasopressin (Lysipressin) in placental mammals: a reinvestigation of the hippopotamidae (*Hippopotamus amphibius*) and Tayassuidae (*Taayassu angulatus*) families. *Gen. Comp. Endocrinol.* 71, 475–483.

Russchen, F. T., Smeets, W. J. A. J., and Hoogland, P. V. (1987). Histochemical identification of pallidal and striatal structures in the lizard *Gekko gecko*: evidence for compartmentalization. *J. Comp. Neurol.* 256, 329–341.

Ryan, S. M., Arnold, A. P., and Elde, R. P. (1981). Enkephalin-like immunoreactivity in vocal control regions of the zebra finch brain. *Brain Res.* 229, 236–240.

Rzasa, P., Kavoustian, K. V., and Prokop, E. K. (1984). Immunochemical evidence for met-enkephalin-like and leu-enkephalin-like peptides in tissues of the earthworm, *Lumbricus terrestris*. *Comp. Biochem. Physiol.* 77C, 345–350.

Saayman, H. S., Naude, R. J., and Oelofsen, W. (1985). Isolation and characterization of a neurophysin from ostrich neurohypophyses. *Int. J. Peptide Protein Res.* 26, 416–424.

Saffrey, M. J., Polak, J. M., and Burnstock, G. (1982). Distribution of vasoactive intestinal polypeptide-, substance P-, enkephalin- and neurotensin-like immunoreactive nerves in the chicken gut during development. *Neuroscience* 7, 279–293.

Said, S. I. (1984). Vasoactive intestinal polypeptide (VIP): current status. *Peptides* 5, 143–150.

Said, S. I., and Mutt, V. (1970). Polypeptide with broad biological activity: isolation from small intestine. *Science* (N.Y.) 169, 1217–1218.

Said, S. I., and Mutt. V. (1972). Isolation from porcine intestinal wall of a vasoactive octacosapeptide related to secretin and to glucagon. *Eur. J. Pharmacol.* 28, 199–204.

Saint Girons, H. (1963). Données histophysiologiques sur le cycle annuel des glandes endocrines et de leurs effecteurs chez l'orvet, *Anguis fragilis* (L.). *Arch. Anat. Morphol. Exp.* 52, 1–51.

Saint Girons, H. (1967). Le cycle sexuel et les correlations hypophysogenitales des males chez *Agama bibroni* Duméril au Maroc. *Bull. Biol. Stockh.* 101, 321–344.

Salin, K. (1988). Gastrin/CCK-like immunoreactivity in Hatschek's groove of *Branchiostoma lanceolatum* (Cephalochordata). *Gen. Comp. Endocrinol.* 70, 436–441.

Sanger, F. (1959). Chemistry of insulin. *Science* (N.Y.) 129, 1340–1344.

Sar, M., Stumpf, W. E., Miller, R. J., Chang, K. J., and Cuatrecasas, P. (1978). Immunohistochemical localization of enkephalin in rat brain and spinal cord. *J. Comp. Neurol.* 182, 17–38.

Sawin, C. T., Bacharach, P., and Lance, V. (1981). Thyrotropin-releasing hormone and thyrotropin in the control of thyroid function in the turtle, *Chrysemys picta. Gen. Comp. Endocrinol.* 45, 7–11.

Sawyer, W. H. (1961). Comparative physiology and pharmacology of the neurohypophysis. *Rec. Prog. Horm. Res.* 17, 437–465.

Scanes, C. G. (1974). Some in vitro effects of synthetic thyrotropin-releasing factor on secretion of thyroid stimulating hormone from the anterior gland of the domestic fowl. *Neuroendocrinology* 15, 1–9.

Schally, A. V., Arimura, A., and Kastin, A. V. (1973). Hypothalamic regulatory hormones. *Science* (N.Y.) 179, 341–350.

Scharrer, E., and Scharrer, B. (1945). Neurosecretion. *Physiol. Rev.* 25, 171–181.

Schoofs, L., Danger, J. M., Jelou, S., Pelletier, G., Huybrechts, R., Vaudry, H., and De Loof, A. (1988). NPY-like peptides occur in the nervous system and midgut of the migratory locust, *Locusta migratoria* and in the brain of the grey fleshfly, *Sarcophaga bullata. Peptides* 9, 1027–1036.

Schot, L. P. C., Boer, H. H., Swaab, D. F., and Van Noorden, S. (1981). Immunocytochemical demonstration of peptidergic neurons in the central nervous system of the pond snail *Cymnaea stagnalis* with antisera raised to biologically active peptides of vertebrates. *Cell Tiss. Res.* 216, 273–291.

Schulman, J. A., Finger, T. E., Brecha, N. C., and Karten, H. J. (1981). Enkephalin immunoreactivity in Golgi cells and mossy fibers of mammalian, avian, amphibian and teleostean cerebellum. *Neuroscience* 6, 2407–2416.

Schwanzel-Fukuda, M., and Silverman, A.-J. (1980). The nervus terminalis of the guinea pig: a new luteinizing hormone-releasing hormone (LHRH) neuronal system. *J. Comp. Neurol.* 191, 213–225.

Sei, C. A., Richard, R., and Dores, R. M. (1989). Steady-state levels of prodynorphin-related end-products from the brain of the amphibian, *Xenopus laevis. Brain Res.* 479, 162–166.

Seki, T., Nakai, Y., Shioda, S., Mitsuma, T., and Kikuyama, S. (1983). Distribution of immunoreactive thyrotropin-releasing hormone in the forebrain and hypophysis of the bullfrog, *Rana catesbeiana. Cell Tiss. Res.* 233, 507–516.

Senn, D. G. (1971). Structure and development of the optic tectum of the snapping turtle (*Chelydra serpentina* L.). *Acta Anat.* 80, 46–57.

Seroogy, K., Tsuruo, Y., Hokfelt, T., Walsh, J., Fahrenkrug, J., Emson, P. C., and Goldstein, M. (1988). Further analysis of presence of peptides in dopamine neurons: cholecystokinin, peptide histidine-isoleucine/vasoactive intestinal polypeptide and substance P in rat supramammillary region and mesencephalon. *Exp. Brain Res.* 72, 523–534.

Setalo, G., Vigh, S., Schally, A. V., Arimura, A., and Flerko, B. (1976). Immunohistochemical study of the origin of LHRH-containing nerve fibers of the rat hypothalamus. *Brain Res.* 103, 597–602.

Sheela, R., and Pandalai, K. R. (1968). Reaction of the paraventricular nucleus to dehydration in the garden lizard, *Calotes versicolor. Gen. Comp. Endocrinol.* 11, 257–261.

Sherwood, N. M. (1986). Evolution of a neuropeptide family: gonadotropin-releasing hormone. *Am. Zool.* 26, 1041–1054.

Sherwood, N. M., Eiden, L., Brownstein, M., Spiess, J., Rivier, J., and Vale, W. (1983). Characterization of a teleostean gonadotropin-releasing hormone. *Proc. Natl. Acad. Sci. USA* 80, 2794–2798.

Sherwood, N. M., Harvey, B., Brownstein, M. J., and Eiden, L. E. (1984). Gonadotropin-releasing hormone (GnRH) in striped mullet (*Mugil cephalus*), milkfish (*Chanos chanos*) and rainbow trout (*Salmo gairdneri*): comparison with salmon GnRH. *Gen. Comp. Endocrinol.* 55, 174–181.

Sherwood, N. M., and Whittier, J. M. (1988). Gonadotropin-releasing hormone from brains of reptiles: turtles (*Pseudemys scripta*) and snakes (*Thamnophis sirtalis parietalis*). *Gen. Comp. Endocrinol.* 69, 319–327.

Sherwood, N. M., Wingfield, J. C., Ball, G. F., and Dufty, A. M. (1988). Identity of GnRh in passerine birds: comparison of GnRH in song sparrow (*Melosospiza melodia*) and starling (*Sturnus vulgaris*) with five vertebrate GnRHs. *Gen. Comp. Endocrinol.* 69, 341–351.

Sherwood, N. M., Zoeller, R. T., and Moore, F. L. (1986). Multiple forms of gonadotropin-releasing hormone in amphibian brains. *Gen. Comp. Endocrinol.* 61, 313–322.

Shiosaka, S., Takatsuki, K., Inagaki, S., Sakanaka, M., Takagi, H., Senba, E., Matsuzaki, T., and Tohyama, M. (1981). Topographic atlas of somatostatin-containing neuron system in the avian brain in relation to catecholamine-containing neuron system. II. Mesencephalon, rhombencephalon and spinal cord. *J. Comp. Neurol.* 202, 115–124.

Silverman, A.-J., and Krey, L. C. (1978). The luteinizing hormone-releasing hormone (LH-RH) neuronal networks of the guinea pig brain. I: Intra- and extra-hypothalamic projections. *Brain Res.* 157, 233–246.

Simantov, R., Goodman, R., Aposhian, D., and Snyder, S. H. (1976). Phylogenetic distribution of a morphine-like peptide "enkephalin." *Brain Res.* 111, 204–211.

Simantov, R., and Snyder, S. H. (1976). Morphine-like peptides in mammalian brain: isolation, structure, elucidation and interaction with opiate receptor. *Proc. Natl. Acad. Sci. USA* 73, 2515–2519.

Skirboll, L., Hokfelt, T., Rehfeld, J., Cuello, A. C., and Dockray, G. J. (1982). Co-existence of substance P and cholecystokinin-like immunoreactivity in neurons of the mesencephalic gray. *Neurosci. Lett.* 28, 35–39.

Skofitsch, G., and Jacobowitz, D. M. (1985). Calcitonin gene-related peptide: detailed immunohistochemical distribution in the central nervous system. *Peptides* 6, 721–745.

Cimato, W. J. A. J. Jonker A. J., and Hoogland, P. V. (1987). Distribution of dopamine in the forebrain and midbrain of the red-eared turtle, *Pseudemys scripta elegans*, reinvestigated using antibodies against dopamine. *Brain Behav. Evol.* 30, 121–142.

Smith, Y., and Parent, A. (1986). Neuropeptide Y-immunoreactive neurons in the striatum of cat and monkey: morphological characteristics, intrinsic organization and co-localization with somatostatin. *Brain Res.* 372, 241–252.

Snyder, S. H. (1980). Brain peptides as neurotransmitters. *Science* (N.Y.) 209, 976–984.

Spiess, J., Rivier, J., Rivier, C., and Vale, W. (1981). Primary structure of corticotropin-releasing factor from ovine hypothalamus. *Proc. Natl. Acad. Sci. USA* 78, 6517–6521.

Spiess, J., Rivier, J. E., Rodk, J. A., Bennett, C. D., and Vale, W. (1979). Isolation and characterization of somatostatin from pigeon pancreas. *Proc. Natl. Acad. Sci. USA* 76, 2974–2978.

Stell, W. (1984). Luteinizing hormone-releasing hormone (LHRH)- and pancreatic polypeptide (PP)-immunoreactive neurons in the terminal nerve of the spiny dogfish, *Squalus acanthias*. *Anat. Rec.* 208, 173A.

Stephan, Y., Dufour, C., and Falkmer, S. (1978). Mise en evidence par immunofluorescence de cellules a polypeptide pancréatique (PP) dans le pancréase et le tube digestif de Poissons osseux et cartilagineux. *C. r. h. Acad. Sci. Ser. D.* 286, 1073–1075.

Sterling, R. J., and Sharp, P. J. (1982). The localization of LH-RH neurones in the diencephalon of the domestic hen. *Cell Tiss. Res.* 222, 283–298.

Stoll, C. J., and Voorn, P. (1985). The distribution of hypothalamic and extrahypothalamic vasotocinergic cells and fibers in the brain of a lizard, *Gekko gecko*: presence of a sex difference. *J. Comp. Neurol.* 239, 193–204.

Straus, E., Ryder, S. W., Eng, J., and Yalow, R. S. (1981). Immunochemical studies relating cholecystokinin in brain and gut. *Recent Prog. Horm. Res.* 37, 447–475.

Straus, E., Yalow, R. S., and Gainer, H. (1975). Molluscan gastrin: concentration and molecular forms. *Science* (N.Y.) 190, 687–689.

Strutz, J. (1982). The origin of efferent fibers to the inner ear in a turtle (*Terrapene ornata*). *Brain Res.* 244, 165–168.

Su, C.-J., White, J. W., Li, W.-H., Luo, C.-C., Frazier, M. L., Saunders, G. F., and Chen, L. (1988). Structure and evolution of somatostatin genes. *Molec. Endocrinol.* 2, 209–216.

Sundby, F., Frandson, E. K., Thomsen, J., Kristiansen, K., and Brunfeldt, K. (1972). Crystallization and amino acid sequence of duck glucagon. *FEBS Lett.* 26, 289–293.

Sundler, F., Alumets, J., Fahrenkrug, J., Hakanson, R., and Shaffalitzky de

Muckadell, O. B. (1979). Cellular localization and ontogeny of immuno-reactive vasoactive intestinal polypeptide (VIP) in the chicken gut. *Cell Tiss. Res.* 196, 193–201.

Sundler, F., Moghimzadeh, E., Hakanson, T., Ekelund, M., and Emson, P. (1983). Nerve fibers in the gut and pancreas of the rat displaying neuropeptide-Y immunoreactivity. *Cell Tiss. Res.* 230, 487–493.

Swanson, L. W. (1977). Immunohistochemical evidence for a neurophysin-containing autonomic pathway arising in the paraventricular nucleus of the hypothalamus. *Brain Res.* 128, 346–353.

Taban, C. H., and Catheini, M. (1978). Localization of substance P-like immunoreactivity in hydra. *Experientia* 35, 811–812.

Taban, C. H., and Catheini, M. (1983). Distribution of substance P-like immunoreactivity in the brain of the newt (*Tristurus cristatus*). *J. Comp. Neurol.* 216, 453–470.

Tagliafierro, G., Bonini, E., Faraldi, G., Farina, L., and Rossi, G. G. (1988). Distribution and ontogeny of VIP-like immunoreactivity in the gastro-entero-pancreatic system of a cartilaginous fish *Scyliorhinus stellaris*. *Cell Tiss. Res.* 253, 23–28.

Takahashi, T. (1985). Thyrotropin-releasing hormone mimics descending slow synaptic potentials in rat spinal motoneurons. *Proc. R. Soc. Lond.* 225, 391–398.

Takatsuki, K., Shiosaka, S., Inagaki, S., Sakanaka, M., Takagi, H., Senba, E., Matsuzaki, T., and Tohyama, M. (1981). Topographic atlas of somatostatin-containing neuron system in the avian brain in relation to catecholamine-containing neuron system. I. Telencephalon and diencephalon. *J. Comp. Neurol.* 202, 103–113.

Tan, L., and Rousseau, P. (1982). The chemical identity of the immunoreactive LHRH-like peptide biosynthesized in the human placenta. *Biochem. Biophys. Res. Commun.* 109, 1061–1071.

Tannenbaum, G. S., Ling, N., and Brazeau, P. (1982). Somatostatin-28 is no longer acting and more selective than somatostatin-14 on pituitary and pancreatic hormone release. *Endocrinology* 101–107.

Tatemoto, K. (1984). PHI—A new brain-gut peptide. *Peptides* 5, 151–154.

Tatemoto, K., Carlquist, M., and Mutt, V. (1982). Neuropeptide Y—a novel brain peptide with structural similarities to peptide YY and pancreatic-polypeptide. *Nature* (London) 296, 659–662.

Tatemoto, K., Lundberg, J. M., Terenius, L., and Mutt, V. (1982). New gut-brain peptides: peptide YY and neuropeptide Y. Abstract presented at the 4th Int. Symp. Gastrointest. Horm., held in Stockholm, Sweden, June 20–23, 1982.

Tatemoto, K., and Mutt, V. (1980). Isolation of two novel candidate hormones using a chemical method for finding naturally occurring polypeptides. *Nature* (London) 285, 417–418.

Taylor, W. L., Collier, K. J., Descenes, R. J., Weith, H. L., and Dixon, J. E. (1981). Sequence analysis of a cDNA coding for a pancreatic precursor to somatostatin. *Proc. Natl. Acad. Sci. USA* 78, 6694–6698.

ten Donkelaar, H. J., and de Boer-van Huizen, R. (1981). Basal ganglia projections to the brainstem in the lizard *Varanus exanthematicus* as demonstrated by retrograde transport of horseradish peroxidase. *Neuroscience* 6, 1567–1590.

ten Donkelaar, H. J., Kusuma, A., and de Boer-van Huizen, R. (1980). Cells of origin of pathways to the spinal cord in some quadrupedal reptiles. *J. Comp. Neurol.* 192, 827–851.

Terashima, S. (1987). Substance P-like immunoreactive fibers in the trigeminal sensory nuclei of the pit viper, *Trimeresurus flavoviridis*. *Neuroscience* 23, 685–691.

Thepen, Th., Voorn, P., Stoll, C. J., Sluiter, A. A., Pool, C. W., and Lohman, A. H. M. (1987). Mesotocin and vasotocin in the brain of the lizard *Gekko gecko*: an immunocytochemical study. *Cell Tiss. Res.* 250, 649–656.

Thorndyke, M., and Dockray, G. J. (1986). Identification and localization of material with gastrin-like immunoreactivity in the neural ganglion of a protochordate, *Ciona intestinalis*. *Regulatory Peptides* 16, 269–279.

Tohyama, M., Shiosaka, S., Takagi, H., Inagaki, S., Takatsuki, K., Senba, E., Kawai, Y., and Minigawa, H. (1981). Somatostatin-like immunoreactivity in the facial, glossopharyngeal and vagal lobes of the carp. *Neurosci. Lett.* 24, 233–236.

Tominaga, M., Ebifani, I., Marubashi, S., Kamimura, T., Katagiri, T., and Sasaki, H. (1981). Species differences of glucagon-like material in the brain. *Life Sci.* 29, 1577–1581.

Triepel, J., and Grimmelikhuijzen, C. J. P. (1984). A critical examination of the occurrence of FMRFamide immunoreactivity in the brain of guinea pig and rat. *Histochemistry* 80, 63–71.

Turner, C. D., and Bagnara, J. T. (1976). *General Endocrinology*. W. B. Saunders Co., Philadelphia.

Twarog, B. M., and Page, I. H. (1953). Serotonin content of some mammalian tissues and urine and a method for its determination. *Am. J. Physiol.* 175, 157–161.

Twery, M. J., and Moss, R. L. (1985). Calcitonin and calcitonin gene-related peptide alter the excitability of neurons in rat forebrain. *Peptides* 6, 373–378.

Udenfriend, S., and Kilpatrick, D. C. (1983). Biochemistry of the enkephalins and enkephalin-containing peptides. *Arch. Biochem. Biophys.* 221, 309–323.

Ueda, S., Takeushi, Y., and Sano, Y. (1983). Immunohistochemical demonstration of serotonin neurons in the central nervous system of the turtle *Clemmys japonica*. *Anat. Embryol.* 168, 1–19.

Uhl, G. R., Goodman, R. R., and Snyder, S. H. (1979). Neurotensin-containing cell bodies, fibers and nerve terminals in the brainstem of the rat. Immunohistochemical mapping. *Brain Res.* 167, 77–91.

Uhl, G. R., Kuhar, M. J., and Snyder, S. H. (1977). Neurotensin, Immunohistochemical localization in rat central nervous system. *Proc. Natl. Acad. Sci. USA* 74, 4059–5063.

Uhl, G. R., and Snyder, S. H. (1980). Neurotensin. In *The Role of Peptides in Neuronal Function*. (J. L. Barker and T. G. Smith, Jr., eds.). Marcel Dekker, New York, pp. 509–544.

Ulinksi, P. S. (1983). *Dorsal Ventricular Ridge: A Treatise on Forebrain Organization in Reptiles and Birds*. John Wiley and Sons, New York.

Ulinski, P. S. (1988). Functional architecture of turtle visual cortex. In *The*

Forebrain of Reptiles: Current Concepts of Structure and Function (W. K. Schwerdtfeger and W. J. A. J. Smeets, eds.). Karger, Basel, pp. 151–161.

Vaillant, C., Dimaline, R., and Dockray, G. J. (1980). The distribution and cellular origin of vasoactive intestinal polypeptide in the avian gastrointestinal tract and pancreas. *Cell Tiss. Res.* 211, 511–523.

Vale, W., Ling, N., Rivier, J., Villarreal, J., Rivier, C., Douglas, C., and Brown, M. (1976). Anatomic and phylogenetic distribution of somatostatin. *Metabolism* 25, 1491–1494.

Vale, W., Spiess, J., Rivier, C., and Rivier, J. (1981). Characterization of a 41-residue ovine hypothalamic peptide that stimulates secretion of corticotropin and beta-endorphin. *Science* (N.Y.) 213, 1394–1397.

Valentino, K. L., Tatemoto, K., Hunter, J., and Barchas, J. D. (1986). Distribution of neuropeptide K-immunoreactivity in the rat central nervous system. *Peptides* 7, 1043–1059.

Vallarino, M. (1984). Immunocytochemical localization of alpha-melanocyte-stimulating hormone in the brain of the lizard, *Lacerta muralis. Cell Tiss. Res.* 237, 521–524.

Vallarino, M. (1986). β-Endorphin-like immunoreactivity in the brain of the lizard *Lacerta muralis. Gen. Comp. Endocrinol.* 64, 52–59.

Vallarino, M., Danger, J. M., Fasolo, A., Pelletier, G., Saint-Pierre, S., and Vaudry, H. (1988). Distribution and characterization of neuropeptide Y in the brain of an elasmobranch fish. *Brain Res.* 448, 67–76.

Vanderhaeghen, J. J. (1985). Neuronal cholecystokinin. In *Handbook of Chemical Neuroanatomy*, Vol. 4, *GABA and Neuropeptides in the CNS*, Part I (A. Bjorklund and T. Hokfelt, eds.). Elsevier Science Publishers, Amsterdam, pp. 406–435.

Vanderhaeghen, J. J., Lotstra, F., Vandesande, F., and Dierickx, K. (1981). Coexistence of cholecystokinin and oxytocin-neurophysin in some magnocellular hypothalamohypophyseal neurons polypeptides. *Cell Tiss. Res.* 221, 227–231.

Vanderhaeghen, J. J., Signeau, J. C., and Gepts, W. (1975). New peptide in the vertebrate CNS reacting with antigastrin antibodies. *Nature* (London) 257, 604–605.

Vandesande, F., and Dierickx, K. (1976). Immunocytochemical demonstration of separate vasotocinergic and mesotocinergic neurons in the amphibian hypothalamic magnocellular neurosecretory system. *Cell Tiss. Res.* 175, 289–296.

Vandesande, F., and Dierickx, K. (1980). Immunocytochemical localization of somatostatin-containing neurons in the brain of *Rana temporaria. Cell Tiss. Res.* 205, 43–53.

Vandesande, F., Dierickx, K., and De Mey, J. (1977). The origin of the vasopressinergic and oxytocinergic fibers of the external region of the median eminence of the rat hypophysis. *Cell Tiss. Res.* 180, 443–452.

Van Noorden, S., and Falkmer, S. (1980). Gut-islet endocrinology: some evolutionary aspects. *Investigative Cell Pathol.* 3, 21–35.

Van Noorden, S., and Paten, G. J. (1978). Localization of pancreatic polypeptide (PP)-like immunoreactivity in the pancreatic islets of some teleost fishes. *Cell Tiss. Res.* 188, 521–525.

Van Noorden, S., and Polak, J. M. (1979). Hormones of the alimentary tract.

In *Hormones and Evolution*, Vol. 2, (E. J. W. Barrington, ed.) Academic Press, London, pp. 791–828.

Van Vossel-Daeninck, J., Dierickx, K., Vandesande, F., and Van Vossel, A. (1981). Electron-microscopic immunocytochemical demonstration of separate vasotocinergic, mesotocinergic and somatostatinergic neurons in the hypothalamic preoptic nucleus of the frog. *Cell Tiss. Res.* 218, 7–12.

Vaudry, H., Jenks, B. G., and van Overbeeke, A. P. (1984). Biosynthesis, processing and release of pro-opiomelanocortin related peptides in the intermediate lobe of the pituitary gland of the frog (*Rana ridibunda*). *Peptides* 5, 905–912.

Vayse, N., Pradayrol, L., Chayvialle, J. A., Pignal, F., Esteve, J. P., Susini, C., Descos, F., and Ribet, A. (1981). Effects of somatostatin-14 and somatostatin-28 on bombesin-stimulated release of gastrin, insulin and glucagon in the dog. *Endocrinology* 108, 1843–1847.

Verhaert, P., Grimmelikhuijzen, C. J. P., and De Loof, A. (1985). Distinct localization of FMRFamide- and bovine pancreatic polypeptide-like material in the brain, retrocerebral complex and subesophageal ganglion of the cockroach *Periplaneta americana* L. *Brain Res.* 348, 331–338.

Verstappen, A., Van Reeith, O., Vaudry, H., Pelletier, G., and Vanderhaeghen, J. J. (1986). Demonstration of a neuropeptide Y (NPY)-like immunoreactivity in the pigeon retina. *Neurosci. Lett.* 70, 193–197.

Vigna, S. R. (1979). Distinction between cholecystokinin-like and gastrin-like biological activities extracted from gastrointestinal tract tissues of some vertebrates. *Gen. Comp. Endocrinol.* 39, 512–520.

Vigna, S. R. (1985). Cholecystokinin and its receptors in vertebrates and invertebrates. *Peptides* 6, 283–287.

Vigna, S. R. (1986). Evolution of hormone and receptor diversity: cholecystokinin and gastrin. *Am. Zool.* 26, 1033–1040.

Vigna, S. R., Thorndyke, M. C., and Williams, J. A. (1986). Evidence for a common evolutionary origin of brain and pancreas cholecystokinin receptors. *Proc. Natl. Acad. Sci. USA* 83, 4355–4359.

Vincent, S. R., Hokfelt, T., Christensson, I., and Terenius, L. (1982). Immunohistochemical evidence for a dynorphin immunoreactive striato-nigral pathway. *Eur. J. Pharmacol.* 85, 251–252.

Vincent, S. R., Johansson, O., Hokfelt, T., Meyerson, B., Sachs, C., Elde, R. P., Terenius, L., and Kimmel, J. (1982). Neuropeptide co-existence in human cortical neurons. *Nature* (London) 298, 65–67.

Vincent, S. R., Skirboll, L., Hokfelt, T., Johansson, O., Lundberg, J. M., Elde, R. P., Terenius, L., and Kimmel, J. (1982). Co-existence of somatostatin- and avian pancreatic polypeptide (APP)-like immunoreactivity in some forebrain neurons. *Neuroscience* 7, 439–446.

Viveros, O. H., Dilberto, E. J., Hazum, E., and Chang, K. J. (1979). Opiate-like materials in the adrenal medulla: evidence for storage and secretion with catecholamines. *Molec. Pharmacol.* 16, 1101–1108.

Vivien-Roels, B., Guerne, J. M., Holden, F. C., and Schroeder, D. M. (1979). Comparative immunohistochemical, radioimmunological and biological attempt to identify arginine-vasotocin (AVT) in the pineal gland of reptiles and fishes. *Progr. Brain Res.* 52, 459–463.

Vogt, M. (1954). The concentration of sympathin in different parts of the central nervous system under normal conditions and after administration of drugs. *J. Physiol.* 123, 451–481.

von Euler, U. S., and Gaddum, J. H. (1931). An unidentified depressor substance in certain tissue extracts. *J. Physiol.* 72, 74–87.

Walmsey, J. K., Young, W. S., and Kuhar, M. J. (1980). Immunohistochemical localization of enkephalin in rat forebrain. *Brain Res.* 190, 153–174.

Walsh, J. H., Lechago, J., Wong, H. C., and Rosenquist, G. L. (1982). Presence of ranatensin-like and bombesin-like peptides in amphibian brains. *Regulatory Peptides* 3, 1–13.

Watson, S. J., Akil, H., Fischli, W., Goldstein, A., Zimmerman, E. A., and Nilaver, G. (1982a). Dynorphin and vasopressin: common localization in magnocellular neurons. *Science* (N.Y.) 216, 85–87.

Watson, S. J., Khachaturian, H., Akil, H., Coy, D. H., and Goldstein, A. (1982b). Comparison of the distribution of dynorphin and enkephalin systems in brain. *Science* (N.Y.) 218, 1134–1136.

Watson, S. J., Khachaturian, H., Taylor, L., Fischli, W., Goldstein, A., and Akil, H. (1983). Pro-dynorphin peptides are found in the same neurons throughout rat brain: immunocytochemical study. *Proc. Natl. Acad. Sci. USA* 80, 891–894.

Weber, E., Evans, C. J., and Barchas, J. D. (1983). Multiple endogeneous ligands for opioid receptors. *Trends Neurosci.* 6, 333–336.

Weber, E., Evans, C. J., Samuelsson, S. S., and Barchas, J. D. (1981). Novel peptide neuronal system in rat brain and pituitary. *Science* (N.Y.) 214, 1248–1251.

Weber, E., Roth, K. A., and Barchas, J. D. (1982). Immunohistochemical distribution of alpha-neo-endorphin/dynorphin neuronal systems in rat brain: evidence for co-localization. *Proc. Natl. Acad. Sci. USA* 79, 3062–3066.

Weiler, R. (1985). Mesencephalic pathway to the retina exhibits enkephalin-like immunoreactivity. *Neurosci. Lett.* 55, 11–16.

Weindl, A., Triepel, J., and Kuchling, G. (1984). Somatostatin in the brain of the turtle *Testudo hermanni* Gmelin: an immunohistochemical mapping study. *Peptides* 5, 91–100.

Werz, M. A., and MacDonald, R. L. (1983). Opioid peptides selective for mu- and delta-opiate receptors reduce calcium-dependent action potential duration by increasing potassium conductance. *Neurosci. Lett.* 42, 173–178.

Werz, M. A., and MacDonald, R. L. (1984). Dynorphin reduces calcium-dependent action potential duration by decreasing voltage-dependent calcium conductance. *Neurosci. Lett.* 46, 185–190.

White, J. D., Krause, J. E., Karten, H. J., and McKelvy, J. F. (1985). Presence and ontogeny of enkephalin and substance P in the chick ciliary ganglion. *J. Neurochem.* 45, 1319–1322.

Wilczynski, W., and Northcutt, R. G. (1983). Connections of the bullfrog striatum: efferent projections. *J. Comp. Neurol.* 214, 333–343.

Wilson, J. X. (1984a). Co-evolution of the renin-angiotensin system and the nervous control of blood circulation. *Can. J. Zool.* 62, 137–147.

Wilson, J. X. (1984b). The renin-angiotensin system in nonmammalian vertebrates. *Endocrine Rev.* 5, 45–61.

Winokur, A. J., and Utiger, R. D. (1974). Thyrotropin releasing hormone: regional distribution in rat brain. *Science* (N.Y.) 185, 265–267.

Witkin, J. W., Paden, C. M., and Silverman, A.-J. (1982). The luteinizing hormone-releasing hormone (LHRH) systems in the rat brain. *Neuroendocrinology* 35, 429–438.

Witkin, J. W., and Silverman, A.-J. (1983). Luteinizing hormone-releasing hormone (LHRH) in rat olfactory nerves. *J. Comp. Neurol.* 218, 213–225.

Wofsey, A. R., Kuhar, M. J., and Snyder, S. H. (1971). A unique synaptosomal fraction which accumulates glutamic and aspartic acids in brain tissue. *Proc. Natl. Acad. Sci. USA* 68, 1102–1106.

Wolf, G. (1976). Immunohistological identification of neurophysin and neurophysin-like substances in different vertebrates. *Endokrinologie* 68, 288–299.

Wolf, S. S., and Moody, T. W. (1985). Receptors for GRP/Bombesin-like peptides in the rat forebrain. *Peptides* 6, 111–114.

Wolters, J. G., ten Donkelaar, H. J., and Verhofstad, A. A. J. (1986). Distribution of some peptides (Substance P, leu-enkephalin, met-enkephalin) in the brainstem and spinal cord of a lizard, *Varanus exanthematicus*. *Neuroscience* 18, 917–946.

Yalow, R. S., and Berson, S. A. (1960). Immunoassay of endogeneous plasma insulin in man. *J. Clin. Invest.* 39, 1157–1175.

Yamashiro, D., Ferrara, P., and Li, C. H. (1980). Synthesis and radioreceptor binding activity of turkey beta-endorphin and deacetylated salmon endorphin. *Int. J. Pept. Protein Res.* 16, 75–78.

Yanaihara, N., Nokihara, K., Yanaihara, C., Iwanaga, T., and Fujita, T. (1983). Immunocytochemical demonstration of PHI and its co-existence with VIP in intestinal nerves of the rat and pig. *Arch. Histol. Jap.* 46, 575–581.

Yanaihara, N., Yanaihara, C., Mochizukim, T., Iwahara, K., Fujita, T., and Iwanaga, T. (1981). Immunoreactive GRP. *Peptides* 2, Suppl. 2, 185–192.

Yasuda, A., Yamaguchi, K., Papkoff, H., Yokoo, Y., and Kawauchi, H. (1989). The complete amino acid sequence of growth hormone from the sea turtle (*Chelonia midas*). *Gen. Comp. Endocrinol.* 73, 241–252.

Youngs, L. J., Winokur, A., and Selzer, M. E. (1985). Thyrotropin-releasing hormone in lamprey central nervous system. *Brain Res.* 338, 177–180.

Yu, K. L., Sherwood, N. M., and Peter, R. E. (1988). Differential distribution of two molecular forms of gonadotropin-releasing hormone in discrete brain areas of goldfish (*Carassius auratus*). *Peptides* 9, 625–630.

Zaloglu, S. (1973). The hypothalamo-hypophysial neurosecretory system and its relation to the reproductive cycle of the lizard, *Ophisops elegans* Menet. *Sci. Rep. Faculty of Sci. Ege University* N° 151, 3–4.

Zhang, Y., Cao, Q., and Zhang, Y. (1981). The primary structure of snake *Zaocys dhumnades*, Cantor insulin. *Scientia Sinica* 24, 1585–1589.

Zipser, B. (1980). Identification of specific leech neurons immunoreactive to enkephalin. *Nature* (London) 283, 857–858.

Zukin, R. S., and Zukin, S. R. (1984). The case for multiple opiate receptors. *Trends Neurosci.* 7, 160–164.

Contributors

Gesineke C. Bangma
Hubrecht Laboratory
Netherlands Institute of Developmental Biology
Uppsalalaan 8
3504 CT Utrecht, The Netherlands

Alan Crowe
Department of Medical and Physiological Physics
Rijksuniversiteit Utrecht
3508 TA Utrecht, The Netherlands

Dennis M. Dacey
Department of Ophthalmology
University of Washington
Seattle, Washington 98195

Alan M. Granda
School of Life Sciences
University of Delaware
Newark, Delaware 19711

Stéphane Hergueta
Laboratoire d'Anatomie Comparée
Muséum National d'Histoire Naturelle
55 rue de Buffon
75005 Paris, France

Michel Lemire
Laboratoire d'Anatomie Comparée
Muséum National d'Histoire Naturelle
55 rue de Buffon
75005 Paris, France

Dom Miceli
Laboratoire de Neuropsychologie Expérimentale
Université de Québec
Trois-Rivières
Québec, Canada G9A 5H7

Gerard J. Molenaar
Rijksuniversiteit Utrecht
Fakulteit der Diergeneeskunde
Vakgroep funktionele Morfologie
3508 TD Utrecht, The Netherlands

Ellengene H. Peterson
Department of Zoologic and Biomedical Sciences
Ohio University
Athens, Ohio 45701

Anton J. Reiner
Department of Anatomy and Neurobiology
Health Science Center
University of Tennessee
875 Monroe Avenue
Memphis, Tennessee 38163

Jacques Repérant
Laboratoire d'Anatomie
  Comparée
Muséum National d'Histoire
  Naturelle
55 rue de Buffon
75005 Paris, France

Jean Paul Rio
Laboratoire de Neuromor-
  phologie
INSERM U-106
Hôpital de la Salpêtrière
75651 Paris, France

Martin I. Sereno
Department of Cognitive
  Sciences
University of California, San
  Diego
La Jolla, California 92093

David F. Sisson
School of Life Sciences
University of Delaware
Newark, Delaware 19716

Hans J. ten Donkelaar
Department of Anatomy and
  Embryology
University of Nijmegen
6500 HB Nijmegen, The Nether-
  lands

Philip S. Ulinski
Department of Organismal Biol
  ogy and Anatomy
University of Chicago
Chicago, Illinois 60637

Roger Ward
Laboratoire de Neuropsycholo-
  gie Expérimentale
Université de Québec
Trois-Rivières
Québec, Canada G9A 5H7

Author Index

Subject Index